Antioxidant Properties of
Natural Products

Antioxidant Properties of Natural Products

Editor

Lillian Barros

MDPI • Basel • Beijing • Wuhan • Barcelona • Belgrade • Manchester • Tokyo • Cluj • Tianjin

Editor
Lillian Barros
Centro de Investigação de Montanha (CIMO),
Instituto Politécnico de Bragança
Portugal

Editorial Office
MDPI
St. Alban-Anlage 66
4052 Basel, Switzerland

This is a reprint of articles from the Special Issue published online in the open access journal *Antioxidants* (ISSN 2076-3921) (available at: https://www.mdpi.com/journal/antioxidants/special_issues/Isabel_Ferreira).

For citation purposes, cite each article independently as indicated on the article page online and as indicated below:

LastName, A.A.; LastName, B.B.; LastName, C.C. Article Title. *Journal Name* **Year**, *Article Number*, Page Range.

ISBN 978-3-03936-718-4 (Pbk)
ISBN 978-3-03936-719-1 (PDF)

Contents

About the Editor

Lillian Barros is an assistant researcher at the Centro de Investigação de Montanha (CIMO), Instituto Politécnico de Bragança and vice-coordinator of CIMO. She obtained her degree in biotechnology engineering (2002) from the Institute Polytechnic of Bragança, and obtained her Ph.D. "Doctor Europeus" in pharmacy (nutrition and bromatology) at the University of Salamanca (2008). She has published more than 448 indexed papers, within the highest impact factor journals in the fields of food science and technology and has an h-index of 54. She has edited several books, written book chapters, registered national and international patents, and is the principal investigator of several national and international research projects. Her scientific work has received interest and has led to the supervision of several post-doc, Ph.D., master, and bachelor students. Her scientific production has been relevant and has reached top positions in the world rankings (Highly Cited Researcher in 2016, 2017, and 2018, Clarivate) and has received awards from several different organizations. She is an associate editor of *Food & Function*, is a currently an active Editorial Board Member for several journals (Elsevier, MDPI, Hindawy, and Bentham Science), and has open Special Issues in these journals. Her research targets are mainly the identification, separation, and recovery of functional molecules from different natural products, as well as their implementation as ingredients and bioactive compounds in food, with the ultimate goal of extracting high-added-value molecules for reuse in the food chain (from agriculture to the consumer).

 *antioxidants*

Editorial

Antioxidant Properties of Natural Products

Lillian Barros

Centro de Investigação de Montanha (CIMO), Instituto Politécnico de Bragança, Campus de Santa Apolónia, 5300-253 Bragança, Portugal; lillian@ipb.pt

Received: 25 March 2020; Accepted: 26 March 2020; Published: 28 March 2020

Professor Isabel C.F.R. Ferreira has worked on the Chemistry of Natural Products, specifically the extraction, purification, and stabilization of myco and phytochemicals; development of natural ingredients, nutraceuticals, and innovative functional food formulations. She has developed numerous methodologies for the extraction, refinement, and chemical characterization of bioactive molecules from mushrooms and plants, and also improved the analytical methodologies of different biomolecules. She has prospected more than 300 species and discovered natural sources of ingredients with different capacities: bioactive, preservative, and coloring. In 2007, she led the research leading to the discovery that Portuguese wild mushrooms are powerful sources of compounds with bioactive properties. Subsequently, she showed that not only wild, but also commercial mushrooms are sources of bioactive compounds and equilibrated nutrients, leading to the first studies on species from Serbia, Ghana, Brazil, China, and Argentina, among others. In 2012, Isabel C.F.R. Ferreira discovered that specific mushroom species have significant antitumor, antimicrobial, and anti-inflammatory activities that can be improved modifying chemical substituents in specific molecules identified in those species.

In 2009, she began prospecting plant species and discovered specific parts of vegetables, flowers, and medicinal plants that are unique sources of bioactive molecules, particularly in the class of phenolic compounds. Subsequent to this, she showed for the first time that synergistic effects can be achieved by mixing defined species and that in vitro culture under optimized conditions can increase the concentration of these compounds. She has also developed novel cheminformatic tools and models to predict bioactive properties and organize molecule databases, and has carried out docking studies on bioactive cellular targets. She has been working with many mushroom and plant producing companies, and in 2011 she started to study physical conservation methods, discovering that gamma rays and electron-beams can decontaminate the mentioned raw materials without changing their nutritional/chemical profile, and also improve the extraction of bioactive molecules. She also discovered that modified atmospheres can recover the use of traditional plant species.

More recently, Professor Isabel C.F.R. Ferreira developed, improved, and optimized several extraction methodologies for different natural molecules with bioactive, coloring, and preservative capacities to be used as natural food ingredients/additives, and also their stabilization. With these stabilized ingredients, she has developed innovative food and cosmetic products (registered patents) with plant and mushroom-origin preservatives, colorants, and bioactives. Isabel F.C.F. Ferreira revision articles on bioactive molecules/properties of mushrooms and plants, natural preservatives, colorants, and conservation technologies, which are some of the most highly cited papers in the field.

The research performed by Professor Isabel C.F.R. Ferreira has placed her in top positions, being listed as a Highly Cited Researcher since 2015. She has published more than 675 articles and several innovative patents, supervised many students, and has been the principal researcher of numerous national and international projects. She has also gained a range of awards for her impact in Food Chemistry, for her advancement of research in Portugal, and for her outstanding work as one of the country's top women in science. Now, she is working for the Portuguese Government as Secretary of State for Interior Valorization.

The special issue in honor of Professor Isabel C.F.R. Ferreira, with the thematic "Antioxidants Properties of Natural Products" has twenty-three original research or review articles. Within the reviews performed, Rusu et al. [1] revised the clinical trials and cohort studies published in the last decade on the influence of antioxidant diets on nuts and peanuts in preventing or delaying age-related illnesses in middle-aged and elderly individuals. Meanwhile, Afonso et al. [2] reviewed the use of brown macroalgae, considered as safe for consumption in Europe, as a valuable ingredient namely for the development pf phaeophyta-enriched food products. Gomez-Zavaglia et al. [3] reviewed the up to date literature on the chemical composition and bioactive effects of seaweeds, and reported a critical discussion about their potential as natural sources of new functional ingredients. Root vegetables as an alternative source of natural red–purple color ingredients for pharmaceutical or food applications were extensively revised by Spyridon and contributors [4]. Furthermore, a discussion of the main bioactivities of these potential sources of natural colorants were also explored by the authors. Additionally, Di Gioia et al. [5] continued this study (part 2) and revised the health effects of red–purple vegetables, including leafy vegetables, fruits, and others, in which the chemical composition and bioactivities of the investigated matrices were discussed. Finally, the revision performed by Balakrishman et al. [6] provides insight into the major components of *Murraya koenigii* (L.) Sprengel, and their pharmacological activities against different pathological conditions.

Regarding the original research works performed, Reyes-Fermín et al. [7] evaluated α-mangostin, a phytochemical found in *Garcinia mangostana* L., regarding its antioxidant action in the preservation of mitochondrial function against *cis*-dichlorodiammineplatinum II induced damage using in vitro assays. Moreover, the antioxidant activity during the seven stages of development of *Amaranthus caudatus* was determined by Kamarać et al. [8], through several assays (ABTS$\bullet^+$, DPPH$\bullet$, O2$\bullet^-$ scavenging activity, FRAP, and Fe^{2+} chelating ability). In addition, the phenolic profile of the plant in each stage was also described by the authors. Petropoulos et al. [9] performed the nutritional and chemical composition evaluation of different parts of *Portulaca oleracea* L., namely stems and leaves, harvested in three growth stages. Additionally, the cytotoxicity of these samples was also assessed. In the research performed by Molina and contributors [10], the *Lonicera caerulea* L. berry was assessed in terms of nutritional and chemical composition and bioactive properties, namely antioxidant and antimicrobial activities. Also, the stabilization of the anthocyanins from the fruit juice was accomplished by microencapsulation.

Tanase et al. [11] characterized *Fagus sylvatica* L. bark in terms of phytochemical profile and bioactive potential. In addition, the improvement of the phenolic compounds extraction by microwave assisted extraction was carried out using a design experiment approach. Hou et al. [12] optimized the extraction of flavonoids from *Pteris cretica* L. by ultrasound-assisted extraction through a Response Surface Methodology. Furthermore, the identification of the major compounds was performed by chromatographic analysis, and the optimal extract was evaluated in terms of its antioxidant potential. Añibarro-Ortega et al. [13] characterized the nutritional composition, organic acids, and phenolic compounds of *Aloe vera* leaf (fillet, mucilage, and rind) and flowers, and their bioactive features, in order to enhance the exploitation of this plant as a valuable source of compounds for food and pharmaceutical industries application. Biosynthesis of silver nanoparticles (AgNP) using *Fagus sylvatica* L. bark polyphenolic extract was accomplished by Tanase et al. [14] and the obtained AgNPs were explored in terms of their antioxidant and antimicrobial activities. Rusu et al. [15] obtained bioactive rich molecules from the by-products of *Corylus avellana* L. (hazelnut involucre) using optimized conditions. The extract was evaluated through in vitro antioxidant tests and enzymatic inhibitory assays. Additionally, the cytotoxic and antioxidant effects of this extract was evaluated on two tumor cell lines and one normal cell line. Medrano-Padial et al. [16] evaluated the cytotoxic activity of a stilbenes-rich extract obtained from grapevine shoots on two tumor cell lines, with the aim of providing more evidence on the use of this kind of phytochemicals in wine preservation. Pereira-Maróstica and coauthors [17] studied the capability of methyl jasmonate to reduce articular and hepatic inflammation and oxidative stress in rats with adjuvant-induced arthritis. In the report performed by Lall et al. [18], the nutritional profile of *Sideritis perfoliata* L. was described. In addition, interesting bioactivities, such as anti-tyrosinase, antibacterial, and anti-elastase potential were evaluated to support the use of this plant as a potential cosmetic ingredient. The influence of cultivars and maturation stage on the phenolic composition of *Lycium barbarum* L. fruits were determined by Mocan and contributors [19]. Additionally, the authors correlated the phytochemical profiles obtained from the fruit extracts with the antioxidant and anti-fungal activities, as also, with tyrosinase inhibition. Molecular mechanisms involved in the biological effects of quercetin were studied by Ayuda-Durán et al. [20], using *Caenorhabditis elegans* as a model organism with different methodological approaches. Lee et al. [21] assessed the 3,3′-Diindolylmethane, a metabolite of indole-3-carbinol present in Brassicaceae vegetables, regarding its potential for prevention and therapy of neurodegenerative diseases. The authors evaluated in vitro effects of this compound against oxidative stress-induced issues. In order to provide a means for the disposal and valorization of walnut leaves, a phenolic-rich extract with high antioxidant activity from this matrix was obtained by Fernández-Agulló et al. [22], under optimized conditions for a maceration extraction system. Finally, Cásedas et al. [23] surveyed the neuroprotective action of urolithin A, as a gastrointestinal metabolite from ellagitannins, against oxidative-stress provoked in neuro-2a cells.

Overall, this special issue includes research related to the work that Professor Isabel C.F.R. Ferreira has been performing across her career, which is to exploit natural products, including natural compounds obtained from plants, mushrooms, marine, bee products, among others, focusing on the structural characterization and/or isolation of compounds with bioactivity; chemical, biological, or biochemical methods for the evaluation of *in vivo* and in vitro bioactivity; clinical and nutritional trials focused on the bioactive properties of compounds, and elucidation of the mechanisms of action of bioactive compounds.

May this themed issue be able to honor someone who, one day, proudly stated: "Never giving up is my helm, and the pleasure of discovery is my anchor". A motivation to keep in mind in science, as in life.

Acknowledgments: The conclusion of this special issue would not have been possible without the generous support and dedication of all researchers in this field of expertise; therefore, I would like to thank all the authors for their contribution in this special issue. I would like to thank Isabel C.F.R. Ferreira for all the knowledge and support she has given me across my research career, and also for the woman she is. It has been a pleasure to be part of your path in valorizing natural mountain products, and as you mention, a trail that has been conducted with enthusiasm and determination.

Conflicts of Interest: The authors declare no conflict of interest.

References

1. Rusu, M.; Fizeșan, I.; Pop, A.; Gheldiu, A.; Mocan, A.; Crișan, G.; Vlase, L.; Loghin, F.; Popa, D.; Tomuta, I. Health Benefits of nut consumption in middle-aged and elderly population. *Antioxidants* **2019**, *8*, 302. [CrossRef] [PubMed]

2. Afonso, N.C.; Catarino, M.D.; Silva, A.M.S.; Cardoso, S.M. Brown macroalgae as valuable food ingredients. *Antioxidants* **2019**, *8*, 365. [CrossRef] [PubMed]

3. Gomez-Zavaglia, A.; Prieto Lage, M.A.; Jimenez-Lopez, C.; Mejuto, J.C.; Simal-Gandara, J. The Potential of Seaweeds as a Source of Functional Ingredients of prebiotic and antioxidant value. *Antioxidants* **2019**, *8*, 406. [CrossRef] [PubMed]

4. Petropoulos, S.A.; Sampaio, S.L.; Di Gioia, F.; Tzortzakis, N.; Rouphael, Y.; Kyriacou, M.C.; Ferreira, I. Grown to be blue—Antioxidant properties and health effects of colored vegetables. Part I: Root vegetables. *Antioxidants* **2019**, *8*, 617. [CrossRef]

5. Di Gioia, F.; Tzortzakis, N.; Rouphael, Y.; Kyriacou, M.C.; Sampaio, S.L.; C.F.R. Ferreira, I.; Petropoulos, S.A. Grown to be blue—Antioxidant properties and health effects of colored vegetables. Part II: Leafy, fruit, and other vegetables. *Antioxidants* **2020**, *9*, 97. [CrossRef]

6. Balakrishnan, R.; Vijayraja, D.; Jo, S.; Ganesan, P.; Su-Kim, I.; Choi, D. Medicinal profile, phytochemistry, and pharmacological activities of *Murraya koenigii* and its primary bioactive compounds. *Antioxidants* **2020**, *9*, 101. [CrossRef]

7. Reyes-Fermín, L.M.; Avila-Rojas, S.H.; Aparicio-Trejo, O.E.; Tapia, E.; Rivero, I.; Pedraza-Chaverri, J. The protective effect of alpha-mangostin against cisplatin-induced cell death in LLC-PK1 cells is associated to mitochondrial function preservation. *Antioxidants* **2019**, *8*, 133. [CrossRef]

8. Karamać, M.; Gai, F.; Longato, E.; Meineri, G.; Janiak, M.A.; Amarowicz, R.; Peiretti, P.G. Antioxidant activity and phenolic composition of amaranth (*Amaranthus caudatus*) during plant growth. *Antioxidants* **2019**, *8*, 173. [CrossRef]

9. Petropoulos, S.A.; Fernandes, Â.; Dias, M.I.; Vasilakoglou, I.B.; Petrotos, K.; Barros, L.; Ferreira, I.C.F.R. Nutritional value, chemical composition and cytotoxic properties of common purslane (*Portulaca oleracea* L.) in relation to harvesting stage and plant part. *Antioxidants* **2019**, *8*, 293. [CrossRef]

10. Molina, A.K.; Vega, E.N.; Pereira, C.; Dias, M.I.; Heleno, S.A.; Rodrigues, P.; Fernandes, I.P.; Barreiro, M.F.; Kostić, M.; Soković, M.; et al. Promising antioxidant and antimicrobial food colourants from *Lonicera caerulea* L. var. Kamtschatica. *Antioxidants* **2019**, *8*, 394. [CrossRef]

11. Tanase, C.; Mocan, A.; Coșarcă, S.; Gavan, A.; Nicolescu, A.; Gheldiu, A.; Vodnar, D.; Muntean, D.; Crișan, O. Biological and chemical insights of beech (*Fagus sylvatica* L.) bark: A source of bioactive compounds with functional properties. *Antioxidants* **2019**, *8*, 417. [CrossRef] [PubMed]

12. Hou, M.; Hu, W.; Wang, A.; Xiu, Z.; Shi, Y.; Hao, K.; Sun, X.; Cao, D.; Lu, R.; Sun, J. Ultrasound-assisted extraction of total flavonoids from *Pteris cretica* L.: process optimization, HPLC analysis, and evaluation of antioxidant activity. *Antioxidants* **2019**, *8*, 425. [CrossRef] [PubMed]

13. Añibarro-Ortega, M.; Pinela, J.; Barros, L.; Ćirić, A.; Silva, S.; Coelho, E.; Mocan, A.; Calhelha, R.; Soković, M.; Coimbra, M.; et al. Compositional features and bioactive properties of *Aloe vera* Leaf (fillet, mucilage, and rind) and flower. *Antioxidants* **2019**, *8*, 444. [CrossRef] [PubMed]

14. Tanase, C.; Berta, L.; Coman, N.; Roșca, I.; Man, A.; Toma, F.; Mocan, A.; Jakab-Farkas, L.; Biró, D.; Mare, A. Investigation of in vitro antioxidant and antibacterial potential of silver nanoparticles obtained by biosynthesis using beech bark extract. *Antioxidants* **2019**, *8*, 459. [CrossRef] [PubMed]

15. Rusu, M.; Fizeșan, I.; Pop, A.; Gheldiu, A.; Mocan, A.; Crișan, G.; Vlase, L.; Loghin, F.; Popa, D.; Tomuta, I. Enhanced recovery of antioxidant compounds from hazelnut (*Corylus avellana* L.) involucre based on extraction optimization: Phytochemical profile and biological activities. *Antioxidants* **2019**, *8*, 460. [CrossRef] [PubMed]

16. Medrano-Padial, C.; Puerto, M.; Moreno, F.J.; Richard, T.; Cantos-Villar, E.; Pichardo, S. In vitro toxicity assessment of stilbene extract for its potential use as antioxidant in the wine industry. *Antioxidants* **2019**, *8*, 467. [CrossRef]

17. Pereira-Maróstica, H.; Castro, L.; Gonçalves, G.; Silva, F.; Bracht, L.; Bersani-Amado, C.; Peralta, R.; Comar, J.; Bracht, A.; Sá-Nakanishi, A. Methyl jasmonate reduces inflammation and oxidative stress in the brain of arthritic rats. *Antioxidants* **2019**, *8*, 485. [CrossRef]

18. Lall, N.; Chrysargyris, A.; Lambrechts, I.; Fibrich, B.; Blom Van Staden, A.; Twilley, D.; de Canha, M.N.; Oosthuizen, C.B.; Bodiba, D.; Tzortzakis, N. Sideritis perfoliata (subsp. *Perfoliata*) nutritive value and its potential medicinal properties. *Antioxidants* **2019**, *8*, 521. [CrossRef]

19. Mocan, A.; Cairone, F.; Locatelli, M.; Cacciagrano, F.; Carradori, S.; Vodnar, D.C.; Crișan, G.; Simonetti, G.; Cesa, S. Polyphenols from *Lycium barbarum* (Goji) Fruit european cultivars at different maturation steps: Extraction, HPLC-DAD analyses, and biological evaluation. *Antioxidants* **2019**, *8*, 562. [CrossRef]

20. Ayuda-Durán, B.; González-Manzano, S.; Miranda-Vizuete, A.; Sánchez-Hernández, E.; R. Romero, M.; Dueñas, M.; Santos-Buelga, C.; González-Paramás, A.M. Exploring target genes involved in the effect of quercetin on the response to oxidative stress in Caenorhabditis *elegans*. *Antioxidants* **2019**, *8*, 585. [CrossRef]

21. Lee, B.D.; Yoo, J.-M.; Baek, S.Y.; Li, F.Y.; Sok, D.-E.; Kim, M.R. 3,3′-Diindolylmethane promotes BDNF and antioxidant enzyme formation via TrkB/Akt pathway activation for neuroprotection against oxidative stress-induced apoptosis in hippocampal neuronal cells. *Antioxidants* **2019**, *9*, 3. [CrossRef] [PubMed]

22. Fernández-Agulló, A.; Castro-Iglesias, A.; Freire, M.S.; González-Álvarez, J. Optimization of the extraction of bioactive compounds from walnut (*Juglans major* 209 x *Juglans regia*) leaves: Antioxidant capacity and phenolic profile. *Antioxidants* **2019**, *9*, 18. [CrossRef] [PubMed]

23. Cásedas, G.; Les, F.; Choya-Foces, C.; Hugo, M.; López, V. The metabolite urolithin-A ameliorates oxidative stress in Neuro-2a Cells, becoming a potential neuroprotective agent. *Antioxidants* **2020**, *9*, 177. [CrossRef] [PubMed]

 *antioxidants*

Review

Health Benefits of Nut Consumption in Middle-Aged and Elderly Population

Marius Emil Rusu [1], Andrei Mocan [2,3], Isabel C. F. R. Ferreira [4,* and Daniela-Saveta Popa [5]

[1] Department of Pharmaceutical Technology and Biopharmaceutics, Faculty of Pharmacy, "Iuliu Hatieganu" University of Medicine and Pharmacy, 8 Victor Babes, 400012 Cluj-Napoca, Romania
[2] Department of Pharmaceutical Botany, Faculty of Pharmacy, "Iuliu Hatieganu" University of Medicine and Pharmacy, 8 Victor Babes, 400012 Cluj-Napoca, Romania
[3] Laboratory of Chromatography, ICHAT, University of Agricultural Sciences and Veterinary Medicine, 400372 Cluj-Napoca, Romania
[4] Centro de Investigação de Montanha (CIMO), Instituto Politécnico de Bragança (IPB), Campus de Santa Apolónia, 5300-253 Bragança, Portugal
[5] Department of Toxicology, Faculty of Pharmacy, "Iuliu Hatieganu" University of Medicine and Pharmacy, 8 Victor Babes, 400012 Cluj-Napoca, Romania
* Correspondence: iferreira@ipb.pt

Received: 23 June 2019; Accepted: 31 July 2019; Published: 12 August 2019

Abstract: Aging is considered the major risk factor for most chronic disorders. Oxidative stress and chronic inflammation are two major contributors for cellular senescence, downregulation of stress response pathways with a decrease of protective cellular activity and accumulation of cellular damage, leading in time to age-related diseases. This review investigated the most recent clinical trials and cohort studies published in the last ten years, which presented the influence of tree nut and peanut antioxidant diets in preventing or delaying age-related diseases in middle-aged and elderly subjects (≥55 years old). Tree nut and peanut ingestion has the possibility to influence blood lipid count, biochemical and anthropometric parameters, endothelial function and inflammatory biomarkers, thereby positively affecting cardiometabolic morbidity and mortality, cancers, and cognitive disorders, mainly through the nuts' healthy lipid profile and antioxidant and anti-inflammatory mechanisms of actions. Clinical evidence and scientific findings demonstrate the importance of diets characterized by a high intake of nuts and emphasize their potential in preventing age-related diseases, validating the addition of tree nuts and peanuts in the diet of older adults. Therefore, increased consumption of bioactive antioxidant compounds from nuts clearly impacts many risk factors related to aging and can extend health span and lifespan.

Keywords: tree nuts; peanuts; middle-aged population; antioxidants; age-related diseases; elderly; nut-enriched Mediterranean diet; inflammation; oxidative stress

---

1. Introduction

Healthy diet, along with physical and cognitive activity, is a modifiable lifestyle factor that has been associated with overall health in old age. Aging is a collective, multifactorial progression of physiological dysfunctions at cellular and tissue levels, as well as the deregulation of normal cell pathways and lost synchronicity among defense mechanisms in the body, leading to the development of age-related diseases [1].

A number of studies explored the phytochemical composition, bioactivity, and antioxidant potential of nuts [2,3]. Scientific evidence demonstrated the potential benefits of a higher intake of nuts or nut-enriched Mediterranean diets (MDs) against risk factors associated with pathological conditions linked with aging, such as cardiometabolic diseases, cancer, or cognitive disorders [4–8]. This review focused on middle-aged and older adults, a constantly growing group, with particular attention given to how these individuals can benefit from healthy dietary plans enhanced with tree nuts and peanuts, and how the extra bioactive antioxidant compounds may add quality to the increased life expectancy.

The aim of this study was to present up-to-date evidence confirming that nut intake and nut-enriched MDs [9] can be beneficial in delaying aging and can support modern medicine in treating age-related diseases. We investigated human studies that assessed the effects of nut consumption on the health of a middle-aged and elderly population 55 years of age and older. We searched PubMed/MEDLINE and EMBASE databases for peer-reviewed articles published in the decade from 2009 to June 2019. We also examined studies found in the references of meta-analysis that were in accordance with our inclusion criteria: subject age ≥55 years and nuts with antioxidant and anti-inflammatory activities added in the diet. In this review, the word "nuts" includes tree nuts, such as walnuts (*Juglans regia*), almonds (*Prunus dulcis*), hazelnuts (*Corylus avellana*), pistachios (*Pistacia vera*), pecans (*Carya illinoinensis*), cashews (*Anacardium occidentale*), pine nuts (*Pinus pinea*), Brazil nuts (*Bertholletia excelsa*), macadamias (*Macadamia integrifolia*), and also peanuts (*Arachis hypogaea*), botanically classified as legumes but with a nutrient profile comparable to tree nuts.

In addition, we summarized several nut phytochemicals with biological antioxidant activity, their possible mechanisms of actions, as well as the influence of nuts on gut microbiota (GM).

2. Association between Nut Consumption and Cardiometabolic Disorders

Globally, cardiometabolic diseases including type 2 diabetes mellitus (T2DM), coronary heart disease (CHD), coronary artery disease (CAD), and stroke are leading causes of morbidity and mortality [10]. Most of these diseases could be prevented by changing behavioral risk factors such as suboptimal diet [11].

Tree nuts and peanuts, because of their healthy antioxidant biochemical profile, can improve the lipid profile, increase insulin sensitivity and metabolism, and favorably influence other cardiometabolic risk factors [12].

2.1. Nut Consumption in Cardiometabolic Morbidity and Mortality

Clinical trials and lengthy prospective studies, focusing on men and women over the age of 55, showed that higher weekly nut intake can lower all causes and cause-specific morbidity and mortality [13–21] (Table 1).

Subjects at high cardiovascular (CV) risk, who supplemented their MD with 30 grams of tree nuts per day, at any point in time during the 4 years study period, had a 53% lower diabetes incidence compared to the control group [13]. Pan et al. [14] observed that women in the highest nut serving group were 26% less likely to develop diabetes than participants in the very low nut intake group, with 95% confidence that the true value is lying between 16%–35%. Similarly, when comparing high to very low nut intake, the prevalence of diabetes, obesity, and MS were 13%, 39%, and 26% less likely, respectively [15].

A single meta-analysis [22] confirmed that subjects in the highest total nut intake group had 32% less risk of dying from diabetes than those in the lowest group (relative risk, RR = 0.68, 95% confidence interval, CI: 0.52–0.90).

Table 1. Nut consumption and cardiometabolic morbidity and mortality.

Author, Year, Country [Ref]	Design	Subjects (F:M) Mean Age (Range)	Length of Study	Comparison Group	Intake of Nuts	Findings
Salas-Salvadó et al., 2011,2018 Spain [13]	RCT	418 (293:125) 67 (55–80) y	4 y	Control (low-fat diet *)	MD + 30 g/d nuts (15 g W, 7.5 g H, 7.5 g A)	↓ diabetes incidence, HR 0.47 (95% CI: 0.23–0.98) (vs. control)
Estruch et al., 2018, Spain [20]	RCT, Parallel, multicenter	2454 (1326:1128) 66.7 ± 6.1	4.8 y	Control (nut free diet)	MD + 30 g/d mixed nuts (15 g W, 7.5 g A, 7.5 g H)	↓ incidence of CV events (myocardial infarction, stroke, death from CV causes) (vs. control) HR 0.64 (95% CI: 0.47–0.88)
Pan et al., 2013, USA [14]	Prospective (NHS)	58,063 F 52–77 y	22 y		Tree nuts and peanuts (1) Never/rarely (2) <1 serving/wk (3) 1 serving/wk (4) 2–4 servings/wk (5) ≥5 servings/wk	↓ T2DM risk (*p*-trend < 0.001) for tree nuts and peanuts HR 0.80 (95% CI: 0.71–0.90) for (2) to (4) vs. (1) HR 0.74 (95% CI: 0.65–0.84) for (5) vs. (1) ↓ T2DM risk (*p*-trend = 0.002) for walnuts HR 0.76 (95% CI: 0.62–0.94) for (4),(5) vs. (1)
Ibarrola-Jurado et al., 2013, Spain [15]	Cross-sectional (PREDIMED)	7210 (4143:3067) 67 (55–80) y			Tree nuts and peanuts (1) <1 serving/wk (2) 1–3 servings/wk (3) >3 servings/wk	↓ prevalence of diabetes (3) vs. (1): OR 0.87 (95% CI: 0.78–0.99), *p*-trend = 0.043 ↓ prevalence of MS (3) vs. (1): OR 0.74 (95% CI: 0.65–0.85), *p*-trend < 0.001 ↓ prevalence of obesity (3) vs. (1): OR 0.61 (95% CI: 0.54–0.68), *p*-trend < 0.001
Guasch-Ferré et al., 2013, Spain [16]	Prospective (PREDIMED)	7216 (4145:3071) 67 y	4.8 y		Tree nuts and peanuts (1) none (2) 1–3 servings/wk (3) >3 servings/wk	↓ CV mortality (3) vs. (1) for walnuts HR 0.53 (95% CI: 0.29–0.98), *p*-trend = 0.047 ↓ CV mortality (3) vs. (1) for tree nuts (no walnuts) and peanuts HR 0.42 (95% CI: 0.20–0.89), *p*-trend = 0.031 ↓ total mortality risk (3) vs. (1) for tree nuts (walnuts included) and peanuts HR 0.61 (95% CI: 0.45–0.83), *p*-trend = 0.01
Hshieh et al., 2015, USA [17]	Prospective	20,742 M 67 y	9.6 y		Tree nuts and peanuts (1) <1 serving/mo (2) 1–3 servings/mo (3) 1 serving/wk (4) 2–4 servings/wk (5) ≥5 servings/wk	↓ CVD deaths (5) vs. (1) HR 0.74 (95% CI: 0.55–1.02), *p*-trend = 0.015

Table 1. *Cont.*

Author, Year, Country [Ref]	Design	Subjects (F:M) Mean Age (Range)	Length of Study	Comparison Group	Intake of Nuts	Findings
Guasch-Ferré et al., 2017, USA [18]	Prospective (a) NHS (b) NHS II (c) HPFS	(a) 76,364 F (b) 92,946 F (c) 41,526 M 56 y	28.7 y 21.5 y 22.5 y		Tree nuts and peanuts (1) Never/almost never (2) <1 time/wk (3) 1 time/wk (4) 2–4 times/wk (5) ≥5 times/wk	(5) vs. (1) for tree nuts and peanuts ↓ CVD-HR 0.86 (95% CI: 0.79–0.93, p-trend < 0.001) ↓ CHD-HR 0.80 (95% CI: 0.72–0.89, p-trend < 0.001) ≥2 times/wk tree nuts and peanuts ↓ 13–19% CVD risk ↓ 15–23% CHD risk
Larsson et al., 2018, Sweden [19]	Prospective	61,364 (28,453:32,911) 58 (45–83) y	17 y		Tree nuts and peanuts (1) none (2) 1–3 times/mo (3) 1–2 times/wk (4) ≥3 times/wk	↓ risk of atrial fibrillation for (4) vs. (1) (linear association) HR 0.82 (95% CI: 0.68–0.99), p-trend = 0.004 ↓ risk of heart failure for (3) vs. (1) (non-linear association) HR 0.80 (95% CI: 0.67–0.97), p-trend = 0.003 (fully adjusted models)
Liu et al., 2019, USA [21]	Prospective (NHS, HPFS)	16,217 (12,006:4211) 64.7–69.4 y	34 y 28 y		Tree nuts and peanuts (1) <1 serving/mo (2) <1 serving/wk (3) 1 serving/wk (4) 2–4 servings/wk (5) ≥5 servings/wk	(5) vs. (1) HRs ↓ CVD incidence, 0.83 (0.71–0.98), p-trend = 0.01 ↓ CHD incidence, 0.80 (0.67–0.96), p-trend = 0.005 ↓ CVD mortality, 0.66 (0.52–0.84), p-trend < 0.001 ↓ all-cause mortality 0.69 (0.61–0.77), p-trend < 0.001

* low-fat diet—all types of fat, from both animal and vegetable sources, reduced, but no fat-free foods. A—almonds; CHD—coronary heart disease; CI—confidence interval; CV—cardiovascular; CVD—cardiovascular diseases; F—women; H—hazelnuts; HR—hazard ratio; HPFS—Health Professionals Follow-Up Study; M—men; MD—Mediterranean diet; MS—metabolic syndrome; NHS—Nurses' Health Study; OR—odds ratio; RCT—randomized controlled trial; RR—risk ratio; T2DM—type 2 diabetes mellitus; W—walnuts.

One prospective study showed a nonsignificant reduction in cardiovascular disease (CVD) mortality when comparing the highest with the lowest nut intake population groups [17]. However, two others concluded that participants who were fed nut-enhanced MDs had a significantly lower risk for CVD and CHD, as well as lower CVD, CHD, and all-cause mortality compared to the no-nut group [16,18]. Participants who consumed walnuts ≥1 serving per week had 19%, 21%, and 17% lower risk for CVD, CHD, and stroke, respectively, while those who consumed walnuts >3 servings per week had 47% lower risk of CV mortality compared with subjects who did not eat walnuts [18]. A very recent prospective study demonstrated lower CVD incidence (hazard ratio, HR = 0.83, 95% CI: 0.71–0.98) and CHD incidence (HR = 0.80, 95% CI: 0.67–0.96), 31% lower all-cause mortality and 34% lower CVD mortality for at least 5 servings of nuts per week compared to less than one per month [21]. In the same study, total nut intake was not significantly associated with stroke incidence. Moreover, a recent intervention study showed that individuals at high CV risk had a lower incidence of major CV events (myocardial infarction, stroke, death from CV causes) when nuts were added daily to MD compared with a low-fat, nut-free control diet [20]. Also, three or more servings of nuts per week, as compared to none, can even lower the risk of atrial fibrillation and heart failure [19].

Peripheral arterial disease (PAD), often caused by atherosclerosis, can lead to heart attack and stroke. A large cross-sectional study on mature adults indicated that daily nut consumption was associated with lower odds of PAD (OR = 0.79; 95% CI: 0.77–0.80, $p < 0.001$) compared to participants with the lowest intake of nuts [23].

A very recent meta-analysis of 11 observational studies confirmed that tree nut (but not peanut) consumption was negatively associated with metabolic syndrome (MS) ($p = 0.04$) [24]. Another meta-analysis showed a significantly reduced risk for CVD and CHD for higher nut consumption (including peanuts) [22]. Luu et al. [24] showed that in different ethnic groups, nut consumption was associated with decreased overall mortality and CVD mortality and that a higher versus lower quintile of nut intake had a statistically significant inverse association for ischemic heart disease (IHD). Mayhew et al. [25], after reviewing several large prospective cohort studies, concluded that nut consumption was inversely associated with total CVD, CVD mortality, total CHD, CHD mortality, and sudden cardiac death. While one meta-analysis established that daily nut intake can reduce the risk of stroke [26], another study did not find a significant association with stroke, but indicated an inverse association with IHD, overall CVD, and all-cause mortality for nut consumption [27]. A meta-analysis of randomized controlled trials (RCTs) and observational studies identified that nut feeding was inversely linked with nonfatal and fatal IHD and diabetes, but not stroke [28]. However, a meta-analysis of prospective cohorts to evaluate the relation between nuts and stroke risk and mortality, reported that nut consumption was inversely associated with stroke risk (RR 0.90, 95% CI: 0.83–0.98) and stroke mortality, when comparing the highest with the lowest nut intake [29].

High blood pressure (BP) is a major risk factor for CVD. The previous feeding studies showed that tree nuts or peanuts had no effect on BP. However, a very recent RCT on an elderly cohort (age, 69 years) proved that walnuts (42.5 grams per day) reduced systolic BP in the walnut group (−4.61 mm Hg, 95% CI: −7.43 to −1.79 mm Hg) compared to the no-walnut group (−0.59 mm Hg, 95% CI: −3.38 to 2.21 mm Hg) ($p = 0.051$) [30]. As no changes in diastolic BP were noticed, it is safe to say that walnut intake can reduce systolic BP in mature subjects, mainly in those with mild hypertension.

2.2. Nut Consumption and Blood Lipids

The effects of tree nut and peanut consumption on lipid profiles from intervention studies published in the last ten years [31–43] are summarized in Table 2. The results of these clinical trials in a middle-aged population indicate a causal association between higher nut intake and lower levels of total cholesterol (T-C), low density lipoprotein-cholesterol (LDL-C), non-high density lipoprotein-cholesterol (non-HDL-C), triglycerides (TG), and apolipoprotein B (apoB), all markers of CV morbidity and mortality.

Table 2. Nut consumption and blood lipid levels in intervention studies.

Author, Year, Country [Ref]	Design	Subjects (F:M) Mean Age (±SD)	Length of Study	COMPARISON GROUP	Intake of Nuts	Findings
Li et al., 2011, Taiwan [31]	RCT, Crossover	20 (11:9) T2DM patients 58 y	12 wk	Control (diet without A)	56 g/d A	↓ T-C 6.0% (95% CI: 1.6–9.4), $p \leq 0.0025$ ↓ LDL-C 11.6% (95% CI: 2.8–19.1), $p \leq 0.0117$ ↓ LDL-C/HDL-C ratio 9.7% (95% CI: 0.3–20.9), $p \leq 0.0128$ (vs. control)
Wu et al., 2014, Germany [32]	RCT, Crossover	40 (30:10) 60 ± 1 y	8 wk	Control (nut-free Western-type diet)	43 g/d W	↓ non-HDL-C (-10 ± 3 mg/dL, $p = 0.025$) ↓ apoB (-5.0 ± 1.3 mg/dL, $p = 0.009$) (vs. baseline)
Hernández-Alonso et al., 2015, Spain [33]	RCT, Crossover	54 (25:29) prediabetics 55 y	9 mo	Control (diet without pistachios)	57 g/d pistachio	↓ LDL-C (P) -28.07 nM (95% CI: -60.43 to 4.29) vs. baseline; $p = 0.02$ ↓ Non-HDL-C (P) -36.02 nM (95% CI: -77.56 to 5.52) vs. baseline; $p = 0.04$
Ruisinger et al., 2015, USA [34]	RCT, Parallel	48 (24:24) on statin therapy 60 y	4 wk	Control (diet without A)	100 g/d A	↓ non-HDL-C (113.4 ± 24.5 vs. 124.7 ± 20.8 mg/dL, $p = 0.024$) ↓ LDL-C (95.6 ± 23.9 vs. 104.3 ± 18.7 mg/dL, $p = 0.068$) ↓ TG (102.1 ± 38.3 vs. 115.0 ± 42.6 mg/dL, $p = 0.068$) (vs. control)
Jamshed et al., 2015, Pakistan [35]	RCT	150 (37:113) CAD patients 60 (32–86) y	12 wk	Control (diet without A)	10 g/d A before breakfast	↑ HDL-C (40 ± 6 vs. 33 ± 5 mg/dL) ↓ TG (118 ± 18 vs. 130 ± 20 mg/dL) (vs. baseline; p all < 0.05)
Njike et al., 2015, USA [36]	RCT, Parallel	112 (81:31) 55 y	6 mo	Control (diet without W)	56 g/d W	↓ T-C (-16.04 ± 27.34 mg/dL vs. baseline, $p < 0.0001$) ↓ LDL-C (-14.52 ± 24.11 mg/dL vs. baseline, $p < 0.0001$)
Huguenin et al., 2015, Brazil, [37]	RCT, Crossover	91 (44:47) hypertensive 62 y	12 wk	Control (nut-free diet)	13 g/d Granulated Brazil nut	↓ Ox LDL-C (60.68 ± 20.88 from 66.31 ± 23.59 U/L) (vs. baseline, $p < 0.05$)
Sauder et al., 2015, USA [38]	RCT, Crossover	30 (15:15) 56.1 ± 7.8 y	4 wk	Control (diet without pistachios)	pistachios (20% of total energy)	↓ T-C (4.00 ± 0.06 vs. 4.15 ± 0.06 mmol/L, $p = 0.048$) ↓ T-C/HDL-C (4.06 ± 0.08 vs. 4.37 ± 0.08, $p = 0.0004$) ↓ TG (1.56 ± 0.10 vs. 1.84 ± 0.10, $p = 0.003$) (vs. control)
Mah et al., 2017, USA [39]	RCT, Crossover	51 (31:20) 56 y	4 wk	Control (diet without cashews)	28–64 g/d cashews	↓ T-C 23.9% (95% CI: 29.3–1.7) vs. 0.8% (95% CI: 21.5–4.5) ↓ LDL-C 24.8% (95% CI: 212.6–3.1) vs. 1.2% (95% CI: 22.3–7.8) ↓ non-HDL-C 25.3% (95% CI: 28.6–2.1) vs. 1.7% (95% CI: 20.9–5.6%)(vs. baseline compared with control, $p < 0.05$)

Table 2. *Cont.*

Author, Year, Country [Ref]	Design	Subjects (F:M) Mean Age (±SD)	Length of Study	COMPARISON GROUP	Intake of Nuts	Findings
Bamberger et al., 2017, Germany [40]	RCT, Crossover	194 (134:60) 63 ± 0.54 y	24 wk	Control (diet without W)	43 g/d W	↓ T-C (−9.5 vs. −2.2 mg/dL, $p = 0.0003$) ↓ LDL-C (−7.3 vs. −1.9 mg/dL, $p = 0.0009$) ↓ non-HDL-C (−9.4 vs. −1.5 mg/dL, $p < 0.001$) ↓ TG (−5.5 vs. 3.4 mg/dL, $p = 0.0943$) ↓ apoB (−6.8 vs. −0.9 mg/dL, $p < 0.0001$) (vs. control)
McKay et al., 2018, USA [41]	RCT, Crossover	26 (5:21) 59.7 (57–70) y	12 wk	Control (isocaloric, no pecans)	42.5 g/d pecans	↓ T-C (−8.89 ± 4.41, $p = 0.056$) ↓ LDL-C (−7.41 ± 3.85, $p = 0.067$)
Jenkins et al., 2018, Canada [42]	RCT, Parallel	117 (39:78) diabetics 62 y	3 mo	Controls (isocaloric (1) carbs diet and (2) carbs and nut diet)	75 g/d mixed nuts (tree nuts and peanuts)	↓ T-C −0.06 mmol/L (95% CI: −0.12 to −0.01), $p = 0.026$ ↓ non HDL-C −0.09 mmol/L (95% CI: −0.17 to −0.01), $p = 0.026$ ↓ apoB −0.09 g/L (95% CI: −0.16 to −0.02), $p = 0.015$ vs. control (1)
Bowen et al., 2019, Australia [43]	RCT	76 (31:45) 61 y	8 wk	Control (nut free diet)	56 g/d A	↓ TC/HDL-C ratio (in women, but not in men)

A—almonds; apoB—apolipoprotein B; CI—confidence interval; F—women; HDL-C—high density lipoprotein-cholesterol; LDL-C (P)—low density lipoprotein-cholesterol (particle); M—men; RCT—randomized controlled trial; SD—standard deviation; T2DM—type 2 diabetes mellitus; T-C—total cholesterol; TG—triglycerides; W—walnuts.

After a 12 week trial and daily intake of 56 g almonds, compared with a no-nut control diet, plasma apoB, apoB/apoA-1 ratio, T-C, LDL-C, and LDL-C/HDL-C ratio decreased significantly by 15.6%, 17.4%, 6.0%, 11.6%, and 9.7%, respectively, in patients with T2DM [31]. Similarly, the inclusion of almonds in the diet of patients on chronic statin therapy revealed a 4.9% reduction in non–HDL-C compared with the no-nut group, and non-statistical significance decreases in LDL-C and TG [34]. Consumed before breakfast, 10 g almonds proved to increase serum HDL-C by 15% after 12 weeks in CAD patients with low initial HDL-C [35]. A new trial, comparing almond snacks with isocaloric carbohydrate snacks, demonstrated that almond snacks can improve the serum T-C/HDL-C ratio in women but not in men, with no change in body weight (BW) or inflammatory biomarkers in overweight and obese adults with high T2DM risk [43]. The short 8 weeks trial might be the cause for the differential gender results.

Our findings are aligned with data from a meta-analysis conducted by Musa-Veloso et al. [44] in which almond consumption was confirmed to significantly decrease T-C ($p < 0.001$), LDL-C ($p = 0.001$), and TG ($p = 0.042$), with no modification in HDL-C ($p = 0.207$). Also, Nishi et al. [45] showed that the daily consumption of almonds by middle-aged adults can improve the blood lipid profile, and a 3.5% decrease in the 10-year CHD risk was noticed for every 30 g increase in almond intake.

Consistent with the effects of other nuts, Brazil nuts and cashews can also improve lipid profiles. In a group of hypertensive and dyslipidemic subjects, the intake of partially defatted Brazil nuts significantly increased plasma selenium and the antioxidant activity of the glutathione peroxidase enzyme, and reduced oxidation in LDL-C compared to the baseline [37]. Mah et al. [39] demonstrated that adding cashews into the diet of a population with high LDL-C risk could lower the T-C, LDL-C, and LDL-C/HDL-C ratio. In agreement with these results, a very recent trial showed a significant decrease for the LDL-C/HDL-C ratio in a cashew diet group compared with a no-cashew control group [46].

Pistachio, a biochemically-rich tree nut, proved to have a lowering effect on CV risk factors. Daily pistachio intake significantly decreased the T-C and T-C/HDL-C ratio ($p < 0.05$), and TG levels ($p = 0.003$) compared to the control in T2DM adults [38]. Also, after 4 months of 57 g pistachio daily, small LDL particles and non-HDL particles significantly decreased compared to the nut-free diet [33]. This change of lipoprotein particle size may explain the decrease of CVD risk. Kay et al. [47] showed that the consumption of a pistachio-enriched diet, when compared to the control, significantly increased serum concentration of antioxidants, including beta-carotene, gamma-tocopherol, and lutein, and significantly decreased serum oxidized-LDL, an important factor in CVD.

Walnut, one of the most versatile tree nuts with its higher content of polyunsaturated fatty acids (PUFAs) including α-linolenic acid, and high antioxidant activity, may influence CVD risks via its lipid-lowering impact. Compared with a control diet without walnuts, a walnut-included diet for 6 months significantly decreased T-C and LDL-C and improved diet quality [36]. In a shorter cross-over trial, a walnut-enriched diet significantly reduced non-HDL-C ($p = 0.025$) and apoB ($p = 0.009$) compared with a control diet, while T-C displayed a tendency toward reduction ($p = 0.073$) [32]. Bamberger et al. [40] indicated that a walnut-enriched diet versus a control diet caused a significant decrease in fasting cholesterol ($p = 0.002$), LDL-C ($p = 0.029$), non-HDL-C ($p \leq 0.001$), TG ($p = 0.015$), and apoB ($p \leq 0.001$) in healthy mature adults. Also, 15 mL walnut oil daily (corresponding to ~28 g walnuts) added for 90 days to the diet of hyperlipidemic T2DM patients significantly decreased the T-C, LDL-C, T-C/HDL-C ratio ($p < 0.001$ for all), and TG level ($p = 0.021$), compared with the control group, while the HDL level showed an increased trend ($p = 0.06$) [48]. Similarly, Austel et al. [49] noticed beneficial changes in blood lipids after replacing saturated fats with walnut oil.

A meta-analysis of 26 trials confirmed that walnut-enriched diets compared with control groups significantly reduced T-C ($p < 0.001$), LDL-C ($p < 0.001$), TG ($p = 0.03$), and apoB ($p = 0.008$), with no significant modifications in BW or blood pressure [50].

Integrated into a typical Western diet, pecans, another tree nut with higher contents of PUFAs, showed a borderline significant lowering of T-C and LDL-C as compared to a nut-free control diet in overweight or obese adults with central adiposity [41].

A new trial confirmed that, compared to a carbohydrate control diet, adding 75 g per day of mixed nuts (tree nuts and peanuts) to healthy diets could significantly lower small LDL-C ($p = 0.018$), with a trend towards reduction for T-C ($p = 0.066$) and non-HDL-C ($p = 0.067$) in T2DM elderly patients [42].

These outcomes are validated by a meta-analysis of 61 trials which concluded that nut intake significantly lowered the levels of T-C, LDL-C, apoB, and TG, with the key factor of changing lipid profile appearing to be nut dose rather than nut type [51].

As many strategies for reducing T-C and LDL-C levels could lower HDL-C levels, all the dietary plans for lowering LDL-C levels should aim to maintain or even increase HDL-C. In their 2017 guidelines, the American Association of Clinical Endocrinologists and American College of Endocrinology recommend a minimum blood HDL-C level of 40 mg/dL in CVD risk individuals [52]. However, data [53] showed that small HDL particles present only a weak defense, the strong protection against CVD risks coming from large HDL units. Equally, small LDL particles, due to their proneness to oxidation compared with larger ones, are responsible for atherosclerosis progress and CVD, while large LDL components are only weakly linked with CVD [53]. Based on these many results, nut consumption can evidently aid in the management of a healthy lipid profile in mature adults.

2.3. Nut Consumption and Biochemical and Anthropometric Parameters

The outcome of tree nut and peanut consumption on biochemical and anthropometric parameters found in clinical trials published in the last ten years [31,41,42,54–60] are summarized in Table 3.

The trial of Li et al. [31] revealed that in Chinese T2DM patients, almond intake, compared with a control diet had a beneficial effect on glycemic control, lowering fasting insulin, fasting glucose, and Homeostatic Model Assessment—Insulin Resistance (HOMA-IR) by 4.1%, 0.8%, and 9.2%, respectively. The ingestion of a single serving (28 g) of almonds before a high-starch meal significantly reduced hemoglobin A1c (HbA1c) ($p = 0.045$) and postprandial glycemia ($p = 0.043$) in individuals with uncomplicated T2DM [54]. Another intervention study showed that almonds significantly decreased post-interventional fasting glucose by 5.9% ($p = 0.01$) and HbA1c by 3.0% ($p = 0.04$) as compared with that of control in T2DM subjects [59]. In the study of Hou et al. [60], the T2DM patients who replaced part of their starchy food with almonds or peanuts had decreased values for fasting blood glucose and 2-h postprandial blood glucose ($p < 0.05$) compared with the baseline. In addition, in the almond group a decrease in the HbA1c level from the baseline was found ($p < 0.05$) and none of the nut groups showed an increase of body mass index (BMI).

Several dietary interventions within the frame of the PREDIMED study tried to establish a relation between MDs supplemented with mixed nuts (walnuts, hazelnuts, and almonds) and biochemical or anthropometric parameters. In a clinical trial with high CV risk participants but no CVD at enrolment, data showed that MD supplemented with 30 g per day mixed nuts for 1 year significantly decreased waist circumference (Wc) compared to baseline, and also lowered LDL-C and shifted LDL size to a less atherogenic pattern [55]. Lasa et al. [56] indicated that a daily quantity of 30 g nuts (walnuts, hazelnuts, and almonds) added to MDs was associated with a significant reduction in BW ($p = 0.021$) compared with the control, with improved glucose metabolism in both the nut group and low-fat diet group. Rodríguez-Rejón et al. [58] concluded that the nut-enhanced MD lowered glycemic index and glycemic load, two indices that can be associated with T2DM and CHD.

An RCT conducted on patients with diabetes established that 75 g mixed nuts (tree nuts and peanuts) added to the diet for 3 months, besides improving the lipid profile, also significantly lowered HbA1c compared with a carbohydrate diet ($p = 0.026$) [42]. In another trial conducted on T2DM subjects, the addition of 15 mL walnut oil to the diet for three month significantly decreased HbA1c level by 7.86% ($p = 0.005$) and fasting blood glucose by 8.24% ($p = 0.001$) compared to control [61].

Table 3. Nut consumption and changes in biochemical and anthropometric parameters in intervention studies.

Author, Year, Country [Ref]	Design	Subjects (F:M) Mean Age (Range)	Length of Study	Comparison Group	Intake of Nuts	Findings
Li et al., 2011, Taiwan [31]	RCT, Crossover	20 (11:9) diabetics 58 y	12 wk	Control (diet without A)	60 g/d A	↓ fasting insulin 4.1% (95% CI: 0.9–12.5), $p \le 0.0184$ ↓ fasting glucose 0.8% (95% CI: 0.4–6.3), $p \le 0.0238$ ↓ HOMA-IR 9.2% (95% CI: 4.4–13.2), $p \le 0.0039$ (vs. control)
Cohen & Johnston, 2011, USA [54]	RCT	13 (6:7) diabetics 66 y	12 wk	Control (diet without A)	28 g/d A	↓ HbA1c (−4% vs. +1% for almond group vs. control), $p = 0.045$ ↓ BMI (−4% vs. 0% for almond group vs. control), $p = 0.047$
Damasceno et al., 2013, Spain [55]	RCT	169 (95:74) 67 (55–80) y	1 y	Baseline and Control (low-fat diet *)	MD + 30 g/d nuts (15 g W, 7.5 g H, 7.5 g A)	↓ Wc −5.1cm (95% CI: −6.8 to −3.4) vs. baseline; $p \le 0.006$ ↓VLDL-C −111 nmol/l (95% CI: −180 to −42) vs. baseline
Lasa et al., 2014, Spain [56]	RCT	191 (114:77) diabetics 67 (55–80) y	1 y	Baseline and Control (low-fat diet)	30 g/d nuts (15 g W, 7.5 g H, 7.5 g A)	↓ BW (−0.71 ± 2.41 kg vs. baseline of 75.2 ± 11.5 kg, $p = 0.021$) ↓ Wc (−4.84 ± 7.50 cm vs. baseline of 99.1 ± 8.96 cm, $p = 0.001$ for women)
Hernández-Alonso et al., 2014, Spain [57]	RCT, Crossover	54 (25:29) prediabetics 55 y	9 mo	Control (diet without pistachios)	57 g/d pistachio	↓ fasting glucose −5.17 mg/dL (95% CI: −8.14 to −2.19) vs. baseline; $p < 0.001$ ↓ fasting insulin −2.04 mU/mL (95% CI: −3.17 to −0.92) vs. baseline; $p < 0.001$ ↓HOMA-IR −0.69 (95% CI: −1.07 to −0.31) vs. baseline; $p < 0.001$ ↑ GLP−1 4.09 pg/mL (95% CI: 1.25–6.94) vs. baseline; $p = 0.01$
Rodríguez-Rejón et al., 2014, Spain [58]	RCT	2866 (1781:1085) non-diabetics 67 (55–80) y	1 y	Control (low-fat diet)	MD + 30 g/d nuts (15 g W, 7.5 g H, 7.5 g A)	↓ dietary GL −10.34 (95% CI: −12.69 to −8.00) ↓dietary GI −1.06 (95% CI: −1.51 to −0.62)

Table 3. *Cont.*

Author, Year, Country [Ref]	Design	Subjects (F:M) Mean Age (Range)	Length of Study	Comparison Group	Intake of Nuts	Findings
Chen et al., 2017, China [59]	RCT, Crossover	33 (20:13) diabetics 55 y	12 wk	Control (isocaloric diet no A)	60 g/d A	↓ fasting glucose 129.3 ± 25.6 (fast) vs. 132.8 ± 24.8 (pre) p = 0.011 ↓ HbA1c (%) 7.01 ± 0.62 (fast) vs. 7.18 ± 0.64 (pre) p = 0.043
Hou et al., 2018, China [60]	RCT	25 (10:15) diabetics 70 (40–80) y	3 mo	Compared to baseline	(1) Peanuts (60 g/d men, 50 g/d women) (2) A (55 g/d men, 45 g/d women)	↓ fasting glucose–in (1) and (2) ↓ postprandial 2-h blood glucose–in (1) and (2) (compared to baseline) (p < 0.05)
Jenkins et al., 2018, Canada [42]	RCT, Parallel	117 (39:78) diabetics 62 y	3 mo	Controls (isocaloric (1) carbs diet and (2) carbs and nut diet)	75 g/d mixed nuts (tree nuts and peanuts)	↓ HbA1c −2.0 mmol/mol (95% CI: −3.8 to −0.3), p = 0.026 vs. control (1)
McKay et al., 2018, USA [41]	RCT, Crossover	26 (5:21) 59.7 (57–70) y	12 wk	Control (isocaloric, no pecans)	42.5 g/d pecans	↓ fasting insulin (−2.00 ± 0.83 μIU/mL, p = 0.024) ↓ HOMA-IR (−0.51 ± 0.23, p = 0.037)

* low-fat diet—all types of fat, from both animal and vegetable sources, reduced, but no fat-free foods. A—almonds; BMI—body mass index; BW—body weight; CI—confidence interval; F—women; GI—glycemic index; GL—glycemic load; GLP-1—glucagon-like peptide-1; H—hazelnuts; HbA1c—hemoglobin A1c; HOMA-IR—Homeostasis Model of Assessment for insulin resistance; HR—hazard ratio; M—men; MD—Mediterranean Diet; RCT—randomized controlled trial; VLDL-C—very low density lipoprotein cholesterol; W—walnuts; Wc—waist circumference.

The supplementation of the diet with pistachios for 9 months in prediabetic participants had a significant lowering effect in fasting plasma glucose, fasting plasma insulin, and HOMA-IR ($p < 0.001$ for all), but only borderline significance in decreasing Wc, BW, and BMI [57]. Also, a short 4 week trial with a pecan-rich diet showed significant reductions in serum insulin and HOMA-IR compared to the control diet ($p < 0.05$) [41].

Similar to our findings, two meta-analysis of RCTs pointed out that nut consumption was related to a significant decrease in BW, BMI, and Wc [62,63].

One reason for these outcomes can be the fact that the metabolizable energy content, or energy available to the body, of several nuts is less than predicted by the Atwater factors: 5% less for pistachios [64], 16% less for cashews [65], 21% less for walnuts [66], and as much as 32% less in the case of almonds [67]. Also, despite 10% higher energy intake from peanuts in another trial, a less than expected increase in BW compared to the control was found [68].

The meta-analysis of Viguiliouk et al. [69] established that tree nut intake significantly lowered fasting blood glucose ($p = 0.03$) and HbA1c ($p = 0.0003$) compared with control diets, but no significant effects were detected for fasting insulin and HOMA-IR. The very recent meta-analysis from Tindall et al. [70] observed that consumption of tree nuts or peanuts significantly decreased fasting insulin and HOMA-IR with no significant effect on fasting blood glucose or HbA1c.

Positive associations between nut consumption and cardiometabolic biomarkers were seen in two cross-sectional studies reported in the literature in the last ten years, although the strength of the association is weaker in these types of studies, which look at a single point in time, compared to longitudinal and intervention studies. A cross-sectional study conducted on an elderly population at high CV risk, showed that nut intake decreased both Wc and BMI (p-trend < 0.005 in both) [71]. In the study of Jaceldo-Siegl et al. [72] obesity was 11% less likely in the high peanut/low tree nut group (odds ratio, OR = 0.89; 95% CI: 0.53–1.48), 37% less likely in the high tree nut/high peanut group (OR = 0.63; 95% CI: 0.40–0.99), and 46% less likely in the high tree nut/low peanut group (OR = 0.54; 95% CI: 0.34–0.88) (p-trend = 0.006) compared to low tree nut/low peanut group. Moreover, in a large US population cross-sectional survey, walnut consumers had a significantly lower prevalence of diabetes with lower levels of fasting blood glucose and HbA1c [73].

2.4. Nut Consumption Effect on Endothelial Function and Inflammation Markers

The studies, summarized in Table 4, proved that tree nut-enhanced diets can have beneficial effects on endothelial function or inflammatory markers [74–79]. Endothelial dysfunction, characterized by a decreased bioavailability of nitric oxide (NO), an endogenous vasodilator synthesized from the amino acid L-arginine, is linked with a greater risk of CV events [80]. Since tree nuts are a rich source of L-arginine, their intake might potentially improve endothelial dysfunction.

The trial of Ma et al. [74] on T2DM patients revealed that consumption of an ad libitum diet supplemented with 56 g walnuts daily for a period of 8 weeks significantly improved flow mediated dilation (FMD), a measure of endothelial function, as compared to a nut-free ad libitum diet. In a larger trial on overweight adults with visceral adiposity, the same research team found again that after daily ingestion of walnuts, FMD improved significantly ($p = 0.019$) as compared with the control diet, suggesting the potential for overall cardiac risk reduction [75]. Similarly, in hypercholesterolemic subjects, a diet with walnuts and walnut oil amended FMD (+34%) and reduced total peripheral resistance [81]. The study of Chen et al. [78] in patients with CAD established that almonds can also increase FMD, but did not significantly change C-reactive protein (CRP), tumor necrosis factor-alpha (TNF-α), or E-selectin compared to control. These results align with a meta-analysis published by Neale et al. [82] where nut intake had a significant effect on FMD ($p = 0.0004$). Equally, a recent meta-analysis of RCTs showed that nut consumption significantly increased FMD ($p = 0.001$) [83]. Bhardwaj et al. [84] noticed that walnut and almond diets, besides positively effecting FMD, can also decrease soluble vascular cell adhesion molecules.

Table 4. Nut consumption and endothelial function and inflammatory biomarkers.

Author, Year, Country [Ref]	Design	Subjects (F:M)Mean Age (±SD)	Length of Study	Comparison Group	Intake of Nuts	Findings
Ma et al., 2010, USA [74]	RCT, Crossover	24 (14:10) 58.1 ± 9.2 y	8 wk	Control (diet without W)	56 g/d W	↑ FMD (2.2 ± 1.7 vs. 1.2 ± 1.6%, $p = 0.04$) (vs. control)
Katz et al., 2012, USA [75]	RCT, Crossover	46 (28:18) 57.4 ± 11.9 y	8 wk	Control (diet without nuts)	56 g/d W	↑ FMD 1.1% (95% CI: 0.2–2.0), $p = 0.019$ (vs. control)
Liu et al., 2013, Taiwan [76]	RCT, Crossover	20 (11:9) diabetics 58 y	12 wk	Control (diet without A)	56 g/d A	↓ IL-6 10.3% (95% CI: 5.2–12.6) ↓ TNF-α 15.7 % (95% CI: −0.3 to 29.9) ↓ CRP 10.3% (95% CI: −24.1 to 40.5), $p = 0.0455$ (vs. control)
Sweazea et al., 2014, USA [77]	RCT, Parallel	21 (12:9) 56.2 y	12 wk	Control (diet without A)	43 g/d A	↓ CRP in almond group vs. control (−1.2 vs. +4.33 mg/L, $p = 0.029$)
Chen et al. 2015, USA [78]	RCT Crossover	45 (26:18) 61.8 ± 8.6 y CAD patients	22 wk	Control (diet without A)	85 g/d A	↑ FMD, % (7.7 ± 3.3 vs. 8.3 ± 3.8%) (vs. control)
Yu et al., 2016, USA [79]	Cross-sectional (a)NHS (b)HPFS	(a) 3654 F; 59 y (b) 1359 M; 65 y	4 y		Tree nuts and peanuts (1) Never or almost never (2) <1 time/wk (3) 1 time/wk (4) 2–4 times/wk (5) ≥5 times/wk	↓ CRP–RR 0.90 (0.84–0.97) for (4) vs. (1); RR 0.84 (0.74–0.95) for (5) vs. (1) (p-trend = 0.006) ↓ IL-6–RR 0.88 (0.83–0.94) for (4) vs. (1); RR 0.88 (0.79–0.99) for (5) vs. (1) (p-trend = 0.016)

A—almonds; CI—confidence interval; CAD—coronary artery disease; CRP—C-reactive protein; F—women; FMD—flow-mediated dilation; NHS—Nurses' Health Study; HPFS—Health Professionals Follow-Up Study; IL-6—interleukin-6; M—men; RCT—randomized controlled trial; RR—relative risk; SBP—systolic blood pressure; TNF-α—tumor necrosis factor-α; W—walnuts.

While walnut and almond diets increased FMD, pistachios had no effect on this measure [38]. This fact was confirmed by the meta-analysis of Fogacci et al. [85] which did not notice a significant change in FMD, but the brachial artery diameter (BAD) significantly improved ($p < 0.001$) following pistachio consumption.

A very recent review [86] established that tree nut and peanut intake had the potential to improve vascular function, which was linked with increased risk of CVD.

In regard to inflammatory markers, the results of Liu et al. [76] proposed that the integration of almonds into a healthy diet in diabetic Chinese patients, as compared to the control diet, can decrease interleukin-6 (IL-6), TNF-α, and CRP. Both IL-6 and TNF-α are mediators of CRP synthesis in the liver. An increased IL-6 level was associated with insulin resistance, hyperglycemia, and T2DM [76]. The trial conducted by Sweazea et al. [77] concluded that almond intake, compared with control, significantly lowered CRP ($p = 0.029$) in diabetic U.S. patients, as well. In a recent study, Borkowski et al. [87] showed that walnuts can reduce the TNFα-mediated production of pro-inflammatory cytokines IL-6 and IL-8 in human adipocytes.

These findings are consistent with the results of a cross-sectional analysis which showed that higher tree nut and peanut intake was associated with decreased amounts of CRP and IL-6 [79]. However, the meta-analysis of Mazidi et al. [88] disclosed that nut consumption had no significant effect on CRP, IL6, TNF-α, and adiponectin, but significantly decreased leptin.

A low level of adiponectin, a protein with anti-inflammatory and anti-atherogenic properties secreted by adipocytes, is perceived as risk factor for CV events. An addition of 9 g of walnut oil or mixed plant oil in the diet increased adiponectin levels in T2DM subjects (+6.84% and +4.47%, respectively, compared to baseline) with borderline significance ($p = 0.051$) [89].

The favorable effects on inflammatory markers may be attributed to tree nuts, as a 12-week RCT on healthy mature adults with peanut intake failed to find any changes in inflammatory biomarkers [68].

The conclusion from these cohort and intervention studies is that the consumption of tree nuts and peanuts, with their high content of biologically active antioxidant compounds, should be encouraged for maintaining lower blood lipid counts and better biochemical and anthropometric parameters, in order to improve the prospect of cardiometabolic morbidity and mortality.

3. Association between Nuts and Cancer

Nutrition was demonstrated to have a causal and protective role in the progress of several types of cancer, the second leading cause of death worldwide [90].

A number of studies published in the last ten years have demonstrated the influence of nut consumption on cancer [91–107] (Table 5). Despite the fact that, in the observational studies, no causality could be proven, there were still several obvious strengths: prospective design for the majority of the studies, large population size, high retention rates with long-term follow-up, and adjustments for a multitude of other potential risk factors.

Table 5. Association between nut consumption and cancer.

Author, Year, Country [Ref]	Design	Subjects (F:M) Mean Age (Range)	Length of Intervention	Intake of Nuts	Findings
Hardman et al., 2019, USA [105]	RCT	10 women 55 y	2 to 3 wk	56 g/d walnuts	↓ growth and survival of breast cancer cells in walnut-diet group vs. control (no walnut)
Raimondi et al., 2010, Canada, [91]	Case-control study	394 men 69 y		Total nuts (g/d) (1) 0 (2) 0.1–1.2 (3) 1.3–3.0 (4) >3	↓ prostate cancer risk (4) vs. (1): OR 0.43 (95% CI: 0.22–0.85), p-trend = 0.01
Ibiebele et al., 2012, Australia [92]	Case-control study	2780 women 57 y		n-6 fatty acid (g) from total nuts (1) 0.13 (0.0–0.29) (2) 0.45 (0.29–0.68) (3) 1.48 (0.73–2.59) (4) 3.35 (2.59–25.9)	↓ ovarian cancer risk (4) vs. (1) OR 0.72 (95% CI: 0.57–0.92), p-trend = 0.02
Guasch-Ferré et al., 2013, Spain [16]	Prospective	7216 (4145:3071) high CV risk 67 y	4.8 y	Tree nuts and peanuts (1) none (2) 1–3 servings/wk (3) >3 servings/wk	↓ cancer mortality (3) vs. (1) for walnuts HR 0.46 (0.27–0.79), p-trend = 0.005 ↓ cancer mortality (3) vs. (1) for all nuts HR 0.60 (0.37–0.98), p-trend = 0.064
Bao et al., 2013, USA [93]	Prospective	75,680 women 59 y	30 y	Tree nuts and peanuts (1) never/almost never (2) 1–3 times/mo (3) 1 time/wk (4) ≥2 times/wk	↓ pancreatic cancer risk (p-trend = 0.01) RR 0.71 (95% CI: 0.51–0.99) for (3) vs. (1) RR 0.68 (95% CI: 0.48–0.96) for (4) vs. (1)
van den Brandt and Schouten, 2015, the Netherlands [94]	Prospective	120,852 (62,573:58,279) 61 y	10 y	Tree nuts and peanuts (1) 0 g/d (2) 0.1–5 g/d (3) 5–10 g/d (4) 10+ g/d	↓ cancer mortality (p-trend = 0.002) HR 0.82 (95% CI: 0.68–0.98) for (3) vs. (1) HR 0.79 (95% CI: 0.67–0.93) for (4) vs. (1)
Yang et al., 2016, USA [95]	Prospective	75,680 women 59 y	30 y	Tree nuts and peanuts (1) never/almost never (2) 1–3 times/mo (3) once/wk (4) ≥2 times/wk	↓ colorectal cancer risk (p-trend = 0.06), RR 0.87 (95% CI: 0.72–1.05) for (4) vs. (1)
Wang et al., 2016, USA [96]	Prospective	47,299 men 65 y	26 y	Tree nuts and peanuts (1) Never or almost never (2) <1 time/wk (3) 1 time/wk (4) 2–4 times/wk (5) ≥5 times/wk	↓ overall mortality after being diagnosed with non-metastatic prostate cancer (5) vs. (1) HR 0.66 (95% CI: 0.52–0.83), p-trend = 0.0005

Table 5. *Cont.*

Author, Year, Country [Ref]	Design	Subjects (F:M) Mean Age (Range)	Length of Intervention	Intake of Nuts	Findings
Lee et al., 2017, Italy/USA [97]	EAGLE case-control study; NIH-AARP Diet and Health cohort study	3639—65 y 495,785—62 y	16 y	Tree nuts and peanuts 10 categories, ranging from "never" to ≥2 times per day; 3 categories for portion size	↓ lung cancer risk (highest vs. lowest nut intake) OR_{EAGLE} 0.74 (95% CI: 0.57–0.95), *p*-trend = 0.017 HR_{AARP} 0.86 (95% CI: 0.81–0.91), *p*-trend < 0.001
Hashemian et al., 2017, USA [98]	Prospective	566,407 (59.6% men) 63 (50–71) y	15.5 y	Tree nuts, peanuts, peanut butter Expressed in g/1000 kcal: (C0) 0 (C1) 0.11 (0.05, 0.16) (C2) 0.51 (0.36, 0.68) (C3) 2.20 (1.35, 4.12)	↓ gastric noncardia adenocarcinoma risk (C3) vs. (C0): HR 0.73 (95% CI: 0.57–0.94, *p*-trend 0.004) for tree nuts and peanuts HR 0.75 (95% CI: 0.60–0.94, *p*-trend 0.02) for peanut butter
Nieuwenhuis and van den Brandt 2018, the Netherlands [99]	Prospective	120,852 (62,573:58,279) 62 (55–69) y	20.3 y	Tree nuts and peanuts: (1) non-consumers (2) 0.1–5 g/d (3) 5–10 g/d (4) >10 g/d	↓ pancreatic cancer risk in men—(3), (4) vs. (1) HR 0.53 (95% CI: 0.28–1.00), *p*-trend = 0.047
Nieuwenhuis and van den Brandt 2018, the Netherlands [100]	Prospective	120,852 (62,573:58,279) 62 (55–69) y	20.3 y	Tree nuts and peanuts: (1) non-consumers (2) 0.1–5 g/d (3) 5–10 g/d (4) >10 g/d	↓ esophageal squamous cell carcinoma risk(4) vs. (1) HR 0.54 (95% CI: 0.30–0.96), *p*-trend = 0.050
Fadelu et al., 2018, USA [101]	Prospective	826 patients with stage III colon cancer	6.5 y	Tree nuts and peanuts (1) none (2) ≥2 servings/wk	↑ disease-free survival (2) vs. (1) HR 0.58 (95% CI: 0.37–0.92), *p*-trend = 0.03 ↑ overall survival (2) vs. (1) HR 0.43 (95% CI: 0.25–0.74), *p*-trend = 0.01 ↓ cancer recurrence and mortality
van den Brandt and Nieuwenhuis 2018, the Netherlands [102]	Prospective	62,573 women 61 (55–69) y	20.3 y	Tree nuts and peanuts: (1) non-consumers (2) 0.1–5 g/d (3) 5–10 g/d (4) >10 g/d	↓ (ER -) breast cancer risk (4) vs. (1) HR 0.55 (95% CI: 0.33–0.93), *p*-trend = 0.025 ↓ ER–PR breast cancer risk (4) vs. (1) HR 0.53 (95% CI: 0.29–0.99), *p*-trend = 0.037
Zhao et al., 2018, China [103]	Case-control study	444 (152:292) 59 (40–69) y		Peanuts: (1) <1/mo (2) 1–3 times/mo (3) 1–3 times /wk (4) 4–6 times/wk	↓ esophageal squamous cell carcinoma risk(4) vs. (1) OR 0.31 (95% CI: 0.16–0.59), *p*-trend < 0.001

Table 5. *Cont.*

Author, Year, Country [Ref]	Design	Subjects (F:M) Mean Age (Range)	Length of Intervention	Intake of Nuts	Findings
Lee et al., 2018, Korea [104]	Case-control study	2769 (894:1875) 57 (48–66) y		Tree nuts and peanuts (1) None (2) <1 serving (15g)/wk (3) 1–3 servings/wk (4) ≥3 servings/wk	↓ colorectal cancer risk (F,M) (4) vs. (1) OR 0.30 (95% CI: 0.20–0.45), p-trend < 0.001 ↓ distal colon cancer risk (4) vs. (1) OR 0.13 (95% CI: 0.04–0.48), p < 0.001 for F OR 0.39 (95% CI: 0.19–0.80), p = 0.004 for M ↓ rectal cancer risk (4) vs. (1) OR 0.40 (95% CI: 0.17–0.95), p = 0.006 for F OR 0.23 (95% CI: 0.12–0.46), p < 0.001 for M
Sui et al., 2019, USA [106]	Prospective, NHS and HPFS	88,783 women 51,492 men 63 y	27.9 y	Tree nuts, servings/wk (1) 0.01 (2) 0.23 (3) 1.25	↓ hepatocellular carcinoma (3) vs. (1) HR 0.64 (95% CI: 0.43–0.95), p-trend = 0.07
Nieuwenhuis and van den Brandt 2019, the Netherlands [107]	Prospective	120,852 (62,573:58,279) 62 (55–69) y	20.3 y	Tree nuts and peanuts: (1) non-consumers (2) 0.1–5 g/d (3) 5–10 g/d (4) >10 g/d	↓ small cell carcinoma (lung cancer subtype) in men—(4) vs. (1) HR 0.62 (95% CI: 0.43–0.89), p-trend = 0.024 ↓ lung cancer risk in men (non-significantly)

AARP—American Association of Retired Persons; CI—confidence interval; CV—cardiovascular; EAGLE—the Environment And Genetics in Lung cancer Etiology; (ER -)—estrogen receptor negative; F—women; HPFS—Health Professionals Follow-up Study; HR—hazard ratio; M—men; n-6—omega 6; NHS—Nurses' Health Study; NIH—National Institutes of Health; OR—odds ratio; PR—progesterone receptor; RCT—randomized controlled trial; RR—relative risk.

Several studies indicated that patients in the highest tree nut and peanut-intake group compared to the lowest intake group at any point in time during the study period were: 40%, 25%, and 14% less likely to die from total cancer, gastric noncardia adenocarcinoma, and lung cancer, respectively [16,97,98]. Also, they were 46%, 45%, and 47% less likely to die from esophageal squamous cell carcinoma, estrogen receptor negative breast cancer, and estrogen-progesterone receptor breast cancer, respectively [100,102]. Subjects having at least two servings of nuts per week had 0.68 times the risk of pancreatic cancer compared with subjects having nuts never or almost never [93]. The oil extracted from walnuts exhibited in vitro ability to reduce the viability of esophageal cancer cells, induced necrosis and cell cycle arrest, and displayed anticarcinogenic effect, thus it may present favorable effects in esophageal cancer in humans [108].

A recent prospective study showed that nut intake was not strongly associated with hepatocellular carcinoma but a significant inverse association with tree nut consumption was noted [106]. Higher intake of tree nuts and peanuts, but not peanut butter, was associated with a significantly reduced risk for small cell carcinoma (lung cancer subtype), after adjusting for smoking frequency and duration, and a non-significant decrease in lung cancer risk for men, results which have not been replicated in women [107].

The following case control studies also reported inverse associations between nut consumption and different types of cancer. In the highest intake group the outcomes were 57% and 28% less likely for prostate cancer and ovarian cancer, respectively [91,92]. Similarly, peanut consumption was linked with a lower risk of esophageal squamous cell carcinoma [103]. Lee et al. [104] noticed that odd ratios were 70% less likely for colorectal cancer in women and men, 60% (in women) and 77% (in men) less likely for rectal cancer, and 87% (in women) and 61% (in men) less likely for distal colon cancer, for the highest nut-intake group. Also, the results from a very recently published study suggest that, particularly among women, moderate to high nut intakes (2 to 5.5 servings/week) may be associated with a lower risk of colorectal adenomas, the precursor to most colorectal cancers [109]. Yang et al. [95] showed that the colorectal cancer risk for women who consumed tree nuts and peanuts $\geq$2 times per week was 13% lower compared to non-consumers, with a borderline statistical significance (p-trend = 0.06). Fadelu et al. [101] proved that higher total nut intake was linked with a significantly reduced incidence of cancer relapse and death in subjects with stage III colon cancer, but the benefit was limited to tree nut consumption.

These results confirm those of Casari and Falasca [110], who linked nut intake with a positive effect against cancer, and Aune et al. [22], who noticed a 15% decreased cancer risk in subjects eating 28 g of nuts daily compared to subjects who did not have nuts.

A new clinical trial showed that walnuts could alter tumor gene expression in women with confirmed breast cancer in ways expected to decrease cancer growth, delay proliferation, reduce metastasis, and increase cancer cell death [105]. Toledo et al. [111] indicated also that walnut-enriched diets could modulate breast cancer growth. These results are in accordance with those reported by Soriano-Hernandez et al. [112] where, in the group that consumed higher amounts of walnuts, almonds and peanuts, the breast cancer risk was between 55 and 67% less likely.

Higher nut consumption might influence cancer risk through its association with lower circulating levels of total cholesterol, CRP, IL-6, and insulin. Different compounds, such as flavonoids from nuts or their by-products might lower the risk of head and neck cancer [113] and could present chemopreventive properties and anticancer potential [114–116]. These bioactive antioxidant molecules can act alone or most likely in synergism to regulate the inflammatory response and immunological activity, prevent the development of prostaglandins or pro-inflammatory cytokines, and potentially inhibit cancer risk.

4. Association between Nuts and Cognitive Disorders

Inflammation-associated chronic pathologies, such as dementia, Parkinson's disease (PD), or Alzheimer's disease (AD), lead to one of the most unfavorable health problems in the elderly:

age-related cognitive deterioration, a condition which may be prevented or delayed by modifiable lifestyle factors, including antioxidant diets [117].

Quite a few studies have examined the association between diets supplemented with nuts and cognitive performance [118–124] (Table 6).

In an elderly population, consumption of walnuts was related to better cognitive performance, mainly working memory, although the causality could not be inferred [120]. These results were consistent with another cross-sectional study that indicated a positive association between nut consumption and cognitive function in mature Chinese adults. Patients with mild cognition impairment symptoms had less tree nuts and peanuts in their diet compared to healthy subjects ($p = 0.031$) [125]. Similarly, positive relations between cognitive functions and nut intake were shown in the US population. Significant improvements in almost all cognitive test scores were noted among older adults who added walnuts in their diet [126].

The scores from two neuropsychological tests, the Mini-Mental State Examination (an indicator of cognitive impairment) and the Clock Drawing Test (a neuropsychological test which evaluates cognitive decline and dementia), were higher for subjects allocated to the nut-enhanced MD compared to the low-fat, nut-free diet group [121]. Comparable results were obtained by Valls-Pedret et al. [123]; in an older population, MD supplemented with tree nuts (walnuts, almonds, hazelnuts) was associated with improved cognitive functions. Also, a high consumption of tree nuts and peanuts was linked to better cognitive function at baseline and might reduce cognitive decline in mature adults [119]. Equally, O'Brien et al. [122] suggested that long-term tree nut and peanut intake was related to overall level of cognition but had no effect on cognitive decline.

A recent RCT demonstrated that peanut consumption could improve vascular and cognitive functions in overweight middle-aged subjects. Small artery elasticity, cerebrovascular reactivity, as well as measures of verbal fluency, processing speed, and short-term memory were all greater after higher intake of roasted, unsalted peanuts with skin [124]. Also, the addition of walnuts (15% of energy) to an ad libitum diet confirmed that regular nut consumption can delay the onset of age-related neurodegenerative disorders. Compared with the control, individuals in the walnut group reported significantly lower intake of animal protein, total carbohydrates, saturated fatty acids, and sodium, but significantly higher ingestion of vegetable protein, antioxidant *n*-3 and *n*-6 PUFAs [127].

Brain-derived neurotrophic factor (BDNF), a protein belonging to the neurotrophic family, controls axonal elongation, neurotransmitter release, growth, differentiation, and survival of presynaptic structure. While low plasma levels of BDNF could lead to the atrophy of specific brain areas in mammals such as the hippocampus and frontal cortex. Higher concentrations of BDNF provided by enhanced-nut MD were likely to prevent depression, memory loss, and cognitive decline [118]. It seems that *n*-3 PUFA, with its powerful antioxidant potential, is responsible for the increased levels of the BDNF signaling factor [128]. Blondeau et al. [129] noticed that alpha-linolenic acid (ALA), the plant-based *n*-3 PUFA, may increase BDNF, thus walnut intake can have a role in neuroprotection, neuroplasticity, and vasodilation of brain arteries.

Major depressive disorder (MDD) is a chronic disease where healthy dietary practices in combination with current treatments may prevent or delay its evolution. Increased consumption of nuts, seeds, vegetables, fruits, and legumes, with proven antioxidant and anti-inflammatory capacities, is the principal nutritional recommendation, with a key reminder that the beneficial effects are possible to come from wholesome nutritious diets rather than from individual nutrients [130]. Ali-Sisto et al. [131] showed that MDD is characterized by a decreased arginine level, an amino acid found in nuts and a precursor of NO, which is needed to modulate neuronal and vasodilation functions, to prevent oxidation of LDL-C, aggregation of platelets, or vascular inflammation, and inhibit oxidation in the central nervous system (CNS). MDD is associated with increased CV events, and the biological mechanism connecting MDD and CVD is apparently a chronic inflammation induced by a low level or bioavailability of arginine [131].

Table 6. Association between nut consumption and cognitive disorders.

Author, Year, Country [Ref]	Design	Subjects (F:M) Mean Age (Range)	Length of Intervention	Comparison Group	Intake of Nuts	Findings
Sánchez-Villegas et al., 2011, Spain [118]	RCT	152 (76:76) 68 y	3 y	Control (low-fat diet *)	MD + 30 g/d nuts (15 g W + 15 g A)	↓ risk for low plasma BDNF values OR 0.22 (95% CI: 0.05–0.90, $p = 0.04$) vs. control
Martínez-Lapiscina et al., 2013, Spain [121]	RCT, multicenter	522 (289:233) 67 y	6.5 y	Control (low-fat diet *)	MD + 30 g/d nuts (15 g W, 7.5 g H, 7.5 g A)	↑ cognition ↑ MMSE 0.57 (95% CI: 0.11–1.03, $p = 0.015$) vs. control ↑ CDT 0.33 (95% CI: 0.003–0.67, $p = 0.048$) vs. control
Valls-Pedret et al., 2015, Spain [123]	RCT	334 (170:164) 67 (55–80) y	4.1 y	Control (low-fat diet *)	MD + 30 g/d nuts (15 g W, 7.5 g H, 7.5 g A)	↓ age-related cognitive decline ↑ memory composite 0.09 (95% CI: −0.05 to 0.23, $p = 0.04$) vs. control ↑ frontal cognition composite 0.03 (95% CI: −0.25 to 0.31, $p = 0.03$) vs. control
Barbour et al., 2017, Australia [124]	RCT, Crossover	61 (32:29) 65 y	12 wk	Control (nut free diet)	56–84 g peanut/d	↑ cognitive functions (short-term memory, verbal fluency, processing speed) vs. control
Nooyens et al., 2011, the Netherlands [119]	Prospective	2613 (1325:1288) 55 (43–70) y	Ongoing since 1995		Tree nuts and peanuts 5 quintiles of nut consumption	↑ cognitive function at baseline ↓ cognitive decline: memory (highest vs. lowest nut intake, $p = 0.03$); global cognitive function (highest vs. lowest nut intake, $p = 0.02$)
Valls-Pedret et al., 2012, Spain [120]	Cross-sectional	447 (233:214) 67 (55–80) y			30 g W/d	↑ cognitive function (working memory, $p = 0.039$)
O'Brien et al., 2014, USA [122]	Prospective	15,467 women 74 y	6 y		Tree nuts and peanuts (1) never, <1/mo (2) 1–3/mo (3) 1/wk (4) 2–4/wk (5) 5/wk	↑ cognitive performance ↑ cognition (4), (5) vs. (1)

* low-fat diet—all types of fat, from both animal and vegetable sources, reduced, but no fat-free foods. A—almonds; BDNF—brain-derived neurotrophic factor; CDT—Clock Drawing Test; CI—confidence interval; F—women; H—hazelnuts; M—men; MD—Mediterranean diet; MMSE—Mini-Mental State Examination; OR—odds ratio; RCT—randomized controlled trial; W—walnuts.

A study on Chinese adults demonstrated that, even after adjusting for potential confounders, the frequency of tree nut and peanut intake (more than 4 times per week) might be associated with a lower incidence of depressive symptoms [132]. Also, Arab et al. [133] reported that U.S. nut consumers, especially walnut eaters, had significantly lower depression scores as compared to no-nut consumers, and the difference was strongest among women.

Optimal dietary choices, such as increasing bioactive antioxidant compound intake through higher nut, fruit, and vegetable consumption, can improve endothelial function, decrease inflammatory biomarkers, protect neuronal and cell-signaling function, increase cognitive performance, and prevent or delay the onset of cognitive dysfunction during aging [134]. In anxiety-based psychopathology, replacing pro-inflammatory saturated fats with anti-inflammatory walnut oil might result in faster, more profound elimination of fear-based learning [135].

5. Other Possible Beneficial Association of Nuts

As the population of the world is getting older, a global priority for the aging population should be the maintenance of independence and freedom of movement. One of the hallmarks of aging and a major public health concern is the loss of mobility due to sarcopenia and dynapenia, or progressive loss of skeletal muscle mass and muscle strength, respectively. Another gerontological condition, described by physical inactivity, slow walking speed, fatigue, and weakness, is frailty [136].

In the elderly, both conditions, sarcopenia and frailty, are characterized by increased levels of pro-inflammatory cytokines, including TNF-α, IL-6, and CRP. Therefore, knowing its antioxidant and anti-inflammatory potential, dietary intake of food rich in monounsaturated fatty acids (MUFAs) and PUFAs is of particular interest for this age group [137]. One recent study mentioned that quantities of 2 to 5 g per day of marine *n*-3 PUFA, corresponding to approximately 20 to 50 g walnuts, is shown to reduce muscle wasting and augment the intracellular anabolic signaling, thus having beneficial effects for the prevention and treatment of sarcopenia in mature adults [138]. Also in this age group, malnutrition and sarcopenia frequently overlap. In order to overcome the loss of lean mass and meet the increased energy requirements, the recommended protein intake is higher (1.2–1.4 g/kg/day) than that of healthy adults [139]. Because the protein level is between 15 to 21% in most tree nuts and around 26% in peanuts, nuts should be included in healthy diets as plant food sources of protein [140].

Nutrition is a factor that could influence osteoarthritis (OA), the most dominant form of arthritis with limited treatment, mainly through symptom management [141]. As food impacts systemic lipid levels, high consumption of saturated fat is linked with higher levels of pro-inflammatory fatty acids, while diets rich in less-inflammatory MUFAs and PUFAs, lipids also found in tree nuts and peanuts (Table 7) may reduce cartilage degradation and OA progression [142].

Table 7. Average fat composition of nuts [3,139].

Mean Value (g/100 g)	Almond	Brazil Nut	Cashew	Hazelnut	Macadamia	Pecan	Pine Nuts	Pistachio	Walnut	Peanut
Total fat	49.9	67.1	43.8	60.7	75.8	72.0	68.4	45.3	65.2	49.2
SFA	3.8	16.1	7.8	4.5	12.1	6.2	4.9	5.9	6.1	6.8
MUFA	31.6	23.9	23.8	45.7	58.9	40.8	18.8	23.3	8.9	24.4
PUFA	12.3	24.4	7.8	7.9	1.5	21.6	34.1	14.4	47.2	15.6
(MUFA + PUFA)/SFA	11.6	3.0	4.1	11.9	5.0	10.1	8.8	6.4	9.2	5.9

MUFA—monounsaturated fatty acids; PUFA—polyunsaturated fatty acids; SFA—saturated fatty acids.

In a group of postmenopausal women, MD enhanced with up to 20 g/day nuts was significantly associated with bone mineral density ($p = 0.045$), indicating that nuts may be beneficial in osteoporosis prevention [143].

Two biomarkers of aging, advanced glycation end products (AGEs) and length of telomeres, can be influenced by nut diets. AGEs, a complex class of compounds, can be formed in vivo, as part of normal metabolism, or acquired exogenously from unhealthy diets or environment. They can accumulate in tissues during aging. Data suggests that AGEs can have damaging effects in CVD, neurodegenerative diseases, T2DM, cancer, or sarcopenia [144–146]. It was shown that nut-enhanced MD may constitute a good instrument towards the inhibition of AGEs formation and absorption [145]. Telomeres, protecting DNA sequences at the end of the chromosomes, present a defensive mechanism against risk factors for a number of age-related diseases, including osteoporosis, CVD, CHD, T2DM, pulmonary fibrosis, AD, and cancer [147–149]. Several studies demonstrated that reactive oxygen species (ROS) can accelerate telomere attrition, induce DNA damage response, and senescence [150]. Sirtuins or the action of telomerase can counter telomere shortening. Among the environmental factors lessening telomere attrition are polyphenols, *n*-3 PUFA, or vitamin E, active antioxidant molecules [150]. A recent study linked regular nut consumption with telomere length and a significant reduction in cellular aging and biologic senescence [151]. Fiber, another valuable compound in nuts, can mediate longer telomeres and reduce biologic aging [152]. A two year trial conducted by Freitas-Simoes et al. [153] in an elderly population confirmed that supplying the diet with walnuts at 15% of daily energy (30 to 60 g/day) postponed leukocyte telomere attrition, potentially influencing the aging process.

Together with physical exercises, long-term daily intake of tree nuts and/or peanuts may contribute to maintaining the health of the skeletal system, muscle mass and strength, as well as the well-being of middle-aged and elderly population.

6. Phytochemicals and Mechanisms Responsible for the Beneficial Activity

Biologically active antioxidant compounds found in nuts can modulate essential physiological processes inside human bodies and influence key mechanisms of actions involved in aging and age-associated diseases [154].

Nut antioxidant polyphenols, the majority of which are found in the pellicle of nuts, can have anti-carcinogenic potential. They retard the initiation, differentiation, and proliferation of cancer cells, modulate signaling pathways related to cell survival, attenuate the growth of tumors, diminish angiogenesis and metastasis, and stimulate the expression of detoxification enzymes and antioxidants [155,156].

Some polyphenols are found in significant amounts in certain types of nuts, giving them specific biological actions. Thus, ellagic acid (EA), physiologically hydrolyzed from ellagitannins (ETs) abundant in walnuts, found also in pecans and pine nuts, could reduce adipocyte expansion and might be beneficial in the management of obesity and the metabolic complications related to obesity [157]. Another example is anacardic acid, a strong antioxidant polyphenol contained in cashew nut shells, which was shown to have anticancer potential, inhibited prostate tumor angiogenesis, cell proliferation, and prompted apoptosis [158–160].

Other polyphenols, found only in very small amounts in nuts, can contribute to beneficial health effects through their hormetic and/or synergistic actions with other polyphenols. In pistachios, the small amounts of genistein, (-) epigallocatechin-3-gallate (EGCG) or resveratrol can act synergistically through common or complementary action pathways with proven antioxidant and anti-aging activity. Thus, the flavonoid (isoflavone) genistein has demonstrated antioxidant, chemopreventive, and chemotherapeutic effects [161]. Growing evidence suggests that EGCG, also present in pecans and hazelnuts, can contribute to the anti-cancer potential [162]. It has an inhibitory proliferation effect on human pancreatic cancer cells [163]. In oral cancer, EGCG exerted an apoptotic therapeutic role, controlling cancer cell proliferation, and in breast cancer showed an anti-angiogenic effect [164]. Resveratrol, another antioxidant phytochemical also found in peanuts besides pistachios, proved to have neurogenesis activity and cancer chemoprevention potential [165,166]. Pterostilbene (PTS), a natural dimethylated analog of resveratrol, had the capability to significantly inhibit secretion of TNF-α and alter the cytokine production in IGROV-1 ovarian cancer cell line [167].

Melatonin, found in walnuts in quantities of 3.5 $\pm$ 1.0 ng/g, could be protective against CV damage and cancer initiation and propagation [168].

Selenium, a trace element supplied mostly by Brazil nuts, is associated with reduced risks for prostate cancer [169] and hepatocellular carcinoma [170].

Nuts are characterized as fatty foods, with total lipid content ranging from 46% in cashew to 76% in macadamia (Table 7). However, the healthy lipid profile of tree nuts and peanuts, mostly MUFAs and PUFAs and low or very low amounts of saturated fatty acids, is a key mechanism of action in the prevention of several age-related diseases [154,171].

Lipophilic bioactive compounds found in nuts can also influence the aging process. Among those compounds, phytosterols can reduce CV risk [172]. Increasing evidence recommends phytosterols for lowering LDL-C [173–175]. Phytosterols, being more hydrophobic than cholesterol, can dislocate cholesterol from intestinal micelles and reduce LDL-C absorption [176]. In combination with *n*-3 PUFAs, phytosterols show both complementary and synergistic lipid-lowering effects in hyperlipidemic mature adults [177].

Lipophilic isomers of vitamin E (tocopherols and tocotrienols), via their antioxidant properties, might inhibit the propagation of free radical damage in biological membranes and enhance immune functions [91]. Oxidation stress and inflammation, processes involved in the decline of cognitive function and neural capacity of the aging brain, can be reduced by tocols through their antioxidant and anti-inflammatory properties [178,179]. It was suggested that dietary intake of tocotrienols could be sufficient to support neuroprotection [180,181].

The lipophilic antioxidant phytochemicals, even in minute amounts, showed increased bioavailability and bioaccessibility, with their intestinal absorption being favored by the presence of lipids in tree nuts and peanuts. Lutein, the most abundant antioxidant carotenoid in the human retina and brain, can be found in pistachios [182]. Age-related macular degeneration (AMD), the primary cause of blindness and vision impairment in old age, can be amended or prevented by lutein [183]. Data indicated the significant impact macular pigment density, a biomarker of brain lutein, might also have on the brain health and cognition in the elderly by improving neurobiological efficiency, neural structure and efficacy, visual perception, and decision-making [184,185]. Recent scientific evidence showed that lutein could stop neuroinflammation, a pathological condition of many neurodegenerative disorders, diminish lipid peroxidation, and, by down-regulation of the NF-κB (the nuclear factor kappa-light-chain-enhancer of activated B cells) pathway, decrease the release of pro-inflammatory cytokines [186]. Compared to other sources, the amount of lutein found in pistachios is low, but as mentioned before, the intestinal absorption of lutein is enhanced by the presence of fatty acids in the tree nut.

Two mechanisms of actions, increased cholesterol efflux and improved endothelial function, favorably affected by whole walnuts and walnut oil, may answer in part the CV benefits of walnut consumption [187]. The favorable effect walnuts have on endothelial function could be credited to ALA, oxylipins (PUFA metabolites with a protective role in CVD and aging), polyphenols, L-arginine, and magnesium [176]. Walnut kernels provide ~9% ALA, while walnut oil provides ~10% ALA [81].

Similarly, ALA might be the factor for the decreased number of atherogenic small and dense LDL-C particles and increased number of large HDL-C particles noticed after walnut intake, as well as the reduction of detrimental lipid classes, such as ceramides and sphingomyelins, associated with CVD risk [188].

The synergistic influence exerted by MUFAs and PUFAs, antioxidant polyphenols and lipophilic compounds, and fiber in the modulation of specific miRNAs, have resulted in the improvement of insulin sensitivity via the PI3K-AKT signaling pathway in pre-diabetic and diabetic population [189].

In pathological conditions, such as AD, there is a diminished expression of glucose transporters, which apparently contributes to a reduced utilization of glucose in cognition-critical brain areas. However, transport and metabolism of ketone bodies (KBs), metabolites produced by the liver as alternative energy sources, are not affected in AD [190]. For that reason, periods of ketogenic diets (KDs) can possibly be effective preventive or treatment measures for neurological disorders [191]. We argue that nuts, due to their phytochemical profile (fat content between 49 and 75%, low amounts of carbohydrates, and high content of ketogenic amino acids including leucine) and strong antioxidant potential, can be part of KDs. Important actions of KDs are related to decrease oxidative stress and inflammatory activity and improve mitochondrial function [192].

As seen in this review, higher nut intake by mature adults was associated with a reduced risk of diabetes, CHD, CVD, several types of cancer, and cognitive disorders. For most of these outcomes, there were indications of nonlinear associations between tree nut and peanut consumption and decreased risk noticed up to an intake of around 15–20 g per day, or 4–5 servings per week, with no further decrease with higher intakes. One study revealed that walnuts had a beneficial effect against diabetes at about 5 grams per day, a little more than one serving per week, with again no additional results for greater intake amount. The intake of both tree nuts and peanuts was linked with reduced risks of diabetes, CHD, CVD, and cancer, as well as increased cognitive function and performance. Although, only tree nuts showed increase in flow-mediated dilation and decrease in LDL-C or certain biochemical and anthropometric parameters (fasting insulin, HOMA-IR, HbA1c, BW, Wc). The results of our study are in line with other analyses that have investigated the relationship between nut intake and chronic diseases [22,94].

7. The Association between Nut Intake and Gastrointestinal Microbiota

The relationship between gut microbiota (GM), diet, and healthy aging has been established in many studies [193–195]. Nutrition is a vital instrument in keeping a friendly microbiome, and this is more important in aging, when increased usage of medication can reduce healthy GM diversity and stability [196]. GM can impact CNS function, via gut-brain axis, and regulate the immune system [197,198]. Also, GM can be involved in several brain disorders (autism, PD, schizophrenia) [199]. Patients having PD, a high incidence neurodegenerative disorder among those over 60 years old, revealed pro-inflammatory bacteria in their gastrointestinal tract. Pathological by-products of these microorganisms could leak from the intestinal lumen in the enteric nervous system and aggregate into insoluble fibrils in the CNS [200,201].

Nut polyphenols were reported to increase the abundance of *Bifidobacterium* and *Lactobacillus* bacteria, probiotic strains related to significant lowering of CRP concentrations and increase in plasma HDL-C, cancer prevention, immune-modulation, as well as reductions of pathogenic *Clostridium* species and enteropathogens *Salmonella typhimurium* or *Staphylococcus aureus* [202,203]. Walnut ingestion increased the abundance of *Lactobacillus* [204], while decreasing microbial derived, proinflammatory LDL-C and secondary bile acids in healthy mature adults [205]. Similar results were achieved in a very recent 8-week long RCT, including 194 healthy individuals (mean age 63 years), where after 43 g/day walnut-enriched diet, the probiotic and butyric acid-producing species (*Ruminococcaceae* and *Bifidobacteria*) significantly increased ($p < 0.02$), while *Clostridium* species significantly decreased ($p < 0.05$) [206]. Also, pistachio and almond consumption via the prebiotic compounds may stimulate the growth of beneficial butyrate-producing bacteria and inhibit the development of pathogenic ones [207]. Holscher et al. [208] demonstrated that daily consumption of around 42 g almonds for at least 3 weeks can increase the abundance of *Roseburia* species, a favorable genus known to be negatively affected by age.

The favorable effect of walnut diets on BP may be linked to changes in the GM. As walnuts are not completely metabolized in the upper gastrointestinal tract, they provide substrate to the gut microbiome and may stimulate the production of short-chain fatty acids, including butyrate, which have been associated with normal BP management [209].

Human diet influences the relative abundance of bacterial communities present in the gastrointestinal tract [210]. A significant diversity and number of bacteria ensure a greater ability to resist metabolic changes and infections and constitute the prerequisite for a healthy status of the gut [202]. Consequently, a nut enhanced diet characterized by high antioxidant and anti-inflammatory activities can delay age-related microbiota changes and positively alter the microbial composition of the human GM with benefits for health.

8. Conclusions

Given that population aging has become a global trend, it is necessary to carefully evaluate age-associated diseases and identify strategies for promoting healthy lifestyle leading to healthy aging. Main goals should be the preservation of physical and cognitive functions, the maintenance of high standards of life quality and independence. Clearly, there is a need to design personalized recommendations to prevent, manage, or treat pathological conditions prevalent in the elderly.

As demonstrated in the present review, nuts, via their numerous biological active compounds (proteins, MUFAs and PUFAs, vitamins, minerals, fiber, polyphenols, phytosterols), have antioxidant and anti-inflammatory properties and might ensure cardioprotective benefits, safeguard against metabolic conditions, lower carcinogenic risk, help in cognitive disorders, or aid in sarcopenia and frailty. Just one bioactive compound cannot explain all these health benefits. It seems that antioxidant phytochemicals act synergistically to decrease the age-related oxidative stress and inflammation. Nuts, as complete functional foods, may positively adjust aging processes and play key roles in the relationship between lifespan and health span. Recent data favor the inclusion of robust antioxidant nuts in healthy diets of middle-aged and elderly population, a category steadily growing worldwide.

The scientific findings of our review stress the beneficial effects tree nut and peanut consumption can have in lowering risk factors related to several age-related diseases and highlight the importance of including nuts in healthy dietary plans. Moreover, the study has the potential to advance the perception of nuts as strong antioxidants between nutritionists, nurses, physicians, or general public and could be helpful in public health and health policy decisions.

Author Contributions: M.E.R., D.-S.P. wrote the first draft of the manuscript, which was further edited by all authors. M.E.R., A.M., I.C.F.R.F., D.-S.P. read, critically revised, and approved the final submitted version of the manuscript.

Funding: This research was funded in part by "Iuliu Hatieganu" University of Medicine and Pharmacy in Cluj-Napoca, Romania through a Ph.D. grant (PCD No. 1529/60/18.01.2019 to M.E.R.).

Acknowledgments: We would like to thank George Rusu (Southern Connecticut University, New Haven, CT, USA) for the language review.

Conflicts of Interest: The authors declare no conflict of interest.

Abbreviations

AD	Alzheimer's disease
AGEs	advanced glycation end products
AMD	age-related macular degeneration
apoB	apolipoprotein B
BDNF	brain-derived neurotrophic factor
BMI	body mass index
BP	blood pressure
BW	body weight
CAD	coronary artery disease
CHD	coronary heart disease
CI	confidence interval
CNS	central nervous system
CRP	C-reactive protein
CVD	cardiovascular diseases
EA	ellagic acid
EGCG	(−)-epigallocatechin-3-gallate
ER	estrogen receptor
ETs	ellagitannins
FMD	flow-mediated dilation
GLUTs	glucose transporters
GM	gut microbiota

HbA1c	hemoglobin A1c
HDL-C	high density lipoprotein-cholesterol
HOMA-IR	Homeostatic Model Assessment—Insulin Resistance
HR	hazard ratio
IHD	ischemic heart disease
IL-6	interleukin 6
KBs	ketone bodies
KDs	ketogenic diets
LDL-C	low density lipoprotein-cholesterol
MDD	major depressive disorder
MD	Mediterranean diet
MS	metabolic syndrome
MUFAs	monounsaturated fatty acids
NO	nitric oxide
OR	odds ratio
PD	Parkinson's disease
PTS	pterostilbene
PUFAs	polyunsaturated fatty acids
RCT	randomized controlled trial
ROS	reactive oxygen species
RR	relative risk
T-C	total cholesterol
T2DM	type 2 diabetes mellitus
TG	triglycerides
TNF-α	tumor necrosis factor-α
VLDL-C	very-low-density lipoprotein cholesterol
Wc	waist circumference

References

1. De Almeida, A.; Ribeiro, T.; de Medeiros, I. Aging: Molecular Pathways and Implications on the Cardiovascular System. *Oxid. Med. Cell. Longev.* **2017**, *2017*, 7941563. [CrossRef] [PubMed]
2. Bolling, B.W.; Chen, C.O.; McKay, D.L.; Blumberg, J.B. Tree nut phytochemicals: Composition, antioxidant capacity, bioactivity, impact factors. A systematic review of almonds, Brazils, cashews, hazelnuts, macadamias, pecans, pine nuts, pistachios and walnuts. *Nutr. Res. Rev.* **2011**, *24*, 244–275. [CrossRef] [PubMed]
3. Ros, E. Nuts and CVD. *Br. J. Nutr.* **2015**, *113*, S111–S120. [CrossRef] [PubMed]
4. Blanco Mejia, S.; Kendall, C.; Viguiliouk, E.; Augustin, L.; Ha, V.; Cozma, A.; Mirrahimi, A.; Maroleanu, A.; Chiavaroli, L.; Leiter, L.; et al. Effect of tree nuts on metabolic syndrome criteria: A systematic review and meta-analysis of randomised controlled trials. *BMJ Open* **2014**, *4*, e004660. [CrossRef] [PubMed]
5. Zhou, D.; Yu, H.; He, F.; Reilly, K.H.; Zhang, J.; Li, S.; Zhang, T.; Wang, B.; Ding, Y.; Xi, B. Nut consumption in relation to cardiovascular disease risk and type 2 diabetes: A systematic review and meta-analysis of prospective studies. *Am. J. Clin. Nutr.* **2014**, *100*, 270–277. [CrossRef] [PubMed]
6. Wu, L.; Wang, Z.; Zhu, J.; Murad, A.L.; Prokop, L.J.; Murad, M.H. Nut consumption and risk of cancer and type 2 diabetes: A systematic review and meta-analysis. *Nutr. Rev.* **2015**, *73*, 409–425. [CrossRef] [PubMed]
7. Grosso, G.; Yang, J.; Marventano, S.; Micek, A.; Galvano, F.; Kales, S. Nut consumption on all-cause, cardiovascular, and cancer mortality risk: A systematic review and meta-analysis of epidemiologic studies. *Am. J. Clin. Nutr.* **2015**, *101*, 783–793. [CrossRef] [PubMed]
8. Kim, Y.; Keogh, J.; Clifton, P.M. Nuts and Cardio-Metabolic Disease: A Review of Meta-Analyses. *Nutrients* **2018**, *10*, 1935. [CrossRef] [PubMed]
9. Bach-Faig, A.; Berry, E.M.; Lairon, D.; Reguant, J.; Trichopoulou, A.; Dernini, S.; Medina, F.X.; Battino, M.; Belahsen, R.; Miranda, G.; et al. Mediterranean diet pyramid today. Science and cultural updates. *Public Health Nutr.* **2011**, *14*, 2274–2284. [CrossRef]
10. Liguori, I.; Russo, G.; Curcio, F.; Bulli, G.; Aran, L.; Della-Morte, D.; Gargiulo, G.; Testa, G.; Cacciatore, F.; Bonaduce, D.; et al. Oxidative stress, aging, and diseases. *Clin. Interv. Aging* **2018**, *13*, 757–772. [CrossRef]

11. Micha, R.; Shulkin, M.L.; Peñalvo, J.L.; Khatibzadeh, S.; Singh, G.M.; Rao, M.; Fahimi, S.; Powles, J.; Mozaffarian, D. Etiologic effects and optimal intakes of foods and nutrients for risk of cardiovascular diseases and diabetes: Systematic reviews and meta-analyses from the nutrition and chronic diseases expert group (NutriCoDE). *PLoS ONE* **2017**, *12*, e0175149. [CrossRef] [PubMed]

12. Carughi, A.; Feeney, M.J.; Kris-Etherton, P.; Fulgoni, V., III; Kendall, C.W.C.; Bulló, M.; Webb, D. Pairing nuts and dried fruit for cardiometabolic health. *Nutr. J.* **2016**, *15*, 23. [CrossRef] [PubMed]

13. Salas-Salvadó, J.; Bulló, M.; Babio, N.; Martínez-González, M.; Ibarrola-Jurado, N.; Basora, J.; Estruch, R.; Covas, M.; Corella, D.; Arós, F.; et al. Reduction in the Incidence of Type 2 Diabetes With the Mediterranean Diet: Results of the PREDIMED-Reus nutrition intervention randomized trial. *Diabetes Care* **2011**, *34*, 14–19, Erratum in *Diabetes Care* **2018**, *41*, 2259–2260. [CrossRef] [PubMed]

14. Pan, A.; Sun, Q.; Manson, J.; Willett, W.; Hu, F. Walnut consumption is associated with lower risk of type 2 diabetes in women. *J. Nutr.* **2013**, *143*, 512–518. [CrossRef] [PubMed]

15. Ibarrola-Jurado, N.; Bulló, M.; Guasch-Ferré, M.; Ros, E.; Martínez-González, M.; Corella, D.; Fiol, M.; Wärnberg, J.; Estruch, R.; Román, P.; et al. Cross-sectional assessment of nut consumption and obesity, metabolic syndrome and other cardiometabolic risk factors: The PREDIMED study. *PLoS ONE* **2013**, *8*, e57367. [CrossRef] [PubMed]

16. Guasch-Ferré, M.; Bulló, M.; Martínez-González, M.; Ros, E.; Corella, D.; Estruch, R.; Fitó, M.; Arós, F.; Wärnberg, J.; Fiol, M.; et al. Frequency of nut consumption and mortality risk in the PREDIMED nutrition intervention trial. *BMC Med.* **2013**, *11*, 164. [CrossRef] [PubMed]

17. Hshieh, T.; Petrone, A.; Gaziano, J.; Djoussé, L. Nut consumption and risk of mortality in the Physicians' Health Study. *Am. J. Clin. Nutr.* **2015**, *101*, 407–412. [CrossRef] [PubMed]

18. Guasch-Ferré, M.; Liu, X.; Malik, V.; Sun, Q.; Willett, W.; Manson, J.; Rexrode, K.; Li, Y.; Hu, F.; Bhupathiraju, S. Nut Consumption and Risk of Cardiovascular Disease. *J. Am. Coll. Cardiol.* **2017**, *70*, 2519–2532. [CrossRef]

19. Larsson, S.C.; Drca, N.; Björck, M.; Bäck, M.; Wolk, A. Nut consumption and incidence of seven cardiovascular diseases. *Heart* **2018**, *104*, 1615–1620. [CrossRef]

20. Estruch, R.; Ros, E.; Salas-Salvadó, J.; Covas, M.-I.; Corella, D.; Arós, F.; Gómez-Gracia, E.; Ruiz-Gutiérrez, V.; Fiol, M.; Lapetra, J.; et al. Primary Prevention of Cardiovascular Disease with a Mediterranean Diet Supplemented with Extra-Virgin Olive Oil or Nuts. *N. Engl. J. Med.* **2018**, *378*, e34. [CrossRef]

21. Liu, G.; Guasch-Ferré, M.; Hu, Y.; Li, Y.; Hu, F.B.; Rimm, E.B.; Manson, J.E.; Rexrode, K.; Sun, Q. Nut Consumption in Relation to Cardiovascular Disease Incidence and Mortality among Patients with Diabetes Mellitus. *Circ. Res.* **2019**, *124*, 920–929. [CrossRef] [PubMed]

22. Aune, D.; Keum, N.; Giovannucci, E.; Fadnes, L.; Boffetta, P.; Greenwood, D.; Tonstad, S.; Vatten, L.; Riboli, E.; Norat, T. Nut consumption and risk of cardiovascular disease, total cancer, all-cause and cause-specific mortality: A systematic review and dose-response meta-analysis of prospective studies. *BMC Med.* **2016**, *14*, 207. [CrossRef] [PubMed]

23. Heffron, S.; Rockman, C.; Gianos, E.; Guo, Y.; Berger, J. Greater Frequency of Nut Consumption is Associated with Lower Prevalence of Peripheral Arterial Disease. *Prev. Med.* **2015**, *72*, 15–18. [CrossRef] [PubMed]

24. Luu, H.; Blot, W.; Xiang, Y.; Cai, H.; Hargreaves, M.; Li, H.; Yang, G.; Signorello, L.; Gao, Y.; Zheng, W.; et al. Prospective Evaluation of the Association of Nut/Peanut Consumption With Total and Cause-Specific Mortality. *JAMA Intern. Med.* **2015**, *175*, 755–766. [CrossRef] [PubMed]

25. Mayhew, A.; de Souza, R.; Meyre, D.; Anand, S.; Mente, A. A systematic review and meta-analysis of nut consumption and incident risk of cardiovascular disease and all-cause mortality. *Br. J. Nutr.* **2016**, *115*, 212–225. [CrossRef] [PubMed]

26. Shao, C.; Tang, H.; Zhao, W.; He, J. Nut intake and stroke risk: A dose-response meta-analysis of prospective cohort studies. *Sci Rep.* **2016**, *6*, 30394. [CrossRef] [PubMed]

27. Luo, C.; Zhang, Y.; Ding, Y.; Shan, Z.; Chen, S.; Yu, M.; Hu, F.B.; Liu, L. Nut consumption and risk of type 2 diabetes, cardiovascular disease, and all-cause mortality: A systematic review and meta-analysis. *Am. J. Clin. Nutr.* **2014**, *100*, 256–269. [CrossRef]

28. Afshin, A.; Micha, R.; Khatibzadeh, S.; Mozaffarian, D. Consumption of nuts and legumes and risk of incident ischemic heart disease, stroke, and diabetes: A systematic review and meta-analysis. *Am. J. Clin. Nutr.* **2014**, *100*, 278–288. [CrossRef]

29. Zhang, Z.; Xu, G.; Wei, Y.; Zhu, W.; Liu, X. Nut consumption and risk of stroke. *Eur. J. Epidemiol.* **2015**, *30*, 189–196. [CrossRef]

30. Domènech, M.; Serra-Mir, M.; Roth, I.; Freitas-Simoes, T.; Valls-Pedret, C.; Cofán, M.; López, A.; Sala-Vila, A.; Calvo, C.; Rajaram, S.; et al. Effect of a Walnut Diet on Office and 24-Hour Ambulatory Blood Pressure in Elderly Individuals. *Hypertension* **2019**, *73*, 1049–1057. [CrossRef]

31. Li, S.; Liu, Y.; Liu, J.; Chang, W.; Chen, C.; Chen, C. Almond consumption improved glycemic control and lipid profiles in patients with type 2 diabetes mellitus. *Metabolism* **2011**, *60*, 474–479. [CrossRef] [PubMed]

32. Wu, L.; Piotrowski, K.; Rau, T.; Waldmann, E.; Broedl, U.; Demmelmair, H.; Koletzko, B.; Stark, R.G.; Nagel, J.M.; Mantzoros, C.S.; et al. Walnut-enriched diet reduces fasting non-HDL-cholesterol and apolipoprotein B in healthy Caucasian subjects: A randomized controlled cross-over clinical trial. *Metabolism* **2014**, *63*, 382–391. [CrossRef] [PubMed]

33. Hernández-Alonso, P.; Salas-Salvadó, J.; Baldrich-Mora, M.; Mallol, R.; Correig, X.; Bulló, M. Effect of pistachio consumption on plasma lipoprotein subclasses in pre-diabetic subjects. *Nutr. Metab. Cardiovasc. Dis.* **2015**, *25*, 396–402. [CrossRef] [PubMed]

34. Ruisinger, J.F.; Gibson, C.A.; Backes, J.M.; Smith, B.K.; Sullivan, D.K.; Moriarty, P.M.; Kris-Etherton, P. Statins and almonds to lower lipoproteins (the STALL Study). *J. Clin. Lipidol.* **2015**, *9*, 58–64. [CrossRef] [PubMed]

35. Jamshed, H.; Sultan, F.; Iqbal, R.; Gilani, A. Dietary Almonds Increase Serum HDL Cholesterol in Coronary Artery Disease Patients in a Randomized Controlled Trial. *J. Nutr.* **2015**, *145*, 2287–2292. [CrossRef] [PubMed]

36. Njike, V.Y.; Ayettey, R.; Petraro, P.; Treu, J.A.; Katz, D.L. Walnut ingestion in adults at risk for diabetes: Effects on body composition, diet quality, and cardiac risk measures. *BMJ Open Diabetes Res. Care* **2015**, *3*, e000115. [CrossRef] [PubMed]

37. Huguenin, G.V.; Oliveira, G.M.; Moreira, A.S.; Saint'Pierre, T.D.; Gonçalves, R.A.; Pinheiro-Mulder, A.R.; Teodoro, A.J.; Luiz, R.R.; Rosa, G. Improvement of antioxidant status after Brazil nut intake in hypertensive and dyslipidemic subjects. *Nutr. J.* **2015**, *14*, 54. [CrossRef] [PubMed]

38. Sauder, K.; McCrea, C.; Ulbrecht, J.; Kris-Etherton, P.; West, S. Effects of pistachios on the lipid/lipoprotein profile, glycemic control, inflammation, and endothelial function in type 2 diabetes: A randomized trial. *Metabolism* **2015**, *64*, 1521–1529. [CrossRef] [PubMed]

39. Mah, E.; Schulz, J.A.; Kaden, V.N.; Lawless, A.L.; Rotor, J.; Mantilla, L.B.; Liska, D.J. Cashew consumption reduces total and LDL cholesterol: A randomized, crossover, controlled-feeding trial. *Am. J. Clin. Nutr.* **2017**, *105*, 1070–1078. [CrossRef] [PubMed]

40. Bamberger, C.; Rossmeier, A.; Lechner, K.; Wu, L.; Waldmann, E.; Stark, R.G.; Altenhofer, J.; Henze, K.; Parhofer, K.G. A walnut-enriched diet reduces lipids in healthy caucasian subjects, independent of recommended macronutrient replacement and time point of consumption: A prospective, randomized, controlled trial. *Nutrients* **2017**, *9*, 1097. [CrossRef]

41. McKay, D.L.; Eliasziw, M.; Chen, O.C.; Blumberg, J.B. A pecan-rich diet improves cardiometabolic risk factors in overweight and obese adults: A randomized controlled trial. *Nutrients* **2018**, *10*, 339. [CrossRef] [PubMed]

42. Jenkins, D.J.; Kendall, C.W.; Lamarche, B.; Banach, M.S.; Srichaikul, K.; Vidgen, E.; Mitchell, S.; Parker, T.; Nishi, S.; Bashyam, B.; et al. Nuts as a replacement for carbohydrates in the diabetic diet: A reanalysis of a randomised controlled trial. *Diabetologia* **2018**, *61*, 1734–1747. [CrossRef] [PubMed]

43. Bowen, J.; Luscombe-Marsh, N.D.; Stonehouse, W.; Tran, C.; Rogers, G.B.; Johnson, N.; Thompson, C.H.; Brinkworth, G.D. Effects of almond consumption on metabolic function and liver fat in overweight and obese adults with elevated fasting blood glucose: A randomised controlled trial. *Clin. Nutr. ESPEN* **2019**, *30*, 10–18. [CrossRef] [PubMed]

44. Musa-Veloso, K.; Paulionis, L.; Poon, T.; Lee, H.Y. The effects of almond consumption on fasting blood lipid levels: A systematic review and meta-analysis of randomised controlled trials. *J. Nutr. Sci.* **2016**, *5*, 1–15. [CrossRef]

45. Nishi, S.; Kendall, C.W.C.; Gascoyne, A.-M.; Bazinet, R.P.; Bashyam, B.; Lapsley, K.G.; Augustin, L.S.A.; Sievenpiper, J.L.; Jenkins, D.J.A. Effect of almond consumption on the serum fatty acid profile: A dose–response study. *Br. J. Nutr.* **2014**, *112*, 1137–1146. [CrossRef]

46. Damavandi, D.R.; Mousavi, S.N.; Shidfar, F.; Mohammadi, V.; Rajab, A.; Hosseini, S.; Heshmati, J. Effects of Daily Consumption of Cashews on Oxidative Stress and Atherogenic Indices in Patients with Type 2 Diabetes: A Randomized, Controlled-Feeding Trial. *Int. J. Endocrinol. Metab.* **2019**, *17*, e70744. [CrossRef]

47. Kay, C.D.; Gebauer, S.K.; West, S.G.; Kris-Etherton, P.M. Pistachios increase serum antioxidants and lower serum oxidized-LDL in hypercholesterolemic adults. *J. Nutr.* **2010**, *140*, 1093–1098. [CrossRef]

48. Zibaeenezhad, M.; Farhadi, P.; Attar, A.; Mosleh, A.; Amirmoezi, F.; Azimi, A. Effects of walnut oil on lipid profiles in hyperlipidemic type 2 diabetic patients: A randomized, double-blind, placebo-controlled trial. *Nutr. Diabetes* **2017**, *7*, e259. [CrossRef]
49. Austel, A.; Ranke, C.; Wagner, N.; Görge, J.; Ellrott, T. Weight loss with a modified Mediterranean-type diet using fat modification: A randomized controlled trial. *Eur. J. Clin. Nutr.* **2015**, *69*, 878–884. [CrossRef]
50. Guasch-Ferré, M.; Li, J.; Hu, F.B.; Salas-Salvadó, J.; Tobias, D.K. Effects of walnut consumption on blood lipids and other cardiovascular risk factors: An updated meta-analysis and systematic review of controlled trials. *Am. J. Clin. Nutr.* **2018**, *108*, 174–187. [CrossRef]
51. Del Gobbo, L.; Falk, M.; Feldman, R.; Lewis, K.; Mozaffarian, D. Effects of tree nuts on blood lipids, apolipoproteins, and blood pressure: Systematic review, meta-analysis, and dose-response of 61 controlled intervention trials. *Am. J. Clin. Nutr.* **2015**, *102*, 1347–1356. [CrossRef]
52. Jellinger, P.; Handelsman, Y.; Rosenblit, P.; Bloomgarden, Z.; Fonseca, V.; Garber, A.; Grunberger, G.; Guerin, C.; Bell, D.; Mechanick, J.; et al. American Association of Clinical Endocrinologists and American College of Endocrinology Guidelines for Management of Dyslipidemia and Prevention of Cardiovascular Disease. *Endocr. Pr.* **2017**, *23*, 1–87. [CrossRef] [PubMed]
53. Blesso, C.; Fernandez, M. Dietary Cholesterol, Serum Lipids, and Heart Disease: Are Eggs Working for or Against You? *Nutrients* **2018**, *10*, 426. [CrossRef] [PubMed]
54. Cohen, A.E.; Johnston, C.S. Almond ingestion at mealtime reduces postprandial glycemia and chronic ingestion reduces hemoglobin A(1c) in individuals with well-controlled type 2 diabetes mellitus. *Metabolism* **2011**, *60*, 1312–1317. [CrossRef] [PubMed]
55. Damasceno, N.; Sala-Vila, A.; Cofán, M.; Pérez-Heras, A.; Fitó, M.; Ruiz-Gutiérrez, V.; Martínez-González, M.; Corella, D.; Arós, F.; Estruch, R.; et al. Mediterranean diet supplemented with nuts reduces waist circumference and shifts lipoprotein subfractions to a less atherogenic pattern in subjects at high cardiovascular risk. *Atherosclerosis* **2013**, *230*, 347–353. [CrossRef] [PubMed]
56. Lasa, A.; Miranda, J.; Bulló, M.; Casas, R.; Salas-Salvadó, J.; Larretxi, I.; Estruch, R.; Ruiz-Gutiérrez, V.; Portillo, M. Comparative effect of two Mediterranean diets versus a low-fat diet on glycaemic control in individuals with type 2 diabetes. *Eur. J. Clin. Nutr.* **2014**, *68*, 767–772. [CrossRef] [PubMed]
57. Hernández-Alonso, P.; Salas-Salvadó, J.; Baldrich-Mora, M.; Juanola-Falgarona, M.; Bulló, M. Beneficial effect of pistachio consumption on glucose metabolism, insulin resistance, inflammation, and related metabolic risk markers: A randomized clinical trial. *Diabetes Care* **2014**, *37*, 3098–3105. [CrossRef]
58. Rodríguez-Rejón, A.I.; Castro-Quezada, I.; Ruano-Rodríguez, C.; Ruiz-López, M.D.; Sánchez-Villegas, A.; Toledo, E.; Artacho, R.; Estruch, R.; Salas-Salvadó, J.; Covas, M.I.; et al. Effect of a Mediterranean Diet Intervention on Dietary Glycemic Load and Dietary Glycemic Index: The PREDIMED Study. *J. Nutr Metab.* **2014**, *2014*, 985373. [CrossRef]
59. Chen, C.; Liu, J.; Li, S.; Huang, C.; Hsirh, A.; Weng, S.; Chang, M.; Li, H.; Mohn, E.; Chen, C. Almonds ameliorate glycemic control in Chinese patients with better controlled type 2 diabetes: A randomized, crossover, controlled feeding trial. *Nutr. Metab.* **2017**, *14*, 51. [CrossRef]
60. Hou, Y.-Y.; Ojo, O.; Wang, L.-L.; Wang, Q.; Jiang, Q.; Shao, X.-Y.; Wang, X.-H. A Randomized Controlled Trial to Compare the Effect of Peanuts and Almonds on the Cardio-Metabolic and Inflammatory Parameters in Patients with Type 2 Diabetes Mellitus. *Nutrients* **2018**, *10*, 1565. [CrossRef]
61. Nezhad, M.; Aghasadeghi, K.; Hakimi, H.; Yarmohammadi, H.; Nikaein, F. The Effect of Walnut Oil Consumption on Blood Sugar in Patients With Diabetes Mellitus Type 2. *Int. J. Endocrinol. Metab.* **2016**, *14*, e34889.
62. Li, H.; Li, X.; Yuan, S.; Jin, Y.; Lu, J. Nut consumption and risk of metabolic syndrome and overweight/obesity: A meta-analysis of prospective cohort studies and randomized trials. *Nutr. Metab.* **2018**, *15*, 46. [CrossRef] [PubMed]
63. Flores-Mateo, G.; Rojas-Rueda, D.; Basora, J.; Ros, E.; Salas-Salvadó, J. Nut intake and adiposity: Meta-analysis of clinical trials. *Am. J. Clin. Nutr.* **2013**, *97*, 1346–1355. [CrossRef] [PubMed]
64. Baer, D.J.; Gebauer, S.K.; Novotny, J.A. Measured energy value of pistachios in the human diet. *Br. J. Nutr.* **2012**, *107*, 120–125. [CrossRef] [PubMed]
65. Baer, D.; Novotny, J. Metabolizable Energy from Cashew Nuts is Less than that Predicted by Atwater Factors. *Nutrients* **2019**, *11*, 33. [CrossRef] [PubMed]

66. Baer, D.J.; Gebauer, S.K.; Novotny, J.A. Walnuts Consumed by Healthy Adults Provide Less Available Energy than Predicted by the Atwater Factors. *J. Nutr.* **2016**, *146*, 9–13. [CrossRef] [PubMed]

67. Novotny, J.A.; Gebauer, S.K.; Baer, D.J. Discrepancy between the Atwater factor predicted and empirically measured energy values of almonds in human diets. *Am. J. Clin. Nutr.* **2012**, *96*, 296–301. [CrossRef] [PubMed]

68. Barbour, J.; Howe, P.; Buckley, J.; Bryan, J.; Coates, A. Effect of 12 Weeks High Oleic Peanut Consumption on Cardio-Metabolic Risk Factors and Body Composition. *Nutrients* **2015**, *7*, 7381–7398. [CrossRef] [PubMed]

69. Viguiliouk, E.; Kendall, C.W.C.; Blanco Mejia, S.; Cozma, A.I.; Ha, V.; Mirrahimi, A.; Jayalath, V.H.; Augustin, L.S.A.; Chiavaroli, L.; Leiter, L.A.; et al. Effect of tree nuts on glycemic control in diabetes: A systematic review and meta-analysis of randomized controlled dietary trials. *PLoS ONE* **2014**, *9*, e103376. [CrossRef] [PubMed]

70. Tindall, A.M.; Johnston, E.A.; Kris-Etherton, P.M.; Petersen, K.S. The effect of nuts on markers of glycemic control: A systematic review and meta-analysis of randomized controlled trials. *Am. J. Clin. Nutr.* **2019**, *109*, 297–314. [CrossRef] [PubMed]

71. Casas-Agustench, P.; Bulló, M.; Ros, E.; Basora, J.; Salas-Salvadó, J. Cross-sectional association of nut intake with adiposity in a Mediterranean population. *Nutr. Metab. Cardiovasc. Dis.* **2011**, *21*, 518–525. [CrossRef] [PubMed]

72. Jaceldo-Siegl, K.; Haddad, E.; Oda, K.; Fraser, G.E.; Sabate, J. Tree Nuts Are Inversely Associated with Metabolic Syndrome and Obesity: The Adventist Health Study-2. *PLoS ONE* **2014**, *9*, e85133. [CrossRef] [PubMed]

73. Arab, L.; Dhaliwal, S.K.; Martin, C.J.; Larios, A.D.; Jackson, N.J.; Elashoff, D. Association between walnut consumption and diabetes risk in NHANES. *Diabetes Metab. Res. Rev.* **2018**, *34*, e3031. [CrossRef] [PubMed]

74. Ma, Y.; Njike, V.; Millet, J.; Dutta, S.; Doughty, K.; Treu, J.; Katz, D. Effects of Walnut Consumption on Endothelial Function in Type 2 Diabetic. *Diabetes Care* **2010**, *33*, 227–232. [CrossRef] [PubMed]

75. Katz, D.L.; Davidhi, A.; Ma, Y.; Kavak, Y.; Bifulco, L.; Njike, V.Y. Effects of walnuts on endothelial function in overweight adults with visceral obesity: A randomized, controlled, crossover trial. *J. Am. Coll. Nutr.* **2012**, *31*, 415–423. [CrossRef] [PubMed]

76. Liu, J.F.; Liu, Y.H.; Chen, C.M.; Chang, W.H.; Chen, C.Y.O. The effect of almonds on inflammation and oxidative stress in Chinese patients with type 2 diabetes mellitus: A randomized crossover controlled feeding trial. *Eur. J. Nutr.* **2013**, *52*, 927–935. [CrossRef] [PubMed]

77. Sweazea, K.L.; Johnston, C.S.; Ricklefs, K.D.; Petersen, K.N. Almond supplementation in the absence of dietary advice significantly reduces C-reactive protein in subjects with type 2 diabetes. *J. Funct. Foods* **2014**, *10*, 252–259. [CrossRef]

78. Chen, C.; Holbrook, M.; Duess, M.; Dohadwala, M.; Hamburg, N.; Asztalos, B.; Milbury, P.; Blumberg, J.; Vita, J. Effect of almond consumption on vascular function in patients with coronary artery disease: A randomized, controlled, cross-over trial. *Nutr. J.* **2015**, *14*, 61. [CrossRef] [PubMed]

79. Yu, Z.; Malik, V.; Keum, N.; Hu, F.; Giovannucci, E.; Stampfer, M.; Willett, W.; Fuchs, C.; Bao, Y. Associations between nut consumption and inflammatory biomarkers. *Am. J. Clin. Nutr.* **2016**, *104*, 722–728. [CrossRef] [PubMed]

80. Bitok, E.; Sabaté, J. Nuts and Cardiovascular Disease. *Prog. Cardiovasc. Dis.* **2018**, *61*, 33–37. [CrossRef] [PubMed]

81. West, S.G.; Krick, A.L.; Klein, L.C.; Zhao, G.; Wojtowicz, T.F.; McGuiness, M.; Bagshaw, D.; Wagner, P.; Ceballos, R.M.; Holub, B.J.; et al. Effects of diets high in walnuts and flax oil on hemodynamic responses to stress and vascular endothelial function. *J. Am. Coll. Nutr.* **2010**, *29*, 595–603. [CrossRef]

82. Neale, E.P.; Tapsell, L.C.; Guan, V.; Batterham, M.J. The effect of nut consumption on markers of inflammation and endothelial function: A systematic review and meta-analysis of randomised controlled trials. *BMJ Open.* **2017**, *7*, e016863. [CrossRef] [PubMed]

83. Xiao, Y.; Huang, W.; Peng, C.; Zhang, J.; Wong, C.; Kim, J.H.; Yeoh, E.; Su, X. Effect of nut consumption on vascular endothelial function: A systematic review and meta-analysis of randomized controlled trials. *Clin. Nutr.* **2018**, *37*, 831–839. [CrossRef] [PubMed]

84. Bhardwaj, R.; Dod, H.; Sandhu, M.S.; Bedi, R.; Dod, S.; Konat, G.; Chopra, H.; Sharma, R.; Jain, A.C.; Nanda, N. Acute effects of diets rich in almonds and walnuts on endothelial function. *Indian Heart J.* **2018**, *70*, 497–501. [CrossRef] [PubMed]

85. Fogacci, F.; Cicero, A.F.G.; Derosa, G.; Rizzo, M.; Veronesi, M.; Borghi, C. Effect of pistachio on brachial artery diameter and flow-mediated dilatation: A systematic review and meta-analysis of randomized, controlled-feeding clinical studies. *Crit. Rev. Food Sci. Nutr.* **2019**, *59*, 328–335. [CrossRef] [PubMed]

86. Morgillo, S.; Hill, A.M.; Coates, A.M. The Effects of Nut Consumption on Vascular Function. *Nutrients* **2019**, *11*, 116. [CrossRef] [PubMed]

87. Borkowski, K.; Yim, S.J.; Holt, R.R.; Hackman, R.M.; Keen, C.L.; Newman, J.W.; Shearer, G.C. Walnuts change lipoprotein composition suppressing TNFα-stimulated cytokine production by diabetic adipocyte. *J. Nutr. Biochem.* **2019**, *68*, 51–58. [CrossRef] [PubMed]

88. Mazidi, M.; Rezaie, P.; Ferns, G.A.; Gao, H. Impact of different types of tree nut, peanut, and soy nut consumption on serum C-reactive protein (CRP). *Medicine* **2016**, *95*, e5165. [CrossRef]

89. Müllner, E.; Plasser, E.; Brath, H.; Waldschütz, W.; Forster, E.; Kundi, M.; Wagner, K. Impact of polyunsaturated vegetable oils on adiponectin levels, glycaemia and blood lipids in individuals with type 2 diabetes: A randomised, double-blind intervention study. *J. Hum. Nutr. Diet.* **2014**, *27*, 468–478. [CrossRef]

90. Thanikachalam, K.; Khan, G. Colorectal Cancer and Nutrition. *Nutrients* **2019**, *11*, 164. [CrossRef] [PubMed]

91. Raimondi, S.; Mabrouk, J.B.; Shatenstein, B.; Maisonneuve, P.; Ghadirian, P. Diet and prostate cancer risk with specific focus on dairy products and dietary calcium: A case-control study. *Prostate* **2010**, *70*, 1054–1065. [CrossRef] [PubMed]

92. Ibiebele, T.; Nagle, C.; Bain, C.; Webb, P. Intake of omega-3 and omega-6 fatty acids and risk of ovarian cancer. *Cancer Causes Control.* **2012**, *23*, 1775–1783. [CrossRef] [PubMed]

93. Bao, Y.; Hu, F.; Giovannucci, E.; Wolpin, B.; Stampfer, M.; Willett, W.; Fuchs, C. Nut consumption and risk of pancreatic cancer in women. *Br. J. Cancer* **2013**, *109*, 2911–2916. [CrossRef] [PubMed]

94. Van den Brandt, P.; Schouten, L. Relationship of tree nut, peanut and peanut butter intake with total and cause-specific mortality: A cohort study and meta-analysis. *Int. J. Epidemiol.* **2015**, *44*, 1038–1049. [CrossRef] [PubMed]

95. Yang, M.; Hu, F.; Giovannucci, E.; Stampfer, M.; Willett, W.; Fuchs, C.; Wu, K.; Bao, Y. Nut consumption and risk of colorectal cancer in women. *Eur. J. Clin. Nutr.* **2016**, *70*, 333–337. [CrossRef] [PubMed]

96. Wang, W.; Yang, M.; Kenfield, S.A.; Hu, F.B.; Stampfer, M.J.; Willett, W.C.; Fuchs, C.S.; Giovannucci, E.L.; Bao, Y. Nut consumption and prostate cancer risk and mortality. *Br. J. Cancer* **2016**, *115*, 371–374. [CrossRef] [PubMed]

97. Lee, J.; Lai, G.; Liao, L.; Subar, A.; Bertazzi, P.; Pesatori, A.; Freedman, N.; Landi, M.; Lam, T. Nut Consumption and Lung Cancer Risk: Results from Two Large Observational Studies. *Cancer Epidemiol. Biomarkers Prev.* **2017**, *26*, 826–836. [CrossRef] [PubMed]

98. Hashemian, M.; Murphy, G.; Etemadi, A.; Dawsey, S.M.; Liao, L.M.; Abnet, C.C. Nut and peanut butter consumption and the risk of esophageal and gastric cancer subtypes. *Am. J. Clin. Nutr.* **2017**, *106*, 858–864. [CrossRef]

99. Nieuwenhuis, L.; van den Brandt, P.A. Total nut, tree nut, peanut, and peanut butter consumption and the risk of pancreatic cancer in the Netherlands Cohort Study. *Cancer Epidemiol. Biomarkers Prev.* **2018**, *27*, 274–284. [CrossRef]

100. Nieuwenhuis, L.; van den Brandt, P.A. Tree nut, peanut, and peanut butter consumption and the risk of gastric and esophageal cancer subtypes: The Netherlands Cohort Study. *Gastric Cancer.* **2018**, *21*, 900–912. [CrossRef]

101. Fadelu, T.; Zhang, S.; Niedzwiecki, D.; Ye, X.; Saltz, L.; Mayer, R.; Mowat, R.; Whittom, R.; Hantel, A.; Benson, A.; et al. Nut Consumption and Survival in Patients With Stage III Colon Cancer: Results From CALGB 89803 (Alliance). *J. Clin. Oncol.* **2018**, *36*, 1112–1120. [CrossRef] [PubMed]

102. Van den Brandt, P.A.; Nieuwenhuis, L. Tree nut, peanut, and peanut butter intake and risk of postmenopausal breast cancer: The Netherlands Cohort Study. *Cancer Causes Control* **2018**, *29*, 63–75. [CrossRef] [PubMed]

103. Zhao, Y.; Zhao, L.; Hu, Z.; Wu, J.; Li, J.; Qu, C.; He, Y.; Song, Q. Peanut consumption associated with a reduced risk of esophageal squamous cell carcinoma: A case–control study in a high-risk area in China. *Thorac Cancer* **2018**, *9*, 30–36. [CrossRef] [PubMed]

104. Lee, J.; Shin, A.; Oh, J.; Kim, J. The relationship between nut intake and risk of colorectal cancer: A case control study. *Nutr. J.* **2018**, *17*, 37. [CrossRef] [PubMed]

105. Hardman, W.E.; Primerano, D.A.; Legenza, M.T.; Morgan, J.; Fan, J.; Denvir, J. Dietary walnut altered gene expressions related to tumor growth, survival, and metastasis in breast cancer patients: A pilot clinical trial. *Nutr. Res.* **2019**, *66*, 82–94. [CrossRef] [PubMed]

106. Sui, J.; Yang, W.; Ma, Y.; Li, T.Y.; Simon, T.G.; Meyerhardt, J.A.; Liang, G.; Giovannucci, E.L.; Chan, A.T.; Zhang, X. A prospective study of nut consumption and risk of primary hepatocellular carcinoma in the U.S. women and men. *Cancer Prev. Res.* **2019**, *12*, 367–374. [CrossRef] [PubMed]

107. Nieuwenhuis, L.; van den Brandt, P.A. Nut and peanut butter consumption and the risk of lung cancer and its subtypes: A prospective cohort study. *Lung Cancer* **2019**, *128*, 57–66. [CrossRef] [PubMed]

108. Batirel, S.; Yilmaz, A.M.; Sahin, A.; Perakakis, N.; Kartal Ozer, N.; Mantzoros, C.S. Antitumor and antimetastatic effects of walnut oil in esophageal adenocarcinoma cells. *Clin. Nutr.* **2018**, *37*, 2166–2171. [CrossRef]

109. Yin, X.; Bostick, R.M. Associations of Nut Intakes with Incident Sporadic Colorectal Adenoma: A Pooled Case-Control Study. *Nutr. Cancer* **2019**, *71*, 731–738. [CrossRef]

110. Casari, I.; Falasca, M. Diet and Pancreatic Cancer Prevention. *Cancers* **2015**, *7*, 2309–2317. [CrossRef]

111. Toledo, E.; Salas-Salvadó, J.; Donat-Vargas, C.; Buil-Cosiales, P.; Estruch, R.; Ros, E.; Corella, D.; Fitó, M.; Hu, F.; Arós, F.; et al. Mediterranean Diet and Invasive Breast Cancer Risk Among Women at High Cardiovascular Risk in the PREDIMED Trial: A Randomized Clinical Trial. *JAMA Intern. Med.* **2015**, *175*, 1752–1760. [CrossRef] [PubMed]

112. Soriano-Hernandez, A.; Madrigal-Perez, D.; Galvan-Salazar, H.; Arreola-Cruz, A.; Briseño-Gomez, L.; Guzmán-Esquivel, J.; Dobrovinskaya, O.; Lara-Esqueda, A.; Rodríguez-Sanchez, I.; Baltazar-Rodriguez, L.; et al. The Protective Effect of Peanut, Walnut, and Almond Consumption on the Development of Breast Cancer. *Gynecol. Obs. Investig.* **2015**, *80*, 89–92. [CrossRef] [PubMed]

113. Sun, L.; Subar, A.F.; Bosire, C.; Dawsey, S.M.; Kahle, L.L.; Zimmerman, T.P.; Abnet, C.C.; Heller, R.; Graubard, B.I.; Cook, M.B.; et al. Dietary Flavonoid Intake Reduces the Risk of Head and Neck but Not Esophageal or Gastric Cancer in US Men and Women. *J. Nutr.* **2017**, *147*, 1729–1738. [PubMed]

114. Rusu, M.E.; Mocan, A.; Fizesan, I.; Popa, D.S.; Vlase, L.; Pop, A. Bioactive compounds from walnut (*Juglans regia* L.) septum extracts: Antioxidant and cytotoxic activity. In Proceedings of the Functional and Medical Foods for Chronic Diseases: Bioactive Compounds and Biomarkers, Boston, MA, USA, 22–23 September 2017.

115. Rusu, M.E.; Gheldiu, A.-M.; Mocan, A.; Moldovan, C.; Popa, D.-S.; Tomuta, I.; Vlase, L. Process Optimization for Improved Phenolic Compounds Recovery from Walnut (*Juglans regia* L.) Septum: Phytochemical Profile and Biological Activities. *Molecules* **2018**, *23*, 2814. [CrossRef] [PubMed]

116. Shah, U.N.; Mir, J.I.; Ahmed, N.; Jan, S.; Fazili, K.M. Bioefficacy potential of different genotypes of walnut *Juglans regia* L. *J. Food Sci. Technol.* **2018**, *55*, 605–618. [CrossRef]

117. Frankish, H.; Horton, R. Prevention and management of dementia: A priority for public health. *Lancet* **2017**, *390*, 2614–2615. [CrossRef]

118. Sánchez-Villegas, A.; Galbete, C.; Martinez-González, M.A.; Martinez, J.A.; Razquin, C.; Salas-Salvadó, J.; Estruch, R.; Buil-Cosiales, P.; Martí, A. The effect of the Mediterranean diet on plasma brain-derived neurotrophic factor (BDNF) levels: The PREDIMED-NAVARRA randomized trial. *Nutr. Neurosci.* **2011**, *14*, 195–201. [CrossRef]

119. Nooyens, A.; Bueno-De-Mesquita, H.; van Boxtel, M.; van Gelder, B.; Verhagen, H.; Verschuren, W. Fruit and vegetable intake and cognitive decline in middle-aged men and women: The Doetinchem Cohort Study. *Br. J. Nutr.* **2011**, *106*, 752–761. [CrossRef]

120. Valls-Pedret, C.; Lamuela-Raventós, R.; Medina-Remón, A.; Quintana, M.; Corella, D.; Pintó, X.; Martínez-González, M.; Estruch, R.; Ros, E. Polyphenol-rich foods in the Mediterranean diet are associated with better cognitive function in elderly subjects at high cardiovascular risk. *J. Alzheimers Dis.* **2012**, *29*, 773–782. [CrossRef]

121. Martínez-Lapiscina, E.H.; Clavero, P.; Toledo, E.; Estruch, R.; Salas-Salvadó, J.; Julián, B.S.; Sanchez-Tainta, A.; Ros, E.; Valls-Pedret, C.; Martinez-Gonzalez, M.Á. Mediterranean diet improves cognition: The PREDIMED-NAVARRA randomised trial. *J. Neurol Neurosurg. Psychiatry* **2013**, *84*, 1318–1325. [CrossRef]

122. O'Brien, J.; Okereke, O.; Devore, E.; Rosner, B.; Breteler, M.; Grodstein, F. Long-term intake of nuts in relation to cognitive function in older women. *J. Nutr. Health Aging* **2014**, *18*, 496–502. [CrossRef] [PubMed]

123. Valls-Pedret, C.; Sala-Vila, A.; Serra-Mir, M.; Corella, D.; de la Torre, R.; Martínez-González, M.; Martínez-Lapiscina, E.; Fitó, M.; Pérez-Heras, A.; Salas-Salvadó, J.; et al. Mediterranean Diet and Age-Related Cognitive Decline: A Randomized Clinical Trial. *JAMA Intern. Med.* **2015**, *175*, 1094–1103. [CrossRef] [PubMed]

124. Barbour, J.A.; Howe, P.R.C.; Buckley, J.D.; Bryan, J.; Coates, A.M. Cerebrovascular and cognitive benefits of high-oleic peanut consumption in healthy overweight middle-aged adults. *Nutr. Neurosci.* **2017**, *20*, 555–562. [CrossRef] [PubMed]

125. Dong, L.; Xiao, R.; Cai, C.; Xu, Z.; Wang, S.; Pan, L.; Yuan, L. Diet, lifestyle and cognitive function in old Chinese adults. *Arch. Gerontol. Geriatr.* **2016**, *63*, 36–42. [CrossRef] [PubMed]

126. Arab, L.; Ang, A. A cross sectional study of the association between walnut consumption and cognitive function among adult US populations represented in NHANES. *J. Nutr. Health Aging* **2015**, *19*, 284–290. [CrossRef] [PubMed]

127. Rajaram, S.; Valls-Pedret, C.; Cofan, M.; Sabaté, J.; Serra-Mir, M.; Perez-Heras, A.M.; Arechiga, A.; Casaroli-Marano, R.P.; Alforja, S.; Sala-Vila, A.; et al. The Walnuts and Healthy Aging Study (WAHA): Protocol for a Nutritional Intervention Trial with Walnuts on Brain Aging. *Front. Aging Neurosci.* **2017**, *8*, 333. [CrossRef] [PubMed]

128. Cutuli, D. Functional and Structural Benefits Induced by Omega-3 Polyunsaturated Fatty Acids During Aging. *Curr. Neuropharmacol.* **2017**, *15*, 534–542. [CrossRef] [PubMed]

129. Blondeau, N.; Lipsky, R.H.; Bourourou, M.; Duncan, M.W.; Gorelick, P.B.; Marini, A.M. Alpha-Linolenic Acid: An Omega-3 Fatty Acid with Neuroprotective Properties—Ready for Use in the Stroke Clinic? *Biomed. Res. Int.* **2015**, *2015*, 519830. [CrossRef] [PubMed]

130. Opie, R.; Itsiopoulos, C.; Parletta, N.; Sanchez-Villegas, A.; Akbaraly, T.; Ruusunen, A.; Jacka, F. Dietary recommendations for the prevention of depression. *Nutr. Neurosci.* **2017**, *20*, 161–171. [CrossRef] [PubMed]

131. Ali-Sisto, T.; Tolmunen, T.; Viinamäki, H.; Mäntyselkä, P.; Valkonen-Korhonen, M.; Koivumaa-Honkanen, H.; Honkalampi, K.; Ruusunen, A.; Nandania, J.; Velagapudi, V.; et al. Global arginine bioavailability ratio is decreased in patients with major depressive disorder. *J. Affect. Disord.* **2018**, *229*, 145–151. [CrossRef] [PubMed]

132. Su, Q.; Yu, B.; He, H.; Zhang, Q.; Meng, G.; Wu, H.; Du, H.; Liu, L.; Shi, H.; Xia, Y.; et al. Nut Consumption Is Associated With Depressive Symptoms Among Chinese Adults. *Depress Anxiety* **2016**, *33*, 1065–1072. [CrossRef] [PubMed]

133. Arab, L.; Guo, R.; Elashoff, D. Lower Depression Scores among Walnut Consumers in NHANES. *Nutrients* **2019**, *11*, 275. [CrossRef] [PubMed]

134. Miller, M.G.; Thangthaeng, N.; Poulose, S.M.; Shukitt-Hale, B. Role of fruits, nuts, and vegetables in maintaining cognitive health. *Exp. Gerontol.* **2017**, *94*, 24–28. [CrossRef] [PubMed]

135. Miller, H.C.; Struyf, D.; Baptist, P.; Dalile, B.; Van Oudenhove, L.; Van Diest, I. A mind cleared by walnut oil: The effects of polyunsaturated and saturated fat on extinction learning. *Appetite* **2018**, *126*, 147–155. [CrossRef] [PubMed]

136. Robinson, S.; Reginster, J.; Rizzoli, R.; Shaw, S.; Kanis, J.; Bautmans, I.; Bischoff-Ferrari, H.; Bruyère, O.; Cesari, M.; Dawson-Hughes, B.; et al. Does nutrition play a role in the prevention and management of sarcopenia? *Clin. Nutr.* **2018**, *37*, 1121–1132. [CrossRef] [PubMed]

137. Stocks, J.; Valdes, A.M. Effect of dietary omega-3 fatty acid supplementation on frailty-related phenotypes in older adults: A systematic review and meta-analysis protocol. *BMJ Open* **2018**, *8*, e021344. [CrossRef] [PubMed]

138. Tachtsis, B.; Camera, D.; Lacham-Kaplan, O. Potential Roles of n-3 PUFAs during Skeletal Muscle Growth and Regeneration. *Nutrients* **2018**, *10*, 309. [CrossRef] [PubMed]

139. Rusu, M.E.; Gheldiu, A.-M.; Mocan, A.; Vlase, L.; Popa, D.-S. Anti-aging potential of tree nuts with a focus on phytochemical composition, molecular mechanisms and thermal stability of major bioactive compounds. *Food Funct.* **2018**, *9*, 2554–2575. [CrossRef] [PubMed]

140. Ros, E. Health benefits of nut consumption. *Nutrients* **2010**, *2*, 652–682. [CrossRef] [PubMed]

141. Thomas, S.; Browne, H.; Mobasheri, A.; Rayman, M.P. What is the evidence for a role for diet and nutrition in osteoarthritis? *Rheumatology* **2018**, *57* (Suppl. 4), iv61–iv74. [CrossRef] [PubMed]

142. Lu, B.; Driban, J.B.; Xu, C.; Lapane, K.L.; McAlindon, T.E.; Eaton, C.B. Dietary Fat Intake and Radiographic Progression of Knee Osteoarthritis: Data From the Osteoarthritis Initiative. *Arthritis Care Res.* **2017**, *69*, 368–375. [CrossRef] [PubMed]

143. Rivas, A.; Romero, A.; Mariscal-Arcas, M.; Monteagudo, C.; Feriche, B.; Lorenzo, M.L.; Olea, F. Mediterranean diet and bone mineral density in two age groups of women. *Int. J. Food Sci. Nutr.* **2013**, *64*, 155–161. [CrossRef] [PubMed]

144. Papagrigoraki, A.; Maurelli, M.; del Giglio, M.; Gisondi, P.; Girolomoni, G. Advanced Glycation End Products in the Pathogenesis of Psoriasis. *Int. J. Mol. Sci.* **2017**, *18*, 2471. [CrossRef] [PubMed]

145. Chen, J.; Lin, X.; Bu, C.; Zhang, X. Role of advanced glycation end products in mobility and considerations in possible dietary and nutritional intervention strategies. *Nutr. Metab.* **2018**, *15*, 72. [CrossRef] [PubMed]

146. Wei, Q.; Liu, T.; Sun, D. Advanced glycation end-products (AGEs) in foods and their detecting techniques and methods: A review. *Trends Food Sci. Technol.* **2018**, *82*, 32–45. [CrossRef]

147. Bonfigli, A.; Spazzafumo, L.; Prattichizzo, F.; Bonafè, M.; Mensà, E.; Micolucci, L.; Giuliani, A.; Fabbietti, P.; Testa, R.; Boemi, M.; et al. Leukocyte telomere length and mortality risk in patients with type 2 diabetes. *Oncotarget* **2016**, *7*, 50835–50844. [CrossRef] [PubMed]

148. Stefler, D.; Malyutina, S.; Maximov, V.; Orlov, P.; Ivanoschuk, D.; Nikitin, Y.; Gafarov, V.; Ryabikov, A.; Voevoda, M.; Bobak, M.; et al. Leukocyte telomere length and risk of coronary heart disease and stroke mortality: Prospective evidence from a Russian cohort. *Sci. Rep.* **2018**, *8*, 16627. [CrossRef] [PubMed]

149. Song, S.; Johnson, F. Epigenetic Mechanisms Impacting Aging: A Focus on Histone Levels and Telomeres. *Genes* **2018**, *9*, 201. [CrossRef] [PubMed]

150. Sack, M.; Fyhrquist, F.; Saijonmaa, O.; Fuster, V.; Kovacic, J. Basic Biology of Oxidative Stress and the Cardiovascular System: Part 1 of a 3-Part Series. *J. Am. Coll. Cardiol.* **2017**, *70*, 196–211. [CrossRef] [PubMed]

151. Tucker, L. Consumption of Nuts and Seeds and Telomere Length in 5,582 Men and Women of the National Health and Nutrition Examination Survey (NHANES). *J. Nutr. Health Aging* **2017**, *21*, 233–240. [CrossRef] [PubMed]

152. Tucker, L. Dietary Fiber and Telomere Length in 5674 U.S. Adults: An NHANES Study of Biological Aging. *Nutrients* **2018**, *10*, 400. [CrossRef] [PubMed]

153. Freitas-Simoes, T.; Cofán, M.; Blasco, M.; Soberón, N.; Foronda, M.; Serra-Mir, M.; Roth, I.; Valls-Pedret, C.; Doménech, M.; Ponferrada-Ariza, E.; et al. Walnut Consumption for Two Years and Leukocyte Telomere Attrition in Mediterranean Elders: Results of a Randomized Controlled Trial. *Nutrients* **2018**, *10*, 1907. [CrossRef] [PubMed]

154. Rusu, M.E.; Simedrea, R.; Gheldiu, A.-M.; Mocan, A.; Vlase, L.; Popa, D.-S.; Ferreira, I.C.F.R. Benefits of tree nut consumption on aging and age-related diseases: Mechanisms of actions. *Trends Food Sci. Technol.* **2019**, *88*, 104–120. [CrossRef]

155. Zhou, Y.; Zheng, J.; Li, Y.; Xu, D.-P.; Li, S.; Chen, Y.-M.; Li, H.-B. Natural Polyphenols for Prevention and Treatment of Cancer. *Nutrients* **2016**, *8*, 515. [CrossRef] [PubMed]

156. Shahidi, F.; Yeo, J. Bioactivities of Phenolics by Focusing on Suppression of Chronic Diseases: A Review. *Int. J. Mol. Sci.* **2018**, *19*, 1573. [CrossRef] [PubMed]

157. Kang, I.; Buckner, T.; Shay, N.F.; Gu, L.; Chung, S. Improvements in Metabolic Health with Consumption of Ellagic Acid and Subsequent Conversion into Urolithins: Evidence and Mechanisms. *Adv. Nutr.* **2016**, *7*, 961–972. [CrossRef] [PubMed]

158. Seong, Y.; Shin, P.; Kim, G. Anacardic acid induces mitochondrial-mediated apoptosis in the A549 human lung adenocarcinoma cells. *Int. J. Oncol.* **2013**, *42*, 1045–1051. [CrossRef] [PubMed]

159. Hamad, F.; Mubofu, E. Potential Biological Applications of Bio-Based Anacardic Acids and Their Derivatives. *Int. J. Mol. Sci.* **2015**, *16*, 8569–8590. [CrossRef] [PubMed]

160. Lall, R.; Syed, D.; Adhami, V.; Khan, M.; Mukhtar, H. Dietary Polyphenols in Prevention and Treatment of Prostate Cancer. *Int. J. Mol. Sci.* **2015**, *16*, 3350–3376. [CrossRef] [PubMed]

161. Popa, D.-S.; Rusu, M.E. Isoflavones: Vegetable Sources, Biological Activity, and Analytical Methods for Their Assessment. In *Superfood and Functional Food—The Development of Superfoods and Their Roles as Medicine*; Shiomi, N., Waisundara, V., Eds.; InTech: London, UK, 2017; ISBN 978-953-51-2942-4.

162. Mayr, C.; Wagner, A.; Neureiter, D.; Pichler, M.; Jakab, M.; Illig, R.; Berr, F.; Kiesslich, T. The green tea catechin epigallocatechin gallate induces cell cycle arrest and shows potential synergism with cisplatin in biliary tract cancer cells. *BMC Complement. Altern. Med.* **2015**, *15*, 194. [CrossRef] [PubMed]

163. Bimonte, S.; Leongito, M.; Barbieri, A.; Vecchio, V.D.; Barbieri, M.; Albino, V.; Piccirillo, M.; Amore, A.; Giacomo, R.D.; Nasto, A.; et al. Inhibitory effect of (−)-epigallocatechin-3-gallate and bleomycin on human pancreatic cancer MiaPaca-2 cell growth. *Infect. Agent Cancer* **2015**, *10*, 22. [CrossRef] [PubMed]

164. Budisan, L.; Gulei, D.; Zanoaga, O.M.; Irimie, A.I.; Chira, S.; Braicu, C.; Gherman, C.D.; Berindan-Neagoe, I. Dietary Intervention by Phytochemicals and Their Role in Modulating Coding and Non-Coding Genes in Cancer. *Int. J. Mol. Sci.* **2017**, *18*, 1178. [CrossRef] [PubMed]

165. Varoni, E.; Lo Faro, A.; Sharifi-Rad, J.; Iriti, M. Anticancer Molecular Mechanisms of Resveratrol. *Front. Nutr.* **2016**, *3*, 8. [CrossRef] [PubMed]

166. Poulose, S.M.; Miller, M.G.; Scott, T.; Shukitt-Hale, B. Nutritional Factors Affecting Adult Neurogenesis and Cognitive Function. *Adv. Nutr.* **2017**, *8*, 804–811. [CrossRef] [PubMed]

167. Pei, H.L.; Mu, D.M.; Zhang, B. Anticancer Activity of Pterostilbene in Human Ovarian Cancer Cell Lines. *Med. Sci. Monit.* **2017**, *23*, 3192–3199. [CrossRef]

168. Reiter, R.J.; Manchester, L.; Tan, D.X. Melatonin in walnuts: Influence on levels of melatonin and total antioxidant capacity of blood. *Nutrition* **2005**, *21*, 920–924. [CrossRef] [PubMed]

169. Cui, Z.; Liu, D.; Liu, C.; Liu, G. Serum selenium levels and prostate cancer risk: A MOOSE-compliant meta-analysis. *Medicine* **2017**, *96*, e5944. [CrossRef] [PubMed]

170. Hughes, D.J.; Duarte-Salles, T.; Hybsier, S.; Trichopoulou, A.; Stepien, M.; Aleksandrova, K.; Overvad, K.; Tjønneland, A.; Olsen, A.; Affret, A.; et al. Prediagnostic selenium status and hepatobiliary cancer risk in the European Prospective Investigation into Cancer and Nutrition cohort. *Am. J. Clin. Nutr.* **2016**, *104*, 406–414. [CrossRef]

171. Kris-Etherton, P.M.; Hu, F.B.; Ros, E.; Sabaté, J. The Role of Tree Nuts and Peanuts in the Prevention of Coronary Heart Disease: Multiple Potential Mechanisms. *J. Nutr.* **2008**, *138*, 1746S–1751S. [CrossRef]

172. Gylling, H.; Plat, J.; Turley, S.; Ginsberg, H.N.; Ellegård, L.; Jessup, W.; Jones, P.J.; Lütjohann, D.; Maerz, W.; Masana, L.; et al. Plant sterols and plant stanols in the management of dyslipidaemia and prevention of cardiovascular disease. *Atherosclerosis* **2014**, *232*, 346–360. [CrossRef]

173. Malinowski, J.; Gehret, M. Phytosterols for dyslipidemia. *Am. J. Health Syst. Pharm.* **2010**, *67*, 1165–1173. [CrossRef]

174. Marangoni, F.; Poli, A. Phytosterols and cardiovascular health. *Pharmacol. Res.* **2010**, *61*, 193–199. [CrossRef] [PubMed]

175. Eussen, S.R.B.M.; Rompelberg, C.J.M.; Klungel, O.H.; van Eijkeren, J.C.H. Modelling approach to simulate reductions in LDL cholesterol levels after combined intake of statins and phytosterols/-stanols in humans. *Lipids Health Dis.* **2011**, *10*, 187. [CrossRef] [PubMed]

176. Ros, E.; Izquierdo-Pulido, M.; Sala-Vila, A. Beneficial effects of walnut consumption on human health: Role of micronutrients. *Curr. Opin. Clin. Nutr. Metab. Care* **2018**, *21*, 498–504. [CrossRef] [PubMed]

177. Micallef, M.A.; Garg, M.L. The lipid-lowering effects of phytosterols and (n-3) polyunsaturated fatty acids are synergistic and complementary in hyperlipidemic men and women. *J. Nutr.* **2008**, *138*, 1086–1090. [CrossRef] [PubMed]

178. Mangialasche, F.; Kivipelto, M.; Mecocci, P.; Rizzuto, D.; Palmer, K.; Winblad, B.; Fratiglioni, L. High plasma levels of vitamin E forms and reduced Alzheimer's disease risk in advanced age. *J. Alzheimers Dis.* **2010**, *20*, 1029–1037. [CrossRef] [PubMed]

179. Mangialasche, F.; Westman, E.; Kivipelto, M.; Muehlboeck, J.; Cecchetti, R.; Baglioni, M.; Tarducci, R.; Gobbi, G.; Floridi, P.; Soininen, H.; et al. Classification and prediction of clinical diagnosis of Alzheimer's disease based on MRI and plasma measures of α-/γ-tocotrienols and γ-tocopherol. *J. Intern. Med.* **2013**, *273*, 602–621. [CrossRef]

180. Khanna, S.; Parinandi, N.; Kotha, S.; Roy, S.; Rink, C.; Bibus, D.; Sen, C. Nanomolar vitamin E alpha-tocotrienol inhibits glutamate-induced activation of phospholipase A2 and causes neuroprotection. *J. Neurochem.* **2010**, *112*, 1249–1260. [CrossRef]

181. Park, H.; Kubicki, N.; Gnyawali, S.; Chan, Y.; Roy, S.; Khanna, S.; Sen, C. Natural vitamin E α-tocotrienol protects against ischemic stroke by induction of multidrug resistance-associated protein 1. *Stroke* **2011**, *42*, 2308–2314. [CrossRef]

182. Eisenhauer, B.; Natoli, S.; Liew, G.; Flood, V. Lutein and zeaxanthin—Food sources, bioavailability and dietary variety in age-related macular degeneration protection. *Nutrients* **2017**, *9*, 120. [CrossRef]

183. Buscemi, S.; Corleo, D.; Di Pace, F.; Petroni, M.L.; Satriano, A.; Marchesini, G. The Effect of Lutein on Eye and Extra-Eye Health. *Nutrients* **2018**, *10*, 1321. [CrossRef] [PubMed]

184. Vishwanathan, R.; Iannaccone, A.; Scott, T.; Kritchevsky, S.; Jennings, B.; Carboni, G.; Forma, G.; Satterfield, S.; Harris, T.; Johnson, K.; et al. Macular pigment optical density is related to cognitive function in older people. *Age Ageing* **2014**, *43*, 271–275. [CrossRef] [PubMed]

185. Mewborn, C.M.; Lindbergh, C.A.; Robinson, T.L.; Gogniat, M.A.; Terry, D.P.; Jean, K.R.; Hammond, B.R.; Renzi-Hammond, L.M.; Miller, L.S. Lutein and Zeaxanthin Are Positively Associated with Visual–Spatial Functioning in Older Adults: An fMRI Study. *Nutrients* **2018**, *10*, 458. [CrossRef] [PubMed]

186. Cho, K.; Shin, M.; Kim, S.; Lee, S. Recent Advances in Studies on the Therapeutic Potential of Dietary Carotenoids in Neurodegenerative Diseases. *Oxid. Med. Cell. Longev.* **2018**, *2018*, 4120458. [CrossRef] [PubMed]

187. Berryman, C.E.; Grieger, J.A.; West, S.G.; Chen, C.-Y.O.; Blumberg, J.B.; Rothblat, G.H.; Sankaranarayanan, S.; Kris-Etherton, P.M. Acute consumption of walnuts and walnut components differentially affect postprandial lipemia, endothelial function, oxidative stress, and cholesterol efflux in humans with mild hypercholesterolemia. *J. Nutr.* **2013**, *143*, 788–794. [CrossRef] [PubMed]

188. Tuccinardi, D.; Farr, O.; Upadhyay, J.; Oussaada, S.; Klapa, M.; Candela, M.; Rampelli, S.; Lehoux, S.; Lázaro, I.; Sala-Vila, A.; et al. Mechanisms underlying the cardiometabolic protective effect of walnut consumption in obese people: A cross-over, randomized, double-blind, controlled inpatient physiology study. *Diabetes Obes. Metab.* **2019**. [CrossRef] [PubMed]

189. Ribeiro, P.; Silva, A.; Almeida, A.; Hermsdorff, H.; Alfenas, R. Effect of chronic consumption of pistachios (*Pistacia vera* L.) on glucose metabolism in pre-diabetics and type 2 diabetics: A systematic review. *Crit. Rev. Food Sci. Nutr.* **2019**, *59*, 1115–1123. [CrossRef]

190. Hashim, S.A.; VanItallie, T.B. Ketone body therapy: From the ketogenic diet to the oral administration of ketone ester. *J. Lipid Res.* **2014**, *55*, 1818–1826. [CrossRef] [PubMed]

191. Puchalska, P.; Crawford, P.A. Multi-dimensional Roles of Ketone Bodies in Fuel Metabolism, Signaling, and Therapeutics. *Cell Metab.* **2017**, *25*, 262–284. [CrossRef] [PubMed]

192. Pinto, A.; Bonucci, A.; Maggi, E.; Corsi, M.; Businaro, R. Anti-Oxidant and Anti-Inflammatory Activity of Ketogenic Diet: New Perspectives for Neuroprotection in Alzheimer's Disease. *Antioxidants* **2018**, *7*, 63. [CrossRef] [PubMed]

193. Claesson, M.; Jeffery, I.; Conde, S.; Power, S.; O'Connor, E.; Cusack, S.; Harris, H.; Coakley, M.; Lakshminarayanan, B.; O'Sullivan, O.; et al. Gut microbiota composition correlates with diet and health in the elderly. *Nature* **2012**, *488*, 178–184. [CrossRef] [PubMed]

194. Bian, G.; Gloor, G.B.; Gong, A.; Jia, C.; Zhang, W.; Hu, J.; Zhang, H.; Zhang, Y.; Zhou, Z.; Zhang, J.; et al. The Gut Microbiota of Healthy Aged Chinese Is Similar to That of the Healthy Young. *mSphere* **2017**, *2*, e00327-17. [CrossRef] [PubMed]

195. Kim, S.; Jazwinski, S. The Gut Microbiota and Healthy Aging: A Mini-Review. *Gerontology* **2018**, *64*, 513–520. [CrossRef] [PubMed]

196. Voreades, N.; Kozil, A.; Weir, T.L. Diet and the development of the human intestinal microbiome. *Front. Microbiol.* **2014**, *5*, 494. [CrossRef] [PubMed]

197. Koh, A.; De Vadder, F.; Kovatcheva-Datchary, P.; Bäckhed, F. From Dietary Fiber to Host Physiology: Short-Chain Fatty Acids as Key Bacterial Metabolites. *Cell* **2016**, *165*, 1332–1345. [CrossRef] [PubMed]

198. De Vadder, F.; Kovatcheva-Datchary, P.; Goncalves, D.; Vinera, J.; Zitoun, C.; Duchampt, A.; Bäckhed, F.; Mithieux, G. Microbiota-Generated Metabolites Promote Metabolic Benefits via Gut-Brain Neural Circuits. *Cell* **2014**, *156*, 84–96. [CrossRef] [PubMed]

199. Dinan, T.G.; Cryan, J.F. Gut Instincts: Microbiota as a key regulator of brain development, ageing and neurodegeneration. *J. Physiol.* **2017**, *595*, 489–503. [CrossRef] [PubMed]

200. Perez-Pardo, P.; Hartog, M.; Garssen, J.; Kraneveld, A.D. Microbes Tickling Your Tummy: The Importance of the Gut-Brain Axis in Parkinson's Disease. *Curr. Behav. Neurosci. Rep.* **2017**, *4*, 361–368. [CrossRef] [PubMed]

201. Killinger, B.A.; Labrie, V. Vertebrate food products as a potential source of prion-like α-synuclein. *NPJ Park. Dis.* **2017**, *3*, 33. [CrossRef] [PubMed]

202. Singh, R.K.; Chang, H.W.; Yan, D.; Lee, K.M.; Ucmak, D.; Wong, K.; Abrouk, M.; Farahnik, B.; Nakamura, M.; Zhu, T.H.; et al. Influence of diet on the gut microbiome and implications for human health. *J. Transl. Med.* **2017**, *15*, 73. [CrossRef]

203. Tzounis, X.; Vulevic, J.; Kuhnle, G.G.; George, T.; Leonczak, J.; Gibson, G.R.; Kwik-Uribe, C.; Spencer, J.P. Flavanol monomer-induced changes to the human faecal microflora. *Br. J. Nutr.* **2008**, *99*, 782–792. [CrossRef] [PubMed]
204. Byerley, L.O.; Samuelson, D.; Blanchard, E., 4th; Luo, M.; Lorenzen, B.N.; Banks, S.; Ponder, M.A.; Welsh, D.A.; Taylor, C.M. Changes in the gut microbial communities following addition of walnuts to the diet. *J. Nutr. Biochem.* **2017**, *48*, 94–102. [CrossRef] [PubMed]
205. Holscher, H.D.; Guetterman, H.M.; Swanson, K.S.; An, R.; Matthan, N.R.; Lichtenstein, A.H.; Novotny, J.A.; Baer, D.J. Walnut consumption alters the gastrointestinal microbiota, microbially derived secondary bile acids, and health markers in healthy adults: A randomized controlled trial. *J. Nutr.* **2018**, *148*, 861–867. [CrossRef] [PubMed]
206. Bamberger, C.; Rossmeier, A.; Lechner, K.; Wu, L.; Waldmann, E.; Fischer, S.; Stark, R.G.; Altenhofer, J.; Henze, K.; Parhofer, K.G. A walnut-enriched diet affects gut microbiome in healthy caucasian subjects: A randomized, controlled trial. *Nutrients* **2018**, *10*, 244. [CrossRef] [PubMed]
207. Ukhanova, M.; Wang, X.; Baer, D.J.; Novotny, J.A.; Fredborg, M.; Mai, V. Effects of almond and pistachio consumption on gut microbiota composition in a randomised cross-over human feeding study. *Br. J. Nutr.* **2014**, *111*, 2146–2152. [CrossRef] [PubMed]
208. Holscher, H.D.; Taylor, A.M.; Swanson, K.S.; Novotny, J.A.; Baer, D.J. Almond consumption and processing affects the composition of the gastrointestinal microbiota of healthy adult men and women: A randomized controlled trial. *Nutrients* **2018**, *10*, 126. [CrossRef] [PubMed]
209. Tindall, A.M.; Petersen, K.S.; Skulas-Ray, A.C.; Richter, C.K.; Proctor, D.N.; Kris-Etherton, P.M. Replacing Saturated Fat With Walnuts or Vegetable Oils Improves Central Blood Pressure and Serum Lipids in Adults at Risk for Cardiovascular Disease: A Randomized Controlled-Feeding Trial. *J. Am. Heart Assoc.* **2019**, *8*, e011512. [CrossRef] [PubMed]
210. Di Daniele, N.; Noce, A.; Vidiri, M.; Moriconi, E.; Marrone, G.; Annicchiarico-Petruzzelli, M.; D'Urso, G.; Tesauro, M.; Rovella, V.; De Lorenzo, A. Impact of Mediterranean diet on metabolic syndrome, cancer and longevity. *Oncotarget* **2017**, *8*, 8947–8979. [CrossRef] [PubMed]

Review

Brown Macroalgae as Valuable Food Ingredients

Nuno C. Afonso, Marcelo D. Catarino, Artur M. S. Silva and Susana M. Cardoso *

QOPNA & LAQV-REQUIMTE, Department of Chemistry, University of Aveiro, 3810-193 Aveiro, Portugal
* Correspondence: susanacardoso@ua.pt; Tel.: +351-234-370-360; Fax: +351-234-370-084

Received: 2 August 2019; Accepted: 26 August 2019; Published: 2 September 2019

Abstract: Due to the balanced nutritional value and abundance of bioactive compounds, seaweeds represent great candidates to be used as health-promoting ingredients by the food industry. In this field, Phaeophyta, i.e., brown macroalgae, have been receiving great attention particularly due to their abundance in complex polysaccharides, phlorotannins, fucoxanthin and iodine. In the past decade, brown algae and their extracts have been extensively studied, aiming at the development of well-accepted products with the simultaneous enhancement of nutritional value and/or shelf-life. However, the reports aiming at their bioactivity in in vivo models are still scarce and need additional exploration. Therefore, this manuscript revises the relevant literature data regarding the development of Phaeophyta-enriched food products, namely those focused on species considered as safe for human consumption in Europe. Hopefully, this will create awareness to the need of further studies in order to determine how those benefits can translate to human beings.

Keywords: *Phaeophyceae*; food fortification; algae; fibres; phlorotannins; fucoxanthin; minerals; iodine; nutrition; health-benefits; functional food

---

1. Introduction

Marine macroalgae i.e., seaweeds, have been well recognised for centuries by their importance in the diet of many Far Eastern countries, such as Japan and Korea [1,2]. They are nutritionally very wealthy, being claimed as a great source of complex polysaccharides, minerals, proteins and vitamins, as well as of several phycochemicals [3,4]. Actually, a regular seafood consumption, in which seaweeds are included, has been associated with a myriad of health benefits and a longer life expectancy [5,6] and these combined facts are leading to an increased interest in the manufacture and consumption of high-value macroalgae-derived products in Western cultures. Their consumption is also in line with the increasing awareness of consumers' perceptions towards organic products and of environmentally sustainable products. As a result, according to the Seafood Source report, the global seaweed market is expected to grow to USD 22.1 billion by 2024 [7].

Nowadays, amongst all three types of macroalgae (green, red and brown), brown algae are the most consumed species (66.5%), followed by red (33%) and green (5%) algae [8]. *Phaeophyceae* possess a high content of diverse phycochemicals and have been repeatedly claimed to exert important therapeutic properties, which turn them into great candidates to be used as bioactive agents in many industries, including the functional food market [9–11].

Europe has been recently highlighted as one of the most innovative regions regarding the use of seaweeds as a food ingredient with new products emerging on the European market increasing at exponential rates [12]. In fact, according to the Seafood Source report, the new products containing this new ingredient launched on the European market increased by 147% between 2011 and 2015, making Europe the most innovative region globally after Asia [13]. In this region, algae are considered as novel foods and a limited number of brown macroalgae species are considered to be safe for human consumption, namely *Fucus vesiculosus, Fucus serratus, Himanthalia elongata, Undaria pinnatifida, Ascophyllum nodosum, Laminaria digitata, Laminaria saccharina, Laminaria japonica* and *Alaria esculenta* [14].

The present manuscript reviews the current knowledge on the incorporation of these brown seaweed species and brown seaweed-derived high-value products in common food products.

2. Chemical Particularities of Brown Macroalgae

The health-claims of *Phaeophyceae* are mainly associated with their abundance in specific nutrients and phycochemicals, particularly fibres, phlorotannins, fucoxanthin and minerals. However, their levels are greatly variable according to distinct factors, including the algae genera and species, maturity and the environmental conditions, i.e., the variations to which the natural habitat of algae might be subjected, namely season, temperature, salinity, oceanic currents, waves or even depth of immersion, as well as post-harvesting storage and processing conditions [9,10,15–18]. As such, this section describes their main structural characteristics as well as some of their most relevant bioactivities, highlighting their overall abundance in the targeted macroalgae of this review.

2.1. Polysaccharides

Brown macroalgae are known to produce different types of polysaccharides and/or fibres which, despite their variability, represent major components that can reach up to 70% of their dried weight (DW) [19]. In fact, previous reported data set the polysaccharide contents of relevant species, namely *L. japonica*, *F. vesiculosus*, *A. nodosum*, *Saccharina longicruis*, *U. pinnatifida* and *Sargassum vulgare* at 37.5%, 65.7%, 69.6%, 57.8%, 35.2% and 67.8% DW, respectively [20–23]. Amongst them, alginates, fucoidans and laminarins are the most representative ones.

Alginic acids or alginates, i.e., the salts of alginic acid, are the main polysaccharides in brown seaweeds [24], reaching up to 16.9% DW in *S. vulgare*, 20% DW in *S. longicruris*, 24% DW in *A. nodosum*, 32% DW in *Sargassum carpophyllum*, 40% DW in *Laminaria hyperborean* [25], 41% in *Sargassum siliquosum* and even to 59% DW in *F. vesiculosus* [26]. Within the cell wall, these polysaccharides are known to be partially responsible for the seaweed's flexibility [3] and therefore, expectedly, brown seaweeds grown under turbulent conditions usually have superior alginate contents than those of calm waters. In terms of structure, alginic acids or their corresponding extracted salts consist of α-L-guluronic acid (G) and β-D-mannuronic acid (M) (1→4)-linked residues arranged either in heteropolymeric (MG) and/or homopolymeric (M or G) blocks (Figure 1A–C). Regardless, the variations caused by diverse factors (e.g., algae species, seasonability, parts of the algae) are expected [16]. Noteworthy, alginates are considered one of the most important food colloids, with many applications in several industries such as foods, paper, pharmaceutical or cosmetics [27]. In fact, G-blocks in the presence of ions, such as Ca^{2+} form is the so-called egg-box, thus granting stiffness to the overall structure and conferring gel-forming properties to these polysaccharides [28]. Therefore, they are usually used as thickeners, gels, emulsifiers and stabilizers in order to improve quality parameters, especially in food grade products [29]. In addition to their wide applications, more recently, dietary alginates are being associated with positive health benefits in the gastrointestinal tract and appetite regulation, as well as antihypertensive and anti-diabetic effects [30]. Alginates are also considered great prebiotics as they were demonstrated to significantly promote the growth of several bacteria, including *Bifidobacterium bifidum*, *Bifidobacterium longum* and *Lactobacilli*, alongside with the increase of acetic acid, propionic acid and several short chain fatty acid metabolites, while decreasing deleterious metabolites, including faecal sulphide, phenol, *p*-cresol, indole, ammonia and skatole [31].

Fucoidans i.e., metabolites belonging to the fucans family, also have a structural role in brown algae, mostly preventing dehydration [3]. Their reported content in *Phaeophytae* is variable, ranging from approximately 6–8% DW in *L. japonica*, 3.2–16% DW in *U. pinnatifida*, and 3.4–25.7% DW in *F. vesiculosus* [10,32]. These polysaccharides are mainly composed of fucose and sulphate, although the presence of other types of monosaccharides (glucose, galactose, mannose, xylose and uronic acids), acetyl groups and proteins also occur [33]. Despite being molecules with high structural diversity, the representative backbone of fucoidans consists of (1→3)- and (1→4)-linked α-L-fucopyranose residues, and these polysaccharides are commonly divided in two types, the first being characterized by long chains of (1→3)-linked α-L-fucopyranose residues (mainly present in *L. saccharina*, *L. digitata*, *C. okamuranus*, and *Chorda filum*) and the second consisting of alternating (1→3)- and (1→4)-linked α-L-fucopyranose residues (characteristic from *A. nodosum* and *Fucus* spp.) (Figure 1D,E) [24,34].

Over the last years, extensive biological activities (e.g., antitumor, antioxidant, anticoagulant, antithrombotic, immunoregulatory, antiviral, anti-inflammatory among others) have been demonstrated with promising preclinical results, as recently reviewed [35]. As an example of in vivo studies, the effectiveness of a *F. vesiculosus* fucoidan injection towards oxidative stress in hyperoxaluric rats was demonstrated by Veena et al. [36] to be mediated by the stimulation of antioxidant enzymes, such as superoxide dismutase (SOD), catalase (CAT) and glutathione peroxidase (GPx). Moreover, Huang et al. [37] reported that the ingestion of fucoidans isolated from *L. japonica* reduced the serum levels of total cholesterol, triglycerides, and low-density lipoprotein cholesterol in hyperlipidaemic rats, while increasing the enzymatic activity of lipoprotein lipase, hepatic lipase and lecithin cholesterol acyltransferase. In addition, their relevance in obesity and/or diabetes was also highlighted, in particular, by Xan et al. [38], who reported *F. vesiculosus* fucoidans' ability to inhibit α-glucosidase in vitro and to decrease the fasting blood glucose and glycosylated haemoglobin levels of db/db mice, as well as by Kim et al. [39], when administrating *U. pinnatifida* fucoidans to the same animal model. Although there is limited evidence to implicate a role of fucoidans in the gut microbiota, some works reported that fucoidans from different brown algae species greatly contributed for the increase in the growth of *Bifidobacterium*, *Lactobacillus* and *Ruminococcaceae*, either in mice or human faecal samples [31].

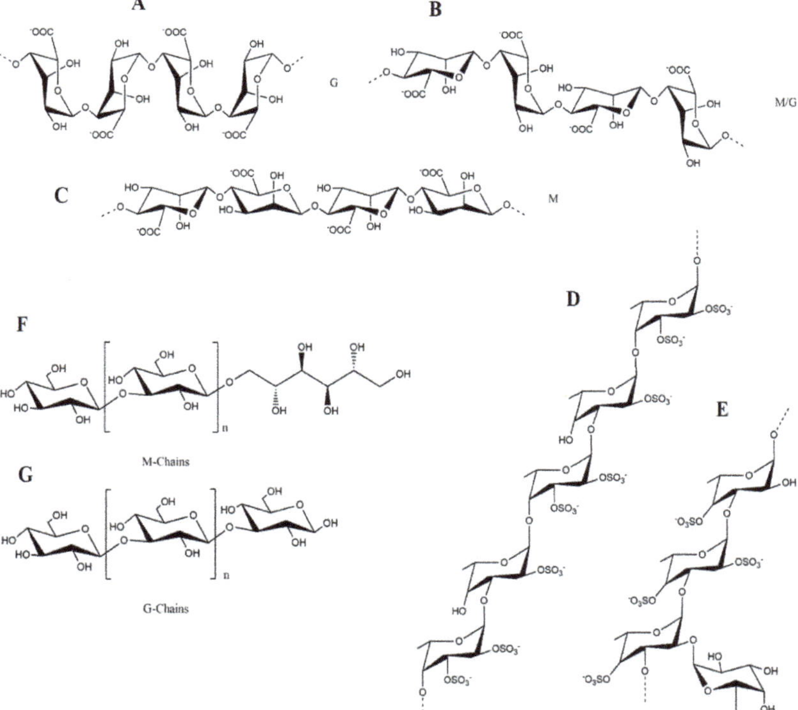

Figure 1. The structure of representative polysaccharides found in brown algae: (**A–C**) alginic acids; (**D–E**) fucoidans from *A. nodosum/F. vesiculosus* and *S. latissima*, respectively; (**F–G**) laminarins M and G chains.

Laminarins, also named laminarans or leucosins, on the other hand, belong to the glucan family and serve as reserve metabolites in brown algae [40]. These are commonly found in the fronds of *Laminaria* and *Saccharina* macroalgae and, to a lesser extent, in *Ascophyllum*, *Fucus* and *Undaria* species [41]. In general, they are relatively small polysaccharides composed of β-(1→3)-linked glucose monomers, containing large amounts of sugars and a low fraction of uronic acids (Figure 1F,G) [42]. Depending on the type of sugar at the reducing end, they are classified in two distinct types, specifically the M chains, which have a terminal 1-O-substituted D-mannitol, and the G chains, possessing a terminal glucose [16]. The content of these polysaccharides is also season-dependent, since seaweeds show no production or very less amounts in the winter and maximum production during summer and autumn [23]. As previously stated, *Laminariales* are known to produce high amounts of laminarins, with contents reaching up to 35% DW, particularly in *L. saccharina* and *L. digitata* [40]. Other reported values of laminarins content comprise those of *A. esculenta*, *U. pinnatifida*, *A. nodosum* and *F. serratus* (11.1%, 3%, 4.5% and up to 19% DW, respectively) [22–24,40].

The bioactivities of laminarins are scarcely exploited, but still they are considered as fibres and therefore can be partially or totally fermented by the endogenous intestinal microflora. This was demonstrated by Devillé et al. [43], when comparing the results from in vitro digestibility tests, where no hydrolysis of this fibre occurred, to those of in vivo tests, for which no traces of laminarin were detected in the faeces of fed Winstar rats after ingestion.

It should be noted that oligosaccharides from brown macroalgae polysaccharides may also exhibit interesting bioactivities, which can differ from those of the original polysaccharides. In this topic, alginate oligosaccharides have been claimed to possess radical scavenging activities with the great potential for application in the food industry [44], and even promising effects on neuro-inflammation, promoting microglial phagocytosis. This could be of great relevance for their application as a nutraceutical agent for neurodegenerative diseases, such as Alzheimer's disease [45]. In turn, in vivo experiments on renovascular hypertensive rats revealed that fucoidan oligosaccharides exhibited anti-hypertensive effects comparable to those of captopril, i.e., an approved drug used for the treatment of hypertension [46].

2.2. Phlorotannins

Phlorotannins are phenolic compounds almost exclusive to *Phaeophytae* and also represent their main phenolic pool. In brown seaweeds, they are associated with a myriad of functions, ranging from structural cell wall components, to biosyinthetic percursors and defensive mediators against natural enemies, acting as herbivore deterrents, inhibitors of digestion and agents against bacteria [11]. Phlorotannins are known to accumulate mostly in physodes (i.e., specialized membrane-bound vesicles of the cell cytoplasm), with levels that might represent up to 25% of seaweed's DW, despite variations which occur depending on distinct factors [47]. For example, the higher levels of phlorotannins in *Fucus* spp. are associated with high salinity waters and solar exposure during summer [10].

Being part of the tannins group, phlorotannins present a polymeric structure derived from several phloroglucinol (1,3,5-trihydroxybenzene) units and possess a high number of hydroxy groups, thus conferring them solubility in water [48]. Depending on the linkage between phloroglucinol monomer units, a wide range of compounds with different molecular weights can be obtained [49], which overall, are divided in four categories for each type of linkage: Fuhalols and phlorethols based on ether linkage, fucols based on C-C linkage, fucophlorethols for a combination of the previous ones, and, finally, eckols and carmalols, based on dibenzodioxin linkage (Figure 2).

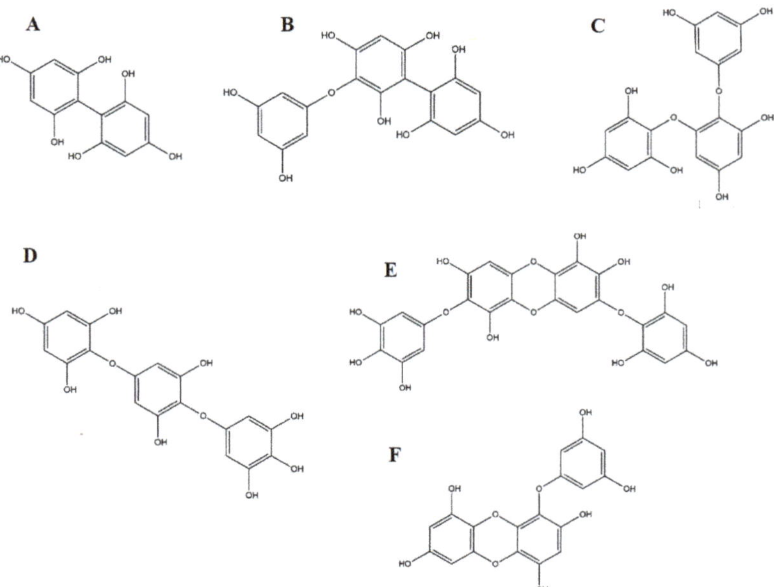

Figure 2. Some representative phlorotannins from brown seaweeds: (**A**) Fucol; (**B**) Fucophlorethol; (**C**) Phlorethol; (**D**) Fuhalol; (**E**) Carmalol; (**F**) Eckol.

Phenolic extracts from brown seaweeds have been demonstrated to exhibit various biological activities, including antioxidant, antidiabetic, anti-inflammatory and others [11,50,51]. In this regard, O'Sullivan et al. [52] observed the augment of glutathione levels in Caco-2 cell models when incubated with *A. nodosum*, *F. vesiculosus* and *F. serratus* phlorotannins extracts, while also highlighting the protective effects of the latter on the same model pretreated with H_2O_2. In vivo experiments have even demonstrated that the oral administration of 200 mg/kg/day of *F. vesiculosus* polyhenol-rich extracts over 4 weeks to Sprague-Dawley rats could increase the blood plasma reducing power, paraoxonase/arylesterase 1 (PON-1) activity and $O_2^{\bullet-}$ scavenging activity by 29%, 33% and 25%, respectively [53]. Likewise, the antidiabetic properties of *A. nodosum* and *F. vesiculosus* phenolic-rich extract were observed in vivo as the postprandial blood glucose levels and insulin peak decreased 90% and 40%, respectively, on rats under hyperglycemic diets supplemented with 7.5 mg/kg compared to the unsupplemented group [54]. In fact, the ingestion of 500 mg of this mixture containing *A. nodosum* and *F. vesiculosus* 30 min prior to the consumption of carbohydrates was shown to reduce the insulin incremental area of the curve and an increase in insulin sensitivity in a human clinical trial [55]. Human trials have also been carried out to evaluate the potential antiobesity effect of polyphenolic-rich extracts of *A. nodosum* (100 mg/day for 8 weeks). Although the treatment did not exhibit any significant benefits (no significant changes in C-reactive protein, antioxidant status or inflammatory cytokines), with the exception of a modest decrease of the DNA damage in the obese group, several phlorotannin metabolites were detected in the subjects plasma and urine, indicating that these compounds are metabolised and absorbed into the systemic circulation [56]. These observations are in line with those reported by Corona et al. [57] who also described the appearance of phlorotannin metabolites in urine and plasma collected from humans after consuming a capsule of *A. nodosum* extract containing about 100 mg of polyphenols.

2.3. Fucoxanthin

In opposition to red and green macroalgae, *Phaophytae* are characterized by the presence of the carotenoid fucoxanthin, which is responsible for their specific coloration. Fucoxanthin is a xanthophyll belonging to the tetraterpenoid family with a structure consisting of an unusual allenic bond and a 5,6-monoepoxide in its molecule (Figure 3). The content of this pigment is highly variable amongst different species, as well as dependent on extrinsic factors, with a large range being even described within the same species. The reported levels comprise in 171 mg/kg (*Fucus spiralis*), 224 mg/kg (*Fucus distichus*), 364 mg/kg (*Fucus evanescens*), 172–660 mg/kg (*A. nodosum*), 178–468 mg/kg (*Laminaria* spp.) [41,58].

Recently, this xanthophyll has earned particular attention mainly because of its promising effects in terms of antidiabetic, anti-obesity and antioxidant activities [59,60], with claims being supported by in vivo studies. For instance, the administration of *U. pinnatifida* lipids rich in fucoxanthin to male diabetic mice were associated with insulin resistance amelioration and the reduction of blood glucose levels [61]. Moreover, fucoxanthin isolated from the same macraolgae species was also shown to inhibit the differentiation of 3T3-L1 preadipocytes into adipocytes by down-regulating peroxisome proliferator-activated receptor gamma (PPARγ) [62]. Furthermore, a diet based on *U. pinnatifida* fucoxanthin was capable of inducing uncoupling protein 1 (UCP1) expression in white adipose tissue (WAT) of obese mice. When added as a supplement to rats fed with a high-fat diet, it prompted a decrement of the mRNA expression of significant enzymes associated with lipid metabolism, such as fatty acid synthase, acyl-CoA cholesterol acyltransferase, hepatic acetyl-CoA carboxylase, glucose-6-phosphate dehydrogenase, hydroxy-3-methylglutaryl coenzyme A and SREBP-1C [63,64].

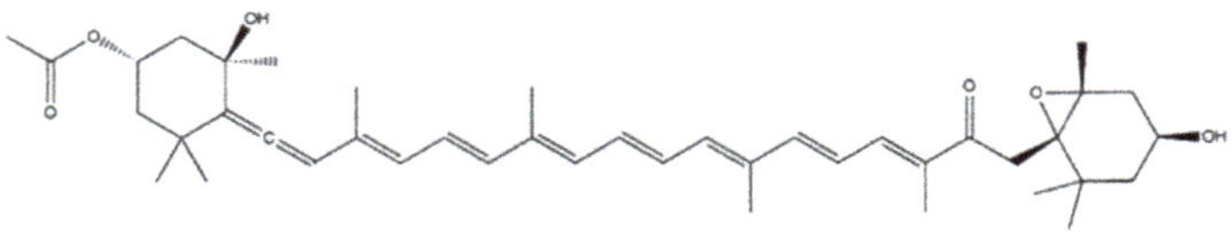

Figure 3. Structure of fucoxanthin.

2.4. Minerals

Due to their structural and physiological features, brown macroalgae are recognized for their superior ability to accumulate minerals. Although the content of minerals like calcium, magnesium, phosphorus, potassium, sodium and iron is usually high within the macroalgae matrix, one of the standout aspects, comparatively to plants in general, are both their low Na/K ratios and high iodine levels [4]. In fact, it is well accepeted that low Na/K ratios are an important aspect for good maintenance of cardiovascular health [65]. Therefore, according to the World Health Organization (WHO), the recommended value for this should be close to one, so consumption of food products with this proportion or below should be considered for healthy cardiovascular purposes [66]. In fact, several studies point to a Na/K ratio ranging between 0.3 and 1.5 in brown seaweeds, with particular interest for *Laminaria* spp. (0.3–0.4) from Spain origins, wich are significantly lower than diverse food products, such as cheddar cheese (8.7), olives (43.6), and sausages (4.9) [4,67–70]. Additionally, *Phaeophyceae* seaweeds, due to the rich composition in alginates and sulphated polysaccharides coupled with the presence of haloperoxidases in the cell walls, allow the accumulation of iodine to more than 30,000 times over its concentration in the surrounding environment which is even higher than any edible plant [71]. The major contents of iodine were documented for *L. digitata*, *A. nodosum*, *H. elogata* and *U. pinnatifida* exhibiting concentrations of 70, 18.2, 10.7 and 3.9 mg/100 g wet weight, respectively [72]. Moreover, other studies also highlight the particular affinity of Laminarales to accumulate iodine, particularly *L. digitata*, in which values are known to reach 9014 and 8122 mg/kg DW, in spring and autumn, respectively [23].

3. Use of Brown Macroalgae as Food Ingredient

Being considered as a rich and balanced source of nutrients and bioactive compounds, consumers and food industries have a growing interest to introduce macroalgae, including *Phaeophytae*, into the dietary habits of the western countries, with new products already being launched in the markets at high rates in Europe. The usage of brown species as food ingredients has, however, to overcome huge challenges, that go from the guarantee of enough biomass to sustain the market development, to the gain of consistent knowledge of their physicochemical features, as well as understanding the extension of their impact when used as ingredients in foods. This section highlights some of the developed foods in the field of seaweed-fortified products, categorized by the respective incorporated algae species, considering the authorized seaweeds for human consumption in France/Europe [14], and finalising with the influence this incorporation has on the foods' chemical, functional and structural behaviour.

3.1. Fucus vesiculosus

F. vesiculosus has found application as a functional ingredient in many different food matrices, mostly as a source of phlorotannins and antioxidant compounds, aiming to prevent food spoilage resultant from oxidative deterioration (Table 1). Fish and fish-derived products are one of the main matrices where several studies with this seaweed have been conducted. In this context, Dellarosa et al. [73] reported that neither aqueous nor 80% ethanol extracts from *F. vesiculosus* had significant effects on the lipid oxidation of fish cakes enriched with omega-3 polyunsaturated fatty acids, throughout a 28-days refrigerate storage. Nevertheless, the authors showed that no off flavour was detected in any samples tested, with low scores of rancid odour and flavour being registered in the sensory analysis. On the other hand, some studies conducted on cod fish muscle and/or protein indicated that the incorporation of *F. vesiculosus* extracts could indeed prevent the lipid peroxidation events and even improve some of their sensorial aspects. In fact, the effects of the incorporation of 1% and 2% of the antioxidant dietary fibre extracted from *F. vesiculosus* into minced horse mackerel revealed a significant reduction of the fish mince lipid oxidation throughout the 5 months of storage at –20 °C. These factors reduced the total drip after thawing and cooking the horse mackerel mince up to 3 months of frozen storage, a fact that could be due to the water holding capacities of the fibre. Furthermore, although the addition of 2% (but not 1%) of antioxidant dietary fibre caused changes in the fish mince flavour compared to the control, these were actually considered positive by the sensory panellists [74].

In a different approach, Wang et al. observed that some oligomeric phlorotannin sub-fractions obtained by Sephadex LH-20 chromatography from an 80% ethanol extract of *F. vesiculosus* were able to completely inhibit the haemoglobin-catalysed lipid oxidation in both washed cod muscle and cod protein isolates systems, during an 8-day period of ice storage. Moreover, with a concentration of 300 mg/kg, the effectiveness of these phlorotannins sub-fractions were comparable to that of 100 mg/kg propyl gallate, i.e., a highly effective synthetic antioxidant in muscle foods, thus evidencing the great potential of oligomeric phlorotannins to be exploited as natural antioxidants in fish and fish-derived products [75]. Similar results were further reported by Jónsdóttir et al. [76], who observed an inhibition of the lipid oxidation in haemoglobin-fortified washed cod mince system after incorporating 300 mg phloroglucinol equivalents/kg of an ethyl acetate fraction obtained from an 80% ethanol extract of *F. vesiculosus*. Other authors also demonstrated that the incorporation of a *F. vesiculosus* phlorotannin-rich fraction (obtained with 80% ethanol and further purified with ethyl acetate) into cod protein hydrolysates, not only prevented the lipid oxidation reactions during storage, but also increased their final antioxidant activity [77,78] and could even improve the bitter, soap, fish oil and rancidity taste of the final protein hydrolysates [77].

Table 1. Selected studies reporting the effects of the incorporation of *F. vesiculosus* or isolates as ingredients in different food matrices.

Functional Food	Functional Ingredient	Results	Ref.
Fish cakes	*F. vesiculosus* extracts: 100% H_2O, 80% EtOH	No off-flavours and lower rancid odour and flavour None of the extracts had influence on lipid oxidation nor quality of the products	[73]
Cod muscle and protein isolates	*F. vesiculosus* 80% EtOH extract and further fractions (EtOAc + Sephadex LH-20)	↓ Lipid oxidation in both fish muscle and protein isolates 300 mg/kg of the oligomeric phlorotannin fractions exhibited an effect comparable to that of 100 mg/kg propyl gallate	[75]
Cod mince	EtOAc fraction of *F. vesiculosus* 80% EtOH extract	↓ Lipid oxidation in fish muscle	[76]
Cod protein hydrolysates	EtOAc fraction of *F. vesiculosus* 80% EtOH extract	↓ Lipid hydroperoxide and TBARS formation during protein hydrolyzation ↑ Antioxidant activity of the final protein hydrolysates	[78]
Cod protein hydrolysates	EtOAc fraction of *F. vesiculosus* 80% EtOH extract	↓ Lipid oxidation during protein hydrolysates freeze drying ↑ Antioxidant activity of the final protein hydrolysates Improved sensorial aspects (bitter, soap, fish oil and rancidity taste)	[77]
Minced horse mackerel	*F. vesiculosus* antioxidant dietary fibre	↓ Lipid oxidation during 5 months of storage at −20 °C ↓ Total dripping after thawing and cooking after up to 3 months of frozen storage Improved fish mince flavour	[74]
Granola bars enriched with fish oil emulsion	*F. vesiculosus* 100% H_2O, 70% acetone and 80% EtOH extracts	↓ Oxidation products after storage ↓ Iron-lipid interactions Acetone and EtOH extracts provided additional lipid oxidation protection ↑ Phenolic content, radical scavenging activity and interfacial affinity of phenolic compounds Possible tocopherol regeneration	[79]
Granola bars enriched with fish oil emulsion	*F. vesiculosus* 100% H_2O, 70% acetone and 80% EtOH extracts	↓ Lipid oxidation during storage ↑ Effectiveness for lower concentrations of EtOH and acetone extracts ↑ Phenolic content, radical scavenging activity and interfacial affinity of phenolic compounds Possible tocopherol regeneration	[80]
Fish-oil-enriched milk and mayonnaise	*F. vesiculosus*: EtOAc fraction from an 80% EtOH extract, 100% H_2O	↑ Lipid stability and ↓ oxidation of EPA and DHA and subsequent secondary degradation products in both foods—H_2O extract at 2.0 g/100 g exerted higher inhibitory effects on mayonnaise's peroxide formation.	[81]
Fish-oil-enriched mayonnaise	*F. vesiculosus* 100% H_2O, 70% acetone, and 80% EtOH extracts	Dose-dependent inhibition of lipid oxidation exhibited by EtOH and acetone extracts H_2O extract increased peroxide formation	[82]
Pork liver pâté	*F. vesiculosus* commercial extract	Decrease in lightness values after storage Redness and yellowness maintained after storage Protection against oxidation comparable to BHT samples ↓ Total volatile compounds	[83]

Table 1. *Cont.*

Functional Food	Functional Ingredient	Results	Ref.
Pork patties	*F. vesiculosus* 50% EtOH extracts	↓ TBARS slightly Did not improve colour, surface discoloration or odour attributes No significant differences between seaweed and control samples in sensory analysis	
Milk	*F. vesiculosus* 60% EtOH extracts	↑ Milk lipid stability and shelf-life characteristics Appearance of greenish colour and fishy taste Overall sensory attributes were worsened	[84]
Yoghurts	*F. vesiculosus* 60% EtOH extracts	No influence on chemical and microbiological characteristics ↑ Yogurts lipid stability and shelf-life characteristics Overall sensory attributes were worsened	[85]
Pasteurized apple beverage	*F. vesiculosus* fucoidan extract	Dose-, time- and temperature-dependent bacteriostatic and bactericidal effects against *L. monocytogenes* and *S. typhimurium* *S. typhimurium* showed higher sensitivity to the extract	[86]
Bread	*F. vesiculosus* powder	↑ Dough viscosity and wheat dough consistency ↓ Porosity ↑ Density, crumb firmness and green colour of crust 4% seaweed powder was considered optimal	[87]

↑: increased; ↓: decreased; BHT: 2,6-di-*tert*-butyl-4-methylphenol; DHA: docosahexaenoic acid; EPA: eicosapentanoic acid; EtOAc: ethyl acetate; EtOH: ethanol; TBARS: Thiobarbituric acid reactive substances.

The fortification of food matrices with fish oils rich in n-3 long chain polyunsaturated fatty acids has been in high demand during recent years due to increasing consumer awareness of the beneficial effects of docosahexaenoic and eicosapentaenoic acids (DHA and EPA, respectively). However, this usually decreases the foods' oxidative stability, leading to the development of undesirable off-flavours and consequent shelf-life reduction [88]. In this field, *F. vesiculosus* extracts were found to be highly promising. According to Karadağ et al., [79] the introduction of 0.5 and 1 g/100 g of both *F. vesiculosus* ethanol and acetone extracts into fish oil-enriched granola bars effectively improved their lipid stability, contributing to an increase of the foods' phenolic content, radical scavenging activity, interfacial affinity of phenolics and eventual regeneration of tocopherol, which consequently cause the reduction of the iron-lipid interactions as well as the lipid oxidation during the storage period. These results agree with previous data demonstrating that addition of both ethanol and acetone *F. vesiculosus* extracts to granola bars enriched with multi-layered fish oil emulsion contributed to the reduction of the formation of primary and secondary oxidation products over the period of storage at 20 °C [80]. Enhancement of lipid stability was also described in two other fish oil-fortified food matrices, namely mayonnaise and milk, after incorporation of 1.0–2.0 g/100 g of an ethyl acetate fraction, obtained from *F. vesiculosus* 80% ethanol extract (rich in phenolics and carotenoids) [81], as well as in fish oil-fortified mayonnaise added with 1.5–2.0 g/kg of both acetone and ethanol extracts of this seaweed species [82]. Interestingly, in the particular case of fish oil-fortified mayonnaise, Hermund et al. [81] found that, despite its lower content of phenolics and carotenoids, *F. vesiculosus* water extracts, at high concentrations, could prevent the peroxides formation more effectively than the ethyl acetate fraction, much likely due to its higher metal chelating capacity resultant from the presence of polysaccharides or other highly polar compounds with strong metal chelating capacities. This outcome was, however, refuted in a latter study that reported an increased peroxide formation in fish oil-enriched mayonnaise also incorporated with *F. vesiculosus* water extracts [82]. The disparity found between these two works might be related to the differences in the trace metal contents of the aqueous extracts performed in each study since the former had much lower iron content than the latter, which might be responsible for the induction of lipid oxidation in the food matrix.

Recently, the fortification of canola oil with 500 ppm of *F. vesiculosus* water extract was reported to reduce approximately 70% of the peroxides formation and 50% of the thiobarbituric acid reactive substances (TBARS) value compared to the control samples, both under accelerated storage conditions (60 °C). This confirms that this extract may in fact hold the potential to be exploited as a food antioxidant agent. Indeed, under similar conditions, butylated hydroxytoluene (BHT) (at 50 ppm) only inhibited peroxides formation and TBARS by 25% and 20%, respectively, thus showing that seaweed extracts could be used as a potential substitute for synthetic antioxidants. In the same line, in a different food matrix, namely low-fat pork liver pâté, the incorporation of 500 mg/kg of a commercial antioxidant extract of *F. vesiculosus* was also shown to be as effective as 50 ppm of BHT at inhibiting the formation of primary and secondary oxidation products over 180 days under storage at 4 °C, as well as in the maintenance of the redness and yellowness which were lowered in the control samples [83]. On the other hand, the fortification of pork patties with *F. vesiculosus* 50% ethanol extracts (250–1000 mg/kg) showed low performances on samples oxidative stability, with modest inhibitory effects on TBARS, compared to the control samples, but very far from that exhibited by BHT. Additionally, regardless the good acceptability in the sensory analysis, the incorporation of these *F. vesiculosus* extracts failed to improve colour, surface discoloration or odour attributes [89]. Therefore, further studies are necessary to conclude whether extracts of this seaweed are suitable for the application as oxidation inhibitors for the long-term storage of meat products.

Further aiming lipid stabilization in dairies, O'Sullivan et al. [84,85] tested the incorporation of 0.25% and 0.5% (*w/w*) of 60% and 40% ethanol extracts from *F. vesiculosus* into milk and yogurt, respectively. Indeed, both products showed a significant reduction of lipid oxidation alongside with improvements on their shelf-life characteristics. However, neither were well accepted in the sensory analysis, even for the lower concentrations, as the panellists reported an unpleasant green/yellowish colour and a fishy taste.

Although the majority of the studies carried out with this seaweed species were focused on their antioxidant activity and capacity to enhance foods' lipid stability, other authors have tried the incorporation of *F. vesiculosus* with different purposes. In a recent work, the incorporation of *F. vesiculosus* fucoidans into a new functional pasteurized apple beverage was found to be useful for controlling the growth of an undesirable microorganism, since strong bacteriostatic and bactericidal effects against *Listeria monocytogenes* and *Salmonella typhimuium* were observed in a dose-, time- and temperature-dependent manner [86]. On the other hand, Arufe et al. [87] studied the influence of the addition of different concentrations (2–8% *w/w*) of *F. vesiculosus* seaweed powder into wheat flour to the final rheological properties of the dough, such as the density and crumb texture. The authors found that for concentrations above 4%, the addition of *F. vesiculosus* powder caused the increase of the elongational dough viscosity and consequent decrease of its porosity, as well as the increase in the bread density, crumb firmness and appearance of a green colour. Therefore, 4% of *F. vesiculosus* powder would be the maximum amount that could be added to the bread without impairing its properties.

3.2. Himanthalia elongata

H. elongata has also been object of many studies comprising the development of seaweed-enriched foods, which, in addition to the improvement stability and/or shelf-life extension, also aimed to provide enhanced nutritional properties to the foods. In this field, many works reporting *H. elongata* fortified-foods were carried out on meat and meat-based products (Table 2). One of the most exploited attributes of this seaweed species is perhaps its wealthy mineral composition, which makes *H. elongata* a good candidate to be used as a salt replacer, contributing to the reduction of salt consumption and related health complications typical of western high-NaCl diets. It also increases the consumption of other elements, such as calcium potassium or iodine, which are usually lacking or below recommended levels in regular diets [4].

Table 2. Selected studies reporting the effects of the incorporation of *H. elongata* or isolates as ingredients in different food matrices.

Functional Food	Functional Ingredient	Results	Ref.
Poultry steaks	3% dry matter *H. elongata*	↑ Purge loss slightly ↓ Cooking loss ↑ Levels of total viable counts, lactic acid bacteria, tyramine and spermidine No important changes observed during chilled storage Positive overall acceptance by a sensory panel	[90]
Pork gel/emulsion systems	2.5% and 5% dry matter *H. elongata*	↑ Water and fat binding properties ↑ Hardness and chewiness of cooked products ↓ Springiness and cohesiveness	[1]
Low-salt pork emulsion systems	5.6% dry matter *H. elongata*	↑ Content of n-3 PUFA ↓ n-6/n-3 PUFA ratio ↓ Thrombogenic index ↑ Concentrations of K, Ca, Mg and Mn	[91]
Pork meat batter	3.4% powder *H. elongata*	↑ Water/oil retention capacity, hardness and elastic modulus. Thermal denaturation of protein fraction was prevented by seaweed alginates Nutritional enhancement	[92]
Restructured meat	5% powder *H. elongata*	Effects in rats: ↓ Total cholesterol ↑ CYP7A1, GPx, SOD, GR expression ↓ CAT expression	[93]
Restructured meat	5% powder *H. elongata*	↓ HSL and FAS and ↑ ACC ($p < 0.05$) expression on rats fed with seaweed fortified meat comparing with rats under hypercholesterolemic diet	[94]
Frankfurters	3.3% *H. elongata* powder	↑ Cooking loss ↓ Emulsion stability Combination of ingredients provided healthier meat products with lower fat and salt contents Worsened physicochemical and sensory characteristics	[95]
Beef patties	10–40% (*w/w*) *H. elongata*	↓ Cooking loss ↑ Tenderness, dietary fibre levels, TPC and antioxidant activity ↓ Microbiological counts and lipid oxidation Patties with 40% seaweed had the highest overall acceptability	[96]
Bread sticks	2.93–17.07% *H. elongata* powder	Highest concentration had higher phycochemical constituents, acceptable edible texture and overall colour	[97]
Bread	8% (*w/w*) *H. elongata*	↑ TPC ↑ Antioxidant activity in DPPH•, ORAC and TEAC	[98]
Yoghurt and Quark	0.25–1% dehydrated *H. elongata*	Alterations in all yoghurt attributes except for buttery odour, and acid and salty flavours Alterations in all quark attributes except yogurt odour, acid flavour and sweet flavour. Sensory characteristics worsened	[99]

↑: increased; ↓: decreased; ACC: acetyl CoA carboxylase; CAT: Catalase; CYP7A1: liver cytochrome P450 7A1; DPPH•: 2,2-diphenyl-1-picrylhydrazyl radical; FAS: fatty acid synthase; GPx: Glutathione peroxidase; GR: Glutathione reductase; HSL: hormone-sensitive lipase; ORAC: oxygen radical absorbance capacity; PUFA: polyunsaturated fatty acids; SOD: superoxide dismutase; TEAC: trolox equivalent antioxidant capacity; TPC: Total phenolic content.

Many of these studies were carried out by the group of Jiménez-Colmenero et al., who have developed several meat products in which the content of sodium chloride was partially replaced by different species of edible seaweeds, including *H. elongata*. Among the seaweed-containing formulations, frankfurters, restructured meats and meat emulsions were shown to have at least 50 to 75% less NaCl than their conventional recipes [1,90,91,95,100,101]. Apart from the NaCl replacement, the fortification of frankfurters and meat emulsions with *H. elongata* also contributed to the increase of K content and subsequent reduction of the Na/K ratio from 3 to values below 1 (i.e., close to those recommended by WHO for maintaining a healthy cardiovascular condition). Additionally, the Ca, Mg and Mn contents in these two meat products increased to >1000%, >300% and >700%, respectively, compared with the conventional formulas, alongside with their water and fat binding properties [1,101]. Other effects resultant from *H. elongata* fortification in these matrices included the reduced cooking loss and increase in the Kramer shear force in restructured poultry meat [90]; increased water and oil retention in pork meat batter [92]; increased dietary fibre content in frankfurters [101]; and increased phenolic content and antioxidant activity in meat emulsions [91]. Overall, these products were well-accepted in the sensory analysis, with exception of frankfurters that were reported unpleasant mainly due to the increase of the dryness feeling and seaweed-like taste.

Cox and Abu-Ghannam [96] also reported that *H. elongata*-fortified beef patties (10–40% *w/w*) were very well accepted in the sensory analysis, particularly those with 40% of seaweed, getting even better scores than the control samples. This was mainly due to the improvements on the samples' texture and overall mouthfeel, which resulted from the decrease in the cooking loss (associated to the incremented fibre content) and the increase in tenderness for more than 50%. Furthermore, a significant enhancement of the phenolic content and antioxidant activity (in a dose-dependent manner), as well as a lowered microbiological count and lipid oxidation before the chilling stage and after 30 days of storage, were observed in all patties containing seaweed. In fact, at the end of the experiment, the samples containing above 20% of *H. elongata*, showed no bacterial growth at all, as well as considerably low levels of the lipid oxidation marker.

In vivo studies on rat models revealed that the introduction of restructured pork meat enriched with 5% *H. elongata* (RPS) in the animals' hypercholesterolemic diet significantly lowered the serum cholesterol levels that were augmented in the group under a non-RPS supplemented hypercholesterolemic diet. Moreover, a significant increase in SOD and GPx, alongside with a decrease of glutathione reductase (GR) expressions, were observed in both groups under hypercholesterolemic and regular RPS-supplemented diets, although increased glutathione reductase activity was also verified. Interestingly, the combined cholesterol and seaweed diet predisposed an increase in the expression of GR, SOD and liver cytochrome P450 7A1 (CYP7A1), i.e., a gene that encodes for the enzyme responsible for the elimination of cholesterol through the production of bile acids, but a decrease in the expression of CAT and GPx, suggesting a possible blocking effect of the hypercholesterolemic agent induced by seaweed incorporation [93]. In a similar study, rats under RPS-supplemented hypercholesterolemic diets, not only exhibited lower plasma cholesterol levels but also lower liver apoptosis markers, namely cellular cycle DNA, caspase-3 and cytochrome c [102]. Supporting these results, González-Torres et al. [94] confirmed that the administration of *H. elongata*-fortified restructured pork meat (at 5%) to rats under cholesterol-rich diets, partially blocked the hypercholesterolemic effect of the dietary pattern while changing the lipogenic/lipolytic enzyme expression (decreasing hormone-sensitive lipase and fatty acid synthase while increasing acetyl CoA carboxylase expressions compared with subjects under hypercholesterolemic diet) and reducing the wasting effect of hypercholesterolemia on adipose tissue in rats.

Apart from meat products, *H. elongata* powder has also been used to enrich breadsticks in order to enhance their nutritional properties. From the 10 formulations tested (with seaweed concentrations of 2.63 to 17.07% *w/w*), the highest was reported to have the most significant influence on the chemical properties of breadsticks. Furthermore, this sample also had higher levels of total dietary fibre, while the total phenolic content and antiradical activity were maximized at 138.25 mg GAE/100 g dry basis and 61.01%, respectively, maintaining an acceptable edible texture and colour of the samples. Therefore, since no significant difference was seen between the control and seaweed enriched breadsticks in terms of sensory analysis, this product could have great acceptability, especially to non-seaweed consumers [97]. The augmented phenolic content as well as the enhanced antioxidant activity were also described on functional breads developed with 8% of *H. elongata* flour [98]. On the other hand, an attempt to supplement yogurt and quark with dehydrated *H. elongata* (0.25–1% *w/w*) turned out to negatively affect almost all the sensory parameters analysed, which makes this seaweed not very suitable for application in these two dairies, at least in these conditions [99].

3.3. Undaria pinnatifida

Similar to *H. elongata*, the applications of *U. pinnatifida* as functional ingredients have mostly been reported in meat and meat-derived products (Table 3). For instance, the incorporation of *U. pinnatifida* (1–4%) into pork beef patties increased their ash content as well as their juiciness due to the lower cooking losses compared to the control [103]. In a similar approach, the reformulation of low-salt (0.5%) and low-fat (<10%) beef patties by the addition of 3% of *U. pinnatifida* and partial or total replacement of pork backfat with olive oil-in-water emulsion, significantly affected the frozen storage characteristics of the products. This presented enhancements in terms of technological, sensory and nutritional properties, as well as improvements in their physiological benefits. These reformulated patties demonstrated less thawing and cooking losses, and were texturally softer than the samples without seaweed, most likely due to the microstructural changes caused by the formation of alginate chains.

Table 3. Selected studies reporting the effects of the incorporation of *U. pinnatifida* or isolates as ingredients in different food matrices.

Functional Food	Functional Ingredient	Results	Ref.
Beef patties	3% dry matter *U. pinnatifida*	↑ Binding properties and cooking retention values of, fat, fatty acids and ash Replacement of animal fat with olive-in-water emulsion and/or seaweed was reportedly healthier. ↓ Thawing and↑ softer texture Changes on the microstructure due to formation of alginate chains Overall acceptable products and fit for consumption	[104,105]
Chicken breast	200 mg/kg *U. pinnatifida*	↑ Redness and yellowness ↓ Lipid oxidation in chilling storage and after cooking Overall appearance and shelf-life were enhanced	[106]
Pork gel/emulsion systems	2.5% and 5% dry matter *U. pinnatifida*	↑ Water and fat binding properties ↑ Hardness and chewiness of cooked products ↓ Springiness and cohesiveness	[1]
Low-salt pork emulsion systems	5.6% dry matter *U. pinnatifida*	↑ Content of n-3 PUFA ↓ n-6/n-3 PUFA ratio ↑ Concentrations of K, Ca, Mg and Mn ↑ Antioxidant capacity	[91]
Pasta	100:0, 95:5, 90:10, 80:20 and 70:30 (semolina/*U. pinnatifida*; *w/w*)	10% *U. pinnatifida* was the most acceptable ↑ Amino acid, fatty acid profile and nutritional value of the product Fucoxanthin was not affected by pasta making and cooking step	[107]
Yoghurt and Quark	0.25–1% dehydrated *U. pinnatifida*	↑ Seaweed flavour with ↓ flavour quality for 0.5% seaweed Alterations in all yoghurt attributes except for buttery odour, and acid and salty flavours Alterations in all quark attributes except yogurt odour, and acid and sweet flavours. Sensory characteristics worsened	[99]
Bread	8% (w:w) *U. pinnatifida*:wheat flour	↑ TPC, ↑ Antioxidant activity in DPPH•, ORAC and TEAC	[98]

↑: increased; ↓: decreased; DPPH•: 2,2-diphenyl-1-picrylhydrazyl radical; ORAC: oxygen radical absorbance capacity; PUFA: polyunsaturated fatty acids; TEAC: trolox equivalent antioxidant capacity; TPC: Total phenolic content.

Moreover, the incorporation of *U. pinnatifida* in the patties' formulation did not hamper their lipid oxidation or microbiological counts, and although the content of Na and K were twice as high as the control samples, the Na/K ratio were still close to 1. Likewise, magnesium and calcium levels were higher in seaweed-fortified samples, corresponding three and six-fold, respectively, to those of the conventional recipe. Interestingly, although a different flavour was pointed out in the sensory analysis, panellists generally described the reformulated patties to be more pleasant and palatable than the control [104]. This reformulation with *U. pinnatifida* also resulted in significant improvements in several parameters on cooked patties, namely in the binding properties and retention values of moisture, ash and particularly fat and fatty acids, the latter parameter being usually the most affected by the cooking process. This means that the incorporation of this seaweed in the patties greatly interfere with the fat and energy content of these food matrices, as well as their fatty acids profile [105]. Identical results were reported on low-salt gel/emulsion meat systems added with 2.5–5% of *U. pinnatifida*, which exhibited better firmness and chewiness due to improvements of the water and fat-biding properties [1]. The incorporation of 5.6% of this species in such systems was also reported to contribute to the increment of the products' phenolic content and antioxidant properties, as well as to improve their mineral profile, increasing the K, Mg, Ca and Mn contents while decreasing the Na content, thus consequently reducing the Na/K ratio from 3.5 in the control samples, to approximately 1. Contrastingly, despite the potential beneficial health effects, increasing the algae was considered a non-satisfactory strategy to achieve healthier lipid meat formulations, since it could affect the food's sensory properties and their lipid content was very low [91]. In turn, Sasaki et al. [106] observed that the addition of 200 mg/kg fucoxanthin extract from *U. pinnatifida* to raw ground chicken breast meat did not prevent the lipid oxidation during their freeze storage period (1 or 6 days). However, it did inhibit TBARS formation of cooked samples stored under the same conditions and improved the products' overall appearance, indicating that fucoxanthin could prevent the oxidation in these products and effectively extend their shelf-life.

Apart from the nutritional stability of the foods, the incorporation of *U. pinnatifida* into foods have also been demonstrated to have great beneficial effects in distinct parameters with impact in the cardiovascular system. According to Moreira et al. [108], the administration of *U. pinnatifida*-fortified restructured pork meat to Wistar rats under a cholesterol-rich diet, not only caused the lowering of the plasma redox index by increasing total and reduced glutathione together with the GR and SOD activity, but also contributed to the decrease of the caspase-3 activity and therefore, hypercholesterolemic-induced apoptotic response of their hepatocytes [102].

Only few studies have focused the use of *U. pinnatifida* in products other than meat. Nevertheless, Prabhasankar et al. [107] reported significantly higher phenolic content and antioxidant activity in the aqueous extracts of uncooked pasta containing different concentrations of *U. pinnatifida* (5–30% w/w) compared to the controls. Although the cooking process caused a loss in these two parameters, they were still significantly higher on seaweed-added pasta compared to the values observed in the conventional pasta. Importantly, the heat processes involved in pasta preparation and cooking did not damaged fucoxanthin. The seaweed incorporation also contributed to the improvement of the pasta amino acid and fatty acid profiles, as well as the increase of bioactive compounds. The pasta incorporated with 10% seaweed, which demonstrated the highest radical scavenging activities, was also the most well accepted in the sensory analysis. The augmented phenolic content and antioxidant activity were also described on functional breads developed with 8% of *U. pinnatifida* flour, although other seaweeds, such as *H. elongata* exhibited better results [98].

The incorporation of *U. pinnatifida*, up to 15% in cottage cheeses, was reported to cause a dose-dependent increment of their Ca, Fe and Mg. However, the textural quality was best for cheeses containing 9% of seaweed [109]. On the other hand, Nuñez and Picon [99] found that, among the 5 different seaweeds used to incorporate in yogurts and quark cheese, dehydrated *U. pinnatifida* at 0.5% (w/w) was the formulation that showed the highest seaweed flavour and the lowest flavour quality in both dairies, worsening almost all of their sensory aspects and making this seaweed unattractive for application in such dairies. To overcome this disadvantage, it would be interesting to explore alternative approaches, such as the application of seaweed in flavoured dairies, the application of algae extracts instead of whole algae or the encapsulation of algae or extracts thereof, in order to assess whether these or other strategies could mask the negative impacts that *U. pinnatifida* has on the sensory aspects of these dairies.

3.4. Ascophyllum Nodosum

Although *A. nodosum* has not been much studied as a functional ingredient for incorporation in foods, some authors have reported promising results in this field (Table 4). For instance, Dierick et al. [110] found that, feeding pigs with 20 g of *A. nodosum*/kg of feed over 21 days caused the levels of iodine in muscle and internal organs to increase 2.7 and 6.8 times, respectively, compared to the pigs fed under a regular diet. This could be a viable approach to increase the daily intake of this mineral which is usually deficient in several European countries [4]. Alternatively, *A. nodosum* extracts applied to low-fat pork liver pâtés (500 mg/kg) was described to increase the protein content by approximately 4% compared to the control samples, without interfering with the chemical composition or microbial characteristics of the samples, throughout 180 days of storage at 4 °C. Furthermore, at the end of the experiment, the oxidative parameters on seaweed-added samples were comparable to those of BHT-added samples, both showing a similar degree of protection against oxidation as well as a significant reduction of volatile compounds after storage [83].

Table 4. Selected studies reporting the effects of the incorporation of *A. nodosum* or isolates as ingredients in different food matrices.

Functional Food	Functional Ingredient	Results	Ref.
Pork	20 g *A. nodosum* /kg feed	↑ I content in piglet's muscles and internal organs	[110]
Pork liver paté	*A. nodosum* extract at 500 mg/kg	↑ Protein content ↑ Redness and yellowness after storage Degree of protection against oxidation comparable to BHT samples ↓ Total volatile compounds	[83]
Milk	*A. nodosum* (100% H_2O and 80% EtOH) extracts (0.25 and 0.5 (*w/w*))	↓ TBARS formation ↑ Radical scavenging and ferrous-ion-chelating activities before and after digestion Supplementation on Caco-2 cells did not affect cellular antioxidant status EtOH extracts had greenish colour and overall sensory attributes were worsened	[84]
Yoghurts	*A. nodosum* (100% H_2O and 80% EtOH) extracts (0.25 and 0.5 (*w/w*))	No influence on chemical characteristics Yoghurts had antioxidant activity before and after digestion Supplementation on Caco-2 cells did not affect cellular antioxidant status Overall sensory attributes were worsened	[85]
Bread	1–4% *A. nodosum* per 400 g loaf	All samples sensorially accepted ↓ Energy intake after 4 h Glucose and cholesterol blood levels not affected	[111]

↑: increased; ↓: decreased; BHT: butylated hydroxytoluene; EtOH: ethanol; TBARS: thiobarbituric acid reactive substances.

On another perspective, *A. nodosum* extracts have proven to be effective in the inhibition of lipid oxidation and the improvement of antioxidant activity in dairies. Indeed, the incorporation of either aqueous or 80% ethanol extracts (0.25% and 0.5%) of this species in milk significantly decreased the TBARS formation and increased the radical scavenging and ferrous-ion-chelating activities either before or after in vitro digestion. However, this did not affect the cellular antioxidant activity or protect against DNA damage in human colon adenocarcinoma Caco-2 cells, suggesting that the fortification with *A. nodosum* extracts could improve certain milk qualities and shelf-life characteristics, but not provide significant biological activity. Interestingly, despite fortified-milk with aqueous extract had good acceptability in the sensory analysis, those formulated with 80% ethanol extract was pointed to have a fishy taste and off flavour, thus having low acceptability by the panellists. Nevertheless, this issue could potentially be addressed by using food flavourings or through micro-encapsulation to camouflage the undesirable flavours [84]. A new set of studies on fortified yogurts with the same *A. nodosum* extracts also revealed the increment of the radical scavenging activity before and after in vitro digestion, which was shown not to affect parameters, such as the product's acidity, microbiology or whey separation. However, as previously stated, the biological activity on cellular models was absent and the sensorial analysis was positive for *A. nodosum* aqueous extracts but not for the 80% ethanol extracts [85]. On another approach, Hall et al. [111] reported that the addition of *A. nodosum* (1–4%) in bread significantly reduced the energy intake after a test meal in a single blind cross trial. Moreover, the same was verified after 24 h of seaweed-enriched bread consumption and no differences were observed in blood glucose and cholesterol levels. The authors highlighted, however, the need of a long-term interventional study to establish the real potential of *A. nodosum*-enriched bread energy intake, in addition to the metabolism of glucose and lipids.

3.5. Laminaria *sp.*

Laminaria is one of the most economically important algae genus since it comprises 31 species, being most widely exploited worldwide as raw materials for alginates production [112]. On the other hand, the studies focusing the use of these seaweeds as functional ingredients in foods are quite limited (Table 5). Nevertheless, due to their high content in iodine, some authors have investigated the use of *Laminaria* sp. as animal feed aiming to increase the iodine content in their muscle before slaughter. Indeed, the work carried out by Schmid et al. [113] demonstrated that feeding chars (*Salvelinus* sp.) with *L. digitata*-fortified fish meal (0.8%) over nine months, contributed to an increase of their total iodine content in approximately four times the levels found in the control fishes. Similar observations were described in other species, such as gilthead seabream (*Sparus aurata*) and rainbow trout, which revealed an increased iodine content in their fillets after *L. digitata* was introduced in their meals as well [114,115]. An identical experiment carried out with pigs also revealed that the supplementation of *L. digitata* in the animal's feed over 3 months resulted in an accumulation of 45% more I in muscle tissue and up to 213% in other internal organs compared to the pigs under a normal diet [116]. In a different approach, four group of pigs were assigned to different diets 35 days pre-slaughter in order to test whether alterations of their diets would affect bacterial count, lipid peroxidation and total antioxidant capacity of fresh meat during storage. Interestingly, the meat excised from the group fed with the *Laminaria* sp.-supplemented diet exhibited the best overall results, showing the highest antioxidant activity, the lowest lipid peroxidation and microbial counts, suggesting that feeding the animals with seaweeds might have a significant impact on the quality and shelf-life of their meat [117].

Alternatively, Moroney et al. [118] tested whether the incorporation of different concentrations (0.01%, 0.1% and 0.5% *w/w*) of *L. digitata* extract, containing laminaran and fucoidan in chopped pork patties would affect their quality and shelf-life period. The results showed that the surface redness of fortified raw patties, upon 14 days under modified atmosphere packages at 4 °C, decreased compared to the control samples, which led to a slight decrease of their quality parameters. Fortification with the extract at 0.5% caused a notable reduction of lipid oxidation in the cooked samples, but the formulated product was not very well accepted in the sensory analysis. A similar work was later conducted with fresh and cooked pork homogenates and commercial horse heart oxymyoglobin incorporated with *L. digitata*-extracted fucoidan, laminaran and a mixture of both. Although fucoidan showed the strongest radical scavenging activity, cooking and digestion of the samples caused a significant decrease of the antioxidant potential in the samples added with this fibre, which could possibly be attributed to its more acidic nature. Interestingly, despite this, polysaccharide was found to reduce lipid oxidation and also was responsible for catalysing the oxidation of oxymyoglobin. Notably, when the digested samples containing the mixture of laminaran and fucoidan were evaluated for their bioaccessibility in a Caco-2 cell model, a decrease in radical scavenging activity of 44.2% and 36.6% was observed after 4 and 20 h of incubation, indicating a theoretical uptake of these polysaccharides. These results highlight the potential use of seaweed extracts as functional ingredients in pork with the advantage of possibly improving the human antioxidant defences [42].

Table 5. Selected studies reporting the effects of the incorporation of *Laminaria* sp. or isolates as ingredients in different food matrices.

Functional Food	Functional Ingredient	Results	Ref.
Chars	*Laminaria digitata* (0.8% in fish meal)	↑ 4 times the I content in fish muscle	[113]
Gilthead seabream	*Laminaria digitata* (10% in fish meal)	↑ I content in fish fillets	[114]
Rainbow trout	*Laminaria digitata* (0.65% in fish meal)	↑ I content in fish fillets	[115]
Pork	*Laminaria digitata* (1.16 and 1.86 g/kg feed)	↑ I content in pigs' muscles by 45% and internal organs by 213%	[116]
Pork	*Laminaria digitata* (5.32 kg/t feed)	↑ Antioxidant activity ↓ Lipid oxidations ↓ Microbial counts	[117]
Pork patties	0.01%, 0.1% and 0.5% (*w/w*) of 9.3% laminarin and 7.8% fucoidan from *L. digitata*	↑ Lipid antioxidant activity for L/F extract (0.5%) No effect in colour, lipid oxidation, texture or sensorial acceptance when adding L/F extract	[118]
Pork homogenates	3 and 6 mg/mL of laminaran, fucoidan and both from *L. digitata*	L had no antioxidant activity The L/F extract had higher antioxidant activity than F, after cooking and post digestion of minced pork. DPPH• antioxidant activity lower in Caco-2 cell model with L/F extracts Seaweed extracts containing F had higher antioxidant activity of the functional cooked meat products.	[42]
Sausages	1–4% *L. japonica* powder	No changes in moisture, protein, and fat contents ↓ Lightness and redness values ↓ Cooking loss ↑ Emulsion stability, hardness, gumminess, and chewiness 1% seaweed had highest overall acceptability	[119]
Pork/chicken patties	*Laminaria japonica* (replacement od 2.25 g of pork/chicken for an equal amount of seaweed)	↓ Increased in postprandial glucose blood levels; ↓ TC and LDL-C	[120]
Yoghurt	*Laminaria* spp. (0.2% or 0.5% *w/w*)	↑ I, Ca, K, Na, Mg, and Fe	[121]

↑: increased; ↓: decreased; DPPH•: 2,2-diphenyl-1-picrylhydrazyl radical; F: fucoidan; L: Laminarin; LDL-C: low density lipoprotein; L/F: Laminarin and Fucoidan; TC: total cholesterol.

In addition to *L. digitata*, other species of this genus have been reported for their positive effects as functional ingredients in foods. This is the case of *Laminaria japonica*, which was incorporated (1–4% *w/w*) in breakfast sausages contributing to a significant dose-dependent increase of their ash content, as well as to the improvements on the emulsion stability and textural parameters such as hardness, gumminess and chewiness. Moreover, the seaweed addition lowered samples' pH, lightness, redness and yellowness, and lowered cooking and water losses, particularly in samples added with 4%. Nevertheless, despite the higher benefits that were observed for higher seaweed powder concentrations, the sensory evaluations determined that the 1% *L. japonica* sausage had the highest overall acceptability [119]. In addition, the incorporation of *L. japonica* in chicken or pork patties was inclusively demonstrated to have positive effects in the post-plasma glucose and lipids profiles in borderline-hyperlipidaemic adults voluntaries. The consumption of fortified-patties with 2.25 g of this species not only lowered the increased post-prandial serum glucose levels compared to the control group, but also the total cholesterol and low density lipoprotein concentrations, while maintaining the same levels of high density lipoprotein [120].

In an alternative to meat products, a new probiotic yogurt containing different concentrations of *Laminaria* sp. was developed with the aim of increasing its iodine content. Indeed, contrarily to the conventional yogurt, the fortified formulation contained not only high levels of I (average of 570 µg I/100 g), but also considerably incremented amounts of Ca, K, Na, Mg, and Fe [121], overall improving their mineral profile.

4. Future Considerations

Previous studies focusing brown seaweed-fortified products have demonstrated that in general, macroalgae can improve the nutritional value of the food products, either by incrementing levels of dietary fibres and/or minerals or their lipidic profiles. Thus, fortified foods with seaweed and/or seaweed extracts come out as possible nutritional alternatives to the original formulations. Moreover, the reported information seems to be solid regarding the fact that the fortification of foods with brown seaweeds and/or their extracts in general have positive impacts on both their oxidative stability and microbial inhibition effects. However, discrepant results are reported regarding the technological properties of the fortified products, namely on the stability of the food's structure. Hence, there is the need to guarantee the compatibility of seaweeds and the overall food matrix, which is not only a result of the seaweed itself, but of the combination of the seaweed with the proper ingredients. In addition, the incorporation of seaweeds in foods frequently comprises their sensorial attributes due to colour changes and the appearance of off-flavours. Nevertheless, some strategies such as fermentation, enzymatic processing or encapsulation of seaweeds or their extracts have already shown interesting effects at cloaking seaweeds' negative sensorial characteristics while maintaining their nutritional properties and stability of bioactive components [122–124]. Nevertheless, further research in this field is necessary to understand if the cost-benefit of the application of such techniques is viable on a larger scale.

Being a crucial factor in nutrition, the bioavailability of relevant nutrients and/or phytochemicals is another critical issue that will require much attention in the following years. This is highly dependent on food components and on individual gastrointestinal conditions. Alginic acid, fucoidans and laminarans are considered dietary fibres, meaning that they may be fermented by colon microflora, therefore surviving the majority of the digestion [125]. The rest of the compounds seem, however, to be absorbed at earlier stages. For instance, in vitro studies suggested that dietary fucoxanthin is metabolized to fucoxanthinol and amarouciaxanthin A [126,127]. In fact, daily administered dietary fucoxanthin (*L. japonica* and *U. pinnatifida* origin) was shown to accumulate as amarouciaxanthin A and fucoxanthinol in several mice tissues [128,129]. In humans, however, the plasma concentrations of fucoxanthin metabolites before and after 1-week dietary interventions with *U. pinnatifida* were shown to be either low (fucoxanthinol) or non-existent (fucoxanthin and amarouciaxanthin A), although a higher subject group would be required in order to confirm these results [130].

As for phlorotannins, to the authors knowledge, only one bioavailability study was made using this particular compound from any of the seaweeds of interest to this review. Recently, in a work developed by Corona et al. (2016), a food-grade phlorotannin extract from *A. nodosum* was submitted to in vitro and in vivo assays, the latter involving the oral administration of a 100 mg capsule with the same extract [57]. The in vitro digestion and fermentation allowed for the identification of 11 compounds including hydroxytrifuhalol A, a C-O-C dimer of phloroglucinol, diphlorethol/difucol and 7-hydroxyeckol, some of which were also detected in the urine and plasma of human participants, thus confirming their absorption into the blood circulation. Moreover, although brown seaweeds are considered a great source of iodine, there is limited information regarding its bioavailability. Domínguez-Gonzaléz et al. [131] found out that despite the high in vitro bioaccessibility of iodine from *U. pinnatifida* and *S. japonica*, only a small percentage was bioavailable using dialysis membranes and an even lower in a biological system model consisting of two major cell types present in the intestine. Nevertheless, more favourable results were demonstrated when iodine-insufficient women were supplemented with encapsulated *A. nodosum*, since one third of the ingested iodine was found to be bioavailable [132].

Hence, it is clear that, not only is there a lack of information regarding the bioavailability of nutrients/phycochemicals in seaweeds and seaweeds-fortified foods in general, but also the relationship between seaweed-fortified products and their potential functionality remains almost unexploited. Indeed, evidences of biological effects of seaweeds-fortified products were barely proven in cellular models and even more rarely in in vivo trials and hence, must still be made to assure the conformity of the results. According to Plaza et al. [133], the principal guideline to follow in the design of a new functional food is to increase as much as possible the benefit/risk ratio, by increasing the benefit to the maximum and reducing the risk to the minimum, considering toxicity studies, for example. Increasing the benefit implies looking for a physiological wide effect, assuring the existing bioavailability and that the mentioned bioavailability is going to be kept along all the useful life of the food. Therefore, since the in vivo biological activity of phycochemicals depends on their bioavailability, in the future, it would be interesting to further access how important properties claimed for brown algae can transpose to human beings through seaweed fortified products. In order to reduce the risk, it is necessary to carry out toxicity studies, to use the functional ingredient in minimal effective doses and to use as functional ingredients, the products naturally found in foods or natural sources.

5. Conclusions

In conclusion, seaweeds are very valuable food sources with reportedly high nutritional value and high in bioactive compounds. In this regard, brown macroalgae are particularly known to accumulate specific metabolites, such as the polysaccharides alginic acids, laminarans and fucoidans, the phenolic compounds phlorotannins, the carotenoid fucoxanthin and exceptionally high levels of iodine with simultaneous low Na/K ratios, which overall confer them great potential to be used as functional ingredients. Thus far, most of the recent studies' main objectives focusing on testing the incorporation of macroalgae in foods, namely those considered as safe for consumption in Europe i.e., *F. vesiculosus*, *F. serratus*, *H. elongata*, *U. pinnatifida*, *A. nodosum*, *L. digitata*, *L. saccharina*, *L. japonica* and *A. esculenta*, were the increment of the product's shelf-life and the sensory properties culminating in the potential commercialization. From an economic point of view, this is rather important, but in terms of the benefits they could bring to the consumer, there is a lot of work to be done. Therefore, in order to effectively exploit this very promising raw material, the focus should be on their bioavailability, especially in humans.

Author Contributions: N.C.A. contribution to writing the original draft. M.C. contribution to writing the original draft. A.M.S.S. contribution to writing—review. S.M.C. contribution to conceptualization, supervision, project managing and writing—review.

Funding: Project CENTRO-01-0145-FEDER-023780 HEPA: Healthier eating of pasta with algae co-financed by the European Regional Development Fund (ERDF), through the partnership agreement Portugal2020–Regional Operation Program CENTRO2020. Foundation for Science and Technology (FCT), the European Union, the National Strategic Reference Framework (QREN), the European Regional Development Fund (FEDER), and Operational Programme Competitiveness Factors (COMPETE), for funding the Organic Chemistry Research Unit (QOPNA) (FCT UID/QUI/00062/2019). Project AgroForWealth (CENTRO-01-0145-FEDER-000001), funded by Centro2020, through FEDER and PT2020, and financed the research contract of Susana M. Cardoso. Foundation for Science and Technology (FCT) funded the PhD grant of Marcelo D. Catarino (PD/BD/114577/2016).

Conflicts of Interest: The authors declare no conflicts of interest.

References and Note

1.   Cofrades, S.; López-Lopez, I.; Solas, M.; Bravo, L.; Jimenez-Colmenero, F. Influence of different types and proportions of added edible seaweeds on characteristics of low-salt gel/emulsion meat systems. *Meat Sci.* **2008**, *79*, 767–776. [CrossRef] [PubMed]
2.   Wijesinghe, W.; Jeon, Y.J. Enzyme-assistant extraction (EAE) of bioactive components: A useful approach for recovery of industrially important metabolites from seaweeds: A review. *Fitoterapia* **2012**, *83*, 6–12. [CrossRef] [PubMed]
3.   Cardoso, S.M.; Carvalho, L.; Silva, P.; Rodrigues, M.; Pereira, O.; Pereira, L.; De Carvalho, L. Bioproducts from Seaweeds: A Review with Special Focus on the Iberian Peninsula. *Curr. Org. Chem.* **2014**, *18*, 896–917. [CrossRef]
4.   Circuncisão, A.R.; Catarino, M.D.; Cardoso, S.M.; Silva, A.M.S. Minerals from Macroalgae Origin: Health Benefits and Risks for Consumers. *Mar. Drugs* **2018**, *16*, 400. [CrossRef] [PubMed]
5.   Cardoso, S.M.; Pereira, O.R.; Seca, A.M.L.; Pinto, D.C.G.A.; Silva, A.M.S. Seaweeds as Preventive Agents for Cardiovascular Diseases: From Nutrients to Functional Foods. *Mar. Drugs* **2015**, *13*, 6838–6865. [CrossRef] [PubMed]
6.   Willcox, D.C.; Willcox, B.J.; Todoriki, H.; Suzuki, M. The Okinawan diet: Health implications of a low-calorie, nutrient-dense, antioxidant-rich dietary pattern low in glycemic load. *J. Am. Coll. Nutr.* **2009**, *28*, 500S–516S. [CrossRef] [PubMed]
7.   Blank, C. The Rise of Seaweed. Available online: https://www.seafoodsource.com/features/the-rise-of-seaweed (accessed on 5 June 2019).
8.   Lorenzo, J.M.; Agregán, R.; Munekata, P.E.S.; Franco, D.; Carballo, J.; Sahin, S.; Lacomba, R.; Barba, F.J. Proximate Composition and Nutritional Value of three macroalgae: *Ascophyllum nodosum*, *Fucus vesiculosus* and *Bifurcaria bifurcata*. *Mar. Drugs* **2017**, *15*, 360. [CrossRef]
9.   Gupta, S.; Abu-Ghannam, N. Bioactive potential and possible health effects of edible brown seaweeds. *Trends Food Sci. Technol.* **2011**, *22*, 315–326. [CrossRef]
10. Catarino, M.D.; Silva, A.M.S.; Cardoso, S.M. Phycochemical Constituents and Biological Activities of *Fucus* spp. *Mar. Drugs* **2018**, *16*, 249. [CrossRef]
11. Catarino, M.D.; Silva, A.M.S.; Cardoso, S.M. Fucaceae: A Source of Bioactive Phlorotannins. *Int. J. Mol. Sci.* **2017**, *18*, 1327. [CrossRef]
12. Holland, J. Seaweed on Track to Become Europe's Next Big Superfood Trend. Available online: https://www.seafoodsource.com/news/food-safety-health/seaweed-on-track-to-become-europe-s-next-big-superfood-trend (accessed on 6 July 2019).
13. Seaweed-Flavoured Food and Drink Launches Increased by 147% in Europe between 2011 and 2015. Available online: http://www.mintel.com/press-centre/food-and-drink/seaweed-flavoured-food-and-drink-launches-increased-by-147-in-europe-between-2011-and-2015 (accessed on 23 Aug 2019).
14. CEVA. *Edible Seaweed and French Regulation*; 2014.
15. Pádua, D.; Rocha, E.; Gargiulo, D.; Ramos, A.; Ramos, A. Bioactive compounds from brown seaweeds: Phloroglucinol, fucoxanthin and fucoidan as promising therapeutic agents against breast cancer. *Phytochem. Lett.* **2015**, *14*, 91–98. [CrossRef]
16. Usman, A.; Khalid, S.; Usman, A.; Hussain, Z.; Wang, Y. Algal Polysaccharides, Novel Application, and Outlook. In *Algae Based Polymers, Blends, and Composites*; Elsevier Inc.: Amsterdam, The Netherlands, 2017; pp. 115–153.
17. Susanto, E.; Fahmi, A.S.; Abe, M.; Hosokawa, M.; Miyashita, K. Lipids, Fatty Acids, and Fucoxanthin Content from Temperate and Tropical Brown Seaweeds. *Aquat. Procedia* **2016**, *7*, 66–75. [CrossRef]
18. Silva, A.F.; Abreu, H.; Silva, A.M.; Cardoso, S.M. Effect of Oven-Drying on the Recovery of Valuable Compounds from *Ulva rigida*, *Gracilaria* sp. and *Fucus vesiculosus*. *Mar. Drugs* **2019**, *17*, 90. [CrossRef] [PubMed]
19. Charoensiddhi, S.; Lorbeer, A.J.; Lahnstein, J.; Bulone, V.; Franco, C.M.; Zhang, W. Enzyme-assisted extraction of carbohydrates from the brown alga Ecklonia radiata: Effect of enzyme type, pH and buffer on sugar yield and molecular weight profiles. *Process. Biochem.* **2016**, *51*, 1503–1510. [CrossRef]
20. Wen, X.; Peng, C.; Zhou, H.; Lin, Z.; Lin, G.; Chen, S.; Li, P. Nutritional Composition and Assessment of *Gracilaria lemaneiformis Bory*. *J. Integr. Plant Biol.* **2006**, *48*, 1047–1053. [CrossRef]

21. Rioux, L.E.; Turgeon, S.; Beaulieu, M. Characterization of polysaccharides extracted from brown seaweeds. *Carbohydr. Polym.* **2007**, *69*, 530–537. [CrossRef]

22. Je, J.Y.; Park, P.J.; Kim, E.K.; Park, J.S.; Yoon, H.D.; Kim, K.R.; Ahn, C.B. Antioxidant activity of enzymatic extracts from the brown seaweed *Undaria pinnatifida* by electron spin resonance spectroscopy. *LWT* **2009**, *42*, 874–878. [CrossRef]

23. Schiener, P.; Black, K.D.; Stanley, M.S.; Green, D.H. The seasonal variation in the chemical composition of the kelp species *Laminaria digitata*, *Laminaria hyperborea*, *Saccharina latissima* and *Alaria esculenta*. *Environ. Biol. Fishes* **2014**, *27*, 363–373. [CrossRef]

24. Bourgougnon, N.; Deslandes, E. Carbohydrates from Seaweeds. In *Seaweed in Health and Disease Prevention*; Elsevier Inc.: Amsterdam, The Netherlands, 2016; pp. 223–274.

25. Horn, S.J.; Moen, E.; Østgaard, K. Direct determination of alginate content in brown algae by near infra-red (NIR) spectroscopy. *J. Appl. Phycol.* **1999**, *11*, 523–527. [CrossRef]

26. Radwanl, A.; Davies, G.; Fataftah, A.; Ghabbour, E.A.; Jansen, S.A.; Willey, R.J. Isolation of humic acid from the brown algae *Ascophyllum nodosum*, *Fucus vesiculosus*, *Laminaria saccharina* and the marine angiosperm *Zostera marina*. *J. Appl. Phycol.* **1997**, *8*, 553–562. [CrossRef]

27. Draget, K.; Skjakbrak, G.; Smidsrod, O. Alginic acid gels: The effect of alginate chemical composition and molecular weight. *Carbohydr. Polym.* **1994**, *25*, 31–38. [CrossRef]

28. Paolucci, M.; Fabbrocini, A.; Grazia, M.; Varricchio, E.; Cocci, E. Development of Biopolymers as Binders for Feed for Farmed Aquatic Organisms. *Aquaculture* **2012**, *1*, 3–34.

29. Brownlee, I.A.; Seal, C.J.; Wilcox, M.; Dettmar, P.W.; Pearson, J.P. Applications of Alginates in Food. In *Alginates: Biology and Applications*; Springer Science & Business Media: Berlin/Heidelberg, Germany, 2009; Volume 13, pp. 211–228.

30. Brown, E.M.; Allsopp, P.J.; Magee, P.J.; Gill, C.I.R.; Nitecki, S.; Strain, C.R. Seaweed and human health. *Nutr. Rev.* **2014**, *72*, 205–216. [CrossRef] [PubMed]

31. Cherry, P.; Yadav, S.; Strain, C.R.; Allsopp, P.J.; McSorley, E.M.; Ross, R.P.; Stanton, C. Prebiotics from Seaweeds: An Ocean of Opportunity? *Mar. Drugs* **2019**, *17*, 327. [CrossRef] [PubMed]

32. Bruhn, A.; Janicek, T.; Manns, D.; Nielsen, M.M.; Balsby, T.J.S.; Meyer, A.S.; Rasmussen, M.B.; Hou, X.; Saake, B.; Göke, C.; et al. Crude fucoidan content in two North Atlantic kelp species, *Saccharina latissima* and *Laminaria digitata*—Seasonal variation and impact of environmental factors. *Environ. Biol. Fishes* **2017**, *29*, 3121–3137. [CrossRef] [PubMed]

33. Ale, M.T.; Mikkelsen, J.D.; Meyer, A.S. Important Determinants for Fucoidan Bioactivity: A Critical Review of Structure-Function Relations and Extraction Methods for Fucose-Containing Sulfated Polysaccharides from Brown Seaweeds. *Mar. Drugs* **2011**, *9*, 2106–2130. [CrossRef] [PubMed]

34. Ale, M.T.; Meyer, A.S. Fucoidans from brown seaweeds: An update on structures, extraction techniques and use of enzymes as tools for structural elucidation. *RSC Adv.* **2013**, *3*, 8131–8141. [CrossRef]

35. Wang, Y.; Xing, M.; Cao, Q.; Ji, A.; Liang, H.; Song, S. Biological Activities of Fucoidan and the Factors Mediating Its Therapeutic Effects: A Review of Recent Studies. *Mar. Drugs* **2019**, *17*, 183. [CrossRef]

36. Veena, C.K.; Josephine, A.; Preetha, S.P.; Varalakshmi, P. Beneficial role of sulfated polysaccharides from edible seaweed *Fucus vesiculosus* in experimental hyperoxaluria. *Food Chem.* **2007**, *100*, 1552–1559. [CrossRef]

37. Huang, L.; Wen, K.; Gao, X.; Liu, Y. Hypolipidemic effect of fucoidan from *Laminaria japonica* in hyperlipidemic rats. *Pharm. Biol.* **2010**, *48*, 422–426. [CrossRef]

38. Shan, X.; Liu, X.; Hao, J.; Cai, C.; Fan, F.; Dun, Y.; Zhao, X.; Liu, X.; Li, C.; Yu, G. In vitro and in vivo hypoglycemic effects of brown algal fucoidans. *Int. J. Biol. Macromol.* **2016**, *82*, 249–255. [CrossRef] [PubMed]

39. Kim, K.J.; Yoon, K.Y.; Lee, B.Y. Fucoidan regulate blood glucose homeostasis in C57BL/KSJ m+/+db and C57BL/KSJ db/db mice. *Fitoterapia* **2012**, *83*, 1105–1109. [CrossRef] [PubMed]

40. Kadam, S.U.; Tiwari, B.K.; Donnell, C.P.O. Extraction, structure and biofunctional activities of laminarin from brown algae. *Int. J. Food Sci. Technol.* **2015**, *50*, 24–31. [CrossRef]

41. Holdt, S.L.; Kraan, S. Bioactive compounds in seaweed: Functional food applications and legislation. *Environ. Biol. Fishes* **2011**, *23*, 543–597. [CrossRef]

42. Moroney, N.C.; O'Grady, M.N.; Lordan, S.; Stanton, C.; Kerry, J.P. Seaweed Polysaccharides (Laminarin and Fucoidan) as Functional Ingredients in Pork Meat: An Evaluation of Anti-Oxidative Potential, Thermal Stability and Bioaccessibility. *Mar. Drugs* **2015**, *13*, 2447–2464. [CrossRef] [PubMed]

43. Deville, C.; Damas, J.; Forget, P.; Dandrifosse, G.; Peulen, O. Laminarin in the dietary fibre concept. *J. Sci. Food Agric.* **2004**, *84*, 1030–1038. [CrossRef]

44. Falkeborg, M.; Cheong, L.Z.; Gianfico, C.; Sztukiel, K.M.; Kristensen, K.; Glasius, M.; Xu, X.; Guo, Z. Alginate oligosaccharides: Enzymatic preparation and antioxidant property evaluation. *Food Chem.* **2014**, *164*, 185–194. [CrossRef]

45. Zhou, R.; Shi, X.Y.; Bi, D.C.; Fang, W.S.; Wei, G.B.; Xu, X. Alginate-Derived Oligosaccharide Inhibits Neuroinflammation and Promotes Microglial Phagocytosis of β-Amyloid. *Mar. Drugs* **2015**, *13*, 5828–5846. [CrossRef]

46. Fu, X.; Xue, C.; Ning, Y.; Li, Z.; Xu, J. Acute antihypertensive effects of fucoidan oligosaccharides prepared from *Laminaria japonica* on renovascular hypertensive rats. *J. Ocean Univ. China(Nat. Sci.)* **2004**, *34*, 560–564.

47. Koivikko, R.; Loponen, J.; Honkanen, T.; Jormalainen, V. Contents of soluble, cell-wall-bound and exuded phlorotannins in the brown alga *Fucus vesiculosus*, with implications on their ecological functions. *J. Chem. Ecol.* **2005**, *31*, 195–212. [CrossRef]

48. Imbs, T.I.; Zvyagintseva, T.N. Phlorotannins are Polyphenolic Metabolites of Brown Algae. *Russ. J. Mar. Biol.* **2018**, *44*, 263–273. [CrossRef]

49. Catarino, M.D.; Silva, A.M.S.; Mateus, N.; Cardoso, S.M. Optimization of Phlorotannins Extraction from *Fucus vesiculosus* and Evaluation of Their Potential to Prevent Metabolic Disorders. *Mar. Drugs* **2019**, *17*, 162. [CrossRef] [PubMed]

50. Wijesekara, I.; Kim, S.-K.; Li, Y.-X.; Li, Y. Phlorotannins as bioactive agents from brown algae. *Process. Biochem.* **2011**, *46*, 2219–2224.

51. Neto, R.T.; Marçal, C.; Queirós, A.S.; Abreu, H.; Silva, A.M.S.; Cardoso, S.M. Screening of *Ulva rigida*, *Gracilaria* sp., *Fucus vesiculosus* and *Saccharina latissima* as Functional Ingredients. *Int. J. Mol. Sci.* **2018**, *19*, 2987. [CrossRef] [PubMed]

52. O'Sullivan, A.; O'Callaghan, Y.; O'Grady, M.; Queguineur, B.; Hanniffy, D.; Troy, D.; Kerry, J.; O'Brien, N. In vitro and cellular antioxidant activities of seaweed extracts prepared from five brown seaweeds harvested in spring from the west coast of Ireland. *Food Chem.* **2011**, *126*, 1064–1070. [CrossRef]

53. Zaragozá, M.C.; López, D.; Sáiz, M.P.; Poquet, M.; Pérez, J.; Puig-Parellada, P.; Màrmol, F.; Simonetti, P.; Gardana, C.; Lerat, Y.; et al. Toxicity and Antioxidant Activity in Vitro and in Vivo of Two *Fucus vesiculosus* Extracts. *J. Agric. Food Chem.* **2008**, *56*, 7773–7780.

54. Roy, M.-C.; Anguenot, R.; Fillion, C.; Beaulieu, M.; Bérubé, J.; Richard, D. Effect of a commercially-available algal phlorotannins extract on digestive enzymes and carbohydrate absorption in vivo. *Food Res. Int.* **2011**, *44*, 3026–3029. [CrossRef]

55. Paradis, M.E.; Couture, P.; Lamarche, B. A randomised crossover placebo-controlled trial investigating the effect of brown seaweed (*Ascophyllum nodosum* and *Fucus vesiculosus*) on postchallenge plasma glucose and insulin levels in men and women. *Appl. Physiol. Nutr. Metab.* **2011**, *36*, 913–919. [CrossRef] [PubMed]

56. Baldrick, F.R.; McFadden, K.; Ibars, M.; Sung, C.; Moffatt, T.; Megarry, K.; Thomas, K.; Mitchell, P.; Wallace, J.M.W.; Pourshahidi, L.K.; et al. Impact of a (poly)phenol-rich extract from the brown algae *Ascophyllum nodosum* on DNA damage and antioxidant activity in an overweight or obese population: A randomized controlled trial. *Am. J. Clin. Nutr.* **2018**, *108*, 688–700. [CrossRef] [PubMed]

57. Corona, G.; Ji, Y.; Anegboonlap, P.; Hotchkiss, S.; Gill, C.; Yaqoob, P.; Spencer, J.P.E.; Rowland, I. Gastrointestinal modifications and bioavailability of brown seaweed phlorotannins and effects on inflammatory markers. *Br. J. Nutr.* **2016**, *115*, 1240–1253. [CrossRef]

58. Collén, J.; Davison, I.R. Reactive oxygen metabolism in intertidal *Fucus* spp. (*Phaeophyceae*). *J. Phycol.* **1999**, *35*, 62–69. [CrossRef]

59. Mikami, K.; Hosokawa, M. Biosynthetic Pathway and Health Benefits of Fucoxanthin, an Algae-Specific Xanthophyll in Brown Seaweeds. *Int. J. Mol. Sci.* **2013**, *14*, 13763–13781. [CrossRef] [PubMed]

60. Peng, J.; Yuan, J.-P.; Wu, C.-F.; Wang, J.-H. Fucoxanthin, a Marine Carotenoid Present in Brown Seaweeds and Diatoms: Metabolism and Bioactivities Relevant to Human Health. *Mar. Drugs* **2011**, *9*, 1806–1828. [CrossRef] [PubMed]

61. Maeda, H.; Hosokawa, M.; Sashima, T.; Murakami-Funayama, K.; Miyashita, K. Anti-obesity and anti-diabetic effects of fucoxanthin on diet-induced obesity conditions in a murine model. *Mol. Med. Rep.* **2009**, *2*, 811–817. [CrossRef] [PubMed]

62. Maeda, H.; Hosokawa, M.; Sashima, T.; Takahashi, N.; Kawada, T.; Miyashita, K. Fucoxanthin and its metabolite, fucoxanthinol, suppress adipocyte differentiation in 3T3-L1 cells. *Int. J. Mol. Med.* **2006**, *18*, 147–152. [CrossRef] [PubMed]

63. Maeda, H.; Hosokawa, M.; Sashima, T.; Funayama, K.; Miyashita, K. Fucoxanthin from edible seaweed, *Undaria pinnatifida*, shows antiobesity effect through UCP1 expression in white adipose tissues. *Biochem. Biophys. Res. Commun.* **2005**, *332*, 392–397. [CrossRef] [PubMed]

64. Ha, A.W.; Kim, W.K. The effect of fucoxanthin rich power on the lipid metabolism in rats with a high fat diet. *Nutr. Res. Pr.* **2013**, *7*, 287–293. [CrossRef] [PubMed]

65. Whelton, P.K. Sodium, Potassium, Blood Pressure, and Cardiovascular Disease in Humans. *Curr. Hypertens. Rep.* **2014**, *16*, 465. [CrossRef]

66. Blaustein, M.P.; Leenen, F.H.H.; Chen, L.; Golovina, V.A.; Hamlyn, J.M.; Pallone, T.L.; Van Huysse, J.W.; Zhang, J.; Gil Wier, W. How NaCl raises blood pressure: A new paradigm for the pathogenesis of salt-dependent hypertension. *Am. J. Physiol. Circ. Physiol.* **2011**, *302*, H1031–H1049. [CrossRef]

67. Larrea-Marín, M.; Pomares-Alfonso, M.; Gómez-Juaristi, M.; Sánchez-Muniz, F.; De La Rocha, S.R. Validation of an ICP-OES method for macro and trace element determination in Laminaria and Porphyra seaweeds from four different countries. *J. Food Compos. Anal.* **2010**, *23*, 814–820. [CrossRef]

68. Moreda-Piñeiro, J.; Alonso-Rodríguez, E.; López-Mahía, P.; Muniategui-Lorenzo, S.; Prada-Rodríguez, D.; Moreda-Piñeiro, A.; Bermejo-Barrera, P. Development of a new sample pre-treatment procedure based on pressurized liquid extraction for the determination of metals in edible seaweed. *Anal. Chim. Acta* **2007**, *598*, 95–102. [CrossRef] [PubMed]

69. Ruperez, P. Mineral content of edible marine seaweeds. *Food Chem.* **2002**, *79*, 23–26. [CrossRef]

70. Paiva, L.; Lima, E.; Neto, A.I.; Marcone, M.; Baptista, J. Health-promoting ingredients from four selected Azorean macroalgae. *Food Res. Int.* **2016**, *89*, 432–438. [CrossRef] [PubMed]

71. Nitschke, U.; Ruth, A.A.; Dixneuf, S.; Stengel, D.B. Molecular iodine emission rates and photosynthetic performance of different thallus parts of *Laminaria digitata* (Phaeophyceae) during emersion. *Planta* **2011**, *233*, 737–748. [CrossRef] [PubMed]

72. MacArtain, P.; Gill, C.I.; Brooks, M.; Campbell, R.; Rowland, I.R. Nutritional Value of Edible Seaweeds. *Nutr. Rev.* **2007**, *65*, 535–543. [CrossRef] [PubMed]

73. Dellarosa, N.; Laghi, L.; Martinsdóttir, E.; Jónsdóttir, R.; Sveinsdóttir, K. Enrichment of convenience seafood with omega-3 and seaweed extracts: Effect on lipid oxidation. *LWT Food Sci. Technol.* **2015**, *62*, 746–752. [CrossRef]

74. Diaz-Rubio, M.E.; Serrano, J.; Borderias, J.; Saura-Calixto, F. Technological Effect and Nutritional Value of Dietary Antioxidant Fucus Fiber Added to Fish Mince. *J. Aquat. Food Prod. Technol.* **2011**, *20*, 295–307. [CrossRef]

75. Wang, T.; Jónsdóttir, R.; Kristinsson, H.G.; Thorkelsson, G.; Jacobsen, C.; Hamaguchi, P.Y.; Ólafsdóttir, G. Inhibition of haemoglobin-mediated lipid oxidation in washed cod muscle and cod protein isolates by *Fucus vesiculosus* extract and fractions. *Food Chem.* **2010**, *123*, 321–330. [CrossRef]

76. Jónsdóttir, R.; Geirsdóttir, M.; Hamaguchi, P.Y.; Jamnik, P.; Kristinsson, H.G.; Undeland, I. The ability of in vitro antioxidant assays to predict the efficiency of a cod protein hydrolysate and brown seaweed extract to prevent oxidation in marine food model systems. *J. Sci. Food Agric.* **2016**, *96*, 2125–2135. [CrossRef]

77. Halldorsdottir, S.M.; Kristinsson, H.G.; Sveinsdóttir, H.; Thorkelsson, G.; Hamaguchi, P.Y. The effect of natural antioxidants on haemoglobin-mediated lipid oxidation during enzymatic hydrolysis of cod protein. *Food Chem.* **2013**, *141*, 914–919. [CrossRef]

78. Halldorsdottir, S.M.; Sveinsdóttir, H.; Gudmundsdóttir, Á.; Thorkelsson, G.; Kristinsson, H.G. High quality fish protein hydrolysates prepared from by-product material with *Fucus vesiculosus* extract. *J. Funct. Foods* **2014**, *9*, 10–17. [CrossRef]

79. Karadağ, A.; Hermund, D.B.; Jensen, L.H.S.; Andersen, U.; Jónsdóttir, R.; Kristinsson, H.G.; Alasalvar, C.; Jacobsen, C. Oxidative stability and microstructure of 5% fish-oil-enriched granola bars added natural antioxidants derived from brown alga *Fucus vesiculosus*. *Eur. J. Lipid Sci. Technol.* **2017**, *119*, 1–12. [CrossRef]

80. Hermund, D.B.; Karadağ, A.; Andersen, U.; Jónsdóttir, R.; Kristinsson, H.G.; Alasalvar, C.; Jacobsen, C. Oxidative Stability of Granola Bars Enriched with Multilayered Fish Oil Emulsion in the Presence of Novel Brown Seaweed Based Antioxidants. *J. Agric. Food Chem.* **2016**, *64*, 8359–8368. [CrossRef] [PubMed]

81. Hermund, D.B.; Yesiltas, B.; Honold, P.; Jónsdóttir, R.; Kristinsson, H.G.; Jacobsen, C. Characterisation and antioxidant evaluation of Icelandic *F. vesiculosus* extracts in vitro and in fish-oil-enriched milk and mayonnaise. *J. Funct. Foods* **2015**, *19*, 828–841. [CrossRef]

82. Honold, P.J.; Jacobsen, C.; Jónsdóttir, R.; Kristinsson, H.G.; Hermund, D.B. Potential seaweed-based food ingredients to inhibit lipid oxidation in fish-oil-enriched mayonnaise. *Eur. Food Res. Technol.* **2016**, *242*, 571–584. [CrossRef]

83. Agregán, R.; Franco, D.; Carballo, J.; Tomasevic, I.; Barba, F.J.; Gómez, B.; Muchenje, V.; Lorenzo, J.M. Shelf life study of healthy pork liver pâté with added seaweed extracts from *Ascophyllum nodosum, Fucus vesiculosus* and *Bifurcaria bifurcata. Food Res. Int.* **2018**, *112*, 400–411. [CrossRef] [PubMed]

84. O'Sullivan, A.M.; O'Callaghan, Y.C.; O'Grady, M.N.; Waldron, D.S.; Smyth, T.J.; O'Brien, N.M.; Kerry, J.P. An examination of the potential of seaweed extracts as functional ingredients in milk. *Int. J. Dairy Technol.* **2014**, *67*, 182–193. [CrossRef]

85. O'Sullivan, A.; O'Grady, M.; O'Callaghan, Y.; Smyth, T.; O'Brien, N.; Kerry, J. Seaweed extracts as potential functional ingredients in yogurt. *Innov. Food Sci. Emerg. Technol.* **2016**, *37*, 293–299. [CrossRef]

86. Poveda-Castillo, G.D.C.; Rodrigo, D.; Martínez, A.; Pina-Pérez, M.C. Bioactivity of Fucoidan as an Antimicrobial Agent in a New Functional Beverage. *Beverages* **2018**, *4*, 64. [CrossRef]

87. Arufe, S.; Della Valle, G.; Chiron, H.; Chenlo, F.; Sineiro, J.; Moreira, R. Effect of brown seaweed powder on physical and textural properties of wheat bread. *Eur. Food Res. Technol.* **2018**, *244*, 1–10. [CrossRef]

88. Jacobsen, C.; Hartvigsen, K.; Lund, P.; Thomsen, M.K.; Skibsted, L.H.; Adler-Nissen, J.; Hølmer, G.; Meyer, A.S. Oxidation in fish oil-enriched mayonnaise3. Assessment of the influence of the emulsion structure on oxidation by discriminant partial least squares regression analysis. *Eur. Food Res. Technol.* **2000**, *211*, 86–98. [CrossRef]

89. Agregán, R.; Barba, F.J.; Gavahian, M.; Franco, D.; Khaneghah, A.M.; Carballo, J.; Ferreira, I.C.F.; Barretto, A.C.S.; Lorenzo, J.M. *Fucus vesiculosus* extracts as natural antioxidants for improvement of physicochemical properties and shelf life of pork patties formulated with oleogels. *J. Sci. Food Agric.* **2019**, *99*, 4561–4570. [CrossRef] [PubMed]

90. Cofrades, S.; López-Lopez, I.; Ruiz-Capillas, C.; Triki, M.; Jimenez-Colmenero, F. Quality characteristics of low-salt restructured poultry with microbial transglutaminase and seaweed. *Meat Sci.* **2011**, *87*, 373–380. [CrossRef] [PubMed]

91. López-Lopez, I.; Bastida, S.; Ruiz-Capillas, C.; Bravo, L.; Larrea, M.T.; Sánchez-Muniz, F.; Cofrades, S.; Jimenez-Colmenero, F. Composition and antioxidant capacity of low-salt meat emulsion model systems containing edible seaweeds. *Meat Sci.* **2009**, *83*, 492–498. [CrossRef] [PubMed]

92. Fernández-Martín, F.; López-López, I.; Cofrades, S.; Colmenero, F.J. Influence of adding Sea Spaghetti seaweed and replacing the animal fat with olive oil or a konjac gel on pork meat batter gelation. Potential protein/alginate association. *Meat Sci.* **2009**, *83*, 209–217. [CrossRef] [PubMed]

93. Moreira, A.R.S.; Benedi, J.; González-Torres, L.; Olivero-David, R.; Bastida, S.; Sánchez-Reus, M.I.; González-Muñoz, M.J.; Sánchez-Muniz, F.J. Effects of diet enriched with restructured meats, containing *Himanthalia elongata*, on hypercholesterolaemic induction, CYP7A1 expression and antioxidant enzyme activity and expression in growing rats. *Food Chem.* **2011**, *129*, 1623–1630. [CrossRef]

94. González-Torres, L.; Churruca, I.; Schultz Moreira, A.R.; Bastida, S.; Benedí, J.; Portillo, M.P.; Sánchez-Muniz, F.J. Effects of restructured pork containing *Himanthalia elongata* on adipose tissue lipogenic and lipolytic enzyme expression of normo- and hypercholesterolemic rats. *J. Nutrigenet. Nutr.* **2012**, *5*, 158–167. [CrossRef]

95. Jimenez-Colmenero, F.; Cofrades, S.; López-Lopez, I.; Ruiz-Capillas, C.; Pintado, T.; Solas, M. Technological and sensory characteristics of reduced/low-fat, low-salt frankfurters as affected by the addition of konjac and seaweed. *Meat Sci.* **2010**, *84*, 356–363. [CrossRef]

96. Cox, S.; Abu-Ghannam, N. Enhancement of the phytochemical and fibre content of beef patties with *Himanthalia elongata* seaweed. *Int. J. Food Sci. Technol.* **2013**, *48*, 2239–2249.

97. Cox, S.; Abu-Ghannam, N. Incorporation of *Himanthalia elongata* seaweed to enhance the phytochemical content of breadsticks using response surface methodology (RSM). *Int. Food Res. J.* **2013**, *20*, 1537–1545.

98. Rico, D.; De Linaje, A.A.; Herrero, A.; Asensio-Vegas, C.; Miranda, J.; Martínez-Villaluenga, C.; De Luis, D.A.; Martin-Diana, A.B. Carob by-products and seaweeds for the development of functional bread. *J. Food Process. Preserv.* **2018**, *42*, e13700. [CrossRef]

99. Nuñez, M.; Picon, A. Seaweeds in yogurt and quark supplementation: Influence of five dehydrated edible seaweeds on sensory characteristics. *Int. J. Food Sci. Technol.* **2017**, *52*, 431–438. [CrossRef]

100. López-Lopez, I.; Cofrades, S.; Jimenez-Colmenero, F. Low-fat frankfurters enriched with n−3 PUFA and edible seaweed: Effects of olive oil and chilled storage on physicochemical, sensory and microbial characteristics. *Meat Sci.* **2009**, *83*, 148–154. [CrossRef] [PubMed]

101. López-Lopez, I.; Cofrades, S.; Ruiz-Capillas, C.; Jimenez-Colmenero, F. Design and nutritional properties of potential functional frankfurters based on lipid formulation, added seaweed and low salt content. *Meat Sci.* **2009**, *83*, 255–262. [CrossRef] [PubMed]

102. Moreira, A.R.S.; Benedi, J.; Bastida, S.; Sánchez-Reus, I.; Sánchez-Muniz, F.J. Nori- and sea spaghetti- but not wakame-restructured pork decrease the hypercholesterolemic and liver proapototic short-term effects of high-dietary cholesterol consumption. *Nutrición Hospitalaria* **2013**, *28*, 1422–1429.

103. Jeon, M.R.; Choi, S.H. Quality Characteristics of Pork Patties Added with Seaweed Powder. *Korean J. Food Sci. Anim. Resour.* **2012**, *32*, 77–83. [CrossRef]

104. López-Lopez, I.; Cofrades, S.; Yakan, A.; Solas, M.; Jimenez-Colmenero, F. Frozen storage characteristics of low-salt and low-fat beef patties as affected by Wakame addition and replacing pork backfat with olive oil-in-water emulsion. *Food Res. Int.* **2010**, *43*, 1244–1254. [CrossRef]

105. López-Lopez, I.; Cofrades, S.; Cañeque, V.; Díaz, M.; Lopez, O.; Jimenez-Colmenero, F. Effect of cooking on the chemical composition of low-salt, low-fat Wakame/olive oil added beef patties with special reference to fatty acid content. *Meat Sci.* **2011**, *89*, 27–34. [CrossRef]

106. Sasaki, K.; Ishihara, K.; Oyamada, C.; Sato, A.; Fukushi, A.; Arakane, T.; Motoyama, M.; Yamazaki, M.; Mitsumoto, M. Effects of Fucoxanthin Addition to Ground Chicken Breast Meat on Lipid and Colour Stability during Chilled Storage, before and after Cooking. *Asian Australas. J. Anim. Sci.* **2008**, *21*, 1067–1072. [CrossRef]

107. Prabhasankar, P.; Ganesan, P.; Bhaskar, N.; Hirose, A.; Stephen, N.; Gowda, L.R.; Hosokawa, M.; Miyashita, K. Edible Japanese seaweed, wakame (*Undaria pinnatifida*) as an ingredient in pasta: Chemical, functional and structural evaluation. *Food Chem.* **2009**, *115*, 501–508. [CrossRef]

108. Moreira, A.S.; González-Torres, L.; Olivero-David, R.; Bastida, S.; Benedi, J.; Sánchez-Muniz, F.J. Wakame and Nori in Restructured Meats Included in Cholesterol-enriched Diets Affect the Antioxidant Enzyme Gene Expressions and Activities in Wistar Rats. *Plant Foods Hum. Nutr.* **2010**, *65*, 290–298. [CrossRef] [PubMed]

109. Lalic, L.M.; Berkovic, K. The influence of algae addition on physicochemical properties of cottage cheese. *Milchwiss. Milk Sci. Int.* **2005**, *60*, 151–154.

110. Dierick, N.; Ovyn, A.; De Smet, S. Effect of feeding intact brown seaweed Ascophyllum nodosum on some digestive parameters and on iodine content in edible tissues in pigs. *J. Sci. Food Agric.* **2009**, *89*, 584–594. [CrossRef]

111. Hall, A.; Fairclough, A.; Mahadevan, K.; Paxman, J. Ascophyllum nodosum enriched bread reduces subsequent energy intake with no effect on post-prandial glucose and cholesterol in healthy, overweight males. A pilot study. *Appetite* **2012**, *58*, 379–386. [CrossRef] [PubMed]

112. Peteiro, C. Alginate Production from Marine Macroalgae, with Emphasis on Kelp Farming. In *Alginates and Their Biomedical Applications*; Rehm, B.H.A., Moradali, F., Eds.; Springer: Berlin/Heidelberg, Germany, 2018; pp. 27–66.

113. Schmid, S.; Ranz, D.; He, M.L.; Burkard, S.; Lukowicz, M.V.; Reiter, R.; Arnold, R.; Le Deit, H.; David, M.; Rambeck, W. A Marine algae as natural source of iodine in the feeding of freshwater fish—A new possibility to improve iodine supply of man. *Rev. Med. Vet. (Toulouse).* **2003**, *154*, 645–648.

114. Ribeiro, A.R.; Gonçalves, A.; Colen, R.; Nunes, M.L.; Dinis, M.T.; Dias, J. Dietary macroalgae is a natural and effective tool to fortify gilthead seabream fillets with iodine: Effects on growth, sensory quality and nutritional value. *Aquaculture* **2015**, *437*, 51–59. [CrossRef]

115. Ribeiro, A.R.; Gonçalves, A.; Bandarra, N.; Nunes, M.L.; Dinis, M.T.; Dias, J.; Rema, P. Natural fortification of trout with dietary macroalgae and selenised-yeast increases the nutritional contribution in iodine and selenium. *Food Res. Int.* **2017**, *99*, 1103–1109. [CrossRef] [PubMed]

116. He, M.L.; Hollwich, W.; Rambeck, W.A. Supplementation of algae to the diet of pigs: A new possibility to improve the iodine content in the meat. *J. Anim. Physiol. Anim. Nutr.* **2002**, *86*, 97–104. [CrossRef]

117. Rajauria, G.; Draper, J.; McDonnell, M.; O'Doherty, J. Effect of dietary seaweed extracts, galactooligosaccharide and vitamin E supplementation on meat quality parameters in finisher pigs. *Innov. Food Sci. Emerg. Technol.* **2016**, *37*, 269–275. [CrossRef]
118. Moroney, N.; O'Grady, M.; O'Doherty, J.; Kerry, J. Effect of a brown seaweed (*Laminaria digitata*) extract containing laminarin and fucoidan on the quality and shelf-life of fresh and cooked minced pork patties. *Meat Sci.* **2013**, *94*, 304–311. [CrossRef]
119. Kim, H.W.; Choi, J.H.; Choi, Y.S.; Han, D.J.; Kim, H.Y.; Lee, M.A.; Kim, S.Y.; Kim, C.J. Effects of Sea Tangle (*Lamina japonica*) Powder on Quality Characteristics of Breakfast Sausages. *Food Sci. Anim. Resour.* **2010**, *30*, 55–61. [CrossRef]
120. Kim, H.H.; Lim, H.-S. Effects of Sea Tangle-added Patty on Postprandial Serum Lipid Profiles and Glucose in Borderline Hypercholesterolemic Adults. *J. Korean Soc. Food Sci. Nutr.* **2014**, *43*, 522–529. [CrossRef]
121. Koval, P.V.; Shulgin, Y.P.; Lazhentseva, L.Y.; Zagorodnaya, G.I. Probiotic drinks containing iodine. *Molochnaya Promyshlennost* **2005**, *6*, 38–39.
122. Indrawati, R.; Sukowijoyo, H.; Indriatmoko; Wijayanti, R.D.E.; Limantara, L. Encapsulation of Brown Seaweed Pigment by Freeze Drying: Characterization and its Stability during Storage. *Procedia Chem.* **2015**, *14*, 353–360. [CrossRef]
123. Einarsdóttir, Á.M. Edible Seaweed for Taste Enhancement and Salt Replacement by Enzymatic Methods. MSc Thesis, University of Iceland, Reykjavík, Iceland, 2014.
124. Seo, Y.S.; Bae, H.N.; Eom, S.H.; Lim, K.S.; Yun, I.H.; Chung, Y.H.; Jeon, J.M.; Kim, H.W.; Lee, M.S.; Lee, Y.B.; et al. Removal of off-flavors from sea tangle (*Laminaria japonica*) extract by fermentation with Aspergillus oryzae. *Bioresour. Technol.* **2012**, *121*, 475–479. [CrossRef] [PubMed]
125. Jimenez-Escrig, A.; Sánchez-Muniz, F.J. Dietary fibre from edible seaweeds: Chemical structure, physicochemical properties and effects on cholesterol metabolism. *Nutr. Res.* **2000**, *20*, 585–598. [CrossRef]
126. Asai, A.; Sugawara, T.; Ono, H.; Nagao, A. Biotransformation of fucoxanthinol into amarouciaxanthin a in mice and hepg2 cells: Formation and cytotoxicity of fucoxanthin metabolites. *Drug Metab. Dispos.* **2004**, *32*, 205–211. [CrossRef]
127. Sugawara, T.; Baskaran, V.; Nagao, A.; Tsuzuki, W. Brown Algae Fucoxanthin Is Hydrolyzed to Fucoxanthinol during Absorption by Caco-2 Human Intestinal Cells and Mice. *J. Nutr.* **2002**, *132*, 946–951. [CrossRef]
128. Hashimoto, T.; Ozaki, Y.; Taminato, M.; Das, S.K.; Mizuno, M.; Yoshimura, K.; Maoka, T.; Kanazawa, K. The distribution and accumulation of fucoxanthin and its metabolites after oral administration in mice. *Br. J. Nutr.* **2009**, *102*, 242–248. [CrossRef]
129. Yonekura, L.; Kobayashi, M.; Terasaki, M.; Nagao, A. Keto-Carotenoids Are the Major Metabolites of Dietary Lutein and Fucoxanthin in Mouse Tissues. *J. Nutr.* **2010**, *140*, 1824–1831. [CrossRef]
130. Asai, A.; Yonekura, L.; Nagao, A. Low bioavailability of dietary epoxyxanthophylls in humans. *Br. J. Nutr.* **2008**, *100*, 273–277. [CrossRef] [PubMed]
131. Domínguez-González, M.R.; Herbello-Hermelo, P.; Vélez, D.; Devesa, V.; Bermejo-Barrera, P.; González, R.D.; Chiocchetti, G.M. Evaluation of iodine bioavailability in seaweed using in vitro methods. *J. Agric. Food Chem.* **2017**, *65*, 8435–8442. [CrossRef] [PubMed]
132. Combet, E.; Ma, Z.F.; Cousins, F.; Thompson, B.; Lean, M.E.J. Low-level seaweed supplementation improves iodine status in iodine-insufficient women. *Br. J. Nutr.* **2014**, *112*, 753–761. [CrossRef] [PubMed]
133. Plaza, M.; Cifuentes, A.; Ibañez, E. In the search of new functional food ingredients from algae. *Trends Food Sci. Technol.* **2008**, *19*, 31–39. [CrossRef]

 *antioxidants*

Review

The Potential of Seaweeds as a Source of Functional Ingredients of Prebiotic and Antioxidant Value

Andrea Gomez-Zavaglia [1], Miguel A. Prieto Lage [2], Cecilia Jimenez-Lopez [2], Juan C. Mejuto [3] and Jesus Simal-Gandara [2,*]

[1]   Center for Research and Development in Food Cryotechnology (CIDCA), CCT-CONICET La Plata, Calle 47 y 116, La Plata, Buenos Aires 1900, Argentina
[2]   Nutrition and Bromatology Group, Department of Analytical and Food Chemistry, Faculty of Science, University of Vigo – Ourense Campus, E32004 Ourense, Spain
[3]   Department of Physical Chemistry, Faculty of Science, University of Vigo – Ourense Campus, E32004 Ourense, Spain
*   Correspondence: jsimal@uvigo.es

Received: 30 June 2019; Accepted: 8 September 2019; Published: 17 September 2019

Abstract: Two thirds of the world is covered by oceans, whose upper layer is inhabited by algae. This means that there is a large extension to obtain these photoautotrophic organisms. Algae have undergone a boom in recent years, with consequent discoveries and advances in this field. Algae are not only of high ecological value but also of great economic importance. Possible applications of algae are very diverse and include anti-biofilm activity, production of biofuels, bioremediation, as fertilizer, as fish feed, as food or food ingredients, in pharmacology (since they show antioxidant or contraceptive activities), in cosmeceutical formulation, and in such other applications as filters or for obtaining minerals. In this context, algae as food can be of help to maintain or even improve human health, and there is a growing interest in new products called functional foods, which can promote such a healthy state. Therefore, in this search, one of the main areas of research is the extraction and characterization of new natural ingredients with biological activity (e.g., prebiotic and antioxidant) that can contribute to consumers' well-being. The present review shows the results of a bibliographic survey on the chemical composition of macroalgae, together with a critical discussion about their potential as natural sources of new functional ingredients.

Keywords: seaweeds; macroalgae; invasive species; prebiotics; antioxidants; functional foods

---

1. Macroalgae Classification

We live on a planet of which ~72% of the surface is water. Since all the necessary elements for life are found in seawater, every form of life emerged from that immense original matrix. These elements and many others, since all the elements of the periodic table are in the sea, have the advantage of being present in quantities that are generally stable and constant along the marine surface, contrary to what happens on earth [1]. However, traces of our marine origins can also be found on earth, since there are similarities between the composition and some properties of the sea and those of biological fluids [2–4].

The oceans contain and give life to approximately 500,000 species, which means that almost three quarters of all known species inhabit seawater. Among them are algae which, although the vast majority inhabit salt water, can also survive in freshwater. These are very peculiar living beings, to whom the development of life on our planet is due, since the algae were pioneers of photosynthesis, thanks to the evolution of chlorophyll function 3200 million years ago [1,5]. Photosynthesis probably began in some blue-green prokaryotic microorganisms that were formerly considered algae, and that currently belong to the *Phylum Cyanobacteria*, which is included in the Monera Kingdom [6].

More than 30,000 species of algae have been described, and their scientific study is called phycology. According to the current definition of algae, the blue-green variety is not considered algae, as they are prokaryotic organisms, and only eukaryotic organisms belong to this category (either unicellular, such as microalgae phytoplankton, or multicellular, such as macroalgae) [3]. The taxonomic classification of algae is complex due to the number of existing varieties and the many applicable classification criteria. They are a polyphyletic group, which means that they belong to different kin groups. Therefore, the classification is not well defined, and may vary according to the authors, but they are currently included in the Protista kingdom, with some exceptions of macroalgae belonging to in the Plantae kingdom [7,8].

The microalgae, Protista microorganisms that are classed as phytoplankton, are important in nature because they represent the first trophic level in the food chain, serving as nutrients to thousands of marine species. They are also essential in chlorophyll function, since they are primary producers, being responsible for 30%–50% of the oxygen contained in the atmosphere [4].

Macroalgae are also a very varied group, with sizes ranging from a few centimeters to specimens that can reach 100 m in length. Approximately 15,000 species of this group have been described [9]. They are also autotrophic and photosynthetic beings, so their habitat is limited to a certain depth, which is usually a maximum of 60 m, always within the intertidal zone, and its growth is usually vertical, looking for sunlight [10,11]. The differences between them and terrestrial plants are that they do not present conductive tissues, but rather they adsorb nutrients throughout their whole surface. They also lack roots, though some present rhizoids or basal discs that allow them to adhere to rocks as a method of restraint, but not to nourish themselves. They form large underwater meadows and are generators of ecosystems in which many different species of bacteria, corals, mollusks, fish, and other marine creatures accumulate and coexist [7,12–14].

Much of the literature agrees that macroalgae can be divided into 3 large groups: The Chlorophytas, commonly known as green algae, the Rhodophytas or red algae, both included in the Plantae kingdom; and the Ochrophytas, mostly classified in the Phaeophyceae class. These are also called brown seaweed and belong to the Protista kingdom, as well as the microalgae kingdom. The classification of this algae macrogroup was made taking into account the pigment that composes it, and through which it manages to perform photosynthesis to carry out autotrophic feeding [13]. There are authors that include some green macroalgae in another differentiated group and at a taxonomic level equivalent to Chlorophytas, which are responsible for life expanding beyond the oceans, and are a precursor of terrestrial plants [15]. Table 1 summarizes the phylogeny of these four large groups of macroalgae. The mentioned differences between algae pigments are collected in Table 2.

Table 1. Phylogenetic summary of the most common macroalgae.

Kingdom	Phylum/Division	Classes	Orders
Chromist	Ochrophyta	Phaeophyceae	Ascoseirales; Desmarestiales; Discosporangiales; Dictyotales Ectocarpales; Fucales; Laminariales; Nemodermatales Ralfsiales.
Plantae	Charophyta	Charophyceae; Chlorokybophyceae; Coleochaetophyceae; Klebsormidiophyceae; Mesostigmatophyceae; Zygnematophyceae.	Bryopsidales; Cladophorales; Dasycladales; Oltmannsiellopsidaes; Trentepohliales; Ulotrichales; Ulvales
	Chlorophyta	Ulvophyceae	
	Rhodophyta	Bangiophyceae; Compsopogonophyceae; Florideophyceae; Porphyridiophyceae; Rhodellophyceae; Stylonematophyceae.	

Table 2. Pigment content of the 3 common groups of macroalgae. Bold pigments represent the predominant ones in each group.

Pigment Class	Green Algae	Brown Algae	Red Algae	Reference
Chlorophylls	Chlorophyll *a* and *b*, and derivatives	Chlorophylls *b* and *c*, and derivatives	Chlorophylls *a* and *d*, and derivatives	[16]
Carotenoids	β-carotene, xanthophylls	Fucoxanthin and xanthophylls, β-carotene	Xanthophylls	[13,16,17]
Phycobiliproteins	-	-	Phycoerythrin and phycocyanin	[13,16]
Example	*Halimeda sp*	*Fucus serratus*	*Palmaria palmata*	

2. The Potential of Invasive Seaweeds

Due to increasing globalization as well as climate change, the arrival of invasive algae to coastal areas of different regions is becoming more common. However, these are not the only causes of the increased proliferation of these non-indigenous marine species. Other causes, such as those related to the marine industry (aquaculture, fisheries, and marine tourism) must also be considered [18–20]. That is why marine biodiversity is seriously threatened by macroalgae invasions. In fact, macroalgae represent between 10% and 40% of the total number of species introduced into the marine ecosystem [21].

These algae not only cause an environmental problem by displacing other native species, causing a loss of biodiversity, damage in the structure and function of the native ecosystem, and a homogenization of the landscape, but also cause losses in the fishing, recreational sector, and many other industrial sectors related with aquatic environment. However, this abundance can also present opportunities. It is therefore of great interest to find possible applications to give added value to these algae. Furthermore, most of the foods humans eat in the modern world come from a small number of domestic animals and plants widely raised and cultivated, most of them having been introduced voluntarily by humans. One of these applications could be the obtainment of compounds of natural origin.

Because invasive algae pose a serious threat in coastal areas, the interest in developing protocols for the control of these exogenous species is booming, although in parallel, researchers are also trying to develop strategies that allow them to be used as a natural source of secondary metabolites. One of the measures proposed is the eradication of the species, which is a promising solution for areas declared as protected, such as Marine Protected Areas. In this sense, the enormous production of biomass by this type of algae can become a benefit, since it offers the possibility of recycling or reuse at the time of eradication. In any case, it should be taken into account that the impact-control studies do not always allow us to reach an accurate conclusion about the impact of a certain invasive algae, since these studies are designed mainly with the intention of evaluating the anthropogenic impacts, when a control-impact evaluation before and after the invasion would be more accurate [22].

The species of marine algae that are in the top five of "potentially invasive", which means that they meet certain characteristics (relating to the mechanisms of reproduction, growth and defense, resistance to pollution, among others), as well as having a high ecological impact, are, from least to greatest: *Grateloupia turuturu* (as *Grateloupia doryphora*), *Asparagopsis armata*, *Undaria pinnatifida*, *Caulerpa taxifolia*, and *Codium fragile* subsp. *tomentosoides* [23]. Some authors [24] classify the factors that favor such invasive attacks of macroalgae in two groups: The abiotic ones, such as salinity and waves, and the biotic ones, such as those related to the competitive abilities of the species. Other authors [25] proposed the Tilman's R* rule, where R refers to the resources available in an area and R* to the balance of available resources, from which one can predict where the invasion is favored, since if the R* of the endogenous species is greater than that of the exogenous species, the invasion is more likely to occur, and may occur in two ways: The invader needs fewer resources than the resident, or the range of acquisition of resources of the invader is greater than that of the endogenous one. On the other hand, there is a study [24] that discusses about the enemy release hypothesis (ERH), based on the fact that the invader has escaped from its habitat due to the presence of enemies and/or herbivores. This can also be the explanation of why invasive algae tend to have greater resistance to herbivores than native ones. For example, *Fucus evanescens*, presents higher amounts of phlorotannins, compounds known to cause animosity to herbivores [24], than the native species [26].

Laminaria sp. (brown algae) show a great adaptability and relocation thanks to their gametophytes, formed by filamentous tufts of approximately 1 or 2 cm. These reproductive structures can be considered as "seed stocks", so their presence is extremely important when colonizing a certain area, as does, for example, the species *Undaria* sp. [27]. Regarding the interactions between the invasive species themselves, information is scarce, so it is an area still to be explored that concerns the entire globe, since invasive algae are not governed by national borders.

Many of these algae contain bioactive compounds that could be contemplated for a wide range of commercial applications like nutrition and pharmaceutical ones. In this regard, invasive algae present great advantages because, according to some authors [20], invasive algae present fast growth rates and biomass accumulation, high levels of repellency against herbivores, and often low levels of epiphytism. Moreover, according to these authors, the low cost of algae farming, along with good economic results and the high demand for products obtained from algae, has led to the intentional introduction of potentially invasive algae in overcoat low-wages countries.

3. Algae Aquaculture

Aquaculture is the science of cultivating aquatic animals, plants, and related organisms like fish, shellfish, seaweed, and microalgae, for human use and consumption, and would be a fast-growing industry. Nowadays, many authors are focused on performing Life Cycle Assessment (LCA) of seafood production to provide new insights into its environmental impacts and therefore to improve environmental sustainability of the aquaculture production systems [28]. Although the technologies related to aquaculture of algae have undergone tremendous development in the last 70 years, especially in Asia, but also in America and Europe, there is still much to improve regarding their science and the social acceptance that entails. One of the main challenges is the development of strains that are thermo-resistant, of rapid growth, with high production of compounds of interest, resistance to morbidity, and antifouling capacity, as well as the development of efficient and economical hatcheries, capable of withstanding storms in the open sea [29]. One fact to keep in mind is that algae aquaculture offers advantages to ecosystems, since it improves the conditions of coastal waters, favoring other species and the environment. Interestingly, although more than 10,000 species of algae are known, aquaculture of algae is mostly made (more than 81% of production) using very few of them [30]: The brown algae, *Saccharina japonica* and *Undaria pinnatifida*, and the red algae *Porphyra sp.*, *Kappaphycus alvarezii* and *Eucheuma striatum* (carrageenophytes), and *Gracilaria/Gracilariopsis* sp. (agarophytes).

One of the priorities when developing aquaculture is to be sustainable, that is, to ensure the minimum possible adverse effects for the environment. One way to achieve this is by developing improved methods for waste treatment. Integrated Multi-Trophic Aquaculture (IMTA) combines the aquaculture of food (for example, of fish) with that of extraction (for example, algae) to create a more balanced ecosystem [31]. However, caution should be exercised, since coastal waters containing high amounts of nutrients may favor the emergence of potentially harmful, invasive, or opportunistic algae, which may have negative consequences for the coastal zone [32–34]. There is another similar concept that excludes the feeding concept, called nutrient bioextraction, which can be applied even to urbanized estuaries, where excess nutrients are currently a problem. In both aquaculture systems described, algae can be used as a solution to eliminate inorganic nutrients (phosphorus compounds, nitrogen, carbon dioxide, and other compounds used for their metabolism), thus decreasing the negative impacts on the environment [35–38]. In this way, while algae are cultivated, the levels of nutrients in the water are reduced, so its acceptance is much greater by users and those who are positioned against aquaculture, since the presence of seaweed provides advantages in the aquaculture system, such as minimal environmental adverse effects and reduction of costs, due to the utilization of wastes to feed other levels [31].

4. General Current Seaweed Industrial Applications

General current seaweed industrial applications are important because these days consumers look for products with a natural origin. This is a new use of algae, which is in high demand. A brief and general bibliographic review allowed us to define the main biological activities (antioxidant, antibacterial, anti-inflammatory, antifungal/antiprotozoal, antiviral, cytotoxicity/antiproliferative, adipogenesis, MAA/UV protection, matrix metalloproteinases and blood fluidity) of algae (summarized next in Table 3).

Table 3. Current marine algae applications.

Applications	Specific	Authors
Anti-biofilm activity		[39–41]
Biofuels		[32,42–45]
Bioremediation		[46–48]
Contraceptive activity		[49,50]
Cosmeceuticals		[51,52]
Fertilizer		[32,43,53]
Fish feed		[32,43]
Food ingredients		[32,43]
Pharmacology/medical	General	[32,54]
	Antibiotic, antiviral, antifungal activity	[55–58]
	Anticancer	[56,59,60]
	Anticoagulant	[61–64]
	Anti-inflammatory	[65–67]
	Antioxidants	[56,68]
Other applications:	Filter	[69]
	Mineralogenic	[43,70]

4.1. Anti-Biofilm Applications

Anti-biofilm activity has been investigated with models based on Gram-positive and -negative bacteria. This is important because biofilms may cause diverse diseases. Several studies concerning anti-biofilm activity can be used as practical examples. For instance, there is a research about how seaweeds can attack *Salmonella enterica* biofilms. That is important because this organism is one of the most prominent causes of bacterial food-borne diseases. In that study, it was observed that brominated furanones obtained by extraction from the red algae *Delisea pulchra*, interfere in this biofilm formation, but more studies are needed to determine a long term efficiency [41]. Another work [40] reported that *Sesbania grandiflora* extract has anti-biofilm and anti-bacterial activity against *Staphylococcus aureus*. Antimicrobial, anti-biofilm, and antifouling properties of sulfated polysaccharides obtained from marine macroalgae were also studied using dental plaque bacteria as model [39].

4.2. Biofuel and Bioremediation Applications

Algae can also be used in the production of biofuels. However, for making the process of obtaining biofuel feasible economically, algae should present a co-production of some other components that could be used biochemically and that have a proven commercial value [42]. This approach would also decrease the cost of algae treatments and should mitigate the eutrophication of lakes. However, using algae as a fuel has several advantages. One of them is that a single algal biomass can be used to produce several different kinds of renewable biofuels depending on the treatment applied (anaerobic digestion and biodiesel derived from algal oil, among others) [32].

Regarding bioremediation, the accumulation of high amounts of organic and inorganic matter is a risk factor for many ecosystems, and therefore, for human health, that is why new solutions are sought to remedy this accumulation. Algae can be one remedy at the aquatic ecosystem level, since they participate actively in the control and biomonitoring of organic pollutants [48].

Algae are a group of great interest for this purpose, since they are ideal for the bioremediation of wastewater thanks to their culture being very economical and easy to achieve on a large scale, since they are capable of capturing a high percentage of metal ions. Furthermore, according to the large amount of biomass that is produced by algae growing in wastewater, a dual-purpose crop could be considered to bioremediate wastewater, and to obtain a production system for other substances of interest, such as compounds with bioactivities [47].

Another study [46] reported that certain algae act as "hyper-accumulators" and "hyper-adsorbents" with a high selectivity for different elements. They also contribute to an alkaline environment, leading heavy metals to precipitate during treatments.

However, in all cases it is necessary to control some parameters such as temperature, pH, nutrients, and availability, among others, to reach the optimum conditions under which algae show the best absorption, removal, and biodegradation of different pollutants [46–48].

4.3. Fish Feed Applications

Most of the fish feed used in aquaculture is made from other fish meat. This has an enormous disadvantage as fishes have to be fed twice and that costs a lot. Moreover, the world's fish stocks are decreasing, and aquaculture is increasing, which makes that system unsustainable.

As for the nutritional aspects of microalgae, they are strongly dependent on culture conditions as well as on factors such as culture phase, temperature, and availability of nutrients [71].

There are evidences of good nutritional properties of algae biomass as a source of micronutrients or as a bulk feedstuff. Moreover, they have a positive effect on the physiological state of the larvae due to, for example, the diversification of bacterial flora [72].

The use of algae as a bulk feedstuff or as a supplement depends on the biomass availability, as well as its composition and cost [73]. However, their use should be limited to a certain concentration due to the amount of toxic metals they may have, such as arsenic, which is one of the main limiting factors.

The consumption of microalgae is carried out directly, in the case of mollusks and crustaceans, and indirectly in urine from previous ingestion by zooplankton species [71].

Algae can be used for animal feed as well. They are used among others in ruminants' nutrition.

4.4. General Food Applications

The main use of algae is the one with direct food applications ("seaweed as a vegetable"). This use represents the main world market for algae. This is mainly due to the great consumption that exists in Asian countries, where they are traditional products of high consumption. The main seaweeds used as human food are [74]: Nori or purple laver (*Porphyra spp.*), aonori or green laver (*Monostroma* spp. and *Enteromorpha spp.*), kombu (*Laminaria japonica*), wakame, (*Undaria pinnatifida*), Hiziki (*Hizikia fusiforme*), mozuku (*Cladosiphon okamuranus*), sea grapes or green caviar (*Caulerpa lentillifera*), dulse (*Palmaria palmata*), Irish moss (*Chondrus crispus*), winged kelp (*Alaria esculenta*), ogo (*Gracilaria spp.*), and *Callophyllis variegate*. Algae products are consumed as food in different ways: Dried, in sushi, in soups, in salads, in tea, in mustard, in pasta, in breads, and many other similar food products.

Algae also contain some compounds that can be used as food ingredients. For example, phycocolloids are a group of natural polymers constituted by polysaccharides with the ability to form gels (hydrocolloid) derived from macroalgae. They are used in almost all industries (food, drugs, paintings, cosmetics, etc.) due to their physical–chemical characteristics. These are the second-most common use of algae. The majority of obtained hydrocolloids are alginate, carrageenan, and agar.

According to a previous study [74], several red and brown algae are used in the production of three hydrocolloids: Agar, alginate, and carrageenan. These compounds are water-soluble carbohydrates mainly used to increase aqueous solutions viscosity, to produce different types of gels, to form water-soluble films, and to stabilize some products. Other functional ingredients present in algae are carotenoids (used as food colorants, feed supplements, and nutraceuticals), lipids, proteins, polysaccharides, and phenolic compounds [32].

4.5. Pharmacology and/or Medical Applications

Algae produce bioactive compounds with rich pharmacological potential. They generate these compounds as a response to environmental conditions or characteristics (competition for space, maintenance of unfolded surfaces, repulsion of predators, etc.) [75]. In recent years, seaweeds have been recognized as producers of an enormous range of biologically active compounds, but the bioactivity of the same species could vary depending on the geographical zone due to environmental and seasonal parameters [75].

4.5.1. Contraceptive Activity Applications

Different studies [49,50] with extracts from red algae were carried out. These extracts were administrated to female rats in the first seven days of their pregnancies. As result, it was demonstrated that these extracts have post-coital contraceptive protection in the rats without showing any relevant adverse effect.

4.5.2. Antibiotic, Antiviral, and Antifungal Activity

Algae's antibiotic activity was demonstrated for several authors through tests of the extracts obtained from algae against Gram-positive and -negative bacteria. Finding these new sources of antibiotics is interesting because microorganisms are getting more resistant to drugs, so providing new drugs is of utmost importance nowadays. That is why novel natural antimicrobial compounds with high potential, good availability, less toxicity, and fewer adverse effects are required [56].

According to a recent article [55], isolated chemical compounds from marine seaweed have been shown to owe bioactivities such as antimicrobial, antioxidant, and anti-inflammatory properties, as well as anticoagulant and apoptotic effects. Viral diseases are still one of the main causes of death in the world. Treatments sometimes fail because of the side effects of infectious diseases or because of drug resistance. Therefore, the study of algae compounds with antiviral activity is of great interest. According to other authors [58,76], algae from the three large groups (green, brown, and red algae) produce interesting polysaccharides that show antibacterial activity against some pathogen bacteria such as *Aeromonas salmonicida* or *Pseudomonas aeruginosa*, both being of importance because of their high pathogenicity, resistance, and incidence in humans. Specifically, some red algae produce sulfated polysaccharides showing antiviral capacity against viruses responsible for different infectious diseases. As an example, red algae *Gracilaria gracilis* extracts have shown antimicrobial activity against *Bacillus subtilis*, an extreme conditions resistant bacteria [77]. One of the main areas of interest is to find new treatments towards herpes simplex virus type 1 (HSV-1), as it is an endemic disease, the virus is developing resistance, and other available drugs have side effects [57].

4.5.3. Anticancer Activity

Actually, cancer is one of the principal causes of death around the world, thus, finding new treatments for this disease is a big challenge. Some studies have linked the anticancer capacity of a certain algae with the content of compounds with antioxidant properties.

Most studies of anticancer properties in algae have been performed on microalga extracts or fractions obtained using low-resolution methods (liquid–liquid extraction, solid–liquid extraction) [59].

Extracts from algae *Bryopsis* sp. contain some compounds (depsipeptides kahalalide A and F) whose future seems to lie with the development of kahalalide F for treatment of lung cancer, tumors, and AIDS. This is due to the fact that kahalalide F has anticancer properties, namely the control of tumors causing colon, lung and prostate cancer [58]. It happens the same with chondriamide A, a compound obtained from *Chondria atropurpurea* that presents antiproliferative activity towards human nasopharyngeal and colorectal cancer cells [60].

Other compounds present in algae, such as several polysaccharides with sulfur groups—e.g., fucoids—have also shown cytotoxic properties [78]. Another example are terpenes. A study carried out using *Chlorella sorokiniana* and *Chaetoceros calcitrans* extracts concluded than they present interesting activities compared to commercially available marine anti-cancer drugs [59]. *L. papillosa* algae was also studied and the conclusion was that it contains some bioactive compounds that could serve as a promising potential antioxidant, antimicrobial, and anticancer agents [56].

4.5.4. Anticoagulant Activity

Phlorotannins and sulfated polysaccharides such as fucoidans in brown algae, carrageenans in red algae, and ulvans in green algae have been recognized as potential anticoagulant agents [61–64]. For example, in a study carried out by [79], they tested the effect of some polysaccharides rich extracts obtained from green algae *Ulva fasciata* and red algae *Agardhiella subulata* in the thromboplastin time and prothrombin time, resulting that both algal extracts showed that prolong those mentioned times during the coagulation cascade, avoiding blood coagulation. However, the anticoagulant activity has been mainly investigated in vitro or in mouse. Hence, and due to the important value of that property, more studies are needed [62].

4.5.5. Anti-Inflammatory Activity

Several studies have demonstrated that some extracts and compounds from a wide variety of algae have potential anti-inflammatory compounds. For example, [66] studied the anti-inflamatory capacity of methanolic extracts obtained from red seaweed *Dichotomaria obtusata*. Other authors [65] reported the enhanced anti-inflamatory ability of brown kelp *Laminaria japonica* by fermentation with *Bacillus subtilis*. Likewise, [67] reported anti-inflammatory properties of several marine algae collected from the Mannar coast gulf (Tuticorin, South India).

4.5.6. Antioxidant Activity

An excessive amount of reactive oxygen species may result in lipid peroxidation, which change the structure of body biomolecules, causing cellular disorders, premature aging, mutations or cell death. Different researches have demonstrated seaweed antioxidant capacity in vitro, attributed to the presence of new antioxidant compounds like carotenoids, certain polysaccharides, and polyphenols, which show scavenger activity, being able to neutralize those reactive oxygen species through their own oxidation, since their affinity to those oxidative compounds is very high [56,68,80–85].

As a curiosity that strengthens the utilization of algae as antioxidant tools, a study carried out by [86] in which fishes were fed with red algae *Gracilaria gracilis* powder, demonstrates that this nutritional input favors genetic expression of some antioxidant enzymes, like superoxide dismutase and catalase, what it also improves the state of the body's immune system.

4.6. General Cosmeceuticals Applications

Another possible use is to obtain compounds with cosmeceutical bioactivity, that is, compounds to be used as ingredients for skin care products. Many of the invasive species have been demonstrated to include these compounds. However, metabolite amounts often vary according to geographical and seasonal conditions, so environmental variability has to be taken into consideration [87]. For that type of product, brown and red marine seaweeds are the most used species. Extracts rich in potential cosmeceutical ingredients, such as phlorotannins, polysaccharides, carotenoids, fatty acids, as well as bioactive proteins, vitamins, and minerals can be obtained from seaweed [52]. These compounds are incorporated into cosmetics to optimize their properties; their capacity to stabilize and preserve products and the bioactive activities of the compounds found (antioxidant, photo-protection, anti-wrinkling, anti-cellulite, moisturizing, and whitening) [88,89]. In vitro studies also demonstrated that these compounds are UV protective, or have an inhibitory effect on melanogenesis [51].

4.7. Other Applications

Traditionally seaweeds have also been collected and used by coastal communities as fertilizers. They are appropriate for that use since they have the characteristic of enriching the soil with N, P, and K, favoring agriculture, and they have the ability to promote seed germination, increase frost resistance, and improve resistance to fungal and insect pests [32,43,53]. The main species that are used for agricultural applications are *Ascophyllum nodosum, Ecklonia maxima, Laminaria sp., Lessonia* species, and *Macrocytis pyriphea* [74]. The most common extractive modality to obtain these interesting compounds is alkaline extraction for which potash is added and heat is applied [90].

Another interesting property of algae is that they are able to take many minerals from the sea, with no substantial differences in the quantity among different types of algae. The most important macroelements they contain are sodium, calcium, magnesium, potassium, chloride, sulfur, and phosphorus. Regarding oligoelements, iodine, iron, zinc, selenium, copper, molybdenum, fluorine, manganese, boron, cobalt, and nickel are the most relevant ones. Iodine is very important as many people suffer from iodine deficiency. The bioavailability depends on the algae species and on the treatment during harvesting and processing [70]. According to this, it can be said that algae possess mineralogenic properties, which represents a useful tool, since these minerals are worthy in both fertilizer and animal feed [43].

Finally, fish farming, as well as intensive livestock production, produces effluents with a high content of inorganic nitrogen and phosphorus, compounds that are needed to produce a biomass of algae and have a high cost. Using these wastes for the production of algae biomass, environmental benefits could be obtained due to the seaweed biofiltration and depuration capacities, favoring the purification of effluents, as well as promoting a reduction of costs thanks to the use of waste waters as a source of necessary compounds [69].

5. Food and Technical Uses of Algae

Although algae extracts can be used in several products, they are mainly used as food ingredients in the formulation of food products (Table 4) [91,92]. In East Asia, as well as in the Pacific Islands, food has been based for centuries on the direct consumption of seaweed, the red seaweed Nori, or laver (*Porphyra*) being the most common commercial ones. In Japan, there are farms in shallow bays and seas comprising approximately 100,000 hectares. The *Porphyra* algae has a life cycle that includes two phases: A small and shell-boring. The first phase can be favored and augmented by humans, by planting on special platforms containing sea beds to which oyster shells are attached by using ropes or nets. In turn, the second phase consists on the germination of the conchospores and their growth along the platforms. It is at this time that the nets or ropes are removed to collect, wash and press the algae so that their drying is accelerated [93].

Table 4. Food and feed applications of the main red, brown and green macroalgae.

Name	Applications	Region/Country	References
Red Alga (Rhodophyta)			
Porphyra (alga nori)	Cultivated for food	Asia	[93]
Saccharina japonica	Cultivated for food	Japan	[93]
Palmaria palmata (Dulse)	Culinary ingredient, flavor-enhancer	USA, Canada, Scotland, Ireland, Iceland	[93]
Gelidium sp., *Gracilaria* sp., *Pterocladia* sp., *Acanthopeltis* sp., *Ahnfeltia* sp.	Instant pie fillings, canned meats or fish, bakery icings, beer and wine clarifiers	Asia	[93]
Eucheuma sp., *Chondria* sp., *Iridaea* sp.	Thickening and stabilizers, imitation of creams, puddings, syrups, canned pet foods.	Philippines, Ireland, Chile, USA, Canada	[93]

Table 4. *Cont.*

Name	Applications	Region/Country	References
Grateloupia sp.	As vegetable	Indo-pacific region	[94]
Brown Alga (Pheophyta)			
Sargassum fusiforme, Sargassum dentifolium	Farmed in small quantities (poultry, improves quality of eggs)	Europe, Asia, North America	[93]
Ascophyllum nodosum	Animal feed (ruminant and poultry diets), human consumption	Norway, UK, Portugal, USA	[93]
Undaria sp., *Hizikia* sp.	Fried in oil, boiled in soup	Japan, Korea, China	[91]
Macrocystis sp., *Laminaria* sp.	Ice-creams, syrups, salad dressings (texturizers, emulsifiers, thickeners)	Europe, USA	[93]
Ascophyllum sp.	Land animal feed (i.e., ruminants)	Iceland	[93]
Laminaria digitata, L. hyperborea, L. latissima	Animal feed	Europe, Asia	[91]
Laminaria japonica	Soup, fried in oil, with soy sauce	Asia	[91]
Fucus vesiculosum	Pigs diet	Sweden	[93]
Enteromorpha prolifera	Poultry diet	Europe	[93]
Pelvetia canaliculata	Pigs diet, human consumption during times of famine	Scotland, Ireland	[93]
Green Alga (Chlorophyta)			
Caulerpa sp.	Farmed in small quantities (poultry, improves quality of eggs), food ("green caviar")	Europe, Asia, Northamerica	[93]
Monostroma sp.	Salads, soups, relishes, meat and fish dishes	Europe, Asia	[93]
Ulva lactuca	Lambs feed, soups, salads	Europe, USA, Asia, Australia, New Zealand	[93]
Ulva intestinalis	Rabbits feed	Egypt, Saudi Arabia	[93]
Chaetomorpha linum	Lambs' feed	Tunisia	[93]

Palmaria palmata, also belonging to the group of red algae, is the most consumed in the north zone of the Atlantic Ocean. It is known by different names, such as Dulse, *duileasg, duileasc*, or *söl*, depending on the region, and its use as a food ingredient is quite widespread. In some areas of the mentioned regions it is also used as a flavor enhancer. Obtaining it is based on the hand collecting of the marine rocks in which they are trapped and accessible when the tide is low [93].

This method of manual collection is also carried out to obtain some brown algae, such as *Laminaria* and *Undaria*, in some Asian countries, such as Japan or Korea. [91], in which the food is supported robustly on the use of algae, using them for the accompaniment of numerous foods such as fish and meat, or for the elaboration of other recipes such as soups [94]. There are green algae whose leaves are similar to those of lettuce, belonging to the species *Monostroma* and *Ulva*, which are used precisely for the preparation of salads, although they are also included in soups and other types of dishes.

From a nutritional viewpoint, seaweeds are an alternative sources of proteins, certain brown species presenting much more protein content than other vegetarian sources, such as soy beans [95]. In the same way, their lipid content includes a concentration of fatty acids within 1 and 6 g/100 g of dry weight. In addition, it should be noted that some species have high concentrations of polyunsaturated fatty acids, namely eicosapentaenoic acid (up to 24%) [95].

One of the most important nutritionally relevant components of seaweeds are polysaccharides, most of them cannot be digested by humans, and thus, they can be considered as dietary fiber (33%–75% of the total composition) [96,97]. In turn, soluble fractions account between 50% and 85% of total dietary fiber content [96]. Some algae polysaccharides with particularly industrial relevance are described below:

Red algae (Rhodophyta) represent a particularly interesting group when considering polysaccharides, as they contain high quantity of sulfated galactans, such as agar or carrageenans. In the past decades, carrageenans have been used as natural ingredients in the elaboration of gels and as thickeners in a wide range of food applications [98]. They can be classified into three groups, according to their industrial use (kappa-, iota-, and lambda-), differing in the quantity and position of their sulfate ester substituents, as well as in the 3,6-anhydro-D-galactose content. The conformation of carrageenans is directly related with their technological properties (i.e., gelification, thickener). Kappa- and iota-carrageenans are gel forming and contain a 3,6-anhydro-galactose unit. In turn, lambda-carrageenans have only galactose residues and are used as thickeners. They lack of 3,6-anhydro-D-galactose ether linkages and, thus, the 4-linked substituent changes to a different conformation, disturbing their helical conformation [96] (Figure 1).

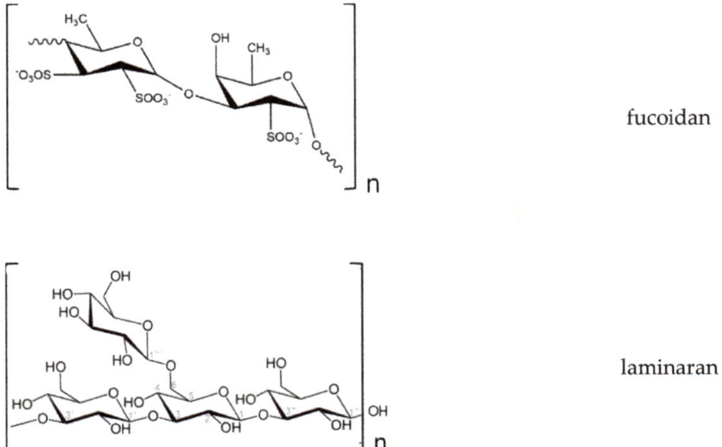

Figure 1. Different carrageenan types from red seaweeds.

Fucans, another kind of polysaccharides, generally present in brown algae, are classified into three main groups: Fucoidans, glycorunogalactofucans, and xylofucoglycuronans. Fucoidans are the leading components of the soluble fiber present in such kind of algae [96]. They are branched polysaccharides sulfate esters, soluble in water and acid, consisting in (1→3) and (1→4)-linked-L-fucose residues, that may be organized as (1→3)-α-fucan chains or as alternating (1→3) and (1→4)-bonded α-L-fucose residues. The L-fucose residues are often substituted with sulfate (SO_3) groups on C-2 or C-4 (rarely on C-3) [99,100]. Besides fucose, fucoidans also contain galactose, mannose, rhamnose, uronic acids, glucose, and xylose [99] (Figure 2).

Figure 2. Structure of fucoidans and laminarans present in brown algae.

Xylofucoglycuronans or ascophyllans consist of a polyuronide backbone, mainly poly-b-(1,4)-D-mannuronic acid branched with 3-O-D-xylosyl-L-fucose-4-sulfate or, sometimes, with uronic acid.

Glycuronogalactofucans composition can be described as linear chains of (1,4)-D-galactose branched with L-fucosyl-3-sulfate or, sometimes, uronic acid at C-5 [101].

Laminarans are other type of polysaccharides present in Pheophytes (*Laminaria* species). They are responsible for the food reserve of brown algae. Although laminaran composition is species dependent, they contain of 20–25 units of glucose on average, and consist in (1,3)-β-D-glucan with β(1,6) branching [102]. Laminaran chains can be classified into two types (M or G): M chains have a mannitol residue at the reducing end, and G chains, a glucose residue [99] (Figure 2).

Other remarkable compounds are alginic acid derivatives, also named alginates, that are constituted by linear polysaccharides with 1,4-linked β-D-mannuronic and α-L-guluronic acid residues distributed unregularly along the chain [99]. Alginates are present in brown seaweeds as sodium and calcium alginate, conforming highly viscous solutions, used for several technological applications (Figure 3).

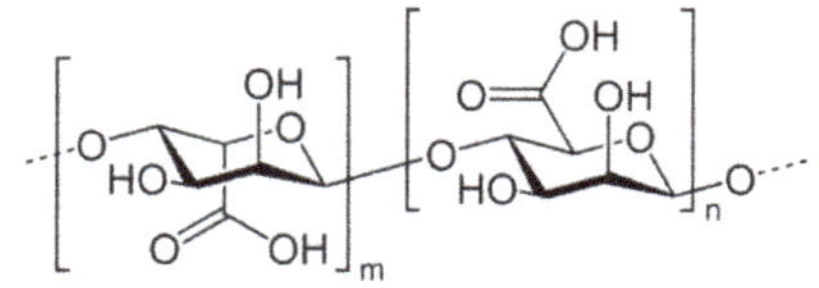

Alginate

Figure 3. Structure of alginate.

The solubility in water of the above-mentioned algae polysaccharides certainly facilitates their commercial extraction. In fact, certain phycocolloids present in the cell walls of many algae can be easily extracted with hot water. Phycocolloids classification includes three major groups: Alginates, carrageenans, and agars, and are interesting because they are safe for humans and animals' consumption. Many of these compounds are currently included in a wide range of processed and ready-to-eat foods, such as instant cakes or synthetic toppings [103]. An example is the use of industrially extracted alginates of brown algae species, such as *Macrocystis*, *Ascophyllum* and *Laminaria*, for the elaboration of ice cream, to avoid the formation of large amounts of ice and give them a creamy and smooth texture. Such alginates have been also used for the elaboration of energetic or sweet bars, and of dressings for salads as satiating agents, or in the preparation of syrups as thickeners and / or emulsifiers [104]. Alginates can also be extracted from red algae, such as *Gelidium*, *Gracilaria*, *Acanthopeltis*, *Pterocladia*, and *Ahnfeltia*, but their food use is reduced to special fillings for baking, canned food products, and as clarifiers in wine or beermaking. However, the use of agar in the R&D industry, as well as in microbiological and analytical laboratories, is really widespread, mainly for the preparation of culture media compatible with a wide variety of cell lines [105]. Carrageenans, also described previously, can be obtained from several species of red marine algae like *Eucheuma* (Philippines), *Chondrus* (United States), and *Iridaea* (Chile). Their main uses are thickening and stabilization of some dairy products like syrups, puddings or canned foods for animals, and they are also used in the elaboration of some medicines and some cosmetics as shampoos and creams [98].

During the 18th century, soda was obtained from brown macroalgae (*Phaeophyceae*) by collecting and scorching. In the following century, the main source of this compound was moved to the soda-rich mines discovered in Austria, among others, so the use of algae for this purpose drastically decreased. However, as the century progressed, they regained importance as a source of salts and iodine, so their exploitation suffered a tremendous boom, until the discovery of iodides and cooking salt, which practically plunged them back into disuse. Later, by the time of the First World War they were revalued for the production of fertilizers, such as potash, and acetone, used for the manufacture of smokeless gunpowder [106].

As mentioned above, algae have been used throughout the world as fertilizers for centuries. It is one of the most widespread uses of algae, as farmers from all coastal areas have collected them since ancient times, both at the oceans and reefs where they were trapped during storms or tides. Once harvested, they were left to dry spread out on the ground, obtaining a dry raw material with a mineral content that can amount up to 50% of the weight, in addition to high amounts of organic nitrogen derivatives. Currently, algae-based fertilizers that are commercialized include a set of micronutrients and macronutrients that help the plants grow properly [107].

6. Prebiotics from Algae

As shown in the previous section, macroalgae are responsible for the production of a large range of primary and secondary metabolites that can be used for different applications in food, cosmetic, pharmaceutical, and other industries [108]. Many of these compounds have well-established antiinflammatory, antimutagenic, antitumor, antidiabetic, hepatoprotective, free radical scavengers, anticoagulants, thrombolytic, and antihyperthensive properties [108,109]. In this section, the properties of non-digestible carbohydrates from algae will be addressed. A brief summary of those health beneficial effects is displayed in Table 5.

Table 5. Prebiotic or prebiotic candidates extracted from marine algae.

Prebiotic/Prebiotic Canditate	Origin/Source	Health Beneficial Effects	References
AGAROS (agaro-oligosaccharides)	Pheophyta (brown algae)	Immunomodulatory (decrease of pro-inflamatory cytokynes) antiinflammatory, carcinostatic, antioxidant, hepatoprotective, and α-glucosidase inhibitory activities	[110]
NAOS (neoagaro-oligosaccharides)	*Gracilaria* sp., *Monostroma* sp.	ROS scavenging, antioxidant and immunomodulatory effects, stimulation of lactobacilli and bifidobacteria populations	[110–113]
COS (carrageenan-oligosaccharides)	*Kappaphycus* sp., *Porphyria* sp., *Gracilaria* sp.	Immunomodulation, skin whitening, and moisturizing, stimulation of lactobacilli and bifidobacteria populations, repair of intestinal damage	[110,113–115]
ALGOS (alginate-oligosaccharides)	*Ascophyllum* sp., *Fucus* sp., *Undaria* sp., *Sargassum* sp., *Laminaria* sp. and *Macrocystis* sp.	Reactive oxygen species (ROS) scavenging, antioxidant and immunomodulatory effects, weight control, reduction of cholesterol, diabetes control (hypoglycemic and hypolipidemic properties), promotion of fecal microbiota metabolism, production of short chain fatty acids by the gut microbiota; decrease of putrefactive compounds and microorganisms, decrease of metabolic syndrome risk	[115,116]
Fucoidans (FUCOS)	*Cladosiphon* (aka Okinawa) *Ascophyllum* (*nodosum*), *Fucus* sp. *Sargassum* sp.	Hypocholesterolaemic, immunomodulatory, anti-obesity, anti-hyperlipidemia, attenuation of hepatic steatosis, anti-diabetes (reduction of insulin resistance), anti-hypertensive, antioxidant	[99,114,116,117]
	Fucus evanescens	Anticoagulant, antioxidant	[109,118]
	Fucus vesciculosus	Anticancer, antimetastatic	[109]

Table 5. *Cont.*

Prebiotic/Prebiotic Canditate	Origin/Source	Health Beneficial Effects	References
Galactofucans	*Laminaria japonica, Sargassum* sp.	Anti-lipidaemic, increases HDL, antiviral, antitumor, immunomodulator, antioxidant, neuroprotective	[114]
	Undaria pinnatifida	Antiviral, anticoagulant, antitumor, anti-proliferative, immunomodulatory, anti-inflammatory induced osteoblastic differentiation	[119]
	Dictyota menstrualis	Peripheral anti-nociceptive, anti-inflammatory, antioxidant; anticoagulant, anti-proliferative	[119]
	Lobophora variegata	Antioxidant, anticoagulant, anti-inflammatory	[119]
	Adenocystis utricularis	Antiviral	[119]
Xylo-galactofucans	*Spatoglossum schröederi*	Anti-thrombotic; Peripheral anti-nociceptive; Anti-proliferative, anti-adhesive, antioxidant	[119]
Arabinoxylans	*Ascophyllum*	Modulation of intestinal microbiota	[120]
Glucans	*Chlorela vulgaris*	Antitumor, infection preventive agent	[119]
Laminarin	*Ascophyllum* sp., *Fucus* sp., *Laminaria* sp., *Saccharina* sp., *Undaria, Enteromorpha* sp.	Antilipidemic, hypocholesterolaemic, fast decrease of blood glucose	[114,116]

According to the last accepted definition, prebiotics are substrates that are selectively used by host microorganisms conferring health benefits [121]. As the target of most prebiotic compounds is the gut, such ingredients must not be hydrolyzed in the upper part of the gastro-intestinal tract. This ensures their safe arrival to the colon, where they stimulate the growth and/or activity of one or a limited number of bacteria, thus positively modulating the intestinal microbiota [121]. Although prebiotic properties must be well-demonstrated both in vivo and in vitro [121], the capacity to safe arrive to the gut converts non-digestible ingredients in potential prebiotic compounds (or prebiotic candidates).

6.1. Chemistry and Obtaining of Prebiotic Compounds from Seaweeds

Some prebiotics (or prebiotic candidates) naturally occurring in seaweeds are saccharides presenting a degree of polymerization (DP) within 2 and 9 (sometimes within 8 and 20). Some researchers consider up to 25 sugar residues [122]. Such oligo/polysaccharides include galacto-oligosaccharides (GOS), agarose-derived oligosaccharides (AGAROS), xylo-oligosaccharides (XOS), neoagaro-oligosaccharides (NAOS), alginate-derived oligosaccharides (ALGOS), arabinoxylans, galactans and glucans [114]. Although not all of them fulfill all the criteria requested for a compound to be considered as prebiotic (i.e., be refracting to hydrolysis and absorption in the upper part of the gastro-intestinal tract, selective promotion of bifidobacteria and/or lactobacilli in the colon, beneficial effect of their fermentation products, stability towards technological processes when incorporated into food products) [121], none of them is degraded by enzymes in the first part of the gastro-intestinal tract. Therefore, they achieve at least one of the requested criteria and take part of the dietary fiber.

Prebiotic oligosaccharides from seaweeds can be obtained by the hydrolysis of naturally occurring polysaccharides. This process has encouraged the development of new extraction techniques, as algae polysaccharides have a wide variety of chemical bonds and conformations, including α or β bonds, *cis* or *trans* configurations, D or L, or *R* or *S* chiral centers. This issue is of great importance since certain bonds can be hydrolyzed with enzymes, whereas others require other techniques as no natural enzymes are able to hydrolyze them [120,123]. Therefore, it is necessary to look for other techniques to obtain oligosaccharides from polysaccharides. Some authors [122] summarized the characteristics of different techniques, as well as their suitability to be used for the hydrolysis of some polysaccharides. Hence, the main non-enzymatic hydrolytic techniques used at an industrial level are ultrasound (in the case of carrageenans, agarose and xylan), microwaves (for hydrolysis of exopolysaccharides obtained from *Porphyridium cruentum*), use of free radicals (for fucoidan and algae galactan), or the application of diluted acids (for example, phosphoric acid) [124–130].

In the case of acids, they can be used as hydrolytic agents only for neutral polysaccharides, namely fucoidans, carrageenans, or galactans. In addition, certain components can be lost during the process and most of the glycosidic bonds are not specifically hydrolyzed, thus giving rise to various low molecular weight derivatives. In spite of that, phosphoric acid is capable of effectively hydrolyzing polysaccharides of some algae species, such as *Chlorella vulgaris* and *Spirulina platensis* [128], mostly composed of uronic acids and giving rise to interesting oligosaccharides with potential prebiotic properties [122].

The use of free radicals is another effective technique for the hydrolysis of polysaccharides, since it does not affect the structure of the compounds when it is correctly performed [122]. Other authors have demonstrated that this technique can be applied for the hydrolysis of fucoidans in low molecular weight compounds (≈8 kDa), although then it is necessary to apply other procedures for the purification of the obtained products [129]. Additionally, the obtained products demonstrated a better anticoagulant capacity with respect to the original polysaccharide [129].

The physical technique of microwaves also gives rise to glucidic compounds of low molecular weight (≈12 kDa) from natural polysaccharides extracted from algae. The obtained products do not undergo structural changes and have a surprisingly enhanced immunomodulatory and anticancer capacities with regard to the original substrate. The higher solubility of the products (with lower molecular weight) was remarked as a possible explanation for such observations [131]. Therefore, the use of this technique offers several clear advantages: it is economical, easy to use, non-toxic and is green, since the energetic and time consumption are not environmentally harmful [132]. All these techniques can be optimized to guarantee the highest results [133].

6.2. Prebiotic Properties of Oligo and Polysaccharides from Seaweeds

Taking into account that some oligo and polysaccharides extracted from algae are not hydrolyzed in the upper part of the gastro-intestinal tract, they represent novel potentially useful raw materials for the obtaining of prebiotics [134]. To determine whether oligo and polysaccharides extracted from seaweeds are prebiotics, their fermentation by microorganisms from the intestinal microbiota is usually assessed [134]. First of all, the stability of algae polysaccharides when exposed to saliva, gastric and intestinal environments is assessed. The activity of intestinal microbiota is commonly evaluated by measuring metabolic end products, such as gases and short-chain fatty acids (i.e., acetic, propionic, butyric acids). Using molecular methods such as fluorescence in situ hybridization (FISH), polymerase chain reaction (PCR), denaturing gradient gel electrophoresis (DGGE) and 16S rRNA gene sequencing provides a complete representation of the in vitro effect of prebiotic candidates from seaweeds on the modulation of intestinal microbiota by increasing the amount of beneficial bacteria and decreasing prejudicial microorganisms [134,135] (Figure 4).

In turn, for in vivo studies, prebiotic candidates are orally administrated to the host (mouse, rats, monogastric, ruminants, and humans), and modifications in the composition of intestinal microbiota (Firmicutes/Bacteroidetes ratio, the two phyla present in the microbiota), and production of short chain fatty acids are determined. Recently, it was reported that polysaccharides obtained from the brown alga *Ascophyllum nodosum* increase the quantity of Bacteroidetes and Firmicutes, suggesting the potential of *Ascophyllum nodosum* polysaccharides to decrease the risk of obesity. Furthermore, the total short chain fatty acids content after fermentation increased significantly. These results suggest that *Ascophyllum nodosum* polysaccharides have potential uses as functional food components to improve human gut health [117]. Other authors evaluated the prebiotic properties of the brown seaweed *Ecklonia radiata* oligosaccharides in vivo, when administered the polysaccharide fraction of seaweed (rich in fucoidan and alginate) to rats. Such fractions lead to a decrease in the levels of toxic protein fermentation products, enhancement of the numbers of butyrate-producing *Faecalibacterium prausnitzii*, and decrease of the numbers of potentially pathogenic *Enterococcus*, thus demonstrating a potential prebiotic effect [136]. Some authors [137] demonstrated an increase in the population of bifidobacteria and lactobacilli both in the cecum and feces of rats fed with diets supplemented with alginate oligosaccharides. Such prebiotic effect was even greater than that of rats fed with a diet containing FOS, a well-established prebiotic. Similar increases were observed by [138] in mice fed with diets supplemented with agarose hydrolysates (NAOS). Moreover, a lower number of *Bacteroides* compared to the controls fed with FOS, were observed [135]. In addition, laminarin supplementation of rats feed also enhanced the cecal population of bifidobacteria, with no significant effect on lactobacilli. Moreover, laminarin also suppressed certain putrefactive compounds considered as risk markers for colon cancer, such as indole, cresol, and sulfide, and had immunomodulatory properties [139].

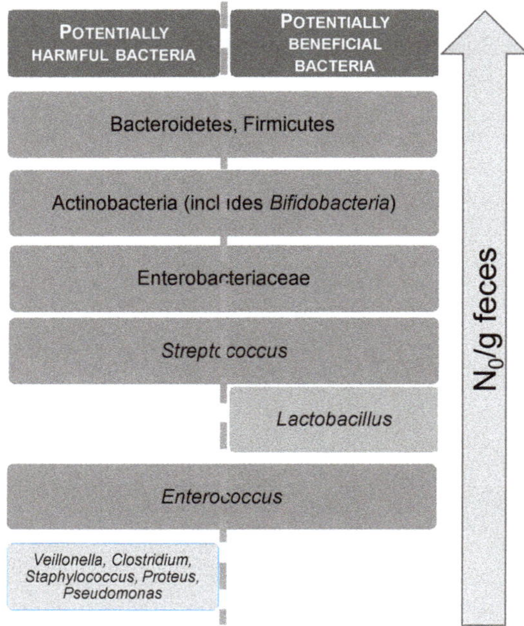

Figure 4. Beneficial bacteria and suppressing pathogenic microorganism.

In turn, green algae, such as *Enteromorpha prolifera* and *Laminaria japonica*, spread all over the Chinese Qingdao coast, can also be fermented by intestinal microbiota [140]. A close relation between the metabolic products of polysaccharides from marine algae and the regulation of enteroendocrine hormone secretion, blood glucose, and lipid metabolism has been recently reported, thus suggesting their effect on alleviation of metabolic syndrome symptoms [116,141] (Figure 5).

The prebiotic effect of polysaccharides from seaweeds has been also evaluated on farm animals [135]. Many studies have evaluated the effects of seaweed polysaccharides (i.e., laminarin) in pigs, lambs, or cattle [142]. However, the differences in digestive physiology and anatomy must be considered when attempting to extrapolate data from ruminants (cattle and sheep) to monogastrics, such as pigs. Some in vitro studies in some animals such as pigs, rabbits, birds, and some ruminants conclude that some algae have the ability to meet the energy and protein requirements for healthy growth, while other algae contain certain compounds that have biological activities, so that both types could be used as prebiotics to favor the breeding of such animals [93].

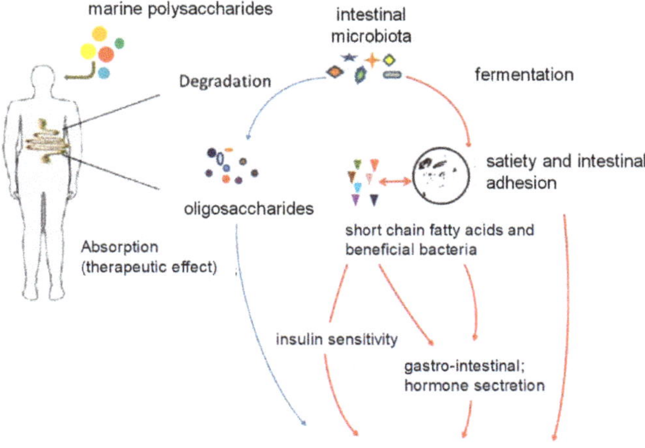

Figure 5. Effects of polysaccharides from marine algae on metabolic syndrome [116].

To summarize this section, algae are an excellent natural source of natural polysaccharides, that can be extracted and hydrolyzed for the obtaining of prebiotic saccharides. Extraction processes based on physical methods are the most efficient ones, since they present minimal or no adverse effects on human health. Besides that, these methods do not alter the original structure of the compounds and are environmentally friendly [132]. Additionally, many of the compounds obtained show good bioactive capacities, so their use as prebiotics is highly recommended [143].

Regarding the functionality of seaweeds, further studies are still necessary to gain information about their intestinal benefits, and if mixtures of polysaccharides can be tailored to improve health benefits. This would be of great importance for the formulation of functional ingredients, improving fermentability by gut microorganisms. In this context, it appears that in the future, human beings could modulate the microbiome through the consumption of drugs and prebiotics, which could be at the cutting edge in the prevention of some diseases [116].

7. Antioxidants from Algae

In spite of having previously named the properties of antioxidants in algae, it is necessary to expand the given information because there are many different compounds, of different natures, that present this quality, whether they are water-soluble (i.e., phenolic compounds or vitamins), or soluble in apolar solvents (i.e., carotenoids or tocopherols) [143].

7.1. Carotenoid Pigments

Carotenoids are one of the most known and widespread natural pigments in the world. Their structure corresponds to that of hydrophobic tetraterpenes with a structure of branched hydrocarbons with a methyl group bound at C-40 [144,145]. The pigmentation produced by carotenoids has to do with the conjugated double bonds that appear in the carbonate skeleton, since they have the ability to capture photons at different wavelengths of the visible spectrum. All organisms that carry out photosynthesis contain carotenoids, which does not mean that they are exclusive compounds of these autotrophic organisms, since they are also present in some bacteria and fungi [146]. More than 600 types of carotenoids are known, which can be divided into two large groups according to their molecular structure: Carotenoids and xanthophylls. The first group is characterized by having linear or cyclic hydrocarbons at one or both ends of the molecule (like β-carotene). In turn, xanthophylls are described as oxygenated derivatives of carotenes (such as violaxanthin), and can be found in green algae, as well as in more developed plants, although there are compounds (i.e., canthaxanthin and loroxanthin) that are exclusive from green algae [147,148]. Similarly, brown algae and some flagellated microalgae are capable of producing another type of xanthophylls, such as diatoxanthin or fucoxanthin [149].

Carotenoid essentially plays a photoprotector role in algae, preventing their photosynthetic system from damage. In addition, they also participate in processes such as phototaxis (movements resulting from the intensity of light) or phototropism (movements in the direction of the light). Some microalgae are capable of generating high amounts of carotenes in response to external stimuli that cause stress, so that they can adapt to new changes [150]. The microalgae species producing the highest amounts of carotenoids are *Dunaliella salina*, which produces β-carotene, and *Haematococcus pluvialis*, which produces astaxanthin. Currently, the use of β-carotene is widespread in the industry because it is a natural ingredient that provides color to matrices, either in food (such as cheeses and butter) or cosmetic formulations [151]. In addition, carotenoids have an associated provitamin A activity. Despite being less known, astaxanthin is also a skin and eye protector, muscle strengthener, immune modulator, and coloring agent. There are studies claiming that the daily ingestion of this astaxanthin slows down the aging process, as it has a regenerative capacity and is a free radical scavenger [152].

7.2. Phycobilin Pigments

Phycobiliproteins are secondary pigments generated by microalgae, which help them to exploit the light energy while protecting them from harmful radiation. It seems that the antioxidant mechanisms within the organisms that create them are very similar to those carried out in food matrices or in the human organism [153]. The main microalgae producing these compounds are cyanobacteria (*Spirulina*, which produces phycocyanin, blue) and cryptomonads, but are also found in red algae (*Porphyridium*, which produces phycoerythrin, red). Despite being a labile molecule, phycocyanin is capable of generating a blue color that other natural dyes do not achieve, so it is used in the manufacture of foods such as ice cream, yogurt, chewing gum, and beverages, as well as in various cosmetic products, mainly in Japan [154].

7.3. Phenolic Compounds

Phenolic compounds are a group of secondary metabolites comprising a wide variety of compounds produced by both terrestrial and aquatic plants, which include, of course, algae [155]. Despite the well-known diversity of structures of phenolic compounds, they must possess a benzene ring having at least one hydrogen substituted with a hydroxyl group. One of their most outstanding features is their antioxidant properties, as they prevent the formation of many free radicals because of their metal ion chelating capacity [156,157]. Phenolic compounds are commonly classified into 5 large groups: Flavonoids (the largest subgroup and associated with different bioactivities, among which, as previously described, the antioxidant, and radical scavenging activity) [158,159], lignans, tannins, tocopherols, and phenolic acids [160]. They are common compounds in algae, especially in brown ones, since some species of brown seaweed have phlorotannins, which are polymers of phloroglucinols (1,3,5-trihydroxybenzene) that can reach up to 15% of the dry weight of these algae [161]. In addition, they are composed of up to eight rings interconnected with each other, which give antioxidant properties much higher than many polyphenols obtained from terrestrial plants, since most of them contain only three or four rings in its structure [155].

7.4. Vitamins and Minerals

Vitamins are well-known for their high antioxidant capacity. However, when attempting to reproduce their synthesis artificially, a decrease in activity with respect to those obtained directly from natural matrices was observed [162]. That is why new natural sources containing essential vitamins are continuously searched. Among them, microalgae biomass contains most of the essential vitamins (e.g., A, B1, B2, B6, B12, C, E, nicotinate, biotin folic acid, and pantothenic acid), as well as an interesting variety of minerals (e.g., Na, K, Ca, Mg, Fe, and Zn) [163]. In fact, some species of microalgae contain high levels of some essential vitamins that are requested in higher quantities by humans. This is the case of B12 vitamin, which can be combined with a mineral content, such as iron, thus converting it in an ideal dietary supplement for vegetarian diets. An example of such combination is Spirulina. However, and as expected, not all algae contain the same amounts of the same vitamins. The vitamin content of each species of algae depends on endogenous factors, such as their life cycles, their growth status and genotype, as well as on exogenous factors, such as the amount of nutrients in their habitat, the intensity of the light reaching them and the processes they suffer since they are collected. Therefore, there are several different factors that can be modulated to achieve the desired amounts of algae vitamins [33].

8. Production and Consumption Statistics and Future Markets

A senior FAO official [31,164] announced in 2017 during the High Level International Meeting on the Global Initiative "Blue Growth" for Latin America and the Caribbean, held in Mexico City, that "aquaculture worldwide is the productive sector with the greatest growth, exceeding as of 2011 the growth rates of bovine cattle". Thus, according to statistics from 2015, the world aquaculture production consisted in 51.9 Mt of fish (68%), 16.4 Mt of mollusks (21%), 7.4 Mt of crustaceans (10%), and 0.9 Mt of other aquatic animal species (1%). In particular, aquaculture in inland waters represents the most important sector of the production of edible organisms (43.6 Mt), which represents 59% of world production. The average annual growth rate of aquaculture production for 2001–2015 was 5.9%, significantly lower than in the last two decades of the 20th century, which stood at 10.8% and 9.5%, respectively. However, the contribution of aquaculture to production has been increasing steadily, reaching 45% in 2015 from 26% in 2000. If we focus on the world production of aquatic plants, mainly marine algae, reached 30.5 Mt in 2015. In this case the extractive processes are merely testimonial since 96% of the production (29.4 Mt) were obtained through aquaculture. Analyzing the distribution of aquatic production at the continental level and according to statistics for 2015, it would be headed by the Asian continent with a production of 68.4 Mtm. America and Europe would be at a distance with 3.3 Mtm and 3.0 Mtm. Africa produced the order of 1.8 and Oceania would scarcely contribute with 0.2 Mtm. The percentage of world production is shown in Figure 6. Figure 7 shows the weight of the top 10 world producers in aquaculture. Table 6 presents the evolution in Mtm of the production of the main producers in aquaculture in recent years.

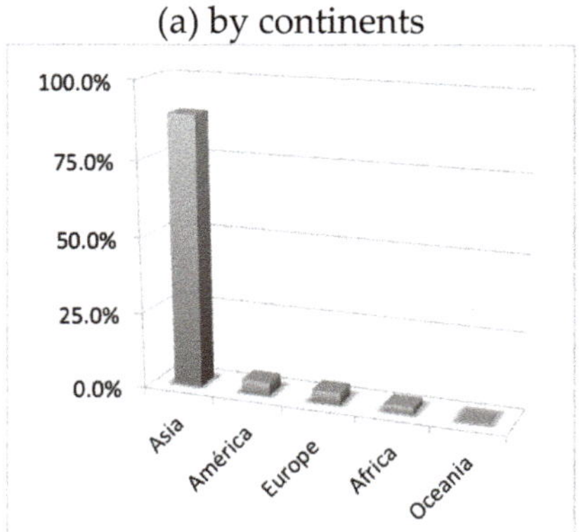

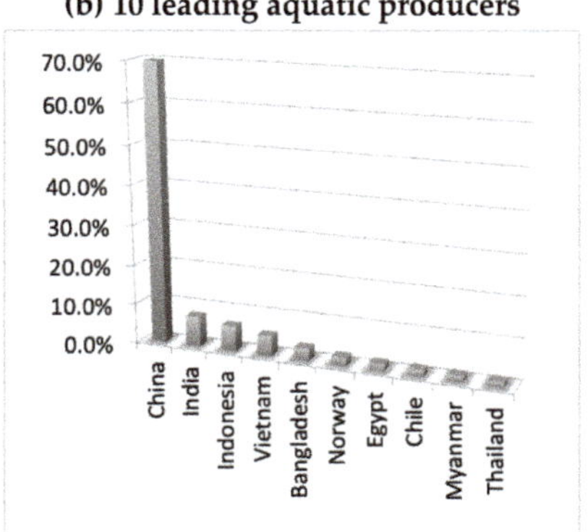

Figure 6. (**a**) Distribution by continents of world aquiculture production (2017) (Source: FAO [31,164–166]) and (**b**) production of the 10 leading aquatic producers in the world (2017) (Source: FAO [31,164–166]).

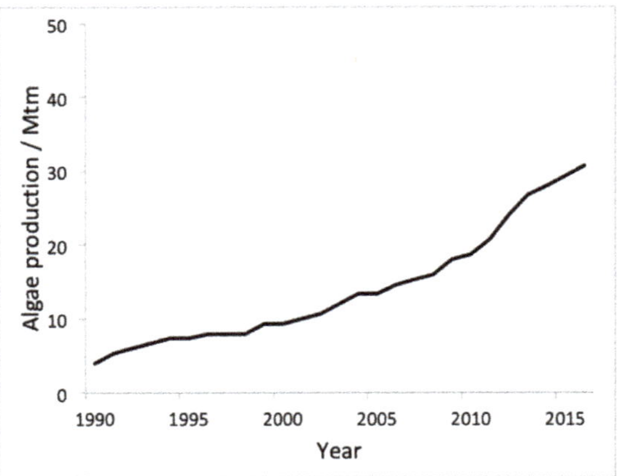

Figure 7. Time evolution of world aquaculture algae production in millions of tons from 1990 to 2016 (source: FAO [31,164–166]).

Table 6. Production in aquaculture (Mtm) of main producers in 2015–2016 (Food and Agriculture Organization of the United Nations [31,164]).

Country	Production/Mtm	
	2016	2015
China	47.6	41.1
India	5.2	4.2
Indonesia	4.3	3.1
Vietnam	3.4	3.1
Bangladesh	2.1	1.7
Norway	1.4	1.3
Egypt	1.2	1
Chile	1.1	1.1
Myanmar	1	0.9
Thailand	0.9	1.2

Access to statistics on seaweed production is largely dependent on annual publications of the Food and Agriculture Organization of the United Nations (FAO), such as 'The State of World's Fisheries and Aquaculture.' The report published in 2018 only comprises production statistics for 2016 [164]. Table 6 shows the production in aquaculture (Mtm) of main producers in 2015–2016.

Meanwhile Figure 6 shows the distribution by continents of world aquiculture production (2017) (Source: FAO) and the production of the 10 leading aquatic producers in the world [31,164–166].

In 2016, aquaculture was the source of 96.5% of the total volume of 31.2 million tons, including wild-collected and farmed plants together. When it comes to aquatic plants cultivation, it mainly refers to seaweeds. Their cultivation almost triplicated in the last ten years. This fact supports the increase in the use of algae, although the demand still exceeds the supply offered by aquaculture. Conversely, the production of algae is decreasing (~5%) because of the collection of natural habitats [164]. Figure 7 shows the world evolution of cultivated aquatic plants.

Macroalgae have a large number of applications being the most relevant the production of human food products [74] and 15% is attributed to algae extracts, like hydrocolloids for their use in fertilizers, animal feeds and for its biological activities [167]. Nowadays, 221 species of algae are collected globally, from which 145 are for human nutrition and 101 for phycocolloid production. These include 125 rhodophytes, 64 phaeophytes and 32 chlorophytes [168]. From these, ten are largely cultivated, specially the brown ones *Laminaria japonica*, *Undaria pinnatifida* and the red ones, *Porphyra* spp., *Porphyra tenera*, *Eucheuma* spp., *Kappaphycus alvarezii* and *Gracilaria* spp. and *Gracilaria verrucosa* [169].

In 2010, 19 million tons of aquatic plants were produced in aquaculture globally, being sewed the largest percentage by far with 99.5% of the total. Of all the seaweed used in the world in 2010, 99.5% was prevenient from aquaculture, being a growing economy with a 7.7% per year increase since the 1970s [170]. The production of Algae is dominated by countries in East and Southeast Asia, reaching to 99.6% of quantity in 2010. China has the major portion of the production, contributing with approximately 58% of worldwide cultivated algae production by quantity. Indonesia, Philippines, South Korea, Japan, and North Korea are some of the other major seaweed producers. Japan has a high value of production, ranking in third position, due to its high-value of *Phorfira* production, accounting for 20% [165,166]. Almost all cultured species produced in East Asia are used for human consumption, with the exception of a small amount of Japanese brown algae that is also used as a source for the extraction of algin and iodine. Contrary, in Southeast Asia, *Eucheuma* algae are the principal cultivated species, used principally as a source for carrageenan obtaining. Aside Asia, a small fraction is also cultivated in Zanzibar (Tanzania) and Chile.

Nowadays, algae have several important uses in the food, cosmetic, and fertilizers industries and in the production of hydrocolloids, such as alginates and agar, being the last one the mainly current commercial exploitation. Furthermore, the production of bioactives and future commercial opportunities for seaweed compounds has increased and more than 15 000 different molecules have been isolated to date [170] hoping to be incorporated in human food, nutraceuticals, pharmaceuticals and animal stock feeds—but also fertilizers, biofuels, and soil conditioners. Opportunities for utilization aquafeeds, in food, feed, and nutraceuticals have presented as offering the most promising results in a reasonable period [171]. The production of seaweed globally has increased over the years as well has consequently the by-products and co-products derived from them.

9. Conclusions on Trends and Challenges for the Sector

The term "functional food" can be defined as foods claiming to have additional function(s) by incorporating new ingredients or greater amounts of the existing ones. Such ingredients, called "functional ingredients" have specific health benefits, including immunomodulation, and reduction of the risk of certain diseases, among others. Seaweeds that can be consumed by humans appear as innovative resources with an interesting potential to be used as functional ingredients. Some of them contain high insoluble dietary fiber. In particular, the high content of non-digestible polysaccharides converts marine algae in a rich source of prebiotics or prebiotic candidates. Further investigations about prebiotic properties both in vitro and in vivo are still necessary to incorporate such ingredients in the formulation of functional foods. On the other hand, certain secondary metabolites from marine algae, including antioxidants, are able to decelerate or prevent oxidation processes, thus being favorable health related compounds. Therefore, there is a need to solve problems such as the determination of the identity of the harvests as fresh materials for functional foods, and the regularization of the products, because seaweeds from different locations can produce different levels of active compounds.

In addition of being a source of novel, beneficial, and natural products, algae and their cultivation offer advantages on a larger scale, since aquaculture is an environmentally friendly process and favors the production of compounds rich in biomass and proteins that could supply or mitigate industries with many more toxic waste. Furthermore, the implementation of this technique in developing countries could entail great financial advantages. However, to do so, it would be first necessary to solve problems, such as the lack of marine spaces, where to carry it out, or the initial economic support. For the development of the aquatic plants cultivation industry (i.e., algae), the transfer of technological improvements (i.e., creation of sea beds not requiring coastal areas) to tropical areas would be very convenient. This way, the problems of lack of cultivable surfaces would be overcome.

In conclusion, governments shRedondelaould become part of this booming industry, supporting sustainable development through investments in infrastructure, personnel, materials, and projects aimed at research, development and innovation, thus committing to a green future for our planet.

Author Contributions: All authors (A.G.-Z., M.A.P.L., C.J.-L., J.S.-G., and J.C.M.) contributed equally to the conceptualization, methodology, validation, formal analysis, investigation, writing—original draft preparation, writing—review and editing, visualization, supervision, and project administration.

Funding: The authors thank CYTED (Ibero-American Program of Science and Technology for Development) for the funds obtained to grant AQUA-CIBUS network (Strengthening aquaculture in Ibero-America: quality, competitiveness and sustainability; Ref. 318RT0549). To MICINN for the financial support for the Ramón and Cajal researcher of M.A. Prieto Lage. To the University of Vigo for the financial support for the pre-doctoral researcher C. Jiménez López. To Xunta de Galicia for the financial support by the Axudas Conecta Peme of the NeuroFood Project (IN852A 2018/58). To the Argentinean Agency for the Promotion of Science and Technology (PICT startup (2016)/4808 and PICT (2017)/1344. A.G.-Z is member of the research career from the Argentinean Research Council (CONICET). The authors are also grateful to the Interreg España-Portugal for financial support through the project 0377_Iberphenol_6_E.

Acknowledgments: This review article is an emotional and well-deserved tribute to Prof Isabel Ferreira (CIMO, IPB) for her scientific career of excellence in a line of research that received a great boost with her research group, the recovery of bioactive ingredients from natural products. We wish her and her group all the best personally and professionally.

Conflicts of Interest: The authors declare no conflicts of interest.

References

1. Sáa, C.F. *Algas de Galicia, Alimento y Salud. Las Verduras del Océano Atlántico: Propiedades, Recetas, Descripción*; Redondela: Algamar, Spain, 2002; ISBN 84-607-4503-1.
2. Rupérez, P.; Saura-Calixto, F. Dietary fibre and physicochemical properties of edible Spanish seaweeds. *Eur. Food Res. Technol.* **2001**, *212*, 349–354. [CrossRef]
3. Santos, S.A.O.; Vilela, C.; Freire, C.S.R.; Abreu, M.H.; Rocha, S.M.; Silvestre, A.J.D. Chlorophyta and Rhodophyta macroalgae: A source of health promoting phytochemicals. *Food Chem.* **2015**, *183*, 122–128. [CrossRef] [PubMed]
4. Sudhakar, K.; Mamat, R.; Samykano, M.; Azmi, W.H.; Ishak, W.F.W.; Yusaf, T. An overview of marine macroalgae as bioresource. *Renew. Sustain. Energy Rev.* **2018**, *91*, 165–179. [CrossRef]
5. Guiry, M.D. How many species of algae are there? *J. Phycol.* **2012**, *48*, 1057–1063. [CrossRef] [PubMed]
6. Adl, S.M.; Simpson, A.G.B.; Farmer, M.A.; Andersen, R.A.; Anderson, R.; Barta, J.R.; Bowser, S.S.; Brugerolle, G.; Fensome, R.A.; Fredericq, S.; et al. The New Higher Level Classification of Eukaryotes with Emphasis on the Taxonomy of Protists. *J. Eukaryot. Mrcrobiol.* **2005**, *52*, 399–451. [CrossRef] [PubMed]
7. Portugal, A.B.; Carvalho, F.L.; de Oliveira Soares, M.; Horta, P.A.; de Castro Nunes, J.M. Structure of macroalgal communities on tropical rocky shores inside and outside a marine protected area. *Mar. Environ. Res.* **2017**, *130*, 150–156. [CrossRef] [PubMed]
8. Draisma, S.G.A.; Prud'homme van Reine, W.F.; Herandarudewi, S.M.C.; Hoeksema, B.W. Macroalgal diversity along an inshore-offshore environmental gradient in the Jakarta Bay–Thousand Islands reef complex, Indonesia. *Estuar. Coast. Shelf Sci.* **2018**, *200*, 258–269. [CrossRef]
9. Vuong, D.; Kaplan, M.; Lacey, H.J.; Crombie, A.; Lacey, E.; Piggott, A.M. A study of the chemical diversity of macroalgae from South Eastern Australia. *Fitoterapia* **2017**, *126*, 53–64. [CrossRef]
10. Ross, A.B.; Jones, J.M.; Kubacki, M.L.; Bridgeman, T. Classification of macroalgae as fuel and its thermochemical behaviour. *Bioresour. Technol.* **2008**, *99*, 6494–6504. [CrossRef]
11. Zhao, F.; Xu, N.; Zhou, R.; Ma, M.; Luo, H.; Wang, H. Community structure and species diversity of intertidal benthic macroalgae in Fengming Island, Dalian. *Acta Ecol. Sin.* **2016**, *36*, 77–84. [CrossRef]
12. Robin, A.; Chavel, P.; Chemodanov, A.; Israel, A.; Golberg, A. Diversity of monosaccharides in marine macroalgae from the Eastern Mediterranean Sea. *Algal Res.* **2017**, *28*, 118–127. [CrossRef]
13. Bonanno, G.; Orlando-Bonaca, M. Chemical elements in Mediterranean macroalgae. A review. *Ecotoxicol. Environ. Saf.* **2018**, *148*, 44–71. [CrossRef]
14. Navarro-Barranco, C.; Florido, M.; Ros, M.; González-Romero, P.; Guerra-García, J.M. Impoverished mobile epifaunal assemblages associated with the invasive macroalga Asparagopsis taxiformis in the Mediterranean Sea. *Mar. Environ. Res.* **2018**, *141*, 44–52. [CrossRef] [PubMed]
15. Timme, R.E.; Bachvaroff, T.R.; Delwiche, C.F. Broad phylogenomic sampling and the sister lineage of land plants. *PLoS ONE* **2012**, *7*. [CrossRef] [PubMed]

16. Werlinger, C.; Alveal, K.; Romo, H. *Biología Marina y Oceanografía: Conceptos y Procesos*; Consejo Nacional del Libro y la Lectura: Santiago, Chile, 2004.

17. Leliaert, F.; Smith, D.R.; Moreau, H.; Herron, M.D.; Verbruggen, H.; Delwiche, C.F.; De Clerck, O. Phylogeny and Molecular Evolution of the Green Algae. *CRC Crit. Rev. Plant Sci.* **2012**, *31*, 1–46. [CrossRef]

18. Hewitt, C.L.; Gollasch, S.; Minchin, D. The Vessel as a Vector–Biofouling, Ballast Water and Sediments. In *Biological Invasions in Marine Ecosystems*; Springer: Berlin, Germany, 2009; pp. 117–131. ISBN 0882-2786.

19. Minchin, D.; Gollasch, S.; Cohen, A.N.; Hewitt, C.L.; Olenin, S. Characterizing Vectors of Marine Invasion. In *Biological Invasions in Marine Ecosystems*; Rilov, G., Crooks, J.A., Eds.; Springer: Berlin, Germany, 2009; pp. 109–116. ISBN 978-3-540-79236-9.

20. Andreakis, N.; Schaffelke, B. Invasive Marine Seaweeds: Pest or Prize? In *Investment Management and Financial Innovations*; Wiencke, C., Bischof, K., Eds.; Springer: Berlin, Germany, 2012; Volume 219, pp. 235–262. ISBN 978-3-642-28450-2.

21. Booth, D.; Provan, J.; Maggs, C.A. Molecular approaches to the study of invasive seaweeds. In *Seaweed Invasions: A Synthesis of Ecological, Economic and Legal Imperatives*; Walter de Gruyter: Berlin, Germany; New York, NY, USA, 2007; Volume 50, pp. 385–396. ISBN 9783110211344. [CrossRef]

22. Forrest, B.M.; Taylor, M.D. Assessing invasion impact: Survey design considerations and implications for management of an invasive marine plant. *Biol. Invasions* **2002**, *4*, 375–386. [CrossRef]

23. Nyberg, C.D.; Wallentinus, I. Can species traits be used to predict marine macroalgal introductions? *Biol. Invasions* **2005**, *7*, 265–279. [CrossRef]

24. Wikström, S.A. Marine Seaweed Invasions—The Ecology of Introduced Fucus Evanescens. Ph.D. Thesis, Botaniska Institutionen, Stockholm, Sweden, 2004.

25. Shea, K.; Chesson, P. Community ecology theory as a framework for biological invasions. *Trends Ecol. Evol.* **2002**, *17*, 170–176. [CrossRef]

26. Pavia, H.; Toth, G. Inducible Chemical Resistance To Herbivory in the Brown Seaweed Ascophyllum Nodosum. *Ecology* **2000**, *81*, 3212–3225. [CrossRef]

27. Thornber, C.S.; Kinlan, B.P.; Graham, M.H.; Stachowicz, J.J. Population ecology of the invasive kelp *Undaria pinnatifida* in California: Environmental and biological controls on demography. *Mar. Ecol. Prog. Ser.* **2004**, *268*, 69–80. [CrossRef]

28. Aitken, D.; Bulboa, C.; Godoy-Faundez, A.; Turrion-Gomez, J.L.; Antizar-Ladislao, B. Life cycle assessment of macroalgae cultivation and processing for biofuel production. *J. Clean. Prod.* **2014**, *75*, 45–56. [CrossRef]

29. Moon, S.M.; Lee, S.A.; Han, S.H.; Park, B.R.; Choi, M.S.; Kim, J.S.; Kim, S.G.; Kim, H.J.; Chun, H.S.; Kim, D.K.; et al. Aqueous extract of Codium fragile alleviates osteoarthritis through the MAPK/NF-κB pathways in IL-1β-induced rat primary chondrocytes and a rat osteoarthritis model. *Biomed. Pharmacother.* **2018**, *97*, 264–270. [CrossRef] [PubMed]

30. FAO. *The State of Fisheries and Aquaculture in the World 2018*; FAO: Rome, Italy, 2018.

31. Ridler, N.; Wowchuk, M.; Robinson, B.; Barrington, K.; Chopin, T.; Robinson, S.; Page, F.; Reid, G.; Szemerda, M.; Sewuster, J.; et al. Integrated multi-trophic aquaculture (IMTA): A potential strategic choice for farmers. *Aquac. Econ. Manag.* **2007**, *11*, 99–110. [CrossRef]

32. Kim, J.K.; Kottuparambil, S.; Moh, S.H.; Lee, T.K.; Kim, Y.J.; Rhee, J.S.; Choi, E.M.; Kim, B.H.; Yu, Y.J.; Yarish, C.; et al. Potential applications of nuisance microalgae blooms. *J. Appl. Phycol.* **2015**, *27*, 1223–1234. [CrossRef]

33. García-Casal, M.N.; Ramírez, J.; Leets, I.; Pereira, A.C.; Quiroga, M.F. Antioxidant capacity, polyphenol content and iron bioavailability from algae (Ulva sp., Sargassum sp. and Porphyra sp.) in human subjects. *Br. J. Nutr.* **2009**, *101*, 79–85. [CrossRef] [PubMed]

34. Abreu, M.H.; Pereira, R.; Yarish, C.; Buschmann, A.H.; Sousa-Pinto, I. IMTA with Gracilaria vermiculophylla: Productivity and nutrient removal performance of the seaweed in a land-based pilot scale system. *Aquaculture* **2011**, *312*, 77–87. [CrossRef]

35. Corey, P.; Kim, J.K.; Duston, J.; Garbary, D.J. Growth and nutrient uptake by Palmaria palmata integrated with Atlantic halibut in a land-based aquaculture system. *ALGAE* **2014**, *29*, 35–45. [CrossRef]

36. Kim, J.K.; Kraemer, G.P.; Yarish, C. Use of sugar kelp aquaculture in Long Island Sound and the Bronx River Estuary for nutrient extraction. *Mar. Ecol. Prog. Ser.* **2015**. [CrossRef]

37. Rose, J.M.; Deonarine, S.; Ferreira, J.G. Nutrient Bioextraction. *Encycl. Sustain. Sci. Technol.* **2015**, 1–33. [CrossRef]

38. Wu, H.; Kim, J.K.; Huo, Y.; Zhang, J.; He, P. Nutrient removal ability of seaweeds on Pyropia yezoensis aquaculture rafts in China's radial sandbanks. *Aquat. Bot.* **2017**, *137*, 72–79. [CrossRef]

39. Jun, J.Y.; Jung, M.J.; Jeong, I.H.; Yamazaki, K.; Kawai, Y.; Kim, B.M. Antimicrobial and antibiofilm activities of sulfated polysaccharides from marine algae against dental plaque bacteria. *Mar. Drugs* **2018**, *16*, 301. [CrossRef]

40. Gandhi, A.D.; Vizhi, D.K.; Lavanya, K.; Kalpana, V.N.; Devi Rajeswari, V.; Babujanarthanam, R. In vitro anti- biofilm and anti-bacterial activity of Sesbania grandiflora extract against Staphylococcus aureus. *Biochem. Biophys. Rep.* **2017**, *12*, 193–197. [CrossRef]

41. Janssens, J.C.A.; Steenackers, H.; Robijns, S.; Gellens, E.; Levin, J.; Zhao, H.; Hermans, K.; De Coster, D.; Verhoeven, T.L.; Marchal, K.; et al. Brominated furanones inhibit biofilm formation by Salmonella enterica serovar Typhimurium. *Appl. Environ. Microbiol.* **2008**, *74*, 6639–6648. [CrossRef]

42. Baghel, R.S.; Trivedi, N.; Gupta, V.; Neori, A.; Reddy, C.R.K.; Lali, A.; Jha, B. Biorefining of marine macroalgal biomass for production of biofuel and commodity chemicals. *Green Chem.* **2015**, *17*, 2436–2443. [CrossRef]

43. Milledge, J.J.; Nielsen, B.V.; Bailey, D. High-value products from macroalgae: The potential uses of the invasive brown seaweed, Sargassum muticum. *Rev. Environ. Sci. Biotechnol.* **2016**, *15*, 67–88. [CrossRef]

44. Li, R.; Zhong, Z.; Jin, B.; Zheng, A. Selection of temperature for bio-oil production from pyrolysis of algae from lake blooms. *Energy Fuels* **2012**, *26*, 2996–3002. [CrossRef]

45. Hu, Z.; Zheng, Y.; Yan, F.; Xiao, B.; Liu, S. Bio-oil production through pyrolysis of blue-green algae blooms (BGAB): Product distribution and bio-oil characterization. *Energy* **2013**, *52*, 119–125. [CrossRef]

46. Bwapwa, J.K.; Jaiyeola, A.T.; Chetty, R. Bioremediation of acid mine drainage using algae strains: A review. *S. Afr. J. Chem. Eng.* **2017**, *24*, 62–70. [CrossRef]

47. Zeraatkar, A.K.; Ahmadzadeh, H.; Talebi, A.F.; Moheimani, N.R.; McHenry, M.P. Potential use of algae for heavy metal bioremediation, a critical review. *J. Environ. Manag.* **2016**, *181*, 817–831. [CrossRef]

48. Buschbaum, C.; Chapman, A.S.; Saier, B. How an introduced seaweed can affect epibiota diversity in different coastal systems. *Mar. Biol.* **2006**, *148*, 743–754. [CrossRef]

49. Ratnasooriya, W.D.; Premakumara, G.A.S.; Tillekeratne, L.M.V. Post-coital contraceptive activity of crude extracts of Sri Lankan marine red algae. *Contraception* **1994**, *50*, 291–299. [CrossRef]

50. de Almeida, C.L.; Falcão, H.D.S.; Lima, G.R.; Montenegro, C.D.A.; Lira, N.S.; de Athayde-Filho, P.F.; Rodrigues, L.C.; de Souza Mde, F.; Barbosa-Filho, J.M.; Batista, L.M. Bioactivities from marine algae of the genus Gracilaria. *Int. J. Mol. Sci.* **2011**, *12*, 4550–4573. [CrossRef]

51. Agatonovic-Kustrin, S.; Morton, D.W. Cosmeceuticals Derived from Bioactive Substances Found in Marine Algae. *J. Oceanogr. Mar. Res.* **2013**, *1*. [CrossRef]

52. Kim, S.K. *Marine Cosmeceuticals: Trends and Prospects*, 1st ed.; Kim, S.K., Ed.; Tayor & Francis Group: Boca Ratón, FL, USA, 2012; ISBN 9781439860281.

53. Uysal, O.; Uysal, F.O.; Ekinci, K. Evaluation of Microalgae as Microbial Fertilizer. *Eur. J. Sustain. Dev.* **2015**, *4*, 77–82. [CrossRef]

54. Rizvi, M.A.; Shameel, M. Pharmaceutical biology of seaweeds from the Karachi coast of Pakistan. *Pharm. Biol.* **2005**, *43*, 97–107. [CrossRef]

55. Chingizova, E.A.; Skriptsova, A.V.; Anisimov, M.M.; Aminin, D.L.; Chingizova, E.A. Antimicrobial activity of marine algal extracts. *Int. J. Phytomed.* **2017**, *9*, 113–122.

56. Omar, H.H.; Al-Judaibiand, A.; El-Gendy, A. Antimicrobial, antioxidant, anticancer activity and phytochemical analysis of the red alga, laurencia papillosa. *Int. J. Pharmacol.* **2018**, *14*, 572–583. [CrossRef]

57. De Souza Barros, C.; Teixeira, V.L.; Paixão, I.C.N.P. Seaweeds with anti-herpes simplex virus type 1 activity. *J. Appl. Phycol.* **2015**, *27*, 1623–1637. [CrossRef]

58. Smit, A.J. Medicinal and pharmaceutical uses of seaweed natural products: A review. *J. Appl. Phycol.* **2004**, *16*, 245–262. [CrossRef]

59. Martínez Andrade, K.A.; Lauritano, C.; Romano, G.; Ianora, A. Marine microalgae with anti-cancer properties. *Mar. Drugs* **2018**, *16*, 165. [CrossRef]

60. Palermo, J.A.; Flower, P.B.; Seldes, A.M. Chondriamides A and B, new indolic metabolites from the red alga Chondria sp. *Tetrahedron Lett.* **1992**, *33*, 3097–3100. [CrossRef]

61. Liu, X.; Wang, S.; Cao, S.; He, X.; Qin, L.; He, M.; Yang, Y.; Hao, J.; Mao, W. Structural characteristics and anticoagulant property in vitro and in vivo of a seaweed Sulfated Rhamnan. *Mar. Drugs* **2018**, *16*, 243. [CrossRef]

62. Kim, S.K.; Wijesekara, I. *Anticoagulant Effect of Marine Algae*, 1st ed.; Elsevier Inc.: Philadelphia, PA, USA, 2011; Volume 64, ISBN 9780123876690.
63. Magalhaes, K.D.; Costa, L.S.; Fidelis, G.P.; Oliveira, R.M.; Nobre, L.T.D.B.; Dantas-Santos, N.; Camara, R.B.G.; Albuquerque, I.R.L.; Cordeiro, S.L.; Sabry, D.A.; et al. Anticoagulant, antioxidant and antitumor activities of heterofucans from the seaweed dictyopteris delicatula. *Int. J. Mol. Sci.* **2011**, *12*, 3352–3365. [CrossRef]
64. Athukorala, Y.; Lee, K.W.; Kim, S.K.; Jeon, Y.J. Anticoagulant activity of marine green and brown algae collected from Jeju Island in Korea. *Bioresour. Technol.* **2007**, *98*, 1711–1716. [CrossRef]
65. Lin, H.T.V.; Lu, W.J.; Tsai, G.J.; Te Chou, C.; Hsiao, H.I.; Hwang, P.A. Enhanced anti-inflammatory activity of brown seaweed Laminaria japonica by fermentation using Bacillus subtilis. *Process Biochem.* **2016**, *51*, 1945–1953. [CrossRef]
66. Delgado, N.G.; Frías Vázquez, A.I.; Sánchez, H.C.; del Valle, R.M.S.; Gómez, Y.S.; Suárez Alfonso, A.M. Anti-inflammatory and antinociceptive activities of methanolic extract from red seaweed Dichotomaria obtusata. *Braz. J. Pharm. Sci.* **2013**, *49*, 65–74. [CrossRef]
67. Radhika, D.; Veerabahu, C.; Prira, R. Anti-inflammatory activities of some seaweed collected from the gulf of mannar coast, tuticorin, south india. *Int. J. Pharma Bio Sci.* **2013**, *4*, 39–44.
68. Surget, G.; Roberto, V.P.; Le Lann, K.; Mira, S.; Guérard, F.; Laizé, V.; Poupart, N.; Cancela, M.L.; Stiger-Pouvreau, V. Marine green macroalgae: A source of natural compounds with mineralogenic and antioxidant activities. *J. Appl. Phycol.* **2017**, *29*, 575–584. [CrossRef]
69. Figueiroa, F.L.; Gil, C.; Rico, R.M.; Moriñigo, M.Á.; Gómez-Pinchetti, J.L.; Abdala Díaz, R. Biofiltración de efluentes mediante algas: Valorización de la biomasa (alimentos funcionales y biodiesel). In *Las Algas Como Recurso: Valorización, Aplicaciones Industriales y Tendencias*; CETMAR: Vigo, Spain, 2011; pp. 209–224. ISBN 9788461535934.
70. Bourgougnon, N.; Bedoux, G.; Sangiardi, A.; Stiger-Pouvreau, V. Las algas: Potencial nutritivo y aplicaciones cosméticas. In *Las Algas Como Recurso. Valorización. Aplicaciones Industriales y Tendencias Aplicaciones*; Centro Tecnológico del Mar-Fundación CETMAR: Vigo, Spain, 2011.
71. Cañavate Hors, J.P. Funciones de las microalgas en acuicultura. In *Las Algas Como Recurso. Valorización. Aplicaciones Industriales y Tendencias Aplicaciones*; CETMAR: Vigo, Spain, 2011.
72. Olsen, A.I.; Olsen, Y.; Attramadal, Y.; Christie, K.; Birkbeck, T.H.; Skjermo, J.; Vadstein, O. Effects of short term feeding of microalgae on the bacterial flora associated with juvenile Artemia franciscana. *Aquaculture* **2000**, *190*, 11–25. [CrossRef]
73. Shields, R.J.; Lupatsch, I. Algae for Aquaculture and Animal Feeds. *Tech. Theor. Prax.* **2012**, *21*, 23–37.
74. Holdt, S.L.; Kraan, S. Bioactive compounds in seaweed: Functional food applications and legislation. *J. Appl. Phycol.* **2011**, *23*, 543–597. [CrossRef]
75. Salvador, N.; Gómez Garreta, A.; Lavelli, L.; Ribera, M.A. Antimicrobial activity of Iberian macroalgae. *Sci. Mar.* **2007**, *71*, 101–114. [CrossRef]
76. Rizzo, C.; Genovese, G.; Morabito, M.; Faggio, C.; Pagano, M.; Spanò, A.; Zammuto, V.; Minicante, S.A.; Manghisi, A.; Cigala, R.M.; et al. Potential antibacterial activity of marine macroalgae against pathogens relevant for aquaculture and human health. *J. Pure Appl. Microbiol.* **2017**, *11*, 1695–1706. [CrossRef]
77. Capillo, G.; Savoca, S.; Costa, R.; Sanfilippo, M.; Rizzo, C.; Giudice, A.L.; Albergamo, A.; Rando, R.; Bartolomeo, G.; Spanò, N.; et al. New insights into the culture method and antibacterial potential of graciliaria gracilis. *Mar. Drugs* **2018**, *16*, 492. [CrossRef]
78. Fedorov, S.N.; Ermakova, S.P.; Zvyagintseva, T.N.; Stonik, V.A. Anticancer and cancer preventive properties of marine polysaccharides: Some results and prospects. *Mar. Drugs* **2013**, *11*, 4876. [CrossRef]
79. Faggio, C.; Pagano, M.; Dottore, A.; Genovese, G.; Morabito, M. Evaluation of anticoagulant activity of two algal polysaccharides. *Nat. Prod. Res.* **2016**, *30*, 1934–1937. [CrossRef]
80. Arive, P.L.C.; Inquimboy, I.H.; No, N.L.I. In Vitro Antioxidant Activity of Selected Seaweeds in the Philippines. *Int. J. Theor. Appl. Sci.* **2017**, *9*, 212–216.
81. Moubayed, N.M.S.; Al Houri, H.J.; Al Khulaifi, M.M.; Al Farraj, D.A. Antimicrobial, antioxidant properties and chemical composition of seaweeds collected from Saudi Arabia (Red Sea and Arabian Gulf). *Saudi J. Biol. Sci.* **2017**, *24*, 162–169. [CrossRef]
82. Lima, R.L.D.; Pires-Cavalcante, M.K.D.S.; De Alencar, D.B.; Viana, A.F.; Sampaio, A.H.; Saker-Sampaio, S. Acta Scientiarum In vitro evaluation of antioxidant activity of methanolic extracts obtained from seaweeds endemic to the coast of Ceará, Brazil. *Acta Sci. Technol.* **2016**, *38*, 247–255. [CrossRef]

83. Foon, T.S.; Ai, L.A.; Kuppusamy, P.; Yusoff, M.M.; Govindan, N. Studies on in vitro antioxidant activity of marine edible seaweed from east coastal region, peninsular Malaysia using different extraction methods. *Res. J. Appl. Sci.* **2013**, *1*, 193–198.

84. Lee, J.H.; Kim, G.H. Evaluation of antioxidant activity of marine algae-extracts from Korea. *J. Aquat. Food Prod. Technol.* **2015**, *24*, 227–240. [CrossRef]

85. Vieira, V.; Prieto, M.A.; Barros, L.; Coutinho, J.A.P.; Ferreira, I.C.F.R.; Ferreira, O. Enhanced extraction of phenolic compounds using choline chloride based deep eutectic solvents from *Juglans regia* L. *Ind. Crop. Prod.* **2018**, *115*, 261–271. [CrossRef]

86. Hoseinifar, S.H.; Yousefi, S.; Capillo, G.; Paknejad, H.; Khalili, M.; Tabarraei, A.; Van Doan, H.; Spanò, N.; Faggio, C. Mucosal immune parameters, immune and antioxidant defence related genes expression and growth performance of zebrafish (Danio rerio) fed on Gracilaria gracilis powder. *Fish Shellfish Immunol.* **2018**, *83*, 232–237. [CrossRef]

87. McReynolds, C. Invasive Marine Macroalgae and Their Current and Potential Use in Cosmetics. 2017. Available online: http://hdl.handle.net/10400.8/2653 (accessed on 12 April 2019).

88. Bedoux, G.; Hardouin, K.; Burlot, A.S.; Bourgougnon, N. *Bioactive Components from Seaweeds: Cosmetic Applications and Future Development*; Elsevier: Amsterdam, The Netherlands, 2014; Volume 71, ISBN 9780124080621.

89. Pinela, J.; Prieto, M.A.; Barros, L.; Carvalho, A.M.; Oliveira, M.B.P.P.; Saraiva, J.A.; Ferreira, I.C.F.R. Cold extraction of phenolic compounds from watercress by high hydrostatic pressure: Process modelling and optimization. *Sep. Purif. Technol.* **2018**, *192*, 501–512. [CrossRef]

90. Hennequart, F. Extractos de algas como bioestimuladores del crecimiento de las plantas. In *Las Algas Como Recurso: Valorización, Aplicaciones Industriales y Tendencias*; CETMAR, Ed.; CETMAR: Vigo, Spain, 2011; pp. 49–62. ISBN 9788461535934.

91. Sohn, C.H. Porphyra, Undaria and Hizikia Cultivation in Korea. *Korean J. Phycol.* **1993**, *8*, 207–216.

92. Oren, A. A hundred years of Dunaliella research: 1905–2005. *Saline Syst.* **2005**, *1*, 2. [CrossRef]

93. Makkar, H.P.S.; Tran, G.; Heuzé, V.; Giger-Reverdin, S.; Lessire, M.; Lebas, F.; Ankers, P. Seaweeds for livestock diets: A review. *Anim. Feed Sci. Technol.* **2016**, *212*, 1–17. [CrossRef]

94. Gavio, B.; Fredericq, S. Grateloupia Turuturu (Halymeniaceae, Rhodophyta): The Correct Identity of the Invasive Species in the Atlantic Known as Grateloupia Doryphora as Inferred From Molecular and Morphological Evidence. *J. Phycol.* **2003**, *38*, 12. [CrossRef]

95. Peinado, I.; Girón, J.; Koutsidis, G.; Ames, J.M. Chemical composition, antioxidant activity and sensory evaluation of five different species of brown edible seaweeds. *Food Res. Int.* **2014**, *66*, 36–44. [CrossRef]

96. Gómez-Ordóñez, E.; Jiménez-Escrig, A.; Rupérez, P. Dietary fibre and physicochemical properties of several edible seaweeds from the northwestern Spanish coast. *Food Res. Int.* **2010**, *43*, 2289–2294. [CrossRef]

97. Ramnani, P.; Chitarrari, R.; Tuohy, K.; Grant, J.; Hotchkiss, S.; Philp, K.; Campbell, R.; Gill, C.; Rowland, I. Invitro fermentation and prebiotic potential of novel low molecular weight polysaccharides derived from agar and alginate seaweeds. *Anaerobe* **2012**, *18*, 1–6. [CrossRef]

98. Delattre, C.; Fenoradosoa, T.A.; Michaud, P. Galactans: An overview of their most important sourcing and applications as natural polysaccharides. *Braz. Arch. Biol. Technol.* **2011**, *54*, 1075–1092. [CrossRef]

99. Gupta, S.; Abu-Ghannam, N. Bioactive potential and possible health effects of edible brown seaweeds. *Trends Food Sci. Technol.* **2011**, *22*, 315–326. [CrossRef]

100. Ale, M.T.; Mikkelsen, J.D.; Meyer, A.S. Important determinants for fucoidan bioactivity: A critical review of structure-function relations and extraction methods for fucose-containing sulfated polysaccharides from brown seaweeds. *Mar. Drugs* **2011**, *9*, 2106. [CrossRef]

101. Jiménez-Escrig, A.; Sánchez-Muniz, F.J. Dietary fibre from edible seaweeds: Chemical structure, physicochemical properties and effects on cholesterol metabolism. *Nutr. Res.* **2000**, *20*, 585–598. [CrossRef]

102. Alderkamp, A.C.; Buma, A.G.J.; Van Rijssel, M. The carbohydrates of Phaeocystis and their degradation in the microbial food web. In *Phaeocystis, Major Link in the Biogeochemical Cycling of Climate-Relevant Elements*; Springer: Berlin/Heidelberg, Germany, 2007; ISBN 9781402062131.

103. Pangestuti, R.; Kim, S. An Overview of Phycocolloids: The Principal Commercial. In *Marine Algae Extracts*; Wiley-VCH Verlag GmbH & Co. KGaA: Weinheim, Germany, 2015; pp. 319–330.

104. Ammar, H.H.; Lajili, S.; Sakly, N.; Cherif, D.; Rihouey, C.; Cerf, D.L.; Bouraoui, A.; Majdoub, H. Influence of the uronic acid composition on the gastroprotective activity of alginates from three different genus of Tunisian brown algae. *Food Chem.* **2017**. [CrossRef]

105. Lee, A.W.; Lim, Y.; Leow, A.T. Biosynthesis of agar in red seaweeds: A review. *Carbohydr. Polym.* **2017**, *164*, 23–30. [CrossRef]

106. Chi, Z.; Fallon, J.V.O.; Chen, S. Bicarbonate produced from carbon capture for algae culture. *Trends Biotechnol.* **2011**, *29*, 537–541. [CrossRef]

107. McLachlan, J. Macroalgae (seaweeds): Industrial resources and their utilization. *Plant Soil* **1985**, *89*, 137–157. [CrossRef]

108. Andrade, P.B.; Barbosa, M.; Pedro, R.; Lopes, G.; Vinholes, J.; Mouga, T.; Valentão, P. Valuable compounds in macroalgae extracts. *Food Chem.* **2013**, *138*, 1819–1828. [CrossRef]

109. Menshova, R.V.; Shevchenko, N.M.; Imbs, T.I.; Zvyagintseva, T.N.; Malyarenko, O.S.; Zaporoshets, T.S.; Besednova, N.N.; Ermakova, S.P. Fucoidans from Brown Alga Fucus evanescens: Structure and Biological Activity. *Front. Mar. Sci.* **2016**, *3*, 1–9. [CrossRef]

110. Yun, E.J.; Yu, S.; Kim, K.H. Current knowledge on agarolytic enzymes and the industrial potential of agar-derived sugars. *Appl. Microbiol. Biotechnol.* **2017**, *101*, 5581–5589. [CrossRef]

111. Seedevi, P.; Moovendhan, M.; Viramani, S.; Shanmugam, A. Bioactive potential and structural chracterization of sulfated polysaccharide from seaweed (Gracilaria corticata). *Carbohydr. Polym.* **2017**, *155*, 516–524. [CrossRef]

112. Souza, B.W.S.; Cerqueira, M.A.; Bourbon, A.I.; Pinheiro, A.C.; Martins, J.T.; Teixeira, J.A.; Coimbra, M.A.; Vicente, A.A. Chemical characterization and antioxidant activity of sulfated polysaccharide from the red seaweed Gracilaria birdiae. *Food Hydrocoll.* **2012**, *27*, 287–292. [CrossRef]

113. Cheong, K.L.; Qiu, H.M.; Du, H.; Liu, Y.; Khan, B.M. Oligosaccharides derived from red seaweed: Production, properties, and potential health and cosmetic applications. *Molecules* **2018**, *23*, 2451. [CrossRef]

114. De Jesus Raposo, M.F.; De Morais, A.M.M.B.; De Morais, R.M.S.C. Emergent sources of prebiotics: Seaweeds and microalgae. *Mar. Drugs* **2016**, *14*, 27. [CrossRef]

115. Peso-Echarri, P.; Frontela-Saseta, C.; González-Bermúdez, C.A.; Ros-Berruezo, G.F.; Martínez-Graciá, C. Polisacáridos de algas como ingredientes funcionales en acuicultura marina: Alginato, carragenato y ulvano. *Rev. Biol. Mar. Oceanogr.* **2012**, *47*, 373–381. [CrossRef]

116. Wang, X.; Wang, X.; Jiang, H.; Cai, C.; Li, G.; Hao, J.; Yu, G. Marine polysaccharides attenuate metabolic syndrome by fermentation products and altering gut microbiota: An overview. *Carbohydr. Polym.* **2018**, *195*, 601–612. [CrossRef]

117. Chen, L.; Xu, W.; Chen, D.; Chen, G.; Liu, J.; Zeng, X.; Shao, R.; Zhu, H. Digestibility of sulfated polysaccharide from the brown seaweed Ascophyllum nodosum and its effect on the human gut microbiota in vitro. *Int. J. Biol. Macromol.* **2018**, *112*, 1055–1061. [CrossRef]

118. Ale, M.T.; Meyer, A.S. Fucoidans from brown seaweeds: An update on structures, extraction techniques and use of enzymes as tools for structural elucidation. *RSC Adv.* **2013**, *3*, 8131–8141. [CrossRef]

119. De Jesus Raposo, M.F.; De Morais, A.M.B.; De Morais, R.M.S.C. Marine polysaccharides from algae with potential biomedical applications. *Mar. Drugs* **2015**, *13*, 2967. [CrossRef] [PubMed]

120. Zaporozhets, T.S.; Besednova, N.N.; Kuznetsova, T.A.; Zvyagintseva, T.N.; Makarenkova, I.D.; Kryzhanovsky, S.P.; Melnikov, V.G. The prebiotic potential of polysaccharides and extracts of seaweeds. *Russ. J. Mar. Biol.* **2014**, *40*, 1–9. [CrossRef]

121. Gibson, G.R.; Hutkins, R.; Sanders, M.E.; Prescott, S.L.; Reimer, R.A.; Salminen, S.J.; Scott, K.; Stanton, C.; Swanson, K.S.; Cani, P.D.; et al. Expert consensus document: The International Scientific Association for Probiotics and Prebiotics (ISAPP) consensus statement on the definition and scope of prebiotics. *Nat. Rev. Gastroenterol. Hepatol.* **2017**, *14*, 491–502. [CrossRef] [PubMed]

122. Courtois, J. Oligosaccharides from land plants and algae: Production and applications in therapeutics and biotechnology. *Curr. Opin. Microbiol.* **2009**, *12*, 261–273. [CrossRef]

123. Klarzynski, O.; Descamps, V.; Plesse, B.; Yvin, J.C.; Kloareg, B.; Fritig, B. Sulfated Fucan Oligosaccharides Elicit Defense Responses in Tobacco and Local and Systemic Resistance Against Tobacco Mosaic Virus. *Mol. Plant Microbe Interact.* **2003**, *16*, 115–122. [CrossRef]

124. Kardos, N.; Luche, J.L. Sonochemistry of carbohydrate compounds. *Carbohydr. Res.* **2001**, *332*, 115–131. [CrossRef]

125. Lii, C.Y.; Chen, C.H.; Yeh, A.I.; Lai, V.M.F. Preliminary study on the degradation kinetics of agarose and carrageenans by ultrasound. *Food Hydrocoll.* **1999**, *13*, 477–481. [CrossRef]

126. Sun, L.; Wang, C.; Shi, Q.; Ma, C. Preparation of different molecular weight polysaccharides from Porphyridium cruentum and their antioxidant activities. *Int. J. Biol. Macromol.* **2009**, *45*, 42–47. [CrossRef]

127. Zúñiga, E.A.; Matsuhiro, B.; Mejías, E. Preparation of a low-molecular weight fraction by free radical depolymerization of the sulfated galactan from Schizymenia binderi (Gigartinales, Rhodophyta) and its anticoagulant activity. *Carbohydr. Polym.* **2006**, *66*, 208–215. [CrossRef]

128. Leal, B. Obtenção de Oligossacarídeos Prebióticos a Partir da Hidrólise Fosfórica da Biomassa de Microalgas Utilizadas na Biomitigação de CO2 de Efluente Gasoso de Churrascaria. Master's Thesis, Universidade Tecnológica Federal do Paraná, Curitiba, Brasil, 2015.

129. Nardella, A.; Chaubet, F.; Boisson-Vidal, C.; Blondin, C.; Durand, P.; Jozefonvicz, J. Anticoagulant low molecular weight fucans produced by radical process and ion exchange chromatography of high molecular weight fucans extracted from the brown seaweed Ascophyllum nodosum. *Carbohydr. Res.* **1996**, *289*, 201–208. [CrossRef]

130. Patel, S.; Goyal, A. The current trends and future perspectives of prebiotics research: A review. *3 Biotech* **2012**, *2*, 115–125. [CrossRef]

131. Zhou, G.; Sun, Y.P.; Xin, H.; Zhang, Y.; Li, Z.; Xu, Z. In vivo antitumor and immunomodulation activities of different molecular weight lambda-carrageenans from Chondrus ocellatus. *Pharmacol. Res.* **2004**, *50*, 47–53. [CrossRef] [PubMed]

132. Sun, L.; Wang, L.; Zhou, Y. Immunomodulation and antitumor activities of different-molecular-weight polysaccharides from Porphyridium cruentum. *Carbohydr. Polym.* **2012**, *87*, 1206–1210. [CrossRef]

133. López, C.J.; Caleja, C.; Prieto, M.A.; Barreiro, M.F.; Barros, L.; Ferreira, I.C.F.R. Optimization and comparison of heat and ultrasound assisted extraction techniques to obtain anthocyanin compounds from Arbutus unedo L. Fruits. *Food Chem.* **2018**, *264*, 81–91. [CrossRef] [PubMed]

134. Di, T.; Chen, G.; Sun, Y.; Ou, S.; Zeng, X.; Ye, H. In vitro digestion by saliva, simulated gastric and small intestinal juices and fermentation by human fecal microbiota of sulfated polysaccharides from Gracilaria rubra. *J. Funct. Foods* **2018**, *40*, 18–27. [CrossRef]

135. Sullivan, L.; Murphy, B.; McLoughlin, P.; Duggan, P.; Lawlor, P.G.; Hughes, H.; Gardiner, G.E. Prebiotics from marine macroalgae for human and animal health applications. *Mar. Drugs* **2010**, *8*, 2038. [CrossRef] [PubMed]

136. Charoensiddhi, S.; Conlon, M.A.; Methacanon, P.; Franco, C.M.M.; Su, P.; Zhang, W. Gut health benefits of brown seaweed Ecklonia radiata and its polysaccharides demonstrated in vivo in a rat model. *J. Funct. Foods* **2017**, *37*, 676–684. [CrossRef]

137. Wang, Y.; Han, F.; Hu, B.; Li, J.; Yu, W. In vivo prebiotic properties of alginate oligosaccharides prepared through enzymatic hydrolysis of alginate. *Nutr. Res.* **2006**, *26*, 597–603. [CrossRef]

138. Hu, B.; Gong, Q.; Wang, Y.; Ma, Y.; Li, J.; Yu, W. Prebiotic effects of neoagaro-oligosaccharides prepared by enzymatic hydrolysis of agarose. *Anaerobe* **2006**, *12*, 260–266. [CrossRef] [PubMed]

139. Kuda, T.; Yano, T.; Matsuda, N.; Nishizawa, M. Inhibitory effects of laminaran and low molecular alginate against the putrefactive compounds produced by intestinal microflora in vitro and in rats. *Food Chem.* **2005**, *91*, 745–749. [CrossRef]

140. Kong, Q.; Dong, S.; Gao, J.; Jiang, C. In vitro fermentation of sulfated polysaccharides from E. prolifera and L. japonica by human fecal microbiota. *Int. J. Biol. Macromol.* **2016**, *91*, 867–871. [CrossRef] [PubMed]

141. Fu, X.; Cao, C.; Ren, B.; Zhang, B.; Huang, Q.; Li, C. Structural characterization and in vitro fermentation of a novel polysaccharide from Sargassum thunbergii and its impact on gut microbiota. *Carbohydr. Polym.* **2018**, *183*, 230–239. [CrossRef] [PubMed]

142. Devillé, C.; Gharbi, M.; Dandrifosse, G.; Peulen, O. Study on the effects of laminarin, a polysaccharide from seaweed, on gut characteristics. *J. Sci. Food Agric.* **2007**, *87*, 1717–1725. [CrossRef]

143. Holck, J.; Hjernø, K.; Lorentzen, A.; Vigsnæs, L.K.; Hemmingsen, L.; Licht, T.R.; Mikkelsen, J.D.; Meyer, A.S. Tailored enzymatic production of oligosaccharides from sugar beet pectin and evidence of differential effects of a single DP chain length difference on human faecal microbiota composition after in vitro fermentation. *Process Biochem.* **2011**, *46*, 1039–1049. [CrossRef]

144. Roriz, C.L.; Barros, L.; Prieto, M.A.; Morales, P.; Ferreira, I.C.F.R. Floral parts of *Gomphrena globosa* L. as a novel alternative source of betacyanins: Optimization of the extraction using response surface methodology. *Food Chem.* **2017**, *229*, 223–234. [CrossRef] [PubMed]

145. Albuquerque, B.R.; Prieto, M.A.; Barreiro, M.F.; Rodrigues, A.; Curran, T.P.; Barros, L.; Ferreira, I.C.F.R. Catechin-based extract optimization obtained from *Arbutus unedo* L. fruits using maceration/microwave/ ultrasound extraction techniques. *Ind. Crops Prod.* **2016**, *95*, 404–415. [CrossRef]

146. Del Campo, J.A.; García-González, M.; Guerrero, M.G. Outdoor cultivation of microalgae for carotenoid production: Current state and perspectives. *Appl. Microbiol. Biotechnol.* **2007**, *74*, 1163–1174. [CrossRef] [PubMed]

147. Baroli, I.; Niyogi, K.K.; Barber, J.; Heifetz, P. Molecular genetics of xanthophyll-dependent photoprotection in green algae and plants. *Philos. Trans. R. Soc. Lond. B Biol. Sci.* **2000**, *355*, 1385–1394. [CrossRef] [PubMed]

148. Grünewald, K.; Hirschberg, J.; Hagen, C. Ketocarotenoid Biosynthesis Outside of Plastids in the Unicellular Green Alga Haematococcus pluvialis. *J. Biol. Chem.* **2001**, *276*, 6023–6029. [CrossRef]

149. Lohr, M.; Wilhelm, C. Xanthophyll synthesis in diatoms: Quantification of putative intermediates and comparison of pigment conversion kinetics with rate constants derived from a model. *Planta* **2001**, *212*, 382–391. [CrossRef] [PubMed]

150. Bhosale, P. Environmental and cultural stimulants in the production of carotenoids from microorganisms. *Appl. Microbiol. Biotechnol.* **2004**, *63*, 351–361. [CrossRef] [PubMed]

151. Raja, R.; Hemaiswarya, S.; Rengasamy, R. Exploitation of Dunaliella for β-carotene production. *Appl. Microbiol. Biotechnol.* **2007**, *74*, 517–523. [CrossRef] [PubMed]

152. Dufossé, L.; Galaup, P.; Yaron, A.; Arad, S.M.; Blanc, P.; Murthy, K.N.C.; Ravishankar, G.A. Microorganisms and microalgae as sources of pigments for food use: A scientific oddity or an industrial reality? *Trends Food Sci. Technol.* **2005**, *16*, 389–406. [CrossRef]

153. Van Den Berg, H.; Faulks, R.; Granado, H.F.; Hirschberg, J.; Olmedilla, B.; Sandmann, G.; Southon, S.; Stahl, W. The potential for the improvement of carotenoid levels in foods and the likely systemic effects. *J. Sci. Food Agric.* **2000**, *80*, 880–912. [CrossRef]

154. Jespersen, L.; Strømdahl, L.D.; Olsen, K.; Skibsted, L.H. Heat and light stability of three natural blue colorants for use in confectionery and beverages. *Eur. Food Res. Technol.* **2005**, *220*, 261–266. [CrossRef]

155. Duan, X.J.; Zhang, W.W.; Li, X.M.; Wang, B.G. Evaluation of antioxidant property of extract and fractions obtained from a red alga, Polysiphonia urceolata. *Food Chem.* **2006**, *95*, 37–43. [CrossRef]

156. Manach, C.; Scalbert, A.; Morand, C.; Rémésy, C.; Jiménez, L. Polyphenols: Food sources and bioavailability. *Am. J. Clin. Nutr.* **2004**, *79*, 727–747. [CrossRef]

157. Pinela, J.; Prieto, M.A.; Barreiro, M.F.; Carvalho, A.M.; Oliveira, M.B.P.P.; Vázquez, J.A.; Ferreira, I.C.F.R. Optimization of microwave-assisted extraction of hydrophilic and lipophilic antioxidants from a surplus tomato crop by response surface methodology. *Food Bioprod. Process.* **2016**, *98*, 283–298. [CrossRef]

158. Kähkönen, M.P.; Hopia, A.I.; Vuorela, H.J.; Rauha, J.P.; Pihlaja, K.; Kujala, T.S.; Heinonen, M. Antioxidant activity of plant extracts containing phenolic compounds. *J. Agric. Food Chem.* **1999**, *47*, 3954–3962. [CrossRef]

159. Pinela, J.; Prieto, M.A.; Barreiro, M.F.; Curran, T.P.; Ferreira, I.C.F.R. Valorisation of tomato wastes for development of nutrient-rich antioxidant ingredients: A sustainable approach towards the needs of the today's society. *Innov. Food Sci. Emerg. Technol.* **2017**, *41*, 160–171. [CrossRef]

160. Thomas, N.V.; Kim, S.K. Potential pharmacological applications of polyphenolic derivatives from marine brown algae. *Environ. Toxicol. Pharmacol.* **2011**, *32*, 325–335. [CrossRef]

161. Wang, T.; Jónsdóttir, R.; Liu, H.; Gu, L.; Kristinsson, H.G.; Raghavan, S.; Ólafsdóttir, G. Antioxidant capacities of phlorotannins extracted from the brown algae Fucus vesiculosus. *J. Agric. Food Chem.* **2012**, *60*, 5874–5883. [CrossRef]

162. Kelman, D.; Posner, E.K.; McDermid, K.J.; Tabandera, N.K.; Wright, P.R.; Wright, A.D. Antioxidant activity of Hawaiian marine algae. *Mar. Drugs* **2012**, *10*, 403. [CrossRef]

163. Becker, W. Handbook of microalgal culture: Biotechnology and applied phycology. In *Handbook of Microalgal Culture: Biotechnology an Applied Phycology*; Richmond, A., Ed.; Blackwell Science Ltd: Hoboken, NJ, USA, 2004; ISBN 0632059532.

164. FAO. *The State of World Fisheries and Aquaculture 2018-Meeting the Sustainable Development Goals*; FAO: Rome, Italy, 2018; Volume 35, ISBN 9789251060292.

165. FAO. *FAO Yearbook. Fishery and Aquaculture Statistics 2011/FAO Annuaire*; FAO: Rome, Italy, 2011.

166. FAO. *The State of World Fisheries and Aquaculture, 2012*; FAO: Rome, Italy, 2013.

167. Barry, L. *Seaweed, Potential as a Marine Vegetable and Other Opportunities*; Rural Industries Research and Development Corporation: Canberra, Australia, 2008.

168. Zemke-White, W.L.; Ohno, M. World seaweed utilisation: An end-of-century summary. *J. Appl. Phycol.* **1999**, *11*, 369–376. [CrossRef]

169. Wikfors, G.H.; Ohno, M. Impact of algal research in aquaculture. *J. Phycol.* **2001**, *37*, 968–974. [CrossRef]

170. FAO. *World Aquaculture 2010*; FAO: Rome, Italy, 2011; ISBN 9789251069974.

171. McHugh, D.J. *A Guide to the Seaweed Industry*; FAO: Rome, Italy, 2003.

 *antioxidants*

Review

Grown to be Blue—Antioxidant Properties and Health Effects of Colored Vegetables. Part I: Root Vegetables

Spyridon A. Petropoulos [1,*], Shirley L. Sampaio [2], Francesco Di Gioia [3], Nikos Tzortzakis [4], Youssef Rouphael [5], Marios C. Kyriacou [6] and Isabel Ferreira [2,*]

[1] Crop Production and Rural Environment, Department of Agriculture, University of Thessaly, 38446 Nea Ionia, Greece
[2] Centro de Investigação de Montanha (CIMO), Instituto Politécnico de Bragança, Campus de Santa Apolónia, 5300-253 Bragança, Portugal; shirleylsampaio@gmail.com
[3] Department of Plant Science, The Pennsylvania State University, University Park, PA 16802, USA; fxd92@psu.edu
[4] Department of Agricultural Sciences, Biotechnology and Food Science, Cyprus University of Technology, 3603 Limassol, Cyprus; nikolaos.tzortzakis@cut.ac.cy
[5] Department of Agricultural Sciences, University of Naples Federico II, 80055 Portici, Italy; youssef.rouphael@unina.it
[6] Department of Vegetable Crops, Agricultural Research Institute, 1516 Nicosia, Cyprus; m.kyriacou@ari.gov.cy
*   Correspondence: spetropoulos@uth.gr (S.A.P.); iferreira@ipb.pt (I.C.F.R.F.)

Received: 16 November 2019; Accepted: 3 December 2019; Published: 4 December 2019

Abstract: During the last few decades, the food and beverage industry faced increasing demand for the design of new functional food products free of synthetic compounds and artificial additives. Anthocyanins are widely used as natural colorants in various food products to replenish blue color losses during processing and to add blue color to colorless products, while other compounds such as carotenoids and betalains are considered as good sources of other shades. Root vegetables are well known for their broad palette of colors, and some species, such as black carrot and beet root, are already widely used as sources of natural colorants in the food and drug industry. Ongoing research aims at identifying alternative vegetable sources with diverse functional and structural features imparting beneficial effects onto human health. The current review provides a systematic description of colored root vegetables based on their belowground edible parts, and it highlights species and/or cultivars that present atypical colors, especially those containing pigment compounds responsible for hues of blue color. Finally, the main health effects and antioxidant properties associated with the presence of coloring compounds are presented, as well as the effects that processing treatments may have on chemical composition and coloring compounds in particular.

Keywords: anthocyanins; antioxidant activity; beet root; betacyanins; cyanidin; blue potatoes; carotenoids; flavonoids; natural colorants; sweet potato

---

1. Introduction

Root vegetables display various colors which usually depend on the presence of three major classes of compounds, namely, flavonoids, betalains, and carotenoids, which they may define their visual appearance and consumer perception [1,2]. Anthocyanins are flavonoids responsible for the different shades of plant epidermal tissues such as purple, blue, red, and pink colors, aiming at attracting pollinators and contributing to the overall plant antioxidant mechanisms under abiotic and biotic stress conditions [3]. They also participate in several physiological processes of the plant, including photosynthesis and plant interactions with the environment [4]. They are produced via the phenylpropanoid pathway and the conversion of leucoanthocyanidins into colored anthocyanidin and glycoside derivatives via anthocyanidin synthase and other enzymes [1,5]. The great number of anthocyanins isolated in nature so far and their high structural variation across plant species raised research interest in these compounds during the last few years in search of novel natural colorants [6,7]. The structural variation of anthocyanins is related to the substitution of hydroxyl and methoxyl groups in the B ring, glycosidic substitution at positions 3 and 5 of the A and C rings, and the possible acylation of glycosidic substitutes with aliphatic and cinnamic acids (Figures 1 and 2) [6]. These structural differences may infer significant variations in the biological activities and antioxidant properties of vegetable products. For example, Oki et al. [8] suggested that antioxidant activities of purple sweet-potato extracts from peonidin-rich cultivars were attributed to anthocyanins, whereas, in those extracts from cyanidin-rich cultivars, the antioxidant capacity was due to the phenolic compounds. Other compounds that transfuse blue color in nature are quinones, quinodes, and various alkaloids which are usually present in fungi, bacteria, and in the animal kingdom [9]. Quinones and quinodes include carbonyl groups within aromatic rings, and they also show a great variation from a structural point of view [9], while alkaloids contain nitrogen atoms and are divided into several distinct classes, including pyridine alkaloids, phenazine alkaloids, and linear tetrapyrrole and indole alkaloids, with different coloring attributes [9].

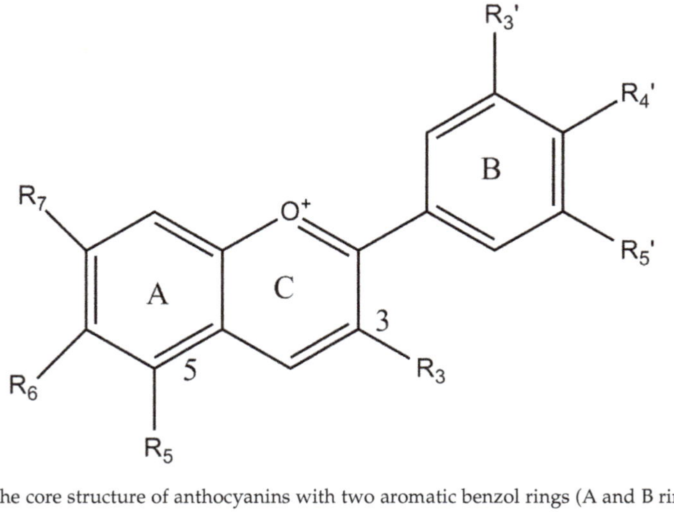

Figure 1. The core structure of anthocyanins with two aromatic benzol rings (A and B rings) and a portion cyclized with oxygen (C ring).

On the other hand, carotenoids are mainly responsible for yellow and orange color with several distinct compounds being detected so far in various vegetables [6,10–12], while betalains such as betacyanins and betaxanthins are also important for the violet and yellow pigmentations, respectively [13]. The main detected carotenoids are β-carotene and lycopene, which are unsaturated hydrocarbons, and they differ in terms of the β-rings, where β-carotene molecules have both ends (Figure 3), and they usually present synergistic effects [14]. Both are fat-soluble, and the number of conjugate double bonds in their structure is closely related to their superoxide inhibitory effect [15,16]. Betacyanins and betaxanthins differ in the moiety derived from betalamic acid, as towel as the fact that betaxanthins are produced from the condensation of betalamic acid with amino acids and they never show glycosidation, whereas betacyanins are the result of condensation of betalamic acid with imino compounds (Figure 4) [17,18]. Further differences are observed within each main class of betalains, namely, betacyanins and betaxanthins, with several structures identified resulting in different individual compounds with different absorption and stability capacity [19]. In particular, the various betacyanins are differentiated through the glycosyl groups attached to the *o*-position of the cyclo-dopa moiety [20], while betaxanthins are differentiated through the conjugated moiety of betalamic acid (amino acids or amines) [20]. The main pigments isolated in the various root vegetables are presented in Table 1.

Figure 2. The main anthocyanins detected in root vegetables.

Figure 3. The main carotenoids detected in root vegetables.

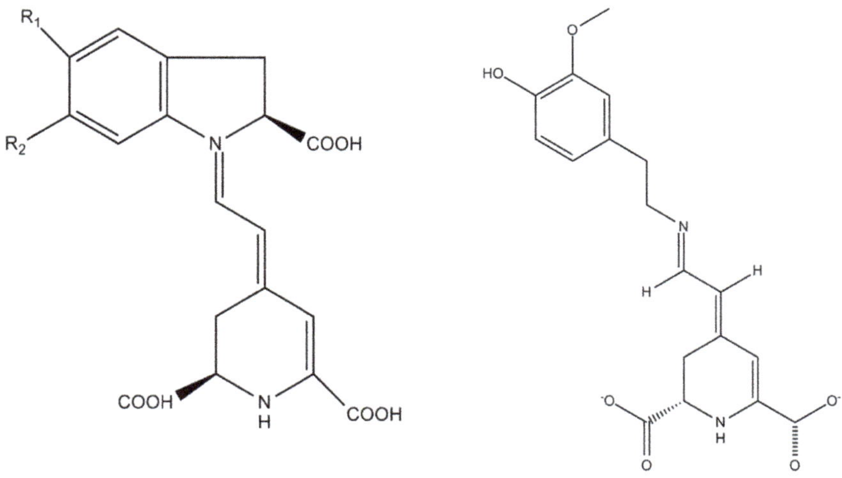

Figure 4. The main betalains detected in root vegetables.

Table 1. The main pigments isolated in various root vegetables.

Species	Edible Part	Color	Class of Compounds	Compounds	References
Potato (*Solanum tuberosum* L.)	Tuber (stem tuber)	Purple	Petunidin derivatives	Petunidin-3-*p*-coumaroylrutinoside-5-glucoside, petunidin-2-*p*-coumarylrutinoside-5-glucoside	[21–24]
		Red	Pelargonidin, delphinidin, cyanidin, peonidin, and malvidin acyl-glycoside derivatives	Pelargonidin-3-*p*-coumaroylrutinoside-5-glucoside	[21,23]
		Purple/red	Carotenoids	Neoxanthin, violaxanthin	[25]
		Yellow	Carotenoids	Antheraxanthin	[25]
Sweet potato (*Ipomoea batatas* L. Lam.)	Tuberous root (root tuber)	Purple	Acylated anthocyanins	Cyanidin, peonidin, and pelargonidin derivatives	[26–30]
Carrot (*Daucus carota* L. ssp. *sativus* Hoffm.)	Taproot (swollen hypocotyl and root)	Purple or black	Cyanidin derivatives	Acylated cyanidin 3-xylosyl(glucosyl)galactosides with sinapic acid, ferulic acid, and coumaric acid;	[31–35]
				Vinylphenol and vinylguaiacol adducts of cyanidin derivatives	[36]
		Red and yellow	Carotenoids	Lycopene and β-carotene	[37,38]
Beet root (*Beta vulgaris* L.)	Root (swollen hypocotyl and root)	Purple	Betalains	Betacyanins	[39,40]
		Yellow	Betalains	Betaxanthins	[41]
				Vulgaxanthin I and betanin	[42]
Yam (*Dioscurea* sp. L.)	Tuber (stem tuber)	Purple	Cyanidin, pelargonidin, and peonidin-type compounds; alatanins A–C	Cyanidin 3-hexoside acylated with two hydroxycinnamic acids, cyanidin 3-glycoside acylated with one hydroxycinnamic acid, cyanidin 3-glycoside acylated with one hydroxycinnamic acid, peonidin 3-glycoside acylated with one hydroxycinnamic acid, alatanin-C	[43–46]
		Yellow	Carotenoids	β-Carotene	[47]
Onion (*Allium cepa* L.)	Bulb (swollen basis of leaves)	Purple	Flavonols and acylated and non-acylated cyanidin glucosides	Dihydroflavonol taxifolin and its 3-, 7-, and 4′-glucosides	[48,49]
Radish (*Raphanus sativus* L.)	Taproot (swollen root and hypocotyl)	Purple	Cyanidin glucosides	Cyanidin 3-(glucosylacyl)acylsophoroside-5-diglucosides, cyanidin 3-sophoroside-5-diglucosides, cyanidin 3-sophoroside-5-glucosides, cyanidin 3-*O*-[2-*O*-(β -glucopyranosyl)-6-*O* -(trans-feruloyl)-β-glucopyranoside]-5-*O*-[6-*O* -(malonyl)-β-glucopyranoside] cyanidin 3-[2-(glucosyl)-6-(*cis*-*p*-coumaroyl)-glucoside]-5-[6-(malonyl)-glucoside]	[50,51]
		Red	Anthocyanins	Pelargonidin 3-sophoroside-5-glucoside, pelargonidin 3-[2-(glucosyl)-6-(*trans*-*p*-coumaroyl)-glucoside]-5-glucoside, pelargonidin 3-[2-(glucosyl)-6-(*trans*-feruloyl)-glucoside]-5-glucoside, pelargonidin 3-[2-(glucosyl)-6-(*trans*-*p*-coumaroyl)-glucoside]-5-(6-malonylglucoside), pelargonidin 3-[2-(glucosyl)-6-(*trans*-feruloyl)-glucoside]-5-(6-malonylglucoside), 3-*O*-[2-*O*-(b-ᴅ-glucopyranosyl)-6-*O*-(trans-caffeoyl)-b-ᴅ -glucopyr-anoside]-5-*O*-(6-*O*-malonyl-b-ᴅ-glucopyranoside), pelargonidin 3-*O*-[2-*O* -(b-ᴅ-glucopyranosyl)-6-*O*-(*cis*-*p*-cou-maroyl) -b-ᴅ-glucopyranoside]-5-*O*-(6-*O*-malonyl-b-ᴅ-glucopyranoside	[52,53]
Kohlrabi (*Brassica oleracea* var. *gongylodes*)	Swollen epicotyl	Purple	Cyanidin and cyanidin glucoside	Cyanidin-3-diglucoside-5-glucoside, cyanidin-3-(sinapoyl)-diglucoside-5-glucoside, cyanidin 3-(feruloyl) (sinapoyl) diglucoside-5-glucoside	[54–57]
Taro (*Colocasia esculenta*)	Corm	Purple		Cyanidin and pelargonidin glucosides	[58]

The first coloring agents used in food products to improve their visual appearance were produced from natural sources; however, the high cost for the production of these coloring agents, the variation in color shades due to the inert variability in natural matrix compositions, and the increasing needs of the market resulted in the use of synthetic compounds originally derived from coal tar and then produced from petroleum and crude oil (e.g., FD&C blue No. 1 and blue No. 2) [9,59]. The consumer concerns about additives and synthetic compounds, amplified by the reports regarding the health risks and the environmental impact associated with these compounds [60–62], necessitated the shift to the root food industry dyes; recently, the food and beverage industry is seeking natural coloring agents that could substitute synthetic dyes and coloring additives [63,64]. The colorant content of root vegetable products is associated with various health benefits including the prevention of modern chronic diseases [65–67]. However, they are often highly concentrated in the epidermal layers and skin tissues which are commonly discarded during domestic processing or in industrial applications [68–70]. For this reason, the research interest in obtaining natural pigments and bioactive compounds from agro-food waste is gaining ground within the context of circular economy and the sustainable use of natural resources [70–74]. There are also several cases where colorants can be found in high concentrations in the flesh due to the presence of pigments in parenchymal cells, increasing the antioxidant capacity and functional value of these products (e.g., potatoes, beets, carrots, and other root vegetables having colored flesh) [3]. Pigment compounds contribute to the overall antioxidant capacity in a dose-dependent and compound-specific manner [75–77], although the bioavailability and the absorption mechanisms within human body still need to be addressed [78]. Notwithstanding the genetic background of each species and/or cultivar, color attributes may be modulated by environmental factors such as the light and temperature conditions, through biotic and abiotic elicitors that may affect chemical composition, hormonal signaling, and enzymatic activities. Although not directly exposed to solar radiation, the pigmentation of root vegetables developing belowground may be indirectly modulated by the level and quality of radiation to which the aboveground plant is exposed [79]. In addition to pre-harvest factors, post-harvest conditions and processing methods may have an impact on bioavailability and biostability of natural matrices and coloring compounds [13,80,81]. Anthocyanins in particular are considered a good option as natural coloring agents due to their low toxicity and the wide range of health effects they present [82]. However, the stability and bioavailability of anthocyanins are affected by several factors (chemical structure and concentration, pH of food matrix, temperature, light, presence of co-pigments, enzymes, and metallic ions, among others), which determine the processing method specificity, and which need to be considered before using these compounds as natural coloring agents in the food industry [83,84]. Moreover, the association of structural differences with biostability and bioavailability is further reflected in the biological activities of these compounds, since, for example, acylated forms are less prone to degradation due to pH variations [14,85]. Therefore, although, for some species, there are already defined protocols for the extraction and processing requirements for obtaining natural colorants (e.g., black carrot, beet root colorants) [71,86–88], there is still a gap in the literature for other colored vegetables which could prove valuable candidates for yielding coloring agents.

The present review aims to present the main colored root vegetable crops, focusing on cultivars with colors atypical for the species. Special attention is given to blue- and purple-colored vegetables since natural colorants of these shades are less common in nature and are highly sought by the food industry, since blue shades are more difficult to replicate in food and beverages due to the susceptibility of coloring compounds to external factors (e.g., pH of the food matrix, extraction conditions). Furthermore, the main compounds responsible for uncommon colors are presented, as well as their antioxidant capacity and health-promoting effects. Finally, the effects of processing treatments on color stability are addressed. The presented information in this review was obtained from worldwide accepted databases such as Scopus, ScienceDirect, PubMed, Google Scholar, and ResearchGate, using the respective names of the studied species (both common and Latin names) and the additional terms of the main colorants and "health effects" as keywords.

2. Main Colored Root Vegetables

2.1. Potato

Potato (*Solanum tuberosum* L., Solanaceae) is the third most important food crop in the world, after wheat and rice [89]. In addition to its nutritional and calorific value, potato varieties also offer bioactive compounds with beneficial effects for human health, such as phenolic compounds and carotenoids, among others [12,23]. Several reports highlighted the beneficial effects of antioxidant-rich potatoes against various diseases, such as cardiovascular diseases [90] and various types of cancer [91,92]. Although yellow- and white-fleshed tubers are the most commonly used ones throughout the world, potato has the highest genetic diversity among cultivated species, with approximately 5000 known varieties with broad variability in terms of flesh and skin color [93]. Red- and blue-fleshed potatoes are particularly rich in phenolic compounds, presenting about three times higher amounts of total polyphenolic content than traditional yellow-fleshed tubers, as well as two to three times higher antioxidant activity [12,23,24,94].

Acylated forms of anthocyanins were reported to be the main compounds responsible for the red and purple flesh color of potatoes [94]. In particular, the deep-purple color of potato flesh and skin is associated with the presence of petunidin derivatives, although studies on metabolite profiling revealed a genotype- and tissue-specific pattern regarding the anthocyanin composition [22]. Petunidin was the major anthocyanidin compound found both in the flesh and the peel of purple potato varieties studied by Yine et al. [21]. In this study, petunidin accounted for 63–66% of the total anthocyanidin content of purple peel and flesh. The same findings were observed by Kita et al. [23] when studying purple- and red-fleshed potato cultivars, where petunidin-3-*p*-coumaroylrutinoside-5-glucoside was the major anthocyanin compound found in the purple-fleshed varieties Salad Blue (29.31 ± 0.73 mg·100 g^{-1} dry weight (dw)), Valfi (43.11 ± 0.37 mg·100 g^{-1} dw), and Blue Congo (36.32 ± 0.33 mg·100 g^{-1} dw). Similarly, Nemś et al. [24] identified petunidin-2-*p*-coumarylrutinoside-5-glucoside as the major anthocyanin present in the cultivars Salad Blue (28.34 ± 9.30 mg·100 g^{-1} dw), Valfi (57.77 ± 28.75 mg·100 g^{-1} dw), and Blue Congo (75.97 ± 12.38 mg·100 g^{-1} dw). On the other hand, in red-fleshed potatoes, pelargonidin acyl-glycoside derivatives appear as the main anthocyanin compounds. Kita et al. [23] found pelargonidin-3-*p*-coumaroylrutinoside-5-glucoside as the major anthocyanin present in red-fleshed varieties, such as Rosalinde (15.14 ± 0.12 mg·100 g^{-1} dw), Herbie 26 (44.46 ± 0.23 mg·100 g^{-1} dw), and Highland Burgundy Red (126.38 ± 0.71 mg·100 g^{-1} dw). Yin et al. [21] carried out an acid hydrolysis of the anthocyanins, studying the composition of the aglycones (anthocyanidins), reporting pelargonidin as the main anthocyanidin present in the red-fleshed cultivar Red Cloud No. 1, with a concentration of 11.73 ± 0.16 mg·100 g^{-1} fresh weight (fw), which corresponded to 82% of the total anthocyanidin content. Other anthocyanin compounds were reported in the literature for red- and purple-fleshed potatoes, including delphinidin, cyanidin, peonidin, and malvidin acyl-glycoside derivatives [21,23]. Moreover, the simulation of domestic cooking processing and gastrointestinal digestion of *Solanum tuberosum* L. cv Vitelotte noire extracts revealed significant antimicrobial and anti-proliferative activities against *Bacillus cereus* and *Escherichia coli* in the first case (domestic cooking processes) and colon (Caco-2 and SW48) and breast cancer (MCF7, MDA-MB-231) cell lines in the latter case (gastrointestinal digestion) [95].

Carotenoids are fat-soluble pigments that can exert antioxidant properties, and they are also present in colored-flesh potatoes. According to Kotíková et al. [25] who compared the carotenoid content of yellow-fleshed, white-fleshed, purple-fleshed, and red-fleshed potato cultivars, significant differences were observed. Interestingly, yellow potatoes showed a much higher average total carotenoid content (26.22 µg·g^{-1} dw) in comparison to the red and purple-fleshed cultivars (5.69 µg·g^{-1} dw), indicating that carotenoid pigments are not highly concentrated in the flesh of purple- and red-fleshed potatoes [25].

Yin et al. [21] investigated 10 colored potato cultivars from China and compared the composition and antioxidant activities of their flesh and peel. The authors found that potato peels were on average 15.34 times richer in anthocyanins than the flesh; the antioxidant activity of the peels extracts was also

5.75 times higher on average than that of the flesh extracts [21]. In the same study, the flesh extracts of cv. Purple Cloud No.1 showed the strongest antioxidant activity among all the tested varieties, along with the highest total content of anthocyanidins (43.38 mg·100 g^{-1} fw), a correlation which indicates anthocyanins as a major contributor to the antioxidant activity of colored potatoes [21].

Recently, there was increasing interest by consumers and food producers in colored potato varieties, due to their attractive organoleptic features (color and taste) and health-promoting chemical composition [12]. The increasing interest of the market for colored potato is stimulating private and public breeding programs to release new specialty potato cultivars such as the red-skin and red-flesh TerraRossa and AmaRosa or the purple-skin yellow-flesh cultivar Huckleberry Gold and Peter Wilcox, marketed as "Purple Sun" or "Blue Gold", which are also characterized by a higher content of anthocyanins, anthocyanidins, and other phenolic compounds [96]. The consumption of anthocyanin-rich food products such as purple-flesh potatoes is associated with the modification of the expression of various genes involved in the metabolism of lipids, inflammation, and energy homeostasis in liver and/or fat tissues [97,98]. Moreover, extracts from purple potato tubers may improve the differentiation of gut epithelia and its barrier function against gut epithelial inflammation through the activation of AMP-activated protein kinase (AMPK) and the increase of CDX2 gene [99]. Color-fleshed potatoes are an excellent source of bioactive compounds that are effective against human colon cancer cell lines (HCT-116 and HT-29); however, prolonged storage may affect their antiproliferative and pro-apoptotic activities [100]. Red- and purple-fleshed potato extracts were also effective against *tert*-butyl hydroperoxide (*t*-BHP)-induced hepatotoxicity through the recovery of serum alanine aminotransferase (ALT) and aspartate aminotransferase (AST) enzyme activities [101]. Therefore, a market niche for colored potato-based food products was created, such as potato chips and crisps. However, the frying process to produce colored potato crisps can cause a 38–70% degradation of anthocyanin compounds, with pelargonidin and malvidin acyl-glycoside derivatives being more stable during the frying process in comparison to petunidin acyl-glycoside derivatives [23]. Nevertheless, despite the reduced contents of anthocyanins in processed compared to raw potatoes, colored potato crisps can present bright attractive colors, in addition to 2–3 times higher antioxidant activities and 40% higher contents of polyphenols than standard snacks made of commonly used yellow potatoes and corn [23,24]. Moreover, in a recent study, Nemś and Pęksa [94] incorporated dried red- and purple-fleshed potatoes into fried snacks and doughs, reporting a beneficial effect on the inhibition of oxidative changes in lipids compared to control material (yellow snacks), particularly when incorporating material from purple-fleshed potato varieties of Blue Congo and Valfi. These effects were attributed to the higher content of colored snacks in polyphenols and anthocyanins than control, with petunidin 2-*p*-coumaroyl-rutinoside-5-glucoside being the major anthocyanin present in both cultivars [94]. Other domestic cooking processes such as boiling, baking, steaming, and microwaving may also affect the anthocyanin content and antioxidant capacity of colored potatoes, with processing (steaming and microwaving) showing the best results in retaining anthocyanin content and antioxidant activity [102–106]. Thermal processing affects not only anthocyanins but also carotenoids which are heat-sensitive and may be degraded, isomerized, or oxidized after domestic cooking processes [25]. According to Qiu et al. [107], anthocyanin content decreased with prolonged drying time and high drying temperatures due to higher degradation rates and shorter half-life values compared to shorter drying procedures with lower temperatures. Therefore, the antioxidant properties of colored potatoes can be beneficial not only to human health but also to the shelf life of processed food products. Another important aspect of processed food products based on processed colored potatoes is that the various types of processing (French fries, chips, and puree) reduce the content of antinutritional factors such as the glykoalkaloids α-chaconine and α-solanine, thus increasing the overall nutritional quality of the semi-processed and final products [108].

2.2. Sweet Potato

Sweet potato (*Ipomoea batatas* (L.) Lam., Convolvulaceae) is a perennial species native to Latin America which is highly appreciated for its fleshly tuberous roots that are widely used in the food and non-food industry depending on starch content and properties [109,110]. In Japan, purple sweet-potato anthocyanins are used as ingredients in several food products and beverages [111,112]. The flesh of the roots is usually white, yellow, or orange, although several cultivars with purple-colored flesh and a high content of anthocyanins also exist [113,114]. It is the fourth most produced vegetable in the world after potato, cassava, and tomato with a total production of 113 million tons in 2017, most of which (63.8%) were produced in China [115]. The nutritional value of the edible roots consists in the richness of carbohydrates, dietary fibers, vitamins, and minerals, while several polyphenolic compounds, peptides, and carotenoids are also present in considerable amounts in the flesh [116] and peels [74] of the tubers. The high calorific value of sweet potato roots makes the species one of the most important food crops in terms of calorific contributors to the human diet [117]. Starch is the main calorific component of sweet-potato tubers with significant variation in its structural and functional properties which depend mostly on the genotype and are not correlated with flesh color [118], although, using a proteomic approach, a recent study revealed that starch degradation may contribute to anthocyanin biosynthesis and accumulation in purple sweet-potato roots [119]. Chlorogenic acid, protocatechuic acid, salicylic acid, and caffeoylquinic acid derivatives are the main phenolic acids detected in purple sweet-potato roots and are responsible for their antioxidant capacity [48,120,121], while orange-fleshed sweet-potato cultivars are rich in provitamin A and also show significant antioxidant activity [113,122,123]. Moreover, in the study of Lebot et al. [124], the antioxidant activity of sweet-potato cultivars with purple, orange, and white flesh was correlated mostly with the presence of caffeoylquinic acid derivatives and less with total anthocyanin content, whereas, according to Oki et al. [8], the contribution of phenolic compounds in radical-scavenging activity is also dependent on the genotype. In contrast, according to the study of Kubow et al. [125], anthocyanins are responsible for the antioxidant capacity of sweet-potato tubers. In the same study, it was reported that the anthocyanin species were detected in the small intestinal and the ascending colonic vessel, depending on the sweet-potato genotype, and the antioxidant activity was increased accordingly [125]. According to the report of Meng et al. [126] who studied the digestion kinetics of sweet-potato polyphenols, the maximum release was recorded 2 h after intestinal digestion and was induced by gastric acid and pepsin [126]. Moreover, acylated anthocyanins from sweet potato are considered as complex and less susceptible to intestinal degradation [127,128], while Sun et al. [129] suggested a prebiotic-like activity of anthocyanins through the modulation of microbiota in the intestine. These results highlight the importance of unraveling the bioavailability and bioaccessibility patterns influencing the antioxidant potential of purple-fleshed sweet potatoes [125].

Acylated anthocyanins are responsible for the intense color of purple-fleshed sweet potatoes [66,130], which renders them good candidates sources for natural colorants with practical application in the food industry [131]. Moreover, peels are also a good source of natural pigments since they contain significant amounts of anthocyanins, and the exploitation of this by-product for obtaining coloring agents would increase the added value of the sweet-potato crop [74]. The total anthocyanin content and compositional profile may differ among the various genotypes, with a total of 39 different anthocyanins isolated so far [132,133]. The main anthocyanins isolated from purple sweet-potato extracts were identified as cyanidin, peonidin, and pelargonidin derivatives [26–30,110,134,135], which were effective against alcohol-induced liver injury in rats when administrated at median doses (100 mg·kg^{-1} body weight), whereas higher doses (300 mg·kg^{-1} body weight) had a pro-oxidant effect and promoted liver injury [136]. Moreover, cyanidin 3-caffeoyl-*p*-hydroxybenzoyl sophoroside-5-glucoside which was isolated from purple-fleshed sweet potatoes was shown to be effective both in vitro and in vivo in inhibiting hepatic glucose secretion and reducing blood glucose [137–139], while peonidin suppressed the excessive expression of the HER2 protein showing anticancer activities [140]. According to Luo et al. [141], cyanidin 3-caffeoyl-feruloyl sophoroside-5-glucoside and peonidin 3-dicaffeoyl sophoroside-5-glucoside were the most effective anthocyanins isolated from the purple sweet-potato cultivar Eshu No. 8. In another study, the oral administration of purple sweet-potato color attenuated cognitive deficits in domoic acid-treated mice through mitochondrial biogenesis signaling and the decrease of p47phox and gp91phox expression [142], while similar results were reported by Zhuang et al. [143], who suggested the regulation of AMPK/autophagy signaling as the mechanism of action. The same pigment was effective against neuroinflammation in mouse brain through the inhibition of mitogen-activated protein kinase (MAPK) and the activation of nuclear factor κB (NF-κB) [144], as well as against bladder cancer through the inhibition of the signaling of phosphatidylinositol-4,5-bisphosphate 3-kinase/Akt or protein kinase B (PI3K/Akt) [145]. In particular, for mitochondrial biogenesis, it was reported that anthocyanins can bind and stimulate estrogen receptor-α and then increase the expression of nuclear respiratory factor-1 (NRF-1) [146]. Anthocyanin-rich extracts from purple sweet potato were moderately effective against human colon cancer cell lines (HCT-116 and HT-29) through the inhibition of tyrosine kinase activity, whereas they showed no effectiveness against the CCD-33Co cell line [67]. Moreover, Yoshimoto et al. reported that the antimutagenic activity of sweet-potato extracts was attributed mainly to cyanidin content (63% inhibition of mutagenicity of Trp-1 against *Salmonella tymphimurium* TA 98 at the dose of 1.5 mM) [147], while Zhao et al. suggested that anthocyanin-rich extracts from sweet potato are potent anti-aging (at the dose of 1000 mg/kg body weight), anti-hyperglycemic (at the dose of 1 g/kg body weight), and anti-tumor agents (68% tumor inhibition at the dose of 1000 mg/kg body weight) [148]. In another study, highly acylated anthocyanins showed effectiveness against hyperuricemia and kidney inflammation in allopurinol-induced hyperuricemic mice [149], while purple sweet-potato color reduced renal damage through the downregulation of vascular endothelial growth factor receptor (VEGFR2) expression [150]. The regular intake of anthocyanins is also highly associated with the prevention of various chronic liver diseases, and it can reduce lipid accumulation in liver tissues and alleviate oxidative stress and hepatic inflammation [25,102,151–156]. Other hepatoprotective effects of purple sweet potatoes include hepatic insulin resistance in high-fat diet-treated mice through the decrease of reactive oxygen species (ROS) production and the inhibition of endoplasmic reticulum (ER) liver stress (administration of purple sweet-potato color at the dose of 700 mg/kg/day) [157], through the decrease in the expression of ionized calcium-binding adapter molecule 1 (Iba1), tumor necrosis factor-α, interleukin-1β, suppressors of cytokine signaling3 (SOCS3), and galectin-3 (administration of purple sweet potato color at the dose of 500 mg/kg/day) [158], or through the inhibition of nucleotide-binding domain, leucine-rich repeat (NLR) family, pyrin domain containing 3 (NLRP3) inflammasome activation (administration of purple sweet potato color at the dose of 700 mg/kg/day) [159]. Moreover, the combinative use of black soybean and purple sweet potato (mixtures of 2:2 for black soybean and purple sweet potato) resulted in improved insulin sensitivity in streptozotocin-induced diabetic rats through the improvement of insulin and insulin receptor substrate-1 (IRS-1) expression, the increase of superoxide dismutase (SOD) levels, and reduced pancreatic necrosis [160]. In a similar study, the mixture of *Curcuma longa* L. and sweet potato (at the dose of 2–5 mg/kg body weight) showed significant immunomodulating properties in murine leukemia retrovirus-infected mice [161]. The administration of purple sweet potato to obese mice fed with a high-fat diet exhibited anti-obesity effects and attenuated gain weight [162]. Other bioactive compounds of purple sweet potatoes include alkali-soluble polysaccharides which presented anti-inflammatory properties in lipopolysaccharide (LPS)-treated macrophages (RAW 264.7) through the inhibition of nitric oxide, interleukin (IL)-6, IL-1β, and tumor necrosis factor alpha (TNF-α) and the increase of IL-10 [163], as well as anti-inflammatory effects against intestinal inflammation on dextran sulfate sodium (DSS)-induced mice [164], hepatoprotective properties [165], and immunomodulatory effects [166–168]. Non-flavonoid compounds and kaempferol derivatives are also present in sweet-potato tuber tissues, and they contribute to the overall bioactive capacity of sweet potato [28].

Processing and storage conditions are important for the chemical composition and the visual quality of sweet-potato tubers, with heating treatments and higher pH having a detrimental effect on anthocyanins and starch content and on flesh color [139,169–172]. Pretreatments such as blanching, osmotic dehydration, ultrasound-assisted dehydration, and ultrasound-assisted osmotic dehydration before microwave drying also had an impact on total phenolic and anthocyanin content of orange- and purple-fleshed sweet-potato slices [173]. Domestic cooking processes may also affect total anthocyanin and total phenolic content, with steaming suggested as the mildest process to retain the highest amount of total anthocyanins compared to fresh samples, while, at the same time, an increase in total phenolic content was observed by Phan et al. [174]. In a similar study, steaming, roasting, and boiling were suggested as the best cooking methods for retaining total phenolic, anthocyanin, and carotenoid content, respectively, in white, yellow, orange, light-purple, and deep-purple sweet-potato tubers [10].

2.3. Carrot

Carrot (*Daucus carota* L. ssp. *sativus* Hoffm.) belongs to the Apiaceae family and is a highly appreciated vegetable consumed for its edible fleshy roots. There are two cultivar groups depending on root color, namely, the carotene or western carrot (*Daucus carota* ssp. *sativus* var. *sativus*) and the eastern or anthocyanin carrot (*Daucus carota* ssp. *sativus* var. *atrorubens* Alef.), which are widely cultivated throughout the world with an annual production of 42.8 million tons including turnips [115]. Although the orange-colored carrots are the most popular ones, a broad genetic basis exists with many other shades of root flesh (red, white, yellow, black, purple, or multi-color) which attract interest due to their nutritional value and associated health effects [175]. Recently, new genetically biofortified cultivars were developed which contain not only α- and β-carotene but also anthocyanins and lycopene [176]. In particular, for black or purple carrots, several research reports highlighted their beneficial health effects on human health, and they are widely used so far as natural sources of blue color and functional ingredients in the food industry [33,40].

Black carrots contain high amounts of mono-acylated anthocyanins which are less prone to thermal degradation, while they can retain their color at various pH values and storage conditions [31,32,34]. These functional and structural characteristics of colored carrot pigments make them good candidates for the extraction of natural colorant agents with practical applications in the food industry, especially in food products with low pH, in beverages and confectioneries [40,131,177]. However, despite their stability under various conditions, Espinosa-Acosta et al. [82] did not suggest their use in food models such as yoghurt and jelly, except for the case of ethanolic extracts of black carrots, which could be incorporated into jellies to increase the antioxidant activity of the final product. Moreover, Assous et al. [86] suggested the use of black-carrot pigments as coloring agents in hard candy and sweet jelly without significant differences in the sensorial profile compared to the control, while the same pigments protected sunflower oil from lipid peroxidation. The use of black-carrot extracts was also proposed for the preparation of jams and marmalades, where the main anthocyanins were slightly affected after gastric ingestion and storage at 4 °C [178], as well in co-pigmentation with other natural colorants (e.g., plum, jamun, strawberry, and pomegranate juices), to increase the color stability to heat treatments and pH variation [179]. On the other hand, red and yellow carrots are rich in carotenoids and lycopene and β-carotene in particular [37,38], which, according to Horvitz et al. [180], are both bioavailable and can provide a significant amount of these carotenoids to human diet.

The main anthocyanins detected are mostly cyanidin derivatives, and, according to Frond et al. [48], the most abundant anthocyanin identified in black-carrot extracts was cyanidin-3-(*p*-coumaroyl)-diglucoside-5-glucoside. In the study of Montilla et al. [33], the main detected anthocyanins in *Daucus carota* subsp. *sativus* var. *atrorubens* Alef. were identified as acylated cyanidin 3-xylosyl(glucosyl)galactosides with sinapic acid, ferulic acid, and coumaric acid, and significant differences were observed between genotypes (Antonina, Beta Sweet, Deep Purple, and Purple Haze) in terms of total and individual anthocyanin content. Similar results were reported in the earlier study of Kammerer et al. [181], with acylated and non-acylated cyanidin derivatives found in the highest amounts, while they also suggested significant differences between 15 different black-carrot cultivars, as well as between roots of the same cultivar. Moreover, in black-carrot juice, two more compounds were identified, namely, cyanidin-3-xylosyl-galactoside and cyanidin-3-xylosyl (feruloylglucosyl)galactoside [35], while Schwarz et al. [36] isolated four more pigments identified as vinylphenol and vinylguaiacol adducts of cyanidin derivatives which are formed during the storage of juice through the reaction of phenolic acids with anthocyanins. Regarding the health effects of anthocyanins, extracts from purple carrot were moderately effective against HCT-116 human colon cancer cell lines through the inhibition of tyrosine kinase activity, whereas they showed no effectiveness against HT-29 and CCD-33Co cell lines [67]. Yet, black-carrot crude extracts exhibited significant antioxidant, cytoprotective, and anti-angiogenic properties, indicating a synergistic effect of the various polyphenols (anthocyanins, phenolic acids, and flavonoids) contained in the root extracts [182]. Although non-digested purple carrot extract is more potent than the digested extract, Olejnik et al. [183] reported that gastrointestinal digested purple-carrot extract had intracellular ROS-inhibitory activity and protected colonic cells against oxidative stress by reducing oxidative DNA damage by 20.7%. According to Blando et al. [184], the anthocyanin-rich extracts from black carrots contained mostly anthocyanins acylated with cinnamic-acid derivatives, which exhibited anti-inflammatory activities through the reduction of the expression of endothelial inflammatory antigens. Apart from anthocyanins, black carrots are a good source of phenolic acids, namely, chlorogenic and caffeic acids, which contribute to the overall antioxidant capacity [48,185].

Processing may affect the chemical composition and antioxidant properties of black-carrot juice, and the use of pectinase during maceration increased the total anthocyanin content, the overall antioxidant capacity, and the juice yield of enzyme-treated compared to normally pressed juice [186,187]. The use of ultrasound and mild heating (50 °C) may increase the extraction yield of anthocyanins from black-carrot pomace, especially the content of cyanidin-3-xyloside-galactoside-glucoside-ferrulic acid and cyanidin-3-xyloside-galactoside, which were detected in the highest amounts [71]. Another processing treatment which could increase the bioavailability and stability of anthocyanins in black-carrot-based food products is the microencapsulation of anthocyanin-rich extracts [83]. Moreover, wounding stress may increase anthocyanin content, chlorogenic acid in particular, thus improving the nutritional and functional value of the obtained food products [188].

2.4. Beet Root

Beet or table beet (*Beta vulgaris* L.) belongs to the Amaranthaceae family and is commonly used for its edible fleshy red roots and tender leaves. Beet roots contain betalains, a class of compounds which is further divided into betacyanins and betaxanthins [39]. The composition of betalains and the ratio of betacyanins to betaxanthins depends on tyrosine production and its conversion to betalains, with significant differences observed between red and yellow beet roots [189]. The most commonly found betacyanins are betanins which are responsible for the red vivid color of beet roots, and they are water-soluble and sensitive to prolonged heating [40]. Betanins are commercially available as color additives (E162) in powder form or as juice concentrates following vacuum evaporation [39]; however, there is a great diversity in flesh color among the beet-root genotypes with variable intensities of red color or other shades ranging from white to orange. Apart from the genotype, color intensity is also affected by growing conditions and maturity stage, storage conditions, and processing treatments [88,190,191]. Beet roots with yellow color are most abundant in betaxanthins, while betacyanins are present in lesser amounts [41]. In the study of Wettasinghe et al. [192], beet-root genotypes with diverse flesh colors exhibited significant differences in antioxidant activity and in phase II enzyme induction capacity, which is associated with cancer chemoprotective effects [192]. Moreover, Lee et al. [42] reported that betanine and betaine extracted from red- and yellow-colored beet roots were effective against HepG2 cell proliferation in a dose-dependent manner. In the same study, the main identified betalains detected in the cultivar with yellow roots (Burpee's Golden Globe) were vulgaxanthin I and betanin [42]. Vulić et al. [193] also reported that the beet-root pomace, a by-product of the beet-root juice extraction, has a high content of betalains and phenolic compounds which exhibited in vitro antiradical activities against 2,2-diphenyl-1-picrylhydrazyl (DPPH) radicals and in vivo antioxidant and hepatoprotective activity, suggesting that it could be used as an excellent nutraceutical resource or an ingredient of functional food products.

2.5. Yam

Yam includes various species of the *Dioscorea* genus (Dioscoreaceae), although sometimes it is confused with other root vegetables such as sweet potatoes, oca, taro, etc., which locally may be referred to as yams [194]. Tuber flesh color can be white, yellow, red, or purple depending on the cultivar, with significant differences in bioactive compound content and antioxidant properties [194,195]. Purple yam or water yam (*Dioscorea alata purpurea*) is usually cultivated in tropical and subtropical regions of the world, and its edible roots are very rich in starch and amylose [196], although a great variation in chemical composition of the edible parts of the species was reported [197]. Resistant starch from purple yam (*D. alata*) was effective against hyperlipidemia in high-fat diet-fed hamsters through the amelioration of lipid metabolism and the modulation of gut microbiota [196,198]. Moreover, extracts from roots significantly reduced blood glucose levels in Wistar rats with alloxan-induced hyperglycemia [199] or cholesterol (total and low-density lipoprotein (LDL)) and triglycerides in hypercholesterolemic rats [200], ameliorated doxorubicin (DOX)-induced cardiotoxicity [201], showed protective effects against aniline-induced spleen toxicity [202] and in vivo anti-inflammatory activities against λ-carrageenan-induced paw edema in mice [203], and could be used as an adjuvant in bone-marrow-derived dendritic cell (DC)-based vaccines for cancer therapy [204]. *D. alata* root extracts may also alleviate cellular fibrosis through the downregulation of the transforming growth factor-beta (TGF-β)/Smad signaling pathway and the modulation of epithelial–mesenchymal transition (EMT) expression in kidneys [205]. On the other hand, according to Chan et al. [206], root extracts are also effective against CCl_4-induced liver injury and hepatic fibrosis. Other health effects include the improvement in function of large bowel and modulation of fecal microflora [207], beneficial effects in gastrointestinal function [208] and cognitive ability [209,210], and the activation of the immune system [211]. The root color of purple yam (*D. alata*) is attributed to the high content of anthocyanins which exhibit significant antibacterial activities [212], anti-inflammatory effects on trinitrobenzenesulfonic acid (TNBS)-induced colitis in mice [213], antiglycative properties [214], and antidiabetic properties [215,216]. The main detected anthocyanins in this species were identified as cyanidin, pelargonidin, and peonidin-type compounds and alatanins A–C [43]; however, the individual compound profile and the overall anthocyanin content are affected by maturity stage and the expression of the concomitant genes [44]. Apart from *D. alata*, which is considered the main purple yam, there are also cultivars of *D. trifida* or cush-cush yam which contain peonidin, cyanidin, and pelargonidin aglycones [45]. Other compounds with bioactive properties are also present, namely, phenolic acids such as ferulic, sinapic, vanillic, caffeic acid, and *p*-coumaric acid, and others, which presented immunomodulatory properties [217,218], proteins with estrogen-stimulating activities that may relieve menopausal syndrome [219], allantoin and dioscin [220], dioscorin [221], or β-sitosterol and ethyl linoleate with anti-atherosclerotic activity [222]. On the other hand, carotenoids and β-carotene in particular are responsible for the root color of yellow yam (*D. cayennensis*) [47]. Yam roots may contain antinutritional factors such as tannins and diosgenin, which also present bioactive properties. For example, antidyslipidemic effects were reported for diosgenin extracts from purple and yellow yams without affecting body weight gain [220,223], while diosgenin and furostanol glycosides and spirostanol glycosides were effective against the proliferation of various cancer cell lines (MCF-7, A 549, and Hep G2) [224].

A very common use of purple yam is the substitution of wheat flour for bakery products and food products in general without affecting the sensorial acceptance of the products by consumers [225–227], while yam flour can be used for gluten-free bakery products [228].

2.6. Onion

Onion (*Allium cepa* L., Alliaceae) is one of the most important species of the *Allium* genus, which is commonly used for its edible bulbs. There is a great number of cultivars available with a great diversity in color, which usually refers to bulb skin color, since, in most cases, the presence of pigments is limited to the outer skins of the bulb [229]. In many countries, onion bulbs are considered the main dietary source of flavonoids, a high proportion of which is attributed to the anthocyanin content [70,230]. However, most of the studies refer to red-onion cultivars which contain various polyphenols including acylated and non-acylated cyanidin glucosides, and less information is available about the profile of anthocyanins in purple onions [48,49]. The biosynthesis of anthocyanins involves the shikimate pathway and the activity of anthocyanidin synthase, which catalyzes the production of anthocyanidins, and, after further enzymatic reactions, the various anthocyanins are produced [5]. Comparing green, yellow, red, and purple onion, Benkeblia [231] observed higher total phenolic content and antioxidant properties in red and purple onion-bulb extracts. Similar results were reported by Zhang et al. [232] in a study comparing white, yellow, and red onion, with the latter showing considerably higher total anthocyanin, flavonoid, and polyphenol content, which was also correlated to high antioxidant activity measured through DPPH, 2,2′-azino-bis(3-ethylbenzothiazoline-6-sulfonic acid (ABTS$^{\bullet+}$), and fluorescence recovery after photobleaching (FRAP) assays. Bulb extracts are potent bioactive natural matrices, and, according to the study of Oboh et al. [233], extracts of purple onion were effective against angiotensin-converting enzyme, α-amylase, and α-glucosidase activity, showing significant antidiabetic and anti-hypertensive effects. Moreover, skins of pearl onion exhibited significant anti-inflammatory properties and inhibitory effects against radical-induced DNA scission [234]. In terms of antioxidant activity, purple onions exhibited higher oxygen radical absorbance capacity (ORAC) values than white onions, which indicates a higher concentration of bioactive compounds [235]. A preliminary study conducted by Khiari et al. [236] suggested that, depending on the quality of the plant residues, onion solid waste, also constituted primarily by the outer dry layers of the bulbs, may be used to extract polyphenols with potential antioxidant activity, and the yield of total polyphenols can be optimized using ethanol extracts, with extraction time up to 6 h, while maintaining relatively low extraction temperature (40 °C gave better results than 60 °C).

2.7. Other Root Vegetables

Radish (*Raphanus sativus* L., Brassicaceae) is a cruciferous species, well known for its normally white edible fleshy hypocotyls which come in different shapes, sizes, and skin colors. Apart from white-fleshed cultivars, there are also genotypes with pink and purple hypocotyls due to the presence of pigments in the xylem [50]. Pigmentation may also change with the hypocotyl development stage [237]. Purple color implies the presence of anthocyanins, and, according to the study of Reference [51], 60 different compounds were detected and identified as cyanidin glucosides. Most of the anthocyanins are present in acylated forms of cyanidin glucosides which increase their stability, and they could be easily used as natural colorants in functional foods [7,50,238,239], while root extracts may also exhibit beneficial health effects against gastric injuries [240].

Purple kohlrabi (*Brassica oleracea* var. *gongylodes*) is another species of the Brassicaceae family with intense purple color, whose edible part is the swollen fleshy meristem. The pigments are located in the meristem skin and consist of cyanidin and cyanidin glucosides which are responsible for the strong antioxidant properties of the species [54–57]. Examining the antioxidant activity of kohlrabi ethanol and water extracts, Pak et al. [241] observed strong DPPH radical-scavenging activity, and purple kohlrabi extracts had higher antioxidant capacity compared to green kohlrabi extracts. Similarly, comparing green and red kohlrabi, Jung et al. [242] observed that the latter had double the total phenolic content, as well as a higher antioxidant (DPPH, ABTS, and peroxynitrite scavenging activity assay (ONOO$^-$)) effect compared to green kohlrabi. In the same study, red kohlrabi methanol extract had stronger anti-inflammatory, antidiabetic, and antioxidant effects than the green kohlrabi methanol extract.

Taro (*Colocasia esculenta* L.) is a root vegetable of the Araceae family with great genetic diversity in plant morphology, including the color of corm flesh, which can be white, purple, brown, or blackish [57,243,244]. The main detected anthocyanins were identified as cyanidin and pelargonidin glucosides, and they exhibit significant antioxidant and anti-inflammatory activities [58].

3. Conclusions

Root vegetables with intense and uncommon colors are very important in the human diet, not only because they increase the overall intake of health-promoting compounds, but also because they diversify the daily diet in terms of color, flavor, and chemical composition, which imparts distinct functional effects on the human body. The inclusion of such root vegetables either raw in fresh salads or in cooked dishes may increase palatability and appeal for healthier food products, although proper marketing is always an issue since consumers are usually reluctant to introduce new flavors and unconventional products that break the mold. Nevertheless, the current trends in the food and beverage market and the increased public demand for substituting synthetic compounds with natural alternatives could boost the establishment of these species and help the crossing over from niche products to widely accepted ones with diverse uses in the food industry. Further research is also needed in order to (i) identify those correlations and mechanisms of action responsible for the antioxidant properties and health effects of the pigmented vegetables, (ii) evaluate agronomic practices that will increase the bioactive capacity of the final products through the improved pigmentation, (iii) study post-harvest and processing treatments that will make these compounds less prone to degradation and easier to use in the design of functional foods and as natural coloring agents, and (iv) define efficient extraction protocols that will allow high yields and high stability and quality of coloring agents extracted from plant sources. Finally, increasing the knowledge about the chemical composition and the health effects of individual compounds of colored root vegetables could be further exploited through breeding programs for the production of elite genotypes with increased content of coloring compounds and tailor-made health effects, as well as through plant in vitro strategies for the production of specific natural secondary metabolites and further use in the pharmaceutical and the food and beverage industries.

Author Contributions: All authors contributed equally to the writing of the original draft and final draft preparation.

Funding: This work was supported by the USDA National Institute of Food and Agriculture and Hatch Appropriations under Project #PEN04723 and Accession #1020664; The authors are grateful to the Foundation for Science and Technology (FCT, Portugal) and FEDER under Program PT2020 for financial support to CIMO (UID/AGR/00690/2019), to FEDER-Interreg España-Portugal programme for financial support through the project 0377_Iberphenol_6_E and TRANSCoLAB 0612_TRANS_CO_LAB_2_P, and to the European Regional Development Fund (ERDF) through the Regional Operational Program North 2020, within the scope of project Mobilizador Norte-01-0247-FEDER-024479: Valor Natural®.

Conflicts of Interest: The authors declare no conflicts of interest.

References

1. Sasaki, N.; Nishizaki, Y.; Ozeki, Y.; Miyahara, T. The role of acyl-glucose in anthocyanin modifications. *Molecules* **2014**, *19*, 18747–18766. [CrossRef] [PubMed]
2. Kyriacou, M.C.; Rouphael, Y. Towards a new definition of quality for fresh fruits and vegetables. *Sci. Hortic. (Amst.)* **2018**, *234*, 463–469. [CrossRef]
3. Chaves-Silva, S.; dos Santos, A.L.; Chalfun-Júnior, A.; Zhao, J.; Peres, L.E.P.; Benedito, V.A. Understanding the genetic regulation of anthocyanin biosynthesis in plants—Tools for breeding purple varieties of fruits and vegetables. *Phytochemistry* **2018**, *153*, 11–27. [CrossRef] [PubMed]
4. Menzies, I.J.; Youard, L.W.; Lord, J.M.; Carpenter, K.L.; van Klink, J.W.; Perry, N.B.; Schaefer, H.M.; Gould, K.S. Leaf colour polymorphisms: A balance between plant defence and photosynthesis. *J. Ecol.* **2016**, *104*, 104–113. [CrossRef]

5. Khandagale, K.; Gawande, S. Genetics of bulb colour variation and flavonoids in onion. *J. Hortic. Sci. Biotechnol.* **2019**, *94*, 522–532. [CrossRef]
6. Wrolstad, R.E.; Culver, C.A. Alternatives to Those Artificial FD&C Food Colorants. *Annu. Rev. Food Sci. Technol.* **2012**, *3*, 59–77.
7. Giusti, M.M.; Wrolstad, R.E. Acylated anthocyanins from edible sources and their applications in food systems. *Biochem. Eng. J.* **2003**, *14*, 217–225. [CrossRef]
8. Oki, T.; Masuda, M.; Furuta, S.; Nishiba, Y.; Terahara, N.; Suda, I. Involvement of anthocyanins and other phenolic compounds in radical-scavenging activity of purple-fleshed sweet potato cultivars. *J. Food Sci.* **2002**, *67*, 1752–1756. [CrossRef]
9. Newsome, A.G.; Culver, C.A.; Van Breemen, R.B. Nature's palette: The search for natural blue colorants. *J. Agric. Food Chem.* **2014**, *62*, 6498–6511. [CrossRef]
10. Tang, Y.; Cai, W.; Xu, B. Profiles of phenolics, carotenoids and antioxidative capacities of thermal processed white, yellow, orange and purple sweet potatoes grown in Guilin, China. *Food Sci. Hum. Wellness* **2015**, *4*, 123–132. [CrossRef]
11. Singh, B.K.; Koley, T.K.; Maurya, A.; Singh, P.M.; Singh, B. Phytochemical and antioxidative potential of orange, red, yellow, rainbow and black coloured tropical carrots (*Daucus carota* subsp. *sativus* Schubl. & Martens). *Physiol. Mol. Biol. Plants* **2018**, *24*, 899–907. [PubMed]
12. Rytel, E.; Tajner-czopek, A.; Kita, A.; Aniołowska, M.; Kucharska, A.Z.; Hamouz, K. Content of polyphenols in coloured and yellow fleshed potatoes during dices processing. *Food Chem.* **2014**, *161*, 224–229. [CrossRef] [PubMed]
13. Khan, M.I.; Giridhar, P. Plant betalains: Chemistry and biochemistry. *Phytochemistry* **2015**, *117*, 267–295. [CrossRef] [PubMed]
14. Khoo, H.E.; Azlan, A.; Tang, S.T.; Lim, S.M. Anthocyanidins and anthocyanins: Colored pigments as food, pharmaceutical ingredients, and the potential health benefits. *Food Nutr. Res.* **2017**, *61*, 1361779. [CrossRef] [PubMed]
15. Merhan, O. The Biochemistry and Antioxidant Properties of Carotenoids. In *Carotenoids*; Cvetkovic, D., Nikolic, G., Eds.; IntechOpen: London, UK, 2016; pp. 51–66.
16. Young, A.J.; Lowe, G.L. Carotenoids—Antioxidant properties. *Antioxidants* **2018**, *7*, 28. [CrossRef] [PubMed]
17. Indra, V.K. *Phytochemicals in Plant Cell Cultures*; Vasil, I.K., Ed.; Elsevier: Amsterdam, The Netherlands, 1988.
18. Rahimi, P.; Abedimanesh, S.; Mesbah-Namin, S.A.; Ostadrahimi, A. Betalains, the nature-inspired pigments, in health and diseases. *Crit. Rev. Food Sci. Nutr.* **2019**, *59*, 2949–2978. [CrossRef]
19. Miguel, M.G. Betalains in some species of the amaranthaceae family: A review. *Antioxidants* **2018**, *7*, 53. [CrossRef]
20. Belhadj Slimen, I.; Najar, T.; Abderrabba, M. Chemical and antioxidant properties of betalains. *J. Agric. Food Chem.* **2017**, *65*, 675–689. [CrossRef]
21. Yin, L.; Chen, T.; Li, Y.; Fu, S.; Li, L.; Xu, M.; Niu, Y. A comparative study on total anthocyanin content, composition of anthocyanidin, total phenolic content and antioxidant activity of pigmented potato peel and flesh. *Food Sci. Technol. Res.* **2016**, *22*, 219–226. [CrossRef]
22. Oertel, A.; Matros, A.; Hartmann, A.; Arapitsas, P.; Dehmer, K.J.; Martens, S.; Mock, H.P. Metabolite profiling of red and blue potatoes revealed cultivar and tissue specific patterns for anthocyanins and other polyphenols. *Planta* **2017**, *246*, 281–297. [CrossRef]
23. Kita, A.; Bąkowska-Barczak, A.; Hamouz, K.; Kułakowska, K.; Lisińska, G. The effect of frying on anthocyanin stability and antioxidant activity of crisps from red- and purple-fleshed potatoes (*Solanum tuberosum* L.). *J. Food Compos. Anal.* **2013**, *32*, 169–175. [CrossRef]
24. Nemś, A.; Pęksa, A.; Kucharska, A.Z.; Sokół-Łętowska, A.; Kita, A.; Drozdz, W.; Hamouz, K. Anthocyanin and antioxidant activity of snacks with coloured potato. *Food Chem.* **2015**, *172*, 175–182. [CrossRef] [PubMed]
25. Kotíková, Z.; Šulc, M.; Lachman, J.; Pivec, V.; Orsák, M.; Hamouz, K. Carotenoid profile and retention in yellow-, purple- and red-fleshed potatoes after thermal processing. *Food Chem.* **2016**, *197*, 992–1001. [CrossRef] [PubMed]
26. Ge, J.; Hu, Y.; Wang, H.; Huang, Y.; Zhang, P.; Liao, Z.; Chen, M. Profiling of anthocyanins in transgenic purple-fleshed sweet potatoes by HPLC-MS/MS. *J. Sci. Food Agric.* **2017**, *97*, 4995–5003. [CrossRef] [PubMed]

27. Hu, Y.; Deng, L.; Chen, J.; Zhou, S.; Liu, S.; Fu, Y.; Yang, C.; Liao, Z.; Chen, M. An analytical pipeline to compare and characterise the anthocyanin antioxidant activities of purple sweet potato cultivars. *Food Chem.* **2016**, *194*, 46–54. [CrossRef]

28. Pacheco, M.T.; Escribano-Bailón, M.T.; Moreno, F.J.; Villamiel, M.; Dueñas, M. Determination by HPLC-DAD-ESI/MSn of phenolic compounds in Andean tubers grown in Ecuador. *J. Food Compos. Anal.* **2019**, *84*, 103258. [CrossRef]

29. Wang, A.; Li, R.; Ren, L.; Gao, X.; Zhang, Y.; Ma, Z.; Ma, D.; Luo, Y. A comparative metabolomics study of flavonoids in sweet potato with different flesh colors (*Ipomoea batatas* (L.) Lam). *Food Chem.* **2018**, *260*, 124–134. [CrossRef]

30. Zhang, J.L.; Luo, C.L.; Zhou, Q.; Zhang, Z.C. Isolation and identification of two major acylated anthocyanins from purple sweet potato (*Ipomoea batatas* L. cultivar Eshu No. 8) by UPLC-QTOF-MS/MS and NMR. *Int. J. Food Sci. Technol.* **2018**, *53*, 1932–1941. [CrossRef]

31. Gizir, A.M.; Turker, N.; Artuvan, E. Pressurized acidified water extraction of black carrot [*Daucus carota* ssp. *sativus* var. *atrorubens* Alef.] anthocyanins. *Eur. Food Res. Technol.* **2008**, *226*, 363–370. [CrossRef]

32. Kirca, A.; Özkan, M.; Cemeroğlu, B. Effects of temperature, solid content and pH on the stability of black carrot anthocyanins. *Food Chem.* **2007**, *101*, 212–218. [CrossRef]

33. Türker, N.; Erdoğdu, F. Effects of pH and temperature of extraction medium on effective diffusion coefficient of anthocynanin pigments of black carrot (*Daucus carota* var. L.). *J. Food Eng.* **2006**, *76*, 579–583. [CrossRef]

34. Montilla, E.C.; Arzaba, M.R.; Hillebrand, S.; Winterhalter, P. Anthocyanin composition of black carrot (*Daucus carota* ssp. *sativus* var. *atrorubens* Alef.) cultivars antonina, beta sweet, deep purple, and purple haze. *J. Agric. Food Chem.* **2011**, *59*, 3385–3390. [CrossRef] [PubMed]

35. Garcia-Herrera, P.; Pérez-Rodríguez, M.L.; Aguilera-Delgado, T.; Labari-Reyes, M.J.; Olmedilla-Alonso, B.; Camara, M.; de Pascual-Teresa, S. Anthocyanin profile of red fruits and black carrot juices, purees and concentrates by HPLC-DAD-ESI/MS-QTOF. *Int. J. Food Sci. Technol.* **2016**, *51*, 2290–2300. [CrossRef]

36. Schwarz, M.; Wray, V.; Winterhalter, P. Isolation and identification of novel pyranoanthocyanins from black carrot (*Daucus carota* L.) juice. *J. Agric. Food Chem.* **2004**, *52*, 5095–5101. [CrossRef] [PubMed]

37. Kaur, A.; Sogi, D.S. Effect of osmotic dehydration on physico-chemical properties and pigment content of carrot (*Daucus carota* L.) during preserve manufacture. *J. Food Process. Preserv.* **2017**, *41*, e13153. [CrossRef]

38. Mayer-Miebach, E.; Behsnilian, D.; Regier, M.; Schuchmann, H.P. Thermal processing of carrots: Lycopene stability and isomerisation with regard to antioxidant potential. *Food Res. Int.* **2005**, *38*, 1103–1108. [CrossRef]

39. Nemzer, B.; Pietrzkowski, Z.; Spórna, A.; Stalica, P.; Thresher, W.; Michałowski, T.; Wybraniec, S. Betalainic and nutritional profiles of pigment-enriched red beet root (*Beta vulgaris* L.) dried extracts. *Food Chem.* **2011**, *127*, 42–53. [CrossRef]

40. Downham, A.; Collins, P. Colouring our foods in the last and next millennium. *Int. J. Food Sci. Technol.* **2000**, *35*, 5–22. [CrossRef]

41. Moreno, D.A.; García-Viguera, C.; Gil, J.I.; Gil-Izquierdo, A. Betalains in the era of global agri-food science, technology and nutritional health. *Phytochem. Rev.* **2008**, *7*, 261–280. [CrossRef]

42. Lee, E.J.; An, D.; Nguyen, C.T.T.; Patil, B.S.; Kim, J.; Yoo, K.S. Betalain and betaine composition of greenhouse- or field-produced beetroot (*Beta vulgaris* L.) and inhibition of HepG2 cell proliferation. *J. Agric. Food Chem.* **2014**, *62*, 1324–1331. [CrossRef]

43. Moriya, C.; Hosoya, T.; Agawa, S.; Sugiyama, Y.; Kozone, I.; Shin-Ya, K.; Terahara, N.; Kumazawa, S. New acylated anthocyanins from purple yam and their antioxidant activity. *Biosci. Biotechnol. Biochem.* **2015**, *79*, 1484–1492. [CrossRef] [PubMed]

44. Yin, J.M.; Yan, R.X.; Zhang, P.T.; Han, X.Y.; Wang, L. Anthocyanin accumulation rate and the biosynthesis related gene expression in *Dioscorea alata*. *Biol. Plant.* **2015**, *59*, 325–330. [CrossRef]

45. Ramos-Escudero, F.; Santos-Buelga, C.; Pérez-Alonso, J.J.; Yáñez, J.A.; Dueñas, M. HPLC-DAD-ESI/MS identification of anthocyanins in *Dioscorea trifida* L. yam tubers (purple sachapapa). *Eur. Food Res. Technol.* **2010**, *230*, 745–752. [CrossRef]

46. Fang, Z.; Wu, D.; Yü, D.; Ye, X.; Liu, D.; Chen, J. Phenolic compounds in Chinese purple yam and changes during vacuum frying. *Food Chem.* **2011**, *128*, 943–948. [CrossRef]

47. Price, E.J.; Bhattacharjee, R.; Lopez-Montes, A.; Fraser, P.D. Carotenoid profiling of yams: Clarity, comparisons and diversity. *Food Chem.* **2018**, *259*, 130–138. [CrossRef] [PubMed]

48. Frond, A.D.; Iuhas, C.I.; Stirbu, I.; Leopold, L.; Socaci, S.; Andreea, S.; Ayvaz, H.; Andreea, S.; Mihai, S.; Diaconeasa, Z.; et al. Phytochemical characterization of five edible purple-reddish vegetables: Anthocyanins, flavonoids, and phenolic acid derivatives. *Molecules* **2019**, *24*, 1536. [CrossRef] [PubMed]

49. Liang, T.; Sun, G.; Cao, L.; Li, J.; Wang, L. Rheological behavior of film-forming solutions and film properties from Artemisia sphaerocephala Krasch. gum and purple onion peel extract. *Food Hydrocoll.* **2018**, *82*, 124–134. [CrossRef]

50. Koley, T.K.; Khan, Z.; Oulkar, D.; Singh, B.K.; Maurya, A.; Singh, B.; Banerjee, K. High resolution LC-MS characterization of phenolic compounds and the evaluation of antioxidant properties of a tropical purple radish genotype. *Arab. J. Chem.* **2017**. [CrossRef]

51. Lin, L.Z.; Sun, J.; Chen, P.; Harnly, J.A. LC-PDA-ESI/MSn identification of new anthocyanins in purple bordeaux radish (*Raphanus sativus* L. Variety). *J. Agric. Food Chem.* **2011**, *59*, 6616–6627. [CrossRef]

52. Park, N.I.; Xu, H.; Li, X.; Jang, I.H.; Park, S.; Ahn, G.H.; Lim, Y.P.; Kim, S.J.; Park, S.U. Anthocyanin accumulation and expression of anthocyanin biosynthetic genes in radish (*Raphanus sativus* L.). *J. Agric. Food Chem.* **2011**, *59*, 6034–6039. [CrossRef]

53. Tatsuzawa, F.; Toki, K.; Saito, N.; Shinoda, K.; Shigihara, A.; Honda, T. Anthocyanin occurrence in the root peels, petioles and flowers of red radish (*Raphanus sativus* L.). *Dye. Pigment.* **2008**, *79*, 83–88. [CrossRef]

54. Park, C.H.; Yeo, H.J.; Kim, N.S.; Eun, P.Y.; Kim, S.-J.; Arasu, M.V.; Al-Dhabi, N.A.; Park, S.-Y.; Kim, J.K.; Park, S.U. Metabolic profiling of pale green and purple kohlrabi (*Brassica oleracea* var. *gongylodes*). *Appl. Biol. Chem.* **2017**, *60*, 249–257. [CrossRef]

55. Zhang, Y.; Hu, Z.; Zhu, M.; Zhu, Z.; Wang, Z.; Tian, S.; Chen, G. Anthocyanin Accumulation and Molecular Analysis of Correlated Genes in Purple Kohlrabi (*Brassica oleracea* var. *gongylodes* L.). *J. Agric. Food Chem.* **2015**, *63*, 4160–4169. [CrossRef] [PubMed]

56. Park, W.T.; Kim, J.K.; Park, S.; Lee, S.W.; Li, X.; Kim, Y.B.; Uddin, M.R.; Park, N.I.; Kim, S.J.; Park, S.U. Metabolic profiling of glucosinolates, anthocyanins, carotenoids, and other secondary metabolites in kohlrabi (*Brassica oleracea* var. *gongylodes*). *J. Agric. Food Chem.* **2012**, *60*, 8111–8116. [CrossRef] [PubMed]

57. Rahim, M.A.; Robin, A.H.K.; Natarajan, S.; Jung, H.J.; Lee, J.; Kim, H.R.; Kim, H.T.; Park, J.I.; Nou, I.S. Identification and characterization of anthocyanin biosynthesis-related genes in kohlrabi. *Appl. Biochem. Biotechnol.* **2018**, *184*, 1120–1141. [CrossRef] [PubMed]

58. Kaushal, P.; Kumar, V.; Sharma, H.K. Utilization of taro (*Colocasia esculenta*): A review. *J. Food Sci. Technol.* **2015**, *52*, 27–40. [CrossRef]

59. Ziarani, M.G.; Moradi, R.; Lashgari, N.; Kruger, H.G. *Metal-Free Synthetic Organic Dyes*; Elsevier: London, UK, 2018; ISBN 978-0-12-815647-6.

60. Pires, T.C.S.P.; Barros, L.; Santos-Buelga, C.; Ferreira, I.C.F.R. Edible flowers: Emerging components in the diet. *Trends Food Sci. Technol.* **2019**, *93*, 244–258. [CrossRef]

61. Amchova, P.; Kotolova, H.; Ruda-Kucerova, J. Health safety issues of synthetic food colorants. *Regul. Toxicol. Pharmacol.* **2015**, *73*, 914–922. [CrossRef]

62. El-Wahab, H.M.F.A.; Moram, G.S.E.D. Toxic effects of some synthetic food colorants and/or flavor additives on male rats. *Toxicol. Ind. Health* **2013**, *29*, 224–232. [CrossRef]

63. Rohrig, B. The chemistry of food coloring: Eating with your eyes. *ChemMatters* **2015**, *2015*, 5–7.

64. Martins, N.; Roriz, C.L.; Morales, P.; Barros, L.; Ferreira, I.C.F.R. Food colorants: Challenges, opportunities and current desires of agro-industries to ensure consumer expectations and regulatory practices. *Trends Food Sci. Technol.* **2016**, *52*, 1–15. [CrossRef]

65. De Pascual-Teresa, S.; Moreno, D.A.; García-Viguera, C. Flavanols and anthocyanins in cardiovascular health: A review of current evidence. *Int. J. Mol. Sci.* **2010**, *11*, 1679–1703. [CrossRef] [PubMed]

66. Horbowicz, M.; Grzesiuk, A.; DĘBski, H.; Kosson, R. Anthocyanins of fruits and vegetables - Their occurrence, analysis and role in human. *Veg. Crop. Res. Bull.* **2008**, *68*, 5–22. [CrossRef]

67. Mazewski, C.; Liang, K.; Gonzalez de Mejia, E. Comparison of the effect of chemical composition of anthocyanin-rich plant extracts on colon cancer cell proliferation and their potential mechanism of action using in vitro, in silico, and biochemical assays. *Food Chem.* **2018**, *242*, 378–388. [CrossRef]

68. Sestari, I.; Zsögön, A.; Rehder, G.G.; de Lira Teixeira, L.; Hassimotto, N.M.A.; Purgatto, E.; Benedito, V.A.; Peres, L.E.P. Near-isogenic lines enhancing ascorbic acid, anthocyanin and carotenoid content in tomato (*Solanum lycopersicum* L. cv Micro-Tom) as a tool to produce nutrient-rich fruits. *Sci. Hortic. (Amst.)* **2014**, *175*, 111–120. [CrossRef]

69. Pérez-Gregorio, M.R.; García-Falcón, M.S.; Simal-Gándara, J. Flavonoids changes in fresh-cut onions during storage in different packaging systems. *Food Chem.* **2011**, *124*, 652–658. [CrossRef]

70. Pérez-Gregorio, M.R.; Regueiro, J.; Simal-Gándara, J.; Rodrigues, A.S.; Almeida, D.P.F. Increasing the added-value of onions as a source of antioxidant flavonoids: A Critical Review. *Crit. Rev. Food Sci. Nutr.* **2014**, *54*, 1050–1062. [CrossRef]

71. Agcam, E.; Akyıldız, A.; Balasubramaniam, V.M. Optimization of anthocyanins extraction from black carrot pomace with thermosonication. *Food Chem.* **2017**, *237*, 461–470. [CrossRef]

72. McGill, A.E.J. The potential effects of demands for natural and safe foods on global food security. *Trends Food Sci. Technol.* **2009**, *20*, 402–406. [CrossRef]

73. Yusuf, M.; Shabbir, M.; Mohammad, F. Natural colorants: historical, processing and sustainable prospects. *Nat. Prod. Bioprospect.* **2017**, *7*, 123–145. [CrossRef]

74. Dusuki, N.J.S.; Abu Bakar, M.F.; Abu Bakar, F.I.; Ismail, N.A.; Azman, M.I. Proximate composition and antioxidant potential of selected tubers peel. *Food Res.* **2019**, *4*, 121–126. [CrossRef]

75. Yahia, E.M.; Barrera, A. Antioxidant capacity and correlation with phenolic compounds and carotenoids in 40 horticultural commodities. *Acta Hortic.* **2010**, *877*, 1215–1220. [CrossRef]

76. Hamouz, K.; Lachman, J.; Pazderů, K.; Tomášek, J.; Hejtmánková, K.; Pivec, V. Differences in anthocyanin content and antioxidant activity of potato tubers with different flesh colour. *Plant Soil Environ.* **2011**, *57*, 478–485. [CrossRef]

77. Sass-Kiss, A.; Kiss, J.; Milotay, P.; Kerek, M.M.; Toth-Markus, M. Differences in anthocyanin and carotenoid content of fruits and vegetables. *Food Res. Int.* **2005**, *38*, 1023–1029. [CrossRef]

78. Oliveira, H.; Roma-Rodrigues, C.; Santos, A.; Veigas, B.; Brás, N.; Faria, A.; Calhau, C.; de Freitas, V.; Baptista, P.V.; Mateus, N.; et al. GLUT1 and GLUT3 involvement in anthocyanin gastric transport-Nanobased targeted approach. *Sci. Rep.* **2019**, *9*, 789. [CrossRef]

79. Reyes, L.E.; Miller, J.C.J.; Cisneros-Zevallos, L. Environmental conditions influence the content and yield of anthocyanins and total phenolics in purple- and red-flesh potatoes during tuber development. *Am. J. Potato Res.* **2004**, *81*, 187–193. [CrossRef]

80. Gu, K.D.; Wang, C.K.; Hu, D.G.; Hao, Y.J. How do anthocyanins paint our horticultural products? *Sci. Hortic. (Amst.)* **2019**, *249*, 257–262. [CrossRef]

81. Leong, S.Y.; Oey, I. Effects of processing on anthocyanins, carotenoids and vitamin C in summer fruits and vegetables. *Food Chem.* **2012**, *133*, 1577–1587. [CrossRef]

82. Espinosa-Acosta, G.; Ramos-Jacques, A.L.; Molina, G.A.; Maya-Cornejo, J.; Esparza, R.; Hernandez-Martinez, A.R.; Sánchez-González, I.; Estevez, M. Stability analysis of anthocyanins using alcoholic extracts from black carrot (*Daucus carota* ssp. *Sativus var. Atrorubens Alef.*). *Molecules* **2018**, *23*, 2744.

83. Ersus, S.; Yurdagel, U. Microencapsulation of anthocyanin pigments of black carrot (*Daucus carota* L.) by spray drier. *J. Food Eng.* **2007**, *80*, 805–812. [CrossRef]

84. Braga, A.R.C.; Murador, D.C.; de Souza Mesquita, L.M.; de Rosso, V.V. Bioavailability of anthocyanins: Gaps in knowledge, challenges and future research. *J. Food Compos. Anal.* **2018**, *68*, 31–40. [CrossRef]

85. Takahata, Y.; Kai, Y.; Tanaka, M.; Nakayama, H.; Yoshinaga, M. Enlargement of the variances in amount and composition of anthocyanin pigments in sweetpotato storage roots and their effect on the differences in DPPH radical-scavenging activity. *Sci. Hortic. (Amst.)* **2011**, *127*, 469–474. [CrossRef]

86. Assous, M.T.M.; Abdel-Hady, M.M.; Medany, G.M. Evaluation of red pigment extracted from purple carrots and its utilization as antioxidant and natural food colorants. *Ann. Agric. Sci.* **2014**, *59*, 1–7. [CrossRef]

87. Chhikara, N.; Kushwaha, K.; Sharma, P.; Gat, Y.; Panghal, A. Bioactive compounds of beetroot and utilization in food processing industry: A critical review. *Food Chem.* **2019**, *272*, 192–200. [CrossRef]

88. Stintzing, F.C.; Carle, R. Betalains-emerging prospects for food scientists. *Trends Food Sci. Technol.* **2007**, *18*, 514–525. [CrossRef]

89. Stokstad, E. The new potato. *Science* **2019**, *363*, 574–577. [CrossRef]

90. Tsang, C.; Smail, N.F.; Almoosawi, S.; McDougall, G.J.M.; Al-Dujaili, E.A.S. Antioxidant Rich Potato Improves Arterial Stiffness in Healthy Adults. *Plant Foods Hum. Nutr.* **2018**, *73*, 203–208. [CrossRef]

91. Tian, J.; Chen, J.; Ye, X.; Chen, S. Health benefits of the potato affected by domestic cooking: A review. *Food Chem.* **2016**, *202*, 165–175. [CrossRef]

92. Camire, M.E.; Kubow, S.; Donelly, D.J. Potatoes and human health. *Crit. Rev. Food Sci. Nutr.* **2009**, *49*, 823–840. [CrossRef]

93. Burlingame, B.; Mouillé, B.; Charrondière, R. Nutrients, bioactive non-nutrients and anti-nutrients in potatoes. *J. Food Compos. Anal.* **2009**, *22*, 494–502. [CrossRef]

94. Nemś, A.; Pęksa, A. Polyphenols of coloured-flesh potatoes as native antioxidants in stored fried snacks. *LWT Food Sci. Technol.* **2018**, *97*, 597–602. [CrossRef]

95. Ombra, M.N.; Fratianni, F.; Granese, T.; Cardinale, F.; Cozzolino, A.; Nazzaro, F. In vitro antioxidant, antimicrobial and anti-proliferative activities of purple potato extracts (*Solanum tuberosum* cv *Vitelotte noire*) following simulated gastro-intestinal digestion. *Nat. Prod. Res.* **2015**, *29*, 1087–1091. [CrossRef] [PubMed]

96. Shock, C.C.; Brown, C.R.; Sathuvalli, V.; Charlton, B.A.; Yilma, S.; Hane, D.C.; Quick, R.; Rykbost, K.A. TerraRossa: A mid-season specialty potato with red flesh and skin and resistance to common scab and golden cyst nematode. *Am. J. Potato Res.* **2018**, *95*, 597–605. [CrossRef]

97. Ayoub, H.M.; McDonald, M.R.; Sullivan, J.A.; Tsao, R.; Meckling, K.A. Proteomic profiles of adipose and liver tissues from an animal model of metabolic syndrome fed purple vegetables. *Nutrients* **2018**, *10*, 456. [CrossRef] [PubMed]

98. Ayoub, H.M.; McDonald, M.R.; Sullivan, J.A.; Tsao, R.; Platt, M.; Simpson, J.; Meckling, K.A. The effect of anthocyanin-rich purple vegetable diets on metabolic syndrome in obese zucker rats. *J. Med. Food* **2017**, *20*, 1240–1249. [CrossRef]

99. Sun, X.; Du, M.; Navarre, D.A.; Zhu, M.J. Purple potato extract promotes intestinal epithelial differentiation and barrier function by activating AMP-activated protein kinase. *Mol. Nutr. Food Res.* **2018**, *62*, 1700536. [CrossRef]

100. Madiwale, G.P.; Reddivari, L.; Holm, D.G.; Vanamala, J. Storage elevates phenolic content and antioxidant activity but suppresses antiproliferative and pro-apoptotic properties of colored-flesh potatoes against human colon cancer cell lines. *J. Agric. Food Chem.* **2011**, *59*, 8155–8166. [CrossRef]

101. Kim, D.H.; Kim, M.; Oh, S.B.; Lee, K.M.; Kim, S.M.; Nho, C.W.; Yoon, W.B.; Kang, K.; Pan, C.H. The Protective Effect of Antioxidant Enriched Fractions from Colored Potatoes Against Hepatotoxic Oxidative Stress in Cultured Hepatocytes and Mice. *J. Food Biochem.* **2017**, *41*, e12315. [CrossRef]

102. Tian, J.; Chen, J.; Lv, F.; Chen, S.; Chen, J.; Liu, D.; Ye, X. Domestic cooking methods affect the phytochemical composition and antioxidant activity of purple-fleshed potatoes. *Food Chem.* **2016**, *197*, 1264–1270. [CrossRef]

103. Tierno, R.; Hornero-Méndez, D.; Gallardo-Guerrero, L.; López-Pardo, R.; de Galarreta, J.I.R. Effect of boiling on the total phenolic, anthocyanin and carotenoid concentrations of potato tubers from selected cultivars and introgressed breeding lines from native potato species. *J. Food Compos. Anal.* **2015**, *41*, 58–65. [CrossRef]

104. Lachman, J.; Hamouz, K.; Orsák, M.; Pivec, V.; Hejtmánková, K.; Pazderů, K.; Dvořák, P.; Čepl, J. Impact of selected factors-Cultivar, storage, cooking and baking on the content of anthocyanins in coloured-flesh potatoes. *Food Chem.* **2012**, *133*, 1107–1116. [CrossRef]

105. Bellumori, M.; Innocenti, M.; Michelozzi, M.; Cerretani, L.; Mulinacci, N. Coloured-fleshed potatoes after boiling: Promising sources of known antioxidant compounds. *J. Food Compos. Anal.* **2017**, *59*, 1–7. [CrossRef]

106. Perla, V.; Holm, D.G.; Jayanty, S.S. Effects of cooking methods on polyphenols, pigments and antioxidant activity in potato tubers. *LWT Food Sci. Technol.* **2012**, *45*, 161–171. [CrossRef]

107. Qiu, G.; Wang, D.; Song, X.; Deng, Y.; Zhao, Y. Degradation kinetics and antioxidant capacity of anthocyanins in air-impingement jet dried purple potato slices. *Food Res. Int.* **2018**, *105*, 121–128. [CrossRef]

108. Rytel, E.; Tajner-Czopek, A.; Kita, A.; Kucharska, A.Z.; Sokół-Łętowska, A.; Hamouz, K. Content of anthocyanins and glycoalkaloids in blue-fleshed potatoes and changes in the content of α-solanine and α-chaconine during manufacture of fried and dried products. *Int. J. Food Sci. Technol.* **2018**, *53*, 719–727. [CrossRef]

109. Zhang, L.; Zhao, L.; Bian, X.; Guo, K.; Zhou, L.; Wei, C. Characterization and comparative study of starches from seven purple sweet potatoes. *Food Hydrocoll.* **2018**, *80*, 168–176. [CrossRef]

110. Li, A.; Xiao, R.; He, S.; An, X.; He, Y.; Wang, C.; Yin, S.; Wang, B.; Shi, X.; He, J. Research advances of purple sweet potato anthocyanins: extraction, identification, stability, bioactivity, application, and biotransformation. *Molecules* **2019**, *24*, 3816. [CrossRef]

111. Suda, I.; Oki, T.; Masuda, M.; Kobayashi, M.; Nishiba, Y.; Furuta, S. Physiological functionality of purple-fleshed sweet potatoes containing anthocyanins and their utilization in foods. *Jpn. Agric. Res. Q.* **2003**, *37*, 167–173. [CrossRef]

112. Bassa, I.A.; Francis, F.J. Stability of anthocyanins from sweet potatoes in a model beverage. *J. Food Sci.* **1987**, *52*, 1753–1754. [CrossRef]

113. Drapal, M.; Rossel, G.; Heider, B.; Fraser, P.D. Metabolic diversity in sweet potato (*Ipomoea batatas*, L.) leaves and storage roots. *Hortic. Res.* **2019**, *6*, 2. [CrossRef]

114. Rosero, A.; Granda, L.; Pérez, J.L.; Rosero, D.; Burgos-Paz, W.; Martínez, R.; Morelo, J.; Pastrana, I.; Burbano, E.; Morales, A. Morphometric and colourimetric tools to dissect morphological diversity: An application in sweet potato [*Ipomoea batatas* (L.) Lam.]. *Genet. Resour. Crop Evol.* **2019**, *66*, 1257–1278. [CrossRef]

115. FAO. *Statistics Division Production and Trade Statistics*; FAO: Rome, Italy, 2017.

116. Cui, R.; Zhu, F. Physicochemical properties and bioactive compounds of different varieties of sweetpotato flour treated with high hydrostatic pressure. *Food Chem.* **2019**, *299*, 125129. [CrossRef] [PubMed]

117. De Albuquerque, T.M.R.; Sampaio, K.B.; de Souza, E.L. Sweet potato roots: Unrevealing an old food as a source of health promoting bioactive compounds–A review. *Trends Food Sci. Technol.* **2019**, *85*, 277–286. [CrossRef]

118. Guo, K.; Liu, T.; Xu, A.; Zhang, L.; Bian, X.; Wei, C. Structural and functional properties of starches from root tubers of white, yellow, and purple sweet potatoes. *Food Hydrocoll.* **2019**, *89*, 829–836. [CrossRef]

119. Wang, S.; Pan, D.; Lv, X.; Song, X.; Qiu, Z.; Huang, C.; Huang, R.; Chen, W. Proteomic approach reveals that starch degradation contributes to anthocyanin accumulation in tuberous root of purple sweet potato. *J. Proteom.* **2016**, *143*, 298–305. [CrossRef]

120. Zhao, J.G.; Yan, Q.Q.; Xue, R.Y.; Zhang, J.; Zhang, Y.Q. Isolation and identification of colourless caffeoyl compounds in purple sweet potato by HPLC-DAD-ESI/MS and their antioxidant activities. *Food Chem.* **2014**, *161*, 22–26. [CrossRef]

121. Kim, M.Y.; Lee, B.W.; Lee, H.; Lee, Y.Y.; Kim, M.H.; Lee, J.Y.; Lee, B.K.; Woo, K.S.; Kim, H. Phenolic compounds and antioxidant activity of sweet potato after heat treatment. *J. Sci. Food Agric.* **2019**. [CrossRef]

122. Moumouni Koala, A.H.; Somé, K.; Palé, E.; Sérémé, A.; Belem, J.; Nacro, M. Evaluation of eight orange fleshed sweetpotato (OFSP) varieties for their total antioxidant, total carotenoid and polyphenolic contents. *Evaluation* **2013**, *3*, 67–73.

123. Neela, S.; Fanta, S.W. Review on nutritional composition of orange-fleshed sweet potato and its role in management of vitamin A deficiency. *Food Sci. Nutr.* **2019**, *7*, 1920–1945. [CrossRef]

124. Lebot, V.; Michalet, S.; Legendre, L. Identification and quantification of phenolic compounds responsible for the antioxidant activity of sweet potatoes with different flesh colours using high performance thin layer chromatography (HPTLC). *J. Food Compos. Anal.* **2016**, *49*, 94–101. [CrossRef]

125. Kubow, S.; Iskandar, M.M.; Sabally, K.; Azadi, B.; Sadeghi, S.; Kumarathasan, P.; Dhar, D.; Prakash, S.; Burgos, G. Biotransformation of anthocyanins from two purple-fleshed sweet potato accessions in a dynamic gastrointestinal system. *Food Chem.* **2016**, *192*, 171–177. [CrossRef] [PubMed]

126. Meng, X.J.; Tan, C.; Feng, Y. Solvent extraction and in vitro simulated gastrointestinal digestion of phenolic compounds from purple sweet potato. *Int. J. Food Sci. Technol.* **2019**, 2887–2896. [CrossRef]

127. Oliveira, H.; Perez-Gregório, R.; de Freitas, V.; Mateus, N.; Fernandes, I. Comparison of the in vitro gastrointestinal bioavailability of acylated and non-acylated anthocyanins: Purple-fleshed sweet potato vs red wine. *Food Chem.* **2019**, *276*, 410–418. [CrossRef] [PubMed]

128. Yang, Z.W.; Tang, C.E.; Zhang, J.L.; Zhou, Q.; Zhang, Z.C. Stability and antioxidant activity of anthocyanins from purple sweet potato (*Ipomoea batatas* L. cultivar Eshu No. 8) subjected to simulated in vitro gastrointestinal digestion. *Int. J. Food Sci. Technol.* **2019**, 2604–2614. [CrossRef]

129. Sun, H.; Zhang, P.; Zhu, Y.; Lou, Q.; He, S. Antioxidant and prebiotic activity of five peonidin-based anthocyanins extracted from purple sweet potato (*Ipomoea batatas* (L.) Lam.). *Sci. Rep.* **2018**, *8*, 5018. [CrossRef] [PubMed]

130. Esatbeyoglu, T.; Rodríguez-werner, M.; Schlösser, A.; Winterhalter, P.; Rimbach, G. Fractionation, enzyme inhibitory and cellular antioxidant activity of bioactives from purple sweet potato (*Ipomoea batatas*). *Food Chem.* **2017**, *221*, 447–456. [CrossRef]

131. Gérard, V.; Ay, E.; Morlet-Savary, F.; Graff, B.; Galopin, C.; Ogren, T.; Mutilangi, W.; Lalevée, J. Thermal and photochemical stability of anthocyanins from black carrot, grape juice, and purple sweet potato in model beverages in the presence of ascorbic acid. *J. Agric. Food Chem.* **2019**, *67*, 5647–5660. [CrossRef]

132. Gras, C.C.; Nemetz, N.; Carle, R.; Schweiggert, R.M. Anthocyanins from purple sweet potato (*Ipomoea batatas* (L.) Lam.) and their color modulation by the addition of phenolic acids and food-grade phenolic plant extracts. *Food Chem.* **2017**, *235*, 265–274. [CrossRef]

133. He, W.; Zeng, M.; Chen, J.; Jiao, Y.; Niu, F.; Tao, G.; Zhang, S.; Qin, F.; He, Z. Identification and quantitation of anthocyanins in purple-fleshed sweet potatoes cultivated in China by UPLC-PDA and UPLC-QTOF-MS/MS. *J. Agric. Food Chem.* **2016**, *64*, 171–177. [CrossRef]

134. Truong, V.D.; Nigel, D.; Thompson, R.T.; Mcfeeters, R.F.; Dean, L.O.; Pecota, K.V.; Yencho, G.C. Characterization of anthocyanins and anthocyanidins in purple-fleshed sweetpotatoes by HPLC-DAD/ESI-MS/MS. *J. Agric. Food Chem.* **2010**, *58*, 404–410. [CrossRef]

135. Li, J.; Li, X.D.; Zhang, Y.; Zheng, Z.D.; Qu, Z.Y.; Liu, M.; Zhu, S.H.; Liu, S.; Wang, M.; Qu, L. Identification and thermal stability of purple-fleshed sweet potato anthocyanins in aqueous solutions with various pH values and fruit juices. *Food Chem.* **2013**, *136*, 1429–1434. [CrossRef] [PubMed]

136. Cai, Z.; Song, L.; Qian, B.; Xu, W.; Ren, J.; Jing, P.; Oey, I. Understanding the effect of anthocyanins extracted from purple sweet potatoes on alcohol-induced liver injury in mice. *Food Chem.* **2018**, *245*, 463–470. [CrossRef] [PubMed]

137. Jang, H.H.; Kim, H.W.; Kim, S.Y.; Kim, S.M.; Kim, J.B.; Lee, Y.M. In vitro and in vivo hypoglycemic effects of cyanidin 3-caffeoyl-p-hydroxybenzoylsophoroside-5-glucoside, an anthocyanin isolated from purple-fleshed sweet potato. *Food Chem.* **2019**, *272*, 688–693. [CrossRef] [PubMed]

138. Wang, L.; Zhao, Y.; Zhou, Q.; Luo, C.L.; Deng, A.P.; Zhang, Z.C.; Zhang, J.L. Characterization and hepatoprotective activity of anthocyanins from purple sweet potato (*Ipomoea batatas* L. cultivar Eshu No. 8). *J. Food Drug Anal.* **2017**, *25*, 607–618. [CrossRef]

139. Xu, J.; Su, X.; Lim, S.; Griffin, J.; Carey, E.; Katz, B.; Tomich, J.; Smith, J.S.; Wang, W. Characterisation and stability of anthocyanins in purple-fleshed sweet potato P40. *Food Chem.* **2015**, *186*, 90–96. [CrossRef]

140. Laksmiani, N.P.L.; Widiastari, M.I.; Reynaldi, K.R. The inhibitory activity of peonidin purple sweet potato in human epidermal receptor-2 receptor (her-2) expression by in silico study. *J. Phys. Conf. Ser.* **2018**, *1040*, 012010. [CrossRef]

141. Luo, C.; Zhou, Q.; Yang, Z.; Wang, R.; Zhang, J. Evaluation of structure and bioprotective activity of key high molecular weight acylated anthocyanin compounds isolated from the purple sweet potato (*Ipomoea batatas* L. cultivar Eshu No. 8). *Food Chem.* **2018**, *241*, 23–31. [CrossRef]

142. Lu, J.; Wu, D.M.; Zheng, Y.L.; Hu, B.; Cheng, W.; Zhang, Z.F. Purple sweet potato color attenuates domoic acid-induced cognitive deficits by promoting estrogen receptor-α-mediated mitochondrial biogenesis signaling in mice. *Free Radic. Biol. Med.* **2012**, *52*, 646–659. [CrossRef]

143. Zhuang, J.; Lu, J.; Wang, X.; Wang, X.; Hu, W.; Hong, F.; Zhao, X.X.; Zheng, Y. lin Purple sweet potato color protects against high-fat diet-induced cognitive deficits through AMPK-mediated autophagy in mouse hippocampus. *J. Nutr. Biochem.* **2019**, *65*, 35–45. [CrossRef]

144. Li, J.; Shi, Z.; Mi, Y. Purple sweet potato color attenuates high fat-induced neuroinflammation in mouse brain by inhibiting mapk and NF-κB activation. *Mol. Med. Rep.* **2018**, *17*, 4823–4831. [CrossRef]

145. Li, W.L.; Yu, H.Y.; Zhang, X.J.; Ke, M.; Hong, T. Purple sweet potato anthocyanin exerts antitumor effect in bladder cancer. *Oncol. Rep.* **2018**, *40*, 73–82. [CrossRef]

146. Schmitt, E.; Stopper, H. Estrogenic activity of naturally occurring anthocyanidins. *Nutr. Cancer* **2001**, *41*, 145–149. [CrossRef]

147. Yoshimoto, M.; Okuno, S.; Yamaguchi, M.; Yamakawa, O. Antimutagenicity of deacylated anthocyanins in purple-fleshed sweetpotato. *Biosci. Biotechnol. Biochem.* **2001**, *65*, 1652–1655. [CrossRef] [PubMed]

148. Zhao, J.G.; Yan, Q.Q.; Lu, L.Z.; Zhang, Y.Q. In vivo antioxidant, hypoglycemic, and anti-tumor activities of anthocyanin extracts from purple sweet potato. *Nutr. Res. Pract.* **2013**, *7*, 359–365. [CrossRef] [PubMed]

149. Zhang, Z.; Zhou, Q.; Yang, Y.; Wang, Y.; Zhang, J. Highly Acylated Anthocyanins from purple sweet potato (*Ipomoea batatas* L.) alleviate hyperuricemia and kidney inflammation in hyperuricemic mice: Possible attenuation effects on allopurinol. *J. Agric. Food Chem.* **2019**, *67*, 6202–6211. [CrossRef] [PubMed]

150. Zheng, G.H.; Shan, Q.; Mu, J.J.; Wang, Y.J.; Zhang, Z.F.; Fan, S.H.; Hu, B.; Li, M.Q.; Xie, J.; Chen, P.; et al. Purple sweet potato color attenuates kidney damage by blocking VEGFR2/ROS/NLRP3 signaling in high-fat diet-treated mice. *Oxid. Med. Cell. Longev.* **2019**, *2019*. [CrossRef]

151. Jiang, Z.; Chen, C.; Xie, W.; Wang, M.; Wang, J.; Zhang, X. Anthocyanins attenuate alcohol-induced hepatic injury by inhibiting pro-inflammation signalling. *Nat. Prod. Res.* **2016**, *30*, 469–473. [CrossRef]

152. Valenti, L.; Riso, P.; Mazzocchi, A.; Porrini, M.; Fargion, S.; Agostoni, C. Dietary anthocyanins as nutritional therapy for nonalcoholic fatty liver disease. *Oxid. Med. Cell. Longev.* **2013**, *2013*. [CrossRef]

153. Del Rio, D.; Rodriguez-Mateos, A.; Spencer, J.P.E.; Tognolini, M.; Borges, G.; Crozier, A. Dietary (poly)phenolics in human health: structures, bioavailability, and evidence of protective effects against chronic diseases. *Antioxid. Redox Signal.* **2012**, *18*, 1818–1892. [CrossRef]

154. Han, K.H.; Shimada, K.I.; Sekikawa, M.; Fukushima, M. Anthocyanin-rich red potato flakes affect serum lipid peroxidation and hepatic SOD mRNA level in rats. *Biosci. Biotechnol. Biochem.* **2007**, *71*, 1356–1359. [CrossRef]

155. Hwang, Y.P.; Choi, J.H.; Choi, J.M.; Chung, Y.C.; Jeong, H.G. Protective mechanisms of anthocyanins from purple sweet potato against tert-butyl hydroperoxide-induced hepatotoxicity. *Food Chem. Toxicol.* **2011**, *49*, 2081–2089. [CrossRef] [PubMed]

156. Suda, I.; Ishikawa, F.; Hatakeyama, M.; Miyawaki, M.; Kudo, T.; Hirano, K.; Ito, A.; Yamakawa, O.; Horiuchi, S. Intake of purple sweet potato beverage affects on serum hepatic biomarker levels of healthy adult men with borderline hepatitis. *Eur. J. Clin. Nutr.* **2008**, *62*, 60–67. [CrossRef] [PubMed]

157. Zhang, Z.F.; Lu, J.; Zheng, Y.L.; Wu, D.M.; Hu, B.; Shan, Q.; Cheng, W.; Li, M.Q.; Sun, Y.Y. Purple sweet potato color attenuates hepatic insulin resistance via blocking oxidative stress and endoplasmic reticulum stress in high-fat-diet-treated mice. *J. Nutr. Biochem.* **2013**, *24*, 1008–1018. [CrossRef] [PubMed]

158. Qin, S.; Sun, D.; Mu, J.; Ma, D.; Tang, R.; Zheng, Y. Purple sweet potato color improves hippocampal insulin resistance via down-regulating SOCS3 and galectin-3 in high-fat diet mice. *Behav. Brain Res.* **2019**, *359*, 370–377. [CrossRef] [PubMed]

159. Wang, X.; Zhang, Z.F.; Zheng, G.H.; Wang, A.M.; Sun, C.H.; Qin, S.P.; Zhuang, J.; Lu, J.; Ma, D.F.; Zheng, Y.L. The inhibitory effects of purple sweet potato color on hepatic inflammation is associated with restoration of NAD^+ levels and attenuation of nlrp3 inflammasome activation in high-fat-diet-treated mice. *Molecules* **2017**, *22*, 1315. [CrossRef] [PubMed]

160. Gofur, A.; Witjoro, A.; Ningtiyas, E.W.; Setyowati, E.; Mukharromah, S.A.; Athoillah, M.F.; Lestari, S.R. The evaluation of dietary black soybean and purple sweet potato on insulin sensitivity in streptozotocin-induced diabetic rats. *Pharmacogn. J.* **2019**, *11*, 639–646. [CrossRef]

161. Park, S.-J.; Lee, D.; Lee, M.; Kwon, H.-O.; Kim, H.; Park, J.; Jeon, W.; Cha, M.; Jun, S.; Park, K.; et al. The Effects of Curcuma longa L., Purple sweet potato, and mixtures of the two on immunomodulation in C57BL/6J mice infected with LP-BM5 murine leukemia retrovirus. *J. Med. Food* **2018**, *21*, 689–700. [CrossRef]

162. Ju, R.; Zheng, S.; Luo, H.; Wang, C.; Duan, L.; Sheng, Y.; Zhao, C.; Xu, W.; Huang, K. Purple Sweet Potato Attenuate Weight Gain in High Fat Diet Induced Obese Mice. *J. Food Sci.* **2017**, *82*, 787–793. [CrossRef]

163. Chen, H.; Sun, J.; Liu, J.; Gou, Y.; Zhang, X.; Wu, X.; Sun, R.; Tang, S.; Kan, J.; Qian, C.; et al. Structural characterization and anti-inflammatory activity of alkali-soluble polysaccharides from purple sweet potato. *Int. J. Biol. Macromol.* **2019**, *131*, 484–494. [CrossRef]

164. Gou, Y.; Sun, J.; Liu, J.; Chen, H.; Kan, J.; Qian, C.; Zhang, N.; Jin, C. Structural characterization of a water-soluble purple sweet potato polysaccharide and its effect on intestinal inflammation in mice. *J. Funct. Foods* **2019**, *61*, 103502. [CrossRef]

165. Sun, J.; Zhou, B.; Tang, C.; Gou, Y.; Chen, H.; Wang, Y.; Jin, C.; Liu, J.; Niu, F.; Kan, J.; et al. Characterization, antioxidant activity and hepatoprotective effect of purple sweetpotato polysaccharides. *Int. J. Biol. Macromol.* **2018**, *115*, 69–76. [CrossRef] [PubMed]

166. Tang, C.; Sun, J.; Liu, J.; Jin, C.; Wu, X.; Zhang, X.; Chen, H.; Gou, Y.; Kan, J.; Qian, C.; et al. Immune-enhancing effects of polysaccharides from purple sweet potato. *Int. J. Biol. Macromol.* **2019**, *123*, 923–930. [CrossRef] [PubMed]

167. Tang, C.; Sun, J.; Zhou, B.; Jin, C.; Liu, J.; Gou, Y.; Chen, H.; Kan, J.; Qian, C.; Zhang, N. Immunomodulatory effects of polysaccharides from purple sweet potato on lipopolysaccharide treated RAW 264.7 macrophages. *J. Food Biochem.* **2018**, *42*, e12535. [CrossRef]

168. Tang, C.; Sun, J.; Zhou, B.; Jin, C.; Liu, J.; Kan, J.; Qian, C.; Zhang, N. Effects of polysaccharides from purple sweet potatoes on immune response and gut microbiota composition in normal and cyclophosphamide treated mice. *Food Funct.* **2018**, *9*, 937–950. [CrossRef]

169. Jiang, T.; Mao, Y.; Sui, L.; Yang, N.; Li, S.; Zhu, Z.; Wang, C.; Yin, S.; He, J.; He, Y. Degradation of anthocyanins and polymeric color formation during heat treatment of purple sweet potato extract at different pH. *Food Chem.* **2019**, *274*, 460–470. [CrossRef]

170. Quan, W.; He, W.; Lu, M.; Yuan, B.; Zeng, M.; Gao, D.; Qin, F.; Chen, J.; He, Z. Anthocyanin composition and storage degradation kinetics of anthocyanins-based natural food colourant from purple-fleshed sweet potato. *Int. J. Food Sci. Technol.* **2019**, *54*, 2529–2539. [CrossRef]

171. Thi Lan Khanh, P.; Chittrakorn, S.; Rutnakornpituk, B.; Phan Tai, H.; Ruttarattanamongkol, K. Processing effects on anthocyanins, phenolic acids, antioxidant activity, and physical characteristics of Vietnamese purple-fleshed sweet potato flours. *J. Food Process. Preserv.* **2018**, *42*, e13722. [CrossRef]

172. Yea, C.S.; Nevara, G.A.; Muhammad, K.; Ghazali, H.M.; Karim, R. Physical properties, resistant starch content and antioxidant profile of purple sweet potato powder after 12 months of storage. *Int. J. Food Prop.* **2019**, *22*, 974–984. [CrossRef]

173. Lagnika, C.; Jiang, N.; Song, J.; Li, D.; Liu, C.; Huang, J.; Wei, Q.; Zhang, M. Effects of pretreatments on properties of microwave-vacuum drying of sweet potato slices. *Dry. Technol.* **2018**, *37*, 1901–1914. [CrossRef]

174. Phan, K.T.L.; Chittrakorn, S.; Tai, H.P.; Ruttarattanamongkol, K. Effects of cooking methods on the changes of total anthocyanins, phenolics content and physical characteristics of purple-fleshed sweet potato (*Ipomoea batatas* L.) grown in Vietnam. *Int. J. Adv. Sci. Eng. Inf. Technol.* **2018**, *8*, 227–233. [CrossRef]

175. Kamiloglu, S.; Van Camp, J.; Capanoglu, E. Black carrot polyphenols: Effect of processing, storage and digestion—An overview. *Phytochem. Rev.* **2018**, *17*, 379–395. [CrossRef]

176. Arscott, S.A.; Tanumihardjo, S.A. Carrots of many colors provide basic nutrition and bioavailable phytochemicals acting as a functional food. *Compr. Rev. Food Sci. Food Saf.* **2010**, *9*, 223–239. [CrossRef]

177. Kamiloglu, S.; Capanoglu, E.; Bilen, F.D.; Gonzales, G.B.; Grootaert, C.; Van de Wiele, T.; Van Camp, J. Bioaccessibility of polyphenols from plant-processing byproducts of black carrot (*Daucus carota* L.). *J. Agric. Food Chem.* **2016**, *64*, 2450–2458. [CrossRef] [PubMed]

178. Kamiloglu, S.; Pasli, A.A.; Ozcelik, B.; Van Camp, J.; Capanoglu, E. Influence of different processing and storage conditions on in vitro bioaccessibility of polyphenols in black carrot jams and marmalades. *Food Chem.* **2015**, *186*, 74–82. [CrossRef] [PubMed]

179. Kumar, M.; Dahuja, A.; Sachdev, A.; Kaur, C.; Varghese, E.; Saha, S.; Sairam, K.V.S.S. *Black Carrot (Daucus carota ssp.) and Black Soybean (Glycine max (L.) Merr.) Anthocyanin Extract: A Remedy to Enhance Stability and Functionality of Fruit Juices by Copigmentation, Waste and Biomass Valorization*; Springer: Berlin/Heidelberg, Germany, 2018.

180. Horvitz, M.A.; Simon, P.W.; Tanumihardjo, S.A. Lycopene and β-carotene are bioavailable from lycopene "red" carrots in humans. *Eur. J. Clin. Nutr.* **2004**, *58*, 803–811. [CrossRef] [PubMed]

181. Kammerer, D.; Carle, R.; Schieber, A. Quantification of anthocyanins in black carrot extracts (*Daucus carota* ssp. *sativus* var. *atrorubens* Alef.) and evaluation of their color properties. *Eur. Food Res. Technol.* **2004**, *219*, 479–486. [CrossRef]

182. Smeriglio, A.; Denaro, M.; Barreca, D.; D'Angelo, V.; Germanò, M.P.; Trombetta, D. Polyphenolic profile and biological activities of black carrot crude extract (*Daucus carota* L. ssp. sativus var. *atrorubens* Alef.). *Fitoterapia* **2018**, *124*, 49–57. [CrossRef]

183. Olejnik, A.; Rychlik, J.; Kidoń, M.; Czapski, J.; Kowalska, K.; Juzwa, W.; Olkowicz, M.; Dembczyński, R.; Moyer, M.P. Antioxidant effects of gastrointestinal digested purple carrot extract on the human cells of colonic mucosa. *Food Chem.* **2016**, *190*, 1069–1077. [CrossRef]

184. Blando, F.; Calabriso, N.; Berland, H.; Maiorano, G.; Gerardi, C.; Carluccio, M.A.; Andersen, Ø.M. Radical scavenging and anti-inflammatory activities of representative anthocyanin groupings from pigment-rich fruits and vegetables. *Int. J. Mol. Sci.* **2018**, *19*, 169. [CrossRef]

185. Kammerer, D.; Carle, R.; Schieber, A. Characterization of phenolic acids in black carrots (*Daucus carota* ssp. *sativus* var. *atrorubens* Alef.) by high-performance liquid chromatography/ electrospray ionization mass spectrometry. *Rapid Commun. Mass Spectrom.* **2004**, *18*, 1331–1340. [CrossRef]

186. Khandare, V.; Walia, S.; Singh, M.; Kaur, C. Black carrot (*Daucus carota* ssp. *sativus*) juice: Processing effects on antioxidant composition and color. *Food Bioprod. Process.* **2011**, *89*, 482–486. [CrossRef]

187. Iliopoulou, I.; Thaeron, D.; Baker, A.; Jones, A.; Robertson, N. Analysis of the thermal degradation of the individual anthocyanin compounds of black carrot (*Daucus carota* L.): A new approach using high-resolution proton nuclear magnetic resonance spectroscopy. *J. Agric. Food Chem.* **2015**, *63*, 7066–7073. [CrossRef] [PubMed]

188. Santana-Gálvez, J.; Pérez-Carrillo, E.; Velázquez-Reyes, H.H.; Cisneros-Zevallos, L.; Jacobo-Velázquez, D.A. Application of wounding stress to produce a nutraceutical-rich carrot powder ingredient and its incorporation to nixtamalized corn flour tortillas. *J. Funct. Foods* **2016**, *27*, 655–666. [CrossRef]

189. Wang, M.; Lopez-Nieves, S.; Goldman, I.L.; Maeda, H.A. Limited tyrosine utilization explains lower betalain contents in yellow than in red table beet genotypes. *J. Agric. Food Chem.* **2017**, *65*, 4305–4313. [CrossRef]

190. Herbach, K.; Stintzing, F.; Carle, R. Impact of thermal treatment on color and pigment pattern of red beet (*Beta vulgaris* L.) preparations. *J. Food Sci.* **2004**, *69*, 419–498. [CrossRef]

191. Montes-Lora, S.; Rodríguez-Pulido, F.J.; Cejudo-Bastante, M.J.; Heredia, F.J. Implications of the red beet ripening on the colour and betalain composition relationships. *Plant Foods Hum. Nutr.* **2018**, *73*, 216–221. [CrossRef]

192. Wettasinghe, M.; Bolling, B.; Plhak, L.; Xiao, H.; Parkin, K. Phase II enzyme-inducing and antioxidant activities of beetroot (*Beta vulgaris* L.) extracts from phenotypes of different pigmentation. *J. Agric. Food Chem.* **2002**, *50*, 6704–6709. [CrossRef]

193. Vulić, J.J.; Ćebović, T.N.; Čanadanović-Brunet, J.M.; Ćetković, G.S.; Čanadanović, V.M.; Djilas, S.M.; Šaponjac, V.T.T. *In vivo* and *in vitro* antioxidant effects of beetroot pomace extracts. *J. Funct. Foods* **2014**, *6*, 168–175. [CrossRef]

194. Leng, M.S.; Tobit, P.; Demasse, A.M.; Wolf, K.; Gouado, I.; Ndjouenkeu, R.; Rawel, H.M.; Schweigert, F.J. Nutritional and anti-oxidant properties of yam (*Dioscorea schimperiana*) based complementary food formulation. *Sci. African* **2019**, *5*, e00132. [CrossRef]

195. Champagne, A.; Hilbert, G.; Legendre, L.; Lebot, V. Diversity of anthocyanins and other phenolic compounds among tropical root crops from Vanuatu, South Pacific. *J. Food Compos. Anal.* **2011**, *24*, 315–325. [CrossRef]

196. Li, T.; Teng, H.; An, F.; Huang, Q.; Chen, L.; Song, H. The beneficial effects of purple yam (*Dioscorea alata* L.) resistant starch on hyperlipidemia in high-fat-fed hamsters. *Food Funct.* **2019**, *10*, 2642–2650. [CrossRef]

197. Patel, K.S.; Karmakar, N.; Desai, K.D.; Narwade, A.V.; Chakravarty, G.; Debnath, M.K. Exploring of greater yam (*Dioscorea alata* L.) genotypes through biochemical screening for better cultivation in south Gujarat zone of India. *Physiol. Mol. Biol. Plants* **2019**, *25*, 1235–1249. [CrossRef]

198. Li, T.; Chen, L.; Xiao, J.; An, F.; Wan, C.; Song, H. Prebiotic effects of resistant starch from purple yam (*Dioscorea alata* L.) on the tolerance and proliferation ability of *Bifidobacterium adolescentis* in vitro. *Food Funct.* **2018**, *9*, 2416–2425. [CrossRef]

199. Estiasih, T.; Umaro, D.; Harijono, H. Hypoglycemic effect of crude water soluble polysaccharide extracted from tubers of purple and yellow water yam (*Dioscorea alata* L.) on alloxan-induced hyperglycemia Wistar rats. *Prog. Nutr.* **2018**, *20*, 59–67.

200. Rosida; Purnawati, A.; Susiloningsih, E.K. Hypocholesterolemic and hypoglycemic effects of autoclaved-cooled water yam (*Dioscorea alata*) on hypercholesterolemia rats. *Int. Food Res. J.* **2018**, *25*, S181–S184.

201. Chen, C.T.; Wang, Z.H.; Hsu, C.C.; Lin, H.H.; Chen, J.H. Taiwanese and Japanese yam (*Dioscorea* spp.) extracts attenuate doxorubicin-induced cardiotoxicity in mice. *J. Food Drug Anal.* **2017**, *25*, 872–880. [CrossRef]

202. Khan, R.; Upaganlawar, A.B.; Upasani, C. Protective effects of *Dioscorea alata* L. in aniline exposure-induced spleen toxicity in rats: A biochemical study. *Toxicol. Int.* **2014**, *21*, 294–299.

203. Chiu, C.S.; Deng, J.S.; Chang, H.Y.; Chen, Y.C.; Lee, M.M.; Hou, W.C.; Lee, C.Y.; Huang, S.S.; Huang, G.J. Antioxidant and anti-inflammatory properties of taiwanese yam (*Dioscorea japonica* Thunb. var. *pseudojaponica* (Hayata) Yamam.) and its reference compounds. *Food Chem.* **2013**, *141*, 1087–1096. [CrossRef]

204. Chang, W.T.; Chen, H.M.; Yin, S.Y.; Chen, Y.H.; Wen, C.C.; Wei, W.C.; Lai, P.; Wang, C.H.; Yang, N.S. Specific dioscorea phytoextracts enhance potency of TCL-loaded DC-based cancer vaccines. *Evidence-Based Complement. Altern. Med.* **2013**, *2013*. [CrossRef]

205. Liu, S.F.; Chang, S.Y.; Lee, T.C.; Chuang, L.Y.; Guh, J.Y.; Hung, C.Y.; Hung, T.J.; Hung, Y.J.; Chen, P.Y.; Hsieh, P.F.; et al. *Dioscorea alata* attenuates renal interstitial cellular fibrosis by regulating smad- and epithelial-mesenchymal transition signaling pathways. *PLoS ONE* **2012**, *7*, e47482. [CrossRef]

206. Chan, Y.C.; Chang, S.C.; Liu, S.Y.; Yang, H.L.; Hseu, Y.C.; Liao, J.W. Beneficial effects of yam on carbon tetrachloride-induced hepatic fibrosis in rats. *J. Sci. Food Agric.* **2010**, *90*, 161–167. [CrossRef]

207. Wang, C.-H.; Tsai, C.-H.; Lin, H.-J.; Wang, T.-C.; Chen, H.-L. Uncooked Taiwanese yam (*Dioscorea alata* L. cv. Tainung No. 2) beneficially modulated the large bowel function and faecal microflora in BALB/c mice. *J. Sci. Food Agric.* **2006**, *87*, 1374–1380. [CrossRef]

208. Hsu, C.C.; Huang, Y.C.; Yin, M.C.; Lin, S.J. Effect of yam (*Dioscorea alata* compared to *Dioscorea japonica*) on gastrointestinal function and antioxidant activity in mice. *J. Food Sci.* **2006**, *71*, 5–8. [CrossRef]

209. Chan, Y.-C.; Hsu, C.-K.; Wang, M.-F.; Liao, J.-W.; Su, T.-Y. Beneficial effect of yam on the amyloid β-protein, monoamine oxidase B and cognitive deficit in mice with accelerated senescence. *J. Sci. Food Agric.* **2006**, *86*, 1517–1525. [CrossRef]

210. Chan, Y.C.; Hsu, C.K.; Wang, M.F.; Su, T.Y. A diet containing yam reduces the cognitive deterioration and brain lipid peroxidation in mice with senescence accelerated. *Int. J. Food Sci. Technol.* **2004**, *39*, 99–107. [CrossRef]

211. Fu, S.L.; Hsu, Y.H.; Lee, P.Y.; Hou, W.C.; Hung, L.C.; Lin, C.H.; Chen, C.M.; Huang, Y.J. Dioscorin isolated from *Dioscorea alata* activates TLR4-signaling pathways and induces cytokine expression in macrophages. *Biochem. Biophys. Res. Commun.* **2006**, *339*, 137–144. [CrossRef]

212. Mahmad, N.; Taha, R.M.; Othman, R.; Abdullah, S.; Anuar, N.; Elias, H.; Rawi, N. Anthocyanin as potential source for antimicrobial activity in *Clitoria ternatea* L. and *Dioscorea alata* L. *Pigment Resin Technol.* **2018**, *47*, 490–495. [CrossRef]

213. Chen, T.; Hu, S.; Zhang, H.; Guan, Q.; Yang, Y.; Wang, X. Anti-inflammatory effects of *Dioscorea alata* L. anthocyanins in a TNBS-induced colitis model. *Food Funct.* **2017**, *8*, 659–669. [CrossRef]

214. Guo, X.; Sha, X.; Cai, S.; Wang, O.; Ji, B. Antiglycative and antioxidative properties of ethyl acetate fraction of Chinese purple yam (*Dioscorea alata* L.) extracts. *Food Sci. Technol. Res.* **2015**, *21*, 563–571. [CrossRef]

215. Guo, X.X.; Sha, X.H.; Liu, J.; Cai, S.B.; Wang, Y.; Ji, B.P. Chinese purple yam (*Dioscorea alata* L.) extracts inhibit diabetes-related enzymes and protect HepG2 cells against oxidative stress and insulin resistance induced by FFA. *Food Sci. Technol. Res.* **2015**, *21*, 677–683. [CrossRef]

216. Mahmoudian-Sani, M.R.; Asadi-Samani, M.; Luther, T.; Saeedi-Boroujeni, A.; Gholamian, N. A new approach for treatment of type 1 diabetes: Phytotherapy and phytopharmacology of regulatory T cells. *J. Ren. Inj. Prev.* **2017**, *6*, 158–163. [CrossRef]

217. Zhang, J.; Tian, H.; Zhan, P.; Du, F.; Zong, A.; Xu, T. Isolation and identification of phenolic compounds in Chinese purple yam and evaluation of antioxidant activity. *LWT Food Sci. Technol.* **2018**, *96*, 161–165. [CrossRef]

218. Dey, P.; Ray, S.; Chaudhuri, T.K. Immunomodulatory activities and phytochemical characterisation of the methanolic extract of *Dioscorea alata* aerial tuber. *J. Funct. Foods* **2016**, *23*, 315–328. [CrossRef]

219. Lu, J.; Wong, R.N.S.; Zhang, L.; Wong, R.Y.L.; Ng, T.B.; Lee, K.F.; Zhang, Y.B.; Lao, L.X.; Liu, J.Y.; Sze, S.C.W. Comparative Analysis of Proteins with Stimulating Activity on Ovarian Estradiol Biosynthesis from Four Different *Dioscorea* Species in vitro Using Both Phenotypic and Target-based Approaches: Implication for Treating Menopause. *Appl. Biochem. Biotechnol.* **2016**, *180*, 79–93. [CrossRef]

220. Wu, Z.G.; Jiang, W.; Nitin, M.; Bao, X.Q.; Chen, S.L.; Tao, Z.M. Characterizing diversity based on nutritional and bioactive compositions of yam germplasm (*Dioscorea* spp.) commonly cultivated in China. *J. Food Drug Anal.* **2016**, *24*, 367–375. [CrossRef]

221. Han, C.H.; Liu, J.C.; Fang, S.U.; Hou, W.C. Antioxidant activities of the synthesized thiol-contained peptides derived from computer-aided pepsin hydrolysis of yam tuber storage protein, dioscorin. *Food Chem.* **2013**, *138*, 923–930. [CrossRef]

222. Koo, H.J.; Park, H.J.; Byeon, H.E.; Kwak, J.H.; Um, S.H.; Kwon, S.T.; Rhee, D.K.; Pyo, S. Chinese yam extracts containing β-sitosterol and ethyl linoleate protect against atherosclerosis in apolipoprotein e-deficient mice and inhibit muscular expression of VCAM-1 in vitro. *J. Food Sci.* **2014**, *79*, H719–H729. [CrossRef]

223. Harijono, T.E.; Ariestiningsih, A.D.; Wardani, N.A.K. The effect of crude diosgenin extract from purple and yellow greater yams (*Dioscorea alata*) on the lipid profile of dyslipidemia rats. *Emirates J. Food Agric.* **2016**, *28*, 506–512. [CrossRef]

224. Lee, Y.C.; Lin, J.T.; Wang, C.K.; Chen, C.H.; Yang, D.J. Antiproliferative effects of fractions of furostanol and spirostanol glycosides from yam (*Dioscorea pseudojaponica* Yamamoto) and diosgenin on cancer and normal cells and their apoptotic effects for MCF-7 cells. *J. Food Biochem.* **2012**, *36*, 75–85. [CrossRef]

225. Jailani, F.; Kusumawardani, S.; Puspitasari, C.; Maula, A.; Purwandari, U. Annealled purple yam (*Dioscorea alata* var. *purpurea*) flour improved gelatinisation profile, but increased glycemic index of substituted bread. *Int. Food Res. J.* **2013**, *20*, 865–871.

226. Hsu, C.L.; Hurang, S.L.; Chen, W.; Weng, Y.M.; Tseng, C.Y. Qualities and antioxidant properties of bread as affected by the incorporation of yam flour in the formulation. *Int. J. Food Sci. Technol.* **2004**, *39*, 231–238. [CrossRef]

227. Li, P.H.; Huang, C.C.; Yang, M.Y.; Wang, C.C.R. Textural and sensory properties of salted noodles containing purple yam flour. *Food Res. Int.* **2012**, *47*, 223–228. [CrossRef]

228. Seguchi, M.; Ozawa, M.; Nakamura, C.; Tabara, A. Development of gluten-free bread baked with yam flour. *Food Sci. Technol. Res.* **2012**, *18*, 543–548. [CrossRef]

229. Brewster, J.L. *Onions and Other Vegetable Alliums*, 2nd ed.; CABI: Wallingford, UK, 2008; ISBN 9781845933999.

230. Slimestad, R.; Fossen, T.; Vågen, I.M. Onions: A source of unique dietary flavonoids. *J. Agric. Food Chem.* **2007**, *55*, 10067–10080. [CrossRef]

231. Benkeblia, N. Free-Radical Scavenging capacity and antioxidant properties of some selected onions (*Allium cepa* L.) and garlic (*Allium sativum* L.) extracts. *Brazilian Arch. Biol. Technol.* **2005**, *48*, 753–759. [CrossRef]

232. Zhang, S.; Deng, P.; Xu, Y.C.; Lü, S.W.; Wang, J.J. Quantification and analysis of anthocyanin and flavonoids compositions, and antioxidant activities in onions with three different colors. *J. Integr. Agric.* **2016**, *15*, 2175–2181. [CrossRef]

233. Oboh, G.; Ademiluyi, A.O.; Agunloye, O.M.; Ademosun, A.O.; Ogunsakin, B.G. Inhibitory Effect of garlic, purple onion, and white onion on key enzymes linked with type 2 diabetes and hypertension. *J. Diet. Suppl.* **2019**, *16*, 105–118. [CrossRef]

234. Albishi, T.; John, J.A.; Al-Khalifa, A.S.; Shahidi, F. Antioxidant, anti-inflammatory and DNA scission inhibitory activities of phenolic compounds in selected onion and potato varieties. *J. Funct. Foods* **2013**, *5*, 930–939. [CrossRef]

235. Ou, B.; Huang, D.; Hampsch-Woodill, M.; Flanagan, J.A.; Deemer, E.K. Analysis of antioxidant activities of common vegetables employing oxygen radical absorbance capacity (ORAC) and ferric reducing antioxidant power (FRAP) assays: A comparative study. *J. Agric. Food Chem.* **2002**, *50*, 3122–3128. [CrossRef]

236. Khiari, Z.; Makris, D.P.; Kefalas, P. An Investigation on the Recovery of Antioxidant Phenolics from Onion Solid Wastes Employing Water/Ethanol-Based Solvent Systems. *Food Process. Technol.* **2009**, *2*, 337–343. [CrossRef]

237. Hanlon, P.R.; Barnes, D.M. Phytochemical composition and biological activity of 8 varieties of radish (*Raphanus sativus* L.) sprouts and mature taproots. *J. Food Sci.* **2011**, *76*, 185–192. [CrossRef]

238. Masukawa, T.; Kadowaki, M.; Matsumoto, T.; Nakatsuka, A.; Cheon, K.S.; Kato, K.; Tatsuzawa, F.; Kobayashi, N. Enhancement of food functionality of a local pungent radish "izumo orochi daikon" 'susanoo' by introduction of a colored root character. *Hortic. J.* **2018**, *87*, 356–363. [CrossRef]

239. Tatsuzawa, F.; Saito, N.; Toki, K.; Shinoda, K.; Shigihara, A.; Honda, T. Acylated cyanidin 3-sophoroside-5-glucosides from the purple roots of red radish (*Raphanus sativus* L.) "Benikanmi.". *J. Jpn. Soc. Hortic. Sci.* **2010**, *79*, 103–107. [CrossRef]

240. Ahn, M.; Koh, R.; Kim, G.O.; Shin, T. Aqueous extract of purple Bordeaux radish, *Raphanus sativus* L. ameliorates ethanol-induced gastric injury in rats. *Orient. Pharm. Exp. Med.* **2013**, *13*, 247–252. [CrossRef]

241. Pak, W.M.; Kim, K.B.W.R.; Kim, M.J.; Kang, B.K.; Bark, S.W.; Kim, B.R.; Ahn, N.K.; Choi, Y.U.; Yoon, S.R.; Ahn, D.H. Antioxidative effect of extracts from different parts of Kohlrabi. *J. Appl. Biol. Chem.* **2014**, *57*, 353–358. [CrossRef]

242. Jung, H.A.; Karki, S.; Ehom, N.; Yoon, M.; Kim, E.J.; Choi, J.S. Anti-diabetic and anti-inflammatory effects of green and red kohlrabi cultivars (*Brassica oleracea* var. *gongylodes*). **2014**, *19*, 281–290. [CrossRef]

243. Paull, R.E.; Uruu, G.; Arakaki, A. Variation in the cooked and chipping quality of taro. *Horttechnology* **2000**, *10*, 823–829. [CrossRef]

244. Beyene, T.M. Morpho–Agronomical Characterization of Taro (*Colocasia esculenta*) Accessions in Ethiopia. *Plant* **2013**, *1*, 1–9. [CrossRef]

 *antioxidants*

Review

Grown to Be Blue—Antioxidant Properties and Health Effects of Colored Vegetables. Part II: Leafy, Fruit, and Other Vegetables

Francesco Di Gioia [1], Nikolaos Tzortzakis [2], Youssef Rouphael [3], Marios C. Kyriacou [4], Shirley L. Sampaio [5], Isabel C.F.R. Ferreira [5] and Spyridon A. Petropoulos [6,*]

[1] Department of Plant Science, The Pennsylvania State University, University Park, PA 16802, USA; fxd92@psu.edu
[2] Department of Agricultural Sciences, Biotechnology and Food Science, Cyprus University of Technology, 3603 Lemesos, Cyprus; nikolaos.tzortzakis@cut.ac.cy
[3] Department of Agricultural Sciences, University of Naples Federico II, 80055 Portici, Italy; youssef.rouphael@unina.it
[4] Department of Vegetable Crops, Agricultural Research Institute, 1516 Nicosia, Cyprus; m.kyriacou@ari.gov.cy
[5] Centro de Investigação de Montanha (CIMO), Instituto Politécnico de Bragança, Campus de Santa Apolónia, 5300-253 Bragança, Portugal; shirleylsampaio@gmail.com (S.L.S.); iferreira@ipb.pt (I.C.F.R.F.)
[6] Department of Agriculture, Crop Production and Rural Environment, University of Thessaly, 38446 N. Ionia, Magnissia, Greece
* Correspondence: spetropoulos@uth.gr; Tel.: +30-242-109-3196

Received: 30 December 2019; Accepted: 20 January 2020; Published: 23 January 2020

Abstract: The current trend for substituting synthetic compounds with natural ones in the design and production of functional and healthy foods has increased the research interest about natural colorants. Although coloring agents from plant origin are already used in the food and beverage industry, the market and consumer demands for novel and diverse food products are increasing and new plant sources are explored. Fresh vegetables are considered a good source of such compounds, especially when considering the great color diversity that exists among the various species or even the cultivars within the same species. In the present review we aim to present the most common species of colored vegetables, focusing on leafy and fruit vegetables, as well as on vegetables where other plant parts are commercially used, with special attention to blue color. The compounds that are responsible for the uncommon colors will be also presented and their beneficial health effects and antioxidant properties will be unraveled.

Keywords: anthocyanins; antioxidants; flavonoids; fruit vegetables; functional quality; leafy vegetables; inflorescence; lettuce; natural colorants; tomato

---

1. Introduction

Vegetables are considered an invaluable ingredient of human diet, since they diversify color of various food products and they also possess beneficial health effects due to their content in various phytochemicals such as flavonoids, betalains and carotenoids and the overall high antioxidant capacity [1–3]. Consumption of purple/blue fresh produce is associated with increased nutrient intake and reduced risk for metabolic syndrome [4]. Based on food intake data from NHANES 2001–2002, the daily intake of anthocyanins was estimated to be 12.5 mg/day/person in the United States [5]. The predominant dietary anthocyanins are malvidin, delphinidin, and peonidin glycosides [6], which can be found in many plant foods, including berries, purple sweet potatoes, grapes, and wine [7].

In comparison with other flavonoids, anthocyanins possess a positive charge on their C-ring, which leads to different colors in response to various pH [8].

Anthocyanins not only have aesthetic importance by generating characteristic purple, bluish, orange, and reddish pigments in various plant tissues [9], but also have biological functions including protective effects against radiation, reactive oxygen species scavenging, defense against pathogens and stress conditions, and attracting seed and pollen dispersers [10–12]. These compounds also have nutritional value and potential health benefits [13].

Leafy vegetables are widely consumed throughout the world and they significantly contribute to the overall recommended daily intake for several nutrients essential for human body [14]. They include several species among which lettuce is considered the most important salad vegetable and several reports highlighted its significance in human nutrition [15,16]. On the other hand, the Solanaceae family has approximately 2700 species and 99 genera and includes some of the most important fruit vegetables consumed globally. *Solanum*, the largest and most complex genus in this family, is of great economic importance with several species used as foods, medicines, and ornamental plants [17]. In other popular vegetables consumed worldwide such as broccoli, cauliflower, and artichoke, the immature inflorescence constitutes the edible portion and represent a rich source of flavonoids, anthocyanins, and other bioactive compounds that are also responsible for their pigmentation [18–21]. In the case of sweet corn, another popular vegetable used for fresh consumption as well as for canned and freezing processing, differing from the regular corn only for a higher accumulation of sugar in the kernels, the ear including cob and kernels constitute a rich source of natural colorants including carotenoids and flavonoids [22]. Yet, in other cases like asparagus the edible portion and source of anthocyanins is constituted by the young stems [23].

Recently we published a review paper regarding colored root vegetable species focusing on the most important coloring compounds and their antioxidant effects. With the present review we aim to continue this work and present the rest of colored vegetable crops, focusing on leafy and fruit vegetables and relevant species where other plant parts are consumed. Having in mind the same context, the main coloring compounds are highlighted, while a special section for each species is allocated to their health effects and antioxidant properties. The information presented in this review was systematically gathered from scientific databases such as Scopus, ScienceDirect, PubMed, Google Scholar, and ResearchGate by using various keywords and key phrases, e.g., the common and Latin of the main species and/or the terms "health effects", "antioxidant compounds", "colored leafy vegetables", "colored fruit vegetables", "blue vegetables", "purple vegetables", and "anthocyanins".

2. Leafy Vegetables

2.1. Lettuce

Lettuce (*Lactuca sativa* L.) belongs to the Asteraceae (Compositae) family and is a very popular vegetable crop used for fresh consumption and as salad ingredient owing to its sensory and health-promoting properties [14,24–26]. Lettuce is widely cultivated throughout the globe and it is rightly considered the most important of leafy greens as it ranks highest in production (27 million tons in 2017; [27]). Lettuce comes in a wide variety of head formations, textures, sizes, leaf shapes, and colors and it is conventionally classified according to Mou [28] into six major groups (i.e., types): (i) Butterhead, (ii) Cos or Romaine, (iii) Crisphead, (iv) Leaf or Cutting, v) Stalk or Stem, and vi) Latin. Compared to several other leafy vegetables, lettuce is an excellent source of vitamin B9 and total flavonoids. Wang et al. [29] and Gan and Azrina [30] reported that total vitamin B9 and flavonoid contents in lettuce were higher by 16% and 220%, respectively, compared to spinach, which is another important and widely consumed leafy green. Quali-quantitative variation in lettuce vitamins and secondary metabolites (i.e., phytochemicals) depends on many pre-harvest factors such as genotype, environmental conditions, harvest maturity, and agricultural practices [31]. However, the genetic material is the predominant pre-harvest factor and the major determinant of the biosynthesis and accumulation of lipophilic (i.e., carotenoids, chlorophylls, and vitamin E) and hydrophilic (i.e., phenolic compounds and vitamin C) antioxidant molecules [32]. Vitamins are essential micronutrients required for human metabolism and functionalities implicated in the reduction of cardiovascular and degenerative diseases [33]. Folate (vitamin B9) and vitamin C (as ascorbic and dehydroascorbic acids) are eminently present in lettuce [14]. In their review article Kim and co-workers reported that folate and vitamin C contents vary with leaf type and particularly leaf coloration, with red leaf, butterhead, and romaine lettuces being particularly good sources of folate [15,29,34], while leaf green lettuce had the highest vitamin C concentration [24,25,34,35].

Carotenoids, which constitute an important group of lipophilic pigments frequently present in yellow-orange vegetables and in dark green leafy vegetables, vary in concentration among lettuce types and colors. Mou [36] assessed the genetic variability in β-carotene and lutein content, the most abundant carotenoids in lettuce, across 52 genotypes (including butterhead, crisphead, Latin, leaf, primitive, romaine, stem lettuce and wild species) that were categorized by type in the following order: Romaine and green leaf > red leaf > butterhead > crisphead. The author also reported that the two target carotenoids were significantly and positively correlated with chlorophyll a and b as well as with total chlorophyll content. Contrarily to the findings of Mou [36], Baslam et al. [37], and Nicolle et al. [25] demonstrated that the content of β-carotene and lutein may not entirely correlate with leaf green pigmentation, since the carotenoids content appeared to be lower in green compared to red-pigmented lettuce plants. The contradiction between these results may indicate that the content in carotenoids may not be consistently related to leaf pigmentation [14]. Nevertheless, the frequent consumption of carotenoids-rich lettuce could be of high importance since several epidemiological studies demonstrated that the onset of chronic diseases such as heart disease, vision impairment, and certain types of cancer (lung, prostate, and colon) could be reduced [38–41].

Several authors reported that the contents of secondary metabolites in lettuce differed greatly among genotypes and depended particularly on leaf color (dark red, red, green/red, and green) [14,42]. According to Mulabagal et al. [43] red lettuce contains a single anthocyanin, namely cyanidin-3-*O*-(6″-malonyl-β-glucopyranoside) which is further converted in two cyanidin derivatives (cyanidin-3-*O*-(6″-malonyl-β-glucopyranoside methyl ester) and cyanidin-3-*O*-β-glucopyranoside), all presenting significant antioxidant activities against lipid peroxidation and cyclooxygenase activity. Kim et al. [42] explored the genetic material of 23 lettuce cultivars belonging to three major lettuce groups (crisphead, oak-leaf, and romaine) in respect to their phytochemical content and antioxidant potential. The authors reported that most phytochemicals varied significantly with genetic material and were associated mainly with leaf color. The red-leaf and to a lesser extent the green/red cultivars exhibited the highest concentration of the following antioxidant molecules: Cyanidin, carotenoids (lutein, violaxanthin and lactucaxanthin), fatty acids such as α-linolenic and linoleic acid, total polyphenols, and antioxidant potential [42]. In the same study, the authors were also able to demonstrate that the methanolic extract of red-pigmented lettuce contained potent scavengers of ABTS (scavenging assay of 2,2-azino-bis(3-ethylbenzthiazoline-6-sulphonic acid) diammonium salt radical) and DPPH (scavenging assay of 2,2-diphenyl-1-picrylhydrazyl radical) radicals. The higher radical-quenching activity of red-pigmented lettuce cultivars irrespective of type renders them more bioactive thus their systematic inclusion in the human diet could be an efficient tool to minimize the impact of oxidative stress-related diseases [25]. Similarly, Hao et al. [44], reported the differences in nutritive quality (e.g., cellulose, protein, starch, sugar, and vitamin C contents) among 74 red/purple and green varieties of leaf lettuce using grey correlation analysis (a methodology of treating and analyzing qualitative data according to grey system procedure). The authors concluded that purple-leaf lettuce "P-S23" exhibited significantly higher grey comprehensive evaluation value (0.8) in comparison to the green counterparts (values < 0.5). In addition, leaf pigmentation significantly correlates to the constitution and concentration of phenolic compounds belonging mainly in the subgroups of phenolic acids, flavonoids, and anthocyanins [14]. Chicoric, caffeic, and chlorogenic acids and their derivatives are the most abundant phenolic acids present in lettuce, whereas the most outstanding flavonoids include anthocyanins, quercetin, kaempferol, and flavone luteolin [35,45–47]. Several authors reported a higher total phenolic content in red butterhead, red leaf and red romaine lettuces compared to their green counterparts [16,35,48–51]. The red color of lettuce has been associated with a higher total phenolics content, known to impart a greater antioxidant activity than vitamins C and E [52], and has been attributed primarily to anthocyanins, an important group of flavonoids responsible for the red/purple coloration [53].

Regarding the health benefits attributed to lettuce, according to an in vivo study carried out by Lee et al. [54] on mice fed with a high-fat diet, supplementation of the diet with 8% red lettuce on a body weight (bw) basis decreased the total cholesterol and the low density lipoprotein (LDL) by 9% and 123%, respectively, thus highlighting the potential effects of red-pigmented lettuce consumption against cardiovascular disease. A putative mechanism behind the cholesterol reduction could be the synergistic effects of lipophilic and hydrophilic antioxidant molecules such as α-tocopherol, anthocyanins, β-carotene, and phenolic compounds. Similarly, Nicolle et al. [24] observed that feeding male rats with 20% of red oak-leaf lettuce decreased significantly the LDL (low density lipoproteins)/HDL (high-density lipoproteins) cholesterol ratio and the liver cholesterol content.

In addition to preclinical trials, clinical studies demonstrated that weekly consumption of lettuce was able to reduce the incidence of colorectal cancer [55]. The protective effect of frequent lettuce consumption against colorectal cancer has been attributed to the presence of β-carotene and vitamin C and not to Ca and vitamins B9 and E contents [55]. Recently, Qin [56] reported that the new cultivar B-2 of red-pigmented lettuce, characterized by high concentration of anthocyanins, flavones, and phenolic acids, can minimize oxidative stress-related diseases, leading to anti-tumor effects against human lung adenocarcinoma, hepatoma, and human cancer colorectal adenoma cell lines. Based on the above considerations, clinical and preclinical studies demonstrated that frequent consumption of fresh lettuce, in particular the red-pigmented varieties, carries potential anti-diabetic, cholesterol lowering, and anti-tumor properties.

2.2. Basil

Basil (*Ocimum basilicum* L.) belongs to the Lamiaceae family and it is one of the most important aromatic herbs cultivated worldwide as it flourishes under a wide range of climatic conditions. Basil cultivated both in open-field and under greenhouse conditions is an essential ingredient of renowned pesto sauce (widely consumed in Italy), while it is also used for fresh consumption and as a culinary spice [57]. In addition, the extraction of essential oils from basil is of high interest to both cosmetic and pharmaceutical industries. Basil comes in a wide variety of types, conventionally classified into seven morphotypes: (i) Large-leafed "Italian" basil, (ii) tall, slender basil, (iii) dwarf "Bush" basil, (iv) compact "Thai" basil, (v) purple basil (with clove-like aroma), (vi) citriodorum basil (flavored types), and (vii) purpurascens basil (sweet purple-colored basil) [58]. The herbs of the Lamiaceae family, such as basil, are characterized by strong antioxidant capacity. Basil in particular is a rich source of phenolic compounds, including phenolic acids such as rosmarinic, caffeic, chicoric and caftaric acids [59–61], vitamin C, and carotenoids such as lutein and β-carotene [62]. Furthermore, certain purple/red cultivars also have important concentrations of the hydrophilic anthocyanins, especially the "Dark Opal", "Purple Ruffles", and "Rubin" cultivars" [61,63].

The main carotenoids detected in basil are mostly lutein and β-carotene; but Calucci et al. [64] ranks basil first among aromatic herbs with respect to the concentrations of xanthophyll carotenoids. In the study of Kopsell et al. [62], the main detected carotenoids in sweet basil were identified as lutein, β-carotene, and zeaxanthin, and significant differences in carotenoid profiles were observed between different growing conditions (open-field versus greenhouse) and cultivars ("Cinnamon", "Genovese", "Italian large leaf", "Nufar", "Osmin purple", "Red Rubin", "Spicy bush", and "Sweet Tai"). According to Marchand et al. [38] and Johnson et al. [40], the frequent consumption of vegetables and herbs was more strongly correlated with reduced risk of certain types of cancer and degenerative ophthalmic diseases, in comparison to the ingestion of monomolecular carotenoid supplements. In addition to carotenoids, basil is considered also an important source of vitamin C. It is well established that vitamin C is crucial for immune and antioxidant functions, and according to the WHO, 80–90 mg of vitamin C should be ingested daily [65]. However, due to the water solubility of ascorbic acid, regular dietary intake is essential to normal metabolic functioning [66]. According to Murarikova and Neugebauerova [65] the ascorbic acid content varied from 34.3 to 220.0 mg/kg fresh weight among the tested varieties ("Dark Green", "Lettuce Leaf", "Mammolo Genovese", "Manes", "Ohre", "Purple Opal", and "Red Rubin") and the different growing seasons.

Rosmarinic acid is noted in the scientific literature as the most abundant phenolic constituent of basil [59,60]. On the other hand, a study carried out by Kwee and Niemeyer [61] showed that 9 over 15 basil cultivars contained other caffeic acid derivatives, such as chicoric acid, in higher concentrations than rosmarinic acid. Moreover, the antioxidant and antimicrobial properties of "Napoletano" green and purple basil were analyzed by Tenore et al. [67], who reported that the main phenolic acids in purple basil were rosmarinic, ferulic and gallic acid; while in green basil, the most abundant phenolic constituents were gallic acid, followed by rosmarinic and ferulic acids. Interestingly, the functional molecule rosmarinic acid detected in "Napoletano" type basil was by far higher than what on average has been reported for other common varieties such as "Sweet basil", "Thai basil", "Genovese Italiano", and "Purple Petra" (112, 128, 117, and 352 per 100 g fresh weight, respectively) [59]. The main anthocyanins detected in purple basil extract were cyanidin-based *p*-coumaril and malonil acids, acting as powerful antioxidants with potential use as medicinal agents [68]. Tenore and co-workers also demonstrated in the same study that extracts of "Napoletano" green and purple basil both had a broad antimicrobial spectrum able to reduce the growth of all human pathogenic and food spoilage bacteria and molds tested [67].

Regarding the health-promoting effects of basil, several preclinical and clinical studies showed that extracts from basil, particularly the purple one, may alleviate hyperglycemia associated with type 2 diabetes [69,70]. The anti-diabetic beneficial effects of basil extract may be due in part to catechin and especially to rosmarinic acid, which has been found to inhibit key enzymes such as α-amylase, α-glucosidase, and aldose reductase [71–73]. In addition to being anti-diabetic, basil extract may also be an efficient tool against hyperlipidemia by effecting lower cholesterol and triglyceride levels in the blood [74,75]. Reduced uptake of lipids and lower values of total cholesterol and low density lipoprotein may reduce the risk of cardiovascular diseases.

2.3. Perilla

Perilla (*Perilla frutescens* L. Briit) belongs to the Lamiaceae family (formely Labiateae) which consists of 235 genera and more than 700 species [76]. Perilla is an edible herb widely consumed in Asian countries such as China, Korea, Japan, and India. Similar to spinach, perilla leaves are also characterized by a high concentration of carotenoids. In fact, according to Müller-Waldeck et al. [77], perilla may contain high contents of carotenoids, especially up to five-fold higher lutein than other carotenoid-rich leafy vegetables. The leaves of perilla contain a range of bioactive phenolic molecules such as caffeic acid, catechin, chrysoeriol, ferrulic acid, luteolin, quercetin, and rosmarinic acid [78,79]. In particular, secondary metabolites such as rosmarinic acid and perillaldehyde (an essential oil constituent) have demonstrated potential to prevent a wide range of diseases particularly owing to their anti-diabetic, anti-depressant, anti-bacterial, anti-cancer, and antimicrobial properties [76]. Thus the concentrations of these two phytochemicals are crucial for their clinical and culinary applications. It is worth noting that these two secondary metabolites are produced in perilla by two different biosynthetic pathways, namely the monoterpene and phenylpropanoid pathways, and may increase independently in relation to the perilla chemotype and abiotic environmental stress conditions [80]. Perilla is present in nature in two main chemical-varietal phenotypes: (i) The red-pigmented cultivar *P. frutescens* var. *crispa*, known as "Zi-So" and widely grown in China where it is used as a spicy herb, leafy vegetable, and medicinal plant, and (ii) the non-pigmented green cultivar *P. frutescens* var. *frutescens* known in Japan as "Shisoyo" or "Shiso" and mainly used as an oil crop but also as ingredient of skin creams and food products [81,82]. According to Meng et al. [78,83], three cinnamic derivates (caffeic acid, coumaroyl tartaric acid, and rosmarinic acid) ranged from 0.1 to 11 mg/g; six flavonoids (apigenin 7-*O*-caffeoylglucoside, apigenin 7-*O*-diglucuronide, luteolin 7-*O*-diglucuronide, luteolin 7-*O*-glucuronide, scutellarein 7-*O*-diglucuronide, and scutellarein 7-*O*-glucuronide) ranged from 3.5 to 18.5 mg/g; and six anthocyanins (0.7–2 mg/g) including cis-shisonin, cyanidin 3-*O*-(*E*)-caffeoylglucoside-5-*O*-malonylglucoside malonylshisonin and shisonin were detected on eight tested cultivars of perilla. Concerning the health effects of perilla extract, Narisawa et al. [84] reported that perilla leaves carry anti-tumor properties. In their work, the authors showed that treatment of female rats with a 12% fat diet based on perilla extract and safflower oil in 1:3 or 1:1 ratio effected better protection against colon cancers as compared to safflower oil alone [84].

Comparing green (Korean cultivar) and red-pigmented (Japanese cultivar) perilla, Rouphael et al. [82] observed that green perilla produced exclusively perilla ketone (PK), whereas the red perilla contained perillaldehyde (PA). Similar results were reported by Martinetti et al. [81] in a study profiling two red-leaf ("Aka Shiso" and "Purple Zi Su") and three green-leaf cultivars ("Ao Shiso", "Qing Su", and "Korean perilla") with the later containing PK instead of PA. The terpenoid component present in green-pigmented perilla has been demonstrated to be toxic for cattle and horses, since PK is considered a potent lung toxin [77,85], but the health-effect as well as the toxic dose/concentration to humans is still controversial, therefore Müller-Waldeck and co-workers [77] concluded that some Korean genotypes (green-pigmented cultivars) are not suitable/recommended for fresh consumption. Interestingly, several authors showed that PA and PK present in red and green perilla can stimulate the TRPA1 (Transient Receptor Potential) cation channels which are actively involved in multiple biological mechanisms such as pain perception and their functional role in the prevention of certain types of tumor has been proved [86,87].

2.4. Swiss Chard

Green chard also known as Swiss chard (*Beta vulgaris* var. *cicla* L.) belongs to the Amaranthaceae-Chenopodiaceae family and is considered an important leafy vegetable grown for its green or reddish leaves and the white, yellow, or red leaf stalk. Green beet belongs to the same family of the root vegetable red beet (*Beta vulgaris* var. *rubra* L.). Traditionally, Swiss chard has been employed for its health-promoting properties as folk remedy for liver/kidney diseases, for triggering the hematopoietic and immune systems and also as a target diet in some tumors treatment [88]. As they grow, Swiss chard leaves accumulate a wide range of macro and micro minerals such as P, K, Ca, Mg, and Fe and several lipophilic vitamins (such as A and E and also carotenoids), as well as hydrophilic vitamins (such as B3, B5, B9, and C) [89]. According to Mzoughi et al. [90] Swiss chard leaves have a nutritional and functional profile catering to modern human diets. In the latter study, Swiss chard leaves were characterized by high concentrations of secondary metabolites such as (myricitrin, *p*-coumaric, and rosmarinic acid), flavonoids, carotenoids (β-carotene, chlorophyll, and lycopene) and some target volatile compounds (decanal, E-anethole, and octanoic acid). Mzoughi and co-workers demonstrated that the high antioxidant capacity on ABTS and DPPH of Swiss chard ethanol extract was accompanied with significant inhibitory effects on α-amylase and α-glucosidase; thus the Swiss chard extract could be explored in the near future as potential functional food with antioxidant and anti-diabetic properties [90].

In a recent review paper, Ninfali et al. [91] reported that green beet extract may regulate the hematic concentration of glucose, decrease lipid peroxidation, lower triglycerides and cholesterol levels, and improve glutathione levels. The health protective secondary metabolites found in *B. vulgaris cicla* have been identified as a class of G-Glycosyl flavonoids including (i) isovitexin, (ii) vitexin, (iii) vitexin-2-*O*-xyloside and iv) vitexin-2-*O*-rhamnoside, which are characterized by high biological activity [91]. According to Lee et al. [92], vitexin is able to reduce drastically the mitochondrial membrane potential in leukemia cell. Similarly, Nifali et al. [93] and Gennari et al. [94] reported that vitexin-2-*O*-xyloside and vitexin-2-*O*-rhamnoside were able to reduce the proliferation rate of

MCF-7 breast and RKO cancer cells. Concerning the anti-inflammatory properties of Swiss chard, Borghi et al. [95] demonstrated that the administration of 10 mg/kg of vitexin is able to decrease the levels of pro-inflammatory cytokines. Overall, in vitro and in vivo experiments carried out on animals and humans demonstrated that the biological activity of vitexin, vitexin-2-*O*-xyloside, and vitexin-2-*O*-rhamnoside can trigger the expression of a wide range of genes associated with inhibition of cancer cell proliferation and anti-inflammation activities.

2.5. Brassica Leafy Vegetables

The group of brassicaceous leafy vegetables, formerly referred to as cruciferous vegetables, includes a wide range of species with potential health-promoting properties such as kale (*Brassica oleracea* var. *sabellica*), pack choi (*Brassica rapa* var. *chinensis*), mizuna (*Brassica rapa* var. *japonica*), watercress (*Nasturtium officinale* R.Br.), wild and salad rocket (*Diplotaxis tenuifolia* [L.] DC and *Eruca cesicaria* [L.] Cav., respectively). According to FAOSTAT [27] database, roughly 12% of the vegetables grown worldwide are members of the Brassicaceae family. Recent reports [96,97] have suggested that brassicaceous leafy vegetables constitute valuable sources of phytochemicals. They contain high levels of vitamins (C, E [as α- and γ-tocopherols] and K [phylloquinone]), carotenoids, and phenolic compounds. In addition to the latter phytochemicals, *Brassica* leafy vegetables are characterized by sulfur-containing glucosinolates and methylcysteinsulfoxide compounds. The genetic factor is the most important and influential one in terms of modulating the biosynthesis and accumulation of phytochemicals in *Brassica* leafy vegetables [97]. In a comparative study of antioxidant molecules in four *Brassica* leafy vegetables (mizuna, salad rocket, watercress, and wild rocket), the authors observed a large variability in phytochemical concentrations [98]. For instance, watercress showed the highest polyphenol and vitamin C content, while salad and wild rocket were characterized by high concentrations of kaempferol and quercetin derivatives and finally mizuna exhibited significant concentrations of isorhamnetin and sinapic acid [98]. The authors highlighted the potential value of salad *Brassica* leafy vegetables as dietary sources of antioxidants conferring a wide range of positive health effects against type 2 diabetes and cardiovascular diseases.

Kale is a leafy *Brassica* species considered a potent source of glucosinolates and isothiocyanates, their main breakdown products. A study from the CDC (center of disease control and prevention) reported kale being ranked 15th among "powerhouse" vegetables and fruits [99]. The most abundant glucosinolates in kale were: 3-(methylsulphinyl)propyl, 2-propenyl and also 4-(methylsulphinyl)butyl glucosinolates. Genotypic variation within eight cultivars of kale ("Starbor", "Beira, "Scarlet", "Premier", "Olympic Red", "Toscano", "Dwarf Siberian", and "Red Russian") revealed that "Beira" and "Olympic Red" were characterized by the highest total concentration of glucosinolates and were proposed as functional foods [100]. In another experiment carried out by Hahn et al. [101] on 25 kale cultivars, the authors observed a great variation in glucosinolate profiles. Similarly, Ferioli et al. [102] reported higher variation in the aliphatic compared to the indole glucosinolates (9- and 5-fold, respectively) across 25 kale cultivars harvested from different European countries (Italy, Portugal, and Turkey). The presence of isothiocyanates in *Brassica* leafy vegetables has been reported to confer anti-diabetic, anti-inflammatory and anti-cancer properties [103–108]. Although less popular, even "ornamental cabbage" or kale (*Brassica oleracea* L. var. *acephala* DC.) have genotypes characterized by the accumulation of different pigments [109–111]. *Brassica* leafy vegetables, in particular kale, are considered additionally as a rich source of carotenoids (lutein and β-carotene), as well as chlorophylls (a and b). Carotenoid concentrations of 33 kale cultivars were analyzed and quantified [112]. Zeaxanthin was the most abundant carotenoid in 21 cultivars. Moreover, American and hybrid cultivars and accessions were characterized by high concentrations of zeaxanthin, whereas, German landraces, German commercial varieties, Italian, and red-colored kale varieties exhibited high concentrations of chlorophyll a and b [112].

Emerging market trends catering to shifting consumer perceptions of quality [32], have resulted in colored *Brassica* leafy vegetables (e.g., violet kale or pack choi) containing anthocyanins garnering the attention of nutritionists and horticultural scientists. Recently, Mageney et al. [112] proposed that the anthocyanin content could be used as a marker to differentiate between varieties/cultivars. Testing green and red-pigmented pack choi, Zheng et al. [113] observed that red pack choi produced higher concentrations of carotenoids, total phenolic compounds, total flavonoids, glucosinolates, and anthocyanins compared to its green counterpart. Importantly, the regular intake of anthocyanins from such colored leafy vegetables has been positively correlated with the prevention of various liver diseases, and also with the reduction of colon cancer, hepatic inflammation and oxidative stress [114].

The main pigments isolated in the studied leafy vegetables are presented in Table 1.

Table 1. The main pigments isolated in various leafy vegetables.

Species	Color	Class of Compounds	Compounds (Content on a Fresh Weight (fw) or Dry Weight (dw) Basis)	References
Lettuce (*Lactuca sativa* L.)	Red	Anthocyanins	cyanidin (1558.0–3656.9 μγ/g dw), cyanidin-3-O-(6″-malonyl-β-glucopyranoside), cyanidin-3-O-(6″-malonyl-β-glucopyranoside methyl ester), cyanidin-3-O-β-glucopyranoside, cyanidin-3-glucoside (1.40–3.07 g/100 g fw)	[42,43,46,48]
		Carotenoids	all-*E*-violaxanthin (23.9–33.4 μg/g fw), 9′-*Z*-neoxanthin (11.3–14.6 μg/g fw), all-*E*-Luteoxanthin, all-*E*-lactucaxanthin (19.7–23.0 μg/g fw), all-*E*-lutein (31.3–38.2 μg/g fw), and all-*E*-β-carotene (9.0–13.3 μg/g fw)	[42]
	Green	Carotenoids	all-*E*-violaxanthin (15.9–37.1 μg/g fw), 9′-*Z*-neoxanthin (5.0–11.4 μg/g fw), all-*E*-Luteoxanthin, all-*E*-lactucaxanthin (7.5–17.4 μg/g fw), all-*E*-lutein (16.4–36.3 μg/g fw), and all-*E*-β-carotene (4.2–12.9 μg/g fw)	[42]
		Chlorophylls	Chlorophyll a and b (6.95–26.92 mg/100 g fw and 4.60–10.30 mg/100 g fw, respectively)	[43,48]
		Anthocyanins	Cyanidin-3-glucoside (0.192–0.260 g/100 g fw)	[48]
Basil (*Ocimum basilicum* L.).	Purple/red	Anthocyanins	Anthocyanin A (0.325–0.423 mg/g dw),-anthocyanin B (0.057–0.641 mg/g dw), anthocyanin C (0.362–0.877 mg/g dw), anthocyanin D (0.063–0.662 mg/g dw), cyanidin-based (1.78–3.18 mg/g dw) pigments- and peonidin-based (19.8% fw) pigments, cyanidin-3-(6,6′-di-p-coumaroyl)-sophoroside-5-glcucoside (7.5 mg/g extract)	[63,68,115,116]
	Green	Carotenoids	Lutein (4.99–6.64 mg/100 g fw or 22.1–24.0 mg/100 g dw), β-carotene 4.42–6.01 mg/100 g fw, zeaxanthin (0.30–0.45 mg/100 g fw or 1.8 mg/100 g dw)	[62,64]
Perilla (*Perilla frutescens* L. Briit)	Red	Anthocyanins	Cyanidin and cyanidin derivatives (6.44 mg/g dw), shisonin (0.126–0.416 mg/g dw), malonylshishonin (0.462–1.116 mg/g dw)	[83,117–120]
Swiss chard (*Beta vulgaris* var. *cicla* L.)	Yellow	Betaxanthins	Vulgaxanthin I, miraxanthin V	[121,122]
	Red/purple	Betacyanins	Betanin, isobetanin, betanidin, and isobetanidin	[123]

3. Fruit Vegetables

3.1. Tomato

Tomato (*Solanum lycopersicum* L.) is an important fruit vegetable, widely consumed all over the world due to its rich nutrient content, special taste, and diverse ways of consumption (fresh, soups, juices, purees, dried, and sauces) [124]. Fruit color and pigments content are two important traits that largely reflect tomato fruit quality, as well as the antioxidant activity which is mainly correlated to the hydrophilic (e.g., soluble phenolic compounds and vitamin C) than to the lipophilic compounds (e.g., carotenoids, vitamin E, and lipophilic phenols) [125]. Fruit color is mainly related to pigments content, such as chlorophyll, carotene, lycopene, phytoene and anthocyanin, and their relative proportions at different maturity stage [126]. The most abundant carotenoid is lycopene, followed by phytoene, phytofluene, ζ-carotene, γ-carotene, ß-carotene, neurosporene, and lutein [127]. Color development is due to the chlorophyll degradation and the synthesis of carotenoids as fruit is developed and ripen. Therefore, the genetic development of tomato fruit color and pigments content is an interesting research area to improve fruit quality and satisfy the diverse consumers' demands [128].

The consumption of tomato and related food products is associated with the decrease of various diseases incidence such as chronic degenerative diseases, cardiovascular disease, and age-related macular degeneration (AMD) in human health [129]. Raiola et al. [124] reported the nutritional importance of tomato phytochemicals against inflammation processes and prevention of chronic non-communicable diseases (e.g., obesity, diabetes, coronary heart disease, and hypertension). Anthocyanins normally are not produced in tomato fruit, however, some wild tomato species, such as *S. chilense*, *S. cheesmaniae*, *S. lycopersicoides*, and *S. habrochaites* biosynthesize anthocyanins in the sub-epidermal tissue of the fruit, and some alleles from those genotypes have been introgressed into cultivated genotypes [130]. Therefore, combining the dominant Anthocyanin fruit (Aft) gene from *S. chilense* and the recessive atroviolacea (atv) gene from *S. cheesmaniae* into a cultivated tomato background, anthocyanins biosynthesis has been achieved [131,132].

Purple tomatoes have antioxidants and phytochemical properties in both flesh and peel, often in superior levels than those found in conventional red tomatoes [133]. A genetically modified (GM) purple tomato was found to have additional health-promoting effects by prolonging the life of cancer-susceptible mice compared to tomatoes with conventional (red) color [134]. Extracts from fruit of purple tomato (breeding line V118) showed significant and dose dependent anti-inflammatory effect against paw edema in an in vivo study with rat models (edema inhibition: 7.48%–13.8%), suggesting that anthocyanins may play a role in the anti-inflammatory effect [135]. Interestingly, during the last 20 years there has been an increasing interest in developing highly consumed food, such as flavonoids-rich tomato fruit. To that direction, transgenic approaches have been applied to modify the biosynthesis of phenylpropanoids, in order to alter the tomato flavonoid biosynthesis [130].

Phenolic content is varied at different developmental stages of tomatoes, as sun black (SB) tomato had 5.8 and 8.6 mg GAE/g dw phenolic content at mature green and red ripening stage, respectively, contents that were 152% and 134% higher than wild type (WT) [130]. Li et al. [136] reported similar total phenolic content (659.11 mg GAE/100 g dw) for purple tomato as has been reported for other tomato varieties (from 290 to 500 mg GAE/100 g dw) [137]. Individual components of phenolics may also vary among purple and red varieties, as the main phenolic compounds content (chlorogenic acid, naringenin, and rutin) was higher (65.56, 12.82, and 52.39 mg/100 g dw, respectively) [136] in purple tomatoes compared with red tomatoes were chlorogenic acid, naringenin, and rutin content was 2.67, 1.84, and 6.61 mg/100 g dw [138] and 16.7, 2.2, and 16.9 mg/100 g dw, respectively [139]. Apart from chlorogenic acid, naringenin, and rutin Li et al. [136] reported other phenolic compounds in purple tomatoes such as p-coumaric acid (15.68±0.74 mg/100 g dw), gentistic acid (15.25 ± 0.76 mg/100 g dw), ferulic acid (14.51 ± 0.99 mg/100 g dw), caffeic acid (13.65 ± 0.83 mg/100 g dw), and protocatechuic acid (8.95 ± 0.16 mg/100 g dw). Indeed, chlorogenic acid content depends on the developmental stage of fruit and varied from 0.5 to 1.3 mg/g dw in sun black (SB) tomato extracts [130]. Moreover, composition of flesh and peel in tomato mutants differed as flesh contained chlorogenic acid (7.96 ± 0.75 mg/100 g fw), quercetin (5.03 ± 1.02 mg/100 g fw), luteolin (0.45 ± 0.01 mg/100 g fw), and total phenolics (1.18 ± 0.06 mg/g fw), while in peel the respective compounds content was: chlorogenic acid (8.43 ± 0.15 mg/100 g fw), quercetin (5276.15 ± 15.10 mg/100 g fw), luteolin (21.28 ± 1.07 mg/100 g fw), and total phenolics (5.95 ± 0.27 mg/g fw) [133].

Among the several phytochemicals identified in plants, *cis*- and *trans*-resveratrol (3,4,5-trihydroxystilbene) are polyphenols that belong to the stilbene class; however, only the *trans* form is biologically active in the human body [140]. Numerous biocidal activities exerted by resveratrol have been reported such as antioxidant, antidiabetic and estrogenic activity, anticancer effects through the preservation of the regular cell cycle, the inhibition of tumor invasion and angiogenesis, and cardiovascular effects through the reduction of the expression of endothelial adhesion cells and the inhibition of cell apoptosis and platelet aggregation [141–145]. However, resveratrol's daily intake by humans has still to be established [143]. Vagula et al. [146] quantified *trans*-resveratrol in *S. americanum* Mill. fruit, which ranged between 1.07 and 0.796 µg/g for fruit pulp and peel, respectively, and these levels were significantly higher when compared to freeze-stored fruit (0.1353 µg of *trans*-resveratrol/g of sample) and to other berries [146].

Vasco et al. [147] reported the higher antioxidant capacity of purple (purple-red variety) tamarillo or tree tomato (*Solanum betaceum* Cav.), compared to the golden-yellow variety and reported 9.3 µmol trolox/g fw, 3.0 µmol trolox/g fw and 40 µmol trolox/g fw for seed-jelly, pulp, and peel tissues of purple fruits compared to 3.8 µmol trolox/g fw, 2.3 µmol trolox/g fw and 22 µmol trolox/g fw for seed-jelly, pulp and peel tissues of yellow fruits, respectively. Similarly, Sestari et al. [133] reported increased antioxidant capacity (DPPH) in peel (38.12 ± 4.27 µmol trolox/g fw) compared to flesh (8.64 ± 0.45 µmol trolox/g fw) in tomato mutants. Interestingly, the oxygen radical absorption capacity assay (ORAC) value for the hydrophilic extracts in purple tomato reported by Li et al. [136] was 323.23 µmol trolox/g dw, which was 2-fold higher than the ORAC value of the traditional tomato cultivar San Marzano (140 µmol trolox/g dw) reported by Ninfali et al. [148]. Similar observations were made by Blando et al. [130] who stated 3-fold higher trolox equivalent antioxidant capacity (TEAC) value (31.6 µmol trolox/g dw) in sun black tomatoes compared with the wild type (10.3 µmol trolox/g dw) at the red-ripening stage. Moreover, antioxidant capacity of purple fruit was higher at the ripe (red ripe stage-RR) fruit compared to the unripe (mature green stage-MG) ones, probably explained by the great increase in polyphenols accumulation during ripening (from 5.8 to 8.6 mg GAE/g dw, in MG and RR, respectively) [130]. In the same study, total ascorbic acid content was higher in sun black than wild type fruits (37.3 ± 1.4 vs. 27.1 ± 1.1 mg 100/g fw, respectively) [130].

The antioxidant capacity of fruit is not only related to phenolics and ascorbic acid but also to the carotenoids, flavonoids, and anthocyanins content. Lycopene, the main phytochemical of tomatoes, is known for its important role in human health related functionalities [149]. Lycopene supplementation in an in vivo study with iodoacetamide-induced colitis rats showed reduced tissue malondialdehyde (MDA) levels, the histological signs of colon injury, and increased superoxide dismutase levels in the red blood cells [150]. In the study Li et al. [136], lycopene was the dominant carotenoid (185.01 µg/g dw) in breeding line V118, followed by β-carotene (47.11 µg/g dw) and lutein (2.66 µg/g dw). About 8.1% of the total carotenoids in V118 were *cis*-carotenoids, a lower value compared to that of most of the tomato varieties studied [136]. Moreover, Li et al. [128] reported that purple fruit (cv "Zi Ying") had increased antioxidant capacity compared to green fruit (cv "Lv Ying"), with lycopene content of 36.51 ± 2.86 mg/kg fw and β-carotene 13.38 ± 1.31 mg/kg fw in purple fruit versus lycopene content of 1.35 ± 0.05 mg/kg fw and β-carotene 6.80 ± 0.32 mg/kg fw in green fruit. In a comparative study, Hazra et al. [151] pointed out the dietary role of purple tomato (AftAft dgdg genotype) due to the increased values in ascorbic acid (31.56 ± 2.41 mg/100 g fw), lycopene (6.13 ± 0.39 mg/100 g fw), β-carotene (0.65 ± 0.14 mg/100 g fw), and anthocyanin (20.73 ± 2.86 mg/100 g fw), compared to the overall mean value of 31 hybrids.

Li et al. [136] reported a total carotenoid content of breeding line V118 of 234.78 µg/g dw, being within the range of the average amounts reported for red tomatoes (132–583 µg/g dw) [152]. At the red ripe stage, total carotenoids did not differ between the sun black (SB) and wild type (WT) tomatoes; however, the β-carotene content was significant higher in the SB sample, whereas the lycopene content was lower [130]. Similarly, Vasco et al. [147] reported higher β-carotene levels in purple tamarillo than in yellow tomato variety. Generated double and triple mutants (Anthocyanin fruit/high pigment 2(Aft/hp2) and Anthocyanin fruit/atroviolacium/high pigment 2 (Aft/atv/hp2)) of purple tomatoes had higher lycopene and β-carotene levels and up to 63% of vitamin C compared to tomato cultivar Micro-Tom, suggesting accumulating trends of relevant phytochemicals in near-isogenic lines [133].

Anthocyanins, the most abundant flavonoid constituents in pigmented fruit and vegetables, posse potential health beneficial effects, such as antioxidant, anti-inflammatory, anticancer, and antidiabetic activities [153,154]. Anthocyanins also had notable effects against inflammation by inhibiting cyclooxygenase-2 (COX-2) expression, inducible nitric oxide protein and mRNA expression [155]. The "Giant" and "New Zealand" purple cultivars, had total anthocyanins content of 102.35 ± 1.46 mg/100 g dw and 168.88 ± 2.65 mg/100 g dw, but also revealed high antioxidant activity which might be related to their overall phenolic composition [156]. Zhang et al. [157] reported the role of anthocyanins in postharvest storage of tomatoes as in purple tomatoes, anthocyanins doubled the self-life of fruit by delaying over-ripening and reducing susceptibility to *Botrytis cinerea*.

Li et al. [136] through an LC-MS study reported three major anthocyanins, which were mainly acylglycosides of petunidin and malvidin. Among these anthocyanidins, petunidin was the predominant aglycone (91.9%), and the rest of the minor aglycones accounted for only 9.1% of the total anthocyanidins [136]. Moreover, petunidin is not usually synthesized in vegetables and fruit, and little is known about its health benefits, however in tomato mutants petunidin revealed considerable amounts (>60 mg/100 mg fw) in fruit peels of the lines combining Aft and hp2 genes [133]. The total anthocyanin content in breeding line V118 was 72.31 mg/100 g dw, including 9.04, 50.18, and 13.09 mg/100 g dw of petunidin-3-*O*-caffeoyl-rutinoside-5-*O*-glucoside, petunidin-3-*O*-(*p*-coumaryl)-rutinoside-5-*O*-glucoside, and malvidin-3-*O*-(*p*-coumaryl)-rutinoside-5-*O*-glucoside, respectively [136]. Moreover, Blando et al. [130] reported that petanin (Petunidin 3-(6-(4-(*E-p-* coumaroyl)rhamnosyl)glucoside)-5-glucoside (petanin)) and negretein (Malvidin 3-(6-(4-(*E-p*-coumaroyl)rhamnosyl)glucoside)-5-glucoside) represented 56.6% and 21.4% of the total anthocyanins content in sun black (SB) fruit peel, respectively, whereas no anthocyanins were detected in wild type (WT) tomato fruit.

The content of anthocyanins in the Del/Ros1 transgenic tomato is equally distributed within fruit, with 5.1 ± 0.5 g/kg dw being detected in the peel and 5.8 ± 0.3 g/kg dw in the flesh, but not detected in seeds [7]. These values are higher than those reported for well-known anthocyanin-rich foods such as red raspberry (3.9 g/kg dw; [158]), strawberry (3.2 g/kg dw; [158]), and mulberry (2.1 g/kg dw; [159]). In a study with transgenic plants, the predominant anthocyanins in the Del/Ros1 transgenic tomato were delphinidin-3-(trans-coumaroyl)-rutinoside-5-glucoside and petunidin-3-(trans-coumaroyl)-rutino side-5-glucoside, which contributed to nearly 86% of the total anthocyanins content, while two new anthocyanins, malvidin-3-(p-coumaroyl)-rutinoside-5-glucoside and malvidin-3-(feruloyl)-rutinoside-5-glucoside making up to 6% of the total anthocyanins content, were also reported [7]. Three mutant genes have been identified that can lead to the production of anthocyanins in the peel of the fruit, namely Anthocyanin fruit (Aft), Aubergine (abg), and atroviolacea (atv), while the Aft gene was also identified in crosses with *Solanum chilense* Dunal [160]. This gene is located in chromosome 10 and its presence in tomato leads to the production of anthocyanin pigments, mainly delphinidin, malvidin and petunidin, as well as to higher levels of the flavonols quercetin (3.6-fold), and kaempferol (2.7-fold), in tomato fruit [160,161].

Tamarillo crop is attracting research interest lately due to the high content in antioxidants and phytochemicals. The tamarillo, a non-climacteric edible fruit, is quite popular in local markets, especially in South America, consumed in juices or fresh and being highly appreciated due to high polyphenols levels [156], β-carotene (provitamin A), vitamin B6, vitamin C (ascorbic acid), vitamin E, and iron contents [162]. Among phenols, the presence of anthocyanins (delphinidin, cyanidin, and pelargonidin glycosides) and hydroxycinnamoyl derivatives (e.g., 3-*O*-caffeoylquinic acids, caffeoyl glucose and feruloyl glucose) have been described in several reports [147,163,164], while recently rosmarinic acid has been also identified [156]. The hydroxycinnamoyl derivatives show antioxidant properties and have been related to protective effects on human health [165]. In particular, the caffeoyl ester of rosmarinic acid has various biocidal activities, such as antiviral, antibacterial, anti-inflammatory, and antioxidant effects [166]. Other compounds were tentatively identified as different rosmarinic acid glucosides, caffeoyl glucoside, feruloyl glucoside, and ferulic acid dehydrodimers. Pelargonidin

3-*O*-rutinoside and delphinidin 3-*O*-rutinoside were the main anthocyanins in purple cultivars of tomato fruit [167]. Vasco et al. [147] reported anthocyanins content of 38 mg/100 g fw in purple tamarillo which was higher than previous reports in yellow fruit (8.5 mg/100 g fw) [167].

Both *Solanum americanum* Mill. and *S. villosum* Mill. are important medicinal plants of the Solanaceae family, however the blackish-purple (*S. americanum*) and reddish-orange (*S. villosum*) colored fruit are mostly consumed in India, Ethiopia, Ghana, China, and Brazil [168]. Mohy-Ud-Din et al. [169] reported the different important steroidal glycoalkaloids like β-Solamargine, α-Solamargine, Solasonine, α-Solanine, solasodine, and Solanidine, with latter being well-recognized for its anticancer activities [170].

3.2. Eggplant

Eggplant (*Solanum melongena* L.) fruit are very popular vegetables grown worldwide in subtropical and tropical regions [171]. They considered as one of the top 10 vegetable in terms of antioxidant capacity [172] and contain a variety of antioxidants and phytochemicals such as, ascorbic acid, phenolics, and flavonoids that provide health benefits [173]. The most abundant phenolic compound is 5-*O*-caffeoylquinic acid, known as chlorogenic acid (ChA), which is considered as the main contributor to the overall antioxidant capacity [174,175]. However, eggplant fruit are poor sources of provitamin A and vitamin E, with average values of 27 IU 100/g fw and 0.30 mg/100 g fw, respectively [173]. American purple fruit are the most commonly marketed type, though white cultivars have gained consumers acceptance in recent years.

Eggplants are quite versatile vegetables and could be subjected to a number of different processing and cooking methods which may further affect fruit antioxidant capacity [176]. Akanitapichat et al. [173] reported that the antioxidant activities of eggplant were correlated (r = 0.531–0.796) with the total amounts of phenolics and flavonoids. Significant correlation was found between hepatoprotective activities and total phenolics/flavonoids content (r = 0.637–0.884) and antioxidant activities (r = 0.585–0.958), indicating the contribution of the polyphenols present in eggplant to its hepatoprotective effect (human hepatoma cell line HepG2) against tert-Butylhydroperoxide (t-BuOOH)-induced toxicity [173]. Akanitapichat et al. [173] reported that total phenolics content in purple fruit was of 1002.67 ± 8.33 mg GAE/100 g extract. Nisha et al. [176] suggested a higher content of total phenolics and anthocyanins in a purple small-sized fruit variety (106.98 mg/100 g fw and 0.756 mg/100 g fw, respectively) than the three other examined varieties (purple moderate-sized fruit (80.31 mg/100 g fw and 0.525 mg 100/g fw, for total phenolics and anthocyanins, respectively), green long-sized (50.79 mg/100 g fw and 0.0475 mg/100 g fw, for total phenolics and anthocyanins, respectively), and purple big-sized (49.02 mg/100 g fw and 0.53 mg/100 g fw, for total phenolics and anthocyanins, respectively)). In the same study, it was reported that purple small-sized variety revealed the greatest antioxidant activity.

Flavonoids represent only about 10–15% of total phenolics [177] and hydroxycinnamic acid derivatives, particularly free chlorogenic acid is the major phenolic antioxidant regardless of the genotype [177,178]. However, the chlorogenic acid content in eggplant is influenced by both genetic and environmental factors, including the fruit developmental stage, cultivar, and crop and postharvest management [179]. For example, the chlorogenic acid content was 16% higher in purple eggplants compared to white eggplant slices [180]. According to Akanitapichat et al. [173] who compared five eggplant varieties, the purple fruit had higher total flavonoids content (3954.2 ± 6.06 mg catechin equivalents-CE/100 g extract) content compared with green and white ones. Sadilova et al. [181] reported that flavonoids isolated from *S. melongena* showed potent antioxidant activity against chromosomal aberrations induced by Doxorubicin. The purple eggplant had high antioxidant activity of DPPH and ABTS with EC$_{50}$ of 66.74 ± 4.60 µg/mL and 53.18 ± 0.71 µg/mL, compared with green and white varieties [173].

In purple-fruited genotypes of pepper and eggplant the abundance of anthocyanin levels is superior in unripe fruits and decrease upon ripening, often to complete disappearance [182]. It is noteworthy that eggplant fruit reaches its commercial maturity long before its physiological ripeness, as practically it is harvested at the immature fruit stage [183]. The anthocyanin concentration in the purple fruit eggplant cultivars is higher in comparison to other deeply colored fruits and vegetables, e.g., 2.34-fold that of grapes, and 7.08-fold that of red onions [5]. In purple pigmented eggplants, the antioxidant anthocyanins (delphinidin derivatives) is limited as found at the peel tissue which represents less than 5% of the total fruit weight [178,184]. Examining the anthocyanins content of wild type (WT), purple-black (S9-1), green (L6-4), and white (U36-1) eggplants, Xi-Ou et al. [185] found that the anthocyanin content of purple-black (S9-1) was higher than that in WT eggplant, while green eggplant (L6-4) had the lowest levels of anthocyanins. In another study, Zhang et al. [186] reported the total anthocyanins content from purple eggplant (cv Zi Chang) skin to be 1.24 mg/g dw. Moreover, nasunin (delphinidin-3-(*p*-coumaroylrutinoside)-5-glucoside), an anthocyanin isolated from the skin of purple eggplant fruit, is associated with both inhibition of hydroxyl radical generation and superoxide scavenging activity [173,187].

Extracts from eggplant fruit skin were demonstrated to possess high capacity in scavenging of superoxide free radicals and inhibiting hydroxyl radical generation by chelating ferrous iron [187]. Additionally, eggplant extract resulted in hypolipidemic activity in rats fed normal as well as high fat diets [188], suppressed tumor growth and metastasis [189], and inhibited inflammation that can lead to atherosclerosis [190]. Not only fruit but various parts of the plant are useful in the treatment of inflammatory conditions, cardiac debility, neuralgia, ulcers of nose, cholera, bronchitis, and asthma, while they possess analgesic and hypolipidemic properties [191]. *S. melongena* is also a natural source of vitamin A affecting the eye health in children [192].

3.3. Pepper

Pepper (*Capsicum* spp.) is one of the oldest domesticated and utilized crops and the genus *Capsicum* consists of approximately 31 species of which the five domesticated species are *C. annuum*, *C. baccatum*, *C. chinense*, *C. frutescens*, and *C. pubescen* [193]. Average world production and cultivated area of dry and green peppers are estimated at 3.9 and 34.5 million tons respectively, harvested from 1.8 and 1.9 million hectares respectively [27]. Pepper fruit have high nutritive value, as they are rich in vitamin C (ascorbic acid), provitamin A (β-carotene), vitamin E (tocopherols), flavonoids and capsaicinoids, and other carotenoid pigments such as lycopene and zeaxanthin [194]. The noticeable level of phenolic compounds and carotenoid pigments also contributes to the antioxidant properties of sweet pepper [195].

Immature fruits are usually colored in white, green, purple, and black shades and gradually color changes to yellow, orange, red and brown as fruit maturity advances [196,197]. The differences in fruit color is mainly due to the differential accumulation of flavonoids and carotenoids [198]. Anthocyanin accumulation in the outer epidermis of immature pepper fruit is responsible for the purple or black color at 30 d after anthesis [199] and turns to red color at 50 d after anthesis [182].

Liu et al. [182] compared the flavonoids biosynthesis at 30 d after anthesis for green, white, and purple varieties, and reported that anthocyanins, flavones, and flavonols content was significantly higher in purple variety than in the other varieties, with delphinidin, luteolin, chrysoeriol and quercetin derivatives being the most abundant polyphenols. Delphinidin, cyanidin, and malvidin derivatives were the major anthocyanins in colored peppers among the 16 anthocyanins detected of which delphinidin 3,5-diglucoside and delphinidin 3-*O*-rutinoside were specifically accumulated in purple peppers [182]. In the same study, it was reported that the purple color of fruit is related to the high accumulation of cyanidin and delphinidin derivatives at 30 d after anthesis.

In chili peppers, anthocyanins' presence and pigmentation of purple or black and magenta is also possible, and is usually found in flowers, fruit, and foliage [200]. Several reports highlighted the presence of anthocyanins in chili pepper fruit, but so far delphinidin is the only anthocyanidin identified [200,201]. Similarly, Sadilova et al. [181] reported delphinidin-3-*trans*-coumaroylrutinoside-5-glucoside (nasunin 89%) and delphinidin-3-*cis* coumaroylrutinoside-5-glucoside (4.6%) as the main anthocyanins (averaged at 320 µg/g fw) in German chili pepper (*C. annuum* L.), while similar results were found in two Mexican chili peppers [202]. Moreover, hydroponically grown dark violet pepper (cv. "Zorro") had the highest concentration of quercetin and catechin when compared to orange, red and yellow fruit cultivars [203].

Capsanthin-capsorubin synthase (CCS), as a unique enzyme in pepper and tiger lily, converts antheraxanthin and violaxanthin into capsanthin and capsorubin, respectively [204]. Although the exact mechanism is under investigation, Liu et al. [182] reported that the highly active CCS drives antheraxanthin to be converted into capsanthin in purple fruit, which reduces the flux to violaxanthin, eventually resulting in significantly lower levels of antheraxanthin and violaxanthin in purple than in green and white fruit varieties.

3.4. Lablab and Common Bean

Lablab (*Lablab purpureus* L.), an ancient legume species, serves as a vegetable and is widely cultivated throughout the tropics, subtropics and temperate zones [205]. Fruit are green pods, 6 cm long by 2 cm wide, flattened, contain 4–5 seeds and turn light brown-purple when mature [206]. Al-Snafi [207] and Momim et al. [208] reviewed the phytochemical properties of lablab and its medicinal importance, exhibiting antidiabetic, anti-inflammatory, analgesic, antioxidant, cytotoxic, hypolipidemic, antimicrobial, insecticidal, hepatoprotective, antilithiatic, antispasmodic effects. Moreover, Momim et al. [208] and Deoda et al. [209] reported that the juice derived from the fruit pods was used as astringent, digestive, stomachic, to expel worms and for the treatment of inflamed ears and throats. Soetan [210] studied the pharmacological potentials of three varieties ("Rongai brown", "Rongai white", and "Highworth black") of *L. purpureus* seeds and showed that raw and aqueous extracts contained various phytochemicals including trypsin inhibitors, hemagglutinin, cyanogenic glycosides, oxalates, phytates, tannins, and saponins, with greater contents in raw material compered to aqueous extracts. Other biocidal effects have been reported including antilithianic activity [209], hepatopreotective effects [211], and inhibited trypsin and plasmin activity [212].

Momim et al. [208] reported significant antioxidant capacity (DPPH) with the lowest IC_{50} found in purple lablab compared to the white one (430.00 µg/mL *vs* 853.13 µg/mL). In the same study, total flavonoids content in purple fruit was 32.09 ± 0.36 mg quercetin equivalent/g fw while in green lablab it was 42.55 ± 5.77 mg quercetin equivalent/g fw. Bhaisare et al. [206] reported vitamin C content of 81.00 ± 0.16 mg/g and vitamin E content of 73.66 ± 0.08 mg/g in fresh bean seeds of *L. purpureus*.

Total content of anthocyanins in purple (cv. "Hong Fu") pods was about 1.58 mg/g, while low amounts were detected in green (cv. "Qing Feng") ones. Compared to green pods, five kinds of anthocyanins (malvidin, delphinidin, and petunidin derivatives) were found in purple pods by HPLC-ESI-MS/MS and the major compounds were identified as delphinidin derivatives [213]. Besides, nine kinds of polyphenol derivatives, namely quercetin, myricetin, kaempferol, and apigenin derivatives were detected by UPLC-ESI-MS/MS and the major components were quercetin and myricetin derivatives [213].

Common bean (*Phaseolus vulgaris* L.) is another legumes species with pods of varied colors, including black, red, blue, and violet [214,215]. Anthocyanins content may vary significantly depending on the genotype, while polyphenols content is highly associated with the antioxidant activities of pods [216]. Tsuda et al. [217] reported that pelargonidin-3-glucoside, cyanidin-3-glucoside, and delphinidin-3-glucoside isolated from *P. vulgaris* (black bean) seed coat, as well as their standard aglycones, have strong antioxidative activity in a liposomal system and reduced formation of malondialdehyde by UVB irradiation [217]. According to Mazewski et al. [218], purple beans contain mostly condensed tannins which are responsible for the antiproliferative activities against human colon cancer cell lines (HCT-116 and HT-29).

3.5. Pepino

Pepino (*Solanum muricatum* Aiton), a close relative to tomato and potato, is an herbaceous Andean domesticated species grown for its juicy, sweet, and aromatic fruit, with increasing commercial and export interest in South America from exotic fruit markets [219]. Fruit color may be white, cream, yellow, maroon, or purplish [220]. Unripe pepino fruit is green while 51 days from fruit set, newly-acquired purple stripes appear, resulting in fruit softening along with decreases in total pectin and hemicellulose content [221].

Various health benefits were revealed for pepino, including treatment of diabetes, stroke, high blood pressure, heartburn (indigestion), cancer, kidney, constipation, and hemorrhoids [220], activities mostly attributed to the significant amounts of vitamin C, carotenoids, and phenolics [222]. Moreover, Hsu et al. [222] reported the antioxidative, anti-inflammatory, and antiglycative effects of pepino extract. In the same study, aqueous and ethanol extracts had similar content of total phenolic acids (averaged at 1145 mg/100 g dw) but aqueous extracts were richer than ethanol extracts in terms of ascorbic acid (43.8 vs. 6.6 mg/100 g dw), total flavonoids (875 vs. 461 mg/100 g dw), cinnamic acid (75.7 vs. 23.0 mg/100 g dw), ferulic acid (82.3 vs. 11.8 mg/100 g dw), rosmarinic acid (47.2 vs. 8.4 mg/100 g dw), quercetin (126.5 vs. 90.3 mg/100 g dw), and naringenin (57.2 vs. 14.7 mg/100 g dw) [222].

The main pigments isolated in the fruit vegetables are presented in Table 2.

Table 2. The main pigments isolated in various fruit vegetables.

Species	Color	Class of Compounds	Compounds	References
Tomato *Solanum lycopersicum* L.	Red tomato	Carotenoids	Lycopene, phytoene, phytofluene, ζ-carotene, γ-carotene, β-carotene, neurosporene, lutein	[127,149]
	Purple tomato	Carotenoids	Lycopene (36.51–61.30 mg/kg fw), β-carotene (6.5–6.8 mg/kg fw), lutein	[128,136]
	Purple tomato	Petunidin and malvidin acylglycosides	petunidin-3-*O*-caffeoyl-rutinoside-5-*O*-glucoside (1.88–15.36 mg/100 g dw), petunidin-3-*O*-(*p*-coumaryl)-rutinoside-5-*O*-glucoside (16.97–50.18 mg/100 g dw)1, malvidin-3-*O*-(*p*-coumaryl)-rutinoside-5-*O*-glucoside (6.17-27.06 mg/100 g dw), delphinidin-3-(trans-coumaroyl)-rutinoside-5-glucoside (114.53–162.43 mg/100 g dw)	[7,136,156]
	Purple/black tomato	Petunidin and malvidin acylglycosides	petunidin 3-(6-(4-(*E*-*p*-coumaroyl)rhamnosyl)glucoside)-5-glucoside (petanin) (2.77 mg/g dw), Malvidin 3-(6-(4-(*E*-*p*-coumaroyl)rhamnosyl)glucoside)-5-glucoside (1.05 mg/g dw)	[130]
Eggplant *Solanum melongena* L.	Purple eggplant	delphinidin derivatives	delphinidin-3 glucoside-5-(coumaryl) dirhamnoside (1.10 mg/g dw), delphinidin-3-(*p*-coumaroylrutinoside)-5-glucoside (1357–3200 mg/kg dw)	[173,186,187,223]
Pepper *Capsicum* spp. L.	Green/red peppers	Carotenoids	β-carotene, lycopene, zeaxanthin	[194]
	Purple/black pepper	Delphinidin, cyanidin and malvidin derivatives	delphinidin 3,5-diglucoside, delphinidin 3-*O*-rutinoside, delphinidin-3-*trans*-coumaroylrutinoside-5-glucoside (284.8 µg/g fw), delphinidin-3-*cis* coumaroylrutinoside-5-glucoside (14.72 µg/g fw)	[181,182,200]
Lablab *Lablab purpureus* L.	purple lablab	Malvidin and petunidin derivatives	malvidin 3-sambubiose, malvidin 3-glucoside, delphinidin 3-glucoside-5-rutinoside, delphinidin 3-glucose-5-rhamnose and petunidin 3-rutinose	[213]
Common bean *Phaseolus vulgaris* L.	Black bean	Pelargonidin, Cyaniding and delphinidin glucosides	pelargonidin-3-glucoside, cyanidin-3-glucoside, delphinidin-3-glucoside	[217]

4. Other Vegetables

4.1. Broccoli and Cauliflower

Broccoli (*Brassica oleracea* L., var. *italica* Plenk) and cauliflower (*B. oleracea* L., var. *botrytis*) are the two most popular vegetable crops belonging to the Brassicaceae family. Native of the Mediterranean Basin, both species are adapted to a wide range of environmental conditions and are cultivated in all five continents, with an annual production that reached about 26 million tons in 2017, from an estimated harvested area of over 1.39 million hectares worldwide [27]. Primarily known and appreciated for their typical organosulfur compounds [224–226], broccoli, cauliflower, and other *Brassica* species are also a rich source of anthocyanins which are responsible of the purple pigmentation of some varieties [18,227]. Typically green, some cultivars and populations of broccoli are characterized by a purple pigmentation of the sepals of the inflorescence [227,228]. Branca et al. [228] found high levels of anthocyanins in a Sicilian broccoli landrace called "Broccolo nero" (Black broccoli) grown around Mount Etna and characterized by a dark violet pigmentation of the inflorescence, stem, and leaf midribs. In a recent work, Yu et al. [227] identified and mapped a major locus and two minor loci associated with the purple sepal trait in broccoli, while the authors hypothesized that the development of purple color may be induced by cold temperatures. In another recent study, Rahim et al. [229] working on the hypocotyl of young green and purple broccoli seedlings identified seven putative candidate genes (BoPAL, BoDFR, BoMYB114, BoTT8, BoMYC1.1, BoMYC1.2, and BoTTG1) responsible for the biosynthesis of anthocyanins. Among those, BoTT8 was expressed considerably more in purple hypocotyl compared to the green ones. Testing the in vitro cytotoxic effect of Sicilian black broccoli stem and leaf extracts at different concentrations (0.05%, 0.1%, 0.5%, 1%, and 5%) against HT29 (colon cancer) and A2058 (melanoma cancer) cells after 24 h treatment in presence or not of the myrosinase enzyme (responsible for the hydrolyzation of glucosinolates), Terzo et al. [230] found that the juice was less toxic in presence of myrosinase especially at higher concentration of the extract (1%–5%), suggesting that factors other than the glucosinolate content, such as polyphenols (including anthocyanins) could be responsible for the cytotoxic effects against HT29 and A2058 cancer cells. Examining the anthocyanin profile of three cultivars of heat-tolerant purple sprouting broccoli, Rodríguez-Hernández et al. [231] found that cyanidin 3-*O*-diglucoside-5-*O*-glucoside derivatives were the major acylated anthocyanins, and each cultivar and plant portion (leaves, inflorescence) had a particular prominent acylated anthocyanin. The same study revealed that compared to green broccoli cv. "Marathon", purple sprouting broccoli was characterized also by higher levels of glucosinolates. Similarly, Verkerk et al. [232] observed exceptionally high glucoiberin content (396.5 µmol/100 g FW) in purple sprouting broccoli cv "Bordeaux" compared to other green broccoli genotypes, which suggests that there is some sort of interaction between purple pigmentation and glucosinolate profile of broccoli. In another study, analyzing the acylated anthocyanin profile of purple sprouting broccoli and three other green broccoli varieties at the sprouting stage, Moreno et al. [233] observed a significantly higher content of anthocyanins in the purple genotype compared to the green ones and observed that the quantity and quality of anthocyanin pigments were highly variable among the tested genotypes. Out of seventeen anthocyanins identified in the four genotypes only three isomers were predominant in all the genotypes examined: Cyanidin 3-*O*-(acyl)diglucoside-5-*O*-glucoside, cyanidin 3-*O*-(acyl1)(acyl2)diglucoside-5-*O*-glucoside, and cyanidin 3-*O*-(acyl1)(acyl2)diglucoside-5-*O*-(malonyl)glucoside. The purple sprouting genotype was characterized by a higher content of cyanidin 3-*O*-(sinapoyl)(sinapoyl)diglucoside-5-*O*-glucoside, cyanidin 3-*O*-(sinapoyl)diglucoside-5-*O*-glucoside, cyanidin 3-*O*-(feruloyl)diglucoside-5-*O*-glucoside, cyanidin 3-*O*-(sinapoyl)(feruloyl)diglucoside-5-*O*-(malonyl)glucoside, and cyanidin 3-*O*-(sinapoyl)(sinapoyl)diglucoside-5-*O*-(malonyl)glucoside (Table 3). In agreement with previous studies, Moreno et al. [233] concluded that broccoli sprouts could be an excellent source of bioactive compounds rich of flavonoids, including acylated anthocyanins, along with glucosinolates, vitamins, and minerals [234–236], and that future studies should evaluate the potential of further enhancing the content of bioactive compounds in broccoli sprouts.

In the case of cauliflower, while most of the cultivars have been traditionally selected for their white curds [237], many local landraces and commercial cultivars are characterized by colored heads with characteristic pigmentation ranging from green to dark violet. In Italy, green cauliflowers are traditionally grown in Lazio and Marche, while dark violet selections are typically grown in Sicily, Puglia, and other Southern regions characterized by high levels of solar radiation which make it more challenging to produce white curds as traditionally required by the European market [228,238]. Lately, the interest for colored cauliflower varieties substantially increased due to the potential health-beneficial properties of the phenolic compounds that provide the pigmentation of plant tissues [239]. Anthocyanins are in fact responsible for the purple–violet pigmentation also in the case of cauliflower and are considered highly beneficial for human health [8]. The biosynthesis of anthocyanins in cauliflowers is regulated mainly at transcriptional level, and in a particular purple cauliflower mutant it has been demonstrated that the tissue-specific activation of the gene BoMYB2 up-regulated the expression of both BobHLH1 and BobHLH2, leading to the formation of a complex regulation network MYB–bHLH–WD40 (MBW), consisting of MYB, basic Helix-Loop-Helix (bHLH), and WD40 proteins, which in turn activates the structural genes responsible for the biosynthesis of anthocyanins [18,240]. β-carotene accumulation has been also observed in cauliflower curds due to a rare carotenoid gene (Or orange) mutation that activates the biosynthesis of carotenoids in tissues that otherwise would be white [241,242]. Nevertheless, such mutation received limited attention at commercial level and is more relevant to advance our understanding of the carotenoid biosynthesis regulation [243,244].

Analyzing by LC–MS/MS nine Sicilian landraces of violet cauliflower, Scalzo et al. [19] identified cyanidin-3-(6-p-coumaryl)-sophoroside-5-glucoside as the main anthocyanin along with p-coumaryl and feruloyl esterified forms of cyanidin-3-sophoroside-5-glucoside. Scalzo et al. [19] and Kapusta et al. [245] examined also the stability of anthocyanins after processing (blanching, microwave-heating, convection steaming and freezing, and conventional water cooking) and observed substantial changes with the formation of isomers from cyanidin-3-sophoroside-5-glucoside rather than the hydrolysis of anthocyanins, suggesting good stability especially after microwave-heating which could be interesting for food processing applications.

4.2. Cabbage and Kale

Among the *Brassica* species, cabbage (*Brassica oleracea* L. var. *capitata*) and Savoy cabbage (*B. oleracea* L. var. *sabauda* L.) are other two popular cole crops grown all over the world for their "heavy" heads constituted by leaves surrounding the terminal buds and that can be green or red-purple [246]. As for broccoli and cauliflower, the pigmentation of red cabbage genotypes is due to the accumulation of anthocyanins. Comparing four green and four red cabbage genotypes, Yuan et al. [247] observed that the structural genes involved in the biosynthesis of anthocyanins (CHS, F3H, F30H, DFR, LDOX, and GST), were steadily up-regulated in red genotypes for the entire growing period. The same authors observed that the expression of the structural genes responsible for the biosynthesis of anthocyanins was up-regulated in correspondence of nitrogen and phosphorous deficiency. Consistently with the mechanism of transcriptional regulation observed in purple cauliflower, in correspondence of the structural gene up-regulation it was observed a simultaneous increase of the transcript levels of the bHLH gene BoTT8, and of the MYB transcription factor BoMYB2. In a recent study analyzing the gene associated with the purple pigmentation of ornamental cabbage characterized by green external leaves and inner purple leaves, Jin et al. [109] found that phytoregulators such as abscisic acid (ABA) and ethylene (ET) play a key role in promoting the biosynthesis of anthocyanins. The same study identified 14 and 19 putative candidate genes involved in the biosynthesis of ABA and ET, respectively, and among those two ABA-biosynthesis related genes (BoNCED2.1, BoNCED2.2) and two ET-biosynthesis related genes (BoACS11, BoACO4) were expressed significantly more in purple leaves than in green leaves and were strongly correlated with the total anthocyanin content of the purple inner leaves.

Analyzing the anthocyanin profile of red cabbage using HPLC/DAD-ESI/Qtrap MS, Arapitsas et al. [248] separated and identified up to 24 anthocyanins all characterized by cyanidin as aglycon, mono- and/or di-glycoside, non-acylated, or acylated with aromatic and aliphatic acids. Similarly, using HPLC-DAD-MS/MS, Wiczkowski et al. [249] identified twenty cyanidin derivatives, with cyanidin-3-diglucoside-5-glucoside as the base structure, and cyanidin-3-diglucoside-5-glucoside, cyanidin-3-(sinapoyl)(sinapoyl)-diglucoside-5-glucoside, and cyanidin-3-(*p*-coumaroyl)-diglucoside-5-glucoside were the most abundant non-acylated anthocyanins. Moreover, Koss-Mikołajczyk et al. [250] identified nineteen different cyanidin derivatives, with cyanidin-3-(feruloyl)-diglucoside-5-glucoside and cyanidin-3-(sinapoyl)(sinapoyl)-diglucoside-5-

glucoside having been the most predominant. Similar results were obtained by other authors, who however using different analytical procedures and equipment identified a lower number of anthocyanins [19,251].

Analyzing raw and pickled red cabbage, consistently with other studies McDougall et al. [252] identified eighteen anthocyanin structures, most of which had cyanidin-3-diglucoside-5-glucoside as the core structure non-acylated, mono-acylated or di-acylated with *p*-coumaric, caffeic, ferulic and sinapic acids, but pelargonidin-3-glucoside and new cyanidin-3-*O*-triglucoside-5-*O*-glucoside di-acylated with hydroxycinnamic acids were also identified. The same authors examining the stability of anthocyanins after simulated gastrointestinal digestion found that anthocyanin structures were quite stable, and acylated structures were markedly more stable than non-acylated anthocyanins, nevertheless the after-digestion total recovery of anthocyanins was about 25%.

Feeding twelve volunteers with increasing doses (100, 200, and 300 g) of steamed red cabbage containing 1.38 μmol of anthocyanins/g (containing 30 acylated and 6 non-acylated anthocyanins), Charron et al. [253] evaluated the red cabbage anthocyanin bioavailability analyzing the excretion of intact and metabolized anthocyanin compounds in the urine. After 24 h, from the excreted urine were recovered 3 non-acylated and 8 acylated intact anthocyanins and 4 glucuronidated and methylated anthocyanin metabolites. Overall, the recovery of anthocyanins in excreted urine was four times higher for non-acylated compared to the acylated anthocyanins.

Comparing the biological activities (antioxidant, cytotoxic, anti-genotoxic, and influence on enzymatic activities) of the extract of green and red cabbage Koss-Mikołajczyk et al. [250] found that the anthocyanin content and profile was highly correlated with the antioxidant capacity of tested plant extracts measured through different spectrophotometric assays (ABTS, FC, DPPH, and FRAP), and by testing the cellular antioxidant activity. Instead, all the other biological activities tested were not correlated with the content of neither anthocyanins nor glucosinolate derivatives, suggesting that the food matrix effect may be more relevant than the biological activity of the single compound. This aspect should be further examined considering that other cabbage-like vegetables may be subject to different processing and anthocyanins and glucosinolates may have different levels of stability depending on the type of thermal or non-thermal processing and even the effect of the food matrix may change [19,254].

4.3. Artichoke

Native of the Mediterranean Basin and domesticated in Southern Italy during the Roman Empire [255,256], artichoke [*Cynara cardunculus* L. var. *scòlymus* (L.) Fiori] or globe artichoke constitutes a rich source of bioactive compounds and an important component of the Mediterranean diet to which are attributed a number of medicinal properties [20,21]. Artichoke is a major vegetable crop gaining popularity as a natural functional food. Grown on about 122,390 ha worldwide, over 1.5 million tons of artichoke heads were produced in 2017 [27]. As traditional producer and consumer of artichokes, Italy (33.1% of globe artichoke harvested area), Spain (13.4%), France (6.2%), and a few other European countries in the Mediterranean area continue to dominate the production of artichoke at global level. Nevertheless, their share is decreasing as the cultivation of this crop and its consumption are gradually expanding in other regions and reached important land investment in Egypt (10,159 ha), Algeria (5532 ha), Tunisia (3687 ha), Turkey (2994 ha), and Morocco (2923 ha) within the Mediterranean Basin, westward in Perú (8646 ha), Argentina (4472 ha), United States (2914 ha), and Chile (1464 ha), as well as eastward, especially in China (11,803 ha) that has the third largest investment on artichoke crops after Italy and Spain [27]. As a perennial crop, artichoke produces a robust stalk and large leaf biomass (about 75–80% of the above-ground plant biomass) from which during the reproductive phase the floral stems emerge sustaining the primary edible portion constituted by the immature inflorescences called heads or capitula. The capitula is constituted by a fleshy receptacle (the base of the inflorescence) with the inner heart made of tender bracts protected by more fibrous external bracts. The immature inflorescences are a rich source of macro- and microminerals, dietary fibers, inulin, vitamins, sesquiterpene lactones, and phenolic compounds including a series of flavonoids compounds and anthocyanins that are responsible for the pigmentation of the bracts characterizing the heads of different genotypes [20,21,257,258]. The pigmentation of the heads is in fact along with other morphology parameters (shape, presence of spines on bracts) and agronomic characteristics (earliness), one of the main factors used to classify artichoke varietal types. Based on the pigmentation of the heads, artichoke landraces and new hybrids are distinguished in two groups, namely green and pigmented artichokes. The pigmentation of the heads is considered an important quality parameter affecting consumer acceptance in different regions [20,259], and often it is anticipated in the name of the varietal type especially for the pigmented varietal type like in the case of the French "Violet de Provence", "Violet de Hyères", and "Violet du Gapeau", or the Italian "Violetto di provenza", "Violetto di Sicilia", "Violetto Toscano", "Violetto di S. Erasmo", and "Violetto di Chioggia" [260]. In a recent study on the inheritance of artichoke bract pigmentation, Portis et al. [261] found a good level of heritability for bract pigmentation. However, previous studies show that there is large variations of the pigmentation of the bracts within the same landrace, and in the case of artichoke the biosynthesis and accumulation of anthocyanins is highly influenced by environmental conditions and especially by temperatures [261,262]. Artichoke pigmentation is considered a complex trait in which several metabolic pathways could be involved. In 1969, Pochard et al. [263] proposed that the anthocyanin pigmentation in artichoke could be genetically determined by one or two major genes with the involvement of several modifiers. Later, based on the segregation pattern observed in populations obtained crossing inbreed lines and clones and self-pollinated clones of different genetic origins, Cravero et al. [264] proposed that the pigmentation of artichoke heads may be controlled genetically by two independent genes P and U with a simple recessive epistasis, where plants of genotype PP or Pp allow the biosynthesis of anthocyanins and result in purple bracts whereas pp genotype inhibits anthocyanin synthesis resulting in green bracts; at the same time genotypes UU or Uu are characterized by a non-uniform pigmentation encoded by the allele P, and uu genotypes develop uniformly pigmented bracts in presence of the allele P. While further studies should confirm this simple model, the variability of the pigmentation observed within the same genotype suggests that the genetic basis for anthocyanin pigmentation of artichoke bracts is more complex and likely the two loci model proposed is regulated by several modifier genes or multiple alleles that determine the expression of the intensity of the pigmentation in artichoke heads [261,262,264]. More recently, De Palma et al. [265] isolated and functionally characterized CcF3'H as the first structural gene of the flavonoid biosynthesis (encoding for flavonoid 3'-hydroxylase) from *C. cardunculus*. While Blanco et al. [266] isolated and functionally characterized CcMYB12, the artichoke putative homologue of the transcription factor R2R3-MYBs expressing proteins that regulate the biosynthesis of flavonoids and anthocyanins in many species.

Analyzing the polyphenolic profile of artichokes in different genotypes, it is possible to identify two main classes of phenolic compounds, namely hydroxicinnamic acids (C3-C6 skeleton) and flavonoids (C6-C3-C6 skeleton). Within the first class chlorogenic, 3,5-O-dicaffeoylquinic, and 1,5-O-dicaffeoylquinic acids are the predominant compounds followed by other minor mono- and di-caffeoylquinic acids [20,21,267]. Among the flavonoids that quantitatively represent about 10% of the total phenolic compound in artichoke [20], the flavones apigenin-7-O-glucuronide, apigenin-7-O-rutinoside, apigenin-7-O-glucoside, luteolin-7-O-glucuronide, luteolin-7-O-rutinoside, luteolin-7-O-glucoside, naringenin-7-O-rutinoside, and naringenin-7-O-glucoside are the predominant compounds, followed by the anthocyanins including cyanidin, peonidin, and delphinidin derivatives [20,21,257,267,268]. Other studies have revealed that specific phenolic compounds are accumulated only in certain genotypes and certain plant portions even within the same capitula [269,270].

The first attempts to identify the anthocyanin profile of green and purple artichokes were conducted in the late 1970s [20]. However, only later Schütz et al. [271] analyzing by HPLC the pigmented bracts of seven different varietal types of artichoke isolated and identified the main anthocyanins as cyanidin 3,5-diglucoside, cyanidin 3-glucoside, cyanidin 3,5-malonyldiglucoside, cyanidin 3-(3''-malonyl)glucoside, and cyanidin 3-(6''-malonyl)glucoside along with some minor compounds such as peonidin 3-glucoside, peonidin 3-(6''-malonyl)glucoside, and delphinidin glycoside. The same authors observed that the anthocyanin profile varies between different genotypes, nevertheless limited information is available on the genotypic variation of these compounds in artichoke. Analyzing new hybrids and local landraces characterized by bracts with different levels of pigmentation, Bonasia et al. [272] observed that "Violetto di Provenza" and "Tempo" characterized by higher pigmentation had also the highest total phenolic content in the heart and external bracts and in the external bracts, respectively. Examining the content of total anthocyanins in bracts, leaves, and floral stems of two genotypes grown in Tunisia, namely "Violet d'Hyéres" and "Blanc d'Oran", Dabbou et al. [273] found that leaves of "Blanc d'Oran" had the highest concentration of anthocyanins (20.5 µg/g DW) while bracts and floral stems had the lowest concentration (8.3 and 5.9 µg/g DW, respectively). Instead, lower variability between different plant portions was observed in the case of "Violet d'Hyéres" that on average had a total anthocyanin concentration of 14.2 µg/g DW). These results suggest that even the leaves of artichoke plants which represent a big portion of plant biomass may constitute a good source of anthocyanins besides being a source of other phenolic compounds [274]. A very limited number of studies have been conducted to evaluate specifically the biological activity of artichoke anthocyanins [257]. The main bioactive effect attributed to anthocyanins, flavonoids, and other phenolic compounds extracted from artichoke plant tissues is their antioxidant activity demonstrated by several in vitro and in vivo studies [20,21,275,276]. Anthocyanins have also shown lipid lowering effects in a placebo-control double-blind study by reducing serum LDL cholesterol by 7.9%, triglycerides by 23.0%, apolipoprotein by 16.5%, and apolipoprotein C-III by 11.0%, and increasing HDL cholesterol by 19.4% compared with placebo after administration of 160 mg of anthocyanins for 24 weeks twice daily [277]. Intake of anthocyanins seems to have positive effects on the cardiovascular system by also reducing arterial stiffness [278]. In another study, it was observed that artichoke leaf extracts and artichoke flavonoids up-regulate the gene expression of endothelial-type nitric oxide synthase (eNOS, a vasoprotective molecule) in human endothelial cells [279]. While in a follow-up study Xia et al. [280] observed that treatment of human coronary artery smooth muscle cells (HCASMC) with artichoke leaf extracts and particularly with cynarin and cyanidin induced a down-regulation of inducible nitrous oxide synthase (iNOS, a pro-inflammatory molecule that can cause vascular dysfunctions), suggesting that artichoke flavonoid compounds may have great therapeutic potential. In recent years, a number of studies contributed to demonstrate that artichoke heads and leaf extracts and, in some cases, specific phenolic compounds have health-beneficial properties including anti-inflammatory activity, anti-bacterial activity, hepatoprotective activity, hypocholesterolemic and low density lipoproteins (LDL) oxidation inhibition effect, hypoglycemic effect, as well as anticancer activity [20,21,257]. Nevertheless, in most of the cases these biological activities cannot be attributed to a single compound, but are determined by the combined synergistic effect of different compounds [20,21]. In this perspective, the matrix effect is fundamental in determining the bioaccessibility, bioavailability, and the effect of polyphenols, especially considering that polyphenols and anthocyanins in particular have a relatively low bioavailability being quickly transformed into derivatives of phenolic acids [21,281–283].

4.4. Asparagus

Used since ancient time for the diuretic and medicinal properties of its spears [284], cultivated asparagus (*Asparagus officinalis* L.) is now a popular vegetable grown worldwide on over 1.5 million hectares producing over 8.9 million tons of spears [27]. Native of Eastern Europe, cultivated asparagus is grown at commercial level mainly in China, followed far behind by Perú, Mexico, Germany, Spain, USA, Italy, Japan, France, and the Netherlands. The edible portion of this herbaceous perennial is constituted by its spears or young stems produced by the underground crown, which may be green, green-purple, or purple when harvested above ground or white when purposely harvested before exposed to sunlight. Asparagus spears are considered a rich source of minerals [285], amino acids and dietary fibers [286], saponins [287,288], vitamins and volatile sulfur organic compounds [289], and especially of phenolic compounds and flavonoids [290], including anthocyanins that are responsible for the purple color of the bracts of green spears or of the whole spears in purple asparagus genotypes [23,154,291]. Being rich of all these bioactive compounds it is difficult to isolate the effect of anthocyanins and specific pigments, nevertheless several studies have demonstrated the antioxidant properties of anthocyanins in the species [284]. Comparing green, white, and purple spears, Maeda et al. [291] found that purple spears had significantly higher levels of rutin compared to green spears, while rutin was not detected in white asparagus spears. The same authors observed a positive correlation between spears total polyphenol content, rutin, and DPPH radical scavenging activities, suggesting that purple asparagus may provide higher levels of antioxidants compared to green and white asparagus [291]. In this perspective, while most of the cultivars have been selected to produce green spears and have a relatively small content of anthocyanins, and even less pigments when etiolated to produce white spears, more recently, as for other crops, specific functional breeding programs have developed purple cultivars characterized by high levels of anthocyanins to satisfy the need of consumers attracted by new colors and seeking richer sources of natural antioxidants [291]. Most of the modern commercial cultivars of green and white asparagus are diploid ($2n = 20$) and were developed from an old population called "Purple Dutch" from which French growers selected two stocks called "Precoce d'Argenteuil" and "Tardive d'Argenteuil" which were subsequently used to develop modern cultivars and hybrids in The Netherlands, France, Germany, United Kingdom, and in the USA [292,293]. Although any green asparagus genotype may be used to produce etiolated white spears, Dutch, French, and German breeding programs have selected cultivars suitable specifically for white spears production to satisfy the European market requirements by reducing as much possible the content of anthocyanins that may develop and accumulate in the tips of white-harvested spears even during storage [292,294]. On the other hand, using "Violetto d'Albenga" a tetraploid local population traditionally grown in Northern Italy, likely derived from interspecific crossing between *A. officinalis* and *A. maritimus* [295] and characterized by the production of larger, sweeter, and less fibrous spears ranging in color from green to dark purple, breeding programs in the US and New Zealand have developed in the early 1990s the first cultivars producing purple spears such as "Purple Passion" [296] and "Purple Pacific" and "Stewarts Purple" [297].

Using HPLC and NMR to analyze the fresh peel of "Purple Passion" asparagus, Sakaguchi et al. [23] isolated two major anthocyanins identified as: cyanidin 3-[3″-(*O*-β-D-glucopyranosyl)-6″-(*O*-α-L-rhamnopyranosyl)-*O*-β-D-glucopyranoside], and cyanidin 3-rutinoside characterized by high antioxidant activity as determined through ORAC (Oxygen Radical Absorbance Capacity) assay. Recently, Dong et al. [298] comparing the anthocyanin profile of three purple asparagus cultivars ("Jing Zi-2", "Purple Passion", and "Pacific Purple") with a green control ("Jing Lv-1"), identified sixteen anthocyanins, with peonidin, cyanidin and their glycoside derivatives being the predominant compounds. In the same study, through transcriptomics and quantitative real-time polymerase chain reaction (qRT-PCR) analysis, several anthocyanin synthesizing genes (PAL, C4h, 4CL, CHS, CHI, F3H, F3′H, DFR, ANS, and 3GT) and transcription factors genes (bHLH137-like, TT2-like, WD40-like, bZIP61-like, and MADS18-like) that regulates the biosynthesis of anthocyanins like reported for other species were identified and showed to be differentially expressed between purple and green cultivars [298]. The expression of the same genes and the content of anthocyanins were examined also comparing spears grown in the presence of light and in the dark. In accordance with previous studies [299,300], results demonstrated that anthocyanin biosynthetic and regulatory genes are considerably down-regulated in absence of light and pigments are not synthesized in dark conditions, suggesting that light conditions are a key factor for anthocyanin biosynthesis and accumulation. This is further corroborated by the findings of Huyskens-Keil et al. [301] who showed the effects of light quality (white, red, blue, UV-C) on the activity of phenylalanine ammonia-lyase (PAL) and peroxidase (POD) and the synthesis and accumulation of anthocyanins in basal and apical segments of white asparagus spears during postharvest storage. Postharvest conditions and handling may in fact considerably affect the stability of anthocyanins. In another recent study, Barberis et al. [302] reported that dipping the spears of "Purple Passion" for 5 min in a 3 mM solution of oxalic acid (pH 2.9) reduced the lightness of the spears after 12 days only by 13% compared to spears treated with water (control pH 8) or with 1 mM oxalic acid (pH 6) solution, suggesting that the low pH treatment enhanced the stability of the anthocyanin pigments. Analyzing the phenolic profile of green and purple asparagus at harvest and during storage, in accordance with Sakaguchi et al. [23] the same authors [302] found that cyanidin glucosyl rutinoside, cyanidin rutinoside and peonidin rutinoside (at harvest 774.2, 125.5, and 84.8 mg/kg, respectively) were the main anthocyanins identified in "Purple Passion" spears and determining the typical purple color. During storage, the first anthocyanin decreased overtime and was negatively affected by oxalic acid treatment, cyanidin rutinoside increased over time, while peonidin rutinoside remained stable.

4.5. Sweet Corn

Native to America, corn (*Zea mays* L.) has been cultivated in Central America by indigenous populations at least since 3500 BC [303]. When the European explorers arrived in North America, Iroquois Indians living in the region corresponding to the current state of Pennsylvania and New York grew a variety of sweet corn (*Zea mays* L. *saccharata* Sturt.) that turned blue upon ripening, along with other multi-colored varieties [304]. The first samples of maize seeds that arrived in Spain were described as characterized by different colors ranging from white to black [305,306]. Initially grown for curiosity, with the arrival of seeds from North America adapted to higher latitudes, and thus more suitable to the European photoperiod and climate conditions, the new cereal crop started to spread from Spain to the rest of Europe. The rapid diffusion of maize in Europe is also testified by the painter Giuseppe Arcimboldo, in his famous painting "Summer" made in 1573 in which a maize ear is visible [305]. With the process of selection initially conducted in Europe the new crop lost its multi-color pigmentation with the exceptions of a few local populations and assumed the typical white or yellow color of the varieties currently cultivated worldwide.

With over 1.134 million metric tons of corn produced at global level in 2017 [27], corn is by far the first cereal produced in the world, second in terms of acreage only to wheat. Among the different types of maize, sweet corn is particularly important as it is used for human consumption worldwide. In the US, first producer of sweet corn, in 2017 were harvested 464,600 acres (about 188.000 ha) of sweet corn generating a crop value of over $892 million [307]. About 75% of the sweet corn is produced for the fresh market and the remaining portion is used for canned and frozen food processing, making sweet corn the second most important processing vegetable after processing tomato.

Sweet corn is derived from a natural genetic mutation of field corn (dent corn) that was first reported in Pennsylvania in 1770, although it was probably cultivated before by native Americans [308]. The mutant genotype *sugary* (*su*) accumulates more sugar in the endosperm (10.2%) than the standard starchy maize (3.5%) resulting in sweeter taste. Years later, several sugary sweet corn varieties have been selected. These traditional varieties characterized mainly by white or yellow kernels, are harvested before complete physiological ripening when the level of sugar is maximum and have short shelf-life because after harvest the sugar is rapidly converted to starch [309]. Subsequently, sweeter hybrid varieties have been developed by selecting *sugar-enhanced* (*se*) mutants characterized by double content of sugar (20–35%), tender kernels, creamy texture and good corn flavor. Yet, more recently *supersweet* varieties have been developed selecting *shrunken-2* (*sh2*) mutants characterized by even higher sugar content at the immature milky stage (29.9%) and extended shelf life, with lighter shrunken kernels. Lately high sugar hybrids are developed combining *su*, *se*, and *sh2* genes to obtain optimal combinations of sugar, texture, taste, and long shelf life [309,310]. Most of the sweet corn hybrids grown and consumed today at commercial level are characterized by yellow (60%), white (20%), or bicolor (20%, yellow and white) kernels. Nevertheless, in recent years, the interest in reviving ancient colored sweet corn varieties or developing new pigmented varieties characterized by high content of carotenoids [311] and especially anthocyanins is increasing [305,312–316] due to the potential functional properties of anthocyanin-rich genotypes [317,318], as well as to the increasing demand of natural colorants [22,319,320]. The biosynthesis of anthocyanins in maize aleurone (external part of the endosperm) and pericarp (external part of the kernel) of the kernels or in plant tissues involve over twenty structural and regulatory genes that have been identified and functionally characterized as reviewed by Petroni et al. [305,321–323]. The wide range of colors observed in maize kernels is mainly due to the biosynthesis and accumulation of carotenoids and flavonoids. Carotenoids such as β-carotene, zeaxanthin and lutein which are lipid-soluble pigments and are responsible for the color of kernels ranging from yellow to deep orange [311,324]. Maize flavonoids include two main classes of pigments: (i) phlobaphenes which are water-insoluble 3-deoxyflavanoid pigments that accumulate in the pericarp of the kernels and the cob and are responsible for the development of kernel colors ranging from orange to brick red [315]; and (ii) anthocyanins which are water-soluble pigments responsible for the development of pink, red, purple, and blue color in the aleurone and pericarp of the kernels as well as in other plant tissues. Within the anthocyanins, the main class of pigments identified are cyanidin, peonidin, and pelargonidin derivatives with the first two providing bluish-red color and the latter being responsible for more orange-red color [22]. The combination of carotenoids and flavonoids in pericarp and aleurone generates an incredible variety of shades [22] and may be influenced also by the condensation with flavanols [319,320,325,326] or other modifications of the anthocyanin compounds. Analyzing the anthocyanin profile of the kernels of six different purple maize genotypes, González-Manzano et al. [326] identified the dimer catechin-(4a-8)-cyanidin-3,5-*O*-diglucoside and other flavanol-anthocyanin condensed pigments along with several related anthocyanin pigments such as cyanidin-3-glucoside, cyanidin-3-malonylglucoside, peonidin-3-glucoside, peonidin-3-malonylglucoside, pelargonidin-3-dimalonylglucoside, and other derivatives. Malonyl- and dimalonyl acylated anthocyanins are particularly interesting for their higher stability as natural colorants [22]. Analyzing the extract of corn cobs and kernels of a Chinese purple corn, Yang and Zhai [327] identified cyanidin-3-glucoside, pelargonidin-3-glucoside and peonidin-3-glucoside including their malonated derivatives. Besides the great potential of maize pigments as natural food colorants [22,319], there is great interest in the development of anthocyanin-rich maize functional food products [305,311,312,321,328]. Several studies have reported the potential health-beneficial properties of anthocyanin-rich maize [305,317,329]. Examining the phenolic and anthocyanin content of eighteen Mexican maize genotypes, Lopez-Martinez et al. found that total anthocyanins ranged from 1.54 to 850.9 mg of cyanidin-glucoside equivalents/100 g of whole grain flour. Purple genotypes rich in anthocyanins exhibited also the greatest antioxidant activity [330]. Andean purple corn had higher antioxidant activity and antiradical kinetics than blueberries and higher or similar level of anthocyanins [331]. Similarly, analyzing 49 lines of waxy corn (*Zea mays* L. var. *ceratina*) characterized by different colors Harakotr et al. [332] found a large variability of phenolic and anthocyanin compounds and a positive correlation coefficient between anthocyanins and 2,2-diphenyl-1-picrylhydrazyl (DPPH) radical scavenging ability ($r = 0.94$) and between anthocyanins and Trolox equivalent antioxidant capacity (TEAC, $r = 0.88$). Introduction in the diet of anthocyanin-rich corn may contribute to prevent also obesity and diabetes. Tsuda et al. [317] found that feeding high fat (HF) mice for 12 weeks with cyanidin 3-glucoside-rich purple corn significantly reduced HF body weight and white and brown adipose tissues compared to HF mice not receiving purple corn supplement. The HF diet induced also hyperglycemia, hyperinsulinemia, and hyperleptinemia, whereas the same physiological conditions were not observed in HF mice receiving the purple corn supplement. Moreover, an increase of tumor necrosis factor (TNF)-α mRNA level was observed in the HF-diet while it was not observed in the HF group receiving the purple corn supplement which suppressed the mRNA of enzymes involved in fatty acid and triacylglycerol synthesis and reduced the level of sterol regulatory element binding protein-1 mRNA in white adipose tissue. Using near-isogenic maize lines which differed only in the presence or not of anthocyanins, and feeding rats with anthocyanin-rich and anthocyanin-deprived maize for 8 weeks it was observed that in the group fed with anthocyanin-rich corn cardiac tissue damaged following ischemic conditions was reduced by about 30% compared to the group fed with anthocyanin-deprived corn [333]. An increase of myocardial glutathione and omega-3 fatty acids levels in blood indicated that diet supplementation with anthocyanins regulated the cardiac antioxidant defense and the conversion of α-linolenic acid to omega-3 fatty acids [333,334]. In an in vivo study, using db/db mice fed with 10 mg/kg of purple corn extract for 8 weeks, Kang et al. [335] found that anthocyanin-rich purple corn extracts reduced glomerular angiogenesis of diabetic kidneys by inhibiting the induction of vascular endothelial growth factor (VEGF) and hypoxia inducible factor (HIF)-1a induced by hyperglycemic condition. Kang et al. results demonstrated that purple corn extract inhibited glomerular angiogenesis caused by chronic hyperglycemia and diabetes by disturbing the Angpt-Tie-2 ligand-receptor system linked to the renal VEGF receptor 2 (VEGFR2), suggesting that purple corn extract could be used to target abnormal angiogenesis in diabetic nephropathy leading to kidney failure [335]. Yet, in a recent study Mazewski et al. [336] found that anthocyanin-rich purple and red corn may potentially contribute to inhibiting human colon cancer cell proliferation by promoting apoptosis and suppressing angiogenesis.

The main pigments isolated in vegetables consumed for various plant parts are presented in Table 3.

Table 3. The main pigments isolated in vegetables consumed for plant parts other than fruit and leaves.

Species	Edible Part	Color	Class of Compounds	Compounds	References
Broccoli (*Brassica oleracea* L., var. *italica* Plenk)	Inflorescence	Purple	Acylated anthocyanins	cyanidin 3-*O*-(sinapoyl)(sinapoyl)diglucoside-5-*O*-glucoside (0.0653–0.3716 mg/100 g fw), cyanidin 3-*O*-(sinapoyl)diglucoside-5-*O*-glucoside (0.0119–0.0158 mg/100 g fw), cyanidin 3-*O*-(feruloyl)diglucoside-5-*O*-glucoside (0.0201–0.0765 mg/100 g fw), cyanidin 3-*O*-(sinapoyl)(feruloyl)diglucoside-5-*O*-(malonyl)glucoside (0.013–0.048 mg/100 g fw), and cyanidin 3-*O*-(sinapoyl)(sinapoyl)diglucoside-5-*O*-(malonyl)glucoside (0.0159–0.1035 mg/100 g fw)	[231,233]
Cauliflower (*Brassica oleracea* L., var. *botrytis*)	Inflorescence	Purple or dark violet	Cyanidin derivatives	cyanidin-3-(6-*p*-coumaryl)-sophoroside-5-glucoside (9.0–29.9 mg/kg fw), cyanidin-3-(6-*p*-coumaryl)-sophoroside-5-(6-sinapyl)-glucoside (27.8–37.8 mg/kg fw)	[19]
Cabbage (*Brassica oleracea* L. var. *capitata*)	Leaves	Purple	Cyanidin derivatives	cyanidin-3-diglucoside-5-glucoside (0.64 mg/g dw), cyanidin-3-(sinapoyl)(sinapoyl)-diglucoside-5-glucoside (0.26-0.58 mg/g dw), cyanidin-3-(sinapoyl)-diglucoside-5-glucoside (0.85 mg/g dw)and cyanidin-3-(*p*-coumaroyl)-diglucoside-5-glucoside (0.19-0.92 mg/g dw), cyanidin-3-(feruloyl)-diglucoside-5-glucoside (0.14–0.55 mg/g dw)	[249,250,252]
Artichoke [*Cynara cardunculus* L. var. *scòlymus* (L.) Fiori]	Immature inflorescence	Purple	Cyanidin, peonidin, and delphinidin derivatives	cyanidin 3,5-diglucoside (0–11.7 mg/kg dw), cyanidin 3-glucoside (0.3–194.1 mg/kg dw), cyanidin 3,5-malonyldiglucoside (0.5-218.0 mg/kg dw), cyanidin 3-(3''-malonyl)glucoside (0–47.3 mg/kg dw), and cyanidin 3-(6''-malonyl)glucoside (7.6-1234.3 mg/kg dw) along with some minor compounds such as peonidin 3-glucoside, peonidin 3-(6''-malonyl)glucoside, and delphinidin glycoside.	[271]
Asparagus (*Asparagus officinalis* L.)	Spears (young stems)	Purple	Cyanidin and peonidin derivatives	Cyanidin 3-[3''-(*O*-β-D-glucopyranosyl)-6''-(*O*-α-L-rhamnopyranosyl)-*O*-β-D-glucopyranoside], cyanidin 3-rutinoside (774.2 mg/kg fw), cyanidin glucosyl rutinoside (125.5 mg/kg fw), peonidin rutinoside (84.8 mg/kg fw)	[23,298,302]

Table 3. *Cont.*

Species	Edible Part	Color	Class of Compounds	Compounds	References
Sweet corn (*Zea mays saccharata* Sturt.)	Maize kernels	Yellow – deep orange	Carotenoids	β-carotene (2.62 µg/g dw), β-cryptoxanthin (3.96 µg/g dw), zeaxanthin (23.74 µg/g fw), zeinoxanthin (0.75 µg/g dw), lutein (8.2 µg/g dw), antheraxanthin (2.5 µg/g dw))	[311,324]
	Maize kernel pericarp and cobs	Orange – red brik	Phlobaphenes	Phlobaphenes (320.24 A_{510}/100 g)	[315]
	Maize cobs and kernel aleurone and pericarp	Pink, red, purple, blue	Anthocyanins (cyanidin, penodin, pelargonidin,	cyanidin-3-glucoside, pelargonidin-3- glucoside, penodin-3-glucoside, cyanidin-3-(6-malonylglucoside), pelargonidin-3-(6-malonylglucoside), peonidin-3-(6-malonylglucoside), peonidin-3-glucoside, peonidin-3-malonylglucoside, pelargonidin-3-dimalonylglucoside	[326,327]

5. Conclusions

Vegetable products are pivotal in human health and their daily consumption is highly recommended due to their high content in phytochemicals with diverse beneficial health effects. The vegetable crops presented in this review, namely leafy and fruit vegetables and species where other plant parts are consumed, are well known and widely consumed throughout the world, although their colored variants are not always accepted by consumers and proper marketing is needed to introduce them to the public. Currently, the food and beverage industry is seeking for natural alternatives of synthetic compounds to satiate market demands for "non-chemical additives" food products. Therefore, the broad color palette the presented vegetables exhibit not only could diversify human diet and make food products more attractive to consumers and increase their appeal, but also could be proved as valuable sources of natural coloring agents. Under the circumstances, future research should focus on the identification of those promising coloring agents and the establishment of efficient and sustainable techniques for their isolation. Other future requirements, include breeding and agronomic protocols that could improve pigments content in the final produce, as well as post-harvest and processing conditions to increase extraction efficiency of natural coloring agents. Finally, the mechanisms of actions related with the health effects and antioxidant activities of coloring compounds have to be revealed and valorized in the pharmaceutical and food industry for the design of novel functional foods and drugs.

Funding: This work was supported by the USDA National Institute of Food and Agriculture and Hatch Appropriations under Project #PEN04723 and Accession #1020664; The authors are grateful to Foundation for Science and Technology (FCT, Portugal) and FEDER under Programme PT2020 for financial support to CIMO (UID/AGR/00690/2019); to FEDER-Interreg España-Portugal programme for financial support through the project 0377_Iberphenol_6_E and TRANSCoLAB 0612_TRANS_CO_LAB_2_P; the European Regional Development Fund (ERDF) through the Regional Operational Program North 2020, within the scope of project Mobilizador Norte-01-0247-FEDER-024479: ValorNatural®.

Conflicts of Interest: The authors declare no conflict of interest.

References

1. Sasaki, N.; Nishizaki, Y.; Ozeki, Y.; Miyahara, T. The role of acyl-glucose in anthocyanin modifications. *Molecules* **2014**, *19*, 18747–18766. [CrossRef]
2. Luo, W.P.; Fang, Y.J.; Lu, M.S.; Zhong, X.; Chen, Y.M.; Zhang, C.X. High consumption of vegetable and fruit colour groups is inversely associated with the risk of colorectal cancer: A case-control study. *Br. J. Nutr.* **2015**, *113*, 1129–1138. [CrossRef] [PubMed]

3. Carlsen, M.H.; Halvorsen, B.L.; Holte, K.; Bøhn, S.K.; Dragland, S.; Sampson, L.; Willey, C.; Senoo, H.; Umezono, Y.; Sanada, C.; et al. The total antioxidant content of more than 3100 foods, beverages, spices, herbs and supplements used worldwide. *Nutr. J.* **2010**, *9*, 3. [CrossRef] [PubMed]

4. Mcgill, C.R.; Wightman, J.D.; Fulgoni, S.A.; Fulgoni, V.L. Consumption of Purple/Blue Produce Is Associated with Increased Nutrient Intake and Reduced Risk for Metabolic Syndrome: Results From the National Health and Nutrition Examination Survey 1999–2002. *Am. J. Lifestyle Med.* **2011**, *5*, 279–290. [CrossRef]

5. Wu, X.; Beecher, G.R.; Holden, J.M.; Haytowitz, D.B.; Gebhardt, S.E.; Prior, R.L. Concentrations of anthocyanins in common foods in the United States and estimation of normal consumption. *J. Agric. Food Chem.* **2006**, *54*, 4069–4075. [CrossRef]

6. Bognar, E.; Sarszegi, Z.; Szabo, A.; Debreceni, B.; Kalman, N.; Tucsek, Z.; Sumegi, B.; Gallyas, F. Antioxidant and Anti-Inflammatory Effects in RAW264.7 Macrophages of Malvidin, a Major Red Wine Polyphenol. *PLoS ONE* **2013**, *8*, e65355. [CrossRef]

7. Su, X.; Xu, J.; Rhodes, D.; Shen, Y.; Song, W.; Katz, B.; Tomich, J.; Wang, W. Identification and quantification of anthocyanins in transgenic purple tomato. *Food Chem.* **2016**, *202*, 184–188. [CrossRef]

8. Wang, L.S.; Stoner, G.D. Anthocyanins and their role in cancer prevention. *Cancer Lett.* **2008**, *269*, 281–290. [CrossRef]

9. Grotewold, E. The genetic and biochemistry of floral pigments. *Annu. Rev. Plant Biol.* **2006**, *57*, 761–780. [CrossRef]

10. Schaefer, H.M.; Schaefer, V.; Levey, D.J. How plant-animal interactions signal new insights in communication. *Trends Ecol. Evol.* **2004**, *19*, 577–584. [CrossRef]

11. Chaves-Silva, S.; dos Santos, A.L.; Chalfun-Júnior, A.; Zhao, J.; Peres, L.E.P.; Benedito, V.A. Understanding the genetic regulation of anthocyanin biosynthesis in plants—Tools for breeding purple varieties of fruits and vegetables. *Phytochemistry* **2018**, *153*, 11–27. [CrossRef] [PubMed]

12. Menzies, I.J.; Youard, L.W.; Lord, J.M.; Carpenter, K.L.; van Klink, J.W.; Perry, N.B.; Schaefer, H.M.; Gould, K.S. Leaf colour polymorphisms: A balance between plant defence and photosynthesis. *J. Ecol.* **2016**, *104*, 104–113. [CrossRef]

13. He, J.; Giusti, M.M. Anthocyanins: Natural Colorants with Health-Promoting Properties. *Annu. Rev. Food Sci. Technol.* **2010**, *1*, 163–187. [CrossRef] [PubMed]

14. Kim, M.J.; Moon, Y.; Tou, J.C.; Mou, B.; Waterland, N.L. Nutritional value, bioactive compounds and health benefits of lettuce (*Lactuca sativa* L.). *J. Food Compos. Anal.* **2016**, *49*, 19–34. [CrossRef]

15. Johansson, M.; Jägerstad, M.; Frølich, W. Folates in lettuce: A pilot study. *Scand. J. Food Nutr.* **2007**, *51*, 22–30. [CrossRef]

16. Giordano, M.; El-Nakhel, C.; Pannico, A.; Kyriacou, M.C.; Stazi, S.R.; De Pascale, S.; Rouphael, Y. Iron biofortification of red and green pigmented lettuce in closed soilless cultivation impacts crop performance and modulates mineral and bioactive composition. *Agronomy* **2019**, *9*, 290. [CrossRef]

17. Morais, M.G.; Da Costa, G.A.F.; Aleixo, Á.A.; de Oliveira, G.T.; Alves, L.F.; Duarte-Almeida, J.M.; Ferreira, J.M.S.; Dos Santos Lima, L.A.R. Antioxidant, antibacterial and cytotoxic potential of the ripe fruits of *Solanum lycocarpum* A. St. Hil. (Solanaceae). *Nat. Prod. Res.* **2015**, *29*, 480–483. [CrossRef]

18. Chiu, L.W.; Zhou, X.; Burke, S.; Wu, X.; Prior, R.L.; Li, L. The purple cauliflower arises from activation of a MYB transcription factor. *Plant Physiol.* **2010**, *154*, 1470–1480. [CrossRef]

19. Lo Scalzo, R.; Genna, A.; Branca, F.; Chedin, M.; Chassaigne, H. Anthocyanin composition of cauliflower (*Brassica oleracea* L. var. *botrytis*) and cabbage (*B. oleracea* L. var. *capitata*) and its stability in relation to thermal treatments. *Food Chem.* **2008**, *107*, 136–144. [CrossRef]

20. Lattanzio, V.; Kroon, P.A.; Linsalata, V.; Cardinali, A. Globe artichoke: A functional food and source of nutraceutical ingredients. *J. Funct. Foods* **2009**, *1*, 131–144. [CrossRef]

21. D'Antuono, I.; Di Gioia, F.; Linsalata, V.; Rosskopf, E.N.; Cardinali, A. Impact on health of artichoke and cardoon bioactive compounds: Content, bioaccessibility, bioavailability, and bioactivity. In *Phytochemicals in Vegetables: A Valuable Source of Bioactive Compounds*; Petropoulos, S.A., Ferreira, I.C.F.R., Barros, L., Eds.; Bentham Science Publishers Ltd.: Sharjah, UAE, 2018; p. 373.

22. Chatham, L.A.; Paulsmeyer, M.; Juvik, J.A. Prospects for economical natural colorants: Insights from maize. *Theor. Appl. Genet.* **2019**, *132*, 2927–2946. [CrossRef] [PubMed]

23. Sakaguchi, Y.; Ozaki, Y.; Miyajima, I.; Yamaguchi, M.; Fukui, Y.; Iwasa, K.; Motoki, S.; Suzuki, T.; Okubo, H. Major anthocyanins from purple asparagus (*Asparagus officinalis*). *Phytochemistry* **2008**, *69*, 1763–1766. [CrossRef] [PubMed]

24. Nicolle, C.; Cardinault, N.; Gueux, E.; Jaffrelo, L.; Rock, E.; Mazur, A.; Amouroux, P.; Rémésy, C. Health effect of vegetable-based diet: Lettuce consumption improves cholesterol metabolism and antioxidant status in the rat. *Clin. Nutr.* **2004**, *23*, 605–614. [CrossRef] [PubMed]

25. Nicolle, C.; Carnat, A.; Fraisse, D.; Lamaison, J.L.; Rock, E.; Michel, H.; Amoureux, P.; Remesy, C. Characterisation and variation of antioxidant micronutrients in lettuce (*Lactuca sativa* folium). *J. Sci. Food Agric.* **2004**, *84*, 2061–2069. [CrossRef]

26. Rouphael, Y.; Kyriacou, M.C.; Vitaglione, P.; Giordano, M.; Pannico, A.; Colantuono, A.; De Pascale, S. Genotypic variation in nutritional and antioxidant profile among iceberg lettuce cultivars. *Acta Sci. Pol. Hortorum Cultus* **2017**, *16*, 37–45. [CrossRef]

27. FAO FAOSTAT, Crops. FAO Statistics Division. Food and Agriculture Organization of the United Nations 2017. Available online: http://www.fao.org/faostat/en/#data/QC (accessed on 30 December 2019).

28. Mou, B. Lettuce. In *Vegetables, I "Asteraceae, Brassicaceae, Chenopodiaceae, and Cucurbitaceae*; Prohens, J., Nuez, F., Eds.; Springer: New York, NY, USA, 2008; pp. 75–116.

29. Wang, C.; Riedl, K.M.; Schwartz, S.J. Fate of folates during vegetable juice processing—Deglutamylation and interconversion. *Food Res. Int.* **2013**, *53*, 440–448. [CrossRef]

30. Gan, Y.Z.; Azrina, A. Antioxidant properties of selected varieties of lettuce (*Lactuca sativa* L.) commercially available in Malaysia. *Int. Food Res. J.* **2016**, *23*, 2357–2362.

31. Rouphael, Y.; Kyriacou, M.C.; Petropoulos, S.A.; De Pascale, S.; Colla, G. Improving vegetable quality in controlled environments. *Sci. Hortic.* **2018**, *234*, 275–289. [CrossRef]

32. Kyriacou, M.C.; Rouphael, Y. Towards a new definition of quality for fresh fruits and vegetables. *Sci. Hortic.* **2018**, *234*, 463–469. [CrossRef]

33. Survase, S.A.; Bajaj, I.B.; Singhal, R.S. Biotechnological production of vitamins. *Food Technol. Biotechnol.* **2006**, *44*, 381–396.

34. López, A.; Javier, G.A.; Fenoll, J.; Hellín, P.; Flores, P. Chemical composition and antioxidant capacity of lettuce: Comparative study of regular-sized (Romaine) and baby-sized (Little Gem and Mini Romaine) types. *J. Food Compos. Anal.* **2014**, *33*, 39–48. [CrossRef]

35. Llorach, R.; Martínez-Sánchez, A.; Tomás-Barberán, F.A.; Gil, M.I.; Ferreres, F. Characterisation of polyphenols and antioxidant properties of five lettuce varieties and escarole. *Food Chem.* **2008**, *108*, 1028–1038. [CrossRef] [PubMed]

36. Mou, B. Genetic variation of beta-carotene and lutein contents in lettuce. *J. Am. Soc. Hortic. Sci.* **2005**, *130*, 870–876. [CrossRef]

37. Baslam, M.; Morales, F.; Garmendia, I.; Goicoechea, N. Nutritional quality of outer and inner leaves of green and red pigmented lettuces (*Lactuca sativa* L.) consumed as salads. *Sci. Hortic.* **2013**, *151*, 103–111. [CrossRef]

38. Le Marchand, L.; Hankin, J.H.; Kolonel, L.N.; Beecher, G.R.; Wilkens, L.R.; Zhao, L.P. Intake of Specific Carotenoids and Lung Cancer Risk. *Cancer Epidemiol. Biomark. Prev.* **1993**, *2*, 183–187.

39. Giovannucci, E. Tomatoes, Tomato-Based Products, Lycopene, and Cancer: Review of the Epidemiologic Literature. *J. Natl. Cancer Inst.* **1999**, *91*, 317–331. [CrossRef]

40. Johnson, E.J.; Hammond, B.R.; Yeum, K.J.; Qin, J.; Wang, X.D.; Castaneda, C.; Snodderly, D.M.; Russell, R.M. Relation among serum and tissue concentrations of lutein and zeaxanthin and macular pigment density. *Am. J. Clin. Nutr.* **2000**, *71*, 1555–1562. [CrossRef]

41. Slattery, M.L.; Benson, J.; Curtin, K.; Ma, K.N.; Schaeffer, D.; Potter, J.D. Carotenoids and colon cancer. *Am. J. Clin. Nutr.* **2000**, *71*, 575–582. [CrossRef]

42. Kim, D.E.; Shang, X.; Assefa, A.D.; Keum, Y.S.; Saini, R.K. Metabolite profiling of green, green/red, and red lettuce cultivars: Variation in health beneficial compounds and antioxidant potential. *Food Res. Int.* **2018**, *105*, 361–370. [CrossRef]

43. Mulabagal, V.; Ngouajio, M.; Nair, A.; Zhang, Y.; Gottumukkala, A.L.; Nair, M.G. In vitro evaluation of red and green lettuce (*Lactuca sativa*) for functional food properties. *Food Chem.* **2010**, *118*, 300–306. [CrossRef]

44. Hao, J.H.; Wang, L.; Liu, C.J.; Zhang, X.D.; Qi, Z.Y.; Li, B.Y.; Xiao, S.; Fan, S.X. Evaluation of nutritional quality in different varieties of green and purple leaf lettuces. In *IOP Conference Series: Earth and Environmental Science*; IOP Publishing: Bristol, UK, 2018; Volume 185.

45. El-Nakhel, C.; Petropoulos, S.A.; Pannico, A.; Kyriacou, M.C.; Giordano, M.; Colla, G.; Dario Troise, A.; Vitaglione, P.; De Pascale, S.; Rouphael, Y. The bioactive profile of lettuce produced in a closed soilless system as configured by combinatorial effects of genotype and macrocation supply composition. *Food Chem.* **2020**, *309*, 125713. [CrossRef] [PubMed]
46. El-Nakhel, C.; Pannico, A.; Kyriacou, M.C.; Giordano, M.; De Pascale, S.; Rouphael, Y. Macronutrient deprivation eustress elicits differential secondary metabolites in red and green-pigmented butterhead lettuce grown in a closed soilless system. *J. Sci. Food Agric.* **2019**, *99*, 6962–6972. [CrossRef] [PubMed]
47. El-Nakhel, C.; Giordano, M.; Pannico, A.; Carillo, P.; Fusco, G.M.; De Pascale, S.; Rouphael, Y. Cultivar-specific performance and qualitative descriptors for butterhead salanova lettuce produced in closed soilless cultivation as a candidate salad crop for human life support in space. *Life* **2019**, *9*, 61. [CrossRef] [PubMed]
48. Mampholo, B.M.; Maboko, M.M.; Soundy, P.; Sivakumar, D. Phytochemicals and overall quality of leafy lettuce (*Lactuca sativa* L.) varieties grown in closed hydroponic system. *J. Food Qual.* **2016**, *39*, 805–815. [CrossRef]
49. Luna, M.C.; Martínez-Sánchez, A.; Selma, M.V.; Tudela, J.A.; Baixauli, C.; Gil, M.I. Influence of nutrient solutions in an open-field soilless system on the quality characteristics and shelf life of fresh-cut red and green lettuces (*Lactuca sativa* L.) in different seasons. *J. Sci. Food Agric.* **2013**, *93*, 415–421. [CrossRef]
50. Rouphael, Y.; Petropoulos, S.A.; El-Nakhel, C.; Pannico, A.; Kyriacou, M.C.; Giordano, M.; Troise, A.D.; Vitaglione, P.; De Pascale, S. Reducing Energy Requirements in Future Bioregenerative Life Support Systems (BLSSs): Performance and Bioactive Composition of Diverse Lettuce Genotypes Grown Under Optimal and Suboptimal Light Conditions. *Front. Plant Sci.* **2019**, *10*, 1305. [CrossRef]
51. Pannico, A.; El-Nakhel, C.; Kyriacou, M.C.; Giordano, M.; Stazi, S.R.; De Pascale, S.; Rouphael, Y. Combating Micronutrient Deficiency and Enhancing Food Functional Quality Through Selenium Fortification of Select Lettuce Genotypes Grown in a Closed Soilless System. *Front. Plant Sci.* **2019**, *10*, 1495. [CrossRef]
52. Rice-Evans, C.A.; Miller, N.J.; Paganga, G. Antioxidant properties of phenolic compounds. *Trends Plant Sci.* **1997**, *2*, 152–159. [CrossRef]
53. Pérez-López, U.; Pinzino, C.; Quartacci, M.F.; Ranieri, A.; Sgherri, C. Phenolic composition and related antioxidant properties in differently colored lettuces: A study by electron paramagnetic resonance (EPR) kinetics. *J. Agric. Food Chem.* **2014**, *62*, 12001–12007. [CrossRef]
54. Lee, J.H.; Felipe, P.; Yang, Y.H.; Kim, M.Y.; Oh, Y.K.; Sok, D.E.; Kim, H.C.; Kim, M.R. Effects of dietary supplementation with red-pigmented leafy lettuce (*Lactuca sativa*) on lipid profiles and antioxidant status in C57BL/6J mice fed a high-fat high-cholesterol diet. *Br. J. Nutr.* **2009**, *101*, 1246–1254. [CrossRef]
55. Fernandez, E.; La Vecchia, C.; D'Avanzo, B.; Negri, E.; Franceschi, S. Risk factors for colorectal cancer in subjects with family history of the disease. *Br. J. Cancer* **1997**, *75*, 1381–1384. [CrossRef] [PubMed]
56. Qin, X.X.; Zhang, M.Y.; Han, Y.Y.; Hao, J.H.; Liu, C.J.; Fan, S.X. Beneficial phytochemicals with anti-tumor potential revealed through metabolic profiling of new red pigmented lettuces (*Lactuca sativa* L.). *Int. J. Mol. Sci.* **2018**, *19*, 1165. [CrossRef] [PubMed]
57. Sgherri, C.; Cecconami, S.; Pinzino, C.; Navari-Izzo, F.; Izzo, R. Levels of antioxidants and nutraceuticals in basil grown in hydroponics and soil. *Food Chem.* **2010**, *123*, 416–422. [CrossRef]
58. Simon, J.; Morales, M.; Phippen, W.; Vieira, R.; Hao, Z. Basil: A source of aroma compounds and a popular culinary and ornamental herb. *Perspect. New Crop. New Uses* **1999**, *16*, 499–505.
59. Lee, J.; Scagel, C.F. Chicoric acid found in basil (*Ocimum basilicum* L.) leaves. *Food Chem.* **2009**, *115*, 650–656. [CrossRef]
60. Javanmardi, J.; Khalighi, A.; Kashi, A.; Bais, H.P.; Vivanco, J.M. Chemical characterization of basil (*Ocimum basilicum* L.) found in local accessions and used in traditional medicines in Iran. *J. Agric. Food Chem.* **2002**, *50*, 5878–5883. [CrossRef]
61. Kwee, E.M.; Niemeyer, E.D. Variations in phenolic composition and antioxidant properties among 15 basil (*Ocimum basilicum* L.) cultivars. *Food Chem.* **2011**, *128*, 1044–1050. [CrossRef]
62. Kopsell, D.A.; Kopsell, D.E.; Curran-Celentano, J. Carotenoid and chlorophyll pigments in sweet basil grown in the field and greenhouse. *HortScience* **2005**, *40*, 1230–1233. [CrossRef]
63. Flanigan, P.M.; Niemeyer, E.D. Effect of cultivar on phenolic levels, anthocyanin composition, and antioxidant properties in purple basil (*Ocimum basilicum* L.). *Food Chem.* **2014**, *164*, 518–526. [CrossRef]

64. Calucci, L.; Pinzino, C.; Zandomeneghi, M.; Capocchi, A.; Ghiringhelli, S.; Saviozzi, F.; Tozzi, S.; Galleschi, L. Effects of γ-irradiation on the free radical and antioxidant contents in nine aromatic herbs and spices. *J. Agric. Food Chem.* **2003**, *51*, 927–934. [CrossRef]

65. Muráriková, A.; Neugebauerová, J. Seasonal variation of ascorbic acid and nitrate levels in selected basil (*Ocimum basilicum* L.) varieties. *Hortic. Sci.* **2018**, *45*, 47–52. [CrossRef]

66. Gallie, D.R. Increasing vitamin C content in plant foods to improve their nutritional value-successes and challenges. *Nutrients* **2013**, *5*, 3424–3446. [CrossRef]

67. Tenore, G.C.; Campiglia, P.; Ciampaglia, R.; Izzo, L.; Novellino, E. Antioxidant and antimicrobial properties of traditional green and purple "Napoletano" basil cultivars (*Ocimum basilicum* L.) from Campania region (Italy). *Nat. Prod. Res.* **2017**, *31*, 2067–2071. [CrossRef] [PubMed]

68. Phippen, W.B.; Simon, J.E. Anthocyanins in Basil (*Ocimum basilicum* L.). *J. Agric. Food Chem.* **1998**, *46*, 1734–1738. [CrossRef]

69. Cazzola, R.; Camerotto, C.; Cestaro, B. Anti-oxidant, anti-glycant, and inhibitory activity against α-amylase and α-glucosidase of selected spices and culinary herbs. *Int. J. Food Sci. Nutr.* **2011**, *62*, 175–184. [CrossRef]

70. El-Beshbishy, H.A.; Bahashwan, S.A. Hypoglycemic effect of basil (*Ocimum basilicum*) aqueous extract is mediated through inhibition of α-glucosidase and α-amylase activities: An in vitro study. *Toxicol. Ind. Health* **2011**, *28*, 42–50. [CrossRef]

71. Kwon, Y.I.I.; Vattem, D.A.; Shetty, K. Evaluation of clonal herbs of Lamiaceae species for management of diabetes and hypertension. *Asia Pac. J. Clin. Nutr.* **2006**, *15*, 107–118.

72. Ha, T.J.; Lee, J.H.; Lee, M.H.; Lee, B.W.; Kwon, H.S.; Park, C.H.; Shim, K.B.; Kim, H.T.; Baek, I.Y.; Jang, D.S. Isolation and identification of phenolic compounds from the seeds of *Perilla frutescens* (L.) and their inhibitory activities against α-glucosidase and aldose reductase. *Food Chem.* **2012**, *135*, 1397–1403. [CrossRef]

73. Yilmazer-Musa, M.; Griffith, A.M.; Michels, A.J.; Schneider, E.; Frei, B. Grape seed and tea extracts and catechin 3-gallates are potent inhibitors of α-amylase and α-glucosidase activity. *J. Agric. Food Chem.* **2012**, *60*, 8924–8929. [CrossRef]

74. Harnafi, H.; Serghini Caid, H.; el Houda Bouanani, N.; Aziz, M.; Amrani, S. Hypolipemic activity of polyphenol-rich extracts from *Ocimum basilicum* in Triton WR-1339-induced hyperlipidemic mice. *Food Chem.* **2008**, *108*, 205–212. [CrossRef]

75. Amrani, S.; Harnafi, H.; Bouanani, N.E.H.; Aziz, M.; Caid, H.S.; Manfredini, S.; Besco, E.; Napolitano, M.; Bravo, E. Hypolipidaemic Activity of Aqueous *Ocimum basilicum* Extract in Acute Hyperlipidaemia Induced by Triton WR-1339 in Rats and its Antioxidant Property. *Phyther. Res.* **2006**, *20*, 1040–1045. [CrossRef] [PubMed]

76. Dhyani, A.; Chopra, R.; Garg, M. A review on nutritional value, functional properties and pharmacological application of perilla (*Perilla frutescens* L.). *Biomed. Pharmacol. J.* **2019**, *12*, 649–660. [CrossRef]

77. Müller-Waldeck, F.; Sitzmann, J.; Schnitzler, W.H.; Graßmann, J. Determination of toxic perilla ketone, secondary plant metabolites and antioxidative capacity in five *Perilla frutescens* L. varieties. *Food Chem. Toxicol.* **2010**, *48*, 264–270. [CrossRef] [PubMed]

78. Meng, L.; Lozano, Y.; Bombarda, I.; Gaydou, E.M.; Li, B. Polyphenol extraction from eight *Perilla frutescens* cultivars. *Comptes Rendus Chim.* **2009**, *12*, 602–611. [CrossRef]

79. Peng, Y.; Ye, J.; Kong, J. Determination of phenolic compounds in *Perilla frutescens* L. by capillary electrophoresis with electrochemical detection. *J. Agric. Food Chem.* **2005**, *53*, 8141–8147. [CrossRef]

80. Lu, N.; Bernardo, E.L.; Tippayadarapanich, C.; Takagaki, M.; Kagawa, N.; Yamori, W. Growth and accumulation of secondary metabolites in perilla as affected by photosynthetic photon flux density and electrical conductivity of the nutrient solution. *Front. Plant Sci.* **2017**, *8*, 708. [CrossRef]

81. Martinetti, L.; Ferrante, A.; Bassoli, A.; Borgonovo, G.; Tosca, A.; Spoleto, P. Characterization of some qualitative traits in different perilla cultivars. *Acta Hortic.* **2012**, *939*, 301–308. [CrossRef]

82. Rouphael, Y.; Kyriacou, M.C.; Carillo, P.; Pizzolongo, F.; Romano, R.; Sifola, M.I. Chemical Eustress Elicits Tailored Responses and Enhances the Functional Quality of Novel Food *Perilla frutescens*. *Molecules* **2019**, *24*, 185. [CrossRef]

83. Meng, L.; Lozano, Y.F.; Gaydou, E.M.; Li, B. Antioxidant activities of polyphenols extracted from *Perilla frutescens* varieties. *Molecules* **2009**, *14*, 133–140. [CrossRef]

84. Narisawa, T.; Fukaura, Y.; Yazawa, K.; Ishikawa, C.; Isoda, Y.; Nishizawa, Y. Colon cancer prevention with a small amount of dietary perilla oil high in alpha-linolenic acid in an animal model. *Cancer* **1994**, *73*, 2069–2075. [CrossRef]

85. Banno, N.; Akihisa, T.; Tokuda, H.; Yasukawa, K.; Higashihara, H.; Ukiya, M.; Watanabe, K.; Kimura, Y.; Hasegawa, J.I.; Nishino, H. Triterpene acids from the leaves of *Perilla frutescens* and their anti-inflammatory and antitumor-promoting effects. *Biosci. Biotechnol. Biochem.* **2004**, *68*, 85–90. [CrossRef] [PubMed]

86. Bassoli, A.; Borgonovo, G.; Caimi, S.; Scaglioni, L.; Morini, G.; Moriello, A.S.; di Marzo, V.; de Petrocellis, L. Taste-guided identification of high potency TRPA1 agonists from *Perilla frutescens*. *Bioorganic Med. Chem.* **2009**, *17*, 1636–1639. [CrossRef] [PubMed]

87. Laureati, M.; Buratti, S.; Bassoli, A.; Borgonovo, G.; Pagliarini, E. Discrimination and characterisation of three cultivars of *Perilla frutescens* by means of sensory descriptors and electronic nose and tongue analysis. *Food Res. Int.* **2010**, *43*, 959–964. [CrossRef]

88. Kanner, J.; Harel, S.; Granit, R. Betalains - A new class of dietary cationized antioxidants. *J. Agric. Food Chem.* **2001**, *49*, 5178–5185. [CrossRef]

89. Daiss, N.; Lobo, M.G.; Gonzalez, M. Changes in postharvest quality of swiss chard grown using 3 organic preharvest treatments. *J. Food Sci.* **2008**, *73*, S314–S320. [CrossRef]

90. Mzoughi, Z.; Chahdoura, H.; Chakroun, Y.; Cámara, M.; Fernández-Ruiz, V.; Morales, P.; Mosbah, H.; Flamini, G.; Snoussi, M.; Majdoub, H. Wild edible Swiss chard leaves (*Beta vulgaris* L. var. *cicla*): Nutritional, phytochemical composition and biological activities. *Food Res. Int.* **2019**, *119*, 612–621. [CrossRef]

91. Ninfali, P.; Antonini, E.; Frati, A.; Scarpa, E.S. C-Glycosyl Flavonoids from *Beta vulgaris Cicla* and Betalains from Beta vulgaris rubra: Antioxidant, Anticancer and Antiinflammatory Activities—A Review. *Phyther. Res.* **2017**, *31*, 871–884. [CrossRef]

92. Lee, C.Y.; Chien, Y.S.; Chiu, T.H.; Huang, W.W.; Lu, C.C.; Chiang, J.H.; Yang, J.S. Apoptosis triggered by vitexin in U937 human leukemia cells via a mitochondrial signaling pathway. *Oncol. Rep.* **2012**, *28*, 1883–1888. [CrossRef]

93. Ninfali, P.; Bacchiocca, M.; Antonelli, A.; Biagiotti, E.; Di Gioacchino, A.M.; Piccoli, G.; Stocchi, V.; Brandi, G. Characterization and biological activity of the main flavonoids from Swiss Chard (*Beta vulgaris* subspecies *cycla*). *Phytomedicine* **2007**, *14*, 216–221. [CrossRef]

94. Gennari, L.; Felletti, M.; Blasa, M.; Angelino, D.; Celeghini, C.; Corallini, A.; Ninfali, P. Total extract of *Beta vulgaris* var. *cicla* seeds versus its purified phenolic components: Antioxidant activities and antiproliferative effects against colon cancer cells. *Phytochem. Anal.* **2011**, *22*, 272–279. [CrossRef]

95. Borghi, S.M.; Carvalho, T.T.; Staurengo-Ferrari, L.; Hohmann, M.S.N.; Pinge-Filho, P.; Casagrande, R.; Verri, W.A. Vitexin Inhibits Inflammatory Pain in Mice by Targeting TRPV1, Oxidative Stress, and Cytokines. *J. Nat. Prod.* **2013**, *76*, 1141–1149. [CrossRef] [PubMed]

96. Khanam, U.K.S.; Oba, S.; Yanase, E.; Murakami, Y. Phenolic acids, flavonoids and total antioxidant capacity of selected leafy vegetables. *J. Funct. Foods* **2012**, *4*, 979–987. [CrossRef]

97. Neugart, S.; Baldermann, S.; Hanschen, F.S.; Klopsch, R.; Wiesner-Reinhold, M.; Schreiner, M. The intrinsic quality of brassicaceous vegetables: How secondary plant metabolites are affected by genetic, environmental, and agronomic factors. *Sci. Hortic.* **2018**, *233*, 460–478. [CrossRef]

98. Martínez-Sánchez, A.; Gil-Izquierdo, A.; Gil, M.I.; Ferreres, F. A comparative study of flavonoid compounds, vitamin C, and antioxidant properties of baby leaf *Brassicaceae* species. *J. Agric. Food Chem.* **2008**, *56*, 2330–2340. [CrossRef]

99. Di Noia, J. Defining powerhouse fruits and vegetables: A nutrient density approach. *Prev. Chronic Dis.* **2014**, *11*, 3–7. [CrossRef]

100. Kim, M.J.; Chiu, Y.C.; Ku, K.M. Glucosinolates, carotenoids, and vitamins E and K Variation from selected kale and collard cultivars. *J. Food Qual.* **2017**, *2017*, 5123572. [CrossRef]

101. Hahn, C.; Müller, A.; Kuhnert, N.; Albach, D. Diversity of Kale (*Brassica oleracea* var. *sabellica*): Glucosinolate Content and Phylogenetic Relationships. *J. Agric. Food Chem.* **2016**, *64*, 3215–3225. [CrossRef]

102. Ferioli, F.; Giambanelli, E.; D'Antuono, L.F.; Costa, H.S.; Albuquerque, T.G.; Silva, A.S.; Hayran, O.; Koçaoglu, B. Comparison of leafy kale populations from italy, portugal, and turkey for their bioactive compound content: Phenolics, glucosinolates, carotenoids, and chlorophylls. *J. Sci. Food Agric.* **2013**, *93*, 3478–3489. [CrossRef]

103. Bentley-Hewitt, K.L.; Chen, R.K.Y.; Lill, R.E.; Hedderley, D.I.; Herath, T.D.; Matich, A.J.; Mckenzie, M.J. Consumption of selenium-enriched broccoli increases cytokine production in human peripheral blood mononuclear cells stimulated ex vivo, a preliminary human intervention study. *Mol. Nutr. Food Res.* **2014**, *58*, 2350–2357. [CrossRef]

104. Lippmann, D.; Lehmann, C.; Florian, S.; Barknowitz, G.; Haack, M.; Mewis, I.; Wiesner, M.; Schreiner, M.; Glatt, H.; Brigelius-Flohé, R.; et al. Glucosinolates from pak choi and broccoli induce enzymes and inhibit inflammation and colon cancer differently. *Food Funct.* **2014**, *5*, 1073–1081. [CrossRef]

105. Herz, C.; Márton, M.R.; Tran, H.T.T.; Gründemann, C.; Schell, J.; Lamy, E. Benzyl isothiocyanate but not benzyl nitrile from Brassicales plants dually blocks the COX and LOX pathway in primary human immune cells. *J. Funct. Foods* **2016**, *23*, 135–143. [CrossRef]

106. Guzmán-Pérez, V.; Bumke-Vogt, C.; Schreiner, M.; Mewis, I.; Borchert, A.; Pfeiffer, A.F.H. Benzylglucosinolate Derived Isothiocyanate from *Tropaeolum majus* Reduces Gluconeogenic Gene and Protein Expression in Human Cells. *PLoS ONE* **2016**, *11*, e0162379. [CrossRef] [PubMed]

107. Abbaoui, B.; Lucas, C.R.; Riedl, K.M.; Clinton, S.K.; Mortazavi, A. Cruciferous Vegetables, Isothiocyanates, and Bladder Cancer Prevention. *Mol. Nutr. Food Res.* **2018**, *62*, 1800079. [CrossRef] [PubMed]

108. Waterman, C.; Rojas-Silva, P.; Tumer, T.B.; Kuhn, P.; Richard, A.J.; Wicks, S.; Stephens, J.M.; Wang, Z.; Mynatt, R.; Cefalu, W.; et al. Isothiocyanate-rich *Moringa oleifera* extract reduces weight gain, insulin resistance, and hepatic gluconeogenesis in mice. *Mol. Nutr. Food Res.* **2015**, *59*, 1013–1024. [CrossRef] [PubMed]

109. Jin, S.W.; Rahim, M.A.; Kim, H.T.; Park, J.I.; Kang, J.G.; Nou, I.S.; Cheng, Z.M. Molecular analysis of anthocyanin-related genes in ornamental cabbage. *Genome* **2018**, *61*, 111–120. [CrossRef]

110. Jin, S.W.; Rahim, M.A.; Jung, H.J.; Afrin, K.S.; Kim, H.T.; Park, J.I.; Kang, J.G.; Nou, I.S. Abscisic acid and ethylene biosynthesis-related genes are associated with anthocyanin accumulation in purple ornamental cabbage (*Brassica oleracea* var. *acephala*). *Genome* **2019**, *62*, 513–526. [CrossRef]

111. Zhang, B.; Hu, Z.; Zhang, Y.; Li, Y.; Zhou, S.; Chen, G. A putative functional MYB transcription factor induced by low temperature regulates anthocyanin biosynthesis in purple kale (*Brassica oleracea* var. *acephala* f. *tricolor*). *Plant Cell Rep.* **2012**, *31*, 281–289. [CrossRef]

112. Mageney, V.; Baldermann, S.; Albach, D.C. Intraspecific Variation in Carotenoids of *Brassica oleracea* var. *sabellica*. *J. Agric. Food Chem.* **2016**, *64*, 3251–3257. [CrossRef]

113. Zheng, Y.J.; Zhang, Y.T.; Liu, H.C.; Li, Y.M.; Liu, Y.L.; Hao, Y.W.; Lei, B.F. Supplemental blue light increases growth and quality of greenhouse pak choi depending on cultivar and supplemental light intensity. *J. Integr. Agric.* **2018**, *17*, 2245–2256. [CrossRef]

114. Khoo, H.E.; Azlan, A.; Tang, S.T.; Lim, S.M. Anthocyanidins and anthocyanins: Colored pigments as food, pharmaceutical ingredients, and the potential health benefits. *Food Nutr. Res.* **2017**, *61*, 1361779. [CrossRef]

115. McCance, K.R.; Flanigan, P.M.; Quick, M.M.; Niemeyer, E.D. Influence of plant maturity on anthocyanin concentrations, phenolic composition, and antioxidant properties of 3 purple basil (*Ocimum basilicum* L.) cultivars. *J. Food Compos. Anal.* **2016**, *53*, 30–39. [CrossRef]

116. Fernandes, F.; Pereira, E.; Ćirić, A.; Soković, M.; Calhelha, R.C.; Barros, L.; Ferreira, I.C.F.R. *Ocimum basilicum* var. *purpurascens* leaves (red rubin basil): A source of bioactive compounds and natural pigments for the food industry. *Food Funct.* **2019**, *10*, 3161–3171. [CrossRef] [PubMed]

117. Fujiwara, Y.; Kono, M.; Ito, A.; Ito, M. Anthocyanins in perilla plants and dried leaves. *Phytochemistry* **2018**, *147*, 158–166. [CrossRef] [PubMed]

118. Lu, N.; Takagaki, M.; Yamori, W.; Kagawa, N. Flavonoid Productivity Optimized for Green and Red Forms of *Perilla frutescens* via Environmental Control Technologies in Plant Factory. *J. Food Qual.* **2018**, *2018*, 4270279. [CrossRef]

119. He, Y.K.; Yao, Y.Y.; Chang, Y.N. Characterization of anthocyanins in *Perilla frutescens* var. *acuta* Extract by Advanced UPLC-ESI-IT-TOF-MSn Method and their anticancer bioactivity. *Molecules* **2015**, *20*, 9155–9169. [CrossRef]

120. Cui, L.; Zhang, Z.; Li, H.; Li, N.; Li, X.; Chen, T. Optimization of ultrasound assisted extraction of phenolic compounds and anthocyanins from perilla leaves using response surface methodology. *Food Sci. Technol. Res.* **2017**, *23*, 535–543. [CrossRef]

121. Stintzing, F.C.; Carle, R. Betalains - emerging prospects for food scientists. *Trends Food Sci. Technol.* **2007**, *18*, 514–525. [CrossRef]

122. Khan, M.I.; Giridhar, P. Plant betalains: Chemistry and biochemistry. *Phytochemistry* **2015**, *117*, 267–295. [CrossRef]

123. Kugler, F.; Stintzing, F.C.; Carle, R. Identification of Betalains from Petioles of Differently Colored Swiss Chard (*Beta vulgaris* L. ssp. *cicla* [L.] Alef. Cv. Bright Lights) by High-Performance Liquid Chromatography-Electrospray Ionization Mass Spectrometry. *J. Agric. Food Chem.* **2004**, *52*, 2975–2981. [CrossRef]

124. Raiola, A.; Rigano, M.M.; Calafiore, R.; Frusciante, L.; Barone, A. Enhancing the health-promoting effects of tomato fruit for biofortified food. *Mediat. Inflamm.* **2014**, *2014*, 139873. [CrossRef]

125. Kotíková, Z.; Lachman, J.; Hejtmánková, A.; Hejtmánková, K. Determination of antioxidant activity and antioxidant content in tomato varieties and evaluation of mutual interactions between antioxidants. *LWT Food Sci. Technol.* **2011**, *44*, 1703–1710. [CrossRef]

126. Borghesi, E.; Ferrante, A.; Gordillo, B.; Rodríguez-Pulido, F.J.; Cocetta, G.; Trivellini, A.; Mensuali-Sodi, A.; Malorgio, F.; Heredia, F.J. Comparative physiology during ripening in tomato rich-anthocyanins fruits. *Plant Growth Regul.* **2016**, *80*, 207–214. [CrossRef]

127. Khachik, F.; Carvalho, L.; Bernstein, P.S.; Muir, G.J.; Zhao, D.Y.; Katz, N.B. Chemistry, distribution, and metabolism of tomato carotenoids and their impact on human health. *Exp. Biol. Med.* **2002**, *227*, 845–851. [CrossRef] [PubMed]

128. Li, F.; Song, X.; Wu, L.; Chen, H.; Liang, Y.; Zhang, Y. Heredities on fruit color and pigment content between green and purple fruits in tomato. *Sci. Hortic.* **2018**, *235*, 391–396. [CrossRef]

129. Rao, A.V.; Rao, L.G. Carotenoids and human health. *Pharmacol. Res.* **2007**, *55*, 207–216. [CrossRef]

130. Blando, F.; Berland, H.; Maiorano, G.; Durante, M.; Mazzucato, A.; Picarella, M.E.; Nicoletti, I.; Gerardi, C.; Mita, G.; Andersen, Ø.M. Nutraceutical Characterization of Anthocyanin-Rich Fruits Produced by "Sun Black" Tomato Line. *Front. Nutr.* **2019**, *6*, 133. [CrossRef]

131. Povero, G.; Gonzali, S.; Bassolino, L.; Mazzucato, A.; Perata, P. Transcriptional analysis in high-anthocyanin tomatoes reveals synergistic effect of Aft and atv genes. *J. Plant Physiol.* **2011**, *168*, 270–279. [CrossRef]

132. Maligeppagol, M.; Sharath Chandra, G.; Navale, P.M.; Deepa, H.; Rajeev, P.R.; Asokan, R.; Prasad Babu, K.; Bujji Babu, C.S.; Keshava Rao, V.; Krishna Kumar, N.K. Anthocyanin enrichment of tomato (*Solanum lycopersicum* L.) fruit by metabolic engineering. *Curr. Sci.* **2013**, *105*, 72–80.

133. Sestari, I.; Zsögön, A.; Rehder, G.G.; de Lira Teixeira, L.; Hassimotto, N.M.A.; Purgatto, E.; Benedito, V.A.; Peres, L.E.P. Near-isogenic lines enhancing ascorbic acid, anthocyanin and carotenoid content in tomato (*Solanum lycopersicum* L. cv Micro-Tom) as a tool to produce nutrient-rich fruits. *Sci. Hortic.* **2014**, *175*, 111–120. [CrossRef]

134. Gonzali, S.; Mazzucato, A.; Perata, P. Purple as a tomato: Towards high anthocyanin tomatoes. *Trends Plant Sci.* **2009**, *14*, 237–241. [CrossRef]

135. Li, H.; Deng, Z.; Liu, R.; Loewen, S.; Tsao, R. Bioaccessibility, in vitro antioxidant activities and in vivo anti-inflammatory activities of a purple tomato (*Solanum lycopersicum* L.). *Food Chem.* **2014**, *159*, 353–360. [CrossRef] [PubMed]

136. Li, H.; Deng, Z.; Liu, R.; Young, J.C.; Zhu, H.; Loewen, S.; Tsao, R. Characterization of phytochemicals and antioxidant activities of a purple tomato (*Solanum lycopersicum* L.). *J. Agric. Food Chem.* **2011**, *59*, 11803–11811. [CrossRef] [PubMed]

137. Zhou, K.; Yu, L. Total phenolic contents and antioxidant properties of commonly consumed vegetables grown in Colorado. *LWT Food Sci. Technol.* **2006**, *39*, 1155–1162. [CrossRef]

138. Raffo, A.; La Malfa, G.; Fogliano, V.; Maiani, G.; Quaglia, G. Seasonal variations in antioxidant components of cherry tomatoes (*Lycopersicon esculentum* cv. Naomi F1). *J. Food Compos. Anal.* **2006**, *19*, 11–19. [CrossRef]

139. Capanoglu, E.; Beekwilder, J.; Boyacioglu, D.; Hall, R.; De Vos, R. Changes in antioxidant and metabolite profiles during production of tomato paste. *J. Agric. Food Chem.* **2008**, *56*, 964–973. [CrossRef]

140. Walle, T. Bioavailability of resveratrol. *Ann. N. Y. Acad. Sci.* **2011**, *1215*, 9–15. [CrossRef]

141. Stewart, Z.A.; Westfall, M.D.; Pietenpol, J.A. Cell-cycle dysregulation and anticancer therapy. *Trends Pharmacol. Sci.* **2003**, *24*, 139–145. [CrossRef]

142. Gupta, S.C.; Kannappan, R.; Reuter, S.; Kim, J.H.; Aggarwal, B.B. Chemosensitization of tumors by resveratrol. *Ann. N. Y. Acad. Sci.* **2011**, *1215*, 150–160. [CrossRef]

143. Lin, H.Y.; Tang, H.Y.; Davis, F.B.; Davis, P.J. Resveratrol and apoptosis. *Ann. N. Y. Acad. Sci.* **2011**, *1215*, 79–88. [CrossRef]

144. Szkudelski, T.; Szkudelska, K. Anti-diabetic effects of resveratrol. *Ann. N. Y. Acad. Sci.* **2011**, *1215*, 34–39. [CrossRef] [PubMed]

145. Fuentes, E.; Palomo, I. Antiplatelet effects of natural bioactive compounds by multiple targets: Food and drug interactions. *J. Funct. Foods* **2014**, *6*, 73–81. [CrossRef]

146. Vagula, J.M.; Bertozzi, J.; Castro, J.C.; de Oliveira, C.C.; Clemente, E.; de Oliveira Santos Júnior, O.; Visentainer, J.V. Determination of trans-resveratrol in *Solanum americanum* Mill. by HPLC. *Nat. Prod. Res.* **2016**, *30*, 2230–2234. [CrossRef] [PubMed]

147. Vasco, C.; Avila, J.; Ruales, J.; Svanberg, U.; Kamal-Eldin, A. Physical and chemical characteristics of golden-yellow and purple-red varieties of tamarillo fruit (*Solanum betaceum* Cav.). *Int. J. Food Sci. Nutr.* **2009**, *60*, 278–288. [CrossRef] [PubMed]

148. Ninfali, P.; Mea, G.; Giorgini, S.; Rocchi, M.; Bacchiocca, M. Antioxidant capacity of vegetables, spices and dressings relevant to nutrition. *Br. J. Nutr.* **2005**, *93*, 257–266. [CrossRef]

149. Riso, P.; Brusamolino, A.; Contino, D.; Martini, D.; Vendrame, S.; Del Bo', C.; Porrini, M. Lycopene absorption in humans after the intake of two different single-dose lycopene formulations. *Pharmacol. Res.* **2010**, *62*, 318–321. [CrossRef]

150. Reifen, R.; Nissenkorn, A.; Matas, Z.; Bujanover, Y. 5-ASA and lycopene decrease the oxidative stress and inflammation induced by iron in rats with colitis. *J. Gastroenterol.* **2004**, *39*, 514–519. [CrossRef]

151. Hazra, P.; Longjam, M.; Chattopadhyay, A. Stacking of mutant genes in the development of "purple tomato" rich in both lycopene and anthocyanin contents. *Sci. Hortic.* **2018**, *239*, 253–258. [CrossRef]

152. Guil-Guerrero, J.L.; Rebolloso-Fuentes, M.M. Nutrient composition and antioxidant activity of eight tomato (*Lycopersicon esculentum*) varieties. *J. Food Compos. Anal.* **2009**, *22*, 123–129. [CrossRef]

153. Castañeda-Ovando, A.; de Louedes Pacheco-Hernández, M.; Páez-Hernández, M.E.; Rodríguez, J.A.; Galán-Vidal, C.A. Chemical studies of anthocyanins: A review. *Food Chem.* **2009**, *113*, 859–871. [CrossRef]

154. Li, H.; Deng, Z.; Zhu, H.; Hu, C.; Liu, R.; Young, J.C.; Tsao, R. Highly pigmented vegetables: Anthocyanin compositions and their role in antioxidant activities. *Food Res. Int.* **2012**, *46*, 250–259. [CrossRef]

155. Gomes, A.; Fernandes, E.; Lima, J.; Mira, L.; Corvo, M. Molecular Mechanisms of Anti-Inflammatory Activity Mediated by Flavonoids. *Curr. Med. Chem.* **2008**, *15*, 1586–1605. [CrossRef] [PubMed]

156. Espin, S.; Gonzalez-Manzano, S.; Taco, V.; Poveda, C.; Ayuda-Durán, B.; Gonzalez-Paramas, A.M.; Santos-Buelga, C. Phenolic composition and antioxidant capacity of yellow and purple-red Ecuadorian cultivars of tree tomato (*Solanum betaceum* Cav.). *Food Chem.* **2016**, *194*, 1073–1080. [CrossRef] [PubMed]

157. Zhang, Y.; Butelli, E.; De Stefano, R.; Schoonbeek, H.J.; Magusin, A.; Pagliarani, C.; Wellner, N.; Hill, L.; Orzaez, D.; Granell, A.; et al. Anthocyanins double the shelf life of tomatoes by delaying overripening and reducing susceptibility to gray mold. *Curr. Biol.* **2013**, *23*, 1094–1100. [CrossRef] [PubMed]

158. Wang, S.Y.; Lin, H.S. Antioxidant activity in fruits and leaves of blackberry, raspberry, and strawberry varies with cultivar and developmental stage. *J. Agric. Food Chem.* **2000**, *48*, 140–146. [CrossRef] [PubMed]

159. Bae, S.H.; Suh, H.J. Antioxidant activities of five different mulberry cultivars in Korea. *LWT Food Sci. Technol.* **2007**, *40*, 955–962. [CrossRef]

160. Jones, C.M.; Mes, P.; Myers, J.R. Characterization and Inheritance of the Anthocyanin fruit (Aft) Tomato. *J. Hered.* **2003**, *94*, 449–456. [CrossRef]

161. Sapir, M.; Oren-Shamir, M.; Ovadia, R.; Reuveni, M.; Evenor, D.; Tadmor, Y.; Nahon, S.; Shlomo, H.; Chen, L.; Meir, A.; et al. Molecular aspects of Anthocyanin fruit tomato in relation to high pigment-1. *J. Hered.* **2008**, *99*, 292–303. [CrossRef]

162. Lister, C.E.; Morrison, S.C.; Kerkhofs, N.S.; Wright, K.M. *The Nutritional Composition and Health Benefits of New Zealand Tamarillos*; Crop & Food Research: Wellington, New Zealand, 2005.

163. Mertz, C.; Gancel, A.L.; Gunata, Z.; Alter, P.; Dhuique-Mayer, C.; Vaillant, F.; Perez, A.M.; Ruales, J.; Brat, P. Phenolic compounds, carotenoids and antioxidant activity of three tropical fruits. *J. Food Compos. Anal.* **2009**, *22*, 381–387. [CrossRef]

164. Osorio, C.; Hurtado, N.; Dawid, C.; Hofmann, T.; Heredia-Mira, F.J.; Morales, A.L. Chemical characterisation of anthocyanins in tamarillo (*Solanum betaceum* Cav.) and Andes berry (*Rubus glaucus* Benth.) fruits. *Food Chem.* **2012**, *132*, 1915–1921. [CrossRef]

165. Crozier, A.; Jaganath, I.B.; Clifford, M.N. Dietary phenolics: Chemistry, bioavailability and effects on health. *Nat. Prod. Rep.* **2009**, *26*, 1001–1043. [CrossRef]

166. Petersen, M. Rosmarinic acid: New aspects. *Phytochem. Rev.* **2013**, *12*, 207–227. [CrossRef]

167. De Rosso, V.V.; Mercadante, A.Z. HPLC-PDA-MS/MS of anthocyanins and carotenoids from dovyalis and tamarillo fruits. *J. Agric. Food Chem.* **2007**, *55*, 9135–9141. [CrossRef]

168. Samuels, J. Biodiversity of food species of the solanaceae family: A preliminary taxonomic inventory of subfamily Solanoideae. *Resources* **2015**, *4*, 277–322. [CrossRef]

169. Mohy-Ud-Din, A.; Khan, Z.U.D.; Ahmad, M.; Kashmiri, M.A. Chemotaxonomic value of alkaloids in *Solanum nigrum* complex. *Pak. J. Bot.* **2010**, *42*, 653–660.

170. Xu, X.; Conrad, C.; Doktor, D. Optimising phenological metrics extraction for different crop types in Germany using the moderate resolution imaging Spectrometer (MODIS). *Remote Sens.* **2017**, *9*, 254. [CrossRef]

171. Gramazio, P.; Prohens, J.; Plazas, M.; Andjar, I.; Herraiz, F.J.; Castillo, E.; Knapp, S.; Meyer, R.S.; Vilanova, S. Location of chlorogenic acid biosynthesis pathway and polyphenol oxidase genes in a new interspecific anchored linkage map of eggplant. *BMC Plant Biol.* **2014**, *14*, 350. [CrossRef]

172. Cao, G.; Sofic, E.; Prior, R.L. Antioxidant Capacity of Tea and Common Vegetables. *J. Agric. Food Chem.* **1996**, *44*, 3426–3431. [CrossRef]

173. Akanitapichat, P.; Phraibung, K.; Nuchklang, K.; Prompitakkul, S. Antioxidant and hepatoprotective activities of five eggplant varieties. *Food Chem. Toxicol.* **2010**, *48*, 3017–3021. [CrossRef]

174. Whitaker, B.D.; Stommel, J.R. Distribution of hydroxycinnamic acid conjugates in fruit of commercial eggplant (*Solanum melongena* L.) cultivars. *J. Agric. Food Chem.* **2003**, *51*, 3448–3454. [CrossRef]

175. Gajewski, M.; Katarzyna, K.; Bajer, M. The influence of postharvest storage on quality characteristics of fruit of eggplant cultivars. *Not. Bot. Horti Agrobot. Cluj-Napoca* **2009**, *37*, 200–205.

176. Nisha, P.; Abdul Nazar, P.; Jayamurthy, P. A comparative study on antioxidant activities of different varieties of *Solanum melongena*. *Food Chem. Toxicol.* **2009**, *47*, 2640–2644. [CrossRef]

177. Zaro, M.J.; Keunchkarian, S.; Chaves, A.R.; Vicente, A.R.; Concellón, A. Changes in bioactive compounds and response to postharvest storage conditions in purple eggplants as affected by fruit developmental stage. *Postharvest Biol. Technol.* **2014**, *96*, 110–117. [CrossRef]

178. Plazas, M.; Andújar, I.; Vilanova, S.; Hurtado, M.; Gramazio, P.; Herraiz, F.J.; Prohens, J. Breeding for Chlorogenic Acid Content in Eggplant: Interest and Prospects. *Not. Bot. Horti Agrobot. Cluj-Napoca* **2013**, *41*, 26–35. [CrossRef]

179. Concellón, A.; Zaro, M.J.; Chaves, A.R.; Vicente, A.R. Changes in quality and phenolic antioxidants in dark purple American eggplant (*Solanum melongena* L. cv. Lucía) as affected by storage at 0 °C and 10 °C. *Postharvest Biol. Technol.* **2012**, *66*, 35–41. [CrossRef]

180. Zaro, M.J.; Ortiz, L.C.; Keunchkarian, S.; Chaves, A.R.; Vicente, A.R.; Concellón, A. Chlorogenic acid retention in white and purple eggplant after processing and cooking. *LWT Food Sci. Technol.* **2015**, *64*, 802–808. [CrossRef]

181. Sadilova, E.; Stintzing, F.C.; Carle, R. Anthocyanins, colour and antioxidant properties of eggplant (*Solanum melongena* L.) and violet pepper (*Capsicum annuum* L.) peel extracts. *Z. Nat. Sect. C J. Biosci.* **2006**, *61*, 527–535. [CrossRef]

182. Liu, Y.; Lv, J.; Liu, Z.; Wang, J.; Yang, B.; Chen, W.; Ou, L.; Dai, X.; Zhang, Z.; Zou, X. Integrative analysis of metabolome and transcriptome reveals the mechanism of color formation in pepper fruit (*Capsicum annuum* L.). *Food Chem.* **2020**, *306*, 125629. [CrossRef]

183. Mennella, G.; Lo Scalzo, R.; Fibiani, M.; DAlessandro, A.; Francese, G.; Toppino, L.; Acciarri, N.; De Almeida, A.E.; Rotino, G.L. Chemical and bioactive quality traits during fruit ripening in eggplant (*S. melongena* L.) and allied species. *J. Agric. Food Chem.* **2012**, *60*, 11821–11831. [CrossRef]

184. Prohens, J.; Whitaker, B.D.; Plazas, M.; Vilanova, S.; Hurtado, M.; Blasco, M.; Gramazio, P.; Stommel, J.R. Genetic diversity in morphological characters and phenolic acids content resulting from an interspecific cross between eggplant, *Solanum melongena*, and its wild ancestor (*S. incanum*). *Ann. Appl. Biol.* **2013**, *162*, 242–257. [CrossRef]

185. Xi-Ou, X.; Wenqiu, L.; Wei, L.; Xiaoming, G.; Lingling, L.; Feiyue, M.; Yuge, L. The analysis of physiological variations in M2 generation of *Solanum melongena* L. mutagenized by ethyl methane sulfonate. *Front. Plant Sci.* **2017**, *8*, 17. [CrossRef]

186. Zhang, Y.; Hu, Z.; Chu, G.; Huang, C.; Tian, S.; Zhao, Z.; Chen, G. Anthocyanin accumulation and molecular analysis of anthocyanin biosynthesis-associated genes in eggplant (*Solanum melongena* L.). *J. Agric. Food Chem.* **2014**, *62*, 2906–2912. [CrossRef]

187. Noda, Y.; Kneyuki, T.; Igarashi, K.; Mori, A.; Packer, L. Antioxidant activity of nasunin, an anthocyanin in eggplant peels. *Toxicology* **2000**, *148*, 119–123. [CrossRef]

188. Sudheesh, S.; Presannakumar, G.; Vijayakumar, S.; Vijayalakshmi, N.R. Hypolipidemic effect of flavonoids from *Solanum melongena*. *Plant Foods Hum. Nutr.* **1997**, *51*, 321–330. [CrossRef]

189. Matsubara, K.; Kaneyuki, T.; Miyake, T.; Mori, M. Antiangiogenic activity of nasunin, an antioxidant anthocyanin, in eggplant peels. *J. Agric. Food Chem.* **2005**, *53*, 6272–6275. [CrossRef]

190. Han, S.W.; Tae, J.; Kim, J.A.; Kim, D.K.; Seo, G.S.; Yun, K.J.; Choi, S.C.; Kim, T.H.; Nah, Y.H.; Lee, Y.M. The aqueous extract of *Solanum melongena* inhibits PAR2 agonist-induced inflammation. *Clin. Chim. Acta* **2003**, *328*, 39–44. [CrossRef]

191. Mutalik, S.; Paridhavi, K.; Rao, C.M.; Udupa, N. Antipyretic and analgesic effect of leaves of *Solanum melongena* Linn. in Rodents. *Indian J. Pharmacol.* **2003**, *35*, 312–315.

192. Igwe, S.A.; Akunyili, D.N.; Ogbogu, C. Effects of *Solanum melongena* (garden egg) on some visual functions of visually active Igbos of Nigeria. *J. Ethnopharmacol.* **2003**, *86*, 135–138. [CrossRef]

193. Nadeem, M.; Anjum, F.M.; Khan, M.R.; Saeed, M.; Riaz, A. Antioxidant Potential of Bell Pepper (*Capsicum annum* L.)—A Review. *Pak. J. Food Sci.* **2011**, *21*, 45–51.

194. Howard, L.R.; Wildman, R.E.C. Antioxidant vitamin and phytochemical content of fresh and processed pepper fruit (*Capsicum annuum*). In *Handbook of Nutraceuticals and Functional Foods*, 2nd ed.; CRC Press: Boca Raton, FL, USA, 2016; pp. 165–191, ISBN 9781420006186.

195. Topuz, A.; Ozdemir, F. Assessment of carotenoids, capsaicinoids and ascorbic acid composition of some selected pepper cultivars (*Capsicum annuum* L.) grown in Turkey. *J. Food Compos. Anal.* **2007**, *20*, 596–602. [CrossRef]

196. Matsufuji, H.; Ishikawa, K.; Nunomura, O.; Chino, M.; Takeda, M. Anti-oxidant content of different coloured sweet peppers, white, green, yellow, orange and red (*Capsicum annuum* L.). *Int. J. Food Sci. Technol.* **2007**, *42*, 1482–1488. [CrossRef]

197. Wahyuni, Y.; Ballester, A.R.; Sudarmonowati, E.; Bino, R.J.; Bovy, A.G. Metabolite biodiversity in pepper (*Capsicum*) fruits of thirty-two diverse accessions: Variation in health-related compounds and implications for breeding. *Phytochemistry* **2011**, *72*, 1358–1370. [CrossRef]

198. Delgado-Vargas, F.; Paredes-Lopez, O. *Natural Colorants for Food and Nutraceutical Uses*; CRC Press: Boca Raton, FL, USA, 2003; ISBN 1587160765.

199. Lightbourn, G.J.; Griesbach, R.J.; Novotny, J.A.; Clevidence, B.A.; Rao, D.D.; Stommel, J.R. Effects of anthocyanin and carotenoid combinations on foliage and immature fruit color of *Capsicum annuum* L. *J. Hered.* **2008**, *99*, 105–111. [CrossRef]

200. Aza-González, C.; Núñez-Palenius, H.G.; Ochoa-Alejo, N. Molecular biology of chili pepper anthocyanin biosynthesis. *J. Mex. Chem. Soc.* **2012**, *56*, 93–98. [CrossRef]

201. Borovsky, Y.; Oren-Shamir, M.; Ovadia, R.; De Jong, W.; Paran, I. The A locus that controls anthocyanin accumulation in pepper encodes a MYB transcription factor homologous to Anthocyanin2 of Petunia. *Theor. Appl. Genet.* **2004**, *109*, 23–29. [CrossRef]

202. Aza-González, C.; Ochoa-Alejo, N. Characterization of anthocyanins from fruits of two Mexican chili peppers (*Capsicum annuum* L.). *J. Mex. Chem. Soc.* **2012**, *56*, 149–151. [CrossRef]

203. Ghasemnezhad, M.; Sherafati, M.; Payvast, G.A. Variation in phenolic compounds, ascorbic acid and antioxidant activity of five coloured bell pepper (*Capsicum annum*) fruits at two different harvest times. *J. Funct. Foods* **2011**, *3*, 44–49. [CrossRef]

204. Yuan, H.; Zhang, J.; Nageswaran, D.; Li, L. Carotenoid metabolism and regulation in horticultural crops. *Hortic. Res.* **2015**, *2*, 1–11. [CrossRef]

205. Subagio, A.; Morita, N. Effects of protein isolate from hyacinth beans (*Lablab purpureus* (L.) Sweet) seeds on cake characteristics. *Food Sci. Technol. Res.* **2008**, *14*, 12–17. [CrossRef]

206. Bhaisare, M. Antioxidant potential in Dolichose lablab L. (*Lablab purpureus*). *Indian J. Appl. Res.* **2019**, *9*, 35–36.

207. Al-Snafi, A.E. The pharmacology and medical importance of Dolichos lablab (*Lablab purpureus*)—A review. *IOSR J. Pharm.* **2017**, *7*, 22–30. [CrossRef]

208. Momim, M.; Habib, M.R.; Hasan, M.R.; Nayeem, J.; Uddon, N.; Rana, M.S. Anti-inflammatory, antioxidant and cytotoxicity potential of methanolic extract of two Bangladeshi mean *Lablab purpureus* (L.) sweet white and purple. *Int. J. Pharm. Sci. Res.* **2012**, *3*, 779–781.

209. Deoda, R.; Pandya, H.; Patel, M.; Yadav, K.; Kadam, P.; Patil, M. Antilithiatic Activity of Leaves, Bulb and Stem Of *Nymphea odorata* and Dolichos Lablab beans. *Res. J. Pharm. Biol. Chem. Sci.* **2011**, *3*, 814–819.

210. Soetan, K. Comparative evaluation of phytochemicals in the raw and aqueous crude extracts from seeds of three *Lablab purpureus* varieties. *Afr. J. Plant Sci.* **2012**, *6*, 410–415.

211. Im, A.R.; Kim, Y.H.; Lee, H.W.; Song, K.H. Water Extract of *Dolichos lablab* Attenuates Hepatic Lipid Accumulation in a Cellular Nonalcoholic Fatty Liver Disease Model. *J. Med. Food* **2016**, *19*, 495–503. [CrossRef] [PubMed]

212. Koo, S.H.; Choi, Y.; Choi, S.K.; Shin, Y.; Kim, B.G.; Lee, B.L. Purification and Characterization of Serine Protease Inhibitors from Dolichos lablab Seeds; Prevention Effects on Pseudomonal Elastase-Induced Septic Hypotension. *BMB Rep.* **2000**, *33*, 112–119.

213. Cui, B.; Hu, Z.; Zhang, Y.; Hu, J.; Yin, W.; Feng, Y.; Xie, Q.; Chen, G. Anthocyanins and flavonols are responsible for purple color of *Lablab purpureus* (L.) sweet pods. *Plant Physiol. Biochem.* **2016**, *103*, 183–190. [CrossRef]

214. Díaz-Batalla, L.; Widholm, J.M.; Fahey, G.C.; Castaño-Tostado, E.; Paredes-López, O. Chemical components with health implications in wild and cultivated Mexican common bean seeds (*Phaseolus vulgaris* L.). *J. Agric. Food Chem.* **2006**, *54*, 2045–2052. [CrossRef]

215. Ombra, M.N.; D'Acierno, A.; Nazzaro, F.; Riccardi, R.; Spigno, P.; Zaccardelli, M.; Pane, C.; Maione, M.; Fratianni, F. Phenolic Composition and Antioxidant and Antiproliferative Activities of the Extracts of Twelve Common Bean (*Phaseolus vulgaris* L.) Endemic Ecotypes of Southern Italy before and after Cooking. *Oxid. Med. Cell. Longev.* **2016**, *2016*, 1398298. [CrossRef]

216. Aquino-Bolaños, E.N.; García-Díaz, Y.D.; Chavez-Servia, J.L.; Carrillo-Rodríguez, J.C.; Vera-Guzmán, A.M.; Heredia-García, E. Anthocyanin, polyphenol, and flavonoid contents and antioxidant activity in Mexican common bean (*Phaseolus vulgaris* L.) landraces. *Emirates J. Food Agric.* **2016**, *28*, 581–588. [CrossRef]

217. Tsuda, T.; Ohshima, K.; Kawakishi, S.; Osawa, T. Inhibition of Lipid Peroxidation and Radical Scavenging Effect of Anthocyanin Pigments Isolated from the Seeds of *Phaseolus vulgaris* L. In *Food Factors for Cancer Prevention*; Springer: Tokyo, Japan, 1997; pp. 318–322.

218. Mazewski, C.; Liang, K.; Gonzalez de Mejia, E. Comparison of the effect of chemical composition of anthocyanin-rich plant extracts on colon cancer cell proliferation and their potential mechanism of action using in vitro, in silico, and biochemical assays. *Food Chem.* **2018**, *242*, 378–388. [CrossRef]

219. Blanca, J.M.; Prohens, J.; Anderson, G.J.; Zuriaga, E.; Cañizares, J.; Nuez, F. AFLP and DNA sequence variation in an Andean domesticate, pepino (*Solanum muricatum*, Solanaceae): Implications for evolution and domestication. *Am. J. Bot.* **2007**, *94*, 1219–1229. [CrossRef] [PubMed]

220. Mahato, S.; Gurung, S.; Chakravarty, S.; Chhetri, B.; Khawas, T. An introduction to Pepino (*Solanum muricatum* Aiton): Review. *Int. J. Environ. Agric. Biotechnol.* **2016**, *1*, 143–148.

221. O'Donoghue, E.M.; Somerfield, S.D.; De Vré, L.A.; Heyes, J.A. Developmental and ripening-related effects on the cell wall of pepino (*Solanum muricatum*) fruit. *J. Sci. Food Agric.* **1997**, *73*, 455–463. [CrossRef]

222. Hsu, C.C.; Guo, Y.R.; Wang, Z.H.; Yin, M.C. Protective effects of an aqueous extract from pepino (*Solanum muricatum* Ait.) in diabetic mice. *J. Sci. Food Agric.* **2011**, *91*, 1517–1522. [CrossRef] [PubMed]

223. Mauro, R.P.; Agnello, M.; Rizzo, V.; Graziani, G.; Fogliano, V.; Leonardi, C.; Giuffrida, F. Recovery of eggplant field waste as a source of phytochemicals. *Sci. Hortic.* **2019**, *261*, 109023. [CrossRef]

224. Petropoulos, S.A.; Di Gioia, F.; Ntatsi, G. Vegetable organosulfur compounds and their health promoting effects. *Curr. Pharm. Des.* **2017**, *23*, 2850–2875. [CrossRef] [PubMed]

225. Di Gioia, F.; Avato, P.; Serio, F.; Argentieri, M.P. Glucosinolate profile of *Eruca sativa*, *Diplotaxis tenuifolia* and *Diplotaxis erucoides* grown in soil and soilless systems. *J. Food Compos. Anal.* **2018**, *69*, 197–204. [CrossRef]

226. Di Gioia, F.; Rosskopf, E.N.; Leonardi, C.; Giuffrida, F. Effects of application timing of saline irrigation water on broccoli production and quality. *Agric. Water Manag.* **2018**, *203*, 97–104. [CrossRef]

227. Yu, H.; Wang, J.; Sheng, X.; Zhao, Z.; Shen, Y.; Branca, F.; Gu, H. Construction of a high-density genetic map and identification of loci controlling purple sepal trait of flower head in *Brassica oleracea* L. italica. *BMC Plant Biol.* **2019**, *19*, 228. [CrossRef]

228. Branca, F.; Chiarenza, G.L.; Cavallaro, C.; Gu, H.; Zhao, Z.; Tribulato, A. Diversity of Sicilian broccoli (*Brassica oleracea* var. *italica*) and cauliflower (*Brassica oleracea* var. *botrytis*) landraces and their distinctive bio-morphological, antioxidant, and genetic traits. *Genet. Resour. Crop Evol.* **2018**, *65*, 485–502. [CrossRef]

229. Rahim, M.A.; Afrin, K.S.; Jung, H.J.; Kim, H.T.; Park, J.I.; Hur, Y.; Nou, I.S.; Deynze, A. Van Molecular analysis of anthocyanin biosynthesis-related genes reveal BoTT8 associated with purple hypocotyl of broccoli (*Brassica oleracea* var. *italica* L.). *Genome* **2019**, *62*, 253–266. [CrossRef] [PubMed]

230. Terzo, M.N.; Pezzino, F.; Amodeo, L.; Catalano, D.; Viola, M.; Tribulato, A.; Travali, S.; Branca, F. Evaluation of a sicilian black broccoli extract on in vitro cell models. *Acta Hortic.* **2018**, *1202*, 135–142. [CrossRef]

231. Rodríguez-Hernández, M.D.C.; Moreno, D.A.; Carvajal, M.; García-Viguera, C.; Martínez-Ballesta, M.D.C. Natural Antioxidants in Purple Sprouting Broccoli under Mediterranean Climate. *J. Food Sci.* **2012**, *77*, C1058–C1063. [CrossRef] [PubMed]

232. Verkerk, R.; Tebbenhoff, S.; Dekker, M. Variation and distribution of glucosinolates in 42 cultivars of *Brassica oleracea* vegetable crops. *Acta Hortic.* **2010**, *856*, 63–69. [CrossRef]

233. Moreno, D.A.; Pérez-Balibrea, S.; Ferreres, F.; Gil-Izquierdo, Á.; García-Viguera, C. Acylated anthocyanins in broccoli sprouts. *Food Chem.* **2010**, *123*, 358–363. [CrossRef]

234. Kyriacou, M.C.; Rouphael, Y.; Di Gioia, F.; Kyratzis, A.; Serio, F.; Renna, M.; De Pascale, S.; Santamaria, P. Micro-scale vegetable production and the rise of microgreens. *Trends Food Sci. Technol.* **2016**, *57*, 103–115. [CrossRef]

235. Di Gioia, F.; Renna, M.; Santamaria, P. *Sprouts, Microgreens and "Baby Leaf" Vegetables*; Springer: Boston, MA, USA, 2017; pp. 403–432.

236. Di Gioia, F.; Petropoulos, S.A.; Ozores-Hampton, M.; Morgan, K.; Rosskopf, E.N. Zinc and Iron Agronomic Biofortification of Brassicaceae Microgreens. *Agronomy* **2019**, *9*, 677. [CrossRef]

237. Crisp, P.; Jewell, P.A.; Gray, A.R. Improved selection against the purple colour defect of cauliflower curds. *Euphytica* **1975**, *24*, 177–180. [CrossRef]

238. Branca, F. Cauliflower and Broccoli. In *Vegetables I*; Springer: New York, NY, USA, 2007; pp. 151–186.

239. Lo Piero, A.R.; Lo Cicero, L.; Ragusa, L.; Branca, F. Change in the expression of anthocyanin pathway genes in developing inflorescences of sicilian landraces of pigmented broccoli and cauliflower. *Acta Hortic.* **2013**, *1005*, 253–260. [CrossRef]

240. Chiu, L.W.; Li, L. Characterization of the regulatory network of BoMYB2 in controlling anthocyanin biosynthesis in purple cauliflower. *Planta* **2012**, *236*, 1153–1164. [CrossRef]

241. Li, L.; Lu, S.; Cosman, K.M.; Earle, E.D.; Garvin, D.F.; O'Neill, J. β-Carotene accumulation induced by the cauliflower Or gene is not due to an increased capacity of biosynthesis. *Phytochemistry* **2006**, *67*, 1177–1184. [CrossRef] [PubMed]

242. Li, L.; Paolillo, D.J.; Parthasarathy, M.V.; DiMuzio, E.M.; Garvin, D.F. A novel gene mutation that confers abnormal patterns of β-carotene accumulation in cauliflower (*Brassica oleracea* var. *botrytis*). *Plant J.* **2001**, *26*, 59–67. [CrossRef] [PubMed]

243. Zhou, X.; Sun, T.H.; Wang, N.; Ling, H.Q.; Lu, S.; Li, L. The cauliflower Orange gene enhances petiole elongation by suppressing expression of eukaryotic release factor 1. *New Phytol.* **2011**, *190*, 89–100. [CrossRef] [PubMed]

244. Lu, S.; Van Eck, J.; Zhou, X.; Lopez, A.B.; O'Halloran, D.M.; Cosman, K.M.; Conlin, B.J.; Paolillo, D.J.; Garvin, D.F.; Vrebalov, J.; et al. The cauliflower Or gene encodes a DnaJ cysteine-rich domain-containing protein that mediates high levels of β-carotene accumulation. *Plant Cell* **2006**, *18*, 3594–3605. [CrossRef] [PubMed]

245. Kapusta-Duch, J.; Szelag-Sikora, A.; Sikora, J.; Niemiec, M.; Gródek-Szostak, Z.; Kuboń, M.; Leszczyńska, T.; Borczak, B. Health-Promoting Properties of Fresh and Processed Purple Cauliflower. *Sustainability* **2019**, *11*, 4008. [CrossRef]

246. Ordás, A.; Cartea, M.E. Cabbage and Kale. In *Vegetables I*; Springer: New York, NY, USA, 2007; pp. 119–149.

247. Yuan, Y.; Chiu, L.W.; Li, L. Transcriptional regulation of anthocyanin biosynthesis in red cabbage. *Planta* **2009**, *230*, 1141–1153. [CrossRef]

248. Arapitsas, P.; Sjöberg, P.J.R.; Turner, C. Characterisation of anthocyanins in red cabbage using high resolution liquid chromatography coupled with photodiode array detection and electrospray ionization-linear ion trap mass spectrometry. *Food Chem.* **2008**, *109*, 219–226. [CrossRef]

249. Wiczkowski, W.; Szawara-Nowak, D.; Topolska, J. Red cabbage anthocyanins: Profile, isolation, identification, and antioxidant activity. *Food Res. Int.* **2013**, *51*, 303–309. [CrossRef]

250. Koss-Mikołajczyk, I.; Kusznierewicz, B.; Wiczkowski, W.; Płatosz, N.; Bartoszek, A. Phytochemical composition and biological activities of differently pigmented cabbage (*Brassica oleracea* var. *capitata*) and cauliflower (*Brassica oleracea* var. *botrytis*) varieties. *J. Sci. Food Agric.* **2019**, *99*, 5499–5507.

251. Pliszka, B.; Huszcza-Ciołkowska, G.; Mieleszko, E.; Czaplicki, S. Stability and antioxidative properties of acylated anthocyanins in three cultivars of red cabbage (*Brassica oleracea* L. var. capitata L. f. rubra). *J. Sci. Food Agric.* **2009**, *89*, 1154–1158. [CrossRef]

252. McDougall, G.J.; Fyffe, S.; Dobson, P.; Stewart, D. Anthocyanins from red cabbage—Stability to simulated gastrointestinal digestion. *Phytochemistry* **2007**, *68*, 1285–1294. [CrossRef]

253. Charron, C.S.; Clevidence, B.A.; Britz, S.J.; Novotny, J.A. Effect of dose size on bioavailability of acylated and nonacylated anthocyanins from red cabbage (*Brassica oleracea* L. Var. *capitata*). *J. Agric. Food Chem.* **2007**, *55*, 5354–5362. [CrossRef] [PubMed]

254. Volden, J.; Borge, G.I.A.; Bengtsson, G.B.; Hansen, M.; Thygesen, I.E.; Wicklund, T. Effect of thermal treatment on glucosinolates and antioxidant-related parameters in red cabbage (*Brassica oleracea* L. ssp. *capitata* f. *rubra*). *Food Chem.* **2008**, *109*, 595–605. [CrossRef]

255. Rottenberg, A.; Zohary, D. Wild genetic resources of cultivated artichoke. *Acta Hortic.* **2005**, *681*, 307–314. [CrossRef]

256. Sonnante, G.; Pignone, D.; Hammer, K. The domestication of artichoke and cardoon: From Roman times to the genomic age. *Ann. Bot.* **2007**, *100*, 1095–1100. [CrossRef]

257. de Falco, B.; Incerti, G.; Amato, M.; Lanzotti, V. Artichoke: Botanical, agronomical, phytochemical, and pharmacological overview. *Phytochem. Rev.* **2015**, *14*, 993–1018. [CrossRef]

258. Petropoulos, S.A.; Pereira, C.; Ntatsi, G.; Danalatos, N.; Barros, L.; Ferreira, I.C.F.R. Nutritional value and chemical composition of Greek artichoke genotypes. *Food Chem.* **2018**, *267*, 296–302. [CrossRef]

259. Segovia, M.S.; Palma, M.A.; Leskovar, D.I. Factors affecting consumer preferences and willingness to pay for artichoke products. *Acta Hortic.* **2016**, *1147*, 271–280. [CrossRef]

260. Lanteri, S.; Portis, E. Globe Artichoke and Cardoon. In *Vegetables I*; Springer: New York, NY, USA, 2007; pp. 49–74.

261. Portis, E.; Mauro, R.P.; Acquadro, A.; Moglia, A.; Mauromicale, G.; Lanteri, S. The inheritance of bract pigmentation and fleshy thorns on the globe artichoke capitulum. *Euphytica* **2015**, *206*, 523–531. [CrossRef]

262. Basnizki, J.; Zohary, D. Breeding of seed planted artichoke. *Plant Breed. Rev.* **1994**, *12*, 253–269.

263. Pochard, E.; Foury, C.; Chambonet, D. Il miglioramento genetico del carciofo. In Proceedings of the I International Congress on Artichoke, Minerva Medica, Bari, Italy, 23–24 November 1969; pp. 117–143.

264. Cravero, V.P.; Picardi, L.A.; Cointry, E.L. An approach for understanding the heredity of two quality traits (head color and tightness) in globe artichoke (*Cynara scolymus* L.). *Genet. Mol. Biol.* **2005**, *28*, 431–434. [CrossRef]

265. De Palma, M.; Fratianni, F.; Nazzaro, F.; Tucci, M. Isolation and functional characterization of a novel gene coding for flavonoid 3'-hydroxylase from globe artichoke. *Biol. Plant.* **2014**, *58*, 445–455. [CrossRef]

266. Blanco, E.; Sabetta, W.; Danzi, D.; Negro, D.; Passeri, V.; de Lisi, A.; Paolocci, F.; Sonnante, G. Isolation and Characterization of the Flavonol Regulator CcMYB12 From the Globe Artichoke [*Cynara cardunculus* var. *scolymus* (L.) Fiori]. *Front. Plant Sci.* **2018**, *9*, 941. [CrossRef] [PubMed]

267. Schütz, K.; Kammerer, D.; Carle, R.; Schieber, A. Identification and quantification of caffeoylquinic acids and flavonoids from artichoke (*Cynara scolymus* L.) heads, juice, and pomace by HPLC-DAD-ESI/MSn. *J. Agric. Food Chem.* **2004**, *52*, 4090–4096. [CrossRef]

268. Pandino, G.; Lombardo, S.; Mauromicale, G.; Williamson, G. Profile of polyphenols and phenolic acids in bracts and receptacles of globe artichoke (*Cynara cardunculus* var. *scolymus*) germplasm. *J. Food Compos. Anal.* **2011**, *24*, 148–153. [CrossRef]

269. Fratianni, F.; Tucci, M.; de Palma, M.; Pepe, R.; Nazzaro, F. Polyphenolic composition in different parts of some cultivars of globe artichoke (*Cynara cardunculus* L. var. *scolymus* (L.) Fiori). *Food Chem.* **2007**, *104*, 1282–1286. [CrossRef]

270. Negro, D.; Montesano, V.; Grieco, S.; Crupi, P.; Sarli, G.; De Lisi, A.; Sonnante, G. Polyphenol Compounds in Artichoke Plant Tissues and Varieties. *J. Food Sci.* **2012**, *77*, C244–C252. [CrossRef]

271. Schütz, K.; Persike, M.; Carle, R.; Schieber, A. Characterization and quantification of anthocyanins in selected artichoke (*Cynara scolymus* L.) cultivars by HPLC-DAD-ESI-MS n. *Anal. Bioanal. Chem.* **2006**, *384*, 1511–1517. [CrossRef]

272. Bonasia, A.; Conversa, G.; Lazzizera, C.; Gambacorta, G.; Elia, A. Morphological and qualitative characterisation of globe artichoke head from new seed-propagated cultivars. *J. Sci. Food Agric.* **2010**, *90*, 2689–2693. [CrossRef]

273. Dabbou, S.; Dabbou, S.; Flamini, G.; Pandino, G.; Gasco, L.; Helal, A.N. Phytochemical Compounds from the Crop Byproducts of Tunisian Globe Artichoke Cultivars. *Chem. Biodivers.* **2016**, *13*, 1475–1483. [CrossRef]

274. Pandino, G.; Lombardo, S.; Mauromicale, G. Globe artichoke leaves and floral stems as a source of bioactive compounds. *Ind. Crops Prod.* **2013**, *44*, 44–49. [CrossRef]

275. Garbetta, A.; Capotorto, I.; Cardinali, A.; D'Antuono, I.; Linsalata, V.; Pizzi, F.; Minervini, F. Antioxidant activity induced by main polyphenols present in edible artichoke heads: Influence of in vitro gastro-intestinal digestion. *J. Funct. Foods* **2014**, *10*, 456–464. [CrossRef]

276. Gebhardt, R. Antioxidative and protective properties of extracts from leaves of the artichoke (*Cynara scolymus* L.) against hydroperoxide-induced oxidative stress in cultured rat hepatocytes. *Toxicol. Appl. Pharmacol.* **1997**, *144*, 279–286. [CrossRef] [PubMed]

277. Li, D.; Zhang, Y.; Liu, Y.; Sun, R.; Xia, M. Purified Anthocyanin Supplementation Reduces Dyslipidemia, Enhances Antioxidant Capacity, and Prevents Insulin Resistance in Diabetic Patients. *J. Nutr.* **2015**, *145*, 742–748. [CrossRef]

278. Jennings, A.; Welch, A.A.; Fairweather-Tait, S.J.; Kay, C.; Minihane, A.M.; Chowienczyk, P.; Jiang, B.; Cecelja, M.; Spector, T.; Macgregor, A.; et al. Higher anthocyanin intake is associated with lower arterial stiffness and central blood pressure in women. *Am. J. Clin. Nutr.* **2012**, *96*, 781–788. [CrossRef]

279. Li, H.; Xia, N.; Brausch, I.; Yao, Y.; Förstermann, U. Flavonoids from artichoke (*Cynara scolymus* L.) up-regulate endothelial-type nitric-oxide synthase gene expression in human endothelial cells. *J. Pharmacol. Exp. Ther.* **2004**, *310*, 926–932. [CrossRef]

280. Xia, N.; Pautz, A.; Wollscheid, U.; Reifenberg, G.; Förstermann, U.; Li, H. Artichoke, cynarin and cyanidin downregulate the expression of inducible nitric oxide synthase in human coronary smooth muscle cells. *Molecules* **2014**, *19*, 3654–3668. [CrossRef]

281. Dufour, C.; Loonis, M.; Delosière, M.; Buffière, C.; Hafnaoui, N.; Santé-Lhoutellier, V.; Rémond, D. The matrix of fruit & vegetables modulates the gastrointestinal bioaccessibility of polyphenols and their impact on dietary protein digestibility. *Food Chem.* **2018**, *240*, 314–322.

282. Cicero, A.F.G.; Colletti, A.; Bajraktari, G.; Descamps, O.; Djuric, D.M.; Ezhov, M.; Fras, Z.; Katsiki, N.; Langlois, M.; Latkovskis, G.; et al. Lipid lowering nutraceuticals in clinical practice: Position paper from an International Lipid Expert Panel. *Arch. Med. Sci.* **2017**, *13*, 965–1005. [CrossRef]

283. D'Antuono, I.; Garbetta, A.; Linsalata, V.; Minervini, F.; Cardinali, A. Polyphenols from artichoke heads (*Cynara cardunculus* (L.) subsp. *scolymus* Hayek): In vitro bio-accessibility, intestinal uptake and bioavailability. *Food Funct.* **2015**, *6*, 1268–1277.

284. Guo, Q.; Wang, N.; Liu, H.; Li, Z.; Lu, L.; Wang, C. The bioactive compounds and biological functions of *Asparagus officinalis* L.—A review. *J. Funct. Foods* **2019**, 103727. [CrossRef]

285. Conversa, G.; Miedico, O.; Chiaravalle, A.E.; Elia, A. Heavy metal contents in green spears of asparagus (*Asparagus officinalis* L.) grown in Southern Italy: Variability among farms, genotypes and effect of soil mycorrhizal inoculation. *Sci. Hortic.* **2019**, *256*, 108559. [CrossRef]

286. Slatnar, A.; Mikulic-Petkovsek, M.; Stampar, F.; Veberic, R.; Horvat, J.; Jakse, M.; Sircelj, H. Game of tones: Sugars, organic acids, and phenolics in green and purple asparagus (*Asparagus officinalis* L.) cultivars. *Turk. J. Agric. For.* **2018**, *42*, 55–66. [CrossRef]

287. Di Gioia, F.; Petropoulos, S.A. Phytoestrogens, Phytosteroids and Saponins in Vegetables: Biosynthesis, Functions, Health Effects And Practical Applications. In *Functional Food Ingredients from Plants*, 1st ed.; Ferreira, I.C.F.R., Barros, L., Eds.; Elsevier Inc.: London, UK, 2019; Volume 90.

288. Dawid, C.; Hofmann, T. Structural and sensory characterization of bitter tasting steroidal saponins from asparagus spears (*Asparagus officinalis* L.). *J. Agric. Food Chem.* **2012**, *60*, 11889–11900. [CrossRef] [PubMed]

289. Pegiou, E.; Mumm, R.; Acharya, P.; de Vos, R.C.H.; Hall, R.D. Green and White Asparagus (*Asparagus officinalis*): A Source of Developmental, Chemical and Urinary Intrigue. *Metabolites* **2020**, *10*, 17. [CrossRef] [PubMed]

290. Sun, T.; Powers, J.R.; Tang, J. Evaluation of the antioxidant activity of asparagus, broccoli and their juices. *Food Chem.* **2007**, *105*, 101–106. [CrossRef]

291. Maeda, T.; Kakuta, H.; Sonoda, T.; Motoki, S.; Ueno, R.; Suzuki, T.; Oosawa, K. Antioxidation capacities of extracts from green, purple, and white asparagus spears related to polyphenol concentration. *HortScience* **2005**, *40*, 1221–1224. [CrossRef]

292. Geoffriau, E.; Denoue, D.; Rameau, C. Assessment of genetic variation among asparagus (*Asparagus officinalis* L.) populations and cultivars: Agromorphological and isozymic data. *Euphytica* **1991**, *61*, 169–179. [CrossRef]

293. López Anido, F.; Cointry, E. Asparagus. In *Handbook of Plant Breeding. Vegetable II: Fabaceae, Liliaceae, Solanaceae and Umbelliferae*; Prohens, J., Nuez, F., Eds.; Springer: New York, NY, USA, 2008; p. 258, ISBN 9780387741086.

294. Siomos, A.S.; Dogras, C.C.; Sfakiotakis, E.M. Color development in harvested white asparagus spears in relation to carbon dioxide and oxygen concentration. *Postharvest Biol. Technol.* **2001**, *23*, 209–214. [CrossRef]

295. Moreno, R.; Espejo, J.A.; Cabrera, A.; Gil, J. Origin of tetraploid cultivated asparagus landraces inferred from nuclear ribosomal DNA internal transcribed spacers' polymorphisms. *Ann. Appl. Biol.* **2008**, *153*, 233–241. [CrossRef]

296. Benson, B.L.; Mullen, R.J.; Dean, B.B. Three new green asparagus cultivars; Apollo, atlas and grande and one purple cultivar, purple passion. *Acta Hortic.* **1996**, *415*, 59–65. [CrossRef]

297. Falloon, P.G.; Andersen, A.M. Breeding purple asparagus from tetraploid "violetto D'Albenga.". *Acta Hortic.* **1999**, *479*, 109–113. [CrossRef]

298. Dong, T.; Han, R.; Yu, J.; Zhu, M.; Zhang, Y.; Gong, Y.; Li, Z. Anthocyanins accumulation and molecular analysis of correlated genes by metabolome and transcriptome in green and purple asparaguses (*Asparagus officinalis*, L.). *Food Chem.* **2019**, *271*, 18–28. [CrossRef] [PubMed]

299. Flores, F.B.; Oosterhaven, J.; Martínez-Madrid, M.C.; Romojaro, F. Possible regulatory role of phenylalanine ammonia-lyase in the production of anthocyanins in asparagus (*Asparagus officinalis* L). *J. Sci. Food Agric.* **2005**, *85*, 925–930. [CrossRef]

300. Mastropasqua, L.; Tanzarella, P.; Paciolla, C. Effects of postharvest light spectra on quality and health-related parameters in green *Asparagus officinalis* L. *Postharvest Biol. Technol.* **2016**, *112*, 143–151. [CrossRef]

301. Huyskens-Keil, S.; Eichholz-Dündar, I.; Hassenberg, K.; Herppich, W.B. Impact of light quality (white, red, blue light and UV-C irradiation) on changes in anthocyanin content and dynamics of PAL and POD activities in apical and basal spear sections of white asparagus after harvest. *Postharvest Biol. Technol.* **2020**, *161*, 111069. [CrossRef]

302. Barberis, A.; Cefola, M.; Pace, B.; Azara, E.; Spissu, Y.; Serra, P.A.; Logrieco, A.F.; D'hallewin, G.; Fadda, A. Postharvest application of oxalic acid to preserve overall appearance and nutritional quality of fresh-cut green and purple asparagus during cold storage: A combined electrochemical and mass-spectrometry analysis approach. *Postharvest Biol. Technol.* **2019**, *148*, 158–167. [CrossRef]

303. Long, A.; Benz, B.F.; Donahue, D.J.; Jull, A.J.T. First direct AMS dates on early maize from Tehuacán Mexico. *Radiocarbon* **1989**, *31*, 1035–1040. [CrossRef]

304. Pleasant, J.M. *Traditional Iroquois Corn: Its History, Cultivation, and Use*; Plant and Life Sciences Publishing: Ithaca, NY, USA, 2010; ISBN 9781933395142 1933395141.

305. Petroni, K.; Pilu, R.; Tonelli, C. Anthocyanins in corn: A wealth of genes for human health. *Planta* **2014**, *240*, 901–911. [CrossRef]

306. Brandolini, A.; Brandolini, A. Maize introduction, evolution and diffusion in Italy. *Maydica* **2009**, *54*, 233–242.

307. United States Department of Agriculture; National Agricultural Statistics Service USDA-NASS, Data and Statistics. 2019. Available online: https://www.nass.usda.gov/Data_and_Statistics/index.php (accessed on 30 December 2019).

308. Erwin, A.T. Sweet Corn—Mutant or historic species? *Econ. Bot.* **1951**, *5*, 302–306. [CrossRef]

309. Lertrat, K.; Pulam, T. Breeding for Increased Sweetness in Sweet Corn. *Int. J. Plant Breed.* **2007**, *1*, 27–30.

310. Juvik, J.A.; Yousef, G.G.; Han, T.H.; Tadmor, Y.; Azanza, F.; Tracy, W.F.; Barzur, A.; Rocheford, T.R. QTL Influencing Kernel Chemical Composition and Seedling Stand Establishment in Sweet Corn with the shrunken2 and sugary enhancerl Endosperm Mutations. *J. Am. Soc. Hortic. Sci.* **2003**, *128*, 864–875. [CrossRef]

311. O'Hare, T.J.; Fanning, K.J.; Martin, I.F. Zeaxanthin biofortification of sweet-corn and factors affecting zeaxanthin accumulation and colour change. *Arch. Biochem. Biophys.* **2015**, *572*, 184–187. [CrossRef] [PubMed]

312. Lago, C.; Landoni, M.; Cassani, E.; Atanassiu, S.; Canta-Luppi, E.; Pilu, R. Development and characterization of a coloured sweet corn line as a new functional food. *Maydica* **2014**, *59*, 191–200.

313. Casas, M.I.; Duarte, S.; Doseff, A.I.; Grotewold, E. Flavone-rich maize: An opportunity to improve the nutritional value of an important commodity crop. *Front. Plant Sci.* **2014**, *5*, 440. [CrossRef]

314. Lago, C.; Landoni, M.; Cassani, E.; Cantaluppi, E.; Doria, E.; Nielsen, E.; Giorgi, A.; Pilu, R. Study and characterization of an ancient European flint white maize rich in anthocyanins: Millo Corvo from Galicia. *PLoS ONE* **2015**, *10*, e0126521.

315. Cassani, E.; Puglisi, D.; Cantaluppi, E.; Landoni, M.; Giupponi, L.; Giorgi, A.; Pilu, R. Genetic studies regarding the control of seed pigmentation of an ancient European pointed maize (*Zea mays* L.) rich in phlobaphenes: The "Nero Spinoso" from the Camonica valley. *Genet. Resour. Crop Evol.* **2017**, *64*, 761–773. [CrossRef]

316. Lago, C.; Landoni, M.; Cassani, E.; Doria, E.; Nielsen, E.; Pilu, R. Study and characterization of a novel functional food: Purple popcorn. *Mol. Breed.* **2013**, *31*, 575–585. [CrossRef]

317. Tsuda, T.; Horio, F.; Uchida, K.; Aoki, H.; Osawa, T. Dietary Cyanidin 3-*O*-β-*D*-Glucoside-Rich Purple Corn Color Prevents Obesity and Ameliorates Hyperglycemia in Mice. *J. Nutr.* **2003**, *133*, 2125–2130. [CrossRef]

318. Tsuda, T. Dietary anthocyanin-rich plants: Biochemical basis and recent progress in health benefits studies. *Mol. Nutr. Food Res.* **2012**, *56*, 159–170. [CrossRef] [PubMed]

319. Chatham, L.A.; Howard, J.E.; Juvik, J.A. A natural colorant system from corn: Flavone-anthocyanin copigmentation for altered hues and improved shelf life. *Food Chem.* **2019**, *310*, 125734. [CrossRef] [PubMed]

320. Luna-Vital, D.; Li, Q.; West, L.; West, M.; Gonzalez de Mejia, E. Anthocyanin condensed forms do not affect color or chemical stability of purple corn pericarp extracts stored under different pHs. *Food Chem.* **2017**, *232*, 639–647. [CrossRef] [PubMed]

321. Petroni, K.; Tonelli, C. Recent advances on the regulation of anthocyanin synthesis in reproductive organs. *Plant Sci.* **2011**, *181*, 219–229. [CrossRef]

322. Sharma, M.; Chai, C.; Morohashi, K.; Grotewold, E.; Snook, M.E.; Chopra, S. Expression of flavonoid 3'-hydroxylase is controlled by P1, the regulator of 3-deoxyflavonoid biosynthesis in maize. *BMC Plant Biol.* **2012**, *12*, 196. [CrossRef]

323. Wittmeyer, K.; Cui, J.; Chatterjee, D.; Lee, T.F.; Tan, Q.; Xue, W.; Jiao, Y.; Wang, P.H.; Gaffoor, I.; Ware, D.; et al. The dominant and poorly penetrant phenotypes of maize Unstable factor for orange1 are caused by DNA methylation changes at a linked transposon. *Plant Cell* **2018**, *30*, 3006–3023. [CrossRef]

324. Calvo-Brenes, P.; O'Hare, T. Effect of freezing and cool storage on carotenoid content and quality of zeaxanthin-biofortified and standard yellow sweet-corn (*Zea mays* L.). *J. Food Compos. Anal.* **2020**, *86*, 103353. [CrossRef]

325. Janovská, D.; Štočková, L.; Stehno, Z. Prehranske lastnosti mladih rastlin ajde. *Acta Agric. Slov.* **2010**, *95*, 157–162.

326. González-Manzano, S.; Pérez-Alonso, J.J.; Salinas-Moreno, Y.; Mateus, N.; Silva, A.M.S.; de Freitas, V.; Santos-Buelga, C. Flavanol-anthocyanin pigments in corn: NMR characterisation and presence in different purple corn varieties. *J. Food Compos. Anal.* **2008**, *21*, 521–526. [CrossRef]

327. Yang, Z.; Zhai, W. Identification and antioxidant activity of anthocyanins extracted from the seed and cob of purple corn (*Zea mays* L.). *Innov. Food Sci. Emerg. Technol.* **2010**, *11*, 169–176. [CrossRef]

328. Lago, C.; Cassani, E.; Zanzi, C.; Landoni, M.; Trovato, R.; Pilu, R. Development and study of a maize cultivar rich in anthocyanins: Coloured polenta, a new functional food. *Plant Breed.* **2014**, *133*, 210–217. [CrossRef]

329. Tsuda, T. Anthocyanins as functional food factors—Chemistry, nutrition and health promotion. *Food Sci. Technol. Res.* **2012**, *18*, 315–324. [CrossRef]

330. Lopez-Martinez, L.X.; Oliart-Ros, R.M.; Valerio-Alfaro, G.; Lee, C.H.; Parkin, K.L.; Garcia, H.S. Antioxidant activity, phenolic compounds and anthocyanins content of eighteen strains of Mexican maize. *LWT Food Sci. Technol.* **2009**, *42*, 1187–1192. [CrossRef]

331. Cevallos-Casals, B.A.; Cisneros-Zevallos, L. Stoichiometric and Kinetic Studies of Phenolic Antioxidants from Andean Purple Corn and Red-Fleshed Sweetpotato. *J. Agric. Food Chem.* **2003**, *51*, 3313–3319. [CrossRef] [PubMed]

332. Harakotr, B.; Suriharn, B.; Scott, M.P.; Lertrat, K. Genotypic variability in anthocyanins, total phenolics, and antioxidant activity among diverse waxy corn germplasm. *Euphytica* **2015**, *203*, 237–248. [CrossRef]

333. Toufektsian, M.C.; de Lorgeril, M.; Nagy, N.; Salen, P.; Donati, M.B.; Giordano, L.; Mock, H.P.; Peterek, S.; Matros, A.; Petroni, K.; et al. Chronic Dietary Intake of Plant-Derived Anthocyanins Protects the Rat Heart against Ischemia-Reperfusion Injury. *J. Nutr.* **2008**, *138*, 747–752. [CrossRef]

334. Toufektsian, M.C.; Salen, P.; Laporte, F.; Tonelli, C.; de Lorgeril, M. Dietary Flavonoids Increase Plasma Very Long-Chain (n-3) Fatty Acids in Rats. *J. Nutr.* **2011**, *141*, 37–41. [CrossRef]
335. Kang, M.K.; Lim, S.S.; Lee, J.Y.; Yeo, K.M.; Kang, Y.H. Anthocyanin-rich purple corn extract inhibit diabetes-associated glomerular angiogenesis. *PLoS ONE* **2013**, *8*, e79823. [CrossRef]
336. Mazewski, C.; Liang, K.; Gonzalez de Mejia, E. Inhibitory potential of anthocyanin-rich purple and red corn extracts on human colorectal cancer cell proliferation in vitro. *J. Funct. Foods* **2017**, *34*, 254–265. [CrossRef]

 *antioxidants*

Review

Medicinal Profile, Phytochemistry, and Pharmacological Activities of *Murraya koenigii* and Its Primary Bioactive Compounds

Rengasamy Balakrishnan [1], Dhanraj Vijayraja [2], Song-Hee Jo [1], Palanivel Ganesan [3], In Su-Kim [1,*] and Dong-Kug Choi [1,3,*]

[1] Department of Applied Life Sciences and Integrated Bioscience, Graduate School, Konkuk University, Chungju 27478, Korea; rmbalabio@gmail.com (R.B.); wowsong333@naver.com (S.-H.J.)

[2] Department of Biochemistry, Rev. Jacob Memorial Christian College, Ambilikkai 624612, Tamilnadu, India; dvijayraja@gmail.com

[3] Department of Integrated Bio Science and Biotechnology, College of Biomedical and Health Science, Nanotechnology Research Center, Konkuk University, Chungju 27478, Korea; palanivel@kku.ac.kr

* Correspondence: kis5497@kku.ac.kr (I.S.-K.); choidk@kku.ac.kr (D.-K.C.)

Received: 13 December 2019; Accepted: 13 January 2020; Published: 24 January 2020

Abstract: The discovery of several revitalizing molecules that can stop or reduce the pathology of a wide range of diseases will be considered a major breakthrough of the present time. Available synthetic compounds may provoke side effects and health issues, which heightens the need for molecules from plants and other natural resources under discovery as potential methods of replacing synthetic compounds. In traditional medicinal therapies, several plant extracts and phytochemicals have been reported to impart remedial effects as better alternatives. *Murraya koenigii* (*M. koenigii*) belongs to the Rutaceae family, which is commonly used as a medicinally important herb of Indian origin in the Ayurvedic system of medicine. Previous reports have demonstrated that the leaves, roots, and bark of this plant are rich sources of carbazole alkaloids, which produce potent biological activities and pharmacological effects. These include antioxidant, antidiabetic, anti-inflammatory, antitumor, and neuroprotective activities. The present review provides insight into the major components of *M. koenigii* and their pharmacological activities against different pathological conditions. The review also emphasizes the need for more research on the molecular basis of such activity in various cellular and animal models to validate the efficacy of *M. koenigii* and its derivatives as potent therapeutic agents.

Keywords: *Murraya koenigii*; antioxidant; bioactive compounds; pharmacological activity; traditional medicine

1. Introduction

Presently, huge numbers of people in developing countries depend on medicinal plants for healthcare, skin care, economic benefits, and cultural development. For centuries, medicinal plants have been widely used in traditional medicine in countries like India, China, Germany, Thailand, etc. [1]. The World Health Organization (WHO) projected that 80% of the population relies on traditional medicine, which is clearly elucidated by the 19.4 billion USD global revenue for herbal remedies in 2010 [2]. Moreover, the demand for traditional medicinal plants is increasing; for instance, the market for medicinal plants is expanding at an annual rate of 20% in India. Likewise, in China, 30% to 50% of the total medicinal consumption is from preparations of traditional medicine [3]. Nearly 76.7% of the citizens of Thailand have reported mainly using traditional herbal medicine for their primary healthcare [4]. Around 90% of the German population uses natural remedies for certain health issues [5]. Therefore, the medicinal plants used in traditional medical treatments are significant in both developing and industrialized countries. This is clearly demonstrated by the worldwide market for traditional medicine. This market continues to gradually increase [6].

Murraya koenigii (*M. koenigii*) (L) Spreng (Family: Rutaceae) is usually known as "curry leaves". The tropical and subtropical regions in the world have large distributions of *M. koenigii* [7]. Among the 14 global species belonging to the genus of *Murraya*, only two, *M. koenigii* and *M. paniculate*, are available in India. *M. koenigii* is more important due to its huge spectrum of traditional medicinal properties. For centuries, this plant has been used in diverse forms and holds a place of pride in Indian Ayurvedic medicine, known as "krishnanimba" [8]. Different parts of *M. koenigii*, such as its leaves, root, bark, and fruit, are known to promote various biological activities. Aromatic bioactive constituents in the leaves of *M. koenigii* retain their flavor and other qualities, even after drying [9–14]. *M. koenigii* leaves are slightly bitter in taste, pungent in smell, and weakly acidic. They are used as antihelminthics, analgesics, digestives, and appetizers in Indian cookery [15,16]. The green leaves of *M. koenigii* are used in treating piles, inflammation, itching, fresh cuts, dysentery, bruises, and edema. The roots are purgative to some extent. They are stimulating and used for common body aches. The bark is helpful in treating snakebites [17–20]. The essential oil extracted from *M. koenigii* leaves is reported to possess anti-oxidative, hepatoprotective [21–24], antimicrobial, antifungal [25–27], anti-inflammatory, and nephroprotective activities in animal models [28–30]. The medicinal properties of *M. koenigii* have been accredited to several chemical constituents of different carbazole alkaloids and other important metabolites, like terpenoids, flavonoids, phenolics, carbohydrates, carotenoids, vitamins, and nicotinic acid from different parts of the *M. koenigii* plant.

In recent years, greater attention has been paid to the use of *M. koenigii* in traditional medicines and home remedies. On the other hand, limited studies have been conducted for evaluating the pharmacological and medicinal efficacy of *M. koenigii* in promoting health benefits and curing disease [31–36]. This review will discuss the traditional medicinal use of *M. koenigii* and its bioactive compounds, highlighting their pharmacological effects. This review aims to present a well-managed summary and possible recommendation on existing studies to provide information regarding the current reports that can direct future research. Therefore, instead of discussing a few selected studies in a specific time interval, the present review will discuss and cover previous and existing major studies on *M. koenigii* related to the topics chosen. The details, like phytochemical screening, identification, and pharmacological activities, will be systematically categorized, compared, and summarized. We hypothesized that, through all of these efforts, a good summary on pharmacological activities that could initiate future perspectives with the utmost clarity could be produced.

The pharmacological activities of *M. koenigii* are discussed in detail in Figure 1.

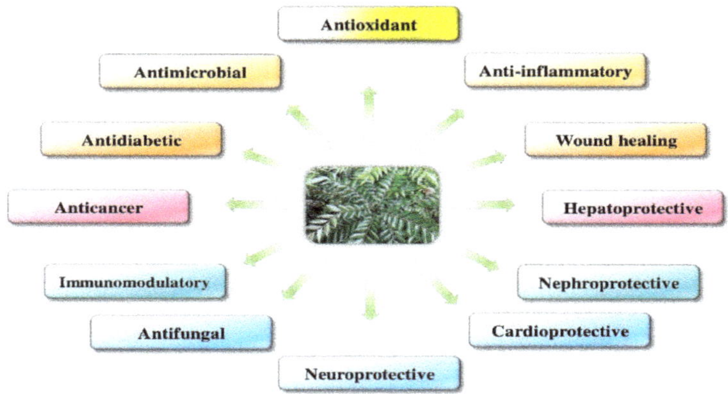

Figure 1. Pharmacological activities of *Murraya koenigii*.

2. *Murraya koenigii* (*M. koenigii*)

2.1. Traditional Uses of M. koenigii

Essential oils and fresh leaf powder of *M. koenigii* are useful in seasoning food items and preparing ready-to-eat foods. Owing to the higher antimicrobial activities of the essential oil from leaf extracts [37,38], this oil can also be used as perfume and flavor agents in traditional practice. Fresh curry leaves are boiled with a coconut oil mixture until they are reduced to a black residue to produce an excellent hair tonic for retaining a normal hair tone and improving hair growth. Curry leaves have a traditional use, either whole or in parts, as antidiarrheal, antifungal, blood purifying, anti-inflammatory, and anti-depressant agents [39–41].

2.2. Medicinal Uses of M. koenigii

M. koenigii has numerous disease remedial activities, for instance, different parts of the plant, such as the leaves, roots, and bark, can be prepared as tonics for inducing digestion and flatulence or as antiemetics [25,42]. After decoction, the leaves become bitter to the taste and are helpful in reducing fever. The juice of the root is given to manage renal pains [43]. The leaves and roots can be given as an anthelmenticku, analgesic, cure for piles, body heat reducer, and thirst quencher and are also helpful in reducing inflammation and itching. They are also useful in managing leucoderma and blood disorders. When eaten raw, the green leaves can offer a cure for dysentery, and when they are boiled in milk, the paste has good application prospects for curing poisonous bites and eruptions [44].

2.3. Phytochemistry of M. koenigii

A wide range of phytochemicals have been isolated from the leaves, roots, and stem bark of *M. koenigii*. *M. koenigii* extracts of leaves, root, stem bark, fruits, and seeds have yielded alkaloids, flavonoids, terpenoids, and polyphenols, as shown in Table 1. The plant leaves contain a substantial amount of proximate composition; the moisture is 63.2%, protein is 8.8%, carbohydrate is 39.4%, total nitrogen is 1.15%, fat is 6.15%, total sugars are 18.92%, starch is 14.6%, and crude fiber is 6.8%. The leaves have been reported as a significant source of several vitamins, such as vitamin A (B-carotene), with a level of 6.04 ± 0.02 mg/100 g; vitamin B3, (niacin), with a level of 2.73 ± 0.02 mg/100 g; vitamin B1 (thiamin), with a level of 0.89 ± 0.01 mg/100 g; calcium, with a level of 19.73 ± 0.02 mg/100 g; magnesium, with a level of 49.06 ± 0.02 mg/100 g; and sodium, with a level of 16.50 ± 0.21 mg/100 g. The alcohol-soluble extract has a value of 1.82%, ash has a value of 13.06%, acid-insoluble ash has a value of 1.35%, cold water (20 °C) extractive has a value of 27.33%, and maximum of hot-water-soluble extractive has a value of 33.45% [15,45]. Wide ranges of carbazole alkaloids, essential oils, terpenoids, and flavonoids have numerous beneficial roles. Table 2 summarizes the major chemical constituents of *M. koenigii* and its pharmacological activities.

Table 1. Phytochemical compounds identified from *M. koenigii*.

Compound	Molecular Formula	Plant Part	References
	Alkaloids		
Mahanine	$C_{23}H_{25}NO_2$	Leaves, stem bark, and seeds	[45–49]
Mahanimbine	$C_{23}H_{25}NO$	Leaves, roots, seeds, and fruits	[47–49]
Murrayanol	$C_{24}H_{29}NO_2$	Leaves, roots, and fruits	[47–49]
Koenimbine	$C_{19}H_{19}NO_2$	Leaves, seeds, and fruits	[46–49]
O-Methylmurrayamine A	$C_{19}H_{20}NO_2$	Leaves	[45–48]
Koenigicine	$C_{20}H_{21}NO_3$	Leaves	[45–48]
Koenigine	$C_{19}H_{19}NO_3$	Leaves and stem bark	[47–49]
Murrayone (Coumarine)	$C_{15}H_{14}O_4$	Leaves	[45–48]
Mahanimbicine	$C_{23}H_{25}NO$	Leaves	[45–48]

Table 1. *Cont.*

Compound	Molecular Formula	Plant Part	References
Bicyclomahanimbicine	$C_{23}H_{25}NO$	Leaves	[45–48]
Phebalosin	$C_{15}H_{14}O_4$	Leaves	[45–48]
Isomahanimbine	$C_{23}H_{25}NO$	Leaves and roots	[45,46,48]
Koenimbidine	$C_{20}H_{21}NO_3$	Leaves and roots	[45,46,48]
Euchrestine B	$C_{24}H_{29}NO_2$	Leaves	[45,46]
Bismurrayafoline E	$C_{48}H_{56}N_2O_4$	Leaves	[45,46]
Isomahanine	$C_{23}H_{25}NO_2$	Leaves, seeds, and fruits	[45,46,49]
Mahanimbinine	$C_{23}H_{27}NO_2$	Leaves and seeds	[45,46,49]
Girinimbilol	$C_{18}H_{19}NO$	Leaves	[45,46]
Pyrayafoline-D	$C_{23}H_{25}NO_2$	Leaves and stem bark	[45,46,49]
Glycozoline	$C_{14}H_{13}NO$	Leaves	[45,46]
Cyclomahanimbine	$C_{23}H_{25}NO$	Leaves	[45,46]
Isomurrayazoline	$C_{23}H_{25}NO$	Leaves	[45,46]
Mahanimboline	$C_{23}H_{25}NO_2$	Leaves	[49]
Mukonicine	$C_{20}H_{21}NO_3$	Leaves	[49]
Isolongifolene	$C_{15}H_{24}$	Leaves	[49]
Mukonal	$C_{13}H_9NO_2$	Stems	[49]
Mukeic acid	$C_{14}H_{11}NO_3$	Stems	[49]
9-Carbethoxy-3-methyl carbazole	$C_{16}H_{15}NO_2$	Roots and stems	[49]
9-Formyl-3-methyl carbazole	$C_{14}H_{11}NO$	Roots and stems	[49]
Murrayazolinol	$C_{23}H_{25}NO_2$	Stems bark	[49,50]
Mahanimbinol	$C_{23}H_{27}NO$	Stems bark	[49,50]
Mukoeic acid	$C_{14}H_{11}NO_3$	Stem bark	[49,50]
Osthol	$C_{15}H_{16}O_3$	Stem bark	[49,50]
Umbelliferone	$C_9H_6O_3$	Stem bark	[49,50]
Murrayanine	$C_{14}H_{11}NO_2$	Stem bark	[49,50]
Mukoenine-A	$C_{18}H_{19}NO$	Roots and stem bark	[49,50]
Mukoenine-B	$C_{23}H_{25}NO_2$	Roots and stem bark	[49,50]
Mukoline	$C_{14}H_{13}NO_2$	Roots	[49,50]
Mukolidine	$C_{14}H_{11}NO_2$	Roots and stem bark	[49,50]
(M)-murrastifoline-F	$C_{28}H_{24}N_2O_2$	Roots and stem bark	[49,50]
3-Methyl-9H-carbazole-9-carbaldehyde	$C_{14}H_{11}NO$	Roots	[49,50]
Bismahanine	$C_{46}H_{48}N_2O_4$	Roots and stem bark	[49,50]
Bikoeniquinone A	$C_{27}H_{20}N_2O_3$	Roots and stem bark	[49,50]
Bismurrayaquinone	$C_{26}H_{16}N_2O_4$	Roots and stem bark	[49,50]
3-Methylcarbazole	$C_{13}H_{11}N$	Roots	[49]
Murrayafoline A	$C_{14}H_{13}NO$	Roots	[49]
Murrayakonine A	$C_{37}H_{36}N_2O_2$	Leaves and stems	[39]
Murrayakonine B	$C_{23}H_{23}NO_2$	Leaves and stems	[39]
Murrayakonine C	$C_{24}H_{25}NO_3$	Leaves and stems	[39]
Murrayakonine D	$C_{23}H_{25}NO_2$	Leaves and stems	[39]
Girinimbine	$C_{18}H_{17}NO$	Roots, stem bark, and seeds	[49,51]
Murrayacine	$C_{18}H_{15}NO_2$	Stem and bark	[49,51]
Murrayazoline	$C_{23}H_{25}NO$	Stem and bark	[49,51]
Flavonoids			
Quercetin	$C_{15}H_{10}O_7$	Leaves	[52]
Apigenin	$C_{15}H_{10}O_5$	Leaves	[52]
Kaempferol	$C_{15}H_{10}O_6$	Leaves	[52]
Rutin	$C_{27}H_{30}O_{16}$	Leaves	[52]
Catechin	$C_{15}H_{14}O_6$	Leaves	[52]
Myricetin	$C_{15}H_{10}O_8$	Leaves	[52]
4-*O*-β-D-Rutinosyl-3-methoxyphenyl-1-propanone	$C_{22}H_{32}O_{12}$	Leaves	[53]
1-*O*-β-D-Rutinosyl-2(*R*)-ethyl-1-pentanol	$C_{19}H_{36}O_{10}$	Leaves	[53]
8-Phenylethyl-*O*-β-D-rutinoside	$C_{20}H_{30}O_{10}$	Leaves	[54]
Terpenoids			
Blumenol A	$C_{13}H_{20}O_3$	Leaves	[53]
Icariside B1	$C_{19}H_{30}O_8$	Leaves	[53]
Loliolide	$C_{11}H_{16}O_3$	Leaves	[53]
Blumenol A	$C_{13}H_{20}O_3$	Leaves	[53]

Table 1. *Cont.*

Compound	Molecular Formula	Plant Part	References
Icariside B1	$C_{19}H_{30}O_8$	Leaves	[53]
(−)-Epiloliolide	$C_{11}H_{16}O_3$	Leaves	[55]
(−)-α-pinene	$C_{10}H_{16}$	Leaves	[55]
(−)-β-pinene	$C_{10}H_{16}$	Leaves	[55]
(+)-β-pinene	$C_{10}H_{16}$	Leaves	[55]
(+)-sabinene	$C_{10}H_{16}$	Leaves	[55]
Squalene	$C_{30}H_{50}$	Leaves and bark	[56]
β-sitosterol	$C_{29}H_{50}O$	Leaves and bark	[56,57]
Polyphenols			
Selin-11-en-4α-ol	$C_{15}H_{26}O$	Leaves and bark	[56]
2-hydroxy-4-methoxy-3,6-dimethylbenzoic acid	$C_{10}H_{12}O_4$	Bark	[56]

Table 2. The major bioactive compounds of *M. koenigii* and its pharmacological activities.

Serial No.	Constituent	Constituent Structure	Activity
1	Mahanine		Cytotoxicity, anti-microbial, and anti-cancer
2	Mahanimbine		Cytotoxicity, anti-oxidant, anti-microbial, anti-diabetic, and hyperlipidemic
3	Isomahanine		Cytotoxicity, anti-oxidant, anti-microbial, anti-diabetic, and hyperlipidemic
4	koenimbine		Cytotoxicity and anti-diarrhea
5	Girinimbine		Anti-tumor

Table 2. *Cont.*

Serial No.	Constituent	Constituent Structure	Activity
6	Isolongifolene	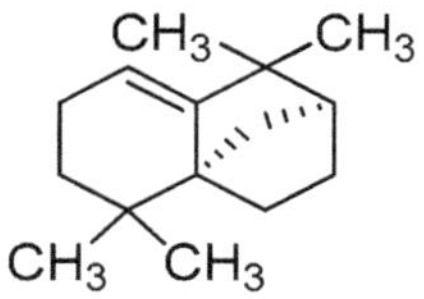	Anti-oxidant and neuroprotective
7	Pyrayafoline D	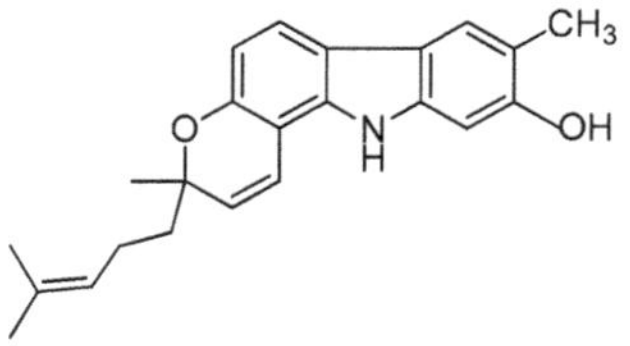	Anti-cancer and anti-bacterial
8	Murrayafoline	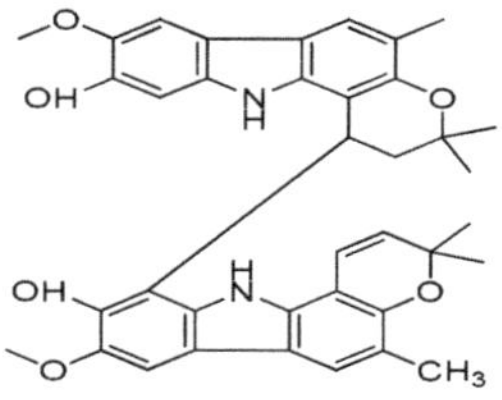	Cytotoxicity and anti-inflammatory
9	Murrayazoline	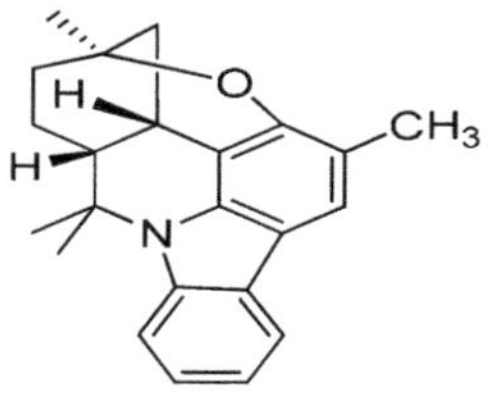	Cytotoxicity and anti-tumor
10	Koenoline	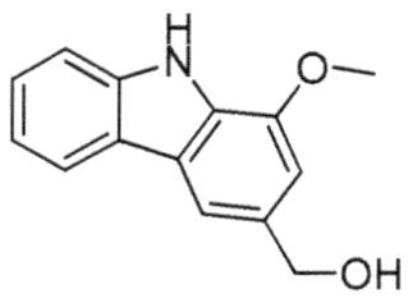	Cytotoxicity
11	9-formyl-3-methyl carbazole	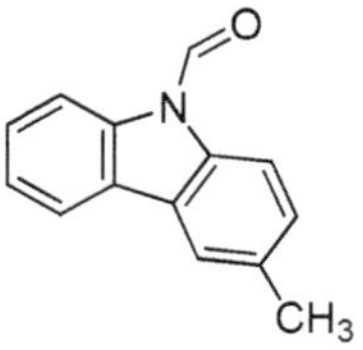	Anti-oxidant
12	*O*-Methylmurrayamine	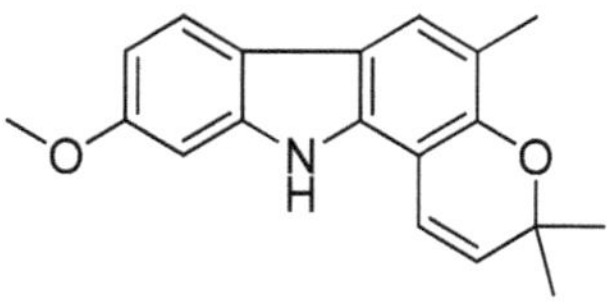	Anti-oxidant and neuroprotective

Table 2. *Cont.*

Serial No.	Constituent	Constituent Structure	Activity
13	Koenine	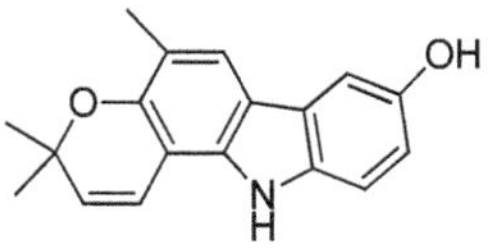	Anti-oxidant
14	Koenigine	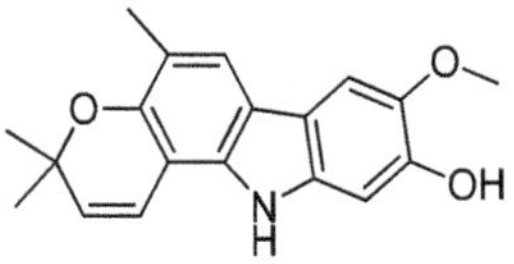	Anti-oxidant
15	Mukonicine	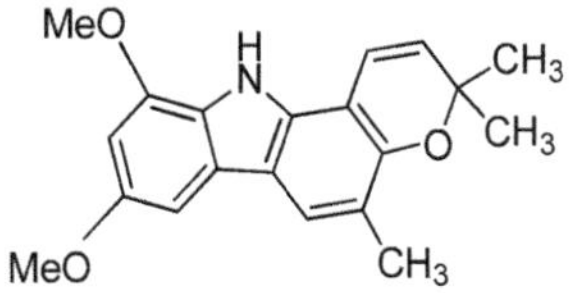	Anti-oxidant
16	Mahanimbinine	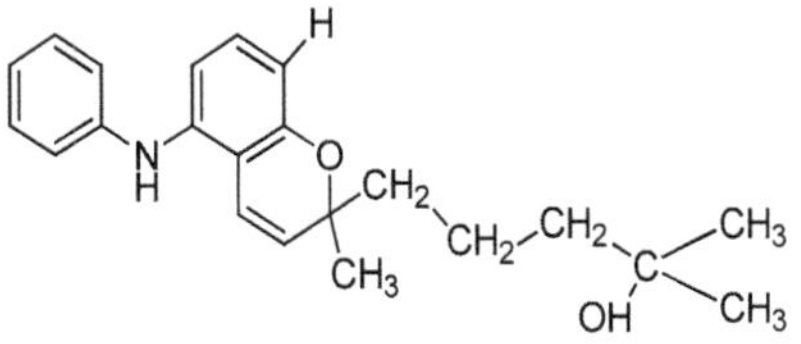	Anti-oxidant, anti-microbial, anti-diabetic, and hyperlipidemic
17	Murrayacinine	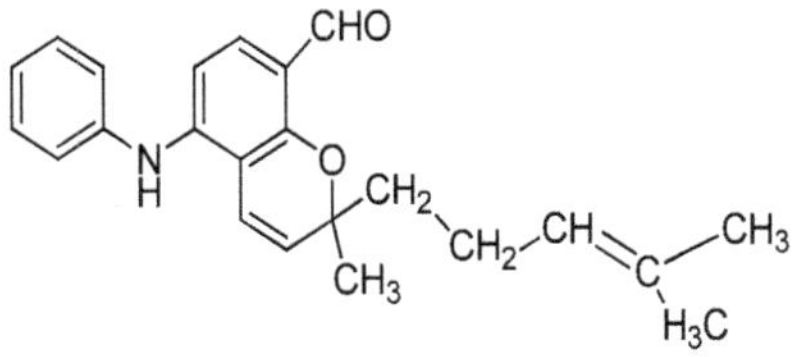	Anti-oxidant, anti-microbial, anti-diabetic, and hyperlipidemic
18	Mahanimboline	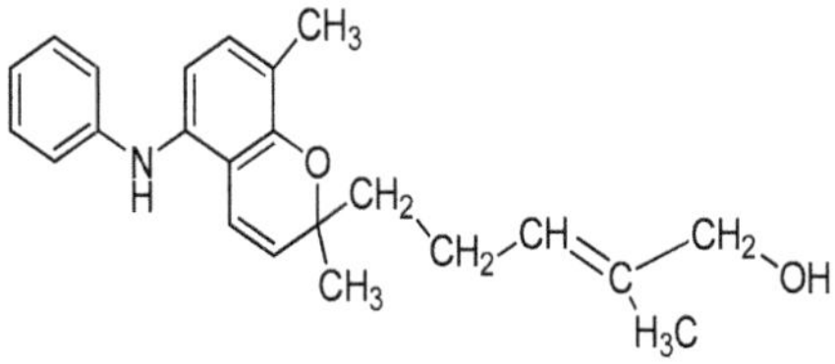	Cytotoxicity, anti-oxidant, anti-microbial, anti-diabetic, and hyperlipidemic
19	Mukoeic acid	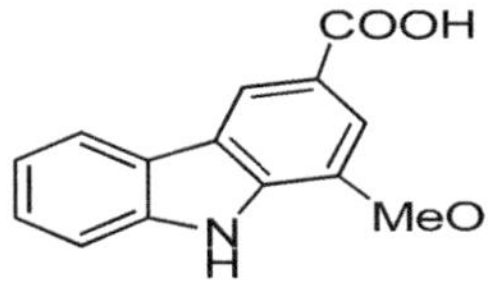	Anti-oxidant
20	Murrayanine	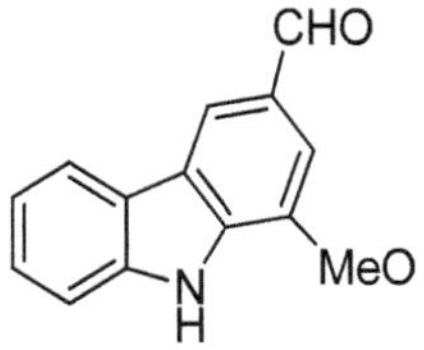	Anti-oxidant

2.4. Bioavailability Study of M. koenigii-Derived Active Constituents

An in vivo pharmacokinetic study revealed that after oral administration of the bioactive compounds at the rate of 0.1 gm/kg body weight (b.w.), the maximum systemic concentration (Cmax) of koenimbine was 1.81 ± 0.55 μM and koenidine was 2.82 ± 0.53 μM, and the time required to reach the maximum concentration was 49.8 ± 8.4 min and 240 ± 0.00 min, respectively [58]. Bhattacharya et al. demonstrated the bioavailability of mahanine—another important bioactive compound of *M. koenigii*—in mice through blood serum estimations based on high-performance liquid chromatography (HPLC) analysis. Mahanine, at a dose of 100 mg/kg of body weight, was found to take 60 min to reach the maximum concentration in blood serum [59].

3. Molecular Mechanism and Activities of *M. koenigii*

3.1. Antioxidants

Reactive oxygen species (ROS), such as singlet oxygen (O_2), hydrogen peroxide (H_2O_2), the superoxide anion ($O_2 \bullet^-$), and the hydroxyl radical ($\bullet OH$), are often generated as byproducts of cellular metabolic reactions and exogenous induction. These ROS create homeostatic imbalances, which lead to the generation of oxidative stress, which in turn, induces cell death and tissue injury [60]. ROS in elevated levels can damage biomolecules such as nucleic acids, proteins, and lipids [61]. Even though the antioxidant defense systems like enzymatic antioxidants and non-enzymatic antioxidants are functioning, uncontrolled ROS accumulation during the life cycle promotes the development of age-dependent diseases, like cancer, atherosclerosis, arthritis etc. [62]. Natural antioxidants from plant sources have been considered a promising therapy for the prevention and treatment of these diseases, especially neurodegenerative disorders, cardiovascular diseases, cancer, and other conditions. Various natural bioactive compounds, such as mahanine, mahanimbine, isolongifolene, koenimbine, girinimbine, isomahanine, koenoline, and O-methylmurrayamine, are present in *M. koenigii* and exhibit remarkable antioxidant properties [63,64].

The leaf extracts of *M. koenigii* have high antioxidant activities [65]. Rao et al. evaluated the antioxidant activities of water and an ethanol extract of *M. koenigii* assessed by the, α-diphenyl-β-picrylhydrazyl (DPPH) free radical scavenging assay, with quercetin as a positive control. The ethanolic extract of *M. koenigii* showed an 80% scavenging activity, which was similar to the activities exhibited by the control antioxidant compound quercetin [66]. Gupta et al. evaluated the antioxidant activities of acetone, alcohol, and aqueous extracts of *M. koenigii* by the DPPH free radical scavenging assay, with ascorbic acid as a positive control. The extracts of *M. koenigii* exhibited activities with an half-maximum effective concentration (EC_{50}) value of acetone of 81.81 ± 19.92 at 4.7 μg/mL, alcohol of 79.80 ± 18.68 at 4.1 μg/mL, and aqueous extract of 62.82 ± 13.62 at 4.4 μg/mL, which was comparable to the EC_{50} value exhibited by ascorbic acid (the positive control), which was 97.13 ± 12.64 at 2.69 μg/mL [67]. Zahin et al. also evaluated the antioxidant activities of both ethyl acetate and petroleum ether fractions of *M. koenigii* through DPPH radical scavenging assay, cupric reducing antioxidant capacity (CUPRAC), and ferric reducing antioxidant power (FRAP) assays, with ascorbic acid as a positive control. The benzene fraction of *M. koenigii* was found to be the most active free radical scavenger, exhibiting an 88.3% decrease at a concentration of 100 μg/mL, followed by ethyl acetate at 79.5% and petroleum ether at 78.7%, while positive controls of ascorbic acid and butylated hydroxytoluene (BHT) at a concentration of 100 μg/mL inhibited 93.1% and 86.5% DPPH absorption, respectively. Similarly, the antioxidant activity created by reducing activity and CUPRAC assays indicated the highest reducing potential in the benzene fraction, followed by petroleum ether and ethyl acetate. The activity was greater than that of ascorbic acid and on par with that of BHT [68].

Yogesh et al. evaluated the antioxidant activity of berry extracts of *M. koenigii* by DPPH free radical scavenging activity and reducing power assays. The results indicated that an *M. koenigii* berry extract is a powerful free radical scavenger compared to known antioxidants, such as butylated hydroxytoluene, ascorbic acid, and tannic acid [69]. Tomar et al. evaluated the total antioxidant activity of acetone and petroleum ether extracts of younger and older *M. koenigii* leaves by estimating the H_2O_2 scavenging activity. The acetone extract of old leaves was found to have a maximum activity at 151.58%, and for young leaves in petroleum ether, the value was 72.23% [70]. Waghmare et al. evaluated the antioxidant property of fruit extracts of *M. koenigii* with DPPH free radical scavenging activity, inhibition of nitric oxide radical (NO) and thiobarbituric acid reactive substances (TBARS) activity, and reducing power assays, and •OH was also estimated, with vitamin C as a positive control. The fruit extract of *M. koenigii* exhibited antioxidant activities, and the EC_{50} value of the extracts for the DPPH assay was 2.6 mg/mL; for the NO radical, was 2.9 mg/mL; for TBARS, was 3.1 mg/mL; for the reducing power assay, was 2.7 mg/mL; and for H_2O_2, was 3.3 mg/mL, which were comparable to the EC_{50} value of 5 mg/mL exhibited by the vitamin C positive control [71]. The antioxidant activities exhibited by the crude extracts of *M. koenigii* were probably due to the presence of flavonoids and phenolic derivatives. The above studies revealed that various extracts of *M. koenigii* display high antioxidant activity. The studies also indicated the potential for this plant to be a natural source of strong antioxidant substances that can be used in therapy for human diseases induced by ROS.

3.2. Oxidative Stress

Chemical species with one or more unpaired electrons are called free radicals. In biological systems, the term "free radicals" refers to reactive oxygen species (ROS). Major ROS include $O_2\bullet^-$, H_2O_2, and •OH [72]. In addition to ROS, reactive nitrogen species (RNS), including peroxynitrite (NO_3^-), NO, and S-nitrosothiols, also contribute to the generation of oxidative stress. Both ROS and RNS arise as intermediates in several metabolic processes and are specifically produced as part of the cellular defense against invasive pathogens. Free radicals also regulate many processes, including cellular growth, glucose metabolism, and proliferation [73].

Apart from certain beneficial effects, free radicals induce various deleterious effects. In a non-specific manner, ROS can react on significant biomolecules, which leads to deleterious effects like a loss of enzyme activity, genetic mutations and permeability alterations in the cell membrane, and RNS-induced protein *S*-nitrosylation [74]. Because DNA is constantly attacked by the free radicals, around 75,000 to 100,000 DNA damage events might occur in each cell per day. •OH is the most reactive species and interacts with all biological molecules, including the C-8 position of guanine to form 8-hydroxyguanine, which is one of the most frequently found oxidized bases in DNA [75].

An increase in the free radical concentration in the body can cause subsequent oxidative and cellular damage to lipids, proteins, RNA, and DNA [76]. The leaf extract of *M. koenigii* has recently been shown to possess potential antioxidant activity and protection against oxidative stress induced in diabetes [77]. The aqueous leaf extract of *M. koenigii* has been shown to reduce lipid peroxidation and decrease cellular damage, thereby protecting the liver from ethanol-induced toxicity [31]. Khan et al. reported the antioxidant effect and preventive effect of curry leaves in cadmium-induced oxidative stress, cardiac tissue damage, and alterations in normal cardiac functions in rats [78]. Mitra et al. reported the heavy metal chelating activity of an *M. koenigii* leaf extract. They found that there was a significant decrease in the tissue cadmium level when the rats were pre-treated with the *M. koenigii* leaf extract before cadmium administration [79].

3.3. Mitochondrial Dysfunction

Mitochondria are the primary source of high-energy metabolism within the cell. Mitochondria are known as the powerhouse of the living cell. Mitochondria also regulate calcium homeostasis and play a role in scavenging free radicals and controlling programmed cell death and/or the apoptosis-signaling pathway [80]. Mitochondrial damage leads to reduced adenosine triphosphate (ATP) production, increased ROS generation, impaired calcium buffering, damage to mitochondrial DNA (mtDNA), an altered mitochondrial morphology, and alterations in mitochondrial fission and fusion. All these events eventually lead to cell death [81]. It is currently believed that the majority of ROS are generated by mitochondrial complexes I and III, likely due to the release of electrons by NADH and dihydroflavine-adenine dinucleotide ($FADH_2$) into the electron transport chain (ETC). Mitochondrial dysfunction is a characteristic of all chronic diseases and aging. It is characterized by a loss of efficiency in the ETC, as well as reductions in the synthesis of high-energy molecules [82]. These diseases include neurodegenerative diseases like Parkinson's disease (PD), Alzheimer's disease (AD), amyotrophic lateral sclerosis, multiple sclerosis, Huntington's disease, cardiovascular diseases, auto-immune diseases, diabetes, and others [83,84].

Mitochondria, as essential organelles, have a noteworthy role in the viability of neuronal cells. Excess ROS formation is due to complex I inhibition inducing impairments in the mitochondrial membrane potential (MMP) and the pro-apoptotic members are believed to permeabilize the outer mitochondrial membrane due to the formation of oligomeric pores, which permits the release of apoptogenic molecules from the intermembrane space. Recent studies have evaluated the neuroprotective activities of isolongifolene and structurally similar compounds, such as girinimbine, murrayazoline, and O-methylmurrayamine A isolated from *M. koenigii*. By using different in vitro assays, a study reported that the above bioactive compounds exhibited the ability to restore the MMP levels and repair the mitochondrial damage [85–87].

3.4. Inflammation

Tissue injury, cell damage, infections due to pathogens, and alterations in biochemicals lead to a biological response called inflammation. In neurological disorders, the important components involved in inflammatory processes are believed to be mast cells, ependymal cells, microglia, astrocytes, and macrophages [88]. Microglia, a type of neuronal support cell acting as resident macrophages located throughout the brain by changing their morphology, actively respond to inflammation and participate in removing damaged neurons and pathogens. An ethanol extract from *M. koenigii* leaves showed significant analgesic and anti-inflammatory activity when explored using carrageenan-induced hind paw edema in albino rats [89]. Another study also confirmed the anti-inflammatory activity of an *M. koenigii* leaf extract in carrageenan-induced paw edema [90]. Additionally, the study recognized the analgesic activity of curry leaves with several experimental models. *M. koenigii* leaf extracts effectively attenuate the pain which is induced by an intraperitoneal injection of acetic acid and subplantar injection of formalin in mice, and the analgesic effect was elucidated with the writhing responses and pain responses in the late phase. Furthermore, it was reported that higher concentrations (20 and 40 mg/kg, per os (p.o.)) reduced the early-phase inflammatory responses induced by formalin [91].

Khurana et al. evaluated the in vitro and in vivo efficacy of a hydroalcoholic extract of *M. koenigii* curry leaves rich in carbazole alkaloids against lipopolysaccharide (LPS)-induced inflammation in RAW 264.7 cells. The activity of inflammatory cytokines interleukin 1 beta (IL-1β), interleukin-6 (IL-6), tumor necrosis factor (TNF-α), and p65-NFκB was significantly reduced by the hydroalcoholic extract of *M. koenigii*. In addition, the hydroalcoholic extract of *M. koenigii* reduced the expression of nitrotyrosine (NT), myeloperoxidase (MPO), IL-1β, intercellular adhesion molecule 1 (ICAM-1), and cyclooxygenase (COX-2), and increased the expression of Nrf2 [92]. Iman et al. evaluated the anti-inflammatory activity of an extract of *M. koenigii* and its bioactive compound girinimbine against lipopolysaccharide/interferon-gamma-induced RAW 264.7 cells. The girinimbine showed reduced levels of NO overproduction and pro-inflammatory cytokine levels IL-1β and TNF-α in the peritoneal fluid. These findings strongly suggest that girinimbine could act as an anti-inflammatory agent by suppressing inflammation [85]. Another study also confirmed the anti-inflammatory activity of an *M. koenigii* leaf extract. Bioactive compounds like murrayakonine A, O-methylmurrayamine A, and mukolidine were reported for their efficiency in inhibiting TNF-α and IL-6 release in LPS-induced inflammation in human peripheral blood mononuclear cells (PBMCs) [39]. The above studies have shed light on the mechanism of the anti-inflammatory activity of *M. koenigii* leaves and their active compounds, which are comparable to nonsteroidal anti-inflammatory drugs (NSAIDs).

3.5. Apoptosis

Apoptosis is a physiological programmed cell death mechanism which is essential for the flawless growth and development of organisms. Furthermore, it is a lively physiological course causing the self-destruction of cells that comprises lethal biochemical and morphological changes in the nucleus and cytoplasm. During cellular strain like oxidative stress and DNA damage, the process of apoptosis can arise, particularly in cells with high proliferation rates and a high expression of pro-apoptotic genes [93]. Intrinsic and extrinsic pathways regulate apoptosis, but both pathways are associated and the molecules involved in those pathways can influence one another.

Previous findings have demonstrated that an *M. koenigii* extract and its primary active compounds regulate multiple signaling pathways, including phosphatidylinositol 3 kinase (PI3K)/protein kinase B (AKT), mammalian target of rapamycin (mTOR) and mitogen-activated protein kinase (MAPK). *M. koenigii* and its primary active compounds exert complementary effects on oxidative stress and the alteration of proteins [94,95]. They are associated with mitochondrial-mediated apoptotic pathways. Murrayazoline and O-methylmurrayamine were shown to induce the downregulation of Akt/mTOR, suggesting downstream targeting of the cell survival pathway and an ability to potentiate the antitumor activity of D.L. Dexter (DLD-1) colon cancer cells; interestingly, this inhibition of the Akt/mTOR pathway could possibly activate the intrinsic apoptotic program [86]. Mahanine and isomahanine derived from *M. koenigii* leaves exert anticancer effects on oral squamous cell carcinoma cells via the induction of microtubule-associated protein 1 light chain 3, type II (LC3B-II), and cleaved caspase-3, suggesting the inhibition of autophagic flux [96]. In human leukemic cells, Mahanine was reported to induce apoptosis by interrupting signal transfer between Apo-1/Fas signaling and the Bid protein and via mitochondrial pathways in humans [59]. Recently, it was reported that girinimbine, a carbazole alkaloid isolated from *M. koenigii*, inhibited the growth of and induced apoptosis in human hepatocellular carcinoma cells (HepG2) [97]. In addition, Xin et al. reported that girinimbine inhibited ovarian cancer cell proliferation in a dose-dependent manner. It also inducted apoptosis and cell cycle arrest due to inhibition of the PI3K/AKT/mTOR and Wnt/b-catenin signaling pathways [98].

The alkaloid koenimbin found in *M. koenigii* was shown to extend pro-apoptotic activities in MCF-7 cancer cells by inhibiting glycogen synthase kinase-3 beta (GSK-3β). Koenimbin induces apoptotic cell death, phosphorylation, the accumulation of β-catenin, and the activation of nuclear factor-κB (NF-kB) in cancers. Moreover, koenimbin suppresses the expression of various anti-apoptotic genes involved in the regulation of cell proliferation and apoptosis [99]. Koenimbine has also been reported to trigger caspase activation, induce the release of cytochrome c, decrease the anti-apoptotic proteins, and increase the pro-apoptotic proteins, and all of these events lead to intrinsic apoptotic pathway activation [99]. Another study demonstrated that pyrayafoline-D and murrafoline-I isolated from *M. koenigii* could induce apoptosis in HL-60 cells. The same study also induced the loss of mitochondrial membrane potential and the subsequent activation of caspase-9/caspase-3, leading to the activation of apoptotic pathways [100].

4. Beneficial Pharmacological Activities of *M. koenigii* and Its Primary Active Derivatives

4.1. Antifungal Activity

The antifungal activity of *M. koenigii* has been reported in various studies. For example, the essential oil of the leaves was reported to possess antifungal activity [18]. The antifungal activity of the leaves of M. koenigii is due to the presence of phytochemical constituents of complex molecular structures and diverse action mechanisms, viz. alkaloids, terpenoids, flavonoids, phenolics, tannins, and saponins, which are known for their antimicrobial properties. Different investigations support the traditional use of the plant as an antifungal agent. In vitro antifungal activity may explain the use of curry leaves for the treatment of diarrhea, dysentery, and skin eruptions in folklore medicines [19]. Bioactive compounds of *M. koenigii* appreciably hold the ability of mycelial growth inhibition and thereby promote antifungal activity. The antifungal activity of *M. koenigii* against a wide range of pathogenic fungi has been studied. *Penicillium notatum, Aspergillus flavus, Aspergillus niger, Fusarium moniliforme, Mucor mucedo, Penicillium funiculosum* etc., were isolated from infected saplings and spoiled foods based on alterations of their growth characteristics, mycelial morphology, and spore morphology (Table 3) [27]. The ethanolic extract of *M. koenigii* exhibited notable effects on the hyphal morphology; namely, an increase in branching potential, which resulted in the development of short slender branches of hyphae with swollen tips. Such effects are usual for any antifungal compound. Bioactive compounds like girinimbine, murrayanine, marmesin-1′-O-beta-D′galactopyranoside, mahanine, murrayacine, mukoeic acid, murrayazolinine, girinimbilol, pyrafoline-D, and murrafoline-I are present in stem bark. Girinimbine, murrayanine, and marmesin-1′-O-beta-D′galactopyranoside have notable anti-fungal activity [20,101].

4.2. Antibacterial Activity

The unsystematic use of antibiotics promotes the development of multiple drug-resistant pathogenic strains of bacteria, which are very harmful, and there is a lack of proper treatment procedures for these ailments. Therefore, the need to search for new antimicrobials remains. Currently, in addition to antibiotics and chemically-synthesized drugs, curiosity for alternative medicines, such as natural or herbal medicines, is increasing. They may have fewer side effects or toxicity owing to their natural sources [102].

Combating microbial infections without side effects is always a tedious process. In this regard, in addition to classical antibiotics and synthetic drugs, there is an ongoing hunt for potent molecules from natural herbal medicines [102]. *M. koenigii* extracts have demonstrated antibacterial effects on a wide variety of microorganisms. Methanol and ethanol extracts of *M. koenigii* leaves were found to be effective against the bacterial strains *Escherichia coli (E. coli), Staphylococcus, Streptococcus,* and *Proteus.* Hence, *M. koenigii* leaves could be efficiently used as a natural remedy in everyday meals for the prevention of several bacterial infections [103]. Pyranocarbazoles isolated from *M. koenigii* exhibited antibacterial activity on bacterial strains of *Staphylococcus aureus* and *Klebsiella pneumonia* [40]. Green synthesized silver nanoparticles (AGNPs) from *M. koenigii* exhibited therapeutic efficacy against multidrug resistant MDR bacteria [103]. *M. koenigii* essential oil showed antibiofilm activity against *Pseudomonas aeruginosa* and it was reported that *M. koenigii* essential oil treatment revealed an 80% reduction in biofilm formation by *P. aeruginosa.* Microscopic analyses confirmed the drop in biofilm formation in *Pseudomonas aeruginosa* when treated with *M. koenigii* essential oil. The presence of antibiofilm substances like spathulenol (5.85%), cinnamaldehyde (0.37%), and linalool (0.04%) was reported in gas chromatography-mass spectrometry (GCMS) studies [104]. *M. koenigii* counteraction on uropathogenic bacteria isolated from clinical samples was reported in a different study [105]. *M. koenigii* was tested for its antibiotic action against *Mycobacterium* species, which was appreciable, like first-line anti-tuberculosis drugs. *M. koenigii* half maximal inhibitory concentration ((IC_{50}) 400 μg/mL) was found to be more effective against *Mycobacterium smegmatis* compared to water extracts and petroleum ether. An *M. koenigii* ethanol extract exhibited significant synergistic antibacterial activity against *Mycobacterium smegmatis* and *Mycobacterium bovis* bacillus calmette-guerin (BCG) in combination with the anti-tuberculosis drug rifampicin [106].

Table 3. An overview of the pharmacological activities of *M. koenigii* and its primary bioactive compounds.

Pharmacological Activities	Plant Parts	Extract	Bioactive Compounds	Model	Main Finding	Reference
				In vitro studies		
Antifungal	Leaves	Essential oil	–	Disc diffusion method	Essential oil extracted from *M. koenigii* exhibited activities with MIC in the range of 25.5 to 75 µg/mL against pathogenic fungi *A. niger*, *F. moniliforme*, *P. notatum*, *M. mucedo*, and *P. funiculosum*	[27]
Antibacterial	Leaves	Solvent-free microwave extraction	–	Soy agar	Minimum inhibitory concentrations (MIC) of solvent-free microwave extraction (SFME) and hydro-distilled oil from *M. koenigii* with values of 400 and 600 µg/mL against *L. innocua* SFME-essential oil at 300 µg/mL provided 92% inhibition, indicating its antibacterial potential	[37]
Antibacterial	Leaves	Methanol	Koenine, koenigine, and mahanine	Broth micro-dilution assay	Koenine, koenigine, and mahanine extracted from *M. koenigii* exhibited activities with MIC values of 3.12–12.5 µg/mL against bacterial strains *S. aureus* and *K. pneumonia*	[40]
Antibacterial	Leaves	Aqueous	–	Agar diffusion assay	*M. koenigii*-AGNPs exhibited inhibitory activity against *E. coli* and *S. aureus*, with a value of 16 mm for *M. koenigii*-AgNPs and 15 mm for $AgNO_3$ solution at 100 µg/well	[18]
Antibacterial	Leaves	Essential oil	–	Microtiter assay	Essential oil extracts of *M. koenigii* treatment resulted in a reduction of biofilm formation in *P. aeruginosa* PAO1. *M. koenigii* essential oil may effectively control *Pseudomonas* biofilms in indwelling medical device	[19]
Antibacterial	Leaves	Petroleum ether, ethanol, and water	–	Colony-forming unit (CFU) assay	Ethanol extracts of *M. koenigii* exhibited activity half maximal inhibitory concentration ((IC_{50}) of 400 µg/mL) against the mycobacterium smegmatis compared to petroleum ether and water extracts	[20]
Hepatoprotective	Leaves	Aqueous	–	Hep G2 cell line	*M. koenigii* leaves preventing alcohol-induced cellular damage	[31]
Antioxidant	Leaves	Ethanol	–	DPPH free radical scavenging assay	exhibited activities with IC_{50} values of 21.4–49.5 µg/mL	[21]
Antioxidant	Leaves	Aqueous	–	TBARS, CAT, SOD, and glutathione (GSH) assay	Carbazole alkaloids from *M. koenigii* extract exhibited activity with IC_{50} values of 120 µg/mL in an ethanol-induced hepatotoxicity in vitro model	[31]
Antioxidant	Leaves	Aqueous/zinc oxide nanoparticles	–	DPPH free radical scavenging assay	Zinc oxide nanoparticle-synthesized *M. koenigii* extract exhibited activity with an IC_{50} value of 36.46 µg/mL	[64]
Antioxidant	Leaves	Aqueous/zinc oxide nanoparticles	–	ABTS radical scavenging assay	Zinc oxide nanoparticle-synthesized *M. koenigii* extract exhibited activity with an IC_{50} value of 11.55 µg/mL	[64]

Table 3. *Cont.*

Pharmacological Activities	Plant Parts	Extract	Bioactive Compounds	Model	Main Finding	Reference
In vitro studies						
Antioxidant	Leaves	Aqueous/zinc oxide nanoparticles	–	Superoxide assay	Zinc oxide nanoparticle-synthesized *M. koenigii* extract exhibited activity with an IC_{50} value of 11.47 µg/mL	[64]
Antioxidant	Leaves	Aqueous/zinc oxide nanoparticles	–	H_2O_2 Assay	Zinc oxide nanoparticle-synthesized *M. koenigii* extract exhibited activity with an IC_{50} value of 54.06 µg/mL	[64]
Antioxidant	Leaves	Ethanoic	–	DPPH free radical scavenging assay	The ethanoic extract of *M. koenigii* showed an 80% scavenging activity, which was similar to the activities exhibited by the control antioxidant compound quercetin	[66]
Antioxidant	Leaves	Aqueous, alcohol, and acetone	–	DPPH free radical scavenging assay	The extracts of *M. koenigii* exhibited activities with an EC_{50} value of acetone of 4.7 µg/mL, alcohol of 4.1 µg/mL, and aqueous of 4.4 µg/mL, which were comparable to the EC_{50} value of 2.6 µg/mL exhibited by ascorbic acid, which was the positive control	[67]
Antioxidant	Leaves	Petroleum ether and ethyl acetate	–	Cupric-reducing antioxidant capacity	CUPRAC assays indicated the highest reducing potential in the benzene fraction, followed by petroleum ether and ethyl acetate	[68]
Antioxidant	Leaves	Benzene, ethyl acetate, acetone, methanol, and ethanol	–	DPPH free radical scavenging assay	Results showed that for 100 µg/mL, the benzene fraction extracted from *M. koenigii* showed 88.3% free radical scavenging activity, followed by ethyl acetate (79.5%), petrol ether (78.7%), acetone (66.1%), methanol (50.7%), and ethanol (53.0%) fractions, respectively, with the positive control being ascorbic acid (93.1%)	[69]
Antioxidant	Fruits	Aqueous	–	DPPH free radical scavenging assay	Fruit extracted from *M. koenigii* exhibited activities with an EC_{50} value of 2.6 mg/mL	[71]
Cytotoxicity	Stem bark and roots	Hexane, chloroform, and methanol	Girinimbine	MTT assay	Girinimbine was shown to significantly inhibit the proliferation of HT-29 cells with an IC_{50} value of 4.79 ± 0.74 µg/mL.	[85]
Cytotoxicity	Leaves	Ethanol	Murrayazoline and O-methylmurrayamine A	MTT assay	Murrayazoline and O-methylmurrayamine A exhibited activities with IC_{50} values of 5.7 and 17.9 mM in both HEK-293 and HaCaT cell lines, respectively	[86]
Cytotoxicity	–	–	Isolongifolene	MTT assay	Isolongifolene exhibited activities at 10 µM, showing a 90% viability in SH-SY5Y cells	[87]
Cytotoxicity	Leaves	Methanol	–	MTT assay	*M. koenigii* methanolic extract exhibited activities with IC_{50} values >400 µg/mL in the CLS-354 cell line	[96]
Cytotoxicity	Leaves	Ethanol	–	MTT assay	*M. koenigii* ethanolic extract exhibited activities with an IC_{50} value of 20 µg/mL in the mouse macrophage RAW 264.7 cell line	[20]

Table 3. *Cont.*

Pharmacological Activities	Plant Parts	Extract	Bioactive Compounds	Model	Main Finding	Reference
Cytotoxicity	Leaves	Hexane, ethyl acetate, and methanol	–	MTT assay	Three extracts of *M. koenigii* exhibited were very active, with values of <1 μg/mL to 2.25 μg/mL, and were thus proved to be potent cytotoxic activity agents against HeLa cancer cells	[11]
Anti-inflammatory	Stems	Methanol	Murrayakonine A, murrayanine, and O-methylmurrayamine-A	Human peripheral blood mononuclear cells	In vitro experiments showed murrayakonine A (IC_{50} 10 μM), murrayanine (IC_{50} 9.4 μM), and O-methylmurrayamine-A (IC_{50} 7 μM) against TNF-α, and murrayanine (IC_{50} 8.4 μM) and methylmurrayamine-A (IC_{50} 8.4 μM) against IL-6, respectively	[39]
Anticancer (Colon)	Leaves	Ethanol	O-methylmurrayamine 5.7–17.9 μM	MCF-7 cells	O-methylmurrayamine A exhibited anti-colon cancer activity through downregulation of the Akt/mTOR survival pathway and activation of the intrinsic pathway of apoptosis	[86]
Anticancer (Oral)	Leaves	Methanol	Mahanine 15 μM	CLS-354 cells	Mahanine increased the expression of LC3B-II, cleaved caspase-3 proteins, and the inhibition of autophagic flux	[96]
Anticancer (Ovarian)	Stem bark	Methanol	Girinimbine 10 μM	Ovarian cancer cell line SKOV3/SV40	Girinimbine was found to be mainly due to the induction of apoptosis and cell cycle arrest due to the inhibition of the PI3K/AKT/mTOR and Wnt/b-catenin signaling pathways	[98]
Anticancer (Breast)	Leaves	Aqueous acetone	Koenimbin 4.89 μg/mL	MCF7 breast cancer stem cells	Koenimbin induced apoptosis in MCF7 cells that was mediated by cell death and regulated the mitochondrial membrane potential by downregulating Bcl2 and upregulating Bax, due to cytochrome *c* release from the mitochondria to the cytosol, and significantly downregulated the Wnt/β-catenin self-renewal pathway	[98]
Anticancer (Prostate)	Leaves	Aqueous acetone	Koenimbin 3.73 μg/mL	Prostate cancer stem cells	Koenimbin induced apoptosis through the intrinsic signaling pathway and suppression of the translocation of cytoplasmic NF-κB into the nucleus, in addition to displaying potential for targeting PCSCs, as affirmed by the prostasphere formation and Aldefluor assay	[99]
Anticancer	Leaves	Methanol	Mahanine 7.5 μM	Glioma HS 683 cells	Mahanine inhibited the cell migration and invasion and inhibited cell growth was simultaneous with the suppression of p-PI3K, p-AKT, and p-mTOR	[96]
Anticancer (Liver)	Leaves	Methanol	Mahanine 25 μM	HepG2, HuCCT1, and KKU-100 cells	Mahanine showed potent cytotoxicity, with increased expression levels of MITF balance between the cellular stresses	[13]
Anticancer (Cervical)	Leaves	Methanol	Mahanine 8.6 μM	HeLa (HPV-18) and SiHa (HPV-16) cell line	Mahanine and cisplatin synergistically displayed growth inhibitory activity in cervical cancer, the inhibition of STAT3 activation, cell migration, and induced apoptosis	[14]
Anticancer (Lung)	Leaves	Methanol	Mahanine 15 μM	NSCLC cancer cell line A549	Mahanine induced the impairment of mTORC2 through rictor inhibition and the destruction of NSCLC cancer cells	[22]

Table 3. *Cont.*

Pharmacological Activities	Plant Parts	Extract	Bioactive Compounds	Model	Main Finding	Reference
Anticancer (Colon)	Leaves	Methanol	Mahanine 0–30 µM	HCT116, HCT116, SW480, and Vero	Mahanine synergistically activated the two tumor suppressors PTEN and p53/p73 and can potentially be used in combination therapy with 5-FU for the treatment of colon carcinoma	[23]
Anticancer (prostate)	Leaves	Methanol	Mahanine 10 µM	PC3 and LNCaP cell line	Mahanine selectively degraded DNMT1 and DNM T3B via the ubiquitin-proteasomal pathway in a dose-dependent manner upon the inactivation of Akt signaling	[24]
Neuroprotective	Leaves	Methanol	Isolongifolene 10 µM	SH-SY5Y cells	Isolongifolene was effectively attenuated in oxidative stress, mitochondrial dysfunction, and apoptosis	[87]
Neuroprotective	Leaves	Methanol	O-methylmurrayamine A	PC12 cells	O-methylmurrayamine A possibly protects against DNA damage, apoptosis, and high levels of cell viability	[35]
In vivo studies						
Antioxidant	Leaves	Aqueous	–	Male albino Wistar rat	The oral administration of an *M. koenigii* leaf extract resulted in a significant reduction in the level of TBARS in both the plasma (3.64 ± 0.13) and pancreas (53.40 ± 2.13) of diabetic rats	[61]
Antioxidant	Leaves	Aqueous	–	Male albino Wistar rat	The oral administration of an *M. koenigii* leaf extract resulted in a significant increase in the level of GSH in both the plasma (24.16 ± 1.30) and pancreas (19.52 ± 1.09) of diabetic rats	[77]
Antioxidant	Leaves	Aqueous	–	Male albino Wistar rat	The oral administration of an *M. koenigii* leaf extract significantly restored the activity of SOD in the pancreas (3.69 ± 0.15) of diabetic rats	[77]
Antioxidant	Leaves	Aqueous	–	Male albino Wistar rat	The oral administration of an *M. koenigii* leaf extract significantly restored the activity of CAT in the pancreas (12.94 ± 0.54) of diabetic rats	[77]
Antioxidant	Leaves	Aqueous	–	Male albino Wistar rat	The oral administration of an *M. koenigii* leaf extract significantly restored the activity of GPx in the pancreas (5.86 ± 0.22) of diabetic rats	[67]
Antioxidant	Leaves	Ethanol	–	Sprague Dawley rats	For 200 and 400 µg/mL b.w, the *M. koenigii* extract showed 80% inhibited free radical generation and 75% restored GSH levels	[10]
Antioxidant	Leaves	Water	–	Male albino Wistar rat	Extract exhibited the potential to reduce lipid peroxidation activity in the liver (2.44 ± 0.029) and kidney (2.34 ± 0.09) in potassium dichromate-induced Wistar rats	[38]
Anti-inflammatory	Stem bark and roots	Hexane, chloroform, and methanol	Girinimbine	Adult zebrafish	Girinimbine treatment significantly suppressed the IL-1β and TNF-α levels induced by peritoneal fluid mice	[85]

Table 3. *Cont.*

Pharmacological Activities	Plant Parts	Extract	Bioactive Compounds	Model	Main Finding	Reference
Anti-inflammatory	Leaves	Ethanol	–	Sprague Dawley rats	Oral administration of an *M. koenigii* extract showed the reduced formation of oedema, with values of 43.28%, 59.67%, and 62.29% induced by carrageenan, histamine, and serotonin in rats	[30]
Hepatoprotective	Leaves	Hydro-ethanolic	–	Male Wistar rats	*M. koenigii* leaves significantly decreased CCl4 -induced hepatotoxic in a time- and dose-dependent manner	[16]
Nephroprotective	Leaves	Aqueous	–	Male Wistar rats	*M. koenigii* extract treatment significantly decreased the renal functional markers, like the blood urea nitrogen and creatinine level	[28]
Anti-Diabetic	Leaves	Ethanol	–	Swiss albino mice	*M. koenigii* possesses antidiabetic activity and has antioxidant effects on STZ-NA-induced diabetes mellitus and particularly significantly decreased the HOMA-IR index	[10]
Anticancer (Colon)	Stem bark and roots	Hexane, chloroform, and methanol	Girinimbine 1.5–100 µg/mL	Zebrafish and Male ICR mice	Girinimbine, supplementation specifically, resulted in the induction of apoptosis, the inhibition of inflammation, and a significant increase in cell numbers in the G0/G1 phase	[85]
Anticancer (Breast)	Leaves	Aqueous	–	Female BALB/c mice	*M. koenigii* aqueous extract has potential for cytotoxicity, anti-inflammatory, and immunomodulatory effects and delays rather than inhibits tumor formation	[12]
Neuroprotective	Leaves	Methanol	–	Male albino mice	*M. koenigii* is effective in attenuating memory impairment and oxidative stress and prevents abnormal oral movements	[36]
Neuroprotective	Leaves	Ethanol	–	Swiss albino mice	*M. koenigii* supplementation resulted in an improvement of acetylcholine (ACh) and reduction in acetylcholinesterase (AChE). In addition, a significant elevation of serum biomarkers, and decline in creatinine, total cholesterol, urea nitrogen, and glucose levels, ameliorated the hepatic and renal functions in the normal ageing process	[30]
Neuroprotective	Leaves	Ethanol	–	Male swiss albino mice	*M. koenigii* leaves elevated the acetylcholine level in the brain and ultimately improved memory impairment. In vitro, it showed BACE1 inhibition and was found to be a non-competitive inhibitor	[33]
Neuroprotective	Leaves	Methanol	Isolongifolene 10 mg/kg b.w.	Male albino Wistar rat	Isolongifolene effectively attenuated behavioral impairment and oxidative stress, acting as an antiaging agent	[34]
Anti-anxiety and anti-depressant	Leaves	Aqueous	–	Swiss albino mice	*M. koenigii* aqueous leaf extract reduced the despair behavior in experimental animal models, suggesting an anti-depressant-like activity and also reduced spontaneous locomotor activity	[41]

4.3. Hepatoprotective Effect

Liver diseases are a worldwide concern, and accessible medical treatments have an inadequate efficacy. Since ancient times, herbs have been used when treating various disease conditions; plant extracts and natural compounds have significant applications as hepatoprotective agents. The liver is the site of drug metabolism and the detoxification site of toxic products, and so it is the organ most exposed to xenobiotics [107]. *M. koenigii* extended hepatoprotective activity when crude aqueous extracts were investigated against ethanol-induced hepatotoxicity in experimental animals. *M. koenigii* was reported to extend a protective effect in liver impairments in chronic alcoholism and was proved to be effective in maintaining the enzymatic oxidant status [108]. Water extracts of carbazole alkaloids and tannin of *M. koenigii* were explored for their hepatoprotective activity against ethanol-induced hepatotoxicity in a HepG2 cell line model. They exhibited excellent hepatoprotective activity, maintaining the enzymatic and non-enzymatic antioxidant level at a near normal value and also maintaining the integrity of the cells [31]. An *M. koenigii* hydro-ethanolic leaf extract was reported to attenuate the CCl_4 hepatotoxic effects in rats. *M. koenigii*-pretreated rats showed a significant decrement in activity levels of hepatic markers and also maintained the level of enzymatic antioxidants [16].

4.4. Immunomodulatory Activity

The immune system makes a network and regulates processes important for maintaining the health of an organism by hindering the entry and invasion of microbes. Impairments in the immune system lead to conditions from chronic inflammation to cancer [109]. In an investigation on the humoral- and cell-mediated immune response to ovalbumin, the immunomodulatory activity of a methanolic extract of *M. koenigii* leaves was evaluated using a carbon clearance test. A considerable increase in the NO production indicated the increased phagocytic activity of macrophages. The *M. koenigii* extract holds promise as an immunomodulatory agent, which acts by stimulating humoral immunity and the phagocytic function [110]. The *M. koenigii* leaf extracts were reported to have certain effects in regulating mice immunology related to oxidative stress metabolism. An *M. koenigii* leaf extract can exhibit an immunomodulatory effect through which it can regulate the oxidative stress metabolism in diabetic mice [111].

4.5. Nephroprotective Activity

M. koenigii has been used as a nephroprotective agent in a diabetic-induced rat model [9]. The *M. koenigii* leaf extract was found to be efficient in maintaining normal levels of serum creatinine, blood urea nitrogen, total serum protein, serum Na^+, urine output, urinary creatinine, urinary urea, total urinary protein, and urinary Na^+. Furthermore, the *M. koenigii* extract maintained the standard pattern in in vivo antioxidants, renal myeloperoxidase (MPO) activity, and histopathology of kidneys against unilateral renal ischemia reperfusion injury. Therefore, the extract of *M. koenigii* was clearly demonstrated to be useful in treating kidney disorders in rats [112]. The nephroprotective activity of *M. koenigii* was elucidated in experimental investigations, which showed decreased levels of blood urea nitrogen (BUN), serum creatinine (Cr), and lipid peroxidation (LPO). An *M. koenigii* extract is efficient against cyclophosphamide-induced nephrotoxicity, which was clearly revealed through the maintenance of high levels of glutathione (GSH) and superoxide dismutase (SOD) compared to the cyclophosphamide-treated group [28]. *M. koenigii* protective activity has been shown to induce significant dose-dependent decreases in serum urea and creatinine levels, as well as marked increases in the levels of plasma antioxidant capacity, in diabetic rats, compared to controls. More noteworthy is the histological integrity of kidneys, which showed comparable tissue regeneration induced by the aqueous extract [9].

4.6. Antidiabetic Activity

Most prominently in developing countries, medicinal plants play a helpful role in managing diabetes mellitus due to their cost effectiveness. Diabetes mellitus, a metabolic disorder, is becoming a serious threat to human health. During the past few years, many phytochemicals responsible for anti-diabetic effects have been isolated from plants. Alkaloids present in the leaves of *M. koenigii* have been explored and reported to have inhibitory effects on the aldose reductase enzyme, glucose utilization, and other enzyme systems for extending anti-diabetic effects [38]. *M. koenigii* was tested for the α-glucosidase inhibitory property and was found to inhibit α glycosidase. Alpha-glucosidase inhibitors are widely used in the treatment of patients with type 2 diabetes [113]. A study reported that an ethanolic extract of *M. koenigii* showed a significant reduction in blood glucose levels, and this effect of reducing blood glucose by *M. koenigii* is mediated by antioxidant properties and insulin mimetic effects. In addition, *M. koenigii* exhibited a profound antioxidant effect by reducing the malondialdehyde (MDA) level, increasing the GSH level, and significantly decreasing the homeostatic model assessment (HOMA)-insulin resistance index. On the whole, it is evident that *M. koenigii* possesses antidiabetic activity and has antioxidant effects in rats [10].

4.7. Anticancer Activity (In Vivo and In Vitro)

M. koenigii possesses potential secondary metabolites that could be developed as anticancer agents. In one study, the cytotoxic activity was evaluated for three extracts: hexane, ethyl acetate, and methanol of *M. koenigii* leaves against the HeLa cell line. The extracts were reported as being potently cytotoxic in nature in HeLa cancer cells. These results established the potential of *M. koenigii* as an anticancer agent in vitro [11]. Additional evidence for the anticancer activity of *M. koenigii* has been obtained from rodent cancer cell lines, as well as different in vivo cancer models [12–14,22–24,114,115]. In an early study, histopathological evidence showed that *M. koenigii* extract treatment generated a decline in neoplasms in the colon [85]. A methanolic extract of *M. koenigii* was reported to have the ability to reduce proliferation in breast cancer cell lines [116]. The total alkaloid extracted from *M. koenigii* leaves has been shown to have promising cytotoxic activity in breast cancer cells, with an IC_{50} of 14.4 µg/mL [117]. The anticancer activity of mahanine and isomahanine in human oral squamous cell carcinoma CLS-354 has also been reported [96].

Girinimbine, another *M. koenigii*-derived carbazole alkaloid, showed growth inhibitory activity in human hepatocellular carcinoma and lung cancer cells in vitro [118]. Rutin, quercetin, kaempferol, and apigenin, present in leaf extracts of *M. koenigii*, showed the dose-dependent inhibition of endogenous 26S proteasome activity in MDA-MB-231 cells [52]. Therefore, *M. koenigii* contains remarkable anticancer compounds, especially mahanine, which has been reported to show anticancer activity targeting different signaling pathways [46]. Girinimbine, a carbazole alkaloid, has been found to have a good role in total leukocyte migration and result in an appreciable reduction in pro-inflammatory cytokine levels. The various activities of *M. koenigii* against different cancer cell lines are shown in Figure 2.

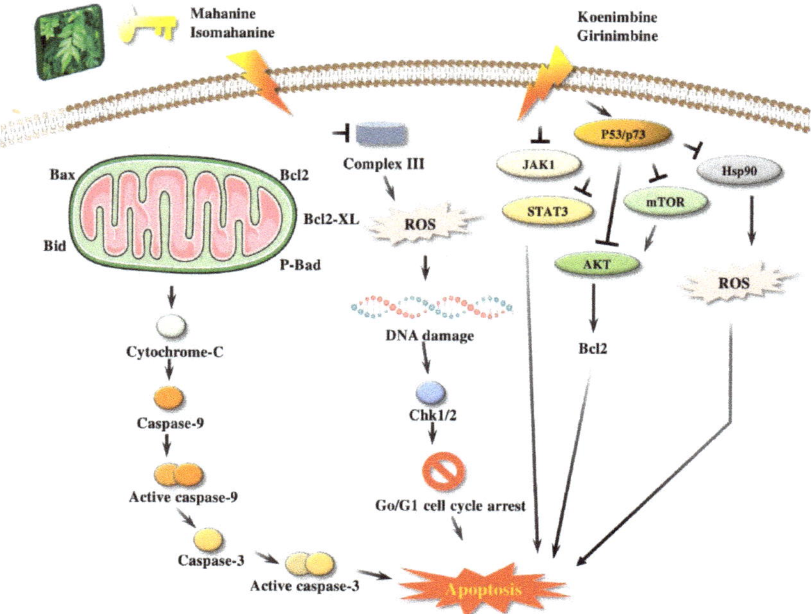

Figure 2. Apoptosis induced by *M. koenigii* bioactive compounds in cancer. Bcl2: B-cell lymphoma 2; Bcl2-XL: B-cell lymphoma-extra-large; P-Bad: P plasmid araB araA araD; ROS: reactive oxygen species; Chk1/2: checkpoint kinase; Go/G1: cell cycle phase; JAK1: janus kinase 1; STAT3: signal transducer and activator of transcription 3; AKT: protein kinase B (also known as AKT); mTOR: mammalian target of rapamycin; P53/p57: tumor protein; Hsp90: heat shock protein.

4.8. Neuroprotective Activity

Supplementation with *M. koenigii* leaf extracts has been reported in the management of a wide spectrum of neurodegenerative diseases, like AD, PD, and others [30,33–35,41]. *M. koenigii* promotes neuroprotective potential against orofacial dyskinesia induced by resperine. Additionally, it stabilizes the levels of protective antioxidant enzymes like SOD, catalase (CAT), and GSH, and inhibits LPO in the forebrain regions of reserpine-treated animals. Furthermore, it has been shown to significantly inhibit reserpine-induced abnormalities in behavior. Similarly, treatment with *M. koenigii* significantly restored the levels of protective antioxidant enzymes (that is, SOD, CAT, and GSH) and inhibited LPO in the forebrain region when compared with reserpine, and it also inhibited catalepsy induced by haloperidol [36]. Isolongifolene (ILF), a tricyclic sesquiterpene of *M. koenigii*, has been reported to render neuroprotective effects against rotenone-induced mitochondrial dysfunction, oxidative stress, and apoptosis in a cellular model. Cytotoxicity, oxidative stress, and mitochondrial dysfunction were also attenuated by ILF in SH-SY5Y cells, which down-regulated Bax and caspases-3, -6, -8, and -9 expression, and up-regulated Bcl-2 expression. IFL was proved to regulate p-P13K, p-AKT, and p-GSK-3 beta expressions [87]. Preclinical studies have reported that *M. koenigii* leaves could enhance memory in rats [119]. The possible neuroprotective potential of amethanolic extract of *M. koenigii* leaves was exhibited in a two-vessel occlusion (2VO) rat model of partial global cerebral ischaemia. The Morris water maze test was implemented to assess the rats' cognitive function postoperatively. Brain samples were histopathologically examined for viable neurons within the CA1 hippocampal region. Test findings showed that *M. koenigii* leaves positively improved memory and learning impairments. *M. koenigii* leaf extracts modestly improved memory in rats with chronic partial global cerebral ischemia [120]. The various activities of *M. koenigii* against neurotoxicity are shown in Figure 3.

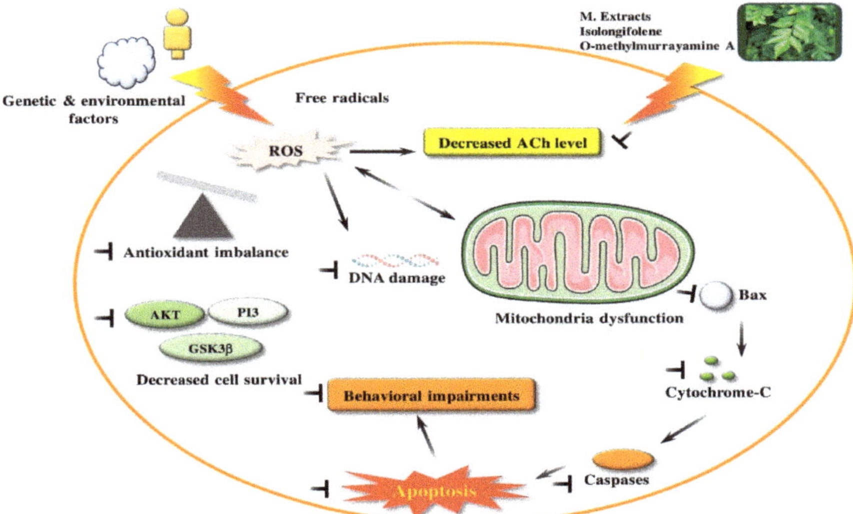

Figure 3. Neuroprotective effect in in vitro and in vivo studies produced by bioactive compounds from *M. koenigii*. PI3: phosphatidylinositol 3 kinase; GSK3β: Glycogen synthase kinase 3 beta; Ach: Acetylcholine; Bax: Bcl2-Associated X protein.

4.9. Radioprotective and Chemoprotective Activity

A methanolic extract of *M. koenigii* was demonstrated to render protection in chromosomal damage against radiation and cyclophasphamide in vivo. Radiation leads to a rise in all types of aberrations, like the fragmentation of chromatids and breakages in chromosomes, rings, and dicentrics. Treatment with a methanolic extract of *M. koenigii* before radiation significantly reduced the aberrations. *M. koenigii* can significantly exert bone marrow protection against radiation and cyclophasphamide [121].

4.10. Wound Healing Effect

Wound healing is a complex and multifactor process involving numerous biochemical and cellular processes which helps in the restoration of functional and anatomical continuity. *M. koenigii* leaves extend wound healing in male albino rats through significantly increased wound contraction and reduced epithelialization, supporting the collagen synthesis which was evident in histopathological studies [122].

5. Conclusions

The current review summarizes the medicinal uses, phytochemistry, and pharmacological properties of *M. koenigii*. *M. koenigii* is a source of several bioactive compounds, including alkaloids, polyphenol, terpenoids, and flavonoids. *M. koenigii* and its derivatives appear to exhibit appreciable pharmacological activities, like anticarcinogenic, proapoptotic, antiangiogenic, antimetastatic, immunomodulatory, and antioxidant properties. The molecular mechanisms underlying these activities of *M. koenigii* and its derivatives are due to their diversified role in combinations of cell signaling pathways at multiple levels in various diseases. *M. koenigii* and its derivatives mitigate oxidative stress, neurotoxicity, neuroinflammation, neuronal loss, and cognitive dysfunctions. However, like other polyphenols, to a certain extent, *M. koenigii* activities are limited by its bioavailability and in such conditions, enhancement of the efficiency should be conducted. Therefore, future studies need to include more experimental studies on bioavailability and efficiency enhancement in clinical investigations.

Author Contributions: R.B., writing—original draft preparation; D.V., writing—review and editing; S.-H.J., table preparation; P.G., validation; I.S.-K., interpreted the figures; D.-K.C., supervision of the review and approval of the final draft. All authors have read and agreed to the published version of the manuscript.

Funding: This work was supported by the Basic Science Research Program through the National Research Foundation of Korea funded by the Ministry of Education, Science and Technology (NRF-2018R1C1B6005129 and 2020R1A2B5B02002032)

Conflicts of Interest: The authors declare no conflicts of interest.

References

1. Handral, H.K.; Pandith, A.; Shruthi, S.D. A review on *Murraya koenigii*: Multipotential medicinal plant. *Asian J. Pharm. Clin. Res.* **2012**, *5*, 5–14.

2. Ujowundu, C.O.; Okafor, O.E.; Agha, N.C.; Nwaogu, L.A.; Igwe, K.O.; Igwe, C.U. Phytochemical and chemical composition of *Combretum zenkeri* leaves. *J. Med. Plants Res.* **2010**, *4*, 965–968.

3. Kang, W.; Wang, Y. China Digital Governance Development Review Over the Past Two Decades. *Int. J. Public Adm. Digit. Age* **2018**, *5*, 92–106. [CrossRef]

4. Zhang, J.; Du, S.; Duan, X.; Zhang, S. Effects of ultrahigh pressure processing on the physicochemical characteristics of Taibai Kudzu starch. *Nongye Gongcheng Xuebao/Trans. Chin. Soc. Agric. Eng.* **2007**, *2017*. [CrossRef]

5. Phumthum, M.; Balslev, H. Thai Ethnomedicinal Plants Used for Diabetes Treatment. *OBM Integr. Complement. Med.* **2018**, *3*, 1–25. [CrossRef]

6. Gunjan, M.; Naing, T.W.; Saini, R.S.; Ahmad, A.; Naidu, J.R.; Kumar, I. Marketing trends & future prospects of herbal medicine in the treatment of various disease. *World J. Pharm. Res.* **2015**, *4*, 132–155.

7. Wojdyło, A.; Oszmiański, J.; Czemerys, R. Antioxidant activity and phenolic compounds in 32 selected herbs. *Food Chem.* **2007**, *105*, 140–149. [CrossRef]

8. Ahluwalia, V.; Sisodia, R.; Walia, S.; Sati, O.P.; Kumar, J.; Kundu, A. Chemical analysis of essential oils of *Eupatorium adenophorum* and their antimicrobial, antioxidant and phytotoxic properties. *J. Pest Sci. (2004)* **2014**, *87*, 341–349. [CrossRef]

9. Yankuzo, H.; Ahmed, Q.U.; Santosa, R.I.; Akter, S.F.U.; Talib, N.A. Beneficial effect of the leaves of *Murraya koenigii* (Linn.) Spreng (Rutaceae) on diabetes-induced renal damage in vivo. *J. Ethnopharmacol.* **2011**, *135*, 88–94. [CrossRef]

10. Husna, F.; Suyatna, F.D.; Arozal, W.; Poerwaningsih, E.H. Anti-Diabetic Potential of *Murraya koenigii* (L) and its Antioxidant Capacity in Nicotinamide-Streptozotocin Induced Diabetic Rats. *Drug Res. (Stuttg)* **2018**, *68*, 631–636. [CrossRef]

11. Amna, U.; Halimatussakdiah, P.W.; Saidi, N.; Nasution, R. Evaluation of cytotoxic activity from Temurui (*Murraya koenigii* [Linn.] Spreng) leaf extracts against HeLa cell line using MTT assay. *J. Adv. Pharm. Technol. Res.* **2019**, *10*, 51–55. [CrossRef] [PubMed]

12. Yeap, S.K.; Abu, N.; Mohamad, N.E.; Beh, B.K.; Ho, W.Y.; Ebrahimi, S.; Yusof, H.M.; Ky, H.; Tan, S.W.; Alitheen, N.B. Chemopreventive and immunomodulatory effects of *Murraya koenigii* aqueous extract on 4T1 breast cancer cell-challenged mice. *BMC Complement. Altern. Med.* **2015**, *4*, 306. [CrossRef] [PubMed]

13. Nooron, N.; Ohba, K.; Takeda, K.; Shibahara, S.; Chiabchalard, A. Dysregulated expression of MITF in subsets of hepatocellular carcinoma and cholangiocarcinoma. *Tohoku J. Exp. Med.* **2017**, *242*, 291–302. [CrossRef] [PubMed]

14. Das, R.; Bhattacharya, K.; Samanta, S.K.; Pal, B.C.; Mandal, C. Improved chemosensitivity in cervical cancer to cisplatin: Synergistic activity of mahanine through STAT3 inhibition. *Cancer Lett.* **2014**, *351*, 81–90. [CrossRef]

15. Bhandari, P. Curry leaf (*Murraya koenigii*) or Cure leaf: Review of its curative properties. *J. Med. Nutr. Nutraceuticals* **2012**, *2*, 92–97. [CrossRef]

16. Desai, S.N.; Patel, D.K.; Devkar, R.V.; Patel, P.V.; Ramachandran, A.V. Hepatoprotective potential of polyphenol rich extract of *Murraya koenigii* L.: An in vivo study. *Food Chem. Toxicol.* **2012**, *50*, 310–314. [CrossRef]

17. Gajaria, T.K.; Patel, D.K.; Devkar, R.V.; Ramachandran, A.V. Flavonoid rich extract of *Murraya koenigii* alleviates in-vitro LDL oxidation and oxidized LDL induced apoptosis in raw 264.7 Murine macrophage cells. *J. Food Sci. Technol.* **2015**, *52*, 3367–3375. [CrossRef]

18. Goutam, M.P.; Purohit, R.M. Antimicrobial activity of the essential oil of the leaves of *Murraya koenigii* (Linn) Spreng (Indian curry leaf). *Indian J. Pharm.* **1974**, *2*, 48–51.

19. Maswada, H.F.; Abdallah, S.A. In vitro antifungal activity of three geophytic plant extracts against three post-harvest pathogenic fungi. *Pakistan J. Biol. Sci.* **2013**, *16*, 1698–1705. [CrossRef]

20. Bonde, S.D.; Nemade, L.S.; Patel, M.R.; Patel, A.A. *Murraya koenigii* (Curry leaf): Ethnobotany, Phytochemistry and Pharmacology—A Review. *Int. J. Pharm. Phytopharm. Res.* **2011**, *1*, 23.

21. Ma, Q.G.; Xu, K.; Sang, Z.P.; Wei, R.R.; Liu, W.M.; Su, Y.L.; Yang, J.B.; Wang, A.G.; Ji, T.F.; Li, L.J. Alkenes with antioxidative activities from *Murraya koenigii* (L.) Spreng. *Bioorg. Med. Chem. Lett.* **2016**, *26*, 799–803. [CrossRef]

22. Chatterjee, P.; Seal, S.; Mukherjee, S.; Kundu, R.; Bhuyan, M.; Barua, N.C.; Baruah, P.K.; Babu, S.P.S.; Bhattacharya, S. A carbazole alkaloid deactivates mTOR through the suppression of rictor and that induces apoptosis in lung cancer cells. *Mol. Cell. Biochem.* **2015**, *405*, 149–158. [CrossRef] [PubMed]

23. Das, R.; Bhattacharya, K.; Sarkar, S.; Samanta, S.K.; Pal, B.C.; Mandal, C. Mahanine synergistically enhances cytotoxicity of 5-fluorouracil through ROS-mediated activation of PTEN and p53/p73 in colon carcinoma. *Apoptosis* **2014**, *19*, 149–164. [CrossRef] [PubMed]

24. Agarwal, S.; Amin, K.S.; Jagadeesh, S.; Baishay, G.; Rao, P.G.; Barua, N.C.; Bhattacharya, S.; Banerjee, P.P. Mahanine restores RASSF1A expression by down-regulating DNMT1 and DNMT3B in prostate cancer cells. *Mol. Cancer* **2013**, *12*, 99. [CrossRef] [PubMed]

25. Mandal, S.; Nayak, A.; Kar, M.; Banerjee, S.K.; Das, A.; Upadhyay, S.N.; Singh, R.K.; Banerji, A.; Banerji, J. Antidiarrhoeal activity of carbazole alkaloids from *Murraya koenigii* Spreng (Rutaceae) seeds. *Fitoterapia* **2010**, *81*, 72–74. [CrossRef] [PubMed]

26. Ningappa, M.B.; Dhananjaya, B.L.; Dinesha, R.; Harsha, R.; Srinivas, L. Potent antibacterial property of APC protein from curry leaves (*Murraya koenigii* L.). *Food Chem.* **2010**, *118*, 747–750. [CrossRef]

27. Tripathi, Y.; Anjum, N.; Rana, A. Chemical Composition and In vitro Antifungal and Antioxidant Activities of Essential Oil from *Murraya koenigii* (L.) Spreng. Leaves. *Asian J. Biomed. Pharm. Sci.* **2018**, *8*, 6–13. [CrossRef]

28. Mahipal, P.; Pawar, R.S. Nephroprotective effect of *Murraya koenigii* on cyclophosphamide induced nephrotoxicity in rats. *Asian Pac. J. Trop. Med.* **2017**, *10*, 808–812. [CrossRef]

29. Rautela, R.; Das, G.K.; Khan, F.A.; Prasad, S.; Kumar, A.; Prasad, J.K.; Ghosh, S.K.; Dhanze, H.; Katiyar, R.; Srivastava, S.K. Antibacterial, anti-inflammatory and antioxidant effects of Aegle marmelos and *Murraya koenigii* in dairy cows with endometritis. *Livest. Sci.* **2018**, *214*, 142–148. [CrossRef]

30. Mani, V.; Ramasamy, K.; Ahmad, A.; Wahab, S.N.; Jaafar, S.M.; Kek, T.L.; Salleh, M.Z.; Majeed, A.B.A. Effects of the total alkaloidal extract of *Murraya koenigii* leaf on oxidative stress and cholinergic transmission in aged mice. *Phyther. Res.* **2013**, *27*, 46–53. [CrossRef]

31. Sathaye, S.; Bagul, Y.; Gupta, S.; Kaur, H.; Redkar, R. Hepatoprotective effects of aqueous leaf extract and crude isolates of *Murraya koenigii* against in vitro ethanol-induced hepatotoxicity model. *Exp. Toxicol. Pathol.* **2011**, *63*, 587–591. [CrossRef] [PubMed]

32. Dar, R.A.; Shahnawaz, M.; Qazi, P.H.; Qazi, H. General overview of medicinal plants: A review. *J. Phytopharm.* **2017**, *6*, 349–351.

33. Mani, V.; Ramasamy, K.; Ahmad, A.; Parle, M.; Shah, S.A.A.; Majeed, A.B.A. Protective effects of total alkaloidal extract from *Murraya koenigii* leaves on experimentally induced dementia. *Food Chem. Toxicol.* **2012**, *50*, 1036–1044. [CrossRef] [PubMed]

34. Balakrishnan, R.; Tamilselvam, K.; Sulthana, A.; Mohankumar, T.; Manimaran, D.; Elangovan, N. Isolongifolene Attenuates Oxidative Stress and Behavioral Impairment in Rotenone-Induced Rat Model of Parkinson's Disease. *Int. J. Nutr. Pharmacol. Neurol. Dis.* **2018**, *8*, 53–58.

35. Zang, Y.D.; Li, C.J.; Song, X.Y.; Ma, J.; Yang, J.Z.; Chen, N.H.; Zhang, D.M. Total synthesis and neuroprotective effect of O-methylmurrayamine A and 7-methoxymurrayacine. *J. Asian Nat. Prod. Res.* **2017**, *19*, 623–629. [CrossRef]

36. Patil, R.; Dhawale, K.; Gound, H.; Gadakh, R. Protective effect of leaves of *Murraya koenigii* on reserpine-induced orofacial dyskinesia. *Iran. J. Pharm. Res.* **2012**, *11*, 635–641.

37. Erkan, N.; Tao, Z.; Vasantha Rupasinghe, H.P.; Uysal, B.; Oksal, B.S. Antibacterial activities of essential oils extracted from leaves of *Murraya koenigii* by solvent-free microwave extraction and hydro-distillation. *Nat. Prod. Commun.* **2012**, *7*, 121–124. [CrossRef]

38. Patel, D.K.; Kumar, R.; Laloo, D.; Hemalatha, S. Natural medicines from plant source used for therapy of diabetes mellitus: An overview of its pharmacological aspects. *Asian Pac. J. Trop. Dis.* **2012**, *2*, 139–150. [CrossRef]

39. Nalli, Y.; Khajuria, V.; Gupta, S.; Arora, P.; Riyaz-Ul-Hassan, S.; Ahmed, Z.; Ali, A. Four new carbazole alkaloids from *Murraya koenigii* that display anti-inflammatory and anti-microbial activities. *Org. Biomol. Chem.* **2016**, *14*, 3322–3332. [CrossRef]

40. Joshi, T.; Jain, T.; Mahar, R.; Singh, S.K.; Srivastava, P.; Shukla, S.K.; Mishra, D.K.; Bhatta, R.S.; Banerjee, D.; Kanojiya, S. Pyranocarbazoles from *Murraya koenigii* (L.) Spreng. as antimicrobial agents. *Nat. Prod. Res.* **2018**, *32*, 430–434. [CrossRef]

41. Sharma, S.; Handu, S.; Dubey, A.; Sharma, P.; Mediratta, P.; Ahmed, Q. Anti-anxiety and anti-depressant like effects of *Murraya koenigii* in experimental models of anxiety and depression. *Anc. Sci. Life* **2017**, *36*, 215–219. [PubMed]

42. Adebajo, A.C.; Ayoola, O.F.; Iwalewa, E.O.; Akindahunsi, A.A.; Omisore, N.O.A.; Adewunmi, C.O.; Adenowo, T.K. Anti-trichomonal, biochemical and toxicological activities of methanolic extract and some carbazole alkaloids isolated from the leaves of *Murraya koenigii* growing in Nigeria. *Phytomedicine* **2006**, *13*, 246–254. [CrossRef] [PubMed]

43. Tembhurne, S.V.; Sakarkar, D.M. Hypoglycemic effects of fruit juice of *Murraya koenigii* (L) in alloxan induced diabetic mice. *Int. J. PharmTech Res.* **2009**, *1*, 1589–1593.

44. Sim, K.M.; Teh, H.M. A new carbazole alkaloid from the leaves of Malayan *Murraya koenigii*. *J. Asian Nat. Prod. Res.* **2011**, *13*, 972–975. [CrossRef] [PubMed]

45. Igara, C.; Omoboyowa, D.; Ahuchaogu, A.; Orji, N.; Ndukwe, M. Phytochemical and nutritional profile of *Murraya koenigii* (Linn) Spreng leaf. *J. Pharmacogn. Phytochem.* **2016**, *5*, 7–9.

46. Samanta, S.K.; Kandimalla, R.; Gogoi, B.; Dutta, K.N.; Choudhury, P.; Deb, P.K.; Devi, R.; Pal, B.C.; Talukdar, N.C. Phytochemical portfolio and anticancer activity of *Murraya koenigii* and its primary active component, mahanine. *Pharmacol. Res.* **2018**, *129*, 227–236. [CrossRef] [PubMed]

47. Ramsewak, R.S.; Nair, M.G.; Strasburg, G.M.; DeWitt, D.L.; Nitiss, J.L. Biologically active carbazole alkaloids from *Murraya koenigii*. *J. Agric. Food Chem.* **1999**, *47*, 444–447. [CrossRef]

48. Joshi, B.S.; Kamat, V.N.; Gawad, D.H. On the structures of girinimbine, mahanimbine, isomahanimbine, koenimbidine and murrayacine. *Tetrahedron* **1970**, *26*, 1475–1482. [CrossRef]

49. Gahlawat, D.K.; Jakhar, S.; Dahiya, P. *Murraya koenigii* (L.) Spreng: An ethnobotanical, phytochemical and pharmacological review. *J. Pharmacogn. Phytochem.* **2014**, *3*, 109–119.

50. Tachibana, Y.; Kikuzaki, H.; Lajis, N.H.; Nakatani, N. Antioxidative activity of carbazoles from *Murraya koenigii* leaves. *J. Agric. Food Chem.* **2001**, *49*, 5589–5594. [CrossRef]

51. Srivastava, S.K.; Srivastava, S.D. New constituents and biological activity of the roots of *Murraya koenigii*. *J. Indian Chem. Soc.* **1993**, *2*, 607–627.

52. Noolu, B.; Gogulothu, R.; Bhat, M.; Qadri, S.S.Y.H.; Sudhakar Reddy, V.; Bhanuprakash Reddy, G.; Ismail, A. In Vivo Inhibition of Proteasome Activity and Tumour Growth by *Murraya koenigii* Leaf Extract in Breast Cancer Xenografts and by Its Active Flavonoids in Breast Cancer Cells. *Anticancer Agents Med. Chem.* **2016**, *16*, 1605–1614. [CrossRef] [PubMed]

53. Ma, Q.; Tian, J.; Yang, J.; Wang, A.; Ji, T.; Wang, Y.; Su, Y. Bioactive carbazole alkaloids from *Murraya koenigii* (L.) Spreng. *Fitoterapia* **2013**, *87*, 1–6. [CrossRef] [PubMed]

54. Umehara, K.; Hattori, I.; Miyase, T.; Ueno, A.; Hara, S.; Kageyama, C. Studies on the Constituents of Leaves of *Citrus unshiu* Marcov. *Chem. Pharm. Bull.* **1988**, *36*, 5004–5008. [CrossRef]

55. Mori, K.; Khlebnikov, V. Carotenoids and Degraded Carotenoids, VIII-Synthesis of (+)-Dihydroactinidiolide, (+)- and (−)-Actinidiolide, (+)- and (−)-Loliolide as well as (+)- and (−)-Epiloliolide. *Liebigs Ann. Chem.* **1993**, *1993*, 77–82. [CrossRef]

56. Tan, S.P.; Ali, A.M.; Nafiah, M.A.; Amna, U.; Ramli, S.A.; Ahmad, K. Terpenes and Phenolic Compounds of *Murraya koenigii*. *Chem. Nat. Compd.* **2017**, *53*, 980–981. [CrossRef]

57. ChV, S. Antioxidant and Biological Activities of Three Morphotypes of *Murraya koenigii* L. from Uttarakhand. *J. Food Process. Technol.* **2013**, *4*, 1–7. [CrossRef]

58. Patel, O.P.S.; Mishra, A.; Maurya, R.; Saini, D.; Pandey, J.; Taneja, I.; Raju, K.S.R.; Kanojiya, S.; Shukla, S.K.; Srivastava, M.N.; et al. Naturally Occurring Carbazole Alkaloids from *Murraya koenigii* as Potential Antidiabetic Agents. *J. Nat. Prod.* **2016**, *79*, 1276–1284. [CrossRef]

59. Bhattacharya, K.; Samanta, S.K.; Tripathi, R.; Mallick, A.; Chandra, S.; Pal, B.C.; Shaha, C.; Mandal, C. Apoptotic effects of mahanine on human leukemic cells are mediated through crosstalk between Apo-1/Fas signaling and the Bid protein and via mitochondrial pathways. *Biochem. Pharmacol.* **2010**, *79*, 361–372. [CrossRef]

60. Brand, M.D.; Affourtit, C.; Esteves, T.C.; Green, K.; Lambert, A.J.; Miwa, S.; Pakay, J.L.; Parker, N. Mitochondrial superoxide: Production, biological effects, and activation of uncoupling proteins. *Free Radic. Biol. Med.* **2004**, *37*, 755–767. [CrossRef]

61. Thannickal, V.J.; Fanburg, B.L. Reactive oxygen species in cell signaling. *Am. J. Physiol. Lung Cell. Mol. Physiol.* **2000**, *279*, L1005–L1028. [CrossRef] [PubMed]

62. Hoidal, J.R. Reactive oxygen species and cell signaling. *Am. J. Respir. Cell Mol. Biol.* **2001**, *25*, 661–663. [CrossRef] [PubMed]

63. Gill, N.S.; Sharma, B. Study on antioxidant potential of *Murraya koenigii* leaves in wistar rats. *Pakistan J. Biol. Sci.* **2013**, *17*, 126–129. [CrossRef]

64. Rehana, D.; Mahendiran, D.; Kumar, R.S.; Rahiman, A.K. In vitro antioxidant and antidiabetic activities of zinc oxide nanoparticles synthesized using different plant extracts. *Bioprocess Biosyst. Eng.* **2017**, *40*, 943–957. [CrossRef] [PubMed]

65. Kusuma, I.W.; Kuspradini, H.; Arung, E.T.; Aryani, F.; Min, Y.H.; Kim, J.S.; Kim, Y.U. Biological Activity and Phytochemical Analysis of Three Indonesian Medicinal Plants, *Murraya koenigii*, *Syzygium polyanthum* and *Zingiber purpurea*. *J. Acupunct. Meridian Stud.* **2011**, *4*, 75–79. [CrossRef]

66. Rao, L.J.M.; Ramalakshmi, K.; Borse, B.B.; Raghavan, B. Antioxidant and radical-scavenging carbazole alkaloids from the oleoresin of curry leaf (*Murraya koenigii Spreng.*). *Food Chem.* **2007**, *100*, 742–747. [CrossRef]

67. Gupta, S.; Paarakh, P.M.; Gavani, U. Antioxidant activity of *Murraya koenigii* linn leaves. *Pharmacologyonline* **2009**, *1*, 474–478.

68. Zahin, M.; Aqil, F.; Husain, F.M.; Ahmad, I. Antioxidant capacity and antimutagenic potential of *Murraya koenigii*. *BioMed Res. Int.* **2013**, *2013*, 263509. [CrossRef]

69. Yogesh, K.; Jha, S.N.; Yadav, D.N. Antioxidant Activities of *Murraya koenigii* (L.) Spreng Berry Extract: Application in Refrigerated (4 ± 1 °C) Stored Meat Homogenates. *Agric. Res.* **2012**, *1*, 183–189. [CrossRef]

70. Rajesh, S.T.; Sharmistha, B.; Shuchi, K. Assessment of antioxidant activity of leaves of *Murraya koenigii* extracts and it's comparative efficacy analysis in different solvents. *J. Pharm. Sci. Res.* **2017**, *9*, 288–291.

71. Waghmare, A.N.; Tembhurne, S.V.; Sakarar, D.M. Phytochemical Analysis and In vitro Antioxidant Properties of *Murraya koenigii* (L.) Fruits. *Am. J. Phytomedicine Clin. Ther.* **2015**, *3*, 403–416.

72. Dexter, D.T.; Wells, F.R.; Lee, A.J.; Agid, F.; Agid, Y.; Jenner, P.; Marsden, C.D. Increased Nigral Iron Content and Alterations in Other Metal Ions Occurring in Brain in Parkinson's Disease. *J. Neurochem.* **1989**, *52*, 1830–1836. [CrossRef] [PubMed]

73. Archer, S.L.; Gomberg-Maitland, M.; Maitland, M.L.; Rich, S.; Garcia, J.G.N.; Weir, E.K. Mitochondrial metabolism, redox signaling, and fusion: A mitochondria-ROS-HIF-1α-Kv1.5 O2-sensing pathway at the intersection of pulmonary hypertension and cancer. *Am. J. Physiol. Heart Circ. Physiol.* **2008**, *294*, H570–H578. [CrossRef] [PubMed]

74. Kowaltowski, A.J.; de Souza-Pinto, N.C.; Castilho, R.F.; Vercesi, A.E. Mitochondria and reactive oxygen species. *Free Radic. Biol. Med.* **2009**, *53*, 885–892. [CrossRef]

75. Matsuzawa, A.; Ichijo, H. Stress-responsive protein kinases in redox-regulated apoptosis signaling. *Antioxid. Redox Signal.* **2005**, *7*, 472–481. [CrossRef]

76. Trachootham, D.; Lu, W.; Ogasawara, M.A.; Del Valle, N.R.; Huang, P. Redox regulation of cell survival. *Antioxid. Redox Signal.* **2008**, *10*, 1343–1374. [CrossRef]

77. Arulselvan, P.; Subramanian, S.P. Beneficial effects of *Murraya koenigii* leaves on antioxidant defense system and ultra structural changes of pancreatic β-cells in experimental diabetes in rats. *Chem. Biol. Interact.* **2007**, *165*, 155–164. [CrossRef]

78. Khan, B.A.; Abraham, A.; Leelamma, S. Anti-oxidant effects of Curry leaf, *Murraya koenigii* and mustard seeds, Brassica juncea in rats fed with high fat diet. *Indian J. Exp. Biol.* **1997**, *35*, 148–150.

79. Mitra, E.; Ghosh, A.K.; Ghosh, D.; Mukherjee, D.; Chattopadhyay, A.; Dutta, S.; Pattari, S.K.; Bandyopadhyay, D. Protective effect of aqueous Curry leaf (*Murraya koenigii*) extract against cadmium-induced oxidative stress in rat heart. *Food Chem. Toxicol.* **2012**, *50*, 1340–1353. [CrossRef]

80. Harman, D. The Biologic Clock: The Mitochondria? *J. Am. Geriatr. Soc.* **1972**, *20*, 145–147. [CrossRef]

81. Beal, M.F. Therapeutic approaches to mitochondrial dysfunction in Parkinson's disease. *Park. Relat. Disord.* **2009**, *3*, S189–S194. [CrossRef]

82. Kushnareva, Y.; Murphy, A.N.; Andreyev, A. Complex I-mediated reactive oxygen species generation: Modulation by cytochrome c and NAD(P)+ oxidation-reduction state. *Biochem. J.* **2002**, *368*, 545–553. [CrossRef] [PubMed]

83. Ischiropoulos, H.; Beckman, J.S. Oxidative stress and nitration in neurodegeneration: Cause, effect, or association? *J. Clin. Investig.* **2003**, *111*, 163–169. [CrossRef] [PubMed]

84. Anderson, S.; Bankier, A.T.; Barrell, B.G.; De Bruijn, M.H.L.; Coulson, A.R.; Drouin, J.; Eperon, I.C.; Nierlich, D.P.; Roe, B.A.; Sanger, F.; et al. Sequence and organization of the human mitochondrial genome. *Nature* **1981**, *290*, 457–465. [CrossRef]

85. Iman, V.; Mohan, S.; Abdelwahab, S.I.; Karimian, H.; Nordin, N.; Fadaeinasab, M.; Noordin, M.I.; Noor, S.M. Anticancer and anti-inflammatory activities of girinimbine isolated from *Murraya koenigii*. *Drug Des. Dev. Ther.* **2017**, *2017*, 103–121. [CrossRef]

86. Arun, A.; Patel, O.P.S.; Saini, D.; Yadav, P.P.; Konwar, R. Anti-colon cancer activity of *Murraya koenigii* leaves is due to constituent murrayazoline and O-methylmurrayamine A induced mTOR/AKT downregulation and mitochondrial apoptosis. *Biomed. Pharmacother.* **2017**, *93*, 510–521. [CrossRef]

87. Balakrishnan, R.; Elangovan, N.; Mohankumar, T.; Nataraj, J.; Manivasagam, T.; Thenmozhi, A.J.; Essa, M.M.; Akbar, M.; Sattar Khan, M.A. Isolongifolene attenuates rotenone-induced mitochondrial dysfunction, oxidative stress and apoptosis. *Front. Biosci. (Sch. Ed.)* **2018**, *10*, 248–261.

88. Bashkatova, V.; Alam, M.; Vanin, A.; Schmidt, W.J. Chronic administration of rotenone increases levels of nitric oxide and lipid peroxidation products in rat brain. *Exp. Neurol.* **2004**, *186*, 235–241. [CrossRef]

89. Gupta, S.; George, M.; Singhal, M.; Sharma, G.; Garg, V. Leaves extract of *Murraya koenigii* linn for anti-inflammatory and analgesic activity in animal models. *J. Adv. Pharm. Technol. Res.* **2010**, *1*, 68–77.

90. Darvekar, V.M.; Patil, V.R.; Choudhari, A.B.; Road, D. Anti-inflammatory Activity of *Murraya koenigii* Spreng on Experimental Animals. *J. Nat. Prod. Plant Resour.* **2011**, *1*, 65–69.

91. Mani, V.; Ramasamy, K.; Abdul Majeed, A.B. Anti-inflammatory, analgesic and anti-ulcerogenic effect of total alkaloidal extract from *Murraya koenigii* leaves in animal models. *Food Funct.* **2013**, *4*, 557–567. [CrossRef] [PubMed]

92. Khurana, A.; Sikha, M.S.; Ramesh, K.; Venkatesh, P.; Godugu, C. Modulation of cerulein-induced pancreatic inflammation by hydroalcoholic extract of curry leaf (*Murraya koenigii*). *Phyther. Res.* **2019**, *33*, 1510–1525. [CrossRef] [PubMed]

93. Oben, K.Z.; Alhakeem, S.S.; McKenna, M.K.; Brandon, J.A.; Mani, R.; Noothi, S.K.; Jinpeng, L.; Akunuru, S.; Dhar, S.K.; Singh, I.P.; et al. Oxidative stress-induced JNK/AP-1 signaling is a major pathway involved in selective apoptosis of myelodysplastic syndrome cells by *Withaferin-A*. *Oncotarget* **2017**, *8*, 77436–77452. [CrossRef] [PubMed]

94. Chen, M.; Yin, X.; Lu, C.; Chen, X.; Ba, H.; Cai, J.; Sun, J. Mahanine induces apoptosis, cell cycle arrest, inhibition of cell migration, invasion and PI3K/AKT/mTOR signalling pathway in glioma cells and inhibits tumor growth in vivo. *Chem. Biol. Interact.* **2019**, *299*, 1–7. [CrossRef]

95. Yu, Y.; Fu, X.; Ran, Q.; Yang, K.; Wen, Y.; Li, H.; Wang, F. Globularifolin exerts anticancer effects on glioma U87 cells through inhibition of Akt/mTOR and MEK/ERK signaling pathways in vitro and inhibits tumor growth in vivo. *Biochimie* **2017**, *141*, 144–151. [CrossRef]

96. Utaipan, T.; Athipornchai, A.; Suksamrarn, A.; Jirachotikoon, C.; Yuan, X.; Lertcanawanichakul, M.; Chunglok, W. Carbazole alkaloids from *Murraya koenigii* trigger apoptosis and autophagic flux inhibition in human oral squamous cell carcinoma cells. *J. Nat. Med.* **2017**, *71*, 158–169. [CrossRef]

97. Syam, S.; Abdul, A.B.; Sukari, M.A.; Mohan, S.; Abdelwahab, S.I.; Wah, T.S. The growth suppressing effects of girinimbine on hepg2 involve induction of apoptosis and cell cycle arrest. *Molecules* **2011**, *16*, 7155–7170. [CrossRef]

98. Xin, Q.; Muer, A. Girinimbine inhibits the proliferation of human ovarian cancer cells in vitro via the phosphatidylinositol-3-kinase (PI3K)/akt and the mammalian target of rapamycin (mTOR) and Wnt/β-catenin signaling pathways. *Med. Sci. Monit.* **2018**, *24*, 5480–5487. [CrossRef]

99. Ahmadipour, F.; Noordin, M.I.; Mohan, S.; Arya, A.; Paydar, M.; Looi, C.Y.; Keong, Y.S.; Siyamak, E.N.; Fani, S.; Firoozi, M.; et al. Koenimbin, a natural dietary compound of *Murraya koenigii* (L) spreng: Inhibition of MCF7 breast cancer cells and targeting of derived MCF7 breast cancer stem cells (CD44+/CD24-/low): An in vitro study. *Drug Des. Dev. Ther.* **2015**, *9*, 1193–1208.

100. Ito, C.; Itoigawa, M.; Nakao, K.; Murata, T.; Tsuboi, M.; Kaneda, N.; Furukawa, H. Induction of apoptosis by carbazole alkaloids isolated from *Murraya koenigii*. *Phytomedicine* **2006**, *13*, 359–365. [CrossRef]

101. Kumar, N.S.; Mukherjee, P.K.; Bhadra, S.; Saha, B.P.; Pal, B.C. Acetylcholinesterase inhibitory potential of a carbazole alkaloid, mahanimbine, from *Murraya koenigii*. *Phyther. Res.* **2010**, *24*, 629–631.

102. Deb, S. Synthesis of silver nano particles using *Murraya koenigii* (Green Curry leaves), Zea mays (Baby corn) and its antimicrobial activity against pathogens. *Int. J. PharmTech Res.* **2014**, *6*, 91–96.

103. Qais, F.A.; Shafiq, A.; Khan, H.M.; Husain, F.M.; Khan, R.A.; Alenazi, B.; Alsalme, A.; Ahmad, I. Antibacterial Effect of Silver Nanoparticles Synthesized Using *Murraya koenigii* (L.) against Multidrug-Resistant Pathogens. *Bioinorg. Chem. Appl.* **2019**, *2019*. [CrossRef] [PubMed]

104. Sankar Ganesh, P.; Rai Vittal, R. In vitro antibiofilm activity of *Murraya koenigii* essential oil extracted using supercritical fluid CO2 method against Pseudomonas aeruginosa PAO1. *Nat. Prod. Res.* **2015**, *29*, 2295–2298. [CrossRef]

105. Rath, S.; Padhy, R.N. Monitoring in vitro antibacterial efficacy of 26 Indian spices against multidrug resistant urinary tract infecting bacteria. *Integr. Med. Res.* **2014**, *13*, 133–141. [CrossRef]

106. Naik, S.K.; Mohanty, S.; Padhi, A.; Pati, R.; Sonawane, A. Evaluation of antibacterial and cytotoxic activity of Artemisia nilagirica and *Murraya koenigii* leaf extracts against mycobacteria and macrophages. *BMC Complement. Altern. Med.* **2014**. [CrossRef]

107. Manfo, F.P.T.; Nantia, E.A.; Kuete, V. Hepatotoxicity and Hepatoprotective Effects of African Medicinal Plants. In *Toxicological Survey of African Medicinal Plants*; Elsevier: Yaoundé, Cameroon, 2014; pp. 223–256.

108. Shah, P.; Singh, S.P.; Kumar, A. Combined effect of hydroethanolic extracts of *Murraya koenigii* and *Phyllanthus niruri* leaves on paracetamol and ethanol-induced toxicity in HepG2 cell line. *Curr. Sci.* **2015**, *109*, 1320. [CrossRef]

109. Kaufmann, T.; Simon, H.U. Targeting disease by immunomodulation. *Cell Death Differ.* **2015**, *22*, 185–186. [CrossRef]

110. Shah, A.S.; Wakade, A.S.; Juvekar, A.R. Immunomodulatory activity of methanolic extract of *Murraya koenigii* (L) Spreng. leaves. *Indian J. Exp. Biol.* **2008**, *46*, 505–509.

111. Paul, S.; Bandyopadhyay, T.K.; Bhattacharyya, A. Immunomodulatory effect of leaf extract of *Murraya koenigii* in diabetic mice. *Immunopharmacol. Immunotoxicol.* **2011**, *33*, 691–699. [CrossRef]

112. Punuru, P.; Sujatha, D.; Kumari, B.P.; Charisma, V.V.L. Evaluation of aqueous extract of *Murraya koenigii* in unilateral renal ischemia reperfusion injury in rats. *Indian J. Pharmacol.* **2014**, *46*, 171–175. [CrossRef] [PubMed]

113. Gul, M.Z.; Attuluri, V.; Qureshi, I.A.; Ghazi, I.A. Antioxidant and α-glucosidase inhibitory activities of *Murraya koenigii* leaf extracts. *Pharmacogn. J.* **2012**, *4*, 65–72. [CrossRef]

114. Sarkar, S.; Dutta, D.; Samanta, S.K.; Bhattacharya, K.; Pal, B.C.; Li, J.; Datta, K.; Mandal, C.; Mandal, C. Oxidative inhibition of Hsp90 disrupts the super-chaperone complex and attenuates pancreatic adenocarcinoma in vitro and in vivo. *Int. J. Cancer* **2013**, *132*, 695–706. [CrossRef] [PubMed]

115. Pei, C.; He, Q.; Liang, S.; Gong, X. Mahanimbine Exerts Anticancer Effects on Human Pancreatic Cancer Cells by Triggering Cell Cycle Arrest, Apoptosis, and Modulation of AKT/Mammalian Target of Rapamycin (mTOR) and Signal Transducer and Activator of Transcription 3 (STAT3) Signalling Pathway. *Med. Sci. Monit.* **2018**, *1*, 6975–6983. [CrossRef] [PubMed]

116. Noolu, B.; Ajumeera, R.; Chauhan, A.; Nagalla, B.; Manchala, R.; Ismail, A. *Murraya koenigii* leaf extract inhibits proteasome activity and induces cell death in breast cancer cells. *BMC Complement. Altern. Med.* **2013**, *13*, 7. [CrossRef]

117. Ismail, A.; Noolu, B.; Gogulothu, R.; Perugu, S.; Rajanna, A.; Babu, S.K. Cytotoxicity and Proteasome Inhibition by Alkaloid Extract from *Murraya koenigii* Leaves in Breast Cancer Cells-Molecular Docking Studies. *J. Med. Food* **2016**, *19*, 1155–1165. [CrossRef]

118. Mohan, S.; Abdelwahab, S.I.; Cheah, S.C.; Sukari, M.A.; Syam, S.; Shamsuddin, N.; Rais Mustafa, M. Apoptosis effect of girinimbine isolated from *Murraya koenigii* on lung cancer cells in vitro. *Evid.-Based Complement. Altern. Med.* **2013**, *2013*, 689865. [CrossRef]

119. Vasudevan, M.; Parle, M. Antiamnesic potential of *Murraya koenigii* leaves. *Phyther. Res.* **2009**, *23*, 308–316. [CrossRef]
120. Mittal, J. Curry Leaf (*Murraya koenigii*): A Spice with Medicinal Property. *MOJ Biol. Med.* **2017**, *2*, 236–256. [CrossRef]
121. Iyer, D.; Uma, D.P. Effect of *Murraya koenigii* (L.) on radiation induced rate of lipid peroxidation in Swiss albino mice. *Indian Drugs* **2009**, *46*, 160–162.
122. Nagappan, T.; Segaran, T.C.; Wahid, M.E.A.; Ramasamy, P.; Vairappan, C.S. Efficacy of carbazole alkaloids, essential oil and extract of *Murraya koenigii* in enhancing subcutaneous wound healing in rats. *Molecules* **2012**, *17*, 14449–14463. [CrossRef] [PubMed]

 *antioxidants*

Article

The Protective Effect of Alpha-Mangostin against Cisplatin-Induced Cell Death in LLC-PK1 Cells is Associated to Mitochondrial Function Preservation

Laura María Reyes-Fermín [1], Sabino Hazael Avila-Rojas [1], Omar Emiliano Aparicio-Trejo [1], Edilia Tapia [2], Isabel Rivero [3] and José Pedraza-Chaverri [1,*]

[1] Department of Biology, Faculty of Chemistry, National Autonomous University of Mexico (UNAM), Mexico City 04510, Mexico; laura_refe@yahoo.com.mx (L.M.R.-F.); shaggy_goose@hotmail.com (S.H.A.-R.); emilianoaparicio91@gmail.com (O.E.A.-T.)

[2] Department of Nephrology and Laboratory of Renal Pathophysiology, National Institute of Cardiology "Ignacio Chávez", Mexico City 14080, Mexico; ediliatapia@hotmail.com

[3] Department of Pharmacy, Faculty of Chemistry, UNAM, Mexico City 04510, Mexico; riveroic@yahoo.com.mx

* Correspondence: pedraza@unam.mx; Tel./Fax: +52-55-5622-3878

Received: 12 April 2019; Accepted: 13 May 2019; Published: 15 May 2019

Abstract: Cis-dichlorodiammineplatinum II (CDDP) is a chemotherapeutic agent that induces nephrotoxicity by different mechanisms, including oxidative stress, mitochondrial dysfunction, autophagy, and endoplasmic reticulum stress. This study aimed to evaluate if the protective effects of the antioxidant alpha-mangostin (αM) in CDDP-induced damage in proximal tubule Lilly laboratory culture porcine kidney (LLC-PK1) cells, are related to mitochondrial function preservation. It was found that αM co-incubation prevented CDDP-induced cell death. Furthermore, αM prevented the CDDP-induced decrease in cell respiratory states, in the maximum capacity of the electron transfer system (E) and in the respiration associated to oxidative phosphorylation (OXPHOS). CDDP also decreased the protein levels of voltage dependence anion channel (VDAC) and mitochondrial complex subunits, which together with the reduction in E, the mitofusin 2 decrease and the mitochondrial network fragmentation observed by MitoTracker Green, suggest the mitochondrial morphology alteration and the decrease in mitochondrial mass induced by CDDP. CDDP also induced the reduction in mitochondrial biogenesis observed by transcription factor A, mitochondria (TFAM) decreased protein-level and the increase in mitophagy. All these changes were prevented by αM. Taken together, our results imply that αM's protective effects in CDDP-induced toxicity in LLC-PK1 cells are associated to mitochondrial function preservation.

Keywords: alpha-mangostin; cisplatin; nephrotoxicity; mitochondria function; mitophagy

---

1. Introduction

Cis-dichlorodiammineplatinum II (CDDP) is a platinum compound frequently used in many types of cancers [1,2]. However, the main side-effect of CDDP is its nephrotoxicity, especially in proximal tubule epithelial cells. It is estimated that 20 to 30% of patients treated with CDDP developed transient episodes of acute kidney injury, which can progress to chronic kidney disease, depending on the dose and individual pharmacokinetics [2]. The CDDP-toxicity mechanisms involve DNA damage, oxidative stress, mitochondrial damage, endoplasmic reticulum (ER) stress, autophagy, and apoptotic cell death [3,4].

In recent years, the use of biomolecules with antioxidant activity has been widely studied to mitigate the nephrotoxic effects induced by CDDP [5]. Alpha-mangostin (αM), a bioactive compound with direct antioxidant properties that can be extracted from the *Garcinia mangostana* tree, has been broadly used in

Asian traditional medicine and its nephroprotective effect has been demonstrated experimentally [6,7]. αM is a prenylated xanthone with antioxidant, anti-inflammatory, and anti-apoptotic properties [8,9]. Additionally, our research group demonstrated αM's protective effects in CDDP-induced nephrotoxicity, related to the prevention of oxidant and nitrate stress increase, glutathione decrease, tumor necrosis factor alpha (TNF-α) and p53 increase, and apoptosis induction [3,7,9]. Recently, the modulation of mitochondrial function, autophagy, and ER stress have been related to αM's protective effects in cancer and diabetic nephropathy models [10–12]. However, the participation of these processes in αM's protection in CDDP-induced nephrotoxicity has not been studied yet.

Mitochondria are double membrane organelles, which regulate important functions related to energetic homeostasis and cell death. Additionally, mitochondria are considered as one of the main reactive oxygen species (ROS) producers in the cell [13]. It is a widely accepted concept that mitochondrial damage is one of the principal mechanisms involved in CDDP-induced nephrotoxicity. CDDP induces mitochondrial membrane potential loss, as well as alterations in bioenergetics, dynamics (balance between fusion and fission), biogenesis and mitophagy, which favor the induction of apoptosis [14–17]. Additionally, the higher dependence of mitochondrial adenosine triphosphate (ATP) production in the proximal tubule compared with other nephron segments, make the proximal tubule more susceptible to CDDP-damage [2,13].

On the other hand, autophagy is a multistep pathway that degrades and recycles damaged macromolecules and organelles, to maintain intracellular homeostasis. This process involves the sequestration of damaged components inside a double membrane vesicle (autophagosome) and their subsequent degradation when the autophagosome fuses with the lysosome (autolysosome) [18]. Although autophagosome and autolysosome formation involves multiple protein complexes and multiple steps, the increase of the lipidated microtubule-associated protein 1 light chain 3 alpha (LC3) form (commonly refer as LC3-II) and the decrease of p62 levels have been widely used to evaluate the induction of autophagy [19]. In the CDDP nephrotoxicity model, it has been suggested that autophagy induction acts as a protective mechanism early on [20–22]. Recent studies also show that mitophagy (a mitochondrial-specific type of autophagy) has a protective role in CDDP nephrotoxicity [22,23]. Under mitochondrial damage or depolarization, the induction of mitophagy helps to maintain the mitochondrial quality control and, therefore, the cellular homeostasis [19]. In CDDP-induced nephrotoxicity, the mitophagy clearance of damaged mitochondria mediated by the phosphatase and tensin homologue (PTEN) induced kinase 1/parkin RBR E3 ubiquitin protein ligase (PINK1/Parkin) pathway has shown protective effects [22,24].

This study aimed to evaluate if the protective effects of αM in CCDP-induced damage in Lilly laboratory culture porcine kidney (LLC-PK1) cells, was related to αM regulation of mitochondrial function and autophagy (especially mitophagy).

2. Materials and Methods

2.1. Reagents

Alpha-mangostin (αM) was obtained as described in our previous studies [25] from Garcinia mangostana (DNP International Inc., Whittier, CA, USA) with 98% purity. The LLC-PK1 (ATTC® CL-101) epithelial cell line, derived from proximal tubular cells of the porcine kidney, was acquired from American Type Culture Collection (ATCC, Rockville, MD, USA). Cis-dichlorodiammineplatinum II (CDDP), trypan blue, fluorescein diacetate (FDA), tris(hydroxymethyl)aminomethane, sodium deoxycholate, 4-nonylphenyl-polyethylene glycol (NP-40), sodium fluoride (NaF), D-(+)-glucose, sodium pyrophosphate ($Na_4P_2O_7$), sodium orthovanadate (Na_3VO_4), Triton X-100, glycerophosphate, 4-(2-hydroxyethyl)piperazine-1-ethanesulfonic acid (HEPES), Tween 20, glycerol, ethylenediaminetetraacetic acid (EDTA), dimethyl sulfoxide (DMSO), antimycin A, ethylene glycol-bis(2-aminoethylether)-N,N,N′,N′-tetraacetic acid (EGTA), dinitrophenol (DNP), magnesium chloride ($MgCl_2$) tetrahydrate, oligomycin, rotenone, sodium L-ascorbate, sodium azide (NaN3) and tetramethyl-p-phenylene diamine (TMPD), sodium phosphate monobasic (NaH_2PO_4), sodium phosphate dibasic (Na_2HPO_4), sodium dodecyl sulfate (SDS), ponceau S, chloroquine diphosphate salt (CQ), wortmannin and antibodies against microtubule-associated protein light chain 3B (LC3, L7543), ubiquitin-binding protein p62 (p62, P0067) and α-tubulin (α-Tub, T9026) were purchased from Sigma Aldrich (St. Louis, MO, USA). Dulbecco's Modified Eagle's Medium (DMEM), fetal bovine serum (FBS), and penicillin/streptomycin were purchased from Biowest (Riverside, MO, USA). Calcium chloride ($CaCl_2$), sodium bicarbonate ($NaHCO_3$), sodium chloride (NaCl), and hydrochloric acid (HCl) were purchased from J.T. Baker (Xalostoc, Edo. Mex, Mexico). Potassium chloride (KCl) was purchased from Mallinckrodt plc. (St. Louis, MO, USA). Protease inhibitor cocktail was purchased from Roche Applied Science (Mannheim, Germany). MitoTrackerTM Green FM was purchased from Thermo Fisher Scientific. Antibodies against malondialdehyde (MDA, ab6463), peroxisome proliferator-activated receptor gamma (PPARγ) coactivator 1α (PGC-1α, ab54481), 4-hydroxynonenal (4HNE, ab5605), parkin RBR E3 ubiquitin protein ligase (Parkin, ab15954) and total oxidative phosphorylation (OXPHOS, ab110413) rodent western blot (WB) antibody cocktail were purchased from Abcam (Cambridge, MA, USA). Antibodies against nuclear respiratory factor 1 (NRF1, sc-33771), voltage dependence anion channel (VDAC, sc-8828), phosphatase and tensin homologue (PTEN) induced kinase 1 (PINK1, sc-33796), transcription factor A, mitochondria (TFAM, sc-19050) were purchased from Santa Cruz Biotechnology (Dallas, TX, USA). Antibody against mitofusin 2 (MFN2, D2D10) was purchased from Cell Signaling Technology (Danvers, MA, USA). All the other reagents were analytical grade.

2.2. Cell Culture and Cell Viability

The LLC-PK1 cells were cultured in DMEM medium, supplemented with 10% FBS and penicillin/streptomycin (100/50 U/mL) in a humidified incubator with 5% CO_2 atmosphere at 37 °C. The experiments were carried out at a density of 50,000 cell/cm^2, passaged 9 to 30 times since the acquired cryovial was opened. The cells were grown in 100 mm culture dishes and when they reached 90% confluence, were planted in the corresponding experimental plates. After 24 h of growth in 10% FBS medium, LLC-PK1 cells were exposed to 0 to 5 μM αM, 5 to 40 μM CDDP or co-incubation of αM and CDDP for 24 h in 1% FBS medium. Stock 10 mM αM was dissolved in DMSO and a posterior dilution of 1:100 in phosphate buffer saline (PBS) (100 μM) was used to prepare each experimental concentration. A stock 0.5 mg/mL (1.670 mM) of CDDP in 0.9% saline solution was used for different CDDP concentrations. Additional treatment consisted of pre-treatment with 30 μM CQ (2 h) or 10 nM wortmannin (1 h previous and during different treatment: αM, CDDP or αM+CDDP) as autophagy inhibitors. After pre-treatment, αM and/or CDDP were added for 24 h. One bright field image was captured in triplicate with a 20× objective after treatments, using a Cytation 5 Cell Imaging Multi-Mode reader (BioTek Instruments, Winooski, VT, USA).

Cell viability was measured by the FDA assay according to a previously described method [26] in 96-well tissue culture microplates, where fluorescence is directly proportional to cell viability.

2.3. Cell Respirometry

The oxygen consumption experiments in intact cells were performed using a high-resolution respirometry equipment O2k meter (Oroboros Instruments, Innsbruck, Austria) according to a method previously described [27]. Cells were pretreated with the corresponding schemes of αM, CDDP or co-incubation for 24 h. Subsequently, cells were washed with PBS, harvested with trypsin, and quantified by the trypan blue assay. Determinations were made using 2 mL of culture medium with 10% FBS. Each experiment was initiated by the addition of approximately 2.5 million cells, the respiratory parameters were defined as: 1. Routine respiration, corresponding to oxygen consumption with presence only of cells. 2. Leak, corresponding to cellular oxygen consumption in the presence of 5 μM oligomycin. 3. Maximum capacity of electron transfer system (E) was achieved by titrations with 5 μM DNP. 4. Respiratory control (RC) corresponding to the ratio basal/leak. 5. Respiration associated with oxidative phosphorylation (P) was calculated by the formula Routine-Leak. All parameters were corrected by subtracting the non-mitochondrial respiration, obtained by the addition of 1 μM rotenone plus 5 μM antimycin A, and normalized by the number of cells.

Complex I (CI)-linked respiration was measured in a different experiment by the rest of routine respiration minus the respiration after inhibition with 1 μM rotenone. The activity of complex IV (CIV) was evaluated in intact cells in culture medium with 10% FBS supplemented with 0.5 μM rotenone plus 2.5 μM antimycin A. Oxygen consumption was stimulated by the addition of 0.5 mM TMPD plus 2 mM ascorbate. CIV activity was corrected by oxygen consumption in the presence of the appropriate inhibitor (100 mM NaN_3). The results were normalized by the number of cells.

2.4. Total Protein Extract and Western Blot

LLC-PK1 cells were grown on 60 mm plates, after treatment were washed twice with cold PBS and lysed by addition of 400 μL of cold radioimmunoprecipitation assay buffer (RIPA) buffer (50 mM Tris-HCl, pH 7.4, 150 mM NaCl, 1 mM EDTA, 0.5% sodium deoxycholate, 1% NP-40, 0.1% SDS) with protease and phosphatase inhibitors (25 mM NaF, 1 mM Na_3VO_4, 1 mM PMSF, 1 mM $Na_4P_2O_7$, 0.5 mM glycerophosphate and 1X protease inhibitors cocktail) for 30 min at 4 °C with stirring. Lysates were centrifuged 15,000 × *g* during 10 min at 4 °C and the supernatant was collected and stored at −70 °C until the experiment was carried out. The protein concentration was quantified by the Lowry assay [28]. Equal amounts of protein (15 μg) were separated on 10% to 15% in sodium dodecyl sulfate–polyacrylamide gel electrophoresis (SDS-PAGE) as appropriate. Proteins were transferred onto Immobilion PVDF membranes for fluorescent application on wet transfer and blocked with 5% non-fat dry milk for 1 h. Then incubated overnight at 4 °C with the appropriate primary antibody: MDA (1:2000), 4HNE (1:3000), PGC-1α (1:100), NRF1 (1:200), TFAM (1:500), VDAC (1:500) OXPHOS cocktail (1:1500), MNF2 (1:1000), PINK1 (1:1000), Parkin (1:1000), LC3 (1:1500), p62 (1:1000) and α-Tub (1:8000). Then the membranes were incubated with the corresponding fluorescent secondary antibodies (1:10000) for 1 h at room temperature in darkness. Protein bands were detected in an Odyssey scanner (Li-COR Biosciences, Lincoln, NE, USA) and protein band intensity was analyzed by Image StudioTM Life Software Li-COR Odyssey, which led us to normalize to correct for loading variations between bands and give us signal counts measured in a single pixel per unit time through Z-Factor Calculation. In all cases, except OXPHOS subunits, α-TUB served as loading control. Thus, once the protein target was detected, we proceeded to detect its loading control. For that, the membranes were rinsed only with Tris-buffered saline, 0.1% Tween 20 (TBS-T 1x) 3 times for 5 min and then the membranes were incubated with primary α-TUB antibody, following the usual steps as for protein target. The OXPHOS subunits were referred to reversible Ponceau staining as loading control as described by Romero-Calvo et al. [29]. Briefly, after transfer, the membranes were incubated in 1% Ponceau S solution for 2 min. Immediately after, membranes were rinsed with PBS to remove staining saturation, clearly visible bands. Membranes were placed between transparent sheets and scanned in conventional equipment at 300 dpi as TIFF document (HP Scanjet G4050). Again, membranes were rinsed until the staining completely disappeared. Finally, membranes were blocked with 5% non-fatty milk in TBS buffer, and the usual steps were continued.

2.5. Mitochondrial Mass by MitoTracker Green

The mitochondrial mass was indirectly evaluated by a fluorometric assay using MitoTracker green (MTG). After treatment, cells were washed with DMEM w/o FBS and then incubated with 0.5 μM MTG for 20 min at 37 °C. After that, the cells were washed two times with mammalian Ringer Krebs Medium. Micrographs were taken with a 20× objective size with numerical aperture 0.45, and zoomed in with Cytation 5 imaging, GFP filter (Ex/Em, 469 nm/525 nm).

2.6. Statistical Analysis

Data are presented as the mean±standard deviation (SD). Data were analyzed by one-way ANOVA followed by the corresponding Dunnett or Tukey post-test. Comparison between time course and protein level as independent variables of all treatment were performed using two-way ANOVA followed by Bonferroni post-test using the software Graph-Pad Prism 5 (San Diego, CA, USA). A p-value lower than 0.05 was considered to deduce representative changes of the behavior studied with respect to each treatment.

3. Results

3.1. αM Prevented CDDP-Induced Cell Death

LLC-PK1 cells were incubated with various CDDP concentrations: 5, 10, 20, 30, and 40 µM (Figure 1A) and it was observed that at 30 µM CDDP, reduced approximately 50% of cell viability, as we previously reported [26]. Later, cells were incubated with αM concentrations between 1 and 5 µM. As we expected, αM was not toxic up to 5 µM (Figure 1B). To evaluate protection by αM, αM and 30 µM CDDP were co-incubated. We observed that αM protected the cells against CDDP-induced cytotoxicity in a concentration-dependent manner and the protection with the least concentration was found at 4 µM (Figure 1C). So, the subsequent experiments were carried out with 4 µM αM and 30 µM CDDP.

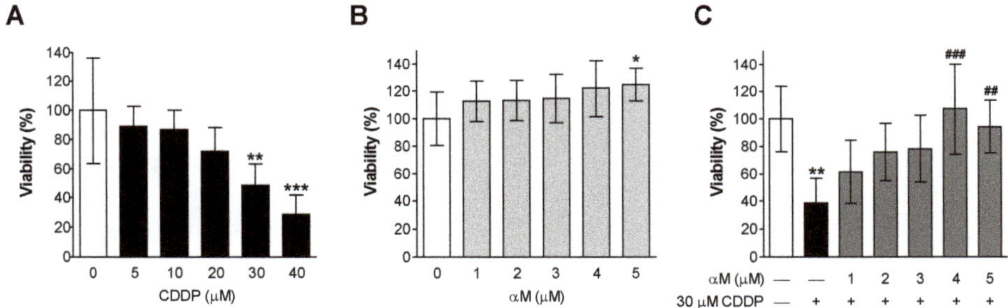

Figure 1. Protective effect of alpha mangostin (αM) against cis-dichlorodiammineplatinum II (CDDP) induced cell death. Lilly laboratory culture porcine kidney (LLC-PK1) cells were treated with (**A**) 0–40 µM cisplatin, (**B**) 0–5 µM αM or (**C**) co incubated with 30 µM CDDP and increasing concentrations of αM for 24 h. After treatment, cell viability was measured by the fluorescein diacetate (FDA) assay. The data are presented as mean ± SD, n = 5–7. *** p < 0.001, ** p < 0.01 and * p < 0.05 vs. control (0). ### p < 0.001 and ## p < 0.01 vs. CDDP.

CDDP's toxicity is partially associated with oxidative stress. To evaluate the effect of CDDP-induced oxidative stress, we evaluated the lipid peroxidation markers 4HNE and MDA by Western blot (WB). As we show in Figure 2, CDDP increased the level of MDA (Figure 2A,B) but not those of 4HNE (Figure 2C,D). αM was unable to prevent CDDP-induced MDA increase.

3.2. αM's Protection is Associated to the Preservation of Mitochondrial Bioenergetics

The main nephrotoxic mechanism of CDDP is mitochondrial damage [13,14]. Therefore, we determined if the preservation of mitochondrial bioenergetics was related to protection by αM. So, we evaluated the respiratory states in the whole cells. As shown in Figure 3, CDDP-treatment reduced the Routine respiration, the Leak of respiration, the maximum capacity of the electron transfer system (E) and the respiration associated to oxidative phosphorylation (P) at 24 h. Interestingly, due to both Routine and Leak respiration reduced, the respiratory control (RC) did not change with CDDP-treatment (Figure 3E). All respiratory state alterations were reversed by αM co-treatment. Importantly αM alone did not alter any respiratory parameters at this concentration (Figure 3).

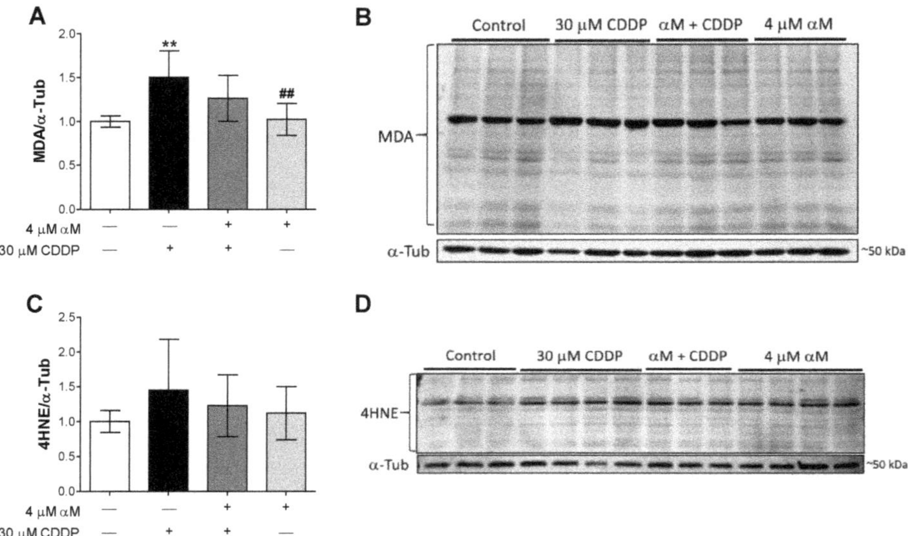

Figure 2. Changes in oxidative stress markers in LLC-PK1 cells treated with αM, CDDP or both. After 24 h of αM-, CDDP-treatment or both, the oxidative stress was evaluated by the lipoperoxidative markers (**A**) malondialdehyde (MDA) and (**C**) 4-hydroxynonenal (4HNE). (**A**) and (**C**) show the densitometry data and (**B**) and (**C**) show the representative blots. The data are presented as mean ± SD, $n = 6$–7. ** $p < 0.01$ vs. control. ## $p < 0.01$ vs. CDDP. α-Tub = alpha-tubulin.

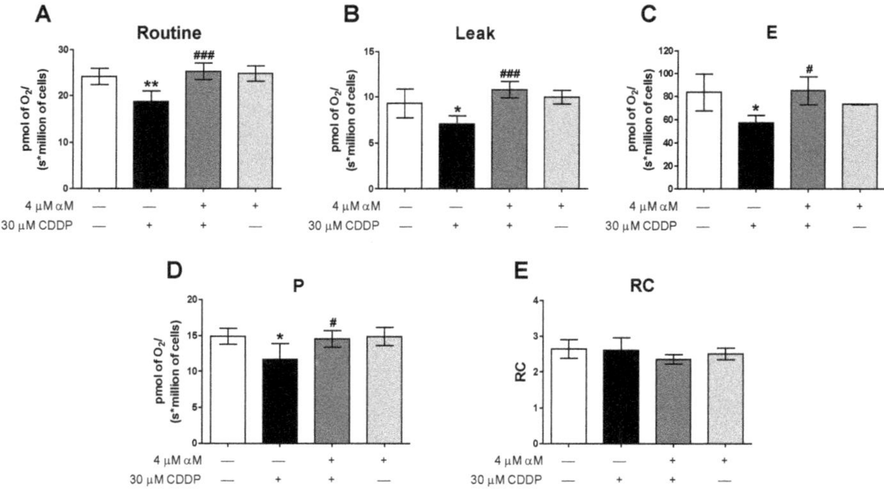

Figure 3. αM co-treatment prevents CDDP induced a decrease in respiratory parameters at 24 h. (**A**) Cellular routine respiration = Routine, (**B**) leak of respiration, (**C**) maximum capacity of electron transfer system = E, (**D**) respiration associated to oxidative phosphorylation = P, (**E**) respiratory control = RC. The data are presented as mean ± SD, $n = 3$–5. ** $p < 0.01$ and * $p < 0.05$ vs. control. ### $p < 0.001$ and # $p < 0.05$ vs. CDDP.

The decrease in all the respiratory states without changes in RC, suggested that the mitochondrial bioenergetics alterations were related to a reduction in the activity of the whole electron transport system (ETS). So, we evaluated the CI-linked respiration, the CIV activity and the percentage of routine respiration attributable to CI. The CDDP-treatment diminished CI-linked respiration but had no effect on CIV activity and the percentage of basal respiration attributable to CI was not reduced (Figure 4). These results, together with the decrease in E, imply a reduction in whole ETS activity induced by CDDP, which could be caused by a reduction in mitochondrial mass.

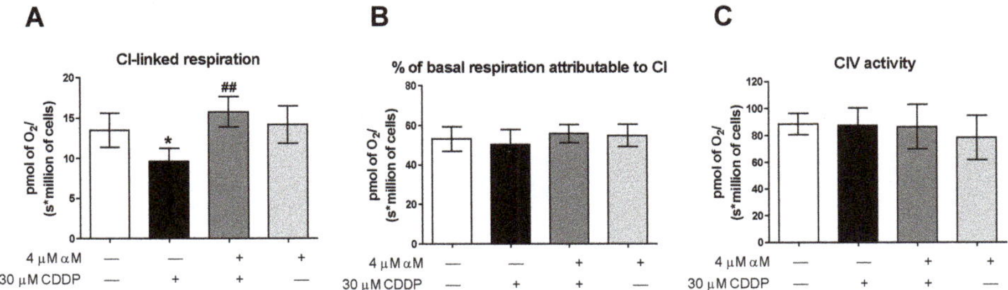

Figure 4. Determination of respiratory parameters at 24 h in LLC-PK1 cells treated with αM, CDDP or both. (**A**) Complex I (CI)-linked respiration, (**B**) percentage of respiration attributable to CI and (**C**) activity of complex IV (CIV). The data are presented as mean ± SD, n = 4–6. * $p < 0.05$ vs. control. ## $p < 0.01$ vs. CDDP.

3.3. Mitochondrial Respirometry Alterations are Associated to Mitochondrial Mass Decrease Related to Mitochondrial Biogenesis Reduction

To determine if the reduction in the respiration states was linked to a decrease in mitochondrial mass, the protein levels of VDAC and some mitochondrial complex subunits were evaluated by WB. We evaluated mitochondrial subunits using an cocktail contained antibodies to labile mitochondrial subunits. So, they allow to evaluate the changes in the mitochondrial proteins of the ETS in the membrane only. Furthermore, several models of acute and chronic kidney damage have previously reported a reduction in protein levels of these subunits as well as in levels of their messengers, which has been related to a reduction in mitochondrial biogenesis and bioenergetics in such models [30–35]. Figure 5 shows that CI-NDUFB8 (Figure 5A), CII-SDHB (Figure 5B) and CIV-MTCO1 (Figure 5C) subunits were decreased with CDDP-treatment, whereas the CV-ATP5A subunit was unchanged (Figure 5D). Moreover, we found a decrease in VDAC protein levels with CDDP-treatment (Figure 5F,G), which was partially prevented by αM-treatment. To corroborate these results, we obtained representative images of the mitochondrial mass using MTG, a selective mitochondrial fluorescent label [36] and also MFN2, fusion marker. Figure 6A shows the images of CDDP treated cells, where a fragmented mitochondrial network was observed, with loss of continuity and fluorescence focal points. In addition, we found the MNF2 level in CDDP treatment that was avoided by αM co-treatment suggests fusion diminishing (Figure 6B,C). Together, these results support the idea that CDDP increases mitochondrial mass reduction and fragmentation, while αM can partially preserve it.

To determine if the observed reduction in mitochondrial mass was related to alterations in mitochondrial biogenesis, we evaluated the levels of mitochondrial biogenesis proteins PCG-1α, NRF1, and TFAM. Interestingly NRF1 levels were upregulated in all treatments (Figure 7B), although we expected a reduction. The implications of these changes are deliberated in the discussion section. Furthermore, although we did not observe changes in PGC-1α (Figure 7A) CDDP-treatment downregulated TFAM levels (Figure 7C) and αM co-treatment prevented it. These results suggest that mitochondrial mass reduction is partially attributable to the decrease in mitochondrial biogenesis induced by CDDP.

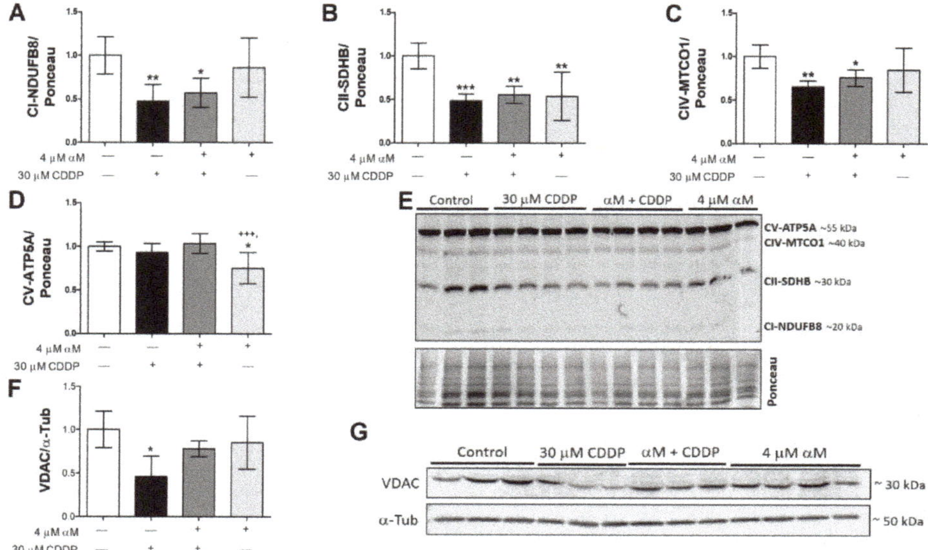

Figure 5. CDDP-induced toxicity is related to a decrease in mitochondrial proteins. LLC-PK1 cells were treated with αM, CDDP or both for 24 h. After treatment, western blotting of total oxidative phosphorylation (OXPHOS) cocktail in the whole cell (**A–D,F**) voltage-dependent anion channel (VDAC) as mitochondrial mass markers were quantified and (**E,G**) representative protein expression is shown. Densitometry values were normalized by Ponceau red staining. The data are presented as mean ± SD, n = 4–8. *** $p < 0.001$, ** $p < 0.01$ and * $p < 0.05$ vs. control. +++ $p < 0.001$ vs. (αM + CDDP). NDUFB8 = NADH: ubiquinone oxidoreductase subunit B8, SDHB = succinate dehydrogenase complex iron–sulfur subunit B, MTCO1 = mitochondrial cytochrome c oxidase I catalytic subunit, ATP5A = ATP synthase F subunit alpha, α-Tub = alpha-tubulin.

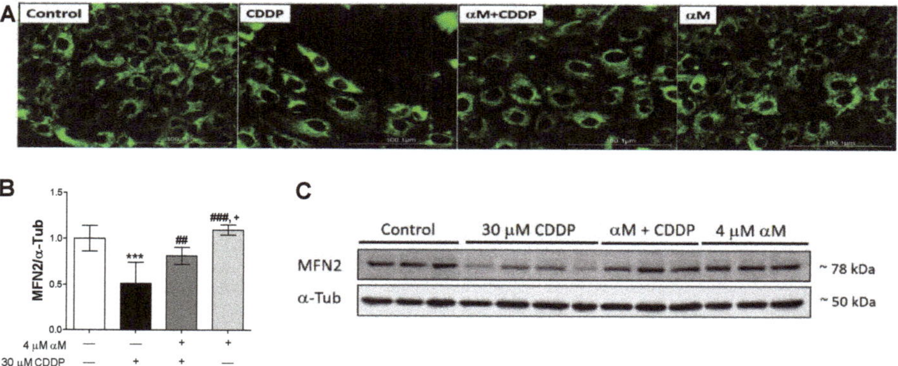

Figure 6. CDDP-induced toxicity is related to a decrease of mitochondrial fusion. LLC-PK1 cells were treated with αM, CDDP or both for 24 h. After treatment, (**A**) representative micrographs of MitoTracker green, mitochondrial mass marker, at 20x and posterior zoom 1:1 and western blotting of (**B**) Fusion marker, mitofusin 2 (MNF2) level and representative blot in (**C**). The data are presented as mean ± SD, n = 6–8. *** $p < 0.001$ vs. control. ### $p < 0.001$, ## $p < 0.01$ vs. CDDP. + $p < 0.05$ vs. (αM+CDDP). α-Tub = alpha-tubulin.

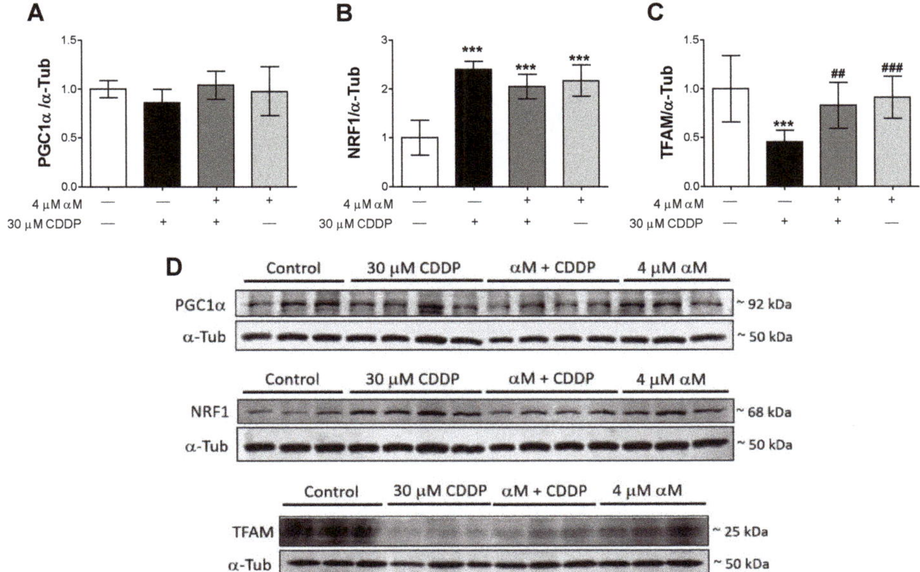

Figure 7. Modulation of mitochondrial biogenesis in LLC-PK1 cells treated with αM, CDDP or both. After 24 h of treatment, protein levels of (**A**) peroxisome proliferator-activated receptor gamma (PPARγ) coactivator 1-alpha (PGC1α), (**B**) nuclear respiratory factor 1 (NRF1), (**C**) mitochondrial transcription factor A (TFAM) were quantified. Representative blots are shown in (**D**). The data are presented as mean ± SD, n = 5–7. *** $p < 0.001$. ### $p < 0.001$ and ## $p < 0.01$ vs. CDDP. α-Tub = alpha-tubulin.

3.4. The Induction of Mitophagy is Related to Protection by αM

Dysfunctional mitochondria are usually eliminated through diverse processes, such as mitophagy [13], so we evaluated if mitophagy was related to CDDP-induced decrease in mitochondrial mass. Figure 8 shows that CDDP increased both PINK1 (Figure 8A) and Parkin (Figure 8B) protein levels and this increase was prevented by αM co-incubation. As the mitophagy process can be related to general macroautophagy, we pre-incubated LLC-PK1 cells with 30 μM CQ for 2 h before αM or CDDP-treatment, and with 10 nM of wortmannin 1 h before and during treatment, as autophagy inhibitors (Figure 9). Although autophagy inhibition has been related to the worsening of CDDP injury, Figure 9A shows that, apparently, CQ only has an effect on CDDP-treatment and does not have any effect over the protection by αM. CQ pre-treatment did not reduce cell viability in αM+CDDP-treatment. Wortmannin, as well as CQ, did not have any effect over the protection by αM (Figure 9B). Probably protection by αM has a mechanism independent of the induction of autophagy. Figure 9C shows micrographs representative of all treatments. CDDP treated cells showed less density at 24 h, as well as cell swelling and loss of structure compared to the control group. αM co-treatment prevented CDDP-alterations. In addition, in CQ and wortmannin + CDDP. we observed similar damage to the CDDP group. However, CQ or wortmannin did not alter the morphological characteristic compared with the αM+CDDP treatment.

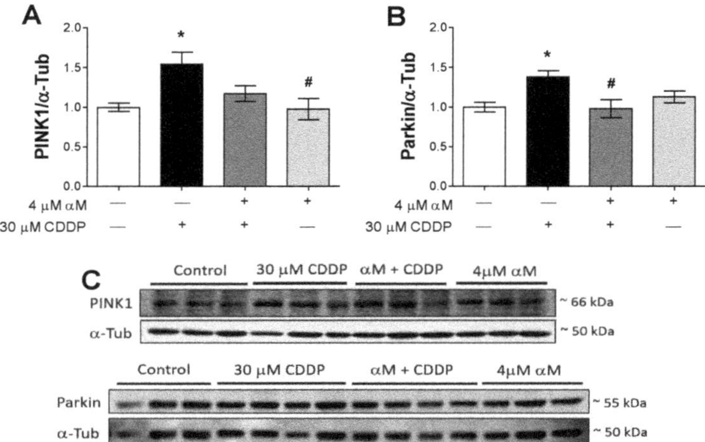

Figure 8. Modulation of mitophagy induction in LLC-PK1 cells treated with αM, CDDP or both. After 24 h of αM-, CDDP-treatment or both, mitophagy proteins (**A**) phosphatase and tensin homologue (PTEN)-induced kinase 1 (PINK1) and (**B**) parkin RBR E3 ubiquitin protein ligase (Parkin) were evaluated. Representative blots are shown in (**C**). The data are presented as mean ± SD, $n = 6$–7. * $p < 0.05$ vs. control. # $p < 0.05$ vs. CDDP. α-Tub = alpha-tubulin.

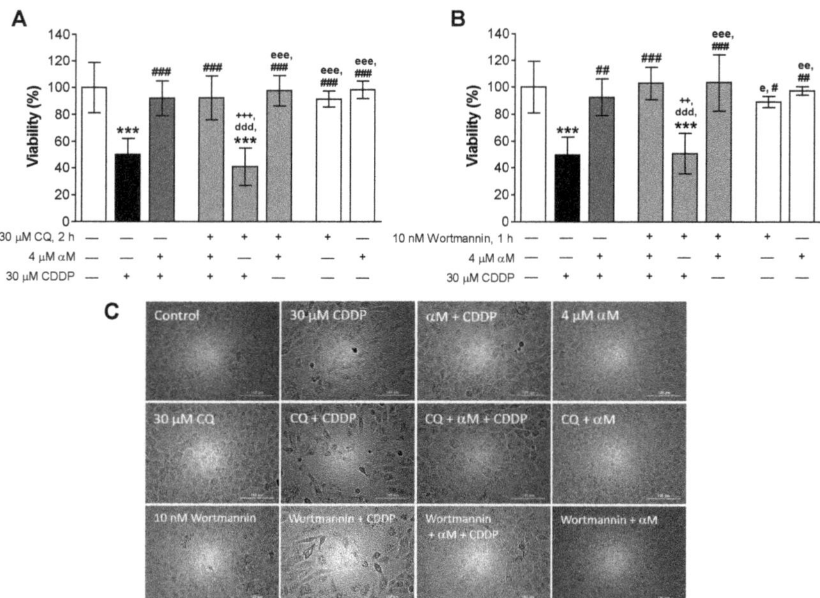

Figure 9. Effect of autophagy inhibition on the protection induced by αM vs. CDDP. LLC-PK1 cells were pretreated with (**A**) 30 μM chloroquine (CQ) for 2 h, as an inhibitor of degradation of cargo in the final autophagy phase, or (**B**) 10 nM wortmannin for 1 h and during the experiment, as an inhibitor of phosphoinositide 3-kinase (PI3K) in the initial autophagy phase induction. After pre-treatment, the cells were co-incubated with 4 μM αM and/or 30 μM CDDP for 24 h. Cell viability was determined, and (**C**) representative micrographs were taken. The data are presented as mean ± SD, $n = 4$–6. *** $p < 0.001$ vs. control. ### $p < 0.001$, ## $p < 0.01$ and # $p < 0.05$ vs. cisplatin. +++ $p < 0.001$ and ++$p < 0.01$ vs. (αM+CDDP). ddd $p < 0.001$ vs. (stressor+αM+CDDP). eee $p < 0.001$, ee $p<0.01$ and e $p < 0.05$ vs. (stressor+CDDP).

Figure 10C shows that CDDP-treatment induced an increase in p62 levels at 24 h, and a slight increase in microtubule-associated protein 1 light chain 3 alpha I (LC3-I) (Figure 10A) and the LC3-II/LC3-I ratio (Figure 10B). The increase in p62 is relevant because of the fact that p62 has been reported as one of receptors involved in PINK1/Parkin-mediated mitophagy balance [37] and aberrant p62 levels may be the result of mitochondrial dysfunction [38], which together with the autophagy results (Figure 9), suggest that mitophagy is involved in the observed mitochondrial mass decrease induced by CDDP.

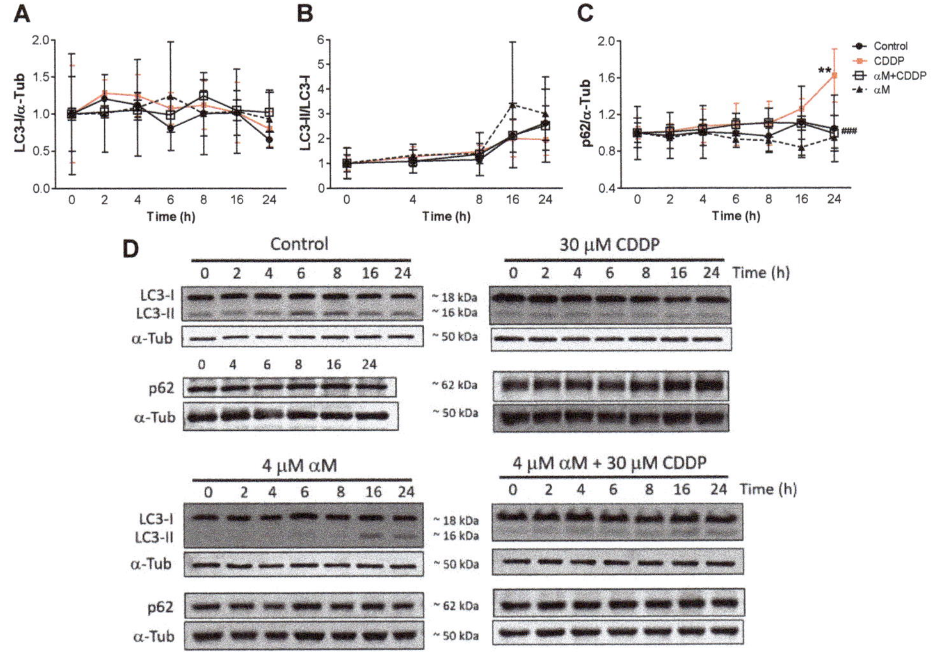

Figure 10. Time-course effect of αM and CDDP on autophagy stress markers. LLC-PK1 cells were incubated with 4 µM αM, 30 µM CDDP or both during 0, 2, 4, 6, 8, 16, and 24 h. After each treatment, (**A**) microtubule-associated protein 1 light chain 3 alpha I (LC3-I), (**B**) lipidated form of LC3 (LC3-II), and (**C**) ubiquitin-binding protein p62 (p62) protein levels were evaluated. Panels A, B, and C show quantitative data, and panel D shows representative blots. Data are presented as mean ± SD, n = 3–5. ** $p < 0.01$ vs. control and ### $p < 0.01$ vs. CDDP. α-Tub = alpha-tubulin.

4. Discussion

Cisplatin (CDDP) is highly deleterious at proximal tubule level given its high density of mitochondria because of its reabsorption function [13]. Its nephrotoxicity is attributable to high CCDP accumulation in the kidneys and adverse impacts on the renal transport system. CDDP entry into the tubular cells by passive diffusion or by a number of cellular transporters including human copper transport protein 1 (Ctr1) and the organic cation transport 2 (OCT2), which are highly expressed on renal tubular cells [39]. Although this transport is expressed in other tissue, diverse strategies have demonstrated its importance in nephrotoxicity development [2]. Inside cells, CDDP metabolism results in its activation to a more potent toxin and glutathione-cisplatin-conjugate derivate and compromises glutathione level and synthesis [40]. CDDP can bind to proteins of cytosol, mitochondria, and ER [2,40,41]. Furthermore, CDDP binding to nuclear and especially mitochondrial DNA, trigger mitochondrial dysfunction [2]. Although hydration and diuretics are used as a strategy to prevent CDDP-induced nephrotoxicity, the optimal strategy to prevent this pathology is still being sought [1].

For this reason, new strategies are being developed that include using phytochemical compounds such as αM [5].

αM has many biological functions [6,8]. There are three studies that demonstrated nephroprotection versus CDDP, without interfering with urinary secretion or antiproliferating activity [3,7,9]. Alpha-mangostin is not specific to tubular cells; however, it preserves the renal function. Pharmacokinetics studies have shown that it can be quantified in the kidney and liver [6] as well as its nephroprotective effects [3,7,9,11]. αM protection in CDDP nephrotoxicity related to its antioxidant properties has been described [3,7,9]. To study αM mechanism we used the LLC-PK1 cell line because it is very stable, undergoing little to no transformation or neoplastic change after numerous passages [42] and it shows tubular cells characteristics [43,44]. We corroborated in LLC-PK1 cells that αM co-incubation protects against CDDP-induced (30 μM) viability decrease (Figure 1C). The protection was found at 4 μM and 5 μM, however, no differences were found between 4 and 5 μM (Figure 1), this time using a much purer αM extract than the previous report [9,25]. It has been previously reported by 3-(4,5-dimethylthiazol-2-yl)-2,5-diphenyltetrazolium bromide (MTT) viability assay 7.5 μM αM reduced viability cell [10], we reproduced this experiments with 1, 2.5, 5, 7.5, and 10 μM and found that αM can reduce near 50% viability at 7.5 μM (data not show). For this reason, we decide to use the minimum αM concentration in the present study. Our group has previously demonstrated that CDDP induces oxidative stress in vitro, and αM was able to prevent the increase in ROS and the decrease in glutathione (GSH) [9]. In our study, αM was unable to prevent MDA increase (Figure 2A). We are tempted to speculate that the use of different markers (MDA versus ROS and GSH) may explain the differences in both studies.

In this study, we demonstrated that CDDP induced mitochondrial bioenergetics alterations, characterized by the reduction in the respiratory parameters: routine, leak, E, and P (Figure 3), as well as CI-linked respiration (Figure 4). The absence of change in RC (Figure 3E), suggests that the CDDP-induced respiratory state alterations were related to a decrease in whole ETS activity rather than a reduction in a specific complex. Which was congruent with the absence of a reduction in the percentage of basal respiration attributable to CI (Figure 4B) and would be responsible for the decrease in OXHPOS capacity (Figure 3C). It was also demonstrated for the first time that αM prevents the bioenergetics alterations induced by CDDP as we can see by the preservation of all evaluated mitochondrial parameters in the co-treatment group (Figures 3 and 4). Mitochondrial dysfunction has been widely recognized as an important factor in renal disease, especially in tubular damage [13]. Furthermore, in the CDDP model, the oxidative stress and the mitochondrial dysfunction favor the activation of apoptosis, which contributes to renal damage [15,17,45]. So, the protection by αM would be, at least in part, attributable to mitochondrial preservation (Figures 3 and 4), which explains the protection of tubular epithelial cells.

Furthermore, the decrease in the levels of VDAC and respiratory complex subunits (Figure 5A–F), implies that the observed mitochondrial bioenergetics alterations are attributable to mitochondrial mass reduction and may also be attributable to mitochondrial network fragmentation. This is congruent with a previous report that demonstrates a shift of mitochondrial dynamics towards fission in CDDP-induced nephropathy [14]. Although the co-treatment with αM had no effect in the evaluation of mitochondrial protein subunit levels (Figure 5A–F), it prevented the alteration in the morphology of the mitochondrial network (Figure 5G), which is congruent with the preservation of mitochondrial dynamics, demonstrated by other antioxidants, such as curcumin in CDDP-induced damage nephropathy [14].

Our results also suggest that the observed reduction in mitochondrial mass is attributable to decreased mitochondrial biogenesis. We found a decrease in TFAM levels in CDDP-treated cells (Figure 6C,D), but PGC1α levels did not change (Figure 6A), suggesting a reduction in mitochondrial biogenesis. We suggest this, first, because although PGC-1α expression did not change, its activity also has to be considered. PGC-1α is regulated via phosphorylation by protein kinase B (AKT) and adenosine monophosphate-activated protein kinase (AMPK), as well as deacetylation by sirtuins [46].

In fact, it has been reported by Ortega-Dominguez [14] that CDDP decreased sirtuin-3 the levels, which triggers the increase in lysine acetylated levels. Therefore, this evidence supports the idea that PGC-1α is inactive in the cisplatin group. Second, as with PGC-1α, NRF1 expression does not necessarily correlate with its activity. NRF1 activity can be regulated by phosphorylation and/or by interactions with PGC-1α, PGC-1β, and PRC [14,46]. In this sense, TFAM levels determination is an indirect proof of its activity. About it, CDDP exacerbated NRF1 levels, whereas TFAM decreased. So, it can be suggested that CDDP-induced NRF1 overexpression is not functional, unlike the co-incubation with αM. Taken together, we can suggest cisplatin suppress the mitochondrial biogenesis pathway. Interestingly, we found increased levels of NRF1 in CDDP-treated cells (Figure 6B). This induction can be explained by the fact that the promoter of the NRF-1 gene possesses antioxidant response elements. Therefore, the nuclear factor erythroid 2-like 2 (Nrf2) can induce its expression [47]. Although we did not evaluate Nrf2 induction in this work, our group has previously reported that CDDP induces the increase in heme oxygenase-1 (HO-1) levels in LLC-PK1 [26], leading to Nrf2 activation, so the observed NRF1 increase can be the consequence of an adaptive response of Nrf2 activation. On the other hand, it was documented that αM can induce Nrf2 activation in retinal cells and liver tissue [48,49], which can also explain the observed increase in NRF1 levels in co-treatment and αM groups (Figure 6B,D), but more experiments are necessary to demonstrate Nrf2 induction.

In view of the fact that mitochondrial mass is regulated by biogenesis and mitophagy [13], we evaluated the mitophagy markers Pink1/Parkin. Co-incubation with αM avoided Parkin (Figure 7B) and PINK1 (Figure 7A) increase induced by CDDP. Recently, mitophagy was described as a response mechanism in CDDP-induced nephrotoxicity, and the PINK/Parkin pathway was suggested as the main route to remove damaged mitochondria [22–24]. The mitophagy machinery can be shared with the macroautophagy, but can also be independently-activated [19]. We show that pharmacological autophagy inhibition with CQ or with wortmannin did not have any effect in the protection by αM and did not increase CDDP toxicity (Figure 8). However, the increase in p62 levels (Figure 9C,D) and the no change of the LC3 ratio in CDDP-treatment at 24 h suggest that increases of PINK/Parkin can use the p62 protein as an adaptor, as suggested in a previous report [37]. However, it is also known that the accumulation of misfolded proteins leads to aberrant p62 expression [38]. Altogether, our results support the idea of a partial preservation of mitochondrial mass and function by αM-treatment.

We found 4 μM of αM did not change the respiratory parameters (Figures 3 and 4) nor the expression of mitochondrial proteins (Figure 5), except CII-SDHB (Figure 5B) and CIV-MTCO1 (Figure 5C). Unfortunately, nowadays, we do not have enough information to explain the origin of these changes. However, it is important to note, that although these subunits decreased in αM treatment, the functional respirometry studies did not show changes in any of the evaluated parameters with respect to the control group (Figures 3 and 4). This implies the observed CII-SDHB and CV-ATP5A decrease did not change the mitochondrial bioenergetics in the αM treatment with respect to the control. And we did not observe changes in the mitochondrial network by MTG (Figure 6A). Additionally, αM alone did not induce mitochondrial biogenesis (Figure 7) or mitophagy activation (Figure 8), implying that at least at this concentration, αM has no negative effects on mitochondrial function. In support of this, studies in heart tissue (200 mg/kg) and hepatic cells (30 μM) describe that αM can avoid mitochondrial dysfunction (preventing the decrease in activity and levels of antioxidant proteins and in RC, as well as ROS increase) induced by isoproterenol or free fatty acids increase [50,51]. By contrast, reports also describe that high αM concentrations (above 25 μM) can induce mitochondrial alterations in mitochondria isolated from kidneys [52], indicating that αM effects are tissue- and concentration-dependent. Although there are no clues regarding αM induced proliferation in normal cells, we evaluated GRP94, an ER chaperon, with 4 μM αM and found can it induce proliferation at 24 h of incubation (data not show). GRP94 is a glycoprotein related to protein folding, stores of calcium and targeting proteins to ER-associated degradation (ERAD), the client protein of GRP94 is related to specialized functions in immunity, grown signaling, and cell adhesion [53]. Probably, GRP94 induction in αM treatment can be related to grown signaling and cell adhesion, and this can explain this result, but more experiments are necessary.

As we show in the integrative scheme (Figure 11) our results suggest that 30 μM CDDP-treatment induces alterations in LCC-PK1 cells principally by reducing mitochondrial function and mass, which is partially attributable to mitochondrial biogenesis reduction and mitophagy activation. αM can prevent mitochondrial bioenergetics alterations and the induction of mitophagy. These mechanisms would be involved in the protective effects observed in LLC-PK tubular cells.

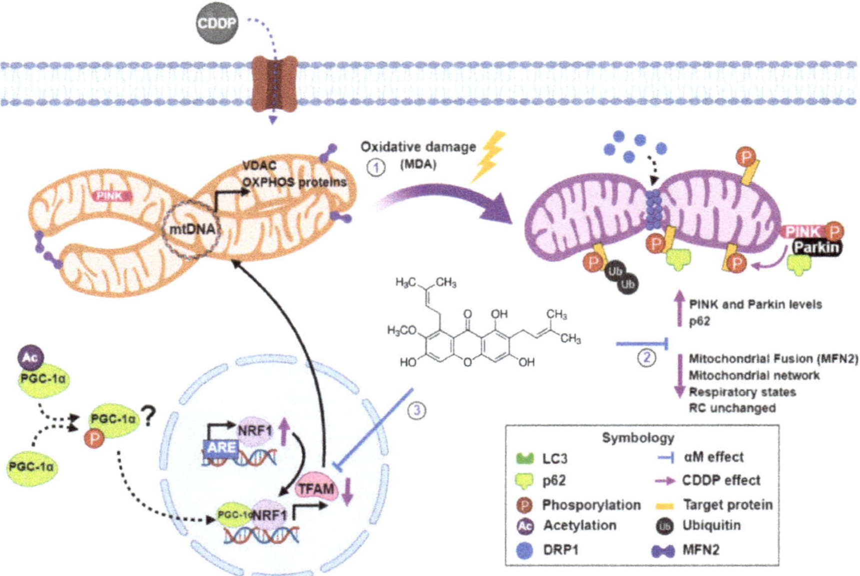

Figure 11. Integrative scheme. CDDP induces (1) the increase MDA level, (2) decrease respiratory states (except respiratory control, RC), the mitochondrial network fragmentation, the mitochondrial fusion protein 2 (MNF2) decrease and induce mitophagy (PINK1/Parkin and p62 increase). Furthermore, CDDP induces (3) TFAM reduction which reduces mitochondrial protein (VDAC and OXPHOS) and probably mitochondrial transcription, and also induces an NRF1 increase. The αM prevented all alterations, except NRF1 induction. Created with BioRender.com.

5. Conclusions

αM's protective effects in CDDP-induced toxicity in LLC-PK1 are mostly attributable to mitochondrial mass and function preservation.

Author Contributions: Conceptualization, L.M.R.-F.; methodology, L.M.R.-F.; formal analysis, L.M.R.-F. and O.E.A.-T.; investigation, L.M.R.-F., O.E.A.-T. and S.H.A.-R.; resources, I.R. and J.P.-C.; writing—original draft preparation, L.M.R.-F.; writing—review and editing L.M.R.-F., O.E.A.-T., S.H.A.-R. and J.P.-C.; supervision, J.P.-C.; funding acquisition, E.T. and J.P.-C.

Funding: This work was supported by Consejo Nacional de Ciencia y Tecnología (CONACyT, México Grant No. AI-S-7495); Programa de Apoyo a Proyecto de Investigación e Innovación Tecnológica (PAPIIT, UNAM, Mexico, Grants IN201316 and IN202219); Programa de Apoyo a la Investigación y el Posgrado (PAIP, Mexico, Grant No. 5000-9105); and Fondos del Gasto Directo autorizado a la Subdirección de Investigación Básica, Instituto Nacional de Cardiología Ignacio Chávez. L.M.R.F is a doctoral student from Programa de Maestría y Doctorado en Ciencias Bioquímicas, Universidad Nacional Autónoma de México (UNAM) and received a fellowship from CONACyT.

Acknowledgments: We thank Elena Martínez-Klimova for her assistance on the review of the present text.

Conflicts of Interest: The authors declare no conflict of interest.

References

1. Duffy, E.A.; Fitzgerald, W.; Boyle, K.; Rohatgi, R. Nephrotoxicity: Evidence in Patients Receiving Cisplatin Therapy. *Clin. J. Oncol. Nurs.* **2018**, *22*, 175–183. [CrossRef] [PubMed]
2. Manohar, S.; Leung, N. Cisplatin nephrotoxicity: A review of the literature. *J. Nephrol.* **2018**, *31*, 15–25. [CrossRef]
3. Pérez-Rojas, J.M.; Cruz, C.; García-López, P.; Sánchez-González, D.J.; Martínez-Martínez, C.M.; Ceballos, G.; Espinosa, M.; Meléndez-Zajgla, J.; Pedraza-Chaverri, J. Renoprotection by alpha-Mangostin is related to the attenuation in renal oxidative/nitrosative stress induced by cisplatin nephrotoxicity. *Free Radic. Res.* **2009**, *43*, 1122–1132. [CrossRef] [PubMed]
4. Sancho-Martínez, S.M.; Prieto-García, L.; Prieto, M.; López-Novoa, J.M.; López-Hernández, F.J. Subcellular targets of cisplatin cytotoxicity: An integrated view. *Pharmacol. Ther.* **2012**, *136*, 35–55. [CrossRef]
5. Heidari-Soreshjani, S.; Asadi-Samani, M.; Yang, Q.; Saeedi-Boroujeni, A. Phytotherapy of nephrotoxicity-induced by cancer drugs: An updated review. *J. Nephropathol.* **2017**, *6*, 254–263. [CrossRef] [PubMed]
6. Ovalle-Magallanes, B.; Eugenio-Pérez, D.; Pedraza-Chaverri, J. Medicinal properties of mangosteen (*Garcinia mangostana* L.): A comprehensive update. *Food Chem. Toxicol.* **2017**, *109*, 102–122. [CrossRef]
7. Pérez-Rojas, J.M.; González-Macías, R.; González-Cortes, J.; Jurado, R.; Pedraza-Chaverri, J.; García-López, P. Synergic Effect of α-Mangostin on the Cytotoxicity of Cisplatin in a Cervical Cancer Model. *Oxid. Med. Cell. Longev.* **2016**, *2016*, 7981397. [CrossRef]
8. Pedraza-Chaverri, J.; Cárdenas-Rodríguez, N.; Orozco-Ibarra, M.; Pérez-Rojas, J.M. Medicinal properties of mangosteen (*Garcinia mangostana*). *Food Chem. Toxicol.* **2008**, *46*, 3227–3239. [CrossRef] [PubMed]
9. Sánchez-Pérez, Y.; Morales-Bárcenas, R.; García-Cuellar, C.M.; López-Marure, R.; Calderon-Oliver, M.; Pedraza-Chaverri, J.; Chirino, Y.I. The alpha-mangostin prevention on cisplatin-induced apoptotic death in LLC-PK1 cells is associated to an inhibition of ROS production and p53 induction. *Chem. Biol. Interact.* **2010**, *188*, 144–150. [CrossRef]
10. Lee, C.-H.; Ying, T.-H.; Chiou, H.-L.; Hsieh, S.-C.; Wen, S.-H.; Chou, R.-H.; Hsieh, Y.-H. Alpha-mangostin induces apoptosis through activation of reactive oxygen species and ASK1/p38 signaling pathway in cervical cancer cells. *Oncotarget* **2017**, *8*, 47425–47439. [CrossRef]
11. Liu, G.; Tang, L.; She, J.; Xu, J.; Gu, Y.; Liu, H.; He, L. Alpha-mangostin attenuates focal segmental glomerulosclerosis of mice induced by adriamycin. *Zhong Nan Da Xue Xue Bao Yi Xue Ban* **2018**, *43*, 1089–1096. [PubMed]
12. Wang, M.-H.; Zhang, K.-J.; Gu, Q.-L.; Bi, X.-L.; Wang, J.-X. Pharmacology of mangostins and their derivatives: A comprehensive review. *Chin. J. Nat. Med.* **2017**, *15*, 81–93. [CrossRef]
13. Ralto, K.M.; Parikh, S.M. Mitochondria in Acute Kidney Injury. *Semin. Nephrol.* **2016**, *36*, 8–16. [CrossRef] [PubMed]
14. Ortega-Domínguez, B.; Aparicio-Trejo, O.E.; García-Arroyo, F.E.; León-Contreras, J.C.; Tapia, E.; Molina-Jijón, E.; Hernández-Pando, R.; Sánchez-Lozada, L.G.; Barrera-Oviedo, D.; Pedraza-Chaverri, J. Curcumin prevents cisplatin-induced renal alterations in mitochondrial bioenergetics and dynamic. *Food Chem. Toxicol.* **2017**, *107*, 373–385. [CrossRef] [PubMed]
15. Pan, H.; Chen, J.; Shen, K.; Wang, X.; Wang, P.; Fu, G.; Meng, H.; Wang, Y.; Jin, B. Mitochondrial modulation by Epigallocatechin 3-Gallate ameliorates cisplatin induced renal injury through decreasing oxidative/nitrative stress, inflammation and NF-kB in mice. *PLoS ONE* **2015**, *10*, e0124775. [CrossRef]
16. Yang, Y.; Liu, H.; Liu, F.; Dong, Z. Mitochondrial dysregulation and protection in cisplatin nephrotoxicity. *Arch. Toxicol.* **2014**, *88*, 1249–1256. [CrossRef] [PubMed]
17. Yu, X.; Meng, X.; Xu, M.; Zhang, X.; Zhang, Y.; Ding, G.; Huang, S.; Zhang, A.; Jia, Z. Celastrol ameliorates cisplatin nephrotoxicity by inhibiting NF-κB and improving mitochondrial function. *EBioMedicine* **2018**, *36*, 266–280. [CrossRef] [PubMed]
18. Kaushal, G.P.; Shah, S.V. Autophagy in acute kidney injury. *Kidney Int.* **2016**, *89*, 779–791. [CrossRef]
19. Galluzzi, L.; Baehrecke, E.H.; Ballabio, A.; Boya, P.; Bravo-San Pedro, J.M.; Cecconi, F.; Choi, A.M.; Chu, C.T.; Codogno, P.; Colombo, M.I.; et al. Molecular definitions of autophagy and related processes. *EMBO J.* **2017**, *36*, 1811–1836. [CrossRef]

20. Takahashi, A.; Kimura, T.; Takabatake, Y.; Namba, T.; Kaimori, J.; Kitamura, H.; Matsui, I.; Niimura, F.; Matsusaka, T.; Fujita, N.; et al. Autophagy guards against cisplatin-induced acute kidney injury. *Am. J. Pathol.* **2012**, *180*, 517–525. [CrossRef] [PubMed]

21. Yang, C.; Kaushal, V.; Shah, S.V.; Kaushal, G.P. Autophagy is associated with apoptosis in cisplatin injury to renal tubular epithelial cells. *Am. J. Physiol. Renal Physiol.* **2008**, *294*, F777–F787. [CrossRef] [PubMed]

22. Zhao, C.; Chen, Z.; Qi, J.; Duan, S.; Huang, Z.; Zhang, C.; Wu, L.; Zeng, M.; Zhang, B.; Wang, N.; et al. Drp1-dependent mitophagy protects against cisplatin-induced apoptosis of renal tubular epithelial cells by improving mitochondrial function. *Oncotarget* **2017**, *8*, 20988–21000. [CrossRef] [PubMed]

23. Ichinomiya, M.; Shimada, A.; Ohta, N.; Inouchi, E.; Ogihara, K.; Naya, Y.; Nagane, M.; Morita, T.; Satoh, M. Demonstration of Mitochondrial Damage and Mitophagy in Cisplatin-Mediated Nephrotoxicity. *Tohoku J. Exp. Med.* **2018**, *246*, 1–8. [CrossRef] [PubMed]

24. Wang, Y.; Tang, C.; Cai, J.; Chen, G.; Zhang, D.; Zhang, Z.; Dong, Z. PINK1/Parkin-mediated mitophagy is activated in cisplatin nephrotoxicity to protect against kidney injury. *Cell Death Dis.* **2018**, *9*, 1113. [CrossRef] [PubMed]

25. Nava Catorce, M.; Acero, G.; Pedraza-Chaverri, J.; Fragoso, G.; Govezensky, T.; Gevorkian, G. Alpha-mangostin attenuates brain inflammation induced by peripheral lipopolysaccharide administration in C57BL/6J mice. *J. Neuroimmunol.* **2016**, *297*, 20–27. [CrossRef] [PubMed]

26. Patricia Moreno-Londoño, A.; Bello-Alvarez, C.; Pedraza-Chaverri, J. Isoliquiritigenin pretreatment attenuates cisplatin induced proximal tubular cells (LLC-PK1) death and enhances the toxicity induced by this drug in bladder cancer T24 cell line. *Food Chem. Toxicol.* **2017**, *109*, 143–154. [CrossRef] [PubMed]

27. Aparicio-Trejo, O.E.; Reyes-Fermín, L.M.; Briones-Herrera, A.; Tapia, E.; León-Contreras, J.C.; Hernández-Pando, R.; Sánchez-Lozada, L.G.; Pedraza-Chaverri, J. Protective effects of N-acetyl-cysteine in mitochondria bioenergetics, oxidative stress, dynamics and S-glutathionylation alterations in acute kidney damage induced by folic acid. *Free Radic. Biol. Med.* **2019**, *130*, 379–396. [CrossRef]

28. Lowry, O.H.; Rosebrough, N.J.; Farr, A.L.; Randall, R.J. Protein measurement with the Folin phenol reagent. *J. Biol. Chem.* **1951**, *193*, 265–275.

29. Romero-Calvo, I.; Ocón, B.; Martínez-Moya, P.; Suárez, M.D.; Zarzuelo, A.; Martínez-Augustin, O.; de Medina, F.S. Reversible Ponceau staining as a loading control alternative to actin in Western blots. *Anal. Biochem.* **2010**, *401*, 318–320. [CrossRef] [PubMed]

30. Zhao, M.; Zhou, Y.; Liu, S.; Li, L.; Chen, Y.; Cheng, J.; Lu, Y.; Liu, J. Control release of mitochondria-targeted antioxidant by injectable self-assembling peptide hydrogel ameliorated persistent mitochondrial dysfunction and inflammation after acute kidney injury. *Drug Deliv.* **2018**, *25*, 546–554. [CrossRef]

31. Wang, C.; Li, L.; Liu, S.; Liao, G.; Li, L.; Chen, Y.; Cheng, J.; Lu, Y.; Liu, J. GLP-1 receptor agonist ameliorates obesity-induced chronic kidney injury via restoring renal metabolism homeostasis. *PLoS ONE* **2018**, *13*, e0193473. [CrossRef]

32. Svensson, K.; Schnyder, S.; Cardel, B.; Handschin, C. Loss of Renal Tubular PGC-1α Exacerbates Diet-Induced Renal Steatosis and Age-Related Urinary Sodium Excretion in Mice. *PLoS ONE* **2016**, *11*, e0158716. [CrossRef] [PubMed]

33. Sun, L.; Yuan, Q.; Xu, T.; Yao, L.; Feng, J.; Ma, J.; Wang, L.; Lu, C.; Wang, D. Pioglitazone Improves Mitochondrial Function in the Remnant Kidney and Protects against Renal Fibrosis in 5/6 Nephrectomized Rats. *Front. Pharmacol.* **2017**, *8*, 545. [CrossRef] [PubMed]

34. Smith, J.A.; Stallons, L.J.; Collier, J.B.; Chavin, K.D.; Schnellmann, R.G. Suppression of mitochondrial biogenesis through toll-like receptor 4-dependent mitogen-activated protein kinase kinase/extracellular signal-regulated kinase signaling in endotoxin-induced acute kidney injury. *J. Pharmacol. Exp. Ther.* **2015**, *352*, 346–357. [CrossRef] [PubMed]

35. Aparicio-Trejo, O.E.; Tapia, E.; Sánchez-Lozada, L.G.; Pedraza-Chaverri, J. Mitochondrial bioenergetics, redox state, dynamics and turnover alterations in renal mass reduction models of chronic kidney diseases and their possible implications in the progression of this illness. *Pharmacol. Res.* **2018**, *135*, 1–11. [CrossRef] [PubMed]

36. Presley, A.D.; Fuller, K.M.; Arriaga, E.A. MitoTracker Green labeling of mitochondrial proteins and their subsequent analysis by capillary electrophoresis with laser-induced fluorescence detection. *J. Chromatogr. B Analyt. Technol. Biomed. Life Sci.* **2003**, *793*, 141–150. [CrossRef]

37. Geisler, S.; Holmström, K.M.; Skujat, D.; Fiesel, F.C.; Rothfuss, O.C.; Kahle, P.J.; Springer, W. PINK1/Parkin-mediated mitophagy is dependent on VDAC1 and p62/SQSTM1. *Nat. Cell Biol.* **2010**, *12*, 119–131. [CrossRef]

38. Liu, H.; Dai, C.; Fan, Y.; Guo, B.; Ren, K.; Sun, T.; Wang, W. From autophagy to mitophagy: The roles of P62 in neurodegenerative diseases. *J. Bioenerg. Biomembr.* **2017**, *49*, 413–422. [CrossRef]

39. Peres, L.A.B.; da Cunha, A.D. Acute nephrotoxicity of cisplatin: Molecular mechanisms. *Braz. J. Nephrol.* **2013**, *35*, 332–340. [CrossRef]

40. Zhang, L.; Cooper, A.J.L.; Krasnikov, B.F.; Xu, H.; Bubber, P.; Pinto, J.T.; Gibson, G.E.; Hanigan, M.H. Cisplatin-induced toxicity is associated with platinum deposition in mouse kidney mitochondria in vivo and with selective inactivation of the alpha-ketoglutarate dehydrogenase complex in LLC-PK1 cells. *Biochemistry* **2006**, *45*, 8959–8971. [CrossRef]

41. Cunningham, R.M.; DeRose, V.J. Platinum Binds Proteins in the Endoplasmic Reticulum of S. cerevisiae and Induces Endoplasmic Reticulum Stress. *ACS Chem. Biol.* **2017**, *12*, 2737–2745. [CrossRef] [PubMed]

42. Hull, R.N.; Cherry, W.R.; Weaver, G.W. The origin and characteristics of a pig kidney cell strain, LLC-PK. *In Vitro* **1976**, *12*, 670–677. [CrossRef]

43. Nielsen, R.; Birn, H.; Moestrup, S.K.; Nielsen, M.; Verroust, P.; Christensen, E.I. Characterization of a kidney proximal tubule cell line, LLC-PK1, expressing endocytotic active megalin. *J. Am. Soc. Nephrol. JASN* **1998**, *9*, 1767–1776. [PubMed]

44. Bergeron, M.; Thiéry, G.; Lenoir, F.; Giocondi, M.C.; Le Grimellec, C. Organization of the endoplasmic reticulum in renal cell lines MDCK and LLC-PK1. *Cell Tissue Res.* **1994**, *277*, 297–307. [CrossRef] [PubMed]

45. Liu, H.; Baliga, R. Endoplasmic reticulum stress-associated caspase 12 mediates cisplatin-induced LLC-PK1 cell apoptosis. *J. Am. Soc. Nephrol. JASN* **2005**, *16*, 1985–1992. [CrossRef] [PubMed]

46. Sanchis-Gomar, F.; García-Giménez, J.L.; Gómez-Cabrera, M.C.; Pallardó, F.V. Mitochondrial biogenesis in health and disease. Molecular and therapeutic approaches. *Curr. Pharm. Des.* **2014**, *20*, 5619–5633. [CrossRef] [PubMed]

47. Piantadosi, C.A.; Carraway, M.S.; Babiker, A.; Suliman, H.B. Heme oxygenase-1 regulates cardiac mitochondrial biogenesis via Nrf2-mediated transcriptional control of nuclear respiratory factor-1. *Circ. Res.* **2008**, *103*, 1232–1240. [CrossRef] [PubMed]

48. Fang, Y.; Su, T.; Qiu, X.; Mao, P.; Xu, Y.; Hu, Z.; Zhang, Y.; Zheng, X.; Xie, P.; Liu, Q. Protective effect of alpha-mangostin against oxidative stress induced-retinal cell death. *Sci. Rep.* **2016**, *6*, 21018. [CrossRef] [PubMed]

49. Fu, T.; Li, H.; Zhao, Y.; Cai, E.; Zhu, H.; Li, P.; Liu, J. Hepatoprotective effect of α-mangostin against lipopolysaccharide/d-galactosamine-induced acute liver failure in mice. *Biomed. Pharmacother. Biomed. Pharmacother.* **2018**, *106*, 896–901. [CrossRef] [PubMed]

50. Sampath, P.D.; Kannan, V. Mitigation of mitochondrial dysfunction and regulation of eNOS expression during experimental myocardial necrosis by alpha-mangostin, a xanthonic derivative from Garcinia mangostana. *Drug Chem. Toxicol.* **2009**, *32*, 344–352. [CrossRef] [PubMed]

51. Tsai, S.-Y.; Chung, P.-C.; Owaga, E.E.; Tsai, I.-J.; Wang, P.-Y.; Tsai, J.-I.; Yeh, T.-S.; Hsieh, R.-H. Alpha-mangostin from mangosteen (*Garcinia mangostana* Linn.) pericarp extract reduces high fat-diet induced hepatic steatosis in rats by regulating mitochondria function and apoptosis. *Nutr. Metab.* **2016**, *13*, 88. [CrossRef] [PubMed]

52. Martínez-Abundis, E.; García, N.; Correa, F.; Hernández-Reséndiz, S.; Pedraza-Chaverri, J.; Zazueta, C. Effects of alpha-mangostin on mitochondrial energetic metabolism. *Mitochondrion* **2010**, *10*, 151–157. [CrossRef] [PubMed]

53. Zhu, G.; Lee, A.S. Role of the unfolded protein response, GRP78 and GRP94 in organ homeostasis. *J. Cell. Physiol.* **2015**, *230*, 1413–1420. [CrossRef] [PubMed]

Article

Antioxidant Activity and Phenolic Composition of Amaranth (*Amaranthus caudatus*) during Plant Growth

Magdalena Karamać [1,*], Francesco Gai [2], Erica Longato [3], Giorgia Meineri [3], Michał A. Janiak [1], Ryszard Amarowicz [1] and Pier Giorgio Peiretti [2]

[1] Institute of Animal Reproduction and Food Research, Polish Academy of Sciences, Tuwima 10, 10-748 Olsztyn, Poland; m.janiak@pan.olsztyn.pl (M.A.J.); r.amarowicz@pan.olsztyn.pl (R.A.)

[2] Institute of Sciences of Food Production, National Research Council, 10095 Grugliasco, Italy; francesco.gai@ispa.cnr.it (F.G.); piergiorgio.peiretti@ispa.cnr.it (P.G.P.)

[3] Department of Veterinary Sciences, University of Turin, 10095 Grugliasco, Italy; erica.longato@unito.it (E.L.); giorgia.meineri@unito.it (G.M.)

* Correspondence: m.karamac@pan.olsztyn.pl; Tel.: +48-895-234-622

Received: 15 May 2019; Accepted: 10 June 2019; Published: 12 June 2019

Abstract: The antioxidant activity and phenolic composition of the aerial part of *Amaranthus caudatus* at seven stages of development were investigated. Total phenolic content, $ABTS^{\bullet+}$, $DPPH^{\bullet}$, and $O_2^{\bullet-}$ scavenging activity, ferric-reducing antioxidant power (FRAP), and Fe^{2+} chelating ability were evaluated. The phenolic profile was characterized by 17 compounds. Rutin was predominant in all growth stages, although its content, similar to the quantity of other phenolics, changed during the growth cycle. Flavonols were most abundant in the plants of early flowering and grain fill stages. In contrast, the highest content of hydroxycinnamic acid derivatives was found in the early vegetative stage. The results of antioxidant assays also showed significant differences among plant stages. Generally, the lowest antioxidant activity was found in the shooting and budding stages. Significantly higher activity was observed in amaranths in earlier (vegetative) and later (early flowering and grain fill) stages, suggesting that plants in these stages are valuable sources of antioxidants.

Keywords: amaranth; morphological stage; scavenging activity; ferrous ions chelating ability; reducing power; phenolic compounds; rutin; growth cycle

---

1. Introduction

Amaranth (*Amaranthus* spp.) is a gluten-free pseudocereal that is cultivated primarily in Mexico and South America, but also thrives in all temperate-tropical areas of the world [1]. Moreover, in certain regions of the world, such as eastern Africa, amaranth leaves are consumed as a vegetable because it is a fast-growing plant available most of the year. There has been a renewed interest in this ancient and highly nutritious food crop due to the excellent nutritional value of seed and leaves [2–4]. Both seeds and leaves are rich sources of proteins, which constitute up to 15–43% and 14–30% of fresh matter (FM), respectively. Amaranth proteins have a well-balanced amino acid composition [4], high bioavailability [3], and good functional properties [5]. Dietary fiber, vitamins and precursors of vitamins (ascorbic acid, riboflavin, tocols, carotenoids), as well as minerals (Ca, Fe, Mg, K, Cu, Zn, and Mn) are other important nutrients present in seeds and leaves of amaranth. Their contents are high compared to these in some cereals and green leafy vegetables [1,2,4]. Nutrient composition causes an increasing interest in amaranth as a food ingredient, especially in the production of gluten-free products [6,7].

In addition to macro- and micronutrients, amaranth contains secondary plant metabolites, which may play a significant role in the human diet due to their potential health beneficial effects [8].

Considerable research has been conducted over the past years on phenolic compound profile of amaranth seeds [8–11] and their functional and bioactive properties [12–15]. However, recent findings confirmed that leaves and other aerial parts of *Amaranthus* also are important sources of phenolic compounds [16–18]. Among phenolic compounds, those belonging to hydroxycinnamic acids, benzoic acids, and flavonols and their glycosides were identified in amaranth leaves, flowers, and stalks [18,19]. Steffensen et al. [20] found additional hydroxycinnamyl amides in aerial parts of young plants. Phenolic compounds are well-known as antioxidants. Conforti et al. [21] showed that amaranth leaf extracts contained phenolics, inhibited nitric oxide production, and scavenged free radicals. The reducing potential and antioxidant activity in lipid systems of various parts of amaranth shoot system were determined as well [16,19,22]. Other phytochemicals with antioxidant activity which occur in amaranth are betalains, especially betacyanins [23]. The contents of these pigments vary among *Amaranthus* species and genotypes [24]. If they are present in the plant, they are accumulated mainly in seedlings, leaves, and inflorescences.

Although the phenolic compound and betacyanin profiles of individual botanical parts of amaranth as well as their antioxidant potential are well-known, information about effects of the growth cycle on these phytochemicals and their activity is still scarce. It is important to identify the changes in antioxidant activity and phenolic compound composition by the growth stage of the amaranth.

In a previous paper [25], we reported the nutritive characteristics of *A. caudatus*, and in particular, we evaluated the effects of plant aging on the chemical composition, gross energy, *in vitro* true digestibility, neutral detergent fiber digestibility and fatty acid profile. However, to the best of our knowledge, no research has been conducted to investigate the change of antioxidant activities and phenolic composition of extracts of *A. caudatus* during plant growth. Barba de la Rosa et al. [26] stated that the knowledge of *Amaranthus* spp. as a source of phytochemicals will increase their importance as a potential source of antioxidant compounds in the human diet.

The aim of this study was to compare the antioxidant activity, TPC, and individual phenolic compounds of *A. caudatus* during plant growth, to identify the ideal growing stage to achieve the maximum antioxidant properties, and to optimize its use as a source of antioxidants for use in food production.

2. Materials and Methods

2.1. Chemicals

The Folin-Ciocalteu phenol reagent (FCR), gallic acid, 2,2-diphenyl-1-picrylhydrazyl (DPPH), 2,2′-azinobis-(3-ethylbenzothiazoline-6-sulfonic acid) (ABTS), 2,4,6-tri(2-pyridyl)-s-triazine (TPTZ), 6-hydroxy-2,5,7,8-tetramethyl-chroman-2-carboxylic acid (Trolox), ferrozine, caffeic acid, quercetin, *p*-coumaric acid, ferulic acid, rutin, and kaempferol-3-*O*-rutinoside were obtained from Sigma-Aldrich (St. Louis, MO, USA). Methanol, acetonitrile, sodium carbonate, ferric chloride, ferrous chloride, ferrous sulfate, and potassium persulfate were acquired from Avantor Performance Materials (Gliwice, Poland). The ACL kit was purchased from Analytik Jena (Jena, Germany).

2.2. Plant Material

A. caudatus seeds were generously donated by Pedon S.p.A. (Molvena, Italy). The amaranth stands were seeded in the spring in field trials (Department of Veterinary Science of the University of Turin) located in Grugliasco, Piedmont, NW Italy (293 m a.s.l., 45°03′57.9″N 7°35′36.9″E), and no irrigation or fertilizers were applied after sowing. The herbage samples were collected in the morning after evaporation of dew and never collected on rainy days. Sampling was performed according to Peiretti et al. [25] at seven progressive morphological stages from early vegetative to grain fill stage from May to July 2014. Plants were cut to a 1–2 cm stubble height with edging shears from two replicate 2 m^2 subplots randomly located in 2 × 7 m^2 plots.

2.3. Extraction

Plant samples were lyophilized (FreeZone Freeze Dry System, Labconco, Kansas City, MO, USA) and ground. Three-gram portions of materials were suspended in 30 mL of 80% (*v/v*) methanol. The closed bottles containing suspensions were shaken for 15 min in a water bath (SW22, Julabo, Seelbach, Germany) at 65 °C. After filtration, the residues were extracted twice more. Combined filtrates for each sample were dried by evaporation of methanol under vacuum at 50 °C (Rotavapor R-200, Büchi Labortechnik, Flawil, Switzerland) and lyophilization of remaining water. Yield of extraction (%) was calculated on the basis of weight of plant sample and dry extract.

2.4. Total Phenolic Content

The colorimetric reaction with FCR was carried out to determinate the total phenolic content (TPC) [27]. The 0.25 mL of extract solution in methanol (1.25 mg/mL), 0.25 mL of FCR, 0.5 mL of saturated sodium carbonate solution, and 4 mL of water were mixed. After 25 min, the reaction mixtures were centrifuged (5 min, 5000× g, MPW-350R, MPW Med. Instruments, Warsaw, Poland), and the absorbance of supernatants was measured at 725 nm (DU-7500 spectrophotometer, Beckman Instruments, Fullerton, CA, USA). TPC was expressed as gallic acid equivalents (GAE) per g of extract or per g of FM of plant.

2.5. Trolox Equivalent Antioxidant Capacity

The Trolox equivalent antioxidant capacity (TEAC) was determined according to the assay reported by Re et al. [28]. ABTS$^{\bullet+}$ was generated and diluted as in the original description. Next, solutions of ABTS$^{\bullet+}$ (2 mL) and amaranth extracts (20 µL with a concentration of 2.5 mg/mL) were vortexed and heated at 30 °C (TH-24 block heater, Meditherm, Warsaw, Poland) within 6 min. After incubation, absorbance was measured at 734 nm (DU-7500 spectrophotometer). The results were expressed as µmol Trolox equivalents (TE) per g of extract or per g of FM of plant.

2.6. Ferric-Reducing Antioxidant Power

The Benzie and Strain method [29] was used to evaluate the ferric-reducing antioxidant power (FRAP) of amaranth extracts. First, FRAP reagent was prepared. Next, the portions of 2.25 mL of this reagent were mixed with 225 µL of water and 75 µL of aqueous solution of amaranth extracts (1 mg/mL). Absorbance of mixtures was recorded at 593 nm after incubation at 37 °C for 30 min. FRAP results were expressed as µmol Fe^{2+} equivalents per g of extract or per g of FM using the calibration curve for $FeSO_4$.

2.7. Photochemiluminescence Assay

In the photochemiluminescence (PCL) assay, superoxide radical anions ($O_2^{\bullet-}$) were generated from luminol [30]. The antiradical activity against these radicals was determined using the ACL (antioxidant capacity of lipid-soluble substances) kit (Analytik Jena, Jena, Germany). The amaranth extracts were dissolved in methanol to a concentration of 0.25 mg/mL. The portions of 10 µL of solutions were mixed with 2.3 mL of methanol (reagent 1), 200 µL of buffer solution (reagent 2), and 25 µL of luminol (reagent 3). The Photochem device (Analytik Jena) supported by PCLsoft software was used to perform reactions and calculate the results, which were expressed as µmol of Trolox equivalents (TE) per g of extract or per g of FM of plant.

2.8. Fe^{2+} Chelating Ability

The ability of amaranth extracts to chelate ferrous ions was determined by a method with ferrozine [31] that was modified to be performed with multi-well plates [32]. Extract solution in water (0.25 mg/mL), 0.4 mM $FeCl_2 \times 4H_2O$ and 5 mM ferrozine were mixed in a ratio of 10:1:2 (*v/v/v*). After 10 min, absorbance was measured at 562 nm using an Infinite M1000 microplate reader (Tecan, Männedorf, Switzerland). The chelating ability was expressed as percentage of Fe^{2+} bound.

2.9. Scavenging of the DPPH Radicals

The DPPH$^\bullet$ scavenging activity of amaranth extracts was evaluated by the method of Brand-Williams et al. [33]. Extract solutions in methanol over a range of concentrations from 2–10 mg/mL were prepared. To 100 µL of these solutions, 0.25 mL of 1 mM DPPH and 2 mL of methanol were added. After 20 min, absorbance of the mixtures was read at 517 nm. The curves of absorbance values vs. concentration of samples (mg/assay) were plotted. Additionally, EC_{50} values defined as concentration of extract (mg/mL reaction mixture) needed to scavenge the 50% of initial DPPH$^\bullet$ were evaluated.

2.10. Analysis of Phenolic Compounds

The HPLC-DAD Shimadzu system (Shimadzu, Kioto, Japan), which consisted of a CBM-20A controller, DGU-20A5R degassing unit, two LC-30AD pumps, SIL-30AC autosampler, SPD-M30A diode array detector (DAD), and CTO-20AC oven, was used to analyze the phenolic compound of the amaranth extracts. Extracts dissolved in 80% (*v/v*) methanol were injected (10 µL) into a Luna C8(2) (4.6 × 150 mm, particle size 3 µm, Phenomenex, Torrance, CA, USA) column [34]. The mobile phase consisted of solvents A (acetonitrile-water-trifluoroacetic acid, 5:95:0.1, *v/v/v*) and B (acetonitrile-trifluoroacetic acid, 100:0.1, *v/v*): From 0–16 min, the eluent composition was changed from 0–24% B. The flow rate was 1 mL/min. The oven temperature was 25 °C and detector wavelength ranged from 200 to 600 nm. The content of individual phenolic compounds in the extracts was expressed on the basis of a calibration curve of the corresponding standards or structurally related substances.

For identification of extract compounds, HPLC-MS/MS analysis was carried out using a liquid chromatography (LC) system coupled with a quadrupole ion trap mass spectrometer (QTRAP® 5500 LC-MS/MS System, AB Sciex, Framingham, MA, USA). Operating MS/MS conditions were the following: Nitrogen curtain gas flow rate of 25 L/min, collision gas glow rate of 9 L/min, ion spray source voltage of −4.5 kV, temperature of 350 °C, nebulizer gas flow rate of 35 L/min, and turbo gas flow rate of 30 L/min. Negative-mode of electrospray ionization (ESI$^-$) was used. Qualification was based on multiple reaction monitoring (MRM) of selected ion pairs in the first quadrupole (Q1) and third quadrupole (Q3). The presence of compounds previously identified in aerial parts of amaranth [10,18–20,35] were verified.

2.11. Statistical Analysisl

The plant growth experiment was carried out in duplicate. Antioxidant assays and HPLC-DAD separations were performed for at least three repetitions. All results were presented as means ± standard deviations (SD). Significance of differences among mean values were estimated by one-way ANOVA and Fisher's LSD test at a level of $p < 0.05$ (GraphPad Prism; GraphPad Software, San Diego, CA, USA). Principal component analysis (PCA) allowed examination of the relationships between TPC, individual phenolic compound contents and values of antioxidant assays obtained for amaranth at different growth stages.

3. Results and Discussion

3.1. Total Phenolic Content

The extraction yield and TPC of the aerial parts of amaranth at different growth stages, expressed on the basis of both extract and fresh matter, are reported in Table 1. The extraction yield did not differ statistically ($p \geq 0.05$) for plants in vegetative, shooting and budding stages, while lower values were recorded for early flowering and grain fill stages of amaranth. The TPC ranged from 18.3 to 33.7 mg GAE/g extract. TPC expressed on the basis of plant fresh matter was in the range of 0.68 to 1.11 mg/g. The highest values were observed in the earliest (early and medium vegetative) and the oldest (early flowering and grain fill) stages of plant. The lowest TPC was noted for extracts obtained from the shooting and budding stages. Li et al. [19] compared phenolic contents of different botanical parts of three *Amaranthus* species (*A. hypochondriacus*, *A. caudatus*, and *A. cruentus*) and found that the leaves had the highest TPC while the seeds and stalks contained the lowest. Therefore, in our study, the high TPC levels found in the amaranth of vegetative stages could be related to the major proportion of leaves with respect to stalks present in these plant growth stages compared to the others. A similar trend was reported in other pseudocereals, such as quinoa [36], as well as soybean [32], where authors found that extracts obtained from the early and late vegetative stages were characterized by the highest TPC. The TPC of amaranth extract of early flowering stage plants presented in Table 1 was in agreement with those reported for methanol-water and acetone extracts from separate leaves (24.8–32.3 mg GAE/g extract) and flowers (27.2–33.3 mg GAE/g extract) collected from a flowering stage of *A. hybridus* [16]. In turn, higher TPC (2.9 mg GAE/g FM) was noted for the five-week-old *A. caudatus* leaves [26] corresponding to an intermediate between early and medium vegetative stages in our study.

Table 1. Yield of extraction and total phenolic content (TPC) of the amaranth extract and fresh matter (FM) in different growth stages.

Growth Stage	Time after Seeding (days)	Extraction Yield (%)	TPC	
			mg GAE/g Extract	mg GAE/g FM
Early vegetative	34	24.3 ± 0.3 ab	33.7 ± 4.5 a	0.94 ± 0.14 abc
Medium vegetative	41	26.3 ± 3.2 a	31.1 ± 3.5 ab	1.04 ± 0.05 ab
Late vegetative	55	27.4 ± 0.6 a	27.9 ± 0.4 b	0.81 ± 0.03 bcd
Shooting	62	26.7 ± 0.5 a	18.6 ± 2.9 c	0.68 ± 0.16 d
Budding	69	26.6 ± 0.4 a	18.3 ± 3.4 c	0.71 ± 0.15 cd
Early flowering	74	22.7 ± 0.5 bc	27.4 ± 0.6 ab	1.11 ± 0.03 a
Grain fill	78	20.3 ± 1.7 c	27.3 ± 3.5 ab	0.88 ± 0.07 abc

GAE, gallic acid equivalents. [abcd] Means with the different lowercase letters in the same column are significantly different ($p < 0.05$).

3.2. Antioxidant Activity

The antioxidant activity of amaranth samples of each growth stage was determined as radical scavenging activity (TEAC, PCL-ACL, and DPPH assay) and ability to reduce Fe^{3+} to Fe^{2+} (FRAP). Additionally, Fe^{2+} chelating ability of extracts was determined because phenolic compounds form stable complexes with ferrous ions, and in this way, decrease the extent of free ions to Fenton's reaction in which highly reactive $^{•}OH$ are generated [31]. The results of antioxidant activity are presented in Table 2 and Figures 1 and 2. The significantly higher TEAC was recorded for amaranth in vegetative, early flowering and grain fill stages as opposed to in shooting and budding stages ($p < 0.05$), both when results were expressed on the basis of extract and fresh matter of plant. The FRAP ranged from 469 to 830 µmol Fe^{2+}/g extract with the highest value observed for the early vegetative stage and the lowest, again, for the shooting and budding stages of the plant. The FRAP expressed on plant fresh matter basis was less varied between amaranth growth stages (17.4–24.6 µmol Fe^{2+}/g). The PCL-ACL ranged from 422 to 858 µmol TE/g extract and from 16.5 to 23.9 µmol TE/g FM, respectively. For both types of PCL-ACL expression, amaranth in vegetative stages had significantly higher activity compared to plants in subsequent morphological states ($p < 0.05$). In turn, the Fe^{2+} chelating ability of extracts ranged from 16.1—19.9% for late vegetative and shooting stages to 37.5% for the medium vegetative stage (Figure 1). The changes of antiradical activity against $DPPH^{•}$ of amaranth extracts with increasing assay content as well as the EC_{50} values, are presented in Figure 2. The $DPPH^{•}$ scavenging activity expressed as EC_{50} showed significant differences between certain growth stages of amaranth ($p < 0.05$). The highest antiradical activity was obtained for extracts from plants in the early flowering and grain fill stages with the lowest EC_{50} value. The highest EC_{50} values (294–317 µg/mL), which did not differ statistically from each other ($p \geq 0.05$), were found for late vegetative, shooting, and budding stages.

To the best of our knowledge, the changes of antioxidant activity of amaranth aerial parts during the growth cycle have not been previously demonstrated, although the differences in the antioxidant potential of various morphological parts of the plant were shown [16,19,37]. In general, FRAP, DPPH•, or ABTS•+ scavenging capacity decreased in this order: Leaves ≥ flowers >> stem > seeds. Among some amaranth species, this relationship was demonstrated for *A. caudatus* [19]. It is known that environmental factors, such as light and temperature, also play an important role in amaranth antioxidant metabolism. Khandaker et al. [38] found that the antioxidant activity of leaves of red amaranth (*A. tricolor*) was higher under full sunlight intensity than under dark conditions. Modi [39] studied the effect of growth temperature on yield, nutritional value, and antioxidant activity of the leaves of five *Amaranthus* spp. (*A. hybridus* var. *cruentus*, *A. hypochondriacus*, *A. tricolor*, *A. thunbergii*, and *A. hybridus*) harvested at 20, 40, and 60 days after sowing and found significant differences between growth temperature and stage of development. Authors concluded that for greater nutritional benefit, *Amaranthus* should be grown under warm conditions and that younger leaves are preferable. The maximum antioxidant activity, independent of the temperature and species considered, was found in leaves after 60 days of sowing, corresponding to an intermediate stage between late vegetative and shooting found in our study.

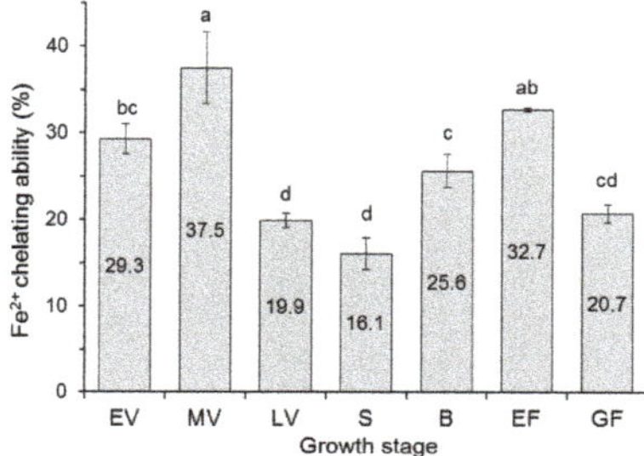

Figure 1. Fe^{2+} chelating ability of the amaranth extracts. EV, early vegetative; MV, medium vegetative; LV, late vegetative; S, shooting; B, budding; EF, early flowering; GF, grain fill. Different letters above bars indicate significant differences among means ($p < 0.05$).

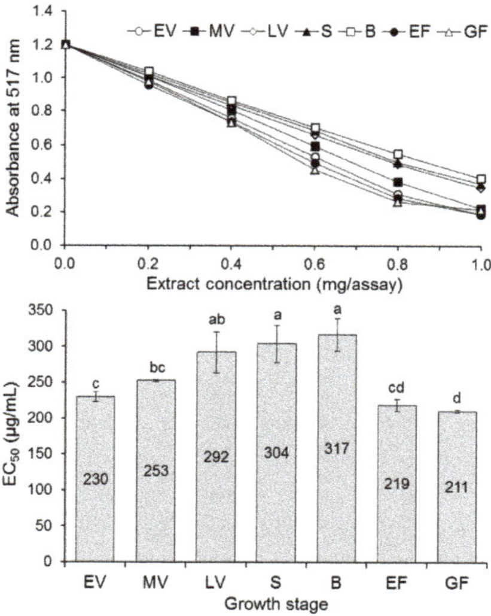

Figure 2. DPPH• scavenging activity of the amaranth extracts. EV, early vegetative; MV, medium vegetative; LV, late vegetative; S, shooting; B, budding; EF, early flowering; GF, grain fill. Different letters above bars indicate significant differences among means ($p < 0.05$).

Table 2. Antioxidant activity of the amaranth extract and fresh matter (FM) in different growth stages.

Growth Stage	TEAC		FRAP		PCL-ACL	
	µmol TE/g Extract	µmol TE/g FM	µmol Fe^{2+}/g Extract	µmol Fe^{2+}/g FM	µmol TE/g Extract	µmol TE/g FM
Early vegetative	259 ± 27 [ab]	7.23 ± 0.91 [bc]	830 ± 27 [a]	23.1 ± 1.3 [ab]	858 ± 24 [a]	23.9 ± 1.2 [a]
Medium vegetative	283 ± 27 [a]	9.52 ± 0.34 [a]	665 ± 83 [b]	22.3 ± 1.5 [ab]	698 ± 43 [b]	23.6 ± 2.8 [a]
Late vegetative	272 ± 18 [ab]	8.58 ± 0.38 [ab]	648 ± 25 [b]	20.4 ± 1.2 [abc]	681 ± 42 [b]	21.5 ± 1.8 [ab]
Shooting	176 ± 6.0 [c]	6.44 ± 0.69 [c]	475 ± 31 [c]	17.4 ± 2.4 [c]	486 ± 3.9 [cd]	17.8 ± 1.4 [bc]
Budding	171 ± 8.1 [c]	6.67 ± 0.53 [c]	469 ± 75 [c]	18.3 ± 3.5 [bc]	422 ± 43 [d]	16.5 ± 2.2 [c]
Early flowering	218 ± 23 [bc]	8.83 ± 0.87 [ab]	606 ± 15 [bc]	24.6 ± 0.4 [a]	483 ± 20 [cd]	19.6 ± 0.7 [bc]
Grain fill	236 ± 45 [ab]	7.62 ± 1.13 [bc]	640 ± 112 [b]	20.7 ± 2.7 [abc]	558 ± 34 [c]	18.1 ± 0.3 [bc]

TEAC, Trolox equivalent (TE) antioxidant capacity; FRAP, ferric-reducing antioxidant power; PCL-ACL, photochemiluminescence-antioxidant capacity of lipid-soluble substances. [abcd] Means with the different lowercase letters in the same column are significantly different ($p < 0.05$).

3.3. Phenolic Compound Profile

The HPLC-DAD analysis showed that the phenolic profile was characterized by 17 compounds (Figure 3). The absorption maxima of UV-Vis spectra of compounds and MRM Q1/Q3 ion pairs used for their identification are listed in Table 3. Compounds **10**, **13**, and **17** were identified as caffeic acid, rutin, and kaempferol-3-*O*-rutinoside, respectively, by comparison with standards. Their presence in the extracts was confirmed by HPLC-MS/MS analysis. This analysis allowed also showing the presence of twelve hydroxycinnamic acids (1–12) and quercetin glucoside (15) in the extracts. Two other compounds (**14 and 16**) were tentatively identified as hydroxycinnamic acid derivatives based on the shape of UV spectra with maxima absorption at 315–329 nm and a shoulder at the shorter wavelength [34]. Most of the identified compounds (caffeic acid, caffeoyl-, coumaroyl- and feruloyl-glucaric isomers, feruloyl- and caffeoyl- quinic acids, rutin, kaempferol-3-*O*-rutinoside, and quercetin glucoside) were detected previously in leaves, seeds, and other aerial parts of *A. caudatus* [10,18–20]. Coumaroylquinic acids were found in stems of *A. spinosus* [35]. In turn, the free phenolic acids, such as, ferulic, *p*-coumaric, *p*-hydroxybenzoic, vanillic, sinapic, gallic, and protocatechuic acids, as well as betacyanins, were not identified in amaranth in the present study, although, according to literature data, they were determined in leaves, flowers, stalks and seeds of *A. caudatus* [10,11,19,40]. The lack of identification of betacyanins in our samples probably results from the selection of the betacyanin-free *A. caudatus* genotype for experiments.

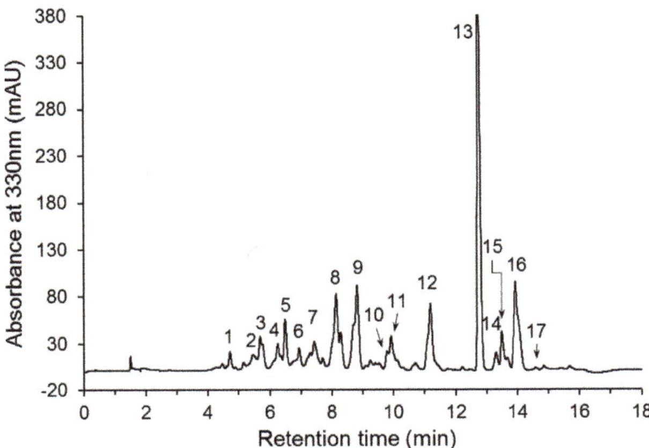

Figure 3. HPLC-DAD chromatogram of the phenolic compounds present in the amaranth extract.

Table 3. Chromatographic and spectral data of the phenolic compounds identified in amaranth extracts.

Compound No [1]	t_R (min) [2]	λ_{max} (nm) [3]	Ion Pair [4] (Q1/Q3)	Compound
1	4.80	301sh;327	371/209; 371/173	Caffeoylglucaric acid 1
2	5.52	301sh;327	371/209; 371/173	Caffeoylglucaric acid 2
3	5.75	302sh; 327	371/209; 371/173	Caffeoylglucaric acid 3
4	6.31	302sh; 328	371/209; 371/173	Caffeoylglucaric acid 4
5	6.54	303sh; 328	371/209; 371/173	Caffeoylglucaric acid 5
6	6.95	304sh; 327	371/191; 371/209	Coumaroylglucaric acid 1
7	7.47	304sh; 327	371/191; 371/209	Coumaroylglucaric acid 2
8	8.18	299sh; 326	385/191; 385/209	Feruloylglucaric acid
9	8.84	303sh; 327	353/191; 353/173	Caffeoylquinic acid
10	9.78	296sh; 324	179/135; 179/107	Caffeic acid
11	9.93	301sh; 314	337/173; 337/155	Coumaroylquinic acid
12	11.21	304sh; 329	367/173; 367/155	Feruloylquinic acid
13	12.75	256, 354	609/301; 609/463	Rutin
14	13.32	302sh; 315	-	Hydroxycinnamic acid derivative
15	13.49	256, 354	463/301	Quercetin glucoside
16	13.95	303sh; 329	-	Hydroxycinnamic acid derivative
17	14.40	265; 348	593/285	Kaempferol-3-*O*-rutinoside

[1] Compound number corresponds to peak number in Figure 3. [2] Retention time (t_R) of HPLC-DAD separation. [3] Maximum absorption (λ_{max}) of UV-Vis spectrum in HPLC-DAD analysis. [4] Ion pair of multiple reaction monitoring (MRM) of HPLC-MS/MS analysis.

The contents of individual phenolic compounds (**1–16**) in different growth stages of amaranth, expressed both per gram of extract and per gram of fresh matter of plant, are presented in Tables 4 and 5, respectively. Compound **17** (kaempferol-3-*O*-rutinoside) was not quantified because its content in the samples was very low (< 1 µg/g FM). The predominant compound in all samples was rutin (15.0–36.2 mg/g extract; 418–1169 µg/g FM), and its content consisted of approximately 95% of the sum of flavonols. Compounds **11** and **16** were the most abundant hydroxycinnamic acid derivatives. The huge amount of rutin in comparison with other phenolic compounds was in line with literature data on aerial components, especially leaves, of different species of amaranth [19,20,41]. In turn, high content of hydroxycinnamic acid derivatives with very high quantity of caffeic acid derivatives was found by Neugart et al. [18] in *A. cruentus* leaves. These authors reported a 3.3–3.4-fold higher content of caffeic acid derivatives than quercetin glycosides in leaves harvested at the 8-12 leaf stage. In our study, the content of hydroxycinnamic acid derivatives predominated over the quantity of flavonols in the early vegetative stage, but with a lower difference between values (718 vs. 439 µg/g FM). Plant growth stage significantly affected the amount of individual phenolic compounds (Tables 4 and 5). The rutin content (and sum of flavonols) increased for subsequent growth stages of plant, and the highest value was measured in early flowering and grain fill stages at the same level ($p \geq 0.05$). Interestingly, on the contrary, the sum of hydroxycinnamic acid derivatives was found to be the highest in amaranth of the early vegetative stage, and this value decreased with age of amaranth for values expressed both on the basis of extract and fresh matter of plant. Our observation regarding the changes of rutin content during the amaranth growth cycle confirmed the trends previous reported by Kalinova and Dadakova [42]. These authors determined the rutin contents in leaves, flowers, stems, and seeds of six *Amaranthus* spp. (*A. caudatus*, *A. hypochondriacus*, *A. hybrid*, *A. retroflexus*, *A. cruentus*, and *A. tricolor*) and found that the rutin content in leaves was related to the developmental stage of the crop and that it usually increased with plant aging.

Table 4. Individual phenolic compound contents in extracts of amaranth in different growth stages (mg/g).

Comp.No	Compound	Early Vegetative	Medium Vegetative	Late Vegetative	Shooting	Budding	Early Flowering	Grain Fill
1 [1]	Caffeoylglucaric acid 1	0.54 ± 0.08 [a]	0.46 ± 0.09 [a]	0.45 ± 0.01 [a]	0.23 ± 0.01 [b]	0.16 ± 0.01 [b]	0.19 ± 0.06 [b]	0.13 ± 0.03 [b]
2 [1]	Caffeoylglucaric acid 2	0.77 ± 0.19 [a]	0.91 ± 0.03 [a]	0.83 ± 0.06 [a]	0.43 ± 0.02 [b]	0.35 ± 0.01 [b]	0.26 ± 0.15 [b]	0.41 ± 0.11 [b]
3 [1]	Caffeoylglucaric acid 3	1.33 ± 0.15 [ab]	1.60 ± 0.33 [a]	1.19 ± 0.06 [b]	0.59 ± 0.07 [c]	0.47 ± 0.05 [c]	0.52 ± 0.07 [c]	0.47 ± 0.12 [c]
4 [1]	Caffeoylglucaric acid 4	1.12 ± 0.22 [a]	0.95 ± 0.37 [ab]	1.13 ± 0.04 [a]	0.60 ± 0.01 [bc]	0.53 ± 0.01 [c]	0.40 ± 0.01 [c]	0.50 ± 0.12 [c]
5 [1]	Caffeoylglucaric acid 5	1.70 ± 0.17 [a]	1.25 ± 0.37 [b]	1.20 ± 0.03 [b]	0.60 ± 0.01 [c]	0.50 ± 0.02 [c]	0.53 ± 0.02 [c]	0.47 ± 0.11 [c]
6 [2]	Coumaroylglucaric acid 1	0.20 ± 0.03 [bc]	0.11 ± 0.01 [c]	0.23 ± 0.02 [b]	0.20 ± 0.04 [bc]	0.22 ± 0.06 [b]	0.39 ± 0.07 [a]	0.08 ± 0.05 [b]
7 [2]	Coumaroylglucaric acid 2	0.61 ± 0.14 [b]	0.72 ± 0.09 [a]	0.63 ± 0.01 [ab]	0.44 ± 0.02 [bc]	0.39 ± 0.05 [c]	0.34 ± 0.00 [c]	0.58 ± 0.14 [ab]
8 [3]	Feruloylglucaric acid	1.00 ± 0.07 [b]	1.16 ± 0.47 [ab]	1.21 ± 0.08 [ab]	1.04 ± 0.09 [b]	1.17 ± 0.21 [ab]	1.67 ± 0.04 [a]	1.74 ± 0.39 [a]
9 [1]	Caffeoylquinic acid	0.79 ± 0.10 [c]	0.73 ± 0.10 [c]	1.08 ± 0.14 [bc]	0.97 ± 0.15 [bc]	1.11 ± 0.21 [bc]	1.60 ± 0.19 [a]	1.42 ± 0.36 [ab]
10	Caffeic acid	0.84 ± 0.03 [a]	0.69 ± 0.22 [b]	0.39 ± 0.02 [c]	0.20 ± 0.01 [cd]	0.12 ± 0.03 [d]	0.09 ± 0.02 [d]	Tr
11 [2]	Coumaroylquinic acid	0.63 ±0.09 [a]	0.59 ± 0.13 [a]	0.57 ± 0.10 [a]	0.53 ± 0.05 [a]	0.44 ± 0.06 [ab]	0.33 ±0.04 [bc]	0.22 ± 0.03 [c]
12 [3]	Feruloylquinic acid	9.89 ± 0.07 [a]	5.17 ± 1.11 [b]	3.54 ± 0.09 [c]	1.27 ± 0.03 [d]	0.46 ± 0.18 [e]	0.39 ± 0.01 [e]	0.34 ± 0.01 [e]
13	Rutin	15.0 ± 1.8 [b]	19.7 ± 5.7 [b]	19.3 ± 4.5 [b]	19.6 ± 2.8 [b]	20 ± 2.7 [b]	31.7 ± 2.7 [a]	36.2 ± 8.2 [a]
14 [2]	Hydroxycinnamic acid derivative	1.55 ± 0.01 [a]	0.69 ± 0.17 [b]	0.59 ± 0.10 [b]	0.17 ± 0.1 [c]	0.23 ± 0.05 [c]	0.09 ± 0.01 [c]	0.18 ± 0.00 [c]
15 [4]	Quercetin glucoside	0.78 ± 0.07 [c]	0.86 ± 0.21 [c]	1.34 ± 0.57 [abc]	1.23 ± 0.15 [bc]	1.28 ± 0.03 [bc]	1.97 ± 0.16 [ab]	2.07 ± 0.60 [a]
16 [2]	Hydroxycinnamic acid derivative	4.80 ± 0.10 [a]	2.51 ± 0.70 [b]	2.83 ± 0.49 [b]	1.59 ± 0.23 [c]	1.11 ± 0.27 [c]	1.26 ± 0.17 [c]	1.17 ± 0.08 [c]
17	Kaempferol-3-O-rutinoside	Tr	Tr	Tr	Tr	Tr	Tr	Tr
	Sum of all compounds	41.5 ± 2.8 [a]	38.1 ± 2.7 [abc]	36.5 ± 6.3 [abc]	29.6 ± 3.3 [bc]	28.5 ± 3.7 [c]	41.8 ± 2.8 [ab]	46.1 ± 9.0 [a]
	Sum of hydroxycinnamic acids	25.8 ± 1.0 [a]	17.5 ± 2.8 [b]	15.9 ± 1.2 [b]	8.86 ± 0.38 [c]	7.27 ± 1.03 [c]	7.99 ± 0.03 [c]	7.91 ± 1.50 [c]
	Sum of flavonols	15.79 ± 1.8 [b]	20.6 ± 6.0 [b]	20.6 ± 5.1 [b]	20.8 ± 2.9 [b]	21.2 ± 2.7 [b]	33.7 ± 2.8 [a]	38.2 ± 8.8 [a]

[1] Expressed as caffeic acid equivalents. [2] Expressed as *p*-coumaric acid equivalents. [3] Expressed as ferulic acid equivalents. [4] Expressed as quercetin equivalents. Tr, traces. [abcd] Means with different lowercase letters in the same row are significantly different ($p < 0.05$). Compound number corresponds to peak number in Figure 3.

Table 5. Individual phenolic compound contents in fresh matter of amaranth in different growth stages (µg/g).

Comp.No	Compound	Early Vegetative	Medium Vegetative	Late Vegetative	Shooting	Budding	Early Flowering	Grain Fill
1 [1]	Caffeoylglucaric acid 1	14.9 ± 1.8 [a]	15.4 ± 4.0 [a]	14.1 ± 0.1 [a]	8.36 ± 0.98 [b]	6.36 ± 0.54 [b]	7.63 ± 2.36 [b]	4.16 ± 0.60 [b]
2 [1]	Caffeoylglucaric acid 2	21.3 ± 4.7 [bc]	30.6 ± 0.9 [a]	26.1 ± 1.3 [ab]	15.6 ± 0.3 [cd]	13.7 ± 0.7 [cd]	10.6 ± 6.4 [d]	13.3 ± 3.0 [d]
3 [1]	Caffeoylglucaric acid 3	37.1 ± 3.4 [b]	53.4 ± 7.9 [a]	37.7 ± 1.0 [b]	21.6 ± 2.5 [c]	18.2 ± 2.4 [c]	21.1 ± 3.0 [c]	15.1 ± 3.3 [c]
4 [1]	Caffeoylglucaric acid 4	31.1 ± 5.5 [ab]	32.2 ± 14 [ab]	35.7 ± 0.5 [a]	21.9 ± 1.8 [abc]	20.6 ± 1.2 [bc]	16.4 ± 0.3 [c]	16.0 ± 3.3 [c]
5 [1]	Caffeoylglucaric acid 5	47.3 ± 3.6 [a]	42.5 ± 11 [a]	37.7 ± 0.1 [a]	21.8 ± 2.0 [b]	19.7 ± 1.2 [b]	21.6 ± 0.9 [b]	15.2 ± 2.9 [b]
6 [2]	Coumaroylglucaric acid 1	5.46 ± 0.07 [bc]	3.59 ± 0.04 [c]	7.22 ± 0.06 [bc]	7.51 ± 0.9 [bc]	8.81 ± 2.5 [b]	16.0 ± 3.2 [a]	9.07 ± 1.24 [b]
7 [2]	Coumaroylglucaric acid 2	16.8 ± 3.6 [bc]	24.3 ± 1.6 [a]	19.9 ± 0.0 [ab]	16.0 ± 2.0 [bc]	15.3 ± 2.4 [bc]	13.7 ± 0.0 [c]	18.6 ± 3.7 [bc]
8 [3]	Feruloylglucaric acid	27.9 ± 1.2 [c]	39.4 ± 15.2 [bc]	38.1 ± 1.7 [bc]	38.0 ± 6.1 [bc]	45.9 ± 9.5 [bc]	67.7 ± 1.3 [a]	56.1 ± 0.10 [ab]
9 [1]	Caffeoylquinic acid	22.0 ± 2.3 [c]	24.8 ± 4.7 [c]	34.1 ± 3.7 [bc]	35.7 ± 8.0 [bc]	43.3 ± 9.5 [b]	65.0 ± 7.2 [a]	45.9 ± 9.6 [b]
10	Caffeic acid	26.2 ± 0.3 [a]	22.7 ± 5.8 [a]	12.2 ± 0.4 [b]	7.24 ± 0.74 [bc]	4.69 ± 1.21 [c]	3.04 ± 0.09 [c]	Tr
11 [2]	Coumaroylquinic acid	17.6 ± 3.1 [a]	20.2 ± 5.7 [a]	17.9 ± 2.8 [a]	19.6 ± 3.2 [a]	17.1 ± 1.7 [a]	13.4 ± 1.8 [ab]	7.28 ± 0.8 [bc]
12 [3]	Feruloylquinic acid	275 ± 4.2 [a]	175 ± 48 [b]	111 ± 1.0 [c]	46.2 ± 2.2 [d]	18.1 ± 7.5 [d]	15.8 ± 0.4 [d]	11.2 ± 0.3 [d]
13	Rutin	418 ± 40 [c]	658 ± 155 [bc]	606 ± 129 [bc]	718 ± 154 [bc]	780 ± 130 [b]	1286 ± 99 [a]	1169 ± 215 [a]
14 [2]	Hydroxycinnamic acid derivative	40.6 ± 0.7 [a]	23.3 ± 7.2 [b]	18.4 ± 2.8 [b]	6.11 ± 3.44 [c]	9.00 ± 2.25 [c]	3.55 ± 0.48 [c]	5.88 ± 0.30 [c]
15 [4]	Quercetin glucoside	21.6 ± 2.5 [c]	29.3 ± 8.7 [c]	41.9 ± 17 [bc]	45.1 ± 8.6 [bc]	50.0 ± 2.9 [b]	79.6 ± 5.8 [a]	66.7 ± 0.17 [ab]
16 [2]	Hydroxycinnamic acid derivative	134 ± 5.7 [a]	85.1 ± 28.7 [b]	89.0 ± 13.8 [b]	58.5 ± 12.7 [bc]	43.4 ± 9.8 [c]	51.1 ± 6.5 [c]	38.0 ± 0.8 [c]
17	Kaempferol-3-*O*-rutinoside	Tr	Tr	Tr	Tr	Tr	Tr	Tr
	Sum of all compounds	1157 ± 51 [bc]	1281 ± 14 [bc]	1147 ± 175 [bc]	1087 ± 196 [c]	1114 ± 181 [c]	1690 ± 101 [a]	1491 ± 270 [ab]
	Sum of hydroxycinnamic acids	718 ± 14 [a]	593 ± 132 [ab]	499 ± 29 [b]	324 ± 38 [c]	264 ± 49 [c]	324 ± 3.9 [c]	256 ± 39 [c]
	Sum of flavonols	439 ± 42 [c]	688 ± 163 [bc]	648 ± 146 [bc]	763 ± 163 [bc]	830 ± 132 [b]	1366 ± 105 [a]	1235 ± 232 [a]

[1] Expressed as caffeic acid equivalents. [2] Expressed as *p*-coumaric acid equivalents. [3] Expressed as ferulic acid equivalents. [4] Expressed as quercetin equivalents. Tr, traces. [abcd] Means with different lowercase letters in the same row are significantly different ($p < 0.05$). Compound number corresponds to peak number in Figure 3.

3.4. Principal Component Analysis

The PCA was carried out to clarify the relationship between the different growth stages of amaranth aerial parts and the evaluated variables. The data set obtained for 16 phenolic compounds, five antioxidant assays, and TPC with respect to seven growth stages was subjected to PCA. The results are presented in Figure 4. The first two principal components (PC1 and PC2) significantly explained most of the variations–87.08% of the total variance. The clustering of growth stages with three separate groups: 1–vegetative stages (early, medium, and late vegetative), 2–shooting and budding stages, and 3–early flowering and grain fill stages was obtained as shown in plot A (Figure 4). The variables that were differentiating distribution of each growth stage were depicted in plot B (Figure 4). The early, medium, and late vegetative stages were associated with the most antioxidant assays: TEAC, FRAP, PCL-ACL, and Fe^{2+} chelating ability. The TPC and content of some hydroxycinnamic acid derivatives (compounds **1–5**, **7**, **10–12**, **14**, and **16**) also affected this group of stages. In turn, the third group of growth stages (early flowering and grain fill) was clearly related to content of flavonols and three remaining hydroxycinnamic acid derivatives (compounds **6**, **8**, and **9**). The DPPH assay was negatively correlated with other antioxidant activity assays because the EC_{50} values were used in PCA. In the plot, the DPPH assay influenced the distribution of shooting and budding stages due to their low DPPH$^{\bullet}$ scavenging activity with high EC_{50} values.

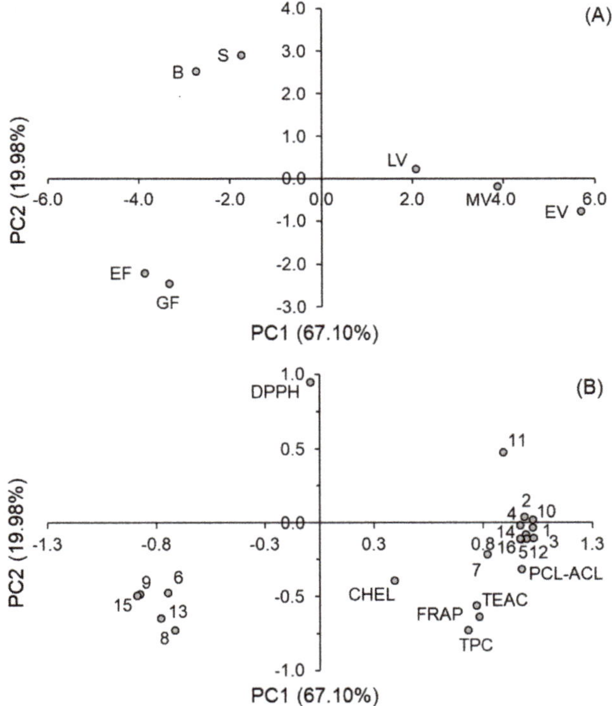

Figure 4. Principal component analysis (PCA) plots of the data set of variables: Total phenolic content (TPC), individual phenolic compound (**1–16**), and antioxidant assays obtained for amaranth in different growth stages. EV, early vegetative; MV, medium vegetative; LV, late vegetative; S, shooting; B, budding; EF, early flowering; GF, grain fill; TEAC, Trolox equivalent antioxidant capacity; FRAP, ferric-reducing antioxidant power; PCL-ACL, antioxidant capacity in photochemiluminescence assay; CHEL, Fe^{2+} chelating ability. (**A**): Distribution of growth stages, (**B**): Distribution of variables.

4. Conclusions

Comparison of antioxidant activity and phenolic composition of the aerial parts of *A. caudatus* showed growth cycle dependent effects. The amaranth in shooting and budding stages had the lowest antioxidant activity and the lowest total phenolic content. The earlier (vegetative) and later (early flowering and grain fill) stages formed two groups, both with high antioxidant activity, though with differing dominant classes of phenolic compounds. The content of flavonols (mainly rutin) increased with growth cycle, and these compounds could be primarily responsible for the antioxidant activity of amaranth in early flowering and grain fill stages. In plants in vegetative stages, especially the early vegetative stage, the participation of hydroxycinnamic acid derivatives in the pool of phenolic compounds was high. The contribution of these compounds to high antioxidant activity of amaranth in vegetative stages seemed significant. It is worth noting the predominance of rutin quantity among individual phenolic compounds in all *A. caudatus* growth states, as well as its increasing content during the growth cycle.

The extracts obtained from amaranth in vegetative, early flowering, and grain fill stages, due to their high content of hydroxycinnamic acid derivatives and rutin, can be a valuable source of antioxidants that can be exploited for the production of nutraceuticals or used as a functional food ingredient.

Author Contributions: Conceptualization, M.K., F.G., R.A. and P.G.P.; methodology, M.K., G.M. and M.A.J.; formal analysis, M.K. and M.A.J.; investigation, M.K., E.L. and M.A.J.; resources, E.L. and G.M.; writing—original draft, M.K., F.G. and P.G.P.; writing—review and editing, M.K., F.G., R.A. and P.G.P.; visualization, M.A.J.; supervision, R.A.

Funding: This research received no external funding.

Acknowledgments: The Italian National Research Council provides a visiting grant to Ryszard Amarowicz in the framework of a free exchange programme between the Polish Academy of Sciences and the Italian National Research Council. The authors thank Wiesław Wiczkowski for assistance in the HPLC-MS/MS analysis.

Conflicts of Interest: The authors declare no conflict of interest.

References

1. Rastogi, A.; Shukla, S. Amaranth: A new millennium crop of nutraceutical values. *Crit. Rev. Food Sci. Nutr.* **2013**, *53*, 109–125. [CrossRef] [PubMed]
2. Repo-Carrasco-Valencia, R.; Peña, J.; Kallio, H.; Salminen, S. Dietary fiber and other functional components in two varieties of crude and extruded kiwicha (*Amaranthus caudatus*). *J. Cereal Sci.* **2009**, *49*, 219–224. [CrossRef]
3. Guzman-Maldonado, S.; Paredes-Lopez, O. Functional products of plant indigenous to Latin America. Amaranth and quinoa, common beans and botanicals. In *Functional Foods. Biochemical and Processing Aspects*; Mazza, G., Ed.; Technomic Publishing Co. Inc.: Lancaster, PA, USA, 1998; pp. 293–328.
4. Venskutonis, P.R.; Kraujalis, P. Nutritional components of amaranth seeds and vegetables: A review on composition, properties, and uses. *Compr. Rev. Food Sci. Food Saf.* **2013**, *12*, 381–412. [CrossRef]
5. López, D.N.; Galante, M.; Raimundo, G.; Spelzini, D.; Boeris, V. Functional properties of amaranth, quinoa and chia proteins and the biological activities of their hydrolyzates. *Food Res. Int.* **2019**, *116*, 419–429. [CrossRef] [PubMed]
6. Alvarez-Jubete, L.; Arendt, E.K.; Gallaghera, E. Nutritive value of pseudocereals and their increasing use as functional gluten-free ingredients. *Trends Food Sci. Technol.* **2010**, *21*, 106–113. [CrossRef]
7. Conte, P.; Fadda, C.; Drabińska, N.; Krupa-Kozak, U. Technological and nutritional challenges, and novelty in gluten-free breadmaking—A review. *Pol. J. Food Nutr. Sci.* **2019**, *69*, 5–21. [CrossRef]
8. Tang, Y.; Tsao, R. Phytochemicals in quinoa and amaranth grains and their antioxidant, anti-inflammatory, and potential health beneficial effects: A review. *Mol. Nutr. Food Res.* **2017**, *61*, 1600767. [CrossRef]
9. Alvarez-Jubete, L.; Wijngaard, H.; Arendt, E.K.; Gallagher, E. Polyphenol composition and in vitro antioxidant activity of amaranth, quinoa buckwheat and wheat as affected by sprouting and baking. *Food Chem.* **2010**, *119*, 770–778. [CrossRef]

10. Paucar-Menacho, M.; Dueñas, M.; Peñas, E.; Frias, J.; Martínez-Villaluenga, C. Effect of dry heat puffing on nutritional composition, fatty acid, amino acid and phenolic profiles of pseudocereals grains. *Pol. J. Food Nutr. Sci.* **2018**, *68*, 289–297. [CrossRef]
11. Repo-Carrasco-Valencia, R.; Hellström, J.K.; Pihlava, J.M.; Mattila, P.H. Flavonoids and other phenolic compounds in Andean indigenous grains: Quinoa (*Chenopodium quinoa*), kañiwa (*Chenopodium pallidicaule*) and kiwicha (*Amaranthus caudatus*). *Food Chem.* **2010**, *120*, 128–133. [CrossRef]
12. Conforti, F.; Statti, G.; Loizzo, M.R.; Sacchetti, G.; Poli, F.; Menichini, F. In vitro antioxidant effect and inhibition of α-amylase of two varieties of *Amaranthus caudatus* seeds. *Biol. Pharm. Bull.* **2005**, *28*, 1098–1102. [CrossRef] [PubMed]
13. Nsimba, R.Y.; Kikuzaki, H.; Konishi, Y. Antioxidant activity of various extracts and fractions of *Chenopodium quinoa* and *Amaranthus* spp. seeds. *Food Chem.* **2008**, *106*, 760–766. [CrossRef]
14. Peiretti, P.G.; Meineri, G.; Gai, F.; Longato, E.; Amarowicz, R. Antioxidative activity and phenolic compounds of pumpkin (*Cucurbita pepo*) seeds and amaranth (*Amaranthus caudatus*) grain extracts. *Nat. Prod. Res.* **2017**, *31*, 2178–2182. [CrossRef] [PubMed]
15. Asao, M.; Watanabe, K. Functional and bioactive properties of quinoa and amaranth. *Food Sci. Technol. Res.* **2010**, *16*, 163–168. [CrossRef]
16. Kraujalis, P.; Venskutonis, P.R.; Kraujalienė, V.; Pukalskas, A. Antioxidant properties and preliminary evaluation of phytochemical composition of different anatomical parts of amaranth. *Plant Foods Hum. Nutr.* **2013**, *68*, 322–328. [CrossRef] [PubMed]
17. Jiménez-Aguilar, D.M.; Grusak, M.A. Minerals, vitamin C, phenolics, flavonoids and antioxidant activity of Amaranthus leafy vegetables. *J. Food Compos. Anal.* **2017**, *58*, 33–39. [CrossRef]
18. Neugart, S.; Baldermann, S.; Ngwene, B.; Wesong, J.; Schreiner, M. Indigenous leafy vegetables of Eastern Africa—A source of extraordinary secondary plant metabolites. *Food Res. Int.* **2017**, *100*, 411–422. [CrossRef]
19. Li, H.; Deng, Z.; Liu, R.; Zhu, H.; Draves, J.; Marcone, M.; Sun, Y.; Tsao, R. Characterization of phenolics, betacyanins and antioxidant activities of the seed, leaf, sprout, flower and stalk extracts of three Amaranthus species. *J. Food Compos. Anal.* **2015**, *37*, 75–81. [CrossRef]
20. Steffensen, S.K.; Pedersen, H.A.; Labouriau, R.; Mortensen, A.G.; Laursen, B.; de Troiani, R.M.; Noellemeyer, E.J.; Janovska, D.; Stavelikova, H.; Taberner, A.; et al. Variation of polyphenols and betaines in aerial parts of young, field-grown *Amaranthus* genotypes. *J. Agric. Food Chem.* **2011**, *59*, 12073–12082. [CrossRef]
21. Conforti, F.; Marrelli, M.; Carmela, C.; Menichini, F.; Valentina, P.; Uzunov, D.; Statti, G.A.; Duez, P.; Menichini, F. Bioactive phytonutrients (omega fatty acids, tocopherols, polyphenols), *in vitro* inhibition of nitric oxide production and free radical scavenging activity of non-cultivated Mediterranean vegetables. *Food Chem.* **2011**, *129*, 1413–1419. [CrossRef]
22. Ozsoy, N.; Yilmaz, T.; Kurt, O.; Can, A.; Yanardag, R. *In vitro* antioxidant activity of *Amaranthus lividus* L. *Food Chem.* **2009**, *116*, 867–872. [CrossRef]
23. Miguel, M.G. Betalains in some species of the Amaranthaceae family: A review. *Antioxidants* **2018**, *7*, 53. [CrossRef] [PubMed]
24. Cai, Y.; Sun, M.; Wu, H.; Huang, R.; Corke, H. Characterization and quantification of betacyanin pigments from diverse *Amaranthus* species. *J. Agric. Food Chem.* **1998**, *46*, 2063–2070. [CrossRef]
25. Peiretti, P.G.; Meineri, G.; Longato, E.; Tassone, S. Chemical composition, *in vitro* digestibility and fatty acid profile of *Amaranthus caudatus* herbage during its growth cycle. *Anim. Nutr. Feed Technol.* **2018**, *18*, 107–116. [CrossRef]
26. Barba de la Rosa, A.P.; Fomsgaard, I.S.; Laursen, B.; Mortensen, A.G.; Olvera-Martínez, L.; Silva-Sánchez, C.; Mendoza-Herrera, A.; González-Castañeda, J.; De León-Rodrígueza, A. Amaranth (*Amaranthus hypochondriacus*) as an alternative crop for sustainable food production: Phenolic acids and flavonoids with potential impact on its nutraceutical quality. *J. Cereal Sci.* **2009**, *49*, 117–121. [CrossRef]
27. Karamać, M.; Kosińska, A.; Estrella, I.; Hernández, T.; Dueñas, M. Antioxidant activity of phenolic compounds identified in sunflower seeds. *Eur. Food Res. Technol.* **2012**, *235*, 221–230. [CrossRef]
28. Re, R.; Pellegrini, N.; Proteggente, A.; Pannala, A.; Yang, M.; Rice-Evans, C. Antioxidant activity applying an improved ABTS radical cation decolorization assay. *Free Radic. Biol. Med.* **1999**, *26*, 1231–1237. [CrossRef]
29. Benzie, I.F.F.; Strain, J.J. The ferric reducing ability of plasma (FRAP) as a measure of "antioxidant power": The FRAP assay. *Anal. Biochem.* **1996**, *239*, 70–76. [CrossRef]

30. Popov, I.; Lewin, G. Oxidants and antioxidants part B—Antioxidative homeostasis: Characterization by means of chemiluminescent technique. *Methods Enzymol.* **1999**, *300*, 437–456.

31. Karamać, M.; Pegg, R.B. Limitations of the tetramethylmurexide assay for investigating the Fe(II) chelation activity of phenolic compounds. *J. Agric. Food Chem.* **2009**, *57*, 6425–6431. [CrossRef]

32. Peiretti, P.G.; Karamać, M.; Janiak, M.; Longato, E.; Meineri, G.; Amarowicz, R.; Gai, F. Phenolic composition and antioxidant activities of soybean (*Glycine max* (L.) Merr.) plant during growth cycle. *Agronomy* **2019**, *9*, 153. [CrossRef]

33. Brand-Williams, W.; Cuvelier, M.E.; Berset, C. Use of a free-radical method to evaluate antioxidant activity. *LWT Food Sci. Technol.* **1995**, *28*, 25–30. [CrossRef]

34. Janiak, M.A.; Slavova-Kazakova, A.; Kancheva, V.D.; Ivanova, M.; Tsrunchev, T.; Karamać, M. Effects of γ-irradiation of wild thyme (*Thymus serpyllum* L.) on the phenolic compounds profile of its ethanolic extract. *Pol. J. Food Nutr. Sci.* **2017**, *67*, 309–316. [CrossRef]

35. Stintzing, F.C.; Kammerer, D.; Schieber, A.; Adama, H.; Nacoulma, O.G.; Carle, R. Betacyanins and phenolic compounds from *Amaranthus spinosus* L. and *Boerhavia erecta* L. *Z. Naturforsch. C* **2004**, *59*, 1–8. [CrossRef] [PubMed]

36. Gai, F.; Peiretti, P.G.; Karamać, M.; Amarowicz, R. Changes in antioxidative capacity and phenolic compounds in quinoa (*Chenopodium quinoa* Willd.) plant extracts during growth cycle. In *Quinoa: Cultivation, Nutritional Properties and Effects on Health*; Peiretti, P.G., Gai, F., Eds.; Nova Science Publishers Inc.: Hauppage, NY, USA, 2019; pp. 63–81.

37. López-Mejía, O.A.; López-Malo, A.; Palou, E. Antioxidant capacity of extracts from amaranth (*Amaranthus hypochondriacus* L.) seeds or leaves. *Ind. Crops Prod.* **2014**, *53*, 55–59. [CrossRef]

38. Khandaker, L.; Ali, M.B.; Oba, S. Total polyphenol and antioxidant activity of red amaranth (*Amaranthus tricolor* L.) as affected by different sunlight level. *J. Jpn. Soc. Hortic. Sci.* **2008**, *77*, 395–401. [CrossRef]

39. Modi, T.A. Growth temperature and plant age influence on nutritional quality of *Amaranthus* leaves and seed germination capacity. *Water SA* **2007**, *33*, 369–378. [CrossRef]

40. Klimczak, I.; Małecka, M.; Pachołek, B. Antioxidant activity of ethanolic extracts of amaranth seeds. *Nahrung* **2002**, *46*, 184–186. [CrossRef]

41. Niveyro, S.L.; Mortensen, A.G.; Fomsgaard, I.S.; Salvo, A. Differences among five amaranth varieties (*Amaranthus* spp.) regarding secondary metabolites and foliar herbivory by chewing insects in the field. *Arthropod-Plant Interact.* **2013**, *7*, 235–245. [CrossRef]

42. Kalinova, J.; Dadakova, E. Rutin and total quercetin content in amaranth (*Amaranthus* spp.). *Plant Foods Hum. Nutr.* **2009**, *64*, 68–74. [CrossRef]

 *antioxidants*

Article

Nutritional Value, Chemical Composition and Cytotoxic Properties of Common Purslane (*Portulaca oleracea* L.) in Relation to Harvesting Stage and Plant Part

Spyridon A. Petropoulos [1],*, Ângela Fernandes [2], Maria Inês Dias [2], Ioannis B. Vasilakoglou [3], Konstantinos Petrotos [3], Lillian Barros [2] and Isabel C. F. R. Ferreira [2],*

[1] Department of Agriculture Crop Production and Rural Environment, University of Thessaly, 38446 N. Ionia, Magnissia, Greece
[2] Centro de Investigação de Montanha (CIMO), Instituto Politécnico de Bragança, Campus de Santa Apolónia, 5300-253 Bragança, Portugal
[3] Department of Crop Production-Agrotechnology, University of Thessaly, 41110 Larissa, Greece
* Correspondence: spetropoulos@uth.gr (S.A.P.); iferreira@ipb.pt (I.C.F.R.F.);
   Tel.: +30-2421-093-196 (S.A.P.); +351-273303219 (I.C.F.R.F.)

Received: 21 July 2019; Accepted: 6 August 2019; Published: 8 August 2019

Abstract: Purslane (*Portulaca oleraceae* L.) is a widespread weed, which is highly appreciated for its high nutritional value with particular reference to the content in omega-3 fatty acids. In the present study, the nutritional value and chemical composition of purslane plants in relation to plant part and harvesting stage were evaluated. Plants were harvested at three growth stages (29, 43 and 52 days after sowing (DAS)), while the edible aerial parts were separated into stems and leaves. Leaves contained higher amounts of macronutrients than stems, especially at 52 DAS. α-tocopherol was the main isoform, which increased at 52 DAS, as well total tocopherols (values were in the ranges of 197–327 µg/100 g fresh weight (fw) and 302–481 µg/100 g fw, for α-tocopherol and total tocopherols, respectively). Glucose and fructose were the main free sugars in stems and leaves, respectively, whereas stems contained higher amounts of total sugars (values were ranged between 0.83 g and 1.28 g/100 g fw). Oxalic and total organic acid content was higher in leaves, especially at the last harvesting stage (52 DAS; 8.6 g and 30.3 g/100 g fw for oxalic acid and total organic acids, respectively). Regarding the fatty acid content, stems contained mainly palmitic (20.2–21.8%) and linoleic acid (23.02–27.11%), while leaves were abundant in α-linolenic acid (35.4–54.92%). Oleracein A and C were the major oleracein derivatives in leaves, regardless of the harvesting stage (values were in the ranges of 8.2–103.0 mg and 21.2–143 mg/100 g dried weight (dw) for oleraceins A and C, respectively). Cytotoxicity assays showed no hepatotoxicity, with GI_{50} values being higher than 400 µg/mL for all the harvesting stages and plant parts. In conclusion, early harvesting and the separation of plant parts could increase the nutritional value of the final product through increasing the content of valuable compounds, such as omega-3 fatty acids, phenolic compounds and oleracein derivatives, while at the same time, the contents of anti-nutritional compounds such as oxalic acid are reduced.

Keywords: α-linolenic acid; fatty acids; hepatotoxicity; omega-3 fatty acids; *Portulaca oleraceae*; phenolic compounds; oleracein derivatives; purslane; tocopherols

---

1. Introduction

Purslane (*Portulaca oleracea* L.) is a widespread weed belonging to the Portulacaceae family, with extensive distribution throughout the world [1]. It is also a basic component of the so-called Mediterranean diet and an ingredient of many salad dishes [2]. Although it is considered a very invasive weed [3], it is also highly appreciated for the high nutritional value of its edible plant parts, due to their high content in omega-3 fatty acids and, particularly, in α-linolenic acid [4]. Other valuable components of purslane edible parts include minerals, such as calcium, potassium and phosphorus, macronutrients such as proteins and carbohydrates [5], as well as tocopherols, carotenoids and ascorbic acid [6,7]. Moreover, phenolic compounds and oleracein derivatives of extracts of purslane leaves have been attributed with antioxidant properties [8], while their content may be affected by cooking and food additives [9]. *P. oleracea* is tolerant under stress conditions, such as heat, drought and salinity stress, a trait that could be useful within the ongoing climate change context and provide alternative solutions to farmers in climate-affected regions [10].

Several studies have reported the effects of cultivation practices, growing conditions and genetic factors on the chemical composition of purslane. In particular, Rahdari et al. [11], Uddin et al. [12] and Teixeira and Carvalho [13] reported that salinity stress may significantly affect proximate and mineral composition, while salinity stress also has an impact on polyphenol and carotenoid content and antioxidant activity [14]. Genetic material is also important for obtaining edible parts of high nutritional value, since many studies have suggested significant differences in the chemical compositions of purslane accessions and genotypes, especially regarding fatty acid and oxalic acid content [6,15] and bioactive properties [16]. Cultivation practices such as planting date or fertilization rates can also be proven as useful means towards the modulation of the chemical composition of purslane edible parts [17,18]. In addition, cultivation systems may affect the nutritional value of purslane aerial parts, with soilless culture showing promising results in regards to improving fatty acid composition and decreasing oxalic acid content [19]. High oxalic acid content is considered as an antinutrient factor, and apart from cultivation systems, the proper harvesting stage and cultivar selection, as well as the modulation of nutrient solution composition may also be helpful in decreasing its content and increasing the overall nutritional value of purslane edible parts [15,20]. Harvesting stage and plant parts also have a significant impact on macronutrients (total solids, proteins, ash, and carbohydrates) and mineral content [5], as well as on phenolic compounds, oleracein derivatives and organic acid profile [4].

The aim of the present work was to study the effect of harvesting stage on the chemical composition of purslane edible parts, as well as the distribution of the main nutrient components and phytochemicals within the aerial parts of the plant (leaves and stems). Considering the importance of omega-3 fatty acids and oxalic acid for purslane's nutritional value, special attention was given to oxalic acid and fatty acid content, along with the plant part selection and harvesting stage, which could be used as simple cultivation practices to increase the nutritional and added value of the final product. Finally, the characterization of phenolic compounds and oleracein derivatives in the aerial plant parts (leaves and stems) in relation to harvesting stage was also performed.

2. Materials and Methods

2.1. Plant Material and Growing Conditions

This trial was carried out at the experimental field of the University of Thessaly, in Larissa (Greece; 39°37′18.6″ N, 22°22′55.1″ E), during the summer of 2016. Seeds of common purslane (*Portulaca oleracea* L.) were obtained from Hortus Sementi Srl. (Budrio, Italy) and were sown directly in soil on 06/06/2016. Prior to sowing, a base dressing of 100 kg/ha with 10-10-10 fertilizer (N-P-K) was applied. Irrigation was applied with sprinklers at regular intervals (once a week, starting on the day of sowing). The soil was sandy clay loam (38% sand, 36% silt, and 26% clay), with pH = 7.4 (1:1 soil/H20) and organic matter content = 1.3%. No pesticides or other agrochemicals were applied during cultivation. Harvesting took place at three different growth stages, namely on 05/07/2016 (29 days after sowing (DAS)), on 19/07/2016 (43 DAS), and on 28/07/2016 (52 DAS).

After each harvesting stage, the aerial plant parts were divided in stems and leaves. Fresh samples of plant tissues were placed in a forced-air oven, and dry weight was recorded after drying the samples at 70 °C until constant weight. Batch samples of fresh plant tissues were stored at −80 °C and were then lyophilized. The lyophilized samples were ground to powder with a pestle and mortar, and were put in plastic air-sealed bags and stored at −80 °C until further analysis.

For phenolic and oleracein composition, as also for cytotoxicity, a hydroethanolic extract was prepared using 1 g of dried sample with 30 mL ethanol/water (80:20 *v/v*), under magnetic stirring for 1 h. After filtration through a Whatman filter paper N°. 4, the plant residue was re-extracted and the combined filtrates were evaporated at 40 °C (rotary evaporator Büchi R-210, Flawil, Switzerland) and subsequently lyophilized to obtain a dry extract.

2.2. Chemical Analyses

2.2.1. Standards and Reagents

Acetonitrile 99.9% of HPLC grade was from Fisher Scientific (Lisbon, Portugal). Ellipticine was purchased from Sigma-Aldrich (St. Louis, MO, USA), as also were acetic acid, sulforhodamine B (SRB), trypan blue, trichloroacetic acid (TCA), and Tris-(hydroxymethyl)aminomethan (TRIS). Phenolic compound standards were from Extrasynthèse (Genay, France). RPMI-1640 medium, fetal bovine serum (FBS), Hank's balanced salt solution (HBSS), L-glutamine, nonessential amino acids solution (2 mM), penicillin/streptomycin solution (100 U/mL and 100 mg/mL, respectively), and trypsin-EDTA (ethylenediaminetetraacetic acid) were from Hyclone (Logan, UT, USA). All other chemicals were of analytical purity and obtained from common suppliers. Water was treated via the purification system Milli-Q water (TGI Pure Water Systems, Greenville, SC, USA).

2.2.2. Nutritional Compounds and Energetic Value

Nutritional compounds of the samples were analyzed (moisture, fat, ash, proteins and carbohydrates) following the Association of Analytical Communities (AOAC) procedures [21]. Briefly, moisture content was determined using an air oven at 105 ± 5 °C until constant weight. Crude protein was evaluated by the macro-Kjeldahl method (N × 6.25) using an automatic distillation and titration unit (model Pro-Nitro-A, JP Selecta, Barcelona, Spain), ash content was determined by incineration at 600 ± 15 °C, and the crude fat was determined by extraction with petroleum ether using a Soxhlet apparatus (Behr Labor Technik, Dusseldorf, Germany). Total carbohydrates were determined by the difference according to the equation: total carbohydrates (g/100 g fresh weight (fw)) = 100 − (g moisture + g fat + g ash + g proteins).

Energy was determined according to the Atwater system following the equation: (kcal/100 g fw) = 4 × (g proteins + g carbohydrates) + 9 × (g fat).

2.2.3. Tocopherols

Tocopherols were determined in the lyophilized samples following a procedure previously described by Dias et al. [22]. The separation of compounds was performed using a high performance liquid chromatography system (Knauer, Smartline system 1000, Berlin, Germany) coupled to a fluorescence detector (FP-2020; Jasco, Easton, PA, USA) programmed for excitation at 290 nm and emission at 330 nm, using the internal standard (IS; tocol, Matreya, Pleasant Gap, PA, USA) method for quantification. After separation, the compounds were identified by comparison with authentic standards. The quantification was based on the fluorescence signal response of each standard, using the internal standard method. The results were expressed in μg per 100 g of fresh weight (fw).

2.2.4. Free Sugars

The lyophilized samples were extracted using a methodology previously described by the authors [23]. The separation of compounds was performed using a HPLC system (Knauer, Smartline system 1000, Berlin, Germany) coupled with a refraction index detector (Knauer Smartline 2300). The mobile phase consisted of an acetonitrile:water mixture (70:30 *v/v*, acetonitrile HPLC-grade,

Lab-Scan, Lisbon, Portugal), and separation was achieved using a Eurospher 100-5 NH2 column (4.6 × 250 mm, 5 μm, Knauer, Berlin, Germany). After separation, the compounds were identified by comparison with standards and quantification was performed by the IS method (melezitose). Results were processed using the Clarity 2.4 software (DataApex, Prague, Czech Republic) and expressed in g per 100 g of fw.

2.2.5. Organic Acids

The lyophilized samples were extracted using a methodology previously described and optimized by the authors [23]. The analysis was performed by Ultra-Fast Liquid Chromatography coupled with a Diode-Array Detector (UFLC-DAD; Shimadzu 20A series UFLC, Shimadzu Corporation, Kyoto, Japan). Compounds were identified and quantified by comparison of the retention time, spectra and peak area, recorded at 215 nm, with those obtained from commercial standards. The results were recorded and processed using LabSolutions Multi LC-PDA software (Shimadzu Corporation, Kyoto, Japan) and were expressed in g/100 g fw.

2.2.6. Fatty Acids

Fatty acids were determined by gas-liquid chromatography with flame ionization detection (GC-FID; DANI1000, Contone, Switzerland), after the extraction and derivatization procedures previously described by Obodai et al. [24]. Fatty acid identification and quantification were performed by comparing the relative retention times of fatty acid methyl ester (FAME) peaks of the tested samples with commercial standards (fatty acid methyl esters (FAMEs) reference standard mixture 37, Sigma-Aldrich, St. Louis, MO, USA). Results were expressed as the relative percentage for each detected fatty acid, using CSW 1.7 software (Data Apex 1.7, Prague, Czech Republic).

2.2.7. Phenolic Compounds and Oleracein Derivatives

Phenolic compounds and oleracein derivatives were evaluated using an ultra-performance liquid chromatography (UPLC) system equipped with a diode array detector (280 nm, 330 nm and 370 nm as preferred wavelengths) coupled to an electrospray ionization mass spectrometry detector (MS) (Dionex Ultimate 3000 UPLC and Linear Ion Trap LTQ XL, Thermo Scientific, San Jose, CA, USA), operating under the conditions described by Bessada et al. [25]. Data acquisition was carried out with the Xcalibur® data system. The identification was made by comparison of retention times, UV-VIS and mass spectra of the sample compounds with those obtained from the available standards, as also with reported data from the literature, and tentatively identified by using the fragmentation pattern. The estimation of phenolic compounds and oleracein derivatives was carried out using the calibration curves obtained from standards, which were constructed based on their UV-VIS signals; the quantification of all compounds was performed at 330 nm. In the case of a non-available standard compound, the most similar structural compound available in the laboratory was used to perform the quantification. A manual integration using the baseline to valley integration mode with baseline projection was performed to obtain the area of the peaks. The results were expressed in mg/100 g dried weight (dw).

2.3. Cytotoxicity

2.3.1. Cytotoxicity in Non-Tumor Liver Cell Primary Culture

Hepatotoxic activity was evaluated following the method described by Abreu et al. [26], using a primary cell culture (PLP2) prepared from a porcine liver. Extracts were tested at a final concentration range from 400 to 1.56 μg/mL. Briefly, a cell culture was prepared from a freshly harvested porcine liver (obtained from a local slaughter house) and designated as PLP2 [26]. Briefly, the liver tissue was rinsed in Hank's balanced salt solution containing 100 U/mL penicillin + 100 μg/mL streptomycin, and was divided into 1×1 mm^3 explants. Some of them were placed into 25 cm^2 tissue flasks containing Dulbecco's Modified Eagle's medium (DMEM, supplemented with 10% fetal bovine serum, 2 mM non-essential amino acids, 100 U/mL penicillin, and 100 mg/mL streptomycin) and incubated at 37 °C under a humidified atmosphere with 5% CO_2. A phase contrast microscope was used for direct monitoring of the cell cultivation every 2 to 3 days. Before reaching the confluence, cells were subcultured and plated in 96-well plates at a density of 1.0×10^4 cells/well, and cultivated in commercial DMEM medium supplemented with 10% FBS, 100 U/mL penicillin and 100 µg/mL streptomycin. The results were measured through the Sulforhodamine B method, where the amount of pigmented cells is directly proportional to the total protein mass and therefore to the number of bounded cells. Results were expressed as GI_{50} values (concentration that inhibits 50% of cell growth) and Ellipticine was used as a positive control.

2.3.2. Cytotoxicity in Human Tumor Cell Lines

Four human tumor cell lines were used: MCF-7 (breast adenocarcinoma), NCI-H460 (non-small cell lung cancer), HeLa (cervical carcinoma) and HepG2 (hepatocellular carcinoma), as previously described by Barros et al. [23]. MCF-7, HeLa and HepG2 were obtained from the European Collection of Cell Cultures (ECACC, Salisbury, UK) and NCI-H460 was kindly provided by the National Cancer Institute (NCI, Bethesda, MD, USA). Extracts were tested at a final concentration range from 400 to 1.56 µg/mL. RPMI-1640 medium containing 10% heat-inactivated FBS and 2 mM glutamine was used to routinely maintain the adherent cell cultures at 37 °C, in a humidified air incubator containing 5% CO_2. For the experiments, each cell line was placed at an appropriate density (1.0×10^4 cells/well) into 96-well plates. The cell growth inhibition was measured using sulforhodamine B, results were expressed as GI_{50} values and Ellipticine was used as a positive control.

2.4. Statistical Analysis

The experimental design was a completely randomized design (CRD), with three replications per treatment (harvesting stage, plant part). For the statistical analysis of chemical composition and bioactivity assays, three samples were analyzed for each treatment, whereas all of the assays were carried out in triplicate. Statistical analysis was conducted with the aid of Statgraphics 5.1.plus (Statpoint Technologies, Inc., VA, USA). Data were evaluated by a two-way ANOVA for the main effects, whereas the means of values were compared by Tukey's honestly significant difference (HSD) test ($p = 0.05$) in the case of harvesting stage effect, and by Student's t-test ($p = 0.05$) in the case of plant part effect.

3. Results and Discussion

The statistical analysis of the data showed a significant interaction between plant parts and harvesting stage for all the recorded parameters. Therefore, the interpretation of the data refers to the combined effect of both factors. Moisture content and proximate analysis results are presented in Table 1. The moisture content of leaves was the highest at 29 and 43 DAS and decreased at 52 DAS with plant maturity, whereas stems contained more water at 43 DAS, followed by the harvesting stages of 52 and 29 DAS. For macronutrient content (fat, proteins, ash, and carbohydrates) and the energetic value of plant tissues, the highest content in leaves was observed at the last harvest (52 DAS), which could be probably attributed to a concentration effect. Moreover, the opposite trend was observed in the case of stems, where harvesting at 29 DAS resulted in the highest content of ash and carbohydrates and the highest energetic value. However, fat content did not differ significantly between the first (29 DAS) and the last harvest (52 DAS), whereas protein content was the highest at the last harvest (52 DAS). Similar trends regarding the macronutrient content at different harvesting stages have been previously reported by Mohamed and Hussein [5], who also suggested an increase in fat, protein and ash content with increasing maturity, whereas they also reported an opposite trend for carbohydrate content. The values for moisture and macronutrient content detected in our study were in the same range as those reported by Ezekwe et al. [17], who studied eight purslane accessions planted at different dates, as well with the study described by Teixeira and Carvalho [13], who evaluated the effect of salinity on the proximate composition of *Portulaca oleracea* cv. "Golden leaf" for two growing seasons. The only difference with the above-mentioned reports was observed in lipid content, which was lower in our study. This finding could be attributed to differences in genotype, harvesting stage and growth conditions. According to Jin et al. [10], stress conditions such as heat, drought and salinity may accelerate protein catabolism and result in lower protein content. In addition, Teixeira and Carvalho [13] suggested that apart from harvesting stage (49 and 57 DAS), growing season (spring and summer) may also have an effect on the proximate composition of purslane leaves.

Significant differences in moisture content were also observed between both plant parts, with leaves having a higher moisture content only at the first harvest stage (29 DAS), since developing plants gradually become succulent and contain more water in stems than leaves (Table 1). Similar results have been reported by Oliveira et al. [4], who also detected higher moisture content in leaves compared to stems, regardless of purslane genotype. Macronutrient content and energetic value also differed between plant parts at the studied harvesting stages. In particular, leaves contained more fat and proteins at the first harvest, whereas stems had a higher content of carbohydrates and ash and a higher energetic value at the same harvesting stage. Moreover, leaves had higher macronutrient content and energetic value than stems at the second and the last harvesting stages (apart from carbohydrate content at the second harvest), which could be partly attributed to the lower moisture content values and the resulting concentration effect. Oliveira et al. [4] also reported differences in fat content between leaves and stems of purslane plants from four different locations in Northern Portugal, whereas there were significant differences in the reported values compared to our study, probably because plants were collected in the wild instead of being cultivated. Moreover, no details regarding the harvesting stage were available in the study of Oliveira et al. [4] in order to make direct comparisons with our study. According to Ezeabara et al. [27], moisture, fat, protein and ash contents were higher in leaves than in stems, when plants were harvested at flowering stage, whereas the opposite trend was observed for carbohydrate content. These results are in agreement with the values of the last harvest of our study, indicating that late harvests increase the nutritional value of leaves in comparison to stems, due to maturation progress and the decrease in moisture content.

The contents of individual and total tocopherols in relation to the harvesting stage and plant part are presented in Table 2. Tocopherols, also known as vitamin E, are significant bioactive compounds with antioxidant properties against lipid peroxidation of biological membranes [28]. α-tocopherol was the main detected isoform in both leaves and stems, followed by γ-, β- and δ-tocopherols. Leaves contained significantly higher amounts of all isoforms and total tocopherols, regardless of the harvesting stage, while harvesting at later stages (52 DAS) had a beneficial effect on the content of most of the individual and total tocopherols. The only exception was observed for γ-tocopherol content, which was the highest at early growth stages (29 DAS). A contrasting trend was observed in the case of stems, where the highest content of individual and total tocopherols was recorded at the stage of 29 DAS, except for β-tocopherol where no significant differences were observed between harvesting at 29 and 43 DAS. Similarly to our study, Szalai et al. [7] reported that α- and γ-tocopherols were the main detected tocopherols in fully mature leaves of three purslane microspecies, although they did not detect the other two vitamin E isoforms (β- and δ-tocopherols) that were present in our study. Moreover, the highest content of α- and total tocopherols at the last harvesting stage (52 DAS) could be attributed to unfavorable growing conditions (namely, high temperatures and high transpiration), which could induce tocopherols' biosynthesis as a means of plants' antioxidant defense [29].

Table 1. Nutritional value (g/100 g fresh weight (fw)) and energetic value (kcal/100 g fw) of purslane stems and leaves in relation to harvesting stage (mean ± SD).

Harvest Stage (DAS) *	Plant Part	Moisture (%)	Fat	Proteins	Ash	Carbohydrates	Energy
29	Stems	88.49 ± 0.41c	0.111 ± 0.002a	1.31 ± 0.01b	2.48 ± 0.09a	7.6 ± 0.1a	47 ± 9a
	Leaves	91.00 ± 0.49a	0.157 ± 0.001b	1.57 ± 0.02c	2.14 ± 0.05b	5.13 ± 0.02c	43.2 ± 0.1c
43	Stems	91.97 ± 0.08a	0.091 ± 0.001b	0.76 ± 0.01c	1.78 ± 0.03b	5.40 ± 0.02b	35.54 ± 0.03c
	Leaves	90.81 ± 0.16a	0.148 ± 0.002b	1.91 ± 0.01b	1.89 ± 0.05c	5.25 ± 0.03b	45.70 ± 0.02b
52	Stems	91.31 ± 0.1b	0.111 ± 0.003a	1.44 ± 0.01a	1.74 ± 0.04b	5.39 ± 0.03b	41.52 ± 0.02b
	Leaves	88.16 ± 0.41b	0.230 ± 0.001a	2.96 ± 0.04a	2.40 ± 0.06a	6.2 ± 0.1a	61.3 ± 0.1a
Student's *t*	Plant Part **	<0.001	<0.001	<0.01	<0.01	<0.01	<0.01

* DAS: days after sowing. Different Latin letters (a–c) in the same column refer to significant differences between harvest stages for the same plant part (stems or leaves) at *p* = 0.05.
** Comparison of means of different plant parts (stems and leaves) from the same harvest was performed with Student's *t*-test at *p* = 0.05.

Table 2. Composition in tocopherols (μg/100 g fw) and sugars (g/100 g fw) of purslane stems and leaves in relation to harvesting stage (mean ± SD).

Harvest Stage (DAS) *	Plant Part	α-Tocopherol	β-Tocopherol	γ-Tocopherol	δ-Tocopherol	Total Tocopherols
29	Stems	26.0 ± 0.2a	3.4 ± 0.3a	14.4 ± 0.1a	0.99 ± 0.03	44.7 ± 0.5a
	Leaves	215 ± 4b	14.0 ± 0.7b	140.7 ± 0.1a	9.6 ± 0.5b	380 ± 4b
43	Stems	19.3 ± 0.5b	3.6 ± 0.6a	8.3 ± 0.1b	nd	31 ± 1b
	Leaves	197 ± 3c	12.4 ± 0.2b	87.7 ± 0.2c	5.1 ± 0.2c	302 ± 2c
52	Stems	10.4 ± 0.2c	1.8 ± 0.1b	5.2 ± 0.2c	nd	17.4 ± 0.6c
	Leaves	327 ± 3a	44 ± 2a	97 ± 8b	13.5 ± 0.5a	481 ± 9a
Student's *t*	Plant part	<0.001	<0.001	<0.001	<0.001	<0.001
Harvest Stage (DAS) *	Plant Part	Fructose	Glucose	Sucrose	Trehalose	Total Sugars
29	Stems	0.308 ± 0.006c	0.358 ± 0.001c	0.135 ± 0.001a	0.024 ± 0.001a	0.830 ± 0.006c
	Leaves	0.11 ± 0.01b	0.041 ± 0.002c	nd	0.012 ± 0.001c	0.160 ± 0.007b
43	Stems	0.44 ± 0.02a	0.53 ± 0.02b	0.051 ± 0.001c	0.012 ± 0.001b	1.03 ± 0.04b
	Leaves	0.183 ± 0.007a	0.113 ± 0.002a	0.009 ± 0.001a	0.026 ± 0.001b	0.330 ± 0.009a
52	Stems	0.39 ± 0.01b	0.74 ± 0.02a	0.118 ± 0.003b	0.022 ± 0.001a	1.28 ± 0.04a
	Leaves	0.179 ± 0.007a	0.100 ± 0.001b	0.014 ± 0.001a	0.041 ± 0.001a	0.330 ± 0.008a
Student's *t*	Plant part **	<0.001	<0.001	<0.001	<0.001	<0.001

nd: not detected; * DAS: days after sowing; Different Latin letters (a–c) in the same column refer to significant differences between harvest stages for the same plant part (stems or leaves) at *p* = 0.05. ** Comparison of means of different plant parts (stems and leaves) from the same harvest was performed with Student's *t*-test at *p* = 0.05.

The free sugar composition of leaves and stems in relation to harvesting stage is presented in Table 2. The main detected sugars were glucose and fructose, followed by sucrose and trehalose, which were detected in lower amounts. Stems contained significantly higher amounts of fructose, glucose, sucrose and total free sugars than leaves, regardless of harvesting stage, whereas trehalose content was higher in leaves compared to stems when harvesting took place at late growth stages (43 and 52 DAS). In addition, an increase in glucose and fructose content was observed at late growth stages (43 and 52 DAS) for both leaves and stems, which was also depicted in the total sugar content of plant parts. Sugar composition was similar to our previously reported study, where six different purslane genotypes were evaluated in terms of chemical composition [15], whereas any differences in individual sugars' content could be explained mostly by genotypic differences and barely by the environmental conditions and cultivation practices, which were identical in both studies. Moreover, Mohamed and Hussein [5] suggested glucose as the main detected sugar, while fluctuating trends were observed in sugar content between plant parts at different growth stages (30, 49 and 59 days after planting). The lower amounts of sucrose in leaves compared to fructose and glucose could be attributed to the use of fixed carbon for the biosynthesis of fructose and glucose and the concurrent export of sucrose from leaves to be used as a biosynthetic substrate [30].

The organic acid content of plant parts in relation to harvesting stage is presented in Table 3. Four organic acids were detected in leaves and stems, namely oxalic, malic, quinic and citric acids, regardless of the harvesting stage. However, the composition of individual organic acids differed significantly between plant parts and harvesting stages. In particular, leaves contained mostly quinic and oxalic acids at all the harvesting stages, although quinic acid content was significantly higher at the last growth stage (52 DAS). On the other hand, stems contained oxalic, quinic and malic acids at the earliest growth stage (29 DAS), whereas quinic acid decreased significantly with plant maturity. Similarly, Oliveira et al. [4], who studied organic acid content in the leaves and stems of different purslane genotypes, detected significant differences between plant parts, although the magnitude of these differences varied depending on the genotype. In contrast to our study, Szalai et al. [7] detected oxalic, malic and ascorbic acids in the leaves of three purslane microspecies, a difference that could be attributed mostly to the effect of harvesting stage. However, it is worth mentioning the effect of genotype on organic acid profile and content, which is also important and has been already confirmed in previous studies [4,15].

Table 3. Composition in organic acids (g/100 g fw) of purslane stems and leaves in relation to harvesting stage (mean ± SD).

Harvest Stage (DAS) *	Plant Part	Oxalic Acid	Quinic Acid	Malic Acid	Citric Acid	Total Organic Acids
29	Stems	7.70 ± 0.07a	6.36 ± 0.04a	6.39 ± 0.07b	5.00 ± 0.03a	25.5 ± 0.2a
	Leaves	6.2 ± 0.1b	6.82 ± 0.01c	3.00 ± 0.03a	3.26 ± 0.01a	19.2 ± 0.1b
43	Stems	4.77 ± 0.01c	1.31 ± 0.01c	6.56 ± 0.07a	1.57 ± 0.04c [¥]	14.22 ± 0.04c
	Leaves	5.7 ± 0.1c	8.4 ± 0.2b	1.90 ± 0.04b	1.53 ± 0.02b [¥]	17.6 ± 0.1c
52	Stems	7.16 ± 0.02b	3.57 ± 0.04b	5.38 ± 0.06c	2.78 ± 0.01b	18.89 ± 0.04b
	Leaves	8.6 ± 0.2a	16.8 ± 0.5a	1.67 ± 0.01c	3.24 ± 0.03a	30.3 ± 0.2a
80	Seeds	0.470 ± 0.005	nd	tr	tr	0.470 ± 0.005
Student's *t*	Plant part **	<0.001	<0.001	<0.001	<0.001	<0.001

nd: not detected; tr: traces; * DAS: days after sowing; [¥]: no significant difference was observed between plant parts. Different Latin letters (a–c) in the same column refer to significant differences between harvest stages for the same plant part (stems or leaves) at $p = 0.05$. ** Comparison of means of different plant parts (stems and leaves) from the same harvest was performed with Student's *t*-test at $p = 0.05$.

Purslane is considered as one of the richest plant sources of omega-3 fatty acids. The composition of leaves and stems in relation to harvesting stage is presented in Table 4. The main detected fatty acids in both stems and leaves were linoleic, palmitic, α-linolenic, behenic, oleic and lignoceric acids, although the abundance of individual compounds varied among the studied plant parts and harvesting stages. The major compound in stems was palmitic acid, followed by linoleic and α-linolenic acids. In addition, the highest content of palmitic and linoleic acids was detected at 29 DAS, whereas α-linolenic acid content was the highest at 43 DAS. In contrast, the fatty acid profile of leaves differed significantly from that of stems, with α-linolenic acid contributing the most to the overall fatty acid content, especially at the first two harvesting stages (29 and 43 DAS). The same fatty acids have been previously reported by Oliveira et al. [4], who also suggested α-linolenic, palmitic, oleic, stearic and behenic acids as the main fatty acids in leaves. However, the reported profile of individual fatty acids differed from that in our study, without additional information regarding harvesting stage being available. Similarly, Guil-Guerrero and Rodríguez-García [31] suggested a different profile of fatty acids in phospholipid and neutral lipid fractions of purslane leaves, which contained higher amounts of omega-6 (n6) than omega-3 (n3) fatty acids. Moreover, in our study the overall α-linolenic acid content (the sum of the contents of stems and leaves) was the highest (0.1 g/100 g fw) at the earliest harvesting stage (29 DAS), despite the significantly higher fat content of leaves at the late harvesting stage (0.1 g/100 g fw at 52 DAS; see Table 1). Similar amounts of α-linolenic acid (1.06 g/100 g dw) were reported by Guil-Guerrero and Rodríguez-García [31], although they also suggested that purslane leaves contain higher amounts of linoleic than α-linolenic acid. The highest polyunsaturated fatty acids (PUFA)/saturated fatty acids (SFA) ratio was observed in stems and leaves at the middle (43 DAS) and late harvesting stages (52 DAS), respectively. In contrast, the lowest n6/n3 fatty acid ratio in stems and leaves was recorded at the earliest harvesting stage (29 DAS). In addition, for all the studied harvesting stages and plant parts, the PUFA/SFA ratio was higher than 0.45, while the n6/n3 ratio was lower than 4.0, indicating a high nutritional value of the edible plant parts [32]. Moreover, the detected values for the above-mentioned ratios of our study were within the same range as the ones recorded in the study of Fontana et al. [19].

The identified phenolic compounds and oleracein derivatives in both plant parts of purslane are presented in Table 5, while quantification data are presented in Table 6 and Figure 1. A total of five compounds were identified in the hydroethanolic extracts of purslane aerial plant parts (Table 5). These compounds included three phenolic acids (caffeic acid derivatives, peaks 2, 3 and 4) and two oleracein derivatives (phenolic alkaloids, peaks 1 and 5). Caffeic acid (peak 2) was identified in comparison with a commercial standard, whereas compound 4 was identified as a caffeic acid derivative, thus its pseudomolecular ion was not clearly identified due to the very small amounts present. Compound 3 ([M − H]⁻ at *m/z* 385) presented a unique MS² fragment at *m/z* 223 (sinapic acid), with a loss of a hexosyl moiety (162 u), being tentatively identified as sinapic acid hexoside. Compounds 1 ([M − H]⁻ at *m/z* 664) and 5 ([M − H]⁻ at *m/z* 502) corresponded to oleracein derivatives, being identified as oleracein C and A, respectively, taking into account previous findings in the literature regarding *P. oleracea* [33,34]. Oleraceins are cyclodopa alkaloids, which have been previously reported by Xiang et al. [35], who identified five oleraceins in dried purslane plants, among other compounds. Moreover, Farag and Shakour [36] suggested that the presence of these compounds in purslane aerial parts may be used as a criterion for the classification of different *Portulaca* taxa. To the best of our knowledge, the rest of the identified compounds have not been previously reported in purslane aerial parts.

Table 4. Fatty acid composition (%) of the studied purslane stems and leaves (mean ± SD) in relation to harvesting stage.

Fatty Acids	29 DAS		43 DAS		52 DAS		Student's *t*
	Stems	Leaves	Stems	Leaves	Stems	Leaves	Plant Parts **
C6:0	0.543 ± 0.001a	0.024 ± 0.001c	0.30 ± 0.03b	0.067 ± 0.001b	0.280 ± 0.006c	0.220 ± 0.001a	<0.001
C8:0	0.098 ± 0.003c	0.032 ± 0.003c	0.12 ± 0.01a	0.039 ± 0.001b	0.115 ± 0.008b	0.095 ± 0.007a	<0.001
C10:0	0.100 ± 0.003a	0.052 ± 0.001b	0.091 ± 0.004b	0.051 ± 0.001b	0.094 ± 0.006b	0.125 ± 0.007a	<0.001
C12:0	0.93 ± 0.03a	0.81 ± 0.02c	0.62 ± 0.04b	0.867 ± 0.001b	0.640 ± 0.006b	1.37 ± 0.04a	<0.001
C14:0	1.42 ± 0.02a	0.736 ± 0.002c	0.97 ± 0.02b	0.77 ± 0.01b	0.85 ± 0.02c	1.24 ± 0.01a	<0.001
C15:0	0.45 ± 0.01b	0.49 ± 0.01b	0.472 ± 0.008b	0.420 ± 0.003c	0.62 ± 0.02a	0.75 ± 0.01a	<0.001
C16:0	21.8 ± 0.2a	9.8 ± 0.1c	21.0 ± 0.1b	10.83 ± 0.01b	20.2 ± 0.2c	12.39 ± 0.03a	<0.001
C16:1	0.236 ± 0.009c	0.52 ± 0.01b	0.36 ± 0.01b	0.48 ± 0.01c	0.401 ± 0.009a	0.730 ± 0.001a	<0.001
C17:0	0.54 ± 0.03b	0.15 ± 0.01c	0.48 ± 0.02c	0.159 ± 0.005b	0.816 ± 0.007a	0.265 ± 0.007a	<0.001
C18:0	4.9 ± 0.1a	2.52 ± 0.05c	4.55 ± 0.05c	2.72 ± 0.01b	4.83 ± 0.07b	3.89 ± 0.06a	<0.001
C18:1n9c+t	9.55 ± 0.04b	5.29 ± 0.05b	11.62 ± 0.01a	4.65 ± 0.04c	6.83 ± 0.14c	6.4 ± 0.1a	<0.001
C18:2n6c	23.02 ± 0.02c	11.40 ± 0.08c	27.11 ± 0.02a	11.63 ± 0.02b	25.4 ± 0.2b	14.81 ± 0.02a	<0.001
C18:3n3	17.31 ± 0.04a	54.92 ± 0.08a	15.03 ± 0.07b	54.34 ± 0.03a	11.64 ± 0.01c	35.4 ± 0.1b	<0.001
C20:0	1.86 ± 0.06c	1.79 ± 0.01b	1.99 ± 0.03c	1.80 ± 0.01b	2.2 ± 0.1a	2.95 ± 0.03a	<0.001
C20:1CIS-11	0.40 ± 0.01a	0.08 ± 0.01c	0.258 ± 0.006b	0.11 ± 0.01b	0.146 ± 0.001c ¥	0.140 ± 0.001a ¥	<0.001
C20:3n3+C21:0	0.67 ± 0.01a	0.155 ± 0.004c	0.63 ± 0.04c	0.195 ± 0.004b	0.646 ± 0.002b	0.32 ± 0.02a	<0.001
C20:5n3	0.17 ± 0.01a	0.051 ± 0.003a	0.17 ± 0.01a	0.042 ± 0.001b	0.146 ± 0.003b	0.040 ± 0.001b	<0.001
C22:0	11.0 ± 0.2b	9.0 ± 0.3b	9.6 ± 0.2c	8.62 ± 0.09c	15.40 ± 0.23a	15.0 ± 0.2a	<0.001
C23:0	0.44 ± 0.02c	0.20 ± 0.01b	0.57 ± 0.01b	0.15 ± 0.01c	0.77 ± 0.02a	0.31 ± 0.01a	<0.001
C24:0	4.5 ± 0.1b	2.04 ± 0.08b	4.1 ± 0.2c	2.05 ± 0.01b	7.97 ± 0.02a	3.61 ± 0.04a	<0.001
Total SFA (% of total FA)	48.65 ± 0.02b	27.58 ± 0.06c	44.82 ± 0.01c	28.5 ± 0.1b	54.8 ± 0.1a	42.2 ± 0.3a	<0.001
Total MUFA (% of total FA)	10.18 ± 0.04b	5.89 ± 0.05b	12.24 ± 0.02a	5.25 ± 0.06c	7.38 ± 0.13c ¥	7.3 ± 0.1a ¥	<0.001
Total PUFA (% of total FA)	41.17 ± 0.02b	66.53 ± 0.01a	42.94 ± 0.02a	66.21 ± 0.04b	37.81 ± 0.25c	50.5 ± 0.2c	<0.001
PUFA/SFA	0.864 ± 0.001b	2.412 ± 0.003a	0.958 ± 0.001a	2.319 ± 0.007b	0.690 ± 0.004c	1.196 ± 0.009c	<0.001
n6/n3	1.269 ± 0.005c	0.207 ± 0.001c	1.71 ± 0.01b	0.213 ± 0.001b	2.04 ± 0.01a	0.414 ± 0.002a	<0.001

* DAS: days after sowing; ¥: no significant difference was observed between plant parts. Caproic acid (C6:0); Caprylic acid (C8:0); Capric acid (C10:0); Lauric acid (C12:0); Myristic acid (C14:0); Pentadecylic acid (C15:0); Palmitic acid (C16:0); Palmitoleic acid (C16:1); Margaric acid (C17:0); Stearic acid (C18:0); Oleic acid (C18:1n9); Linoleic acid (C18:2n6c); α-Linolenic acid (C18:3n3); Arachidic acid (C20:0); Eicosenoic acid (C20:1CIS-11); Eicosatrienoic acid (C20:3n3); Heneicosylic acid (C21:0); Eicosapentaeonic acid (C20:5n3); Behenic acid (C22:0); Tricosylic acid (C23:0); Lignoceric acid (C24:0); SFA: saturated fatty acids; MUFA: monounsaturated fatty acids; PUFA: polyunsaturated fatty acids; n6/n3: omega-6/omega-3 fatty acids. Different Latin letters (a–c) in the same row refer to significant differences between harvest stages for the same plant part (stems or leaves) at $p = 0.05$. ** Comparison of means of different plant parts (stems and leaves) from the same harvest was performed with Student's *t*-test at $p = 0.05$.

Table 5. Retention time (Rt), wavelengths of maximum absorption in the UV-VIS region (λ_{max}), mass spectral data, identification and quantification of phenolic compounds and oleracein derivatives in purslane aerial plant parts (leaves and stems).

Peak	Rt (min)	λ_{max} (nm)	[M−H]⁻ (*m/z*)	MS² (*m/z*)	Tentative Identification
1	6.82	345	664	502(100), 340(20), 296(5), 194(3)	Oleracein C
2	8.77	323	179	161(100), 143(63), 119(40)	Caffeic acid
3	9.29	329	385	223 (100)	Sinapic acid hexoside
4	11.89	318	-	179(100), 161(55), 143(31), 119(18)	Caffeic acid derivative
5	15.54	338	502	340(100), 296(5), 194(3), 145(3)	Oleracein A

Table 6. Quantification of phenolic compounds and oleracein derivatives in purslane stems and leaves (mg/100 g dried weight (dw)) in relation to harvesting stage (mean ± SD).

Peak	Phenolic Compound	29		43		52		Plant Parts **
		Stems	Leaves	Stems	Leaves	Stems	Leaves	
1	Oleracein C [A]	15.2 ± 0.5a	143 ± 5a	6.7 ± 0.1b	21.2 ± 0.3c	3.34 ± 0.07c	102 ± 2b	<0.01
2	Caffeic acid [B]	0.44 ± 0.02a	nd	0.45 ± 0.01a	nd	tr	nd	<0.01
3	Sinapic acid hexoside [C]	4.2 ± 0.1a	22.1 ± 0.7a	4.3 ± 0.2a	nd	1.37 ± 0.03b	nd	<0.01
4	Cafferic acid derivative	nd	nd	nd	nd	tr	nd	-
5	Oleracein A [A]	nd	103 ± 2a	0.75 ± 0.01a	8.2 ± 0.1c	0.28 ± 0.02b	34.9 ± 0.8b	<0.01
	TPCOD	19.8 ± 0.4a	268 ± 6a	12.2 ± 0.1b	29.3 ± 0.4c	4.99 ± 0.08c	137 ± 3b	<0.01

* DAS: days after sowing; nd: not detected; TPCOD: Total phenolic compounds and oleracein derivatives; tr: traces. Calibration curves used: [A]: *p*-coumaric acid ($y = 301{,}950x + 6966.7$; $R^2 = 0.999$); [B]: caffeic acid ($y = 388{,}345x + 406{,}369$; $R^2 = 0.999$); [C]: sinapic acid ($y = 197{,}337x + 30{,}036$; $R^2 = 0.999$). Different Latin letters (a–c) in the same column refer to significant differences between harvest stages for the same plant part (stems and leaves) at $p = 0.05$. ** Comparison of means of different plant parts (stems and leaves) from the same harvest was performed with Student's *t*-test at $p = 0.05$.

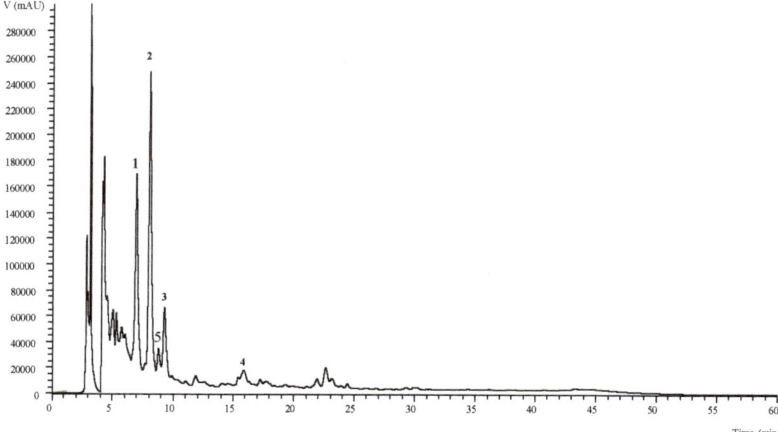

Figure 1. Chromatographic profile of the hydroethanolic extract obtained from purslane stems, recorded at 330 nm. Peak numbers correspond to the compounds already mentioned in Table 5.

The composition of phenolic compounds and oleracein derivatives differed among the tested plant parts and harvesting stages (Table 6). Concerning the quantification of oleraceins, *p*-coumaric acid was the standard applied to perform the quantitative analysis (an available compound with a similar structure); therefore, these compounds were quantified as equivalents to this phenolic acid. Leaves contained significantly higher amounts of individual and total phenolic compounds and oleracein derivatives compared to stems, regardless of harvesting stage. Moreover, harvesting at early stages (29 DAS) resulted in significantly higher contents of phenolic compounds and oleracein derivatives, especially in the case of leaves. Meanwhile, in stems the content of oleracein C was the highest at the same harvesting stage (29 DAS), resulting in the highest total contents of phenolic compounds and oleracein derivatives, accordingly. As mentioned previously for the case of tocopherols (see Table 2), phenolic compounds and oleracein derivatives may also contribute to the overall defense mechanisms of purslane; therefore, the increased content at the earlier stages could be attributed to the protective purposes of developing leaves. Similar results regarding the effect of harvesting stage on the compositions of phenolic compounds and oleracein derivatives have been reported by Lim and Quah [37], who also observed a decrease in total phenolic compounds and oleracein derivatives in leaves with increasing maturity. The differences in the compositions of phenolic compounds and oleracein derivatives could be attributed to differences in the tested genotypes [35,37], to different cultivation regimes and growing conditions [14,18] and also to extraction protocols [38].

The cytotoxic effects on PLP2 non-tumor cell lines of the samples showed no hepatotoxicity, with GI$_{50}$ values being higher than 400 µg/mL for all the harvesting stages and plant parts (data not shown). Concerning the evaluation of in vitro activity against the four tumor cells lines, the extracts did not present any activity at the tested concentrations (data not show), with GI$_{50}$ values being higher than 400 µg/mL.

The hepatoprotective effects of ethanolic extracts of purslane aerial parts have been previously reported by Eidi et al. [39], whose findings are in agreement with the results of our study. Similarly, ethanolic and aqueous extracts of air-dried purslane leaves showed hepatoprotective properties against paracetamol-induced liver damage [40]. Moreover, previous studies have suggested weak to moderate inhibitory effects against two mutagens (benzo[a]pyrene (B[a]P) and 2-Amino-3-methyl-imidazo[4,5-f]quinolone (IQ)) and no mutagenic activity on *Salmonella typhimurium* [41], while Choi and Ryeom [42] and Zhao et al. [43] reported significant antitumor activity against leukaemia and cervical carcinoma cell lines, respectively.

4. Conclusions

The results of the present study showed a significant effect of plant parts and harvesting stages on the nutritional value and chemical composition of purslane. Leaves contained higher amounts of macronutrients than stems, especially at 52 DAS, while α-tocopherol was the main vitamin E isoform, which increased at 52 DAS resulting in the highest overall tocopherol content. Glucose and fructose were the main sugars in stems and leaves, respectively, while stems contained higher amounts of total sugars in comparison to leaves. Regarding oxalic acid as well as total organic acids, the highest contents were recorded in leaves, especially at the last harvesting stage (52 DAS). The edible plant parts contained both omega-6 and omega-3 fatty acids, although leaves were more abundant in α-linolenic acid than stems, which contained mostly palmitic and linoleic acids. Phenolic compounds and oleracein derivatives were also detected in plant parts, with oleraceins A and C being the main compounds detected in leaves, regardless of harvesting stage. In conclusion, early harvesting and the separation of plant parts could increase the nutritional value of the final product through increasing the content of valuable compounds, such as omega-3 fatty acids, phenolic compounds and oleracein derivatives, while at the same time, the contents of anti-nutritional compounds such as oxalic acid are reduced.

Author Contributions: S.A.P. conceived and designed the research, administered and supervised the project, carried out the cultivation, wrote the original draft and reviewed and edited the final manuscript; Â.F. performed chemical analyses, data curation, and methodology; M.I.D performed chemical analyses, data curation, and methodology; I.B.V. carried out the cultivation and prepared the original draft; K.P. carried out the cultivation and prepared the original draft; L.B. performed chemical analyses, data curation, and methodology, wrote the original draft and reviewed and edited the final manuscript; I.C.F.R.R. administered the project, obtained funding, administered and supervised the project, reviewed and edited the final manuscript.

Funding: The authors are grateful to the Foundation for Science and Technology FCT, Portugal and FEDER under Programme PT2020 for financial support to CIMO (UID/AGR/00690/2019). Â.F., M.I.D., and L.B. thank the national funding by FCT, P.I., through the institutional scientific employment program-contract for their contracts. The authors are also grateful to the FEDER-Interreg España-Portugal program for financial support through the project 0377_Iberphenol_6_E.

References

1.   Petropoulos, S.A.; Karkanis, A.; Martins, N.; Ferreira, I.C.F.R. Phytochemical Composition and Bioactive Compounds of Common Purslane (*Portulaca oleracea* L.) as Affected by Crop Management Practices. *Trends Food Sci. Technol.* **2016**, *55*, 1–10. [CrossRef]

2.   Nebel, S.; Heinrich, M. Ta chórta: A Comparative Ethnobotanical-Linguistic Study of Wild Food Plants in a Graecanic Area in Calabria, Southern Italy. *Econ. Bot.* **2009**, *63*, 78–92. [CrossRef]

3.   Mitich, L.W. Common Purslane (*Portulaca oleracea*). *Weed Technol.* **1997**, *11*, 394–397. [CrossRef]

4.   Oliveira, I.; Valentão, P.; Lopes, R.; Andrade, P.B.; Bento, A.; Pereira, J.A. Phytochemical Characterization and Radical Scavenging Activity of *Portulaca oleraceae* L. Leaves and Stems. *Microchem. J.* **2009**, *92*, 129–134. [CrossRef]

5.   Mohamed, A.; Hussein, A. Chemical Composition of Purslane (*Portulaca oleracea*). *Plant Foods Hum. Nutr.* **1994**, *45*, 1–9. [CrossRef] [PubMed]

6.   Liu, L.; Howe, P.; Zhou, Y.F.; Xu, Z.Q.; Hocart, C.; Zhang, R. Fatty Acids and β-Carotene in Australian Purslane (*Portulaca oleracea*) Varieties. *J. Chromatogr. A* **2000**, *893*, 207–213. [CrossRef]

7.   Szalai, G.; Dai, N.; Danin, A.; Dudai, N.; Barazani, O. Effect of Nitrogen Source in the Fertilizing Solution on Nutritional Quality of Three Members of the *Portulaca oleracea* Aggregate. *J. Sci. Food Agric.* **2010**, *90*, 2039–2045. [CrossRef]

8.   Sicari, V.; Loizzo, M.R.; Tundis, R.; Mincione, A.; Pellicanò, T.M. *Portulaca oleracea* L. (Purslane) Extracts Display Antioxidant and Hypoglycaemic Effects. *J. Appl. Bot. Food Qual.* **2018**, *91*, 39–46. [CrossRef]

9.   Naser Aldeen, M.; Mansour, R.; AlJoubbeh, M. The Effect of Food Additives and Cooking on the Antioxidant Properties of Purslane. *Nutr. Food Sci.* **2019**. [CrossRef]

10.   Jin, R.; Wang, Y.; Liu, R.; Gou, J.; Chan, Z. Physiological and Metabolic changes of Purslane (*Portulaca oleracea* L.) in Response to Drought, Heat, and Combined Stresses. *Front. Plant Sci.* **2016**, *6*, 1–11. [CrossRef]

11. Rahdari, P.; Tavakoli, S.; Hosseini, S.M. Studying of Salinity Stress Effect on Germination, Proline, Sugar, Protein, Lipid and Chlorophyll Content in Purslane (*Portulaca oleracea* L.) Leaves. *J. Sress Physiol. Biochem.* **2012**, *8*, 182–193.

12. Uddin, K.; Juraimi, A.S.; Anwar, F.; Hossain, M.A.; Alam, M.A. Effect of Salinity on Proximate Mineral Composition of Purslane (*Portulacaoleracea* L.). *Aust. J. Crop Sci.* **2012**, *6*, 1732–1736.

13. Teixeira, M.; Carvalho, I.S. Effects of Salt Stress on Purslane (*Portulaca oleracea*) Nutrition. *Ann. Appl. Biol.* **2009**, *154*, 77–86. [CrossRef]

14. Alam, M.A.; Juraimi, A.S.; Rafii, M.Y.; Hamid, A.A.; Aslani, F.; Alam, M.Z. Effects of Salinity and Salinity-Induced Augmented Bioactive Compounds in Purslane (*Portulaca oleracea* L.) for Possible Economical Use. *Food Chem.* **2015**, *169*, 439–447. [CrossRef] [PubMed]

15. Petropoulos, S.; Karkanis, A.; Fernandes, Â.; Barros, L.; Ferreira, I.C.F.R.; Ntatsi, G.; Petrotos, K.; Lykas, C.; Khah, E. Chemical Composition and Yield of Six Genotypes of Common Purslane (*Portulaca oleracea* L.): An Alternative Source of Omega-3 Fatty Acids. *Plant Foods Hum. Nutr.* **2015**, *70*, 420–426. [CrossRef] [PubMed]

16. Alam, M.A.; Juraimi, A.S.; Rafii, M.Y.; Abdul Hamid, A.; Aslani, F.; Hasan, M.M.; Mohd Zainudin, M.A.; Uddin, M.K. Evaluation of Antioxidant Compounds, Antioxidant Activities, and Mineral Composition of 13 Collected Purslane (*Portulaca oleracea* L.) Accessions. *Biomed Res. Int.* **2014**, *2014*, 1–10. [CrossRef]

17. Ezekwe, M.O.; Omara-Alwala, T.R.; Membrahtu, T. Nutritive Characterization of Purslane Accessions as Influenced by Planting Date. *Plant Foods Hum. Nutr.* **1999**, *54*, 183–191. [CrossRef]

18. Montoya-García, C.O.; Volke-Haller, V.H.; Trinidad-Santos, A.; Villanueva-Verduzco, C. Change in the Contents of Fatty Acids and Antioxidant Capacity of Purslane in Relation to Fertilization. *Sci. Hortic. (Amst.)* **2018**, *234*, 152–159. [CrossRef]

19. Fontana, E.; Hoeberechts, J.; Nicola, S.; Cros, V.; Palmegiano, G.B.; Peiretti, P.G. Nitrogen Concentration and Nitrate Ammonium Ratio Affect Yield and Change the Oxalic Acid Concentration and Fatty Acid Profile of Purslane (*Portulaca oleracea* L.) Grown in a Soilless Culture System. *J. Sci. Food Agric.* **2006**, *86*, 2417–2424. [CrossRef]

20. Uddin, M.K.; Juraimi, A.S.; Ali, M.E.; Ismail, M.R. Evaluation of Antioxidant Properties and Mineral Composition of Purslane (*Portulaca oleracea* L.) at Different Growth Stages. *Int. J. Mol. Sci.* **2012**, *13*, 10257–10267. [CrossRef]

21. AOAC. *Official Methods of Analysis of AOAC International*; Horwitz, W., Latimer, G., Eds.; AOAC International: Gaithersburg, MD, USA, 2016.

22. Dias, M.I.; Barros, L.; Dueñas, M.; Pereira, E.; Carvalho, A.M.; Alves, R.C.; Oliveira, M.B.P.P.; Santos-Buelga, C.; Ferreira, I.C.F.R. Chemical Composition of Wild and Commercial *Achillea millefolium* L. and Bioactivity of the Methanolic Extract, Infusion and Decoction. *Food Chem.* **2013**, *141*, 4152–4160. [CrossRef] [PubMed]

23. Barros, L.; Pereira, E.; Calhelha, R.C.; Dueñas, M.; Carvalho, A.M.; Santos-Buelga, C.; Ferreira, I.C.F.R. Bioactivity and Chemical Characterization in Hydrophilic and Lipophilic Compounds of *Chenopodium ambrosioides* L. *J. Funct. Foods* **2013**, *5*, 1732–1740. [CrossRef]

24. Obodai, M.; Mensah, D.L.N.; Fernandes, Â.; Kortei, N.K.; Dzomeku, M.; Teegarden, M.; Schwartz, S.J.; Barros, L.; Prempeh, J.; Takli, R.K.; et al. Chemical Characterization and Antioxidant Potential of Wild Ganoderma Species from Ghana. *Molecules* **2017**, *22*, 196. [CrossRef] [PubMed]

25. Bessada, S.M.F.; Barreira, J.C.M.; Barros, L.; Ferreira, I.C.F.R.; Oliveira, M.B.P.P. Phenolic Profile and Antioxidant Activity of *Coleostephus myconis* (L.) Rchb.f.: An Underexploited and Highly Disseminated Species. *Ind. Crops Prod.* **2016**, *89*, 45–51. [CrossRef]

26. Abreu, R.M.V.; Ferreira, I.C.F.R.; Calhelha, R.C.; Lima, R.T.; Vasconcelos, M.H.; Adega, F.; Chaves, R.; Queiroz, M.-J.R.P. Anti-Hepatocellular Carcinoma Activity Using Human HepG2 Cells and Hepatotoxicity of 6-Substituted Methyl 3-Aminothieno[3,2-b]Pyridine-2-Carboxylate Derivatives: In Vitro Evaluation, Cell Cycle Analysis and QSAR Studies. *Eur. J. Med. Chem.* **2011**, *46*, 5800–5806. [CrossRef] [PubMed]

27. Ezeabara, C.A.; Faith, I.C.; Ilodibia, C.V.; Aziagba, B.O.; Okanume, O.E.; Ike, M.E. Comparative Determination of Phytochemical, Proximate and Mineral Compositions in Various Parts of *Portulaca oleracea* L. *J. Plant Sci.* **2014**, *2*, 294–298. [CrossRef]

28. Tucker, J.M.; Townsend, D.M. Alpha-Tocopherol: Roles in Prevention and Therapy of Human Disease. *Biomed. Pharmacother.* **2005**, *59*, 380–387. [CrossRef] [PubMed]

29. Munné-Bosch, S.; Alegre, L. The Function of Tocopherols and Tocotrienols in Plants. *CRC. Crit. Rev. Plant Sci.* **2002**, *21*, 31–57. [CrossRef]

30. Jin, R.; Shi, H.; Han, C.; Zhong, B.; Wang, Q.; Chan, Z. Physiological Changes of Purslane (*Portulaca oleracea* L.) after Progressive Drought Stress and Rehydration. *Sci. Hortic. (Amsterdam)*. **2015**, *194*, 215–221. [CrossRef]
31. Guil-Guerrero, J.L.; Rodríguez-García, I. Lipids Classes, Fatty Acids and Carotenes of the Leaves of Six Edible Wild Plants. *Eur. Food Res. Technol.* **1999**, *209*, 313–316. [CrossRef]
32. Guil, J.L.; Torija, M.E.; Giménez, J.J.; Rodriguez, I. Identification of Fatty Acids in Edible Wild Plants by Gas Chromatography. *J. Chromatogr. A* **1996**, *719*, 229–235. [CrossRef]
33. Xing, J.; Yang, Z.; Lv, B.; Xiang, L. Rapid Screening for Cyclo-Dopa and Diketopiperazine Alkaloids in Crude Extracts of *Portulaca oleracea* L. Using Liquid Chromatography/Tandem Mass Spectrometry. *Rapid Commun. Mass Spectrom.* **2008**, *22*, 1415–1422. [CrossRef] [PubMed]
34. Yang, Z.; Liu, C.; Xiang, L.; Zheng, Y. Phenolic Alkaloids as a New Class of Antioxidants in *Portulaca oleracea*. *Phyther. Res.* **2009**, *23*, 1032–1035. [CrossRef] [PubMed]
35. Xiang, L.; Xing, D.; Wang, W.; Wang, R.; Ding, Y.; Du, L. Alkaloids from *Portulaca oleracea* L. *Phytochemistry* **2005**, *66*, 2595–2601. [CrossRef] [PubMed]
36. Farag, M.A.; Shakour, Z.T.A. Metabolomics Driven Analysis of 11 Portulaca Leaf Taxa as Analysed via UPLC-ESI-MS/MS and Chemometrics. *Phytochemistry* **2019**, *161*, 117–129. [CrossRef] [PubMed]
37. Lim, Y.Y.; Quah, E.P.L. Antioxidant Properties of Different Cultivars of *Portulaca oleracea*. *Food Chem.* **2007**, *103*, 734–740. [CrossRef]
38. Gallo, M.; Conte, E.; Naviglio, D. Analysis and Comparison of the Antioxidant Component of *Portulaca oleracea* Leaves Obtained by Different Solid-Liquid Extraction Techniques. *Antioxidants* **2017**, *6*, 64. [CrossRef] [PubMed]
39. Eidi, A.; Mortazavi, P.; Moghadam, J.Z.; Mardani, P.M. Hepatoprotective Effects of *Portulaca oleracea* Extract against CCl 4 -Induced Damage in Rats. *Pharm. Biol.* **2015**, *53*, 1042–1051. [CrossRef] [PubMed]
40. Mohammed Abdalla, H.; Soad Mohamed, A.G. In vivo Hepato-protective Properties of Purslane Extracts on Paracetamol-Induced Liver Damage. *Mal. J. Nutr.* **2010**, *16*, 161–170.
41. Yen, G.C.; Chen, H.Y.; Peng, H.H. Evaluation of the Cytotoxicity, Mutagenicity and Antimutagenicity of Emerging Edible Plants. *Food Chem. Toxicol.* **2001**, *39*, 1045–1053. [CrossRef]
42. Choi, B.-D.; Ryeom, K. Screening of Antitumor Activity from the Crude Drugs in Korea. *Korean J. Pharmacogn.* **2000**, *31*, 16–22.
43. Zhao, R.; Gao, X.; Cai, Y.; Shao, X.; Jia, G.; Huang, Y.; Qin, X.; Wang, J.; Zheng, X. Antitumor Activity of *Portulaca oleracea* L. Polysaccharides against Cervical Carcinoma in Vitro and in Vivo. *Carbohydr. Polym.* **2013**, *96*, 376–383. [CrossRef] [PubMed]

 *antioxidants*

Article

Promising Antioxidant and Antimicrobial Food Colourants from *Lonicera caerulea* L. var. *Kamtschatica*

Adriana K. Molina [1,†], Erika N. Vega [1,†], Carla Pereira [1,*], Maria Inês Dias [1], Sandrina A. Heleno [1], Paula Rodrigues [1], Isabel P. Fernandes [1], Maria Filomena Barreiro [1,2], Marina Kostić [3], Marina Soković [3], João C.M. Barreira [1], Lillian Barros [1] and Isabel C.F.R. Ferreira [1,*]

[1] Centro de Investigação de Montanha (CIMO), Instituto Politécnico de Bragança, Campus de Santa Apolónia, 5300-253 Bragança, Portugal; akmolinavg@gmail.com (A.K.M.); erimavega@gmail.com (E.N.V.); maria.ines@ipb.pt (M.I.D.); sheleno@ipb.pt (S.A.H.); prodrigues@ipb.pt (P.R.); ipmf@ipb.pt (I.P.F.); barreiro@ipb.pt (M.F.B.); jbarreira@ipb.pt (J.C.M.B.); lillian@ipb.pt (L.B.)

[2] Laboratory of Separation and Reaction Engineering—Laboratory of Catalysis and Materials (LSRE-LCM), Polytechnic Institute of Bragança, Campus Santa Apolónia 1134, 5301-857 Bragança, Portugal

[3] Department of Plant Physiology, Institute for Biological Research "Siniša Stanković", University of Belgrade, Bulevar Despota Stefana 142, 11000 Belgrade, Serbia; kosticmarince89@gmail.com (M.K.); mris@ibiss.bg.ac.rs (M.S.)

* Correspondence: carlap@ipb.pt (C.P.); iferreira@ipb.pt (I.C.F.R.F); Tel.: +351-273-330904 (C.P.); +351-273-303219 (I.C.F.R.F); Fax: +351-273-325405 (C.P.); +351-273-325405 (I.C.F.R.F)

† These authors contributed equally to this work.

Received: 10 August 2019; Accepted: 9 September 2019; Published: 12 September 2019

Abstract: *Lonicera caerulea* L. (haskap) berries are widely known for their richness in anthocyanins. In this study, such fruits were assessed for their nutritional and chemical composition, but also as sources of anthocyanins with great colouring properties to be applied in foodstuff. Haskap presented high levels of water, four free sugars (mainly fructose and glucose), five organic acids (mainly citric, malic, and quinic), α- and γ-tocopherol, twenty fatty acids (with prevalence of linoleic acid), and eight phenolic compounds, among which six were anthocyanins (mainly cyanidin-3-*O*-glucoside). The extract presented great antioxidant properties, evaluated through TBARS and OxHLIA assays, as well as antimicrobial capacity against six bacteria and six fungi. Two colourants were obtained by spray-drying haskap juice with maltodextrin and a mixture of maltodextrin and arabic gum. These formulations were stable over 12 weeks of storage at room and refrigerated temperature, without significant variations in colour parameters and in anthocyanins concentration. They were considered safe for consumption once neither microbial contamination nor cytotoxicity in non-tumour cells were detected. The results obtained allow for the consideration of haskap as a promising source of colourants to be applied not only in the food industry, but also in other fields that rely on artificial colourants.

Keywords: *Lonicera caerulea* L.; anthocyanins; antioxidant; antimicrobial; colouring formulations

---

1. Introduction

Lonicera caerulea L., commonly known as blue honeysuckle, honeyberry, sweet berry honeysuckle, edible honeysuckle, haskap, haskup, hasukappu, and haskappu [1], is a shrub that can be mainly found in Russia (Peninsula of Kamchatka), northeastern Asia, and Japan, but also appears—only rarely—in Europe, the Alps, and Scandinavia [2]. Its dark blue to purple fruits were recognized by the Japanese Aborigines as the "Elixir of Life" due to the fact of its multiple therapeutic properties [3], which has led to its consumption both in fresh and processed forms, namely, in juices, cakes, jams, ice creams, and nuts. In fact, the antioxidant activity of haskap berries has been reported as comparable to that of blackberry, raspberry, blueberry, strawberry, yellow hawthorn, and blackcurrant [4]. Moreover, several biological activities have been associated to this fruit, namely, protection against the incidence and mortality of cancer and ischemic heart disease, reduction in blood pressure, prevention of osteoporosis and anaemia, curative effects in gastrointestinal disorders, and aging process delay, in addition to antitumorigenic properties, antimicrobial, antidiabetic, and antimutagenic properties [2,5].

Haskap fruits are mainly composed of fibre, protein, calcium, and magnesium, but also possess high concentrations of glucose and fructose, presenting traces of sucrose and sorbitol. They are a rich source of polyunsaturated fatty acids, with considerable levels of linoleic acid [5,6], and are also recognized for their high concentration in ascorbic acid. Indeed, haskap is often called a "super fruit" because its ascorbic acid content is three to ten times higher than in blueberries, which are considered one of the richest sources of this organic acid [3,5]. Regarding phenolic compounds, these fruits are particularly rich in phenolic acids such as chlorogenic, neochlorogenic, and caffeic acids, anthocyanins, proanthocyanidins, and other flavonoids such as quercetin (their respective glycosides) and catechins. The richness of haskap in anthocyanin compounds has been reported in different studies, where it has been shown that the main anthocyanin of these fruit is cyanidin 3-glucoside, comprising 79–92% of its total content [5]. Other anthocyanins present in lower quantities are cyanidin 3,5-diglucoside (4.27%), cyanidin 3-rutinoside (2.07%), peonidin 3-glucoside (3.44%), and pelargonidin 3-glucoside (0.83%) [4]. Given this peculiar wealth in anthocyanin compounds, these fruit can also be considered great sources of natural colourants with a colouration range from red to purple, pink, or blue, with application in different sectors such as food, pharmaceuticals, and cosmetics, among others. With that in mind, the present study aimed to provide a complete chemical characterization (phenolic compounds, sugars, organic acids, tocopherols, and fatty acids) of haskap berries, including their nutritional value and bioactive properties (e.g., antioxidant and antimicrobial). Furthermore, it also aimed for the development of colourant formulations obtained from haskap juice and stabilized through a spray-drying technique using different stabilizing agents, which increases their affinity with different food matrices and, more importantly, their shelf life. The stability of the obtained colourants was assessed along 12 weeks of storage, with special focus on its colouring capacity (anthocyanins concentration), bioactivity, and microbiological quality.

2. Materials and Methods

2.1. Samples

Fresh mature fruits of *Lonicera caerulea* var. *Kamtschatica* of Wojtek cultivar, from Poland, were provided by "Ponto Agrícola Unipessoal, Lda." (Portugal). Part of the sample was frozen, lyophilised, and reduced to a fine and homogeneous powder and another part was used to prepare the colouring formulations.

2.2. Nutritional Value and Chemical Composition

2.2.1. Nutritional Value

Haskap samples were evaluated regarding fat, ash, protein, carbohydrates, and energetic value, according to the AOAC procedures [7]. The results were expressed in g per 100 g of fresh fruit and kcal per 100 g of fresh fruit (for energy).

2.2.2. Free Sugars

These sugars were extracted from the lyophilized fruit and were further injected in HPLC equipment (Knauer, Smartline system 1000, Berlin, Germany) coupled to a refractive index detector (RI detector, Knauer Smartline 2300) as described by Barros et al. [8]. Separation was achieved with a Eurospher 100-5 NH2 column (250 mm × 4.6 mm, 5 μm, Knauer), with an isocratic elution using acetonitrile/deionized water (70:30, *v/v*) at a flow rate of 1 mL/min, operating at 35 °C. The internal standard method was applied, and the results were expressed in g per 100 g of fresh fruit.

2.2.3. Organic Acids

Organic acids were extracted from the lyophilized fruit and analyzed via ultra-fast liquid chromatography (UFLC) coupled to photodiode array detector (PDA), employing a Shimadzu 20A series UFLC (Shimadzu Cooperation), according to Barros et al. [8]. Separation was achieved using a SphereClone reverse phase C18 column (250 mm × 4.6 mm, 5 μm, Phenomenex, Torrance, CA, USA), with an isocratic elution using sulphuric acid (3.6 mM) at a flow rate of 0.8 mL/min, operating at 35 °C. The results were expressed in g per 100 g of fresh fruit.

2.2.4. Tocopherols

After an extraction procedure following the recommendations of Pereira et al. [9], the tocopherols were injected in a Knauer Smartline system 1000 (HPLC, Berlin, Germany) with a fluorescence detector (FP-2020; Jasco, Easton, USA). Separation was achieved using a Polyamide II normal-phase column (250 mm × 4.6 mm, 5 μm, YMC Waters, Milford, MA, USA), with an isocratic elution using n-hexane and ethyl acetate (70:30, *v/v*) at a flow rate of 1 mL/min, operating at 35 °C. The internal standard method was applied, and the results are expressed in mg per 100 g of fresh fruit.

2.2.5. Fatty Acids

To determine these compounds, and after performing the extraction and derivatization procedures, the analysis was achieved through a DANI model GC 1000 instrument equipped with a split/splitless injector, a flame ionization detector (FID) at 260 °C and a Zebron–Kame column (30 m × 0.25 mm ID × 0.20 μm d_f, Phenomenex, Torrance, California, USA) [8]. The fatty acids identification was possible by comparison of the relative retention times of fatty acid methyl ester (FAME) peaks from samples with standards. The obtained data was analyzed using the Clarity DataApex 4.0 Software and expressed in relative percentage of each fatty acid.

2.3. Extracts Preparation

2.3.1. Phenolic Compounds (Non-Anthocyanic) and Bioactive Properties Analysis

From the powdered sample, ethanolic extracts were prepared by stirring 1 g of the sample with 30 mL of an hydroethanolic solution (80:20, *v/v*) for 1 h. After the extraction procedure, the samples were filtered through a Whatman No. 4 filter paper and afterwards, the residue was again extracted with an additional 30 mL of the same solvent in the same conditions. The resulting samples were combined and evaporated under reduced pressure at 35 °C (rotary evaporator Büchi R-210, Flawil, Switzerland) to remove the ethanol, and the remaining aqueous mixture was frozen at −20 °C and lyophilised.

2.3.2. Anthocyanins Analysis

The extract was prepared according to the above described conditions, with a slight modification: trifluoroacetic acid (TFA; 0.1%) was added to the extraction solvent.

2.4. Phenolic Compounds

The obtained extracts were re-suspended in the same solvent in a concentration of 5 mg/mL, filtered (0.2 μm), and finally injected in the HPLC equipment (Dionex Ultimate 3000 UPLC, Thermo Scientific, San Jose, CA, USA), using a diode-array detector (280, 330, 370, and 520 nm wavelengths) and linked to an electrospray ionization mass spectrometry (Linear Ion Trap LTQ XL, Thermo Scientific, San Jose, CA, USA) working in negative (non-anthocyanin compounds) and positive (anthocyanin compounds) mode, according to methodology described by Bessada et al. [10] and Gonçalves et al. [11]. Data were collected and analyzed using the Xcalibur® program (Thermo Finnigan, San Jose, CA,

USA). Separation was achieved using a Water Spherisorb S3 ODS-2 reverse phase C_{18} column (3 μm, 4.6 mm × 150 mm, Waters, Milford, MA, USA) for non-anthocyanin compounds and with an AQUA® reverse phase C_{18} column (5 μm, 150 mm × 4.6 mm, Phenomenex, Torrance, CA, USA) for anthocyanin compounds, working at 35 °C, using previously described gradients [10,11].

The compounds present in the samples were determined according to their UV-Vis and mass spectra and retention times in comparison with authentic standards, and also using information present in the literature. For the quantification, a 7 level calibration curve was obtained of different standard compounds. The contents in the individual phenolic compounds were expressed in mg per g of extract and in mg per g of colourant.

2.5. Bioactive Properties

2.5.1. Antioxidant Activity

The lipid peroxidation inhibition was evaluated trough the TBARS assay, which mainly consists of using porcine (*Sus scrofa*) brain homogenates and the antioxidant potential was measured by the decrease in thiobarbituric acid reactive substances (TBARS) as described by Pereira et al. [12]. These results were expressed in μg/mL corresponding to the EC_{50} value (sample concentration providing 50% of antioxidant activity). Another antioxidant assay applied was the anti-haemolytic activity of the obtained extracts, evaluated through the oxidative haemolysis inhibition assay (OxHLIA), according to Lockowandt et al. [13]. These results were expressed in μg/mL corresponding to the IC_{50} value, which is the concentration capable of promoting a Δt haemolysis delay of 60 and 120 min. Trolox was used as a positive control in both of the assays.

2.5.2. Antimicrobial Activity

For the antimicrobial potential, several microorganisms were analyzed including different bacterial strains and fungi. The bacterial strains were Gram-positive (*Bacillus cereus* (food isolate), *Staphylococcus aureus* (ATCC 6538), and *Listeria monocytogenes* (NCTC 7973)) and Gram-negative *Escherichia coli* (ATCC 35210), *Enterobacter cloacae* (human isolate), and *Salmonella typhimurium* (ATCC 13311)); and the fungi were *Aspergillus fumigatus* (ATCC 1022), *Aspergillus versicolor* (ATCC 11730), *Aspergillus niger* (ATCC 6275), *Penicillium funiculosum* (ATCC 36839), *Penicillium ochrochloron* (ATCC 9112), and *Trichoderma viride* (IAM 5061). The antibacterial activity was evaluated according to Soković et al. [14] and the antifungal potential was assessed according to Soković et al. [15]. The minimum inhibitory concentration (MIC) was determined for both bacteria and fungi, as were the minimum bactericidal (MBC) and minimum fungicidal (MFC) concentrations. Streptomycin and ampicillin were used as positive controls for the bacteria and ketoconazole and bifonazole for the fungi.

2.6. Juice Preparation

Fresh *L. caerulea* fruits were blended in order to obtain a juice rich in anthocyanins. Then, the juice was centrifuged, filtered through Whatman No. 4 filter paper, and divided into four equal parts, in plastic waterproof bags, non-temperature sensitive, for further pasteurization and subsequent microbial load analysis and spray-drying at different conditions.

2.7. Pasteurization

To perform the pasteurization, the bags containing the samples were immersed in a water bath at 95 °C until 90 °C was achieved inside the bags. From that moment on, the bags were left to stand inside the bath for 60 s and were then cooled by covering the bags with ice until 3 °C was reached inside the bags. This temperature was chosen due to the fact of its effectiveness in eliminating amicroorganisms usually present in this kind of product, without anthocyanin degradation. One of the bags containing the pasteurized sample was used to assess the microbial load, according to sub-Section 2.9.3, and the remaining three bags were stored at 3 °C for ~15 h for further spray-drying.

2.8. Spay-Drying

The samples were dried by a spray-drying technique according to the work of Moser et al., namely, by using 80% of maltodextrin or 80% of a mixture maltodextrin:arabic gum (1:1, *w/w*), as drying the adjuvants was aimed at analyzing the efficiency of the drying process and the adequacy of the use of maltodextrin and arabic gum materials [16]. The latter's material's percentage of 80% was relative to the total solids content of the prepared samples, and it was selected after an optimization study involving the testing of different percentages of maltodextrin or the mixture of maltodextrin + arabic gum; it was found that the samples with 80% arabic gum materials was the one that led to higher process yields. The solutions of haskap juice with the selected materials were prepared immediately before atomization. Briefly, the haskap juice samples were mixed with the drying adjuvants and thereafter homogenized by stirring for 10 min at room temperature. The used spray-drying equipment was a Mini Spray Dryer B-290 Büchi (Flawil, Switzerland) programmed in the normal operation mode (nozzle diameter: 0.7 mm; atomized volume: 200 mL, solids content < 33%). After an optimization process, the optimal operation conditions were established as having an inlet temperature of 140 °C, outlet temperature of 72 °C, aspiration 90%, and the pump working at 20% (6 mL/min). The yield of the process was calculated as being the ratio between the weight of the obtained powder (dry basis) and the weight of the solid content of the initial atomized solution in dry basis.

After spray-drying, a portion of each powdered sample was used to perform the required analyses immediately after preparation (t0) and the remaining amount was divided into 2 equal portions, in sterile flasks protected from light, stored at room (23 °C) and refrigerated (3 °C) temperatures, respectively.

2.9. Stability of the Colouring Formulations

2.9.1. Anthocyanins Concentration

To verify if the colouring properties of the formulations were maintained over time, the concentration of anthocyanins in the powders was assessed, as described in Section 2.4., after their preparation (t0) and at different stages of storage, more specifically 4, 8, and 12 weeks, both for room and refrigerated temperature stored samples.

2.9.2. Colour Parameters

To measure the colour of the developed colouring formulations, a colourimeter (model CR-400, Konica Minolta Sensing, Inc., Osaka, Japan) equipped with specific tool for granular materials (model CR-A50) was used as reported by the authors Pereira et al. [17]. The analysis of the colour parameters was made in the CIE $L^*a^*b^*$ colour space, through the illuminant C and the diaphragm aperture was 8 mm. The obtained data were analyzed using a "Spectra Magic Nx" (version CM-S100W 2.03.0006 software, Konica Minolta). The colour measurements were achieved after preparing the colorant preparation (t0) and every 4 weeks of storage, until it reached 12 weeks of shelf life.

2.9.3. Microbial Load

After pasteurization and spray-drying, the samples were analyzed; the powder (1 g) was mixed with peptone water (PW, 9 mL) and serial decimal dilutions of the initial suspension in PW were prepared until achieving 10^{-6}. From these solutions, different counts were performed:

Aerobic plate count (total viable count; ISO 4833-2:2013). The dilutions were inoculated in PCA (plate count agar) by the pour plate technique, in duplicate (LOQ = 1 log CFU/g): 1 mL of suspension was pipetted onto the plate and 15 mL of melted PCA (kept at 50 °C in a water bath or incubator) were poured; it was homogenized and left to solidify. It was incubated at 30 °C for 72 h, in reversed position; when the plates had between 15 and 300 colonies, the count was performed.

Coliforms (and E. coli; ISO 4832:2006). The dilutions were inoculated in VRBLA (violet, red bile lactose agar) using the pour plate technique, in duplicate (LOQ = 1 log CFU/g): 1 mL of suspension was pipetted onto the plate and 15 mL of melted VRBLA (kept at 50 °C in a water bath or incubator) was poured; it was homogenized and left to solidify. On top of the medium, a top layer of 4 mL of VRBLA was poured and it was left to solidify. It was incubated at 30 °C for 48 h, in reversed position. When the plates had between 10 and 150 colonies, the count was performed.

Yeasts and Moulds (ISO 21527-1/2:2008). The dilutions were inoculated in DRBC (dichloran rose bengal chloramphenicol) using the spread plate technique, in duplicate (LOQ = 1.7 log CFU/g): 0.2 mL of suspension were pipetted onto a plate containing 15 mL of the medium and were spread with a disposable spreader. It was incubated at 25 °C for 5 days, in the upright position. When the plates had less than 150 colonies, the count of yeast and mould colonies was performed separately after 2 and 5 days of incubation.

Bacillus cereus (ISO 7932:2004). The dilutions were inoculated in MYP (mannitol yolk polymyxin) using the spread plate technique, in duplicate (LOQ = 1.7 log CFU/g): 0.2 mL of suspension were pipetted onto a plate containing 15 mL of the medium and were spread with disposable spreader. It was incubated at 30 °C for 24–48 h, in reversed position. When the plates had between 10 and 150 colonies, the count was performed.

The microbial load of the different colouring formulations was assessed after their preparation (t0) and after 12 weeks of storage at room and refrigerated temperature.

2.9.4. Cytotoxic Activity in Non-Tumour Cells

The cytotoxicity was determined using a primary culture of porcine liver non-tumour cells, PLP2. The cells growth was monitored by the use of a phase contrast microscope. The cells were sub-cultured and plated in 96 well plates (density of 1.0×10^4 cells/well) with the culture medium Dulbecco's modified Eagle's medium (DMEM) supplemented with FBS (10%), penicillin (100 U/mL), and streptomycin (100 µg/mL) according to Corrêa et al. [18]. The results were expressed in µg/mL of the sample concentration able to inhibit 50% of the net cell growth. Ellipticine was used as positive control. The cytotoxicity was assessed after the colourants preparation (t0) and after 12 weeks of storage at room and refrigerated temperature, to guarantee their safety for food application.

2.10. Statistical Analysis

For each analysis, three samples were assessed, and the assays were carried out in triplicate. The results were analyzed using one-way analysis of variance (ANOVA) post-hoc Tukey and are expressed as mean values and standard deviation (SD). This treatment was carried out using the SPSS v.22.0 program (IBM Corp., Armonk, New York, USA).

3. Results and Discussion

3.1. Nutritional and Chemical Composition

3.1.1. Nutritional Composition

The results obtained for the nutritional composition of haskap fruits are presented in Table 1. The fresh fruit presented 82.9 ± 0.1 g/100 g fw of water (17.1 ± 0.1 g/100 g of dry matter), and carbohydrates as major macronutrients (15.87 ± 0.06 g/100 g fw). It also presented 0.87 ± 0.04 g/100 g fw of protein, 0.38 ± 0.02 g/100 g fw of fat, and 0.00127 ± 0.00008 g/100 g fw of ash, which led to 70.37 ± 0.09 kcal/100 g fw of energetic contribution. In a previous study performed by Palíková et al. [19], haskap revealed a similar water content (82.7 g/100 g fw), but significantly higher concentrations of protein and ash (1.6 and 0.5 g/100 g fw, respectively), thus also presenting a higher energy (330 Kcal/100 g fw) [19]. Also, in a work performed with three cultivars of haskap, *Borealis* and *Indigo Gem* cultivars revealed similar contents of dry matter (17.7 ± 0.6 and 15.91 ± 0.04 g/100 g fw, respectively) and carbohydrates (15.6 ± 0.7 and 14.30 ± 0.08 g/100 g fw, respectively), whereas the *Tundra* cultivar showed lower concentrations of both (12.4 ± 0.2 and 10.19 ± 0.2 g/100 g fw of water and carbohydrates, respectively), thus presenting higher levels of protein, fat, and ash [20].

Other cultivars were previously assessed, namely, *Czelabinka, Duet, Jolanta,* and *Wojtek* (focused on in the present study), and presented similar dry matter amounts (16.1 ± 1, 17 ± 1, 15 ± 1, and 14 ± 2 g/100 g fw, respectively), and higher concentrations of ash (0.096 ± 0.003, 0.095 ± 0.007, 0.09 ± 0.02, and 0.067 ± 0.001 g/100 g fw, respectively) [21]. On the other hand, in a study performed with the *Kamtschatica* variety from Khabarovsk (Russia), lower amounts of fat and carbohydrates were found (0.01 ± 0.01 and 0.9 ± 0.1 g/100 g fw), but the sample revealed higher quantities of water, protein, and ash (88 ± 4, 2.1 ± 0.2, and 0.5 ± 0.1 g/100 g fw) [6]. The differences could be explained by several possible factors, such as soil and cultivation conditions, given the different origin of the samples, or the maturation state, which clearly influences fruits composition [1,22], among many others.

Table 1. Nutritional value and free sugars, organic acids, and tocopherols composition of haskap fruits.

Nutritional value	
Moisture (g/100 g fw)	82.9 ± 0.1
Fat (g/100 g fw)	0.38 ± 0.02
Protein (g/100 g fw)	0.87 ± 0.04
Ash (g/100 g fw)	0.00127 ± 0.00008
Carbohydrates (g/100 g fw)	15.87 ± 0.06
Energy (kcal/100 g fw)	70.37 ± 0.09
Free sugars	
Fructose (g/100 g fw)	4.021 ± 0.006
Glucose (g/100 g fw)	3.86 ± 0.05
Sucrose (g/100 g fw)	0.027 ± 0.002
Unknown (g/100 g fw)	0.181 ± 0.007
Total (g/100 g fw)	7.91 ± 0.06
Organic acids	
Oxalic acid (g/100 g fw)	0.041 ± 0.002
Quinic acid (g/100 g fw)	0.37 ± 0.01
Malic acid (g/100 g fw)	0.77 ± 0.01
Ascorbic acid (g/100 g fw)	0.0248 ± 0.0003
Citric acid (g/100 g fw)	2.76 ± 0.01
Total (g/100 g fw)	3.93 ± 0.01
Tocopherols	
α-Tocopherol (mg/100 g fw)	0.77 ± 0.03
γ-Tocopherol (mg/100 g fw)	0.162 ± 0.003
Total (mg/100 g fw)	0.93 ± 0.03

3.1.2. Free Sugars

Regarding free sugars (Table 1), fructose and glucose (4.021 ± 0.006 and 3.86 ± 0.05 g/100 g fw, respectively) were the most abundant ones, but lower concentrations of sucrose and an unknown sugar were also found. The total amount detected in this study (7.91 ± 0.06 g/100 g fw) was very similar to the one reported by Palíková et al. (7.2 g/100 g fw), with fructose (3.2 g/100 g fw) and glucose (2.9 g/100 g fw) as free sugars detected [19]. These latest were also found in other studies, in concentrations of 2.80 and 2.90 g/100 g fw (fructose), and 3.64 and 3.2 g/100 g fw (glucose) [19,23]. As for Wojtek cultivar studied herein, sucrose was also previously detected by Oszmiański et al. (0.04 g/100 g fw) in the same cultivar [23].

3.1.3. Organic Acids

A total of 3.93 ± 0.01 g/100 g fw of organic acids was quantified (Table 1), with citric acid as the most abundant one (2.76 ± 0.01 g/100 g fw), followed by malic and quinic acids in lower amounts (0.77 ± 0.01 and 0.37 ± 0.01 g/100 g fw, respectively), and also oxalic and ascorbic acids (0.041 ± 0.002 and 0.0248 ± 0.0003 g/100 g fw, respectively). Comparing these results with the ones reported by other authors for this cultivar, some variations can be observed, for instance Wojdyło et al. detected 0.154 ± 0.005 g/100 g fw of citric acid, followed by malic, oxalic, quinic, and ascorbic acids (0.039 ± 0.002, 0.0115 ± 0.0003, 0.0098 ± 0.0007, and 0.0017 ± 0.0002 g/100 g fw, respectively), which concentrations were significantly lower than the ones found in the present study [21]. These authors also detected phytic and shikimic acids in concentrations of 0.047 ± 0.005 and 0.0039 ± 0.0002 g/100 g fw, which were not detected in this work.

3.1.4. Tocopherols

In what concerns tocopherols (Table 1), two isoforms were detected, α and γ, with a clear prevalence of α-tocopherol (0.77 ± 0.03 in a total of 0.93 ± 0.03 mg/100 g fw). In a previous investigation, haskap fruits also revealed α-tocopherol as the most abundant isoform (0.42 mg/100 g fw), despite the lower concentration detected, in comparison to the one found in the present study. In the referred work, the authors also found β, γ, and δ-tocopherol [19].

3.1.5. Fatty Acids

A total of 20 fatty acids was found in haskap fruits (Table 2). Linoleic acid was the most abundant one ($71.79 \pm 0.08\%$), which mostly contributed to the prevalence of polyunsaturated fatty acids ($77.69 \pm 0.03\%$) in the sample. Regarding monounsaturated fatty acids ($14.403 \pm 0.006\%$), the one present in the highest concentration was oleic acid ($14.153 \pm 0.006\%$), and in the case of saturated fatty acids ($7.91 \pm 0.02\%$), palmitic acid contributed with the highest percentage ($5.39 \pm 0.01\%$). The same observations were made by Caprioli et al., which also reported linoleic, oleic, and palmitic acids (66.8, 22.8, and 5.3%, respectively) as the most abundant polyunsaturated, monounsaturated, and saturated fatty acids (69.8, 24.2, and 6.1%, respectively) [6]. Given the fact that linoleic acid intake is related to the prevention and control of adverse blood pressure levels in general populations, the results obtained could explain some of the beneficial properties of this fruit [24].

Table 2. Fatty acids composition of haskap fruits.

Fatty Acid	Relative %
Caproic acid (C6:0)	0.078 ± 0.001
Caprolic acid (C8:0)	0.067 ± 0.001
Capric acid (C10:0)	0.012 ± 0.001
Lauric acid (C12:0)	0.366 ± 0.003
Myristic acid (C14:0)	0.304 ± 0.008
Myristoleic acid (C14:1)	0.067 ± 0.001
Pentadecylic acid (C15:0)	0.092 ± 0.002
Palmitic acid (C16:0)	5.39 ± 0.01
Palmitoleic acid (C16:1)	0.098 ± 0.001
Margaric acid (C17:0)	0.082 ± 0.001
Stearic acid (C18:0)	0.986 ± 0.001
Oleic acid (C18:1n9)	14.153 ± 0.006
Linoleic acid (C18:2n6)	71.79 ± 0.08
γ-Linoleic acid (C18:3n6)	1.4 ± 0.1
Linolenic acid (C18:3n3)	4.24 ± 0.01
Arachidic acid (C20:0)	0.158 ± 0.001
Eicosenoic acid (C20:1)	0.086 ± 0.001
Eicosadienoic acid (C20:2)	0.227 ± 0.002
Behenic acid (C22:0)	0.306 ± 0.001
Lignoceric acid (C24:0)	0.069 ± 0.004
SFA	7.91 ± 0.02
MUFA	14.403 ± 0.006
PUFA	77.69 ± 0.03

SFA: saturated fatty acids; MUFA: monounsaturated fatty acids; PUFA: polyunsaturated fatty acids.

3.2. Phenolic Composition

The phenolic composition of haskap berries is presented in Table 3, being identified as constituting one phenolic acid and six anthocyanins. 5-*O*-Caffeoylquinic acid was identified by comparison with a commercial standard and was present in a concentration of 0.589 ± 0.002 mg/g of extract. In other studies involving haskap fruits, other phenolic compounds (non-anthocyanin) were detected, such as protocatechuic, gentisic, ellagic, ferulic, caffeic, and coumaric acids [19]. On the other hand, Oszmiánski et al. [23] found eight phenolic acids, five flavan-3-ols, twelve flavonols, and five flavones. These differences could possibly be explained by the different processing conditions to which the samples were subjected, such as defrosting processes or the maturation stage, climate, harvest, or even storage conditions [2,5,6]. Indeed, several examples can be found for the effect on polyphenolic composition of the different treatments to which the fruits were subjected, for instance, Ochmian et al. [1] packed the berries in polyethylene bags and stored them at −32 °C for 6 months, while Khattab et al. [25] stored them for 6 months at −18 °C; comparing the results obtained, the samples stored at higher temperatures revealed lower concentrations of phenolic compounds. But besides storage temperature, other factors can affect the phenolic composition; in fact, steam bleaching of the fruits prior to freezing clearly increased the retention of these compounds, whereas storage at −32 °C did not reveal such an influence [25]. There are other reports on the pre-treatment of these berries that also report the contribution of the process to the variation of the phenolic composition; for example, in a study performed by Wojdyło et al. [21], the fruits were cut directly into liquid nitrogen and freeze-dried before grounding and storing at −70 °C, which also led to the preservation of such compounds [21]. Besides, the influence of the maturation stage of these fruits also revealed a clear influence on the polyphenolic composition, with late harvest haskap presenting significantly higher concentrations than early-harvested fruits [1]; more specifically, for *Wojtek* cultivar, the early berries presented lower contents (174.95 mg/100 g) than the late-harvest berries, in which an increase of 27% was verified. For the *Brazowa* cultivar, similar observations were made, with an enhancement of 65% of the initial concentration (144.84 mg/100 g) [26,27].

Regarding anthocyanins (Table 3), these were identified according to the peak characteristics, such as retention time, wavelength of maximum absorption and mass spectral data. Three cyanidin derivatives (cyanidin-*O*-hexoside-*O*-hexoside, cyanidin-3-*O*-glucoside, and cyanidin-*O*-rhamnoside- *O*-hexoside), two peonidin derivatives (peonidin-3-*O*-glucoside and peonidin-*O*-rhamnoside-*O*-hexoside), and a pelargonidin derivative (pelargonidin-3-*O*-glucoside) were identified, as shown in Figure 1. The compounds represented by peaks two, four, and five were positively identified by comparison with commercial standard chromatographic characteristics. The compound corresponding to peak 1 ([M]$^+$ at *m/z* 611) presented two MS^2 fragments, presenting two losses of hexosyl units (*m/z* at 287; −162–162 u), being identified as cyanidin-*O*-hexoside-*O*-hexoside. Also, the compounds represented by peaks three ([M]$^+$ at *m/z* 595) and six ([M]$^+$ at *m/z* 609) were identified as cyanidin-*O*-rhamnoside-*O*-hexoside and peonidin-*O*-rhamnoside-*O*-hexoside, respectively, revealing a loss of one hexosyl (−162u) and rhamnosyl (−146u). The most abundant anthocyanin was cyanidin-3-*O*-glucoside, representing 63% of the total anthocyanin concentration (61.7 ± 0.1 of 97.9 ± 0.2 mg/g of extract). The results obtained are in accordance with those reported by Oszmiański et al [23]., but in this study, cyanidin-3-*O*-glucoside represented 88% of the total content, and another seven anthocyanins were found. Ochmian et al. [1] and Wojdyło et al. [21] also reported this anthocyanin as the one present in the highest concentration (83–90% and 71–89% of the total concentration).

Table 3. Phenolic composition of haskap fruits.

Peak	Rt (min)	λ_{max} (nm)	$[M\text{-}H]^-/[M]^+$ m/z	MS^2	Tentative Identification	Concentration (mg/g ext)
				Non-anthocyanic compounds		
1′	7.2	326	353/ -	191(100),179(6),161(5),135(5)	5-*O*-Caffeoylquinic acid [1]	0.589 ± 0.002
				Anthocyanins		
1	11.94	512	- /611	449(100), 287(28)	Cyanidin-*O*-hexoside-*O*-hexoside [2]	6.75 ± 0.01
2	19.72	515	- /449	287(100)	Cyanidin-3-*O*-glucoside [2]	61.7 ± 0.1
3	22.00	517	- /595	449(31), 287(100)	Cyanidin-*O*-rhamnoside-*O*-hexoside [2]	10.1 ± 0.2
4	23.49	508	- /433	271(100)	Pelargonidin-3-*O*-glucoside [3]	10.30 ± 0.04
5	26.29	515	- /463	301(100)	Peonidin-3-*O*-glucoside [4]	4.872 ± 0.002
6	28.33	520	- /609	463(32), 301(100)	Peonidin-*O*-rhamnoside-*O*-hexoside [4]	4.142 ± 0.002
					Total	97.9 ± 0.2

Calibration curves used for quantification: [1] 5-*O*-caffeoylquinic acid ($y = 312503x - 199432$; R^2: 0.9999; LOD: 16.65; LOQ: 50.45); [2] Cyanidin-3-*O*-glucoside ($y = 134578x - 3E+06$; R^2: 0.9986; LOD: 9.94; LOQ: 30.13); [3] Pelargonidin-3-*O*-glucoside ($y = 61493x - 628875$; R^2: 0.9957; LOD: 24.94; LOQ: 75.580.); and [4] Peonidin-3-*O*-glucoside ($y = 151438x - 3E+06$; R^2: 0.9977; LOD: 12.90; LOQ: 39.09).

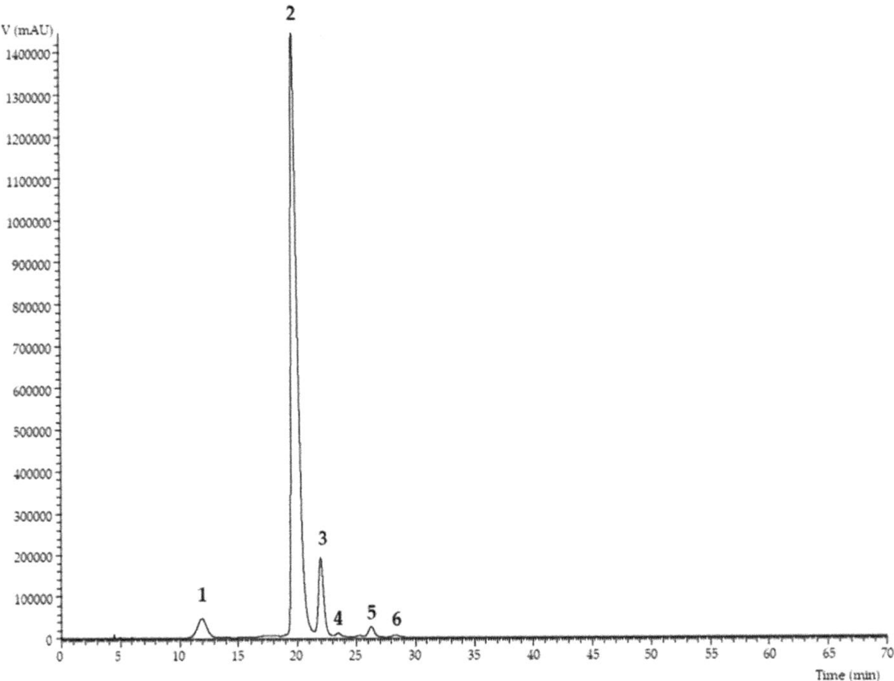

Figure 1. Chromatographic profile of the phenolic compounds found in the haskap fruit hydroethanolic extract, recorded at 520 nm (1: cyanidin-*O*-hexoside-*O*-hexoside; 2: cyanidin-3-*O*-glucoside; 3: cyanidin-*O*-rhamnoside-*O*-hexoside; 4: pelargonidin-3-*O*-glucoside; 5: peonidin-3-*O*-glucoside; and 6: peonidin-*O*-rhamnoside-*O*-hexoside).

3.3. Bioactive Properties

3.3.1. Antioxidant Activity

The antioxidant activity of the haskap fruit extract and colouring formulations was determined by two in vitro assays, the lipid peroxidation inhibition assay (TBARS) and the oxidative haemolysis inhibition assay (OxHLIA), as presented in Table 4. In the TBARS assay, the extract revealed a large amount of antioxidant activity, with an EC_{50} value of 29.9 ± 0.3 µg/mL, which was significantly lower than the one presented by the positive control, Trolox (139 ± 5 µg/mL). Regarding the OxHLIA assay, the extract also revealed strong activity, which was capable of delaying the oxidative haemolysis for 120 min, in a concentration of 938 ± 49 µg/mL, which despite being higher than that needed for Trolox, was still a good result for a natural extract. Regarding the colouring formulations, both samples revealed a lower antioxidant activity in the TBARS assay than the extract, but the fact that they contained only 20% of haskap juice must be taken into consideration. Nevertheless, in the OxHLIA assay, the differences were not so severe, with the formulations containing maltodextrin (80%) and maltodextrin (40%) + arabic gum (40%) revealing, respectively, a higher (737 ± 16 µg/mL) and similar (976 ± 9 µg/mL) capacity to inhibit the oxidative haemolysis, for 120 min, than the extract (938 ± 49 µg/mL). Among formulations, the one prepared with maltodextrin presented higher lipid peroxidation inhibition capacity than that prepared with maltodextrin and arabic gum, but higher concentrations of this formulation were needed to inhibit the oxidative haemolysis.

Table 4. Bioactive properties of haskap fruit extract and colouring formulations.

		Antioxidant Activity (IC_{50} values, µg/mL)				
		Haskap extract	Haskap + M	Haskap + M + AG	Trolox [1]	
TBARS assay		29.9 ± 0.3 [a]	1773 ± 106 [d]	1033 ± 85 [c]	139 ± 5 [b]	
OxHLIA assay	60 min	145 ± 5 [b]	298 ± 13 [c]	394 ± 13 [d]	85 ± 2 [a]	
	120 min	938 ± 49 [c]	737 ± 16 [b]	976 ± 9 [c]	183 ± 4 [a]	
		Antibacterial Activity (MIC and MBC values, mg/mL)				
		Haskap extract	Haskap + M	Haskap + M + AG	Streptomycin [1]	Ampicillin [1]
Bacillus cereus	MIC/MBC	6.81/6.81	5.03/10.06	5.07/10.14	0.10/0.20	0.25/0.40
Staphylococcus aureus	MIC/MBC	6.81/6.81	5.03/10.06	10.14/20.28	0.17/0.25	0.34/0.37
Listeria monocytogenes	MIC/MBC	3.41/3.41	5.03/10.06	7.605/10.14	0.20/0.30	0.40/0.50
Escherichia coli	MIC/MBC	3.41/3.41	2.515/5.03	1.26/2.535	0.20/0.30	0.40/0.50
Enterobacter cloacae	MIC/MBC	3.41/3.41	7.54/10.06	7.60/10.14	0.043/0.25	0.086/0.37
Salmonella typhimurium	MIC/MBC	3.41/3.41	7.54/10.06	7.60/10.14	0.20/0.30	0.75/1.20
		Antifungal Activity (MIC and MFC values, mg/mL)				
		Haskap extract	Haskap + M	Haskap + M + AG	Ketoconazole [1]	Bifonazole [1]
Aspergillus fumigatus	MIC/MFC	13.63/27.27	5.03/10.06	20.28/>20.28	0.38/0.95	0.48/0.64
Aspergillus versicolor	MIC/MFC	6.81/13.63	5.03/10.06	20.28/>20.28	0.20/0.50	0.10/0.20
Aspergillus niger	MIC/MFC	27.27/27.27	20.12/>20.12	20.28/>20.28	0.20/0.50	0.15/0.20
Penicillium funiculosum	MIC/MFC	13.63/27.27	5.03/10.06	>20.28/>20.28	0.20/0.50	0.20/0.25
Penicillium ochrochloron	MIC/MFC	27.27/>27.27	2.52/5.03	20.28/>20.28	3.8/0.48	3.8/0.64
Trichoderma viride	MIC/MFC	2.13/13.63	1.26/2.52	2.54/5.07	4.75/0.64	5.70/0.80

[1] Positive controls; Haskap + M: Colouring formulation containing maltodextrin (80%); Haskap + M + AG: colouring formulation containing maltodextrin (40%) and arabic gum (40%). IC_{50}: extract concentration providing 50% of antioxidant activity; MIC: minimum inhibitory properties; MBC: minimum bactericidal concentration; MFC: minimum fungicidal properties. For the antioxidant activity, different letters in each line mean significant differences ($p < 0.05$).

The results obtained in the present study are not directly comparable to the ones reported in previous studies, as the applied antioxidant assays are different; however, the haskap extracts did reveal a good performance in tests such as TEAC (Trolox-equivalent antioxidant capacity), FRAP (ferric reducing antioxidant power), ORAC (oxygen radical absorbance capacity), DPPH (2,2-diphenyl-1-picrylhydrazyl radical scavenging activity), and ABTS (2,2′-azinobis-(3-ethylbenzothiazoline-6-sulfonic acid) [20,28], notwithstanding the fact that these methods are based on chemical reactions that are not so comparable to those occurring in biological systems, such as the ones used in the present study which were based on porcine brain tissues and sheep blood erythrocytes.

3.3.2. Antimicrobial Activity

The antibacterial and antifungal properties of haskap extract and colouring formulations are presented in Table 4 as MIC (minimum inhibitory concentration), MBC (minimum bactericidal concentration), and MFC (minimum antifungal concentration) values. The extract was able to inhibit the growth of all the studied bacterial strains in concentrations of 3.41 (*L. monocytogenes*, *E. coli*, *E. cloacae*, and *S. typhimurium*) and 6.81 (*B. cereus*, *S. aureus*, and *L. monocytogenes*) mg/mL. Interestingly, these were also the concentrations needed to obtain a bactericidal effect in these microorganisms, which could mean that the compounds responsible for haskap extract antibacterial activity have mainly a bactericidal capacity [29]. In what concerns the antibacterial properties of the colorants, in general, the formulation prepared with maltodextrin revealed lower MIC and MBC values than the other colorant, and in some cases, it was also more active than the extract in inhibiting bacterial growth. Regarding antifungal properties, the extract revealed a stronger activity in *T. viride* and *A. versicolor*, with lower MIC (2.13 and 6.81 mg/mL, respectively) and MFC (13.63 and 13.63 mg/mL, respectively) values. In fact, the MIC value presented by the extract for *T. viride* was even lower than those showed by the positive controls Ketoconazole (4.75 mg/mL) and Bifonazole (5.70 mg/mL). Regarding the formulations, the same tendency could be observed for this fungus, in which both colorants revealed a strongest inhibitory activity than the positive controls (1.26 and 2.54, in the case of haskap + maltodextrin and haskap + maltodextrin+arabic gum, respectively). For the colorant prepared with maltodextrin, the MIC value against *P. ochrochloron* was also lower (2.52 mg/mL) than the ones found for the positive controls (3.8 mg/mL). This formulation was more active against all fungi than the one containing maltodextrin and arabic gum as well as the extract. Raudsepp et al. [30] previously reported the capacity of this fruit extract to inhibit *Bacillus subtilis*, *Kocuria rhizophila*, *E. coli*, *Lactobacillus acidophilus*, and *Campylobacter jejuni*. To the best of our knowledge, this is the first report on haskap fruits' antifungal properties, which limits our possibilities of comparing the results obtained herein with the data in the literature; nevertheless, all the samples (extract and colouring formulations, specially haskap + maltodextrin) revealed promising results in these microorganisms.

3.4. Anthocyanins Composition and Stability of the Colouring Formulations

The main compounds responsible for the colouring properties of haskap fruits are their anthocyanins; thus, an analysis of these compounds in the juice samples was performed and five of the six anthocyanins detected in the previously described hydroethanolic extract were found in the colouring formulations: cyanidin-*O*-hexoside-*O*-hexoside, cyanidin-3-*O*-glucoside, cyanidin-*O*-rhamnoside-*O*-hexoside, peonidin-3-*O*-glucoside, and peonidin-*O*-rhamnoside-*O*-hexoside. The difference in the detected compounds could be explained by the different kind of extract, which for anthocyanins analysis was hydroethanolic and for the colorant's preparation was the centrifuged fruit juice; or even the temperature to which the juice was subjected in the pasteurization and spray-drying processes.

Regarding the spray-drying process, due to the high concentration of sugars in haskap juice, it was not possible to prepare a control sample without the stabilizing materials, thus, two formulations were successfully prepared with good yields: haskap + maltodextrin (80%) and haskap + maltodextrin (40%) + arabic gum (40%), presenting yields of 46 and 51%, respectively.

To evaluate the stability of the prepared formulations, the total concentration of these compounds was considered, alongside with the colouring parameters (L*: lightness, white–black; a*: red colour intensity, red–green; and b*: yellow colour intensity, yellow–blue), both measured at 4, 8, and 12 weeks for samples stored at room and refrigerated temperature. In terms of lightness, the formulation containing maltodextrin + arabic gum allowed higher values (Figure 2a), as well as for the red (Figure 2b) and yellow (Figure 2c) colours' intensity. Even though this formulation presented better results concerning colour parameters, the anthocyanin levels were very similar for both preparations (Figure 2d). Nevertheless, it presented a higher stability over time, with less notorious variations with time and storage temperature than the colorant containing only maltodextrin, as it can be observed in Figure 2a–c. The slight variation observed in the total anthocyanin concentration confirmed the efficient protection of these compounds over time, making spray-drying a suitable methodology for anthocyanin-based colorants development.

In order to guarantee the safety of the prepared colorants, the samples were submitted to microbial load evaluation considering the microorganisms that could represent a risk for the formulation's stability over the time by causing degradation. These formulations were also evaluated in terms of cytotoxicity in non-tumour cells to assure the safety of their consumption. These analyses were performed after samples preparation (t0) and after 12 weeks of storage. According to the obtained results (data not shown), no microbial counts were possible to detect, meaning that the pasteurization process together with application of the spray-drying technique were able to maintain the stability of the colouring formulations. The spray-drying technique allowed the formulation to be in powder form, which drastically decreases the microbial contamination over time. Regarding cytotoxicity, the samples did not reveal any toxic effects on the non-tumour cells at the maximum tested concentration of 400 µg/mL, corroborating its safety for application in foodstuff, which is a very important result, once the objective was to prepare food colourants.

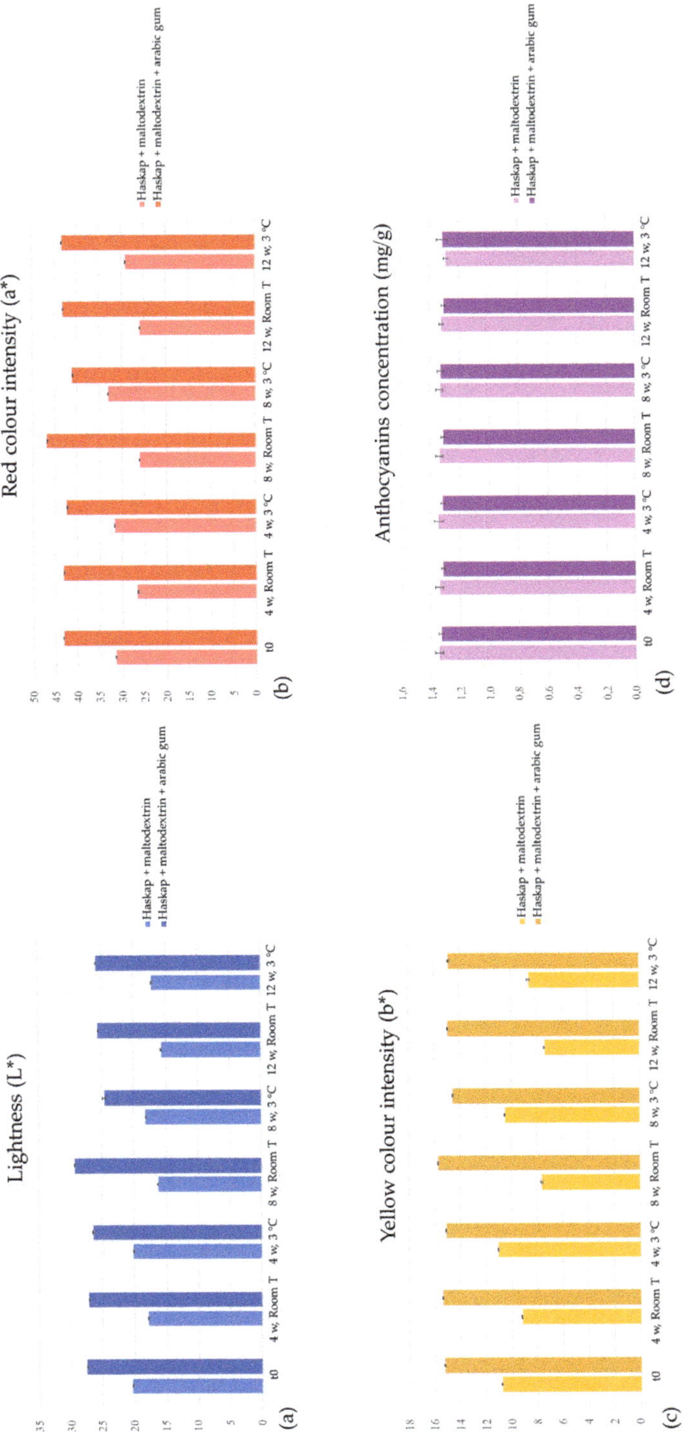

Figure 2. Evolution of colour parameters, L* (**a**), a* (**b**), and b* (**c**), and anthocyanins concentration (**d**) of the colouring formulations. w: weeks, Room T: room temperature.

4. Conclusions

Lonicera caerulea L. was assessed in terms of nutritional and chemical composition and revealed to be mostly composed of water and carbohydrates. It presented four different free sugars, five organic acids, two tocopherol isoforms, and twenty fatty acids, mostly polyunsaturated. Two non-anthocyanic phenolic compounds were detected, with 5-*O*-caffeoylquinic acid as the most abundant one, and six anthocyanins, with a prevalence of cyanidin-3-*O*-glucoside. The fruit hydroethanolic extract revealed excellent antioxidant properties and considerable antibacterial and antifungal activity. From the berries juice, it was possible to obtain two colouring formulations by spray-drying the samples with maltodextrin and a combination of maltodextrin with arabic gum. The developed colorants revealed good antioxidant and antimicrobial properties and did preserve their anthocyanin compounds and related coloration along twelve weeks of storage at room and refrigerated temperature, with neither microbial contamination nor cytotoxicity issues. Thus, food colorants obtained from *L. caerulea* should be considered for food application, in detriment to the most commonly used artificial ones, but also in other industries, such as cosmetics, textiles, or pharmaceuticals, among others. Further studies for the stabilization of these kinds of colourants should also be considered as well as the exploitation of other anthocyanin-rich natural sources.

Author Contributions: Formal analysis, C.P.; Investigation, A.K.M. and E.N.V.; Methodology, M.I.D., J.C.M.B., S.A.H., P.R., I.P.F. and M.K.; Supervision, L.B. and I.C.F.R.F.; Validation, P.R. and M.S.; Writing—original draft, C.P. and L.B.; Writing—review and editing, M.F.B. and I.C.F.R.F.

Funding: The authors are grateful to Foundation for Science and Technology (FCT, Portugal) and FEDER under Programme PT2020 for financial support to CIMO (UID/AGR/00690/2019), and FCT/MCTES (PIDDAC) and FCT for financial support to LSRE-LCM (UID/EQU/50020/2019); national funding by FCT, P.I., through the institutional scientific employment program-contract for M.I. Dias, I.P.F., and L. Barros, and through the individual scientific employment program-contract for S.A. Heleno and J. Barreira; C. Pereira's contract though the celebration of program-contract foreseen in No. 4, 5 and 6 of article 23° of Decree-Law No. 57/2016, of 29th August, amended by Law No. 57/2017, of 19th July; to FEDER-Interreg España-Portugal programme for financial support through the project 0377_Iberphenol_6_E; the European Regional Development Fund (ERDF) through the Regional Operational Program North 2020, within the scope of Project NORTE-01-0145-FEDER-023289: DeCodE and project Mobilizador Norte-01-0247-FEDER-024479: ValorNatural®.

Acknowledgments: The authors are grateful to the "Ponto Agrícola-Unipessoal, Lda." company for the supply of samples.

Conflicts of Interest: The authors declare no conflict of interest.

References

1. Ochmian, I.; Skupień, K.; Grajkowski, J.; Smolik, M.; Ostrowska, K. Chemical composition and physical characteristics of fruits of two cultivars of blue honeysuckle (*Lonicera caerulea* L.) in relation to their degree of maturity and harvest date. *Not. Bot. Horti Agrobot. Cluj Napoca* **2012**, *40*, 155–162. [CrossRef]

2. Svarcova, I.; Heinrich, J.; Valentova, K. Berry fruits as a source of biologically active compounds: The case of *Lonicera caerulea*. *Biomed. Pap. Med. Fac. Palacky Univ. Olomouc* **2007**, *151*, 163–174. [CrossRef]

3. Celli, G.B.; Ghanem, A.; Brooks, M.S.L. Haskap berries (*Lonicera caerulea* L.)—A critical review of antioxidant capacity and health-related studies for potential value-added products. *Food Bioprocess Technol.* **2014**, *7*, 1541–1554. [CrossRef]

4. Khattab, R.; Ghanem, A.; Brooks, M.S.-L. Stability of haskap berry (*Lonicera caerulea* L.) anthocyanins at different storage and processing conditions. *J. Food Res.* **2016**, *5*, 67–79. [CrossRef]

5. Rupasinghe, H.P.V.; Arumuggam, N.; Amararathna, M.; De Silva, A.B.K.H. The potential health benefits of haskap (*Lonicera caerulea* L.): Role of cyanidin-3-*O*-glucoside. *J. Funct. Foods* **2018**, *44*, 24–39. [CrossRef]

6. Caprioli, G.; Iannarelli, R.; Innocenti, M.; Bellumori, M.; Fiorini, D.; Sagratini, G.; Vittori, S.; Buccioni, M.; Santinelli, C.; Bramucci, M.; et al. Blue honeysuckle fruit (*Lonicera caerulea* L.) from eastern Russia: Phenolic composition, nutritional value and biological activities of its polar extracts. *Food Funct.* **2016**, *7*, 1892–1903. [CrossRef]

7. Horwitz, W.; Latimer, G. (Eds.) *AOAC Official Methods of Analysis of AOAC International*; AOAC International: Gaithersburg, MD, USA, 2005; ISBN 9780935584752.

8. Barros, L.; Pereira, E.; Calhelha, R.C.; Dueñas, M.; Carvalho, A.M.; Santos-Buelga, C.; Ferreira, I.C.F.R. Bioactivity and chemical characterization in hydrophilic and lipophilic compounds of *Chenopodium ambrosioides* L. *J. Funct. Foods* **2013**, *5*, 1732–1740. [CrossRef]

9. Pereira, C.; Carvalho, A.M.; Barros, L.; Ferreira, I.C.F.R.F.R. Use of UFLC-PDA for the analysis of organic acids in thirty-five species of food and medicinal plants. *Food Anal. Methods* **2013**, *6*, 1337–1344. [CrossRef]

10. Bessada, S.M.F.; Barreira, J.C.M.; Barros, L.; Ferreira, I.C.F.R.; Oliveira, M.B.P.P. Phenolic profile and antioxidant activity of *Coleostephus myconis* (L.) Rchb.f.: An underexploited and highly disseminated species. *Ind. Crops Prod.* **2016**, *89*, 45–51. [CrossRef]

11. Gonçalves, G.A.; Soares, A.A.; Correa, R.C.G.; Barros, L.; Haminiuk, C.W.I.; Peralta, R.M.; Ferreira, I.C.F.R.; Bracht, A. Merlot grape pomace hydroalcoholic extract improves the oxidative and inflammatory states of rats with adjuvant-induced arthritis. *J. Funct. Foods* **2017**, *33*, 408–418. [CrossRef]

12. Pereira, C.; Calhelha, R.C.; Barros, L.; Queiroz, M.J.R.P.; Ferreira, I.C.F.R. Synergisms in antioxidant and anti-hepatocellular carcinoma activities of artichoke, milk thistle and borututu syrups. *Ind. Crops Prod.* **2014**, *52*, 709–713. [CrossRef]

13. Lockowandt, L.; Pinela, J.; Roriz, C.L.; Pereira, C.; Abreu, R.M.V.; Calhelha, R.C.; Alves, M.J.; Barros, L.; Bredol, M.; Ferreira, I.C.F.R. Chemical features and bioactivities of cornflower (*Centaurea cyanus* L.) capitula: The blue flowers and the unexplored non-edible part. *Ind. Crops Prod.* **2019**, *128*, 496–503. [CrossRef]

14. Soković, M.; Glamočlija, J.; Marin, P.D.; Brkić, D.; Griensven, L.J.L.D. Van Antibacterial effects of the essential oils of commonly consumed medicinal herbs using an in vitro model. *Molecules* **2010**, *15*, 7532–7546. [CrossRef] [PubMed]

15. Soković, M.; Van Griensven, L.J.L.D. Antimicrobial activity of essential oils and their components against the three major pathogens of the cultivated button mushroom, *Agaricus bisporus*. *Eur. J. Plant Pathol.* **2006**, *116*, 211–224. [CrossRef]

16. Moser, P.; De Souza, R.T.; Nicoletti Telis, V.R. Spray drying of grape juice from hybrid CV. BRS Violeta: Microencapsulation of anthocyanins using protein/maltodextrin blends as drying aids. *J. Food Process. Preserv.* **2017**, *41*, e12852. [CrossRef]

17. Pereira, E.; Antonio, A.L.; Barreira, J.C.M.; Barros, L.; Bento, A.; Ferreira, I.C.F.R. Gamma irradiation as a practical alternative to preserve the chemical and bioactive wholesomeness of widely used aromatic plants. *Food Res. Int.* **2015**, *67*, 338–348. [CrossRef]

18. Corrêa, R.C.G.; de Souza, A.H.P.; Calhelha, R.C.; Barros, L.; Glamoclija, J.; Sokovic, M.; Peralta, R.M.; Bracht, A.; Ferreira, I.C.F.R. Bioactive formulations prepared from fruiting bodies and submerged culture mycelia of the Brazilian edible mushroom *Pleurotus ostreatoroseus* Singer. *Food Funct.* **2015**, *6*, 2155–2164. [CrossRef] [PubMed]

19. Palíková, I.; Valentová, K.; Šimánek, V.; Ulrichová, J.; Heinrich, J.; Bednář, P.; Marhol, P.; Křen, V.; Cvak, L.; Růžička, F.; et al. Constituents and antimicrobial properties of blue honeysuckle: A novel source for phenolic antioxidants. *J. Agric. Food Chem.* **2008**, *56*, 11883–11889. [CrossRef]

20. Rupasinghe, H.P.V.; Yu, L.J.; Bhullar, K.S.; Bors, B. Short Communication: Haskap (*Lonicera caerulea*): A new berry crop with high antioxidant capacity. *Can. J. Plant Sci.* **2012**, *92*, 1311–1317. [CrossRef]

21. Wojdyło, A.; Jáuregui, P.N.N.; Carbonell-Barrachina, A.A.; Oszmiański, J.; Golis, T. Variability of phytochemical properties and content of bioactive compounds in Lonicera caerulea L. var. *kamtschatica* berries. *J. Agric. Food Chem.* **2013**, *61*, 12072–12084. [CrossRef]

22. Yamamoto, Y.; Hoshino, Y.; Masago, H.; Kawano, T. Attempt for postharvest ripening of immature fruits of Haskap (*Lonicera caerulea* L. var. *emphyllocalyx* Nakai), an emerging fruit in Northern Japan. *Adv. Hortic. Sci.* **2014**, *28*, 244–249.

23. Oszmiański, J.; Wojdyło, A.; Lachowicz, S. Effect of dried powder preparation process on polyphenolic content and antioxidant activity of blue honeysuckle berries (*Lonicera caerulea* L. var. *kamtschatica*). *LWT Food Sci. Technol.* **2016**, *67*, 214–222. [CrossRef]

24. Miura, K.; Stamler, J.; Nakagawa, H.; Elliott, P.; Ueshima, H.; Chan, Q.; Brown, I.J.; Tzoulaki, I.; Saitoh, S.; Dyer, A.R.; et al. Relationship of dietary linoleic acid to blood pressure: The international study of macro-micronutrients and blood pressure study. *Hypertension* **2008**, *52*, 408–414. [CrossRef] [PubMed]

25. Khattab, R.; Celli, G.B.; Ghanem, A.; Brooks, M.S.L. Effect of frozen storage on polyphenol content and antioxidant activity of haskap berries (*Lonicera caerulea* L.). *J. Berry Res.* **2015**, *5*, 231–242. [CrossRef]

26. Skupien, K.; Oszmianski, J.; Ochmian, I.; Grajkowski, J. Characterization of selected physico-chemical features of blue honeysuckle fruit cultivar Zielona. *Polish J. Nat. Sci. Suppl.* **2007**, *4*, 101–107.

27. Ochmian, I.; Grajkowski, J.; Skupień, K. Yield and chemical composition of blue honeysuckle fruit depending on ripening time. *Bull. Univ. Agric. Sci. Vet. Med. Cluj Napoca Hortic.* **2010**, *67*, 138–147.

28. Bakowska-Barczak, A.M.; Marianchuk, M.; Kolodziejczyk, P. Survey of bioactive components in Western Canadian berries. *Can. J. Physiol. Pharmacol.* **2007**, *85*, 1139–1152. [CrossRef]

29. Radovanović, B.C.; Andelković, A.S.M.; Radovanović, A.B.; Andelković, M.Z. Antioxidant and antimicrobial activity of polyphenol extracts from wild berry fruits grown in Southeast Serbia. *Trop. J. Pharm. Res.* **2013**, *12*, 813–819. [CrossRef]

30. Raudsepp, P.; Anton, D.; Roasto, M.; Meremäe, K.; Pedastsaar, P.; Mäesaar, M.; Raal, A.; Laikoja, K.; Püssa, T. The antioxidative and antimicrobial properties of the blue honeysuckle (*Lonicera caerulea* L.), Siberian rhubarb (*Rheum rhaponticum* L.) and some other plants, compared to ascorbic acid and sodium nitrite. *Food Control* **2013**, *31*, 129–135. [CrossRef]

Article

Biological and Chemical Insights of Beech (*Fagus sylvatica* L.) Bark: A Source of Bioactive Compounds with Functional Properties

Corneliu Tanase [1,*,†], Andrei Mocan [2,*,†], Sanda Coşarcă [1], Alexandru Gavan [3], Alexandru Nicolescu [2], Ana-Maria Gheldiu [2], Dan C. Vodnar [4], Daniela-Lucia Muntean [5] and Ovidiu Crişan [6]

[1] Department of Pharmaceutical Botany, University of Medicine, Pharmacy, Sciences and Technology, 540139 Târgu Mureş, Romania; sandacosarca31@gmail.com
[2] Department of Pharmaceutical Botany, "Iuliu Haţieganu" University of Medicine and Pharmacy, Cluj-Napoca 400337, Romania; alexandru_s_nicolescu@yahoo.com (A.N.); Gheldiu.Ana@umfcluj.ro (A.-M.G.)
[3] Department of Medical Devices, "Iuliu Haţieganu" University of Medicine and Pharmacy, Cluj-Napoca 400439, Romania; gavan.alexandru@umfcluj.ro
[4] Department of Food Science, Faculty of Food Science and Technology, University of Agricultural Sciences and Veterinary Medicine, 400372 Cluj-Napoca, Romania; dan.vodnar@usamvcluj.ro
[5] Department of Analytical chemistry and Drug analysis, University of Medicine, Pharmacy, Sciences and Technology ", 540139 Târgu Mureş, Romania; daniela.muntean@umfst.ro
[6] Department of Organic Chemistry, "Iuliu Haţieganu" University of Medicine and Pharmacy, 400012 Cluj-Napoca, Romania; ovicrisan@yahoo.com
* Correspondence: corneliu.tanase@umfst.ro (C.T.); mocan.andrei@umfcluj.ro (A.M.)
† These authors share the first authorship.

Received: 14 August 2019; Accepted: 9 September 2019; Published: 19 September 2019

Abstract: The present study aimed, on the one hand, to improve the yield of microwave assisted extraction (MAE) of polyphenols from beech bark by using a design of experiments (DoE) approach. On the other hand, beech bark extracts (BBE) were characterized in terms of their phytochemical profile and evaluated for biological potential (antioxidant, antibacterial, antifungal, antimutagen, *anti*-α-glucosidase, and *anti*-tyrosinase). The extraction time varies with the amount of extracted total phenolic content (TPC). The microwave power favors TPC extraction but in different proportions. The optimum conditions which gave the highest TPC (76.57 mg GAE/g dry plant material) were reached when the microwave power was 300 W, extraction time was 4 min, and the solvent was an ethanol–water (50:50) mixture. The practical value of TPC after a controlled experiment was 76.49 mg GAE/g plant material. The identified compounds were vanillic acid, gallic acid, epicatechin, catechin, protocatechuic acid, chlorogenic acid, ferulic acid, and isoquercitrin. The antioxidant potential of BBEs was demonstrated by *in vitro* experiments. The BBEs were active against *Staphylococcus aureus*, *Pseudomonas aeruginosa*, *Salmonella typhimurium*, *Escherichia coli*, and *Candida* species. All extracts were antimutagenic and expressed an inhibition on α-glucosidase and tyrosinase activity. Regarding antimutagen activity, the assayed extracts may be considered to have low or no antimutagen effects.

Keywords: antioxidant; antibacterial; antifungal; antimutagen; *anti*-α-glucosidase; *anti*-tyrosinase; beech bark; microwave assisted extraction; optimization

1. Introduction

Bark plays an important role in protecting woody vascular plants, especially through its content in bioactive compounds with an antimicrobial effect [1]. The bark of woody plants is considered to be a by-product of the forestry and wood industry. This can be an important source of bioactive compounds with a high recovery potential. Numerous studies show the importance and value of bark. Moreover, bark extracts can have biological effects, such as antioxidant, antibacterial, anti-inflammatory, anti-tumor, etc., [2–6]. It has been determined that the bio-activity of bark natural extracts is mainly due to their content in phenolic compounds [7,8].

The beech (*Fagus sylvatica* L.) is one of the most widespread woody vascular plants in Europe and particularly in Romania, with a high economic value [9]. Beech wood is mostly used for fire wood or in the wood processing industry. After processing beech wood, a significant amount of bark is obtained.

Some literature data highlights that beech bark can be a rich source of bioactive compounds [10,11]. Furthermore, in a previous study some of the phenolic compounds obtained from beech bark by hot water extraction were identified (vanillic acid, catechin, taxifolin and syringin) [11,12]. Hofmann et al. identified 37 compounds in beech bark extracts, including catechin, epicatechin, quercetin, taxifolin, procyanidins, syringic acid, and coumaric acid [12]. Regarding the biological activity of beech bark extracts, literature information is quite limited. In previous works, the antibacterial and antitumor activity of beech bark aqueous extracts, obtained by classical extraction, was evaluated [11,13]. Beech bark extracts induced a decrease in A375 melanoma cell viability [13] and antimicrobial activity against *Staphylococcus aureus* including methicillin-resistant strains [11]. In another study, Hofmann et al. [10] found that the most efficient antioxidants in beech bark were the (+)-catechin, procyanidin B dimer 2, (−)-epicatechin, and coniferin isomer 2. Consequently, the research directions are oriented towards the identification and isolation of the bioactive compounds and the description of their mechanisms of action at the level of living organisms, finally with possibility of exploitation and functionalization on an industrial scale.

The main objectives of the current study were: (1) to optimize the extraction yield of phenolic compounds from beech bark (BB) based on an original experimental design; (2) the characterization of the phytochemical profile of optimized beech bark extracts (BBE); (3) the evaluation of the biological potential (antioxidant, antibacterial, antifungal, antimutagen, and enzyme inhibitory activity) of the optimized BBE.

2. Materials and Methods

2.1. Chemicals

The reagents used in this study were acetone, ethanol, methanol, hydrochloric acid (37%), and Folin-Ciocâlteu reagent, all of which were purchased from Merck (Darmstadt, Germany).

The standards used for both spectrophotometric and LC-MS/MS analysis were: quercetin (≥95%), hyperoside (quercetin 3-D-galactoside, ≥97%), isoquercitrin (quercetin 3-D-glucoside, ≥98%), quercitrin (quercetin 3-rhamonoside, ≥78%), (+)-catechin (≥96%), (−)-epicatechin (≥90%), vanillic acid (≥97%), syringic acid (≥95%), protocatechuic acid (3,4-dihydroxybenzoic acid, ≥97%), campesterol (~65%), ergosterol (≥95%), and stigmasterol (~95%) purchased from Sigma-Aldrich, gallic acid (≥98%) purchased from Merck (Darmstadt, Germany), and beta-sitosterol (≥80%) purchased from Carl Roth (Karlsruhe, Germany).

2.2. Plant Sample

The beech (*Fagus sylvatica* L.) bark samples were collected from Gurghiu Mountains, Toplița region, Mureș County, Romania, during November and December 2017. The age of the test trees was about 15–20 years. Only the bark was collected from the stems of the beech trees and (2.0 kg each) splintered manually. The species was identified and authenticated by Dr. Corneliu Tanase from the Department of Pharmaceutical Botany, the first author of the manuscript. The beech bark was air-dried at room temperature (10.5% humidity) and milled in a GRINDOMIX GM 2000 mill to a mean particle size diameter of 0.5 mm. The biomass was directly used without any pre-treatments.

2.3. Extraction

The extraction process was carried out based on a DoE (Design of Experiments) developed using Modde software, version 11.0 (Sartorius Stedim Data Analytics AB, Umeå, Sweden). The D-optimal design type, which is based on the selection of the experiments so that they are spread over the largest area of the variability matrix, allowed the study of two quantitative factors which varied on different levels, namely the microwave power, with 4 levels of variation (300, 450, 600, 800 W) and the extraction time, varied over 3 levels (2, 3, 4 min). The amount of extracted total phenolic content (TPC) using different solvent types (water, ethanol–water 50:50 and 80:20) was introduced in the DoE as 3 separate responses, which allowed a good comparison of the solvents' TPC extraction capacities (Table 1).

Table 1. Variables, symbols, and levels of variation used in the experimental design.

Variables	Symbol	Level of Variation			
Independent Variables (Factors)					
Extraction time (min)	X1		2	3	4
Microwave power (W)	X2	300	450	600	800
Dependent Variables (Responses)					
Extraction solvent-water	BBE 1				
Extraction solvent—50:50 ethanol–water	BBE 2				
Extraction solvent—80:20 ethanol–water	BBE 3				

Notes: mg GAE/g = gallic acid equivalents per gram plant material.

Bark was weighed (2 g) and mixed with 20 mL of solvent (water, 50:50 ethanol–water, 80:20 ethanol–water) in Falcon tubes. The Microwave extraction (MAE) was performed using a domestic microwave oven (Samsung MS23K3513). 7. After extraction, solvent was added to give a final volume of 20 mL. Each tube was centrifuged (Hettich, Micro 22R, Andreas Hettich GmbH & Co., Tuttlingen, Germany) 15 min at 3000 rpm, maintaining the extraction temperature. The supernatant was carefully separated, and the solvent was removed under vacuum at 40 °C using a rotary evaporator (Hei-VAP, Heidolph Instruments GmbH & Co., Schwabach, Germany).

2.4. Quantitative Determinations of Total Phenolic Content

The total phenolic content (TPC) of the BBE extracts was determined by Folin-Ciocâlteu spectrophotometric method according to a method described previously [14]. Gallic acid was used as a reference standard, and the content of phenolics was expressed as gallic acid equivalents (GAE) per gram of dry plant material (mg GAE/g dry plant material).

2.5. Phytochemical Analysis by LC-MS/MS

For the description of bark extracts' phytochemical composition a validated analytical method of liquid chromatography tandem mass spectrometry was employed (LC-MS/MS). The analysis was performed using an Agilent 1100 HPLC Series system (Agilent, Santa Clara, CA, USA) which was equipped with binary gradient pump, degasser, auto sampler, column thermostat, and UV detector. The LC system was coupled with an Agilent Ion Trap 1100 SL mass spectrometer (LC/MSD Ion Trap VL).

The LC-MS/MS analytical method was previously developed and validated [15,16] and was used for the identification of 18 polyphenols in BBE samples: caftaric acid, gentisic acid, caffeic acid, chlorogenic acid, *p*-coumaric acid, ferulic acid, sinapic acid, hyperoside, isoquercitrin, rutin, myricetol, fisetin, quercitrin, quercetin, patuletin, luteolin, kaempferol, and apigenin. For the chromatographic separation of polyphenols, a reverse-phase analytical column was used (Zorbax SB-C18, 100 mm × 3.0 mm i.d., 3.5 μm). The mobile phase was a mixture of methanol: acetic acid 0.1% (v/v) and a binary gradient was used. The elution started with a linear gradient, beginning with 5% methanol and ending at 42% methanol for 35 minutes; then isocratic elution followed for 3 min with 42% methanol. For the chromatographic data processing, the ChemStation and DataAnalysis software (Agilent, Santa Clara, CA, USA) were used.

Moreover, in order to identify other six polyphenols in BBE samples (epicatechin, catechin, syringic acid, gallic acid, protocatechuic acid, and vanillic acid) a second LC-MS method was employed [15]. For the separation of the aforementioned compounds, the same analytical column as previously specified was used. Likewise, the mobile phase consisted of a mixture of methanol: acetic acid 0.1% (v/v) and a binary gradient was used. The elution started with 3% methanol for 3 min, followed by 8% methanol until 8.5 min when 20% methanol was used and kept for the next 10 min, and then the column was rebalanced with 3% methanol. The flow rate was set at 1 mL/min and the sample injection volume was 5 μL. The MS detection mode was selected for detection of the polyphenolic compounds. The ionization on the MS system was in negative mode using an electrospray ion source (capillary +3000 V, nebulizer 60 psi (nitrogen), dry gas nitrogen at 12 L/min, dry gas temperature 360 °C). Each identified polyphenol was further quantified in BBE extracts. The results were expressed as milligrams of phenolic compound per gram of herbal material.

2.6. Antioxidant Activity Assays

The capacity to scavenge the DPPH, monitored according to the method described by Martins et al. [15,17] was performed by using a SPECTRO star Nano microplate reader (BMG Labtech, Offenburg, Germany). The results were expressed as Trolox equivalents (TE) per gram of dry extract (mg TE/g of dry extract).

The radical scavenging activity of the beech bark extracts against ABTS was measured according to Mocan et al. [18]. The results were expressed as milligrams of TE per gram of dry extract (mg TE/g dry extract).

In FRAP assay, the reduction of Fe^{3+}-TPTZ to blue-colored Fe^{2+}-TPTZ complex was monitored [19]. Briefly, the FRAP reagent was prepared by mixing ten volumes of acetate buffer (300 mM, pH 3.6), one volume of TPTZ solution (10 mM TPTZ in 40 mM HCl) and one volume of $FeCl_3$ solution (20 mM $FeCl_3 \cdot 6H_2O$ in 40 mM HCl). The reaction mixture (25 μL sample and 175 μL FRAP reagent) was incubated for 30 min at room temperature (in the dark) and the absorbance was measured at 593 nm (SPECTROstar Nano Multi-Detection Microplate Reader with 96-well plates, BMG Labtech, Ortenberg, Germany). A Trolox™ calibration curve (0.01–0.10 mg/mL) was plotted, and the results were expressed as milligrams of TE per milligram of dry extract (mg TE/mg dry extract).

2.7. Antidiabetic (Glucosidase inhibitory) Assay

The α-glucosidase inhibitory assay was measured according to method described previously [20,21]. In brief, 50 μL of sample diluted in 50 μL 100 mM-phosphate buffer (pH 6.8) in a 96-well microplate, were mixed with 50 μL yeast α-glucosidase for 10 min before 50 μL substrate (5 mM, p-nitrophenyl-α-D-glucopyranoside prepared in same buffer) were added. The release of p-nitrophenol was measured at 405 nm, 20 min after incubation. The blanks for test samples were prepared, and acarbose was used as a standard inhibitor, the results being expressed as IC_{50} or percentage of inhibition (where the sample was not enough active to be able to calculate an IC_{50} value for a tested concentration of 8 mg/mL) using the following formula:

$$\text{Inhibition (\%)} = [(\text{Abs}_{\text{control}} - \text{Abs}_{\text{sample}})/\text{Abs}_{\text{control}}] \times 100 \tag{1}$$

2.8. Tyrosinase Inhibitory Activity

Tyrosinase inhibitory activity of each sample was determined by method previously described by Chen et al. [22] using a SPECTROstar Nano Multi-Detection Microplate Reader with 96-well plates (BMG Labtech, Ortenberg, Germany). Samples were dissolved in water containing 5% DMSO; for each sample four wells were designated as A, B, C, D; each one contained a reaction mixture (200 µL) as follows: (A) 120 µL of 66 mM phosphate buffer solution (pH = 6.8) (PBS), 40 µL of mushroom tyrosinase in PBS (46 U/mL) (Tyr), (B) 160 µL PBS, (C) 80 µL PBS, 40 µL Tyr, 40 µL sample, and (D) 120 µL PBS, 40 µL sample. The plate was then incubated at room temperature for 10 min; after incubation, 40 µL of 2.5 mM L-DOPA in PBS solution were added in each well and the mixtures were incubated again at room temperature for 20 min. The absorbance of each well was measured at 475 nm, and the inhibition percentage of tyrosinase activity was calculated by the following equation, using a kojic acid solution (0.10 mg/mL) as a positive control:

$$\%I = \frac{(A - B) - (C - D)}{(A - B)} \times 100 \tag{2}$$

2.9. Assay of Antimicrobial Activity

2.9.1. Bacteria and Culture Conditions

For bacterial strains were used: one gram positive strain, Staphylococcus aureus (ATCC 49444), and three gram negative strains, namely *Pseudomonas aeruginosa* (ATCC 27853), *Salmonella typhimurium* (ATCC 14028) and *Escherichia coli* (ATCC 25922). The strains were cultured on Muller-Hinton agar stored at 4 °C.

2.9.2. Microdilution Method

The modified microdilution technique was used to evaluate the antimicrobial activity. Cell suspensions were adjusted to a concentration of approximately 2×10^5 CFU/mL in a final volume of 100 µL per well. The inoculum was stored at 4 °C for further use. Minimum inhibitory concentrations (MICs) were performed using 96-well plates. Dilutions of the extracts were carried out in wells containing 100 µL of Muller-Hinton broth and 10 µL of inoculum was added to all the wells. The microplates were incubated for 24 h at 37 °C. The MIC was detected following the addition of 20 µL (0.2 mg/mL) of resazurin solution to each well, and the plates were incubated 2 h at 37 °C. The MIC was identified as the lowest concentration that prevented the color change from blue to pink. The minimum bactericidal concentrations (MBCs) were determined by serial subcultivation of 2 µL into 96-well plates containing 100 µL of broth per well and further incubation for 48 h at 37 °C. The lowest concentration with no visible growth was defined as MBC, indicating 99.5% killing of the original inoculum.

2.9.3. Antifungal Activity

Antifungal activities were investigated by using the following fungi: *Candida albicans* (ATCC 10231), *Candida parapsilosis* (ATCC 22019), and *Candida zeylanoides* (ATCC 20356). Spore suspension (1.5×10^5) was obtained by washing agar plates with sterile solution (0.85% saline, 0.1% Tween 80 (v/v)), then added to each well for a final volume of 100 µL. The minimum inhibitory (MIC) and minimum fungicidal (MFC) concentrations assays were performed using the microdilution method by preparing a serial of dilutions in 96-well plates. The extracts were diluted in 0.85% saline (10 mg/mL), then added to microplates containing Broth Malt medium with inoculum and incubated for 72 h at 28 °C on a rotary shaker. The lowest concentrations without visible growth (at the binocular microscope) were defined as minimal inhibitory concentrations (MICs). The fungicidal concentrations (MFCs) were determined by serial sub-cultivation of 2 µL of tested extracts dissolved in medium and inoculated for 72 h, into microtiter plates containing 100 µL of broth per well and further incubation for 72 h at 28 °C. The lowest concentration with no visible growth was defined as MFC indicating 99.5% killing of the original inoculum. The fungicide fluconazole (Sigma F 8929, Santa Clara, CA, USA) was used as positive control (1–3500 µg/mL). All the experiments were performed in duplicate and repeated thrice. Water was used as a negative control.

2.10. Mutagen and Antimutagen Activity

Mutagen and antimutagenity of samples were examined using the plate incorporation method [23] described in detailed by Sarac and Sen [24]. 4-NPD (4-nitro-o-phenylenediamine) 3 µg/plate and NaN_3 8 µg/plate were used as positive controls for *S. thyphimurium* TA98 and *S. thyphimurium* TA100 (negative control—ethanol:water 1:1, v/v). The concentration of BBE was 5 mg/plate. The antimutagenity of the reference mutagens in the absence of the BBE was defined as 0% inhibition, and the antimutagenity was calculated according to the formula given by Ong et al., [25] as it follows: % Inhibition = [1 − T/M] × 100, where, T is the number of revertants per plate in the presence of mutagen and the BBE and M is the number of revertants per plate (without BBE) in the positive control. The data was presented as mean ± standard deviation (SD). Antimutagenity was recorded as follows: strong: 40% or more inhibition; moderate: 25%–40% inhibition; low/none: 25% or less inhibition [26].

2.11. Statistical Analysis

All samples were analyzed in triplicate (n = 3) and the results were expressed as the mean ± Standard Deviation (SD). The statistical analysis of the obtained experimental data was also performed by using the Modde 11.0 DoE software (Sartorius Stedim Data Analytics AB, Umeå, Sweden), which automatically calculated the statistical parameters necessary for the evaluation of the fitting quality, including the ANOVA test.

3. Results and Discussion

3.1. Fitting of the Experimental Data with the Models

The quantitative factors studied were the microwave power, with 4 levels of variation (300, 450, 600, 800 W) and the extraction time, varied over 3 levels (2, 3, 4 min). The total phenolic contents (TPCs) extracted with different solvent types (water, ethanol–water 50:50 and 80:20) were introduced in the DoE as 3 separate responses, which allowed a good comparison of their TPC extraction capacities.

As presented in Table 2, the design matrix calculated by the DoE software comprised 15 experimental runs, including three replicates that were used to confirm the experimental reproducibility, and to allow the identification of potential errors. Moreover, in order to reduce foreseeable results, the experimental runs were performed in a randomized order, dictated by the software. After performing all experimental runs, the registered response values were centralized and further introduced into the DoE matrix, where the fitting of the data has been accomplished by applying a multiple linear regression (MLR) algorithm. The fitting quality was evaluated by using the recommended, most reliable statistical parameters, i.e., the goodness of fit (R2), prediction capacities (Q2), model reproducibility given by the three performed replicates and model validity represented by the ANOVA test.

Table 2. Matrix of experimental design and experimental results for total phenolic content (TPC).

Sample Code	Run Order	Independent Variables (Factors)		Dependent Variables (Responses)		
		Extraction Time (min)	Microwave Power (W)	BBE 1	BBE 2	BBE 3
				(TPC mg GAE/g Dry Plant Material ± SD)		
N1	1	2	300	51.53	67.27	62.26
N2	6	4	300	**47.44**	76.46	61.87
N3	3	3	300	50.53	**69.59**	64.77
N4	2	2	450	56.79	66.43	66.20
N5	7	4	450	48.19	**77.53**	66.00
N6	5	3	450	54.55	**70.32**	**67.86**
N7	11	4	600	**52.79**	72.43	*70.95*
N8	10	3	600	58.61	72.06	70.20
N9	8	2	800	71.80	73.02	64.12
N10	14	2	800	72.31	72.46	64.27
N11	4	4	800	56.15	73.23	65.97
N12	12	4	800	55.68	73.32	66.07
N13	9	3	600	58.91	72.09	69.81
N14	15	3	600	59.00	71.91	70.33
N15	13	3	600	59.10	71.92	69.93

Notes: values with bold—outliers, TPC—total phenolic content expressed as mg GAE/g—gallic acid equivalents per dry plant material. BBE 1—extracts obtained with water 100%, BBE 2 extracts obtained with ethanol–water 50:50, BBE 3 extracts obtained with ethanol–water 80:20. Each value is the mean ± SD of three independent measurements.

Firstly, in order to improve the raw model by identifying and eliminating potential outliers, two graphs were generated for each response separately; the residuals plot (Figure 1a) which illustrates the residuals on a cumulative normal probability scale and the observed vs. predicted plot (Figure 1b) which allows the investigation of individual points that deviate from the diagonal line. By analyzing the two plots, outliers—marked with gray in Table 3, have been identified and eliminated, providing in this way an excellent fitted model, with high prediction capacities, as shown in the summary of fit chart (Figure 1c). In order to confirm the statistical validity of the model, the ANOVA test was performed (Supplementary file, Table S1). The registered p values were <0.001 for the model, describing a statistically valid design, indicating a significant impact of the factors on the responses. The p values registered for the lack of fit were > 0.099, showing that the model has insignificant lack of fit [27].

The generated coefficient plot (Figure 1d) allows the evaluation of factors' (independent variables) effects over the obtained responses, this type of diagram is used to evaluate the significance and influence of the model terms, each term or interaction being represented as a scaled and centered coefficient. In the present case, the diagram, shows that the extraction time varies inversely (for water 100%) or directly (for ethanol–water 50:50) with the amount of extracted TPC. The microwave power favors TPC extraction but in different proportions.

Table 3. Optimal extraction parameters, DoE predicted and experimental obtained values of the extracted TPC.

Extraction Solvent	Symbol	Extraction Time (min)	Microwave Power (W)	DoE Predicted Values	Experimental Obtained Values	Recovery (%)
				(TPC mg GAE/g Plant Material)		
water	BBE 1	2	800	72.05	69.76	96.82
50:50 ethanol–water	BBE 2	4	300	76.55	76.31	99.69
80:20 ethanol–water	BBE 3	3.1	600	70.09	69.75	99.51

Notes: BBE 1—extracts obtained with water 100%, BBE 2—extracts obtained with ethanol–water 50:50, BBE 3—extracts obtained with ethanol–water 80:20.

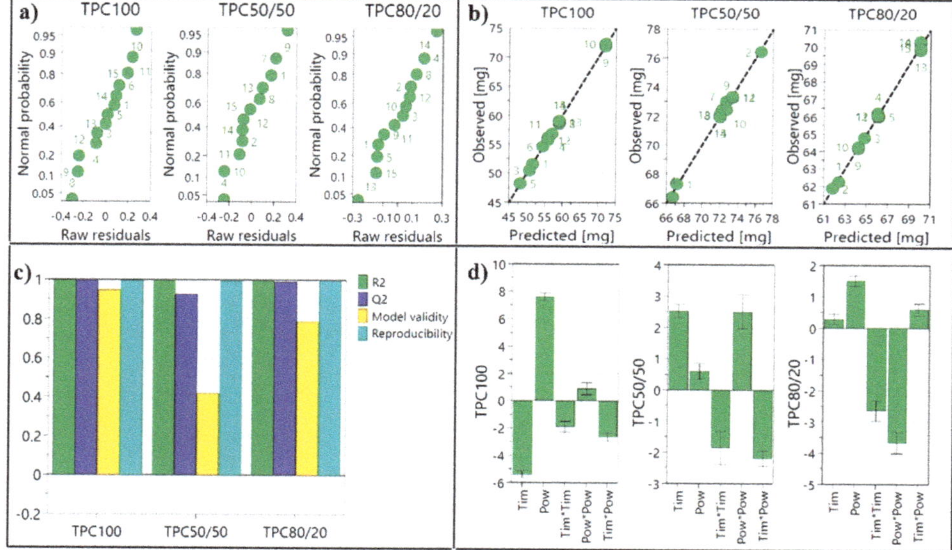

Figure 1. Graphical summary of (**a**) residuals normal probability; (**b**) observed vs. predicted (**c**) summary of fit; (**d**) scaled and centered coefficients (TPC100—total phenolic content in extracts obtained with water 100%, TPC50/50—total phenolic content in extracts obtained with ethanol–water 50:50, TPC80/20—total phenolic content in extracts obtained with ethanol–water 80:20).

3.2. Optimization of the Extraction Parameters

The fitting and analysis of the data obtained after performing all the experimental runs stipulated in the DoE matrix, led to an in depth understanding of the variables' influence over the microwaved extraction process. Further, the optimization module of the DoE software has been used in order to calculate the optimal combination of extraction parameters, having as target the highest extraction yield obtained by using stable parameters. The optimization has been performed separately for each type of extraction solvent, the software also being able to predict not only the optimal combination of extraction parameters, but also the extracted TPC value.

After performing the stipulated optimal runs, results of the experimentally obtained TPC were compared with the DoE predicted ones, the registered recovery percent being >96% for BBE1 and close to 100% for BBE2 and BBE3, highlighting the high prediction capacity of the model. The optimal extraction parameters, DoE predicted- and experimentally-obtained TPC values are presented in Table 3.

3.3. Identification and Quantification of Individual Polyphenols

The optimized extracts obtained from the beech bark (water—BBE1, 50% (v/v) ethanolic—BBE2 and 80% (v/v) ethanolic—BBE3) were investigated in terms of their phytochemical profile. For the individual polyphenols identification and quantification, the previously described LC-MS/MS analytical methods were employed. The results of this analysis are summarized in Table 4 [15].

Table 4. Quantitative (µg/g dry plant material) evaluation of the recovery of main bioactive compounds in samples of BBE.

Sample Code	Bioactive Compounds					
	(−)-Epicatechin	(+)-Catechin	Syringic Acid	Gallic Acid	Protocatechuic Acid	Vanillic Acid
BBE1	22.7 ± 2.72	300.7 ± 44.22	24.2 ± 3.02	1.9 ± 0.14	3.3 ± 0.29	49.9 ± 7.28
BBE2	39.6 ± 3.24	577.4 ± 56.58	7.5 ± 0.78	NF	6.2 ± 0.41	18.0 ± 1.52
BBE3	33.4 ± 3.55	465.1 ± 51.62	5.8 ± 0.68	NF	5.7 ± 0.74	16.1 ± 1.55

Notes: NF—not found, below limit of detection; BBE1—extracts obtained with water 100%, BBE2—extracts obtained with ethanol–water 50:50, BBE3—extracts obtained with ethanol–water 80:20. Data are shown as mean ± standard deviation (SD).

From the 18 phenolic compounds analyzed by the LC-MS/MS method, chlorogenic acid, ferulic acid and isoquercitrin, were identified in all tested BBEs. Quercetin was identified only in beech bark extract obtained with 100%water (BBE1). From the phenolic compounds identified by this method, only ferulic acid found in BBE1, was quantified (43.7 ± 5.03 µg/g dry plant material). The polyphenols (epicatechin, catechin, syringic acid, protocatechuic acid, and vanillic acid) analyzed by the other LC-MS method were identified and quantified in all tested BBEs (Table 4). Gallic acid was identified only in BBE1.

In previous studies some of the phenolic compounds obtained from beech bark were identified, including catechin, epicatechin, quercetin, taxifolin, procyanidins, syringic acid, coumaric acid [11,12].

3.4. Assay of the Antioxidant Activity

The antioxidant activity of optimized beech bark extracts was evaluated by DPPH, FRAP and TEAC assays. The beech bark extracts exhibited scavenging activity against all radicals as shown in Table 5. The strongest antioxidant activity is recorded for BBE2, where the strongest amount of phenols was recorded. Many bark extracts have been evaluated for their antioxidant capacities, commonly associated to their content of phenolic compounds [28–30]. Beech bark contains some of biologically active components, such as catechin, epicatechin, syringic acid, vanillic acid etc., indicating that the antioxidant activity of the extracts can be at least partially ascribed to these bio-components (Table 4). Grzesik et al. [31] studied the antioxidant properties of five catechins, compared with other natural or synthetic compounds. They concluded that catechins showed the strongest ABTS scavenging capacity and the strongest stoichiometry of Fe^{3+} reduction in the FRAP assay [31]. Besides the direct antioxidant properties of catechins, they may show synergistic interaction with endogenous antioxidants and act as indirect antioxidants as well [32,33].

Table 5. Antioxidant activity of optimized beech bark extracts.

Sample Code	TPC mg GAE/g of Dry Extract	DPPH mg TE/g of Dry Extract	TEAC mg TE/g of Dry Extract	FRAP mg TE/g of Dry Extract
BBE 1	69.76 ± 1.54	676.29 ± 19.80	472.08 ± 67.07	625.13 ± 9.62
BBE 2	76.49 ± 2.41	741.43 ± 59.44	619.85 ± 20.75	783.24 ± 31.24
BBE 3	69.86 ± 1.04	505.02 ± 42.02	464.41 ± 37.42	592.84 ± 44.02

TPC—total phenolic content, DPPH—2,2-diphenyl-1-picrylhydrazyl, TEAC—Trolox equivalent antioxidant capacity, FRAP—ferric reducing ability of plasma, BBE1- extracts obtained with water 100%, BBE2—extracts obtained with ethanol–water 50:50, BBE3—extracts obtained with ethanol–water 80:20 ± Standard deviation.

3.5. Assay of the Antimicrobial Activity

The antimicrobial (antibacterial and antifungal) activity of BBEs was tested against four bacteria and three fungi, selected based on their relevance for public health. *Staphylococcus aureus* was the most sensitive strain towards all tested samples, with similar values of MIC (1.56 mg/mL) and MBC

(3.12 mg/mL) (Table 6). Additionally, all tested extracts showed a good antibacterial activity on *E. coli*, *P. aeruginosa*, and *S. typhimurium* strains (MIC—3 mg/mL, and MBC—6 mg/mL).

The antibacterial activity of phenolic compounds has been demonstrated in various studies [4,6]. The previous study of the aqueous beech bark extract, underlined the antimicrobial activity against methicillin-resistant *S. aureus* [11].

Concerning the antifungal activity of the samples (Table 7), all tested *Candida* species, exhibited the highest sensitivity to aqueous extract of beech bark (BBE1) with 25 mg/mL MIC and 50 mg/mL MFC (Table 7). The effect of ethanolic extracts (BBE2, BBE3) on *Candida* species was absent at a concentration of 50 mg/mL.

Thus, the antibacterial activity of BBE could be attributed at least in part to phenolic compounds.

Table 6. Antibacterial activity (mg/mL) of the beech bark extracts (BBE).

Sample Code	Bacteria			
	Staphylococcus aureus (ATCC 49444)	*Escherichia coli* (ATCC 25922)	*Pseudomonas aeruginosa* (ATCC 27853)	*Salmonella typhimurium* (ATCC 14028)
	Minimum Inhibitory Concentration (MIC)			
BBE1	1.56	3	3	3
BBE2	1.56	3	3	3
BBE3	1.56	3	3	3
	Minimum Bactericidal Concentration (MBC)			
BBE1	3.12	6	6	6
BBE2	3.12	6	6	6
BBE3	3.12	6	6	6

Notes: BBE1—extracts obtained with water 100%, BBE2—extracts obtained with ethanol–water 50:50, BBE3—extracts obtained with ethanol–water 80:20.

Table 7. Antifungal activity (mg/mL) of the beech bark extracts (BBE).

Sample Code	Fungi		
	Candida albicans (ATCC 10231)	*Candida parapsilosis* (ATCC 22019)	*Candida zeylanoides* (ATCC 20356)
	Minimum Inhibitory Concentration (MIC)		
BBE1	25	25	25
BBE2	NF	>50	>50
BBE3	NF	>50	>50
	Minimum Fungicidal Concentration (MFC)		
BBE1	50	50	50
BBE2	NF	>50	>50
BBE3	NF	>50	>50

Notes: BBE1—extracts obtained with water 100%, BBE2—extracts obtained with ethanol–water 50:50, BBE3—extracts obtained with ethanol–water 80:20.

3.6. Antimutagenity Activity

The aim was to investigate the mutagen and antimutagen activities of BBE by the bacterial reverse mutation assay in *Salmonella typhimurium* (*S. typhimurium*) strains. The antimutagenity of BBE against *S. typhimurium* TA 98 and TA 100, was tested by comparing the numbers of induced revertants and spontaneous revertants. According with Table 8, TA98 and TA100 strains increase in the number of revertant colonies compared with the negative control when the bacteria was treated with BBE at 5 mg/plate concentration, thereby indicating an antimutagenic activity. Thus, all extracts were antimutagenic to the frameshift TA98 or TA100 *S. typhimurium* strains.

To determine the potential antimutagen activity of the BBE to prevent DNA damage by 4-NPD/NaN3 (positive mutagen/carcinogen), BBE were incubated together with 4-NPD/NaN3. The results are presented in Table 8. The percentage of inhibition of mutagen activity of 4-NPD/NaN3 (antimutagenity) of the beech bark extracts ranged from 3.6 to 17.01% in *S. typhimurium* TA98 and from 15.47 to 16.33% in *S. typhimurium* TA100. The BBE1 had the highest value for antimutagen activity (17% for *S. typhimurium* TA98 and 16.33% for *S. typhimurium* TA100). The antimutagen effect is considered low when the inhibitory effect is less than 25%, moderate when the inhibitory effect is between 25%–40% and strong when the effect is more than 45% [26,34]. Thus, based on the literature data, beech bark extracts assayed in this study may be considered to have low or no antimutagen effects.

Table 8. Antimutagen properties of beech bark extracts (5 mg/plate) on *S. typhimurium* TA98 and TA100 bacterial strains.

| Test Item | Number of Revertants | | | |
| | TA 98 | | TA100 | |
	Mean ± SD	Inhibition %	Mean ± SD	Inhibition %
Negative Control	9.25 ± 3.6 [a]		9.25 ± 2.4 [b]	
BBE1 [a]	161 ± 3.6	17,01	292 ± 6.4	16,33
BBE2 [a]	187 ± 4.4	3.60	295 ± 6.4	15.47
BBE3 [a]	172 ± 3.8	11.34	294 ± 6.2	15,75
4-NPD [c]	193 ± 3.4	-	-	-
NaN$_3$ [c]	-	-	349 ± 15.22	-

Notes: [a] BBE1—extracts obtained with water 100%, BBE2—extracts obtained with ethanol–water 50:50, BBE3—extracts obtained with ethanol–water 80:20; [b] Values expressed are means ± SD of three replications; [c] 4-NPD and NaN$_3$ were used as positive controls for *S. thyphimurium* TA98 and TA100 strains, respectively.

3.7. In Vitro Enzyme Inhibitory Properties of Beech Bark Extracts—α-Glucosidase (Antidiabetic) and Tyrosinase Inhibitory Activity

Diabetes mellitus is a chronic, life-long disorder and is characterized by high blood glucose levels. One of the ways of treating diabetes mellitus is based on glucose absorption delay by inhibiting enzymes such as α-glucosidase in digestive organs [35]. By inhibiting α-glucosidase in the intestine, the rate of oligosaccharides hydrolysis is low, and the carbohydrate digestion process extends into the lower part of the small intestine [36]. Many studies have demonstrated the α-glucosidase inhibitory activity of some bark herbal extracts, proving their strong biochemical potential [37,38]. For example, different extracts of *Canarium tramdenum* bark were evaluated in terms of α-glucosydase inhibition showing a strong inhibitory activity. The different enriched extracts in terpenoids and phenolics appeared as a promising source of natural α-glucosidase inhibitors [38].

As shown in Table 9, the α-glucosidase inhibitory activity was determined for all BBEs and was higher in comparison with the standard acarbose (Table 9). The order of α-glucosidase inhibition is BBE3 > BBE1 > BBE2 > acarbose corresponding to IC$_{50}$ values 38, 92, 168, and 838 μg/mL, respectively.

Table 9. Enzyme inhibitory effects of beech bark extracts.

Nr. crt.	Sample	Glucosidase Inhibition (IC$_{50}$ μg/mL)	Tyrosinase Inhibition (PI—4.025 mg/mL)
1.	BBE1	92	NF
2.	BBE2	168	45.99 ± 5.26%
3.	BBE3	38	NF
4.	Acarbose	838	-
5	Kojic acid (1 mg/mL)	-	97.61 ± 0.24%

Notes: BBE1—extracts obtained with water 100%, BBE2—extracts obtained with ethanol–water 50:50, BBE3—extracts obtained with ethanol–water 80:20, PI—procent of inhibition, NF—not found.

Tyrosinase inhibitors are compounds capable of reducing enzymatic reactions, especially from the skin, which makes them commercially relevant for cosmetic industry [39]. At a concentration of 4.025 mg/mL, inhibitory effects have been shown only for BBE2 sample. However, kojic acid (a standard inhibitor) showed an excellent tyrosinase inhibitory activity of 97.61 ± 0.24%.

4. Conclusions

The results of this study indicate that MAE is an efficient extraction method of phenolic compounds from beech bark. The optimum conditions which gave the highest TPC (76.57 mg GAE/g dry plant material) were reached when the microwave power was 300 W, the extraction time was 4 min, and the solvent was a mixture of ethanol–water (50:50). The practical value of TPC after a control experiment was 76.49 mg GAE/g dry plant material.

The extracts obtained in optimum conditions were characterized by HPLC-MS/MS. The identified compounds in all tested BBEs were: vanillic acid, epicatechin, catechin, protocatechuic acid, chlorogenic acid, ferulic acid, and isoquercitrin. Quercetin and gallic acid were identified only in beech bark extract obtained with 100%water.

The beech bark extracts exhibited free-radical-scavenging activity in all assays, and were active against *S. aureus*, *P. aeruginosa*, *S. typhimurium*, and *E. coli*. All tested *Candida* species were sensitive to beech bark aqueous extract. Additionally, all extracts were active on α-glucosidase. The beech bark extracts assayed may be considered to have low or no antimutagen effects.

This study documented that *Fagus sylvatica* L. bark possesses antioxidant activity and has α-glucosidase inhibitory effects, by several *in vitro* experiments. Ethanol and water extracts can be considered as promising sources of natural antioxidants, antimicrobian agents, and α-glucosidase inhibitors. The isolation of bioactive constituents and investigations on several biomedical properties of *F. sylvatica* bark should be conducted in future experiments.

Supplementary Materials: The following are available online at http://www.mdpi.com/2076-3921/8/9/417/s1, Table S1: ANOVA—complete summary of the regression and residual analysis.

Author Contributions: Conceptualization, C.T. and A.M.; methodology, C.T., A.M., O.C., A.G., S.C., A.N.; software, A.G.; resources, C.T., D.C.V., O.C.; data curation, C.T., A.-M.G., S.C., D.C.V., O.C., A.-M.G.; writing-original draft preparation, C.T.; visualization, A.M., D.-L.M.; supervision, D.-L.M., O.C.; project administration, C.T.; funding acquisition, C.T.

Funding: This work was supported by a grant of Ministry of Research and Innovation, CNCS-UEFISCDI, project number PN-III-P1-1.1-PD-2016-0892, within PNCDI III.

Conflicts of Interest: The authors declare no conflict of interest.

References

1. Alfredsen, G.; Solheim, H.; Slimestad, R. Antifungal effect of bark extracts from some European tree species. *Eur. J. For. Res.* **2008**, *127*, 387. [CrossRef]
2. Tanase, C.; Coșarcă, S.; Muntean, D.-L. A critical review of phenolic compounds extracted from the bark of woody vascular plants and their potential biological activity. *Molecules* **2019**, *24*, 1182. [CrossRef] [PubMed]
3. Bernardo, J.; Ferreres, F.; Gil-Izquierdo, Á.; Videira, R.A.; Valentão, P.; Veiga, F.; Andrade, P.B. In Vitro Multimodal-Effect of *Trichilia catigua* A. Juss. (Meliaceae) Bark Aqueous Extract in CNS Targets. *J. Ethnopharmacol.* **2018**, *211*, 247–255. [CrossRef] [PubMed]
4. Kumar, S.; Pathania, A.S.; Saxena, A.K.; Vishwakarma, R.A.; Ali, A.; Bhushan, S. The Anticancer Potential of Flavonoids Isolated from the Stem Bark of *Erythrina suberosa* through Induction of Apoptosis and Inhibition of STAT Signaling Pathway in Human Leukemia HL-60 Cells. *Chem.-Biol. Interact.* **2013**, *205*, 128–137. [CrossRef]
5. de Souza Santos, C.C.; Guilhon, C.C.; Moreno, D.S.A.; Alviano, C.S.; dos Santos Estevam, C.; Blank, A.F.; Fernandes, P.D. Anti-Inflammatory, Antinociceptive and Antioxidant Properties of *Schinopsis brasiliensis* Bark. *J. Ethnopharmacol.* **2018**, *213*, 176–182. [CrossRef] [PubMed]

6. Salih, E.Y.A.; Kanninen, M.; Sipi, M.; Luukkanen, O.; Hiltunen, R.; Vuorela, H.; Julkunen-Tiitto, R.; Fyhrquist, P. Tannins, Flavonoids and Stilbenes in Extracts of African Savanna Woodland Trees *Terminalia brownii, Terminalia laxiflora* and *Anogeissus leiocarpus* Showing Promising Antibacterial Potential. *S. Afr. J. Bot.* **2017**, *108*, 370–386. [CrossRef]

7. Ambika; Singh, P.P.; Chauhan, S.M.S. Activity-Guided Isolation of Antioxidants from the Leaves of *Terminalia arjuna*. *Nat. Prod. Res.* **2014**, *28*, 760–763. [CrossRef] [PubMed]

8. Ferreres, F.; Gomes, N.G.M.; Valentão, P.; Pereira, D.M.; Gil-Izquierdo, A.; Araújo, L.; Silva, T.C.; Andrade, P.B. Leaves and Stem Bark from *Allophylus africanus* P. Beauv.: An Approach to Anti-Inflammatory Properties and Characterization of Their Flavonoid Profile. *Food Chem. Toxicol.* **2018**, *118*, 430–438. [CrossRef] [PubMed]

9. Bolte, A.; Czajkowski, T.; Kompa, T. The North-Eastern Distribution Range of European Beech—A Review. *For. Int. J. For. Res.* **2007**, *80*, 413–429. [CrossRef]

10. Hofmann, T. Antioxidant Efficiency of Beech (*Fagus sylvatica* L.) Bark Polyphenols Assessed by Chemometric Methods. *Ind. Crops Prod.* **2017**, *108*, 26–35. [CrossRef]

11. Tanase, C.; Cosarca, S.; Toma, F.; Mare, A.; Man, A.; Miklos, A.; Imre, S.; Boz, I. Antibacterial Activities of Beech Bark (*Fagus sylvatica* L.) Polyphenolic Extract. *Environ. Eng. Manag. J.* **2018**, *17*. [CrossRef]

12. Hofmann, T.; Nebehaj, E.; Albert, L. The High-Performance Liquid Chromatography/Multistage Electrospray Mass Spectrometric Investigation and Extraction Optimization of Beech (*Fagus sylvatica* L.) Bark Polyphenols. *J. Chromatogr. A* **2015**, *1393*, 96–105. [CrossRef] [PubMed]

13. Coşarcă, S.-L. Spruce and Beech Bark Aqueous Extracts: Source of Polyphenols, Tannins and Antioxidants Correlated to in Vitro Antitumor Potential on Two Different Cell Lines. *Wood Sci. Technol.* **2019**, *53*, 313–333. [CrossRef]

14. Sánchez-Rangel, J.C.; Benavides, J.; Heredia, J.B.; Cisneros-Zevallos, L.; Jacobo-Velázquez, D.A. The Folin–Ciocalteu Assay Revisited: Improvement of Its Specificity for Total Phenolic Content Determination. *Anal. Methods* **2013**, *5*, 5990–5999. [CrossRef]

15. Rusu, M.E.; Gheldiu, A.-M.; Mocan, A.; Moldovan, C.; Popa, D.-S.; Tomuta, I.; Vlase, L. Process Optimization for Improved Phenolic Compounds Recovery from Walnut (*Juglans regia* L.) Septum: Phytochemical Profile and Biological Activities. *Molecules* **2018**, *23*, 2814. [CrossRef] [PubMed]

16. Toiu, A.; Mocan, A.; Vlase, L.; Pârvu, A.E.; Vodnar, D.C.; Gheldiu, A.-M.; Moldovan, C.; Oniga, I. Phytochemical Composition, Antioxidant, Antimicrobial and in Vivo Anti-Inflammatory Activity of Traditionally Used Romanian *Ajuga laxmannii* (Murray) Benth. ("Nobleman's Beard"—Barba Împăratului). *Front Pharm.* **2018**, *9*, 7. [CrossRef] [PubMed]

17. Martins, N.; Barros, L.; Dueñas, M.; Santos-Buelga, C.; Ferreira, I.C.F.R. Characterization of Phenolic Compounds and Antioxidant Properties of *Glycyrrhiza glabra* L. Rhizomes and Roots. *RSC Adv.* **2015**, *5*, 26991–26997. [CrossRef]

18. Mocan, A.; Schafberg, M.; Crişan, G.; Rohn, S. Determination of Lignans and Phenolic Components of *Schisandra chinensis* (Turcz.) Baill. Using HPLC-ESI-ToF-MS and HPLC-Online TEAC: Contribution of Individual Components to Overall Antioxidant Activity and Comparison with Traditional Antioxidant Assays. *J. Funct. Foods* **2016**, *24*, 579–594. [CrossRef]

19. Damiano, S.; Forino, M.; De, A.; Vitali, L.A.; Lupidi, G.; Taglialatela-Scafati, O. Antioxidant and Antibiofilm Activities of Secondary Metabolites from *Ziziphus jujuba* Leaves Used for Infusion Preparation. *Food Chem.* **2017**, *230*, 24–29. [CrossRef]

20. Les, F.; Venditti, A.; Cásedas, G.; Frezza, C.; Guiso, M.; Sciubba, F.; Serafini, M.; Bianco, A.; Valero, M.S.; López, V. Everlasting Flower (*Helichrysum stoechas* Moench) as a Potential Source of Bioactive Molecules with Antiproliferative, Antioxidant, Antidiabetic and Neuroprotective Properties. *Ind. Crops Prod.* **2017**, *108*, 295–302. [CrossRef]

21. Spínola, V.; Castilho, P.C. Evaluation of Asteraceae Herbal Extracts in the Management of Diabetes and Obesity. Contribution of Caffeoylquinic Acids on the Inhibition of Digestive Enzymes Activity and Formation of Advanced Glycation End-Products (in Vitro). *Phytochemistry* **2017**, *143*, 29–35. [CrossRef] [PubMed]

22. Chen, C.-H.; Chan, H.-C.; Chu, Y.-T.; Ho, H.-Y.; Chen, P.-Y.; Lee, T.-H.; Lee, C.-K. Antioxidant Activity of Some Plant Extracts towards Xanthine Oxidase, Lipoxygenase and Tyrosinase. *Molecules* **2009**, *14*, 2947–2958. [CrossRef] [PubMed]

23. Maron, D.M.; Ames, B.N. Revised Methods for the *Salmonella* Mutagenicity Test. *Mutat. Res.Environ. Mutagen. Relat. Subj.* **1983**, *113*, 173–215. [CrossRef]

24. Saraç, N.; Şen, B. Antioxidant, Mutagenic, Antimutagenic Activities, and Phenolic Compounds of Liquidambar Orientalis Mill. Var. Orientalis. *Ind. Crops Prod.* **2014**, *53*, 60–64. [CrossRef]

25. Ong, T.; Whong, W.-Z.; Stewart, J.; Brockman, H.E. Chlorophyllin: A Potent Antimutagen against Environmental and Dietary Complex Mixtures. *Mutat. Res. Lett.* **1986**, *173*, 111–115. [CrossRef]

26. Evandri, M.G.; Battinelli, L.; Daniele, C.; Mastrangelo, S.; Bolle, P.; Mazzanti, G. The Antimutagenic Activity of *Lavandula angustifolia* (Lavender) Essential Oil in the Bacterial Reverse Mutation Assay. *Food Chem. Toxicol.* **2005**, *43*, 1381–1387. [CrossRef]

27. Eriksson, L.; Johansson, E.; Kettaneh-World, N.; Wold, S. *Design of Experiments. Principles and Applications*; Umetrics Academy: Umeå, Sweden, 2008.

28. Valencia-Avilés, E.; García-Pérez, E.M.; Garnica-Romo, G.M.; Figueroa-Cárdenas, D.J.; Meléndez-Herrera, E.; Salgado-Garciglia, R.; Martínez-Flores, E.H. Antioxidant Properties of Polyphenolic Extracts from *Quercus laurina, Quercus crassifolia*, and *Quercus scytophylla* Bark. *Antioxidants* **2018**, *7*, 81. [CrossRef]

29. Kumar, A.; Anand, V.; Dubey, R.C.; Goel, K.K. Evaluation of Antioxidant Potential of Alcoholic Stem Bark Extracts of *Bauhinia variegata* Linn. *JANS* **2019**, *11*. [CrossRef]

30. Panelli, F.M.; Pierine, T.D.; De Souza, L.S.; Ferron, J.A.; Garcia, L.J.; Santos, C.K.; Belin, A.M.; Lima, P.G.; Borguini, G.M.; Minatel, O.I.; et al. Bark of *Passiflora edulis* Treatment Stimulates Antioxidant Capacity, and Reduces Dyslipidemia and Body Fat in Db/Db Mice. *Antioxidants* **2018**, *7*, 120. [CrossRef]

31. Grzesik, M.; Naparło, K.; Bartosz, G.; Sadowska-Bartosz, I. Antioxidant Properties of Catechins: Comparison with Other Antioxidants. *Food Chem.* **2018**, *241*, 480–492. [CrossRef]

32. Pereira, R.B.; Sousa, C.; Costa, A.; Andrade, P.B.; Valentão, P. Glutathione and the Antioxidant Potential of Binary Mixtures with Flavonoids: Synergisms and Antagonisms. *Molecules* **2013**, *18*, 8858–8872. [CrossRef] [PubMed]

33. Ke, F.; Zhang, M.; Qin, N.; Zhao, G.; Chu, J.; Wan, X. Synergistic Antioxidant Activity and Anticancer Effect of Green Tea Catechin Stabilized on Nanoscale Cyclodextrin-Based Metal–Organic Frameworks. *J. Mater. Sci.* **2019**, *54*, 10420–10429. [CrossRef]

34. Negi, P.S.; Jayaprakasha, G.K.; Jena, B.S. Antioxidant and Antimutagenic Activities of Pomegranate Peel Extracts. *Food Chem.* **2003**, *80*, 393–397. [CrossRef]

35. Kim, K.Y.; Nam, K.A.; Kurihara, H.; Kim, S.M. Potent α-Glucosidase Inhibitors Purified from the Red Alga *Grateloupia elliptica*. *Phytochemistry* **2008**, *69*, 2820–2825. [CrossRef] [PubMed]

36. Kumar, S.; Narwal, S.; Kumar, V.; Prakash, O. α-Glucosidase Inhibitors from Plants: A Natural Approach to Treat Diabetes. *Pharm. Rev.* **2011**, *5*, 19–29. [CrossRef] [PubMed]

37. Adisakwattana, S.; Lerdsuwankij, O.; Poputtachai, U.; Minipun, A.; Suparpprom, C. Inhibitory Activity of *Cinnamon* Bark Species and Their Combination Effect with Acarbose against Intestinal α-Glucosidase and Pancreatic α-Amylase. *Plant Foods Hum. Nutr.* **2011**, *66*, 143–148. [CrossRef] [PubMed]

38. Quan, N.V.; Xuan, T.D.; Tran, H.-D.; Thuy, N.T.D.; Trang, L.T.; Huong, C.T.; Andriana, Y.; Tuyen, P.T. Antioxidant, α-Amylase and α-Glucosidase Inhibitory Activities and Potential Constituents of Canarium Tramdenum Bark. *Molecules* **2019**, *24*, 605. [CrossRef]

39. Fu, R.; Zhang, Y.; Guo, Y.; Chen, F. Antioxidant and Tyrosinase Inhibition Activities of the Ethanol-Insoluble Fraction of Water Extract of *Sapium sebiferum* (L.) Roxb. Leaves. *S. Afr. J. Bot.* **2014**, *93*, 98–104. [CrossRef]

 *antioxidants*

Article

Ultrasound-Assisted Extraction of Total Flavonoids from *Pteris cretica* L.: Process Optimization, HPLC Analysis, and Evaluation of Antioxidant Activity

Mengyang Hou [1,2], Wenzhong Hu [2,3,*], Aosheng Wang [2,3], Zhilong Xiu [1,2], Yusheng Shi [2,3], Kexin Hao [2,3], Xingsheng Sun [2,3], Duo Cao [4], Ruishan Lu [2,3] and Jiao Sun [2,3]

[1] School of Bioengineering, Dalian University of Technology, Dalian 116024, China; mengyanghou@yahoo.com (M.H.); zhlxiu@dlut.edu.cn (Z.X.)
[2] Key Laboratory of Biotechnology and Bioresources Utilization, Ministry of Education, Dalian 116600, China; jnwangaosheng@sina.com (A.W.); shiyusheng@dlun.edu.cn (Y.S.); haokex@sina.com (K.H.); sunxingsheng106@sina.com (X.S.); lrskarma@163.com (R.L.); sunjiao820@sina.com (J.S.)
[3] College of Life Science, Dalian Minzu University, Dalian 116600, China
[4] College of Life Sciences, Yanan University, Yanan 716000, China; caoduo2013@163.com
* Correspondence: wenzhongh@sina.com; Tel.: +86-135-9116-2665

Received: 31 August 2019; Accepted: 17 September 2019; Published: 24 September 2019

Abstract: In the present work, the ultrasonic-assisted extraction (UAE) of total flavonoids (TF) from *Pteris cretica* L. was optimized by response surface methodology (RSM) on the basis of a single-factor experiment. The optimized UAE parameters were as follows: Ethanol concentration 56.74%, extraction time 45.94 min, extraction temperature 74.27 °C, and liquid/solid ratio 33.69 mL/g. Under the optimized conditions, the total flavonoids yield (TFY) was 4.71 ± 0.04%, which was higher than that obtained by heat reflux extraction (HRE). The extracts were further analyzed by HPLC, and five major flavonoids, including rutin, quercitrin, luteolin, apigenin, and luteolin-7-*O*-glucoside, were identified and quantified. Furthermore, the results of the antioxidant test showed that the TF extract obtained under optimized UAE conditions exhibited good 2,2-diphenyl-1-picrylhydrazyl radical (DPPH•) and 2,2-azino-bis(3-ethylbenzothiazoline-6-sulfonic acid) radical (ABTS$^+$•), nitric oxide radical (NO•) scavenging activities, and ferrous ion (Fe^{2+}) chelating capacity, with IC$_{50}$ values of 74.49, 82.92, 89.12, and 713.41 µg/mL, respectively. Results indicated that the UAE technique developed in this work was an efficient, rapid, and simple approach for the extraction of flavonoids with antioxidant activity from *P. cretica*.

Keywords: *Pteris cretica* L.; flavonoids; ultrasonic-assisted extraction; optimization; response surface methodology; HPLC analysis; antioxidant activity

1. Introduction

Pteris cretica L., a perennial evergreen herb, belongs to the genus *Pteris* (Pteridaceae). The genus *Pteris*, geographically distributed over the tropical and subtropical regions of the world, consists of approximately 250 species, many of which have been cultivated for ornamental, culinary, and medicinal purposes [1,2]. Historically, several species of the genus *Pteris* have been used as folk medicine to treat burn injuries, indigestion, diarrhea, furuncles, eczema, apoplexy, jaundice, snakebites, and hemorrhages in China [3,4]. Due to the outstanding medicinal potential, modern investigations on the *Pteris* species have been performed extensively, which reveal that these plants contain various bioactive components, including flavonoids [5–7], sesquiterpenoids [3,8,9], and diterpenoids [9–11]. Flavonoids isolated from plants of this genus, including kaempferol, quercetin, rutin, apigenin, luteolin, and luteolin 7-*O*-sophoroside, especially contribute to various pharmacological effects, such as antibacterial [12], antioxidant [13], and anti-benign prostatic hyperplasia [14] potentials.

As is known to all, extraction is a preliminary step in phytochemical research, which is also a necessary step in pharmaceutical research. Conventional extraction methods, including maceration, decoction, heat reflux, and Soxhlet extraction, are time-consuming, costly, and inefficient [15]. Fortunately, ultrasonic-assisted extraction (UAE) makes up for those shortcomings because the cavitation, vibration, crushing, and mixing effects in media produced by ultrasound can break the cell wall and increase the mass transfer process effectively [16]. In addition, UAE can not only avoid the heat-induced destruction of bioactive ingredients, but also helps to improve the safety of products [17]. Hence, UAE as an alternative technique has been widely used for the extraction of natural bioactive compounds, especially flavonoids, for instance, flavonoids from *Medicago sativa* Linn [18], *Ampelopsis grossedentata* leaves [19], and *Olea europaea* leaves [20]. To the best of our knowledge, research on the extraction of flavonoids from *P. cretica* have not been reported.

Admittedly, extraction yield and pharmacological action depend greatly on the comprehensive effect of several various factors, including the extraction method, solvent type, extraction time, extraction temperature, pH, and liquid/solid ratio [21]. Therefore, to maximize the extraction yield of bioactive substances and pharmacological effects, optimization of the extraction process is quite essential. Recently, response surface methodology (RSM) as an advanced chemometric tool has been frequently applied to the optimization of the extraction process [22]. RSM design can reduce the number of experimental trials, resulting in lower reagent consumption and less laboratory work. Additionally, RSM design yields a mathematical model to account for the reciprocal influence of various independent variables [23]. Box–Behnken design (BBD), one type of RSM, is easier to interpret and perform in comparison with other designs [24].

The present study is an attempt to establish an efficient UAE technique for total flavonoids from *P. cretica*. First, the effects of four independent variables, including ethanol concentration, extraction time, extraction temperature, and liquid/solid ratio, on the yield of total flavonoids were investigated. Second, the reciprocal actions of independent variables were investigated followed by the optimization of the UAE process using RSM. Third, a heat reflux extraction experiment was carried out to verify the efficiency of the UAE method developed. Fourth, a comparison of flavonoid profiles of the extracts obtained by UAE under optimized conditions and heat reflux extraction (HRE) was performed by HPLC. In addition, the antioxidant activity of extracts obtained by optimized UAE was evaluated by 2,2-diphenyl-1-picrylhydrazyl radical (DPPH•), 2,2-azino-bis(3-ethylbenzothiazoline-6-sulfonic acid) radical (ABTS^{+}•), nitric oxide radical (NO•) scavenging activities, and Fe^{2+} chelating activity.

2. Materials and Methods

2.1. Chemicals and Reagents

Rutin, quercetin, luteolin, apigenin, and luteolin-7-*O*-glucoside standards were obtained from Shanghai Yuanye Bio-Technology Co., Ltd. (Shanghai, China). HPLC grade trifluoroacetic acid (TFA) and methanol were obtained from Aladdin Reagent Co., Ltd. (Shanghai, China). Ascrobic acid, 2,2-diphenyl-1-picrylhydrazyl (DPPH) and 2,2-azino-bis(3-ethylbenzothiazoline-6-sulfonic acid) (ABTS), ethylenediaminetetraacetic acid disodium salt (EDTA-2Na), sodium nitroprusside (SNP), potassium persulfate (K$_2$S$_2$O$_8$), and trolox were obtained from Sigma-Aldrich (St. Louis, MO, USA). Other chemicals used (analytical grade) were bought from Kemio Chemical Reagent Co., Ltd. (Tianjin, China).

2.2. Plant Material

P. cretica was collected from Kunming City, Yunnan Province, China. A voucher specimen (FWJ-20180301) was deposited in the College of Life Science, Dalian Minzu University, Dalian, China. The plant material was dried under natural ventilation until reaching constant weight. The plant material was then powdered and stored in an air-tight container for further use.

2.3. Optimization of UAE of TF from P. cretica

2.3.1. Single-Factor Experiments

The UAE of TF from *P. cretica* was performed in a water-bath sonicator (KQ5200DE, 40 kHz frequency and 200 W nominal power, Kunshan Ultrasonic Instrument Co., Jiangsu, China). To study the influence of ethanol concentration on the total flavonoids yield (TFY), 5.0 g of pretreated samples were placed into glass conical flasks (250 mL) and soaked with ethanol solvents (varying from 30–80%, *v/v*), the time for extraction was set at 30 min, the temperature applied for extraction was set at 70 °C, and the liquid/solid ratio was set at 30 mL/g. To select the optimum extraction time, different extraction times (10, 20, 30, 40, 50, and 60 min) were tested under the conditions of ethanol concentration 60%, extraction temperature 70 °C, and liquid/solid ratio 30 mL/g. To choose the best extraction temperature, different extraction temperatures (40, 50, 60, 70, 80, and 90 °C) were studied under the following conditions: Ethanol concentration 60%, extraction time 40 min, and liquid/solid ratio 30 mL/g. Finally, the effect of liquid/solid ratios of 15, 20, 25, 30, 35, and 40 mL/g on the extraction yield was evaluated under the following conditions: Ethanol concentration 60%, extraction time 40 min, and extraction temperature 70 °C. After each UAE, the extraction solution was filtered, and the sample residue was re-extracted twice under the same extraction conditions. The combined extractive solution was then concentrated in vacuo and the extract was stored at −20 °C for further analysis.

2.3.2. RSM Design

Based on the results of single-factor experiments, the UAE of TF from *P. cretica* was further optimized by RSM. In this work, the BBD with four variables (ethanol concentration X_1, extraction time X_2, extraction temperature X_3, and liquid/solid ratio X_4) at three levels (−1, 0, 1) was carried out to evaluate the effect between every two variables on the response value (TFY Y). The BBD procedure of RSM resulted in a total of 29 randomized experiments, and a quadratic model used to analyze experimental data was shown as follows:

$$Y = \alpha_0 + \sum_{i=1}^{4} \alpha_i X_i + \sum_{i=1}^{4} \alpha_{ii} X_i^2 + \sum_{i=1}^{3} \sum_{j=i+1}^{4} \alpha_{ij} X_i X_j \tag{1}$$

where Y is the response value; α_0 refers to the intercept; α_i, α_{ii}, and α_{ij} refer to the linear, quadratic, and interactive coefficients, respectively; X_i and X_j represent the independent variables (i ≠ j).

2.4. Conventional Heat Reflux Extraction (HRE)

A comparative study between the HRE and UAE was carried out to estimate the efficiency of the UAE process established in this work. The HRE of TF from *P. cretica* was performed under the optimized UAE conditions with slight modifications. In brief, 5.0 g of the pre-prepared sample was extracted three times under reflux with 170 mL of 57% ethanol at 75 °C, each time for 2 h. After the HRE, the extract was processed as the method described in Section 2.3.1.

2.5. Measurement of Total Flavonoids Content (TFC)

The total flavonoids content (TFC) was measured using the aluminium chloride colourimetric method reported by with slight modifications [25]. First, to 0.5 mL of the sample solution, 0.15 mL of $NaNO_2$ (5%, *w/v*) was added. After the mixture was stirred for 5 min, 0.15 mL of (10%, *w/v*) $AlCl_3$ was added. After another 5 min, 1 mL of NaOH (1 M) was added. The solution was then made up to a final volume of 5 mL by the addition of ultrapure water. After complete addition, the reaction mixture was further incubated at ambient temperature for 20 min, and the absorbance was immediately recorded at 510 nm with an UV-vis Spectrophotometer (UV-2600, Shimadzu, Kyoto, Japan). Rutin was used as a reference, and the TFC was calculated according to the regression equation: $y = 11.2736x − 0.0028$

($R^2 = 0.9997$, final rutin concentration 18.12–72.48 µg/mL), where y is the absorbance, x is the content (µg/mL), and the TFY was calculated by the equation given below:

$$\text{Extraction yield } (\%) = \frac{C \times V}{W} \times 100 \tag{2}$$

where C represents the TFC (mg/mL); V represents the total volume of extractive solution (mL); and W represents the dry weight of plant material (mg).

2.6. HPLC Analysis

An HPLC system (Shimadzu Co., Kyoto, Japan) equipped with a multi-channel pump (LC-20AD) and a diode array detector (SPD-M20A) was used to analyze the extracts from *P. cretica* obtained by UAE under optimized conditions and HRE. A YMC-Pack ODS-A column (5 µm, 250 mm × 4.6 mm id; YMC Co., Tokyo, Japan) was employed for all separations at 25 °C. The mobile phase was composed of 0.5% TFA in ultrapure water (solvent A) and methanol (solvent B). The elution procedure was as follows: 0–25 min, 25–45% B; 25–40 min, 45–65% B; 40–60 min, 65–90% B; 60–65 min, 90–25% B, with a flow rate of 0.5 mL/min. Detection was carried out at the wavelength of 254 nm. The flavonoids in extracts from *P. cretica* were identified on the basis of comparison of the retention time and ultraviolet spectrum of standards, and the quantification of flavonoids was performed by the external standard method.

2.7. Method Validation

The method was validated for linearity, limit of detection (LOD), limit of quantification (LOQ), precision (inter-day and intra-day precision), stability, and accuracy following the International Conference on Harmonization (ICH) guideline and the previous reports [26,27].

Linearity was examined through the triplicate analysis of mixed standard solutions at six different concentrations, and the calibration curves were constructed by linear regression analysis of the integrated peak areas (y) versus concentrations (x). LOD and LOQ for each analyte under the HPLC conditions were determined at the signal-to-noise ratio (S/N) of 3 and 10, respectively.

Intra-day and inter-day variations were chosen to evaluate the precision of the developed HPLC method. Intra-day precision was assessed by six injections of a mixed standard solution within a single day, and inter-day precision was validated with the mixed standard solution used above once a day for 3 consecutive days. The precision of this method was expressed as relative standard deviation (RSD). Stability was evaluated by analyzing the same sample at 0, 2, 4, 6, 8, and 12 h at room temperature. The accuracy test was performed by the standard addition method. The authentic standards at three different concentration levels (low, medium, and large) were added to the sample, then extracted and quantified based on the established procedures. The recovery was counted according to the following formula:

$$\text{Recovery } (\%) = \frac{\text{total detected amount} - \text{original amount}}{\text{added amount}} \times 100 \tag{3}$$

2.8. Evaluation of Antioxidant Activity

2.8.1. Assay of DPPH• Radical Scavenging Activity

The DPPH• radical scavenging capacity of total flavonoid extracts of *P. cretica* was investigated on the basis of the method reported by Shukla et al. [28] with a minor modification. Briefly, 0.8 mL of the DPPH solution (0.1 mM in ethanol) was mixed with 2.4 mL of total flavonoid extract solution at different concentrations (25–250 µg/mL in ethanol). The reaction mixture was then shaken well and incubated at room temperature for 30 min. The absorbance of the reaction mixture was recorded at 517 nm. Ascrobic acid was used as a positive control in this work. The percentage of DPPH• radical scavenging was calculated using the following equation:

$$\text{Radical scavenging activity } (\%) = \frac{A_0 - A_1}{A_1} \times 100 \qquad (4)$$

where A_0 represents the absorbance of the reaction system without the extract and the A_1 represents the absorbance of the reaction system in the presence of the extract. The test was carried out in triplicate, and the IC_{50} was defined as the concentration of the extract that resulted in a 50% inhibition of DPPH• radical.

2.8.2. Assay of ABTS$^+$• Radical Scavenging Activity

The method reported by Awe et al. [29] was adopted for the assay of ABTS$^+$• radical scavenging activity. First, ABTS was dissolved in phosphate buffered saline (PBS, 0.01 M, pH 7.4) to a 7 mM concentration. The ABTS solution was then mixed with an equal volume of $K_2S_2O_8$ solution (2.45 mM) in the dark at room temperature to produce ABTS$^+$•. After 16 h, the ABTS$^+$• solution was adjusted to a suitable absorbance (0.70 ± 0.02) using water at 734 nm. To 2.0 mL of the prepared ABTS$^+$• solution, 0.5 mL of various concentrations of extract solution (25–150 µg/mL) was added. The reaction mixture was then incubated in the dark at room temperature. A total of 10 mins later, the absorbance of the mixture was immediately recorded at 734 nm. Trolox was used as a positive control in this experiment, and the percentage of ABTS$^+$• radical scavenging was calculated using the Equation (4).

2.8.3. Assay of NO• Radical Scavenging Activity

The assay of NO• radical scavenging activity was carried out following the procedure of Shukla et al. [28] with a minor modification. Briefly, 5 mM of SNP in PBS solution (0.01 M, pH 7.4) was prepared and mixed immediately with 2.0 mL of various concentrations of *P. cretica* extract (25–250 µg/mL). After 150 min of incubation at room temperature, 1.0 mL of Griess reagent (1% sulfanilamide in 2% H_3PO_4 and 0.1% naphthylethylenediamine dihydrochloride) was added. The absorbance of the chromophore formed during the reaction was then recorded at 546 nm. Ascrobic acid was used as a positive control in this work, and the percentage of NO• radical scavenging was calculated using Equation (4).

2.8.4. Assay of Fe^{2+} Chelating Activity

The ferrous ion chelating activity of the extract was measured using the modified method reported by Tohma et al. [30]. In brief, this assay was carried out by mixing 1.0 mL of extract solution (50–1600 µg/mL) with 0.1 mL of $FeCl_2$ solution (2.0 mM), 0.2 mL of ferrozine solution (5.0 mM) and 2.7 mL of deionized water. After 10 min of incubation at room temperature, the absorbance of reaction mixture was recorded at 562 nm, and EDTA-2Na was used as a positive control. The Fe^{2+} chelating activity was calculated according to Equation (4).

2.9. Statistical Analysis

All the experiments were performed in triplicate, and the results were shown as mean ± standard deviation (SD). A variance analysis (ANOVA) is used and the comparison of the averages is carried out by the test of Duncan, and differences were regarded as significant when the $p < 0.05$. The Design-Expert version 10 software (Stat-Ease Inc., Minneapolis, MN, USA) was employed for the RSM design and statistics analysis.

3. Results

3.1. Single-Factor Experiments

3.1.1. Effect of Ethanol Concentration on the TFY

In general, aqueous ethanol is an excellent solvent for the extraction of flavonoids from plants, and the selection of an appropriate ethanol concentration is a crucial step to improve extraction yield.

In the present work, the impact of ethanol concentration on the TFY was studied first. As shown in Figure 1a, the TFY increased dramatically when the ethanol concentration varied from 30% to 60%. However, the extraction yield exhibited a downward trend when the ethanol concentration was above 60%, which was possibly due to the fact that higher ethanol concentration could more easily lead to dissolution of alcohol-soluble impurities [31]. Therefore, 60% ethanol was chosen as the best extraction solvent.

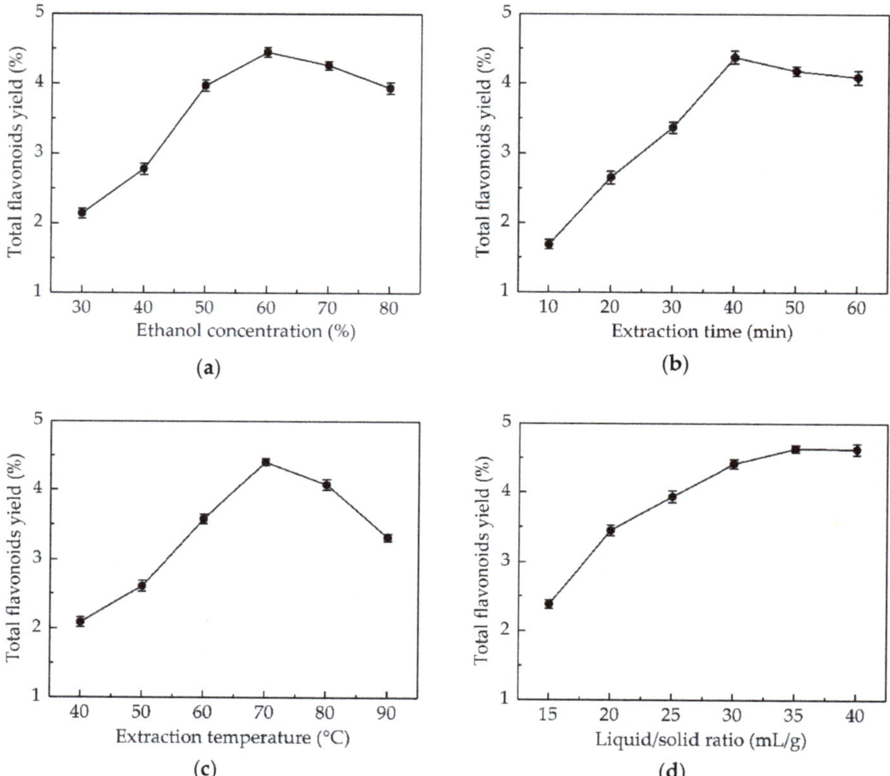

Figure 1. Effects of four independent variables on the total flavonoids yield (TFY): (**a**) Ethanol concentration; (**b**) extraction time; (**c**) extraction temperature; and (**d**) liquid/solid ratio.

3.1.2. Effect of Extraction Time on the TFY

In this work, the influence of extraction time (10–60 min) on the TFY was researched. As shown in Figure 1b, the initial increase of extraction time (10–40 min) resulted in an obvious improvement in the TFY. The extraction yield decreased slightly with the further increase in extraction time, which could be ascribed to the fact that longer extraction time will lead to an increase in the amount of dissolved impurities and the degradation of some flavonoids [32]. Thus, 40 min was regarded as the optimum extraction time.

3.1.3. Effect of Extraction Temperature on the TFY

There is no doubt that temperature is a significant factor influencing the extraction yield. Hence, in order to obtain the optimum extraction temperature, UAE processes were performed at 40, 50, 60, 70, 80, and 90 °C, respectively, and the results were shown in Figure 1c. There was a significant increase in the TFY (varying from 2.09 ± 0.07% to 4.41. ± 0.05%) with increasing extraction temperature from 40 to 60 °C. However, at higher extraction temperature (more than 60 °C), the TFY declined.

One of the possible reasons for this was that the structure of the flavonoids was destroyed under high temperature [33]. Thus, 60 °C was selected as the optimal extraction temperature.

3.1.4. Effect of Liquid/Solid Ratio on the TFY

The effect of liquid/solid ratio (15–40 mL/g) on the TFY was in Figure 1d. The TFY increased with the increasing liquid/solid ratio and reached a maximum (4.64 ± 0.05%) at 35 mL/g. Whereas, extraction yield reached a plateau phase when the liquid/solid ratio continued to rise above 35 mL/g. A high liquid/solid ratio would give rise to an increase in the cost of recovery. Hence, the optimal liquid/solid ratio was set at 35 mL/g.

3.2. Optimization of UAE Process by RSM

3.2.1. Model Fitting and Statistical Analysis

Due to the association between variables, BBD with four factors and three levels was employed to optimize the individual parameters. BBD matrix and response values were listed in Table 1. Subsequently, statistical regression analysis of experimental data was carried out, and a second-order polynomial equation yielded was represented as below:

$$\begin{aligned}
Y = \;& 4.59 - 0.068X_1 + 0.27X_2 + 0.27X_3 + 0.015X_4 \\
& -0.23X_1X_2 - 0.29X_1X_3 + 0.098X_1X_4 + 0.14X_2X_3 \\
& -0.19X_2X_4 - 0.098X_3X_4 - 0.54X_1{}^2 - 0.38X_2{}^2 \\
& -0.55X_3{}^2 - 0.33X_4{}^2
\end{aligned} \tag{4}$$

where X_1, X_2, X_3, and X_4 are the ethanol concentration, extraction time, extraction temperature, and liquid/solid ratio, respectively. Y is the predicted value of TFY.

Table 1. Box–Behnken design matrix and experimental values for the TFY.

Run	X_1 (%)	X_2 (min)	X_3 (°C)	X_4 (mL/g)	Y (%)
1	60 (0)	40 (0)	70 (0)	35 (0)	4.64
2	70 (1)	30 (−1)	70 (0)	35 (0)	3.56
3	60 (0)	30 (−1)	70 (0)	40 (1)	3.87
4	60 (0)	30 (−1)	80 (1)	35 (0)	3.39
5	60 (0)	40 (0)	80 (1)	30 (−1)	4.08
6	50 (−1)	40 (0)	70 (0)	40 (1)	3.60
7	50 (−1)	40 (0)	60 (−1)	35 (0)	2.95
8	60 (0)	40 (0)	70 (0)	35 (0)	4.61
9	50 (−1)	50 (1)	70 (0)	35 (0)	4.27
10	50 (−1)	40 (0)	70 (0)	30 (−1)	3.97
11	70 (1)	40 (0)	70 (0)	40 (1)	3.63
12	60 (0)	40 (0)	80 (1)	40 (1)	4.04
13	60 (0)	40 (0)	60 (−1)	40 (1)	3.58
14	60 (0)	40 (0)	70 (0)	35 (0)	4.61
15	60 (0)	50 (1)	80 (1)	35 (0)	4.23
16	60 (0)	40 (0)	70 (0)	35 (0)	4.62
17	60 (0)	40 (0)	60 (−1)	30 (−1)	3.23
18	60 (0)	50 (1)	60 (−1)	35 (0)	3.61
19	60 (0)	30 (−1)	70 (0)	30 (−1)	3.38
20	70 (1)	40 (0)	60 (−1)	35 (0)	3.46
21	60 (0)	50 (1)	70 (0)	40 (1)	4.01
22	70 (1)	40 (0)	80 (1)	35 (0)	3.48
23	60 (0)	40 (0)	70 (0)	35 (0)	4.48
24	70 (1)	50 (1)	70 (0)	35 (0)	3.64
25	50 (−1)	40 (0)	80 (1)	35 (0)	4.13

Table 1. *Cont.*

Run	X_1 (%)	X_2 (min)	X_3 (°C)	X_4 (mL/g)	Y (%)
26	60 (0)	50 (1)	70 (0)	30 (−1)	4.28
27	70 (1)	40 (0)	70 (0)	30 (−1)	3.61
28	50 (−1)	30 (−1)	70 (0)	35 (0)	3.27
29	60 (0)	30 (−1)	60 (−1)	35 (0)	3.33

The fitness and adequacy of the regression model were evaluated using a variance analysis (ANOVA), and results were given in Table 2. F-value (41.92) and p-value (<0.0001) of the regression model indicated that the established model was very significant. Whereas, F-value (3.28) and p-value (0.1317) of the lack of fit indicated that the lack of fit was not significant as compared with the pure error. In addition, the determination coefficient (R^2) obtained for this model was 0.9767, implying that the model could satisfactorily fit the variability of the TFY. The predicted R^2 of 0.8764 was in reasonable agreement with the adjusted R^2 of 0.9534, implying that the predicted values were highly consistent with the experimental values. In this study, the linear coefficients (X_1, X_2, and X_3), quadratic term coefficients (X_{12}, X_{22}, X_{32}, and X_{42}), and the cross product coefficients (X_1X_2, X_1X_3, X_2X_3, and X_2X_4) had statistically significant effects on the TFY ($p < 0.05$).

Table 2. The analysis of variance for the second-order polynomial model.

Source	Sum of Squares	Df	Mean Square	F-Value	p-Value
Model	6.28	14	0.45	41.92	<0.0001
X_1	0.055	1	0.055	5.11	0.0403
X_2	0.87	1	0.87	81.73	<0.0001
X_3	0.85	1	0.85	79.22	<0.0001
X_4	0.0027	1	0.0027	0.25	0.6233
X_1X_2	0.21	1	0.21	19.77	0.0006
X_1X_3	0.34	1	0.34	31.43	<0.0001
X_1X_4	0.038	1	0.038	3.55	0.0804
X_2X_3	0.078	1	0.078	7.32	0.0170
X_2X_4	0.14	1	0.14	13.49	0.0025
X_3X_4	0.038	1	0.038	3.55	0.0804
X_{12}	1.90	1	1.90	177.64	<0.0001
X_{22}	0.95	1	0.95	88.74	<0.0001
X_{32}	1.95	1	1.95	182.59	<0.0001
X_{42}	0.70	1	0.70	65.06	<0.0001
Residual	0.15	14	0.011		
Lack of fit	0.13	10	0.013	3.28	0.1317
Pure error	0.016	4	0.004		
Cor total	6.43	28			
R^2	0.9767				
Adjusted R^2	0.9534				

3.2.2. Response Surface Analysis

The three-dimensional (3D) response surface images illustrate the mutual influence of any two independent variables on the dependent variable while keeping other variables at the 0 level, and the shape of the 3D response surface plots provides information on the influence degree. To be specific, elliptical or saddle shapes of the 3D surfaces usually represent the interaction between two independent variables on the dependent variable, which is relatively significant, whereas a gently sloping surface indicates that the interaction between two factors on the dependent variable is relatively mild [34,35].

Figure 2a–f offered a visual interpretation of the interactions between two variables (X_1X_2, X_1X_3, X_1X_4, X_2X_3, X_2X_4, and X_3X_4) on the response variable (Y). Figure 2a–c showed the influences of ethanol concentration (X_1) with extraction time (X_2) and extraction temperature (X_3) on the extraction yields of TF, respectively. The initial increase of X_1 (50% to about 57%) led to an increase in TFY and followed by a decline thereafter (about 57% to 70%). Similarly, as could be seen from Figure 2a,d,e, a rapid rise in TFY was obtained when X_2 varied from 30 to about 46 min, then the extraction yield of TF was decreased slowly with increasing extraction time. Figure 2b,d,f indicated that with an increase of X_3 from 60 to about 74 °C, the TFY increased quickly, followed by a slight decline with further increase in X_3. According to Figure 2c,e,f, it could be found that a maximum level of TFY was obtained at a X_4 of 34 mL/g. The graphs also indicated that the influence of X_4 on the TFY is insignificant ($p > 0.05$). In addition, as evident from Figure 2c, the interaction effect of X_1 and X_4 had a insignificant influence on the TFY from *P. cretica* ($p > 0.05$).

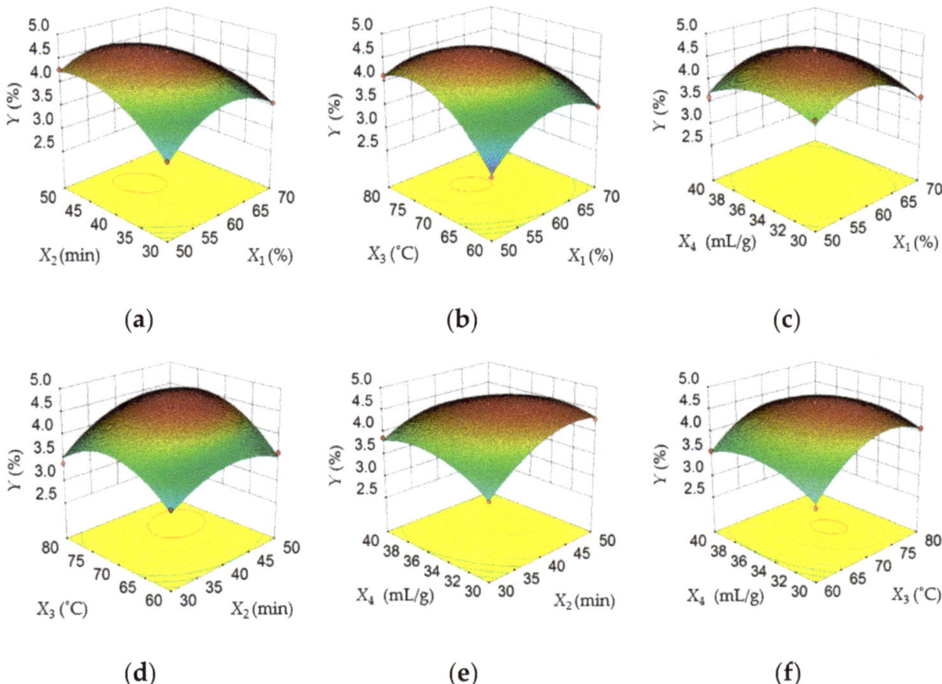

Figure 2. Response surface plots showing interaction between different variables (X_1: Ethanol concentration; X_2: Extraction time; X_3: Extraction temperature; and X_4: Liquid/solid ratio) on the TFY.

3.2.3. Validation of the Optimized Model

The optimal conditions for UAE of TF from *P. cretica* obtained by RSM were as follows: Ethanol concentration (X_1) 56.74%, extraction time (X_2) 45.94 min, extraction temperature (X_3) 74.27 °C, and liquid/solid ratio (X_4) 33.69 mL/g. Under optimized UAE conditions, the theoretical TFY was 4.74%. Subsequently, in order to prove the reliability of the prediction model, the confirmatory experiments were carried out under the optimized extraction conditions. Confirmatory experiments gave a satisfactory result that the TFY was 4.71 ± 0.04%, which was well-matched with theoretical value. Therefore, it could be concluded that the regression model was appropriate for the optimization of the UAE process of TF from *P. cretica*.

In this study, HRE was also performed to confirm the superiority of UAE of TF from *P. cretica*. The results demonstrated that the TFY obtained by UAE under optimized conditions was higher than that obtained by HRE (3.48 ± 0.05%). Additionally, a lot of previous literatures have reported that UAE gave a higher TFY obtained by UAE when compared with HRE. This phenomenon was believed to be caused by cavitation, vibration, crushing, mixing, and other comprehensive effects produced by ultrasound, which enhanced the extractability of flavonoids [36].

3.3. Method Validation for Quantitative Analysis of Five Flavonoids

Table 3 listed the linear equation, correlation coefficient (R^2), linear range, LOD, and LOQ of each compound determined. It could be found that all calibration curves showed good linearity

($R^2 \geq 0.9995$) within the tested concentration range. For these compounds, the LOD values ranged from 0.014 to 0.094 µg/mL, while the LOQ values ranged from 0.16 to 1.31 µg/mL. The results of precision, stability, and accuracy tests were shown in Table 4. All RSD values of the intra-day and inter-day precision ranged from 0.94–2.43% and 1.87–2.51%, respectively, which indicated good precision of the developed method. The results of the stability test showed that the RSD values of peak areas of the five compounds were ≤2.56%, which indicated that the sample was stable for 12 h at room temperature. According to the calculation, the mean recoveries of five compounds ranged from 98.93 to 101.3%, with RSD values ranging from 0.85 to 2.19%, which demonstrated a good accuracy of the developed method. In sum, the verification tests demonstrated that the developed method was feasible for the simultaneous quantification of five compounds from *P. cretica*.

Table 3. Linear regression, limit of detection (LOD), and limit of quantification (LOQ) of the five tested compounds.

Analytes	Linear Equation	R^2	Linear Range (µg/mL)	LOD (µg/mL)	LOQ (µg/mL)
luteolin-7-*O*-glucoside	$y = 73104x - 11077$	0.9995	5.00–100	0.051	0.67
Rutin	$y = 62441x + 21414$	0.9998	5.00–100	0.094	1.31
Quercitrin	$y = 135884x + 32168$	0.9997	2.50–100	0.042	0.44
Luteolin	$y = 146356x - 48920$	0.9998	1.00–100	0.014	0.16
Apigenin	$y = 86719x - 22486$	0.9999	1.00–100	0.026	0.34

Table 4. Precision, stability, and recovery of the five tested compounds.

Analytes	Precision (RSD, %)		Stability (RSD, %)	Recovery	
	Intra-Day	Inter-Day		Mean Recovery (%)	RSD (%)
Luteolin-7-*O*-glucoside	1.34	2.51	2.56	99.67	0.85
Rutin	2.11	1.87	2.48	100.4	1.48
Quercitrin	0.94	2.02	1.89	101.3	2.19
Luteolin	1.52	1.98	2.05	98.93	1.83
Apigenin	2.43	2.16	2.33	99.21	2.07

3.4. Analysis of Flavonoids in the Extracts by HPLC

Without any doubt, pharmacological activity of *P. cretica* depends on its chemical constituents. Therefore, flavonoid profiles in the *P. cretica* extracts were analyzed by HPLC. Figure 3 indicated that the HPLC chromatograms of extracts obtained by UAE and HRE showed highly similar characteristics. A total of five flavonoids, including luteolin-7-*O*-glucoside, rutin, quercetin, luteolin, and apigenin, were identified, and their elution times were 37.340, 39.021, 43.031, 50.568, and 54.306 min, respectively. Based on the results of HPLC quantitative analysis, the contents of the five flavonoids (mg/g plant material, dry weight) were in the following descending order: Luteolin-7-*O*-glucoside (3.237 ± 0.015 mg/g) > rutin (1.483 ± 0.009 mg/g) > apigenin (0.803 ± 0.013 mg/g) > luteolin (0.679 ± 0.007 mg/g) > quercitrin (0.582 ± 0.011 mg/g). However, the contents of five flavonoids extracted by HRE were all lower compared to the UAE: Luteolin-7-*O*-glucoside (2.862 ± 0.005 mg/g) > rutin (1.133 ± 0.007 mg/g) > apigenin (0.659 ± 0.011 mg/g) > luteolin (0.486 ± 0.009 mg/g) > quercitrin (0.574 ± 0.013 mg/g). Thus, it could be concluded that the UAE was a promising extraction method for the separation of bioactive flavonoids from *P. cretica* in the future. To our knowledge, this work was the first quantitative analysis of these five flavonoids in *P. cretica*.

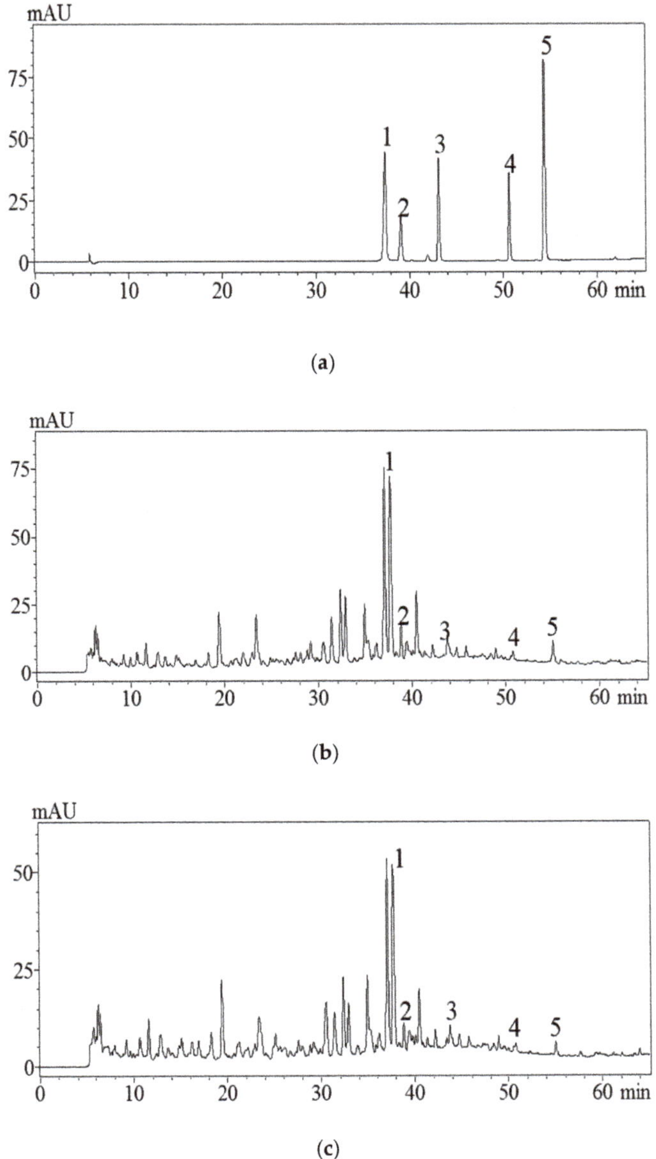

Figure 3. HPLC profiles of standards mixture (**a**) and *P. cretica* extract obtained by optimized ultrasonic-assisted extraction (UAE) (**b**). *P. cretica* extract obtained by heat reflux extraction (HRE) (**c**). Peaks (1): Luteolin-7-*O*-glucoside, (2) rutin, (3) quercitrin, (4) luteolin, and (5) apigenin.

3.5. Antioxidant Activity

3.5.1. DPPH• Radical Scavenging Activity

The DPPH• radical scavenging activity assay is a fairly common method to evaluate the antioxidant activity of natural products [37]. A comparison of the ascorbic acid and the *P. cretica* extract obtained by optimized UAE was performed, and the results are shown in Figure 4. It was observed that the DPPH• radical scavenging effect of the *P. cretica* extract was increased with increasing concentration.

The *P. cretica* extract at the test concentrations of 25, 50, 100, 150, 200, and 250 µg/mL showed 21.23 ± 2.78, 40.45 ± 3.04, 56.17 ± 2.45, 65.56 ± 3.14, 77.13 ± 2.98, and 85.94 ± 2.37% DPPH• radical scavenging, respectively. The IC_{50} value of the *P. cretica* extract was 74.49 µg/mL, while for ascorbic acid it was 35.19 µg/mL. The results indicated that the UAE extract of *P. cretica* had a good potential for scavenging DPPH• radical. It has been reported that DPPH• radical scavenging activity of plant extracts is ascribable to the presence of flavonoids [38].

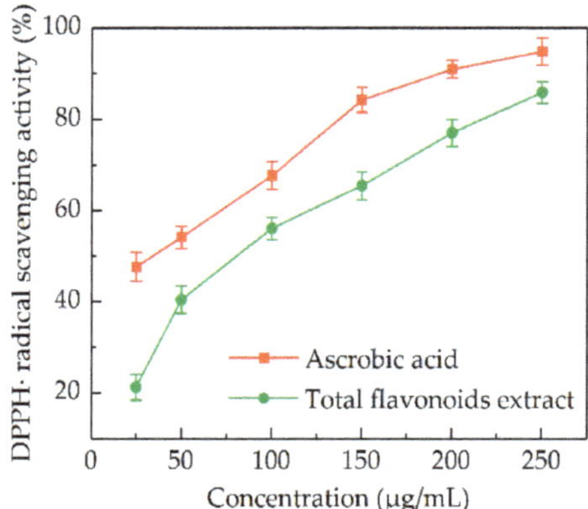

Figure 4. Scavenging activity of *P. cretica* extract obtained by optimized ultrasound-assisted extraction against DPPH• radical compared with ascorbic acid.

3.5.2. ABTS⁺• Radical Scavenging Activity

One of the most widely used organic radicals for the determination of antioxidant activity of natural products is the radical cation derived ABTS [39]. The results of ABTS⁺• radical scavenging activity of the *P. cretica* extract at different concentrations (25–150 µg/mL) were shown in Figure 5. It was found that the *P. cretica* extract had a scavenging activity on the ABTS⁺• radical in a dose dependent manner. The IC_{50} value of standard trolox was 30.10 µg/mL, while the IC_{50} value of the *P. cretica* extract was 82.92 µg/mL. A 150 µg/mL of the *P. cretica* extract exhibited 78.41 ± 2.97% inhibition. From these results, it could be stated that the *P. cretica* extract was a good ABTS⁺• radical scavenger.

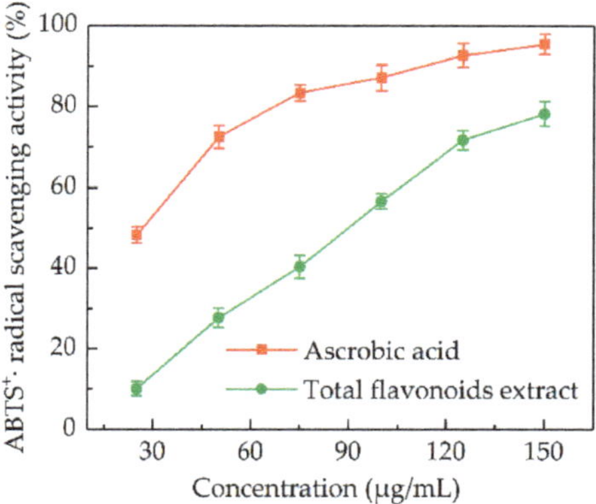

Figure 5. Scavenging activity of the *P. cretica* extract obtained by optimized ultrasound-assisted extraction against ABTS⁺• radical compared with trolox.

3.5.3. NO• Radical Scavenging Activity

Despite the potential health benefits of NO• radical, its role in causing oxidative damage is becoming progressively prominent [40]. In this work, the scavenging effect of the *P. cretica* extract on NO• radical was studied. Figure 6 showed that the *P. cretica* extract within the concentration range tested (25–250 µg/mL) exhibited NO• radical scavenging activity in a dose-dependent manner. Especially, the *P. cretica* extract exhibited a maximum percentage scavenging of 82.71 ± 1.54%, which was comparable to the NO• radical scavenging activity of ascorbic acid. The IC_{50} value of ascorbic acid was 51.60 µg/mL, and for the *P. cretica* extract it was 89.12 µg/mL. Results indicated that *P. cretica* might serve as an excellent NO• radical scavenger due to the high level of flavonoids. It has been reported that the flavonoids are very powerful NO• radical scavengers [41].

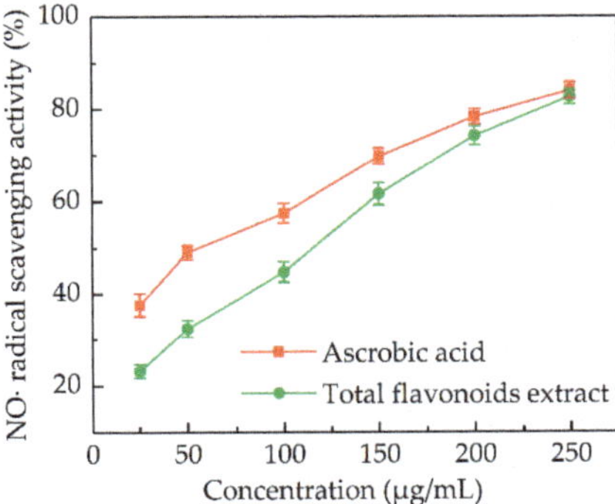

Figure 6. NO• radical scavenging activity of the *P. cretica* extract obtained by optimized ultrasound-assisted compared with ascorbic acid.

3.5.4. Fe^{2+} Chelating Activity

It has been reported that Fe^{2+} is a powerful pro-oxidant among various metal ions and Fe^{2+} chelation may prevent oxidative damage through retarding metal-catalyzed oxidation and inhibiting ROS production [42]. In this work, the Fe^{2+} chelating activity of the *P. cretica* extract was determined and compared with a chelating standard (EDTA-2Na). As shown in Figure 7, at different test doses of 50, 100, 200, 400, 800, and 1600 µg/mL, the Fe^{2+} chelating activity of the *P. cretica* extract was 4.74 ± 1.77, 9.98 ± 1.89, 18.65 ± 1.86, 31.56 ± 2.48, 51.87 ± 2.17, and 74.21 ± 1.83%, respectively, with an IC_{50} value of 713.41 µg/mL. According to these results, it could be concluded that the *P. cretica* extract had certain Fe^{2+} chelating activity.

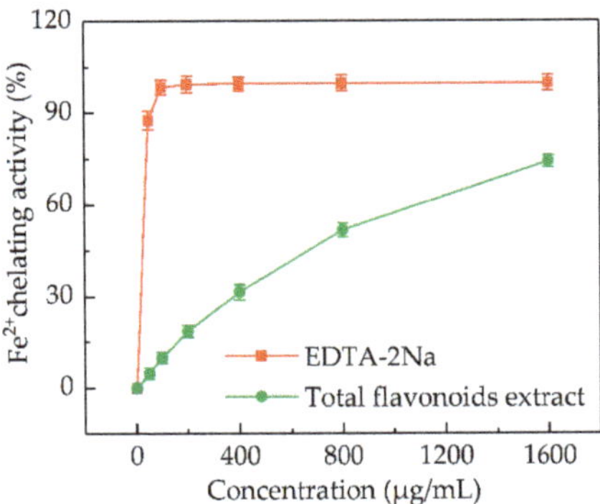

Figure 7. Fe^{2+} chelating activity of the *P. cretica* extract obtained by optimized ultrasound-assisted compared with EDTA-2Na.

4. Conclusions

In this work, the UAE was proved to be a sample, rapid, and effective procedure for the extraction of TF from *P. cretica*, and the process parameters optimized by RSM were as follows: Ethanol concentration 56.74%, extraction time 45.94 min, extraction temperature 74.27 °C, and liquid/solid ratio 33.69 mL/g. Under optimized UAE conditions, the TFY was 4.71 ± 0.04%, which was higher than that obtained by HRE. Moreover, HPLC analysis indicated that luteolin-7-*O*-glucoside, rutin, quercitrin, luteolin, and apigenin were the predominant flavonoids in *P. cretica*. It was also found that UAE is more suitable than HRE for the extraction of these compounds. In addition, the *P. cretica* extract obtained through the optimized UAE method exhibited good DPPH•, ABTS$^+$•, and NO• scavenging activities. The results of the present wok should contribute to the further and deeper investigation of *P. cretica*, which might be used as a novel source of natural antioxidants.

Author Contributions: Conceptualization, W.H.; methodology, W.H. and M.H.; software, A.W., K.H., and X.S.; validation, W.H. and Z.X; formal analysis, D.C.; investigation, M.H. and R.L.; resources, W.H.; data curation, M.H. and J.S.; writing—original draft preparation, M.H.; writing—review and editing, W.H.; supervision, Y.S.; project administration, W.H.; funding acquisition, W.H.

Funding: This research was funded by the "Thirteenth Five-Year Plan" for the National Key Research and Development Program (No. 2016YFD0400903) and National Natural Science Foundation of China (No. 31471923).

Conflicts of Interest: The authors declare no conflict of interest.

References

1. Chao, Y.S.; Rouhan, G.; Amoroso, V.B.; Chiou, W.L. Molecular phylogeny and biogeography of the fern genus *Pteris* (Pteridaceae). *Ann. Bot.* **2014**, *114*, 109–124. [CrossRef] [PubMed]
2. Testo, W.L.; Watkins, J.E.; Pittermann, J.; Momin, R. *Pteris × caridadiae* (Pteridaceae), a new hybrid fern from Costa Rica. *Brittonia* **2015**, *67*, 138–143. [CrossRef]
3. Harinantenaina, L.; Matsunami, K.; Otsuka, H. Chemical constituents of *Pteris cretica* Linn. (Pteridaceae). *Biochem. Syst. Ecol.* **2009**, *37*, 133–137. [CrossRef]
4. Qiu, M.; Yang, B.; Cao, D.; Zhu, J.; Jin, J.; Chen, Y.; Zhou, L.; Luo, X.; Zhao, Z. Two new hydroxylated *ent*-kauranoic acids from *Pteris semipinnata*. *Phytochem. Lett.* **2016**, *16*, 156–162. [CrossRef]
5. Harinantenaina, L.; Matsunami, K.; Otsuka, H. Chemical and biologically active constituents of *Pteris multifida*. *J. Nat. Med.* **2008**, *62*, 452–455. [CrossRef] [PubMed]
6. Wang, H.B.; Wong, M.H.; Lan, C.Y.; Qin, Y.R.; Shu, W.S.; Qiu, R.L.; Ye, Z.H. Effect of arsenic on flavonoid contents in *Pteris* species. *Biochem. Syst. Ecol.* **2010**, *38*, 529–537. [CrossRef]
7. Imperato, F.; Nazzaro, R. Luteolin 7-*O*-sophoroside from *Pteris cretica*. *Phytochemistry* **1996**, *41*, 337–338. [CrossRef]
8. Luo, X.; Li, C.; Luo, P.; Lin, X.; Ma, H.; Seeram, N.P.; Song, C.; Xu, J.; Gu, Q. Pterosin sesquiterpenoids from *Pteris cretica* as hypolipidemic agents via activating liver X receptors. *J. Nat. Prod.* **2016**, *79*, 3014–3021. [CrossRef]
9. Ge, X.; Ye, G.; Li, P.; Tang, W.J.; Gao, J.L.; Zhao, W.M. Cytotoxic diterpenoids and sesquiterpenoids from *Pteris multifida*. *J. Nat. Prod.* **2008**, *71*, 227–231. [CrossRef]
10. Shi, Y.S.; Zhang, Y.; Hu, W.Z.; Zhang, X.F.; Fu, X.; Lv, X. Dihydrochalcones and diterpenoids from *Pteris ensiformis* and their bioactivities. *Molecules* **2017**, *22*, 1413. [CrossRef]
11. Kim, J.W.; Seo, J.Y.; Oh, W.K.; Sung, S.H. Anti-neuroinflammatory *ent*-kaurane diterpenoids from *Pteris multifida* roots. *Molecules* **2017**, *22*, 27. [CrossRef] [PubMed]
12. Singh, M.; Govindarajan, R.; Rawat, A.K.S.; Khare, P.B. Antimicrobial flavonoid rutin from *Pteris vittata* L. against pathogenic gastrointestinal microflora. *Am. Fern J.* **2008**, *98*, 98–104. [CrossRef]
13. Chen, Y.H.; Chang, F.R.; Lin, Y.J.; Wang, L.; Chen, J.F.; Wu, Y.C.; Wu, M.J. Identification of phenolic antioxidants from Sword Brake fern (*Pteris ensiformis* Burm.). *Food Chem.* **2007**, *105*, 48–56. [CrossRef]
14. Dai, G.C.; Hu, B.; Zhang, W.F.; Peng, F.; Wang, R.; Liu, Z.Y.; Xue, B.X.; Liu, J.Y.; Shan, Y.X. Chemical characterization, anti-benign prostatic hyperplasia effect and subchronic toxicity study of total flavonoid extract of *Pteris multifida*. *Food Chem. Toxicol.* **2017**, *108*, 524–531. [CrossRef] [PubMed]

15. Yang, R.F.; Geng, L.L.; Lu, H.Q.; Fan, X.D. Ultrasound-synergized electrostatic field extraction of total flavonoids from *Hemerocallis citrina baroni*. *Ultrason. Sonochem.* **2017**, *34*, 571–579. [CrossRef] [PubMed]

16. Delgado-Povedano, M.M.; de Castro, M.D.L. A review on enzyme and ultrasound: A controversial but fruitful relationship. *Anal. Chim. Acta* **2015**, *889*, 1–21. [CrossRef] [PubMed]

17. Wen, C.; Zhang, J.; Zhang, H.; Dzah, C.S.; Zandile, M.; Duan, Y.; Ma, H.; Luo, X. Advances in ultrasound assisted extraction of bioactive compounds from cash crops—A review. *Ultrason. Sonochem.* **2018**, *48*, 538–549. [CrossRef]

18. Jing, C.L.; Dong, X.F.; Tong, J.M. Optimization of ultrasonic-assisted extraction of flavonoid compounds and antioxidants from alfalfa using response surface method. *Molecules* **2015**, *20*, 15550–15571. [CrossRef]

19. Zhang, H.; Xie, G.; Tian, M.; Pu, Q.; Qin, M. Optimization of the ultrasonic-assisted extraction of bioactive flavonoids from *Ampelopsis grossedentata* and subsequent separation and purification of two flavonoid aglycones by high-speed counter-current chromatography. *Molecules* **2016**, *21*, 1096. [CrossRef]

20. Wang, B.; Qu, J.; Luo, S.; Feng, S.; Li, T.; Yuan, M.; Huang, Y.; Liao, J.; Yang, R.; Ding, C. Optimization of ultrasound-assisted extraction of flavonoids from olive (*Olea europaea*) leaves, and evaluation of their antioxidant and anticancer activities. *Molecules* **2018**, *23*, 2513. [CrossRef]

21. Majd, M.H.; Rajaei, A.; Bashi, D.S.; Mortazavi, S.A.; Bolourian, S. Optimization of ultrasonic-assisted extraction of phenolic compounds from bovine pennyroyal (*Phlomidoschema parviflorum*) leaves using response surface methodology. *Ind. Crop. Prod.* **2014**, *57*, 195–202. [CrossRef]

22. Balavigneswaran, C.K.; Kumar, T.S.J.; Packiaraj, R.M.; Veeraraj, A.; Prakash, S. Anti-oxidant activity of polysaccharides extracted from *Isocrysis galbana* using RSM optimized conditions. *Int. J. Biol. Macromol.* **2013**, *60*, 100–108. [CrossRef] [PubMed]

23. Sun, Y.; Li, T.; Yan, J.; Liu, J. Technology optimization for polysaccharides (POP) extraction from the fruiting bodies of *Pleurotus ostreatus* by Box–Behnken statistical design. *Carbohydr. Polym.* **2010**, *80*, 242–247. [CrossRef]

24. Wang, W.; Li, Q.; Liu, Y.; Chen, B. Ionic liquid-aqueous solution ultrasonic-assisted extraction of three kinds of alkaloids from *Phellodendron amurense Rupr* and optimize conditions use response surface. *Ultrason. Sonochem.* **2015**, *24*, 13–18. [CrossRef] [PubMed]

25. Choi, J.; An, X.; Lee, B.H.; Lee, J.S.; Heo, H.J.; Kim, T.; Ahn, J.W.; Kim, D.O. Protective effects of bioactive phenolics from jujube (*Ziziphus jujuba*) seeds against H_2O_2–induced oxidative stress in neuronal PC-12 cells. *Food Sci. Biotechnol.* **2015**, *24*, 2219–2227. [CrossRef]

26. Bae, I.K.; Ham, H.M.; Jeong, M.H.; Kim, D.H.; Kim, H.J. Simultaneous determination of 15 phenolic compounds and caffeine in teas and mate using RP-HPLC/UV detection: Method development and optimization of extraction process. *Food Chem.* **2015**, *172*, 469–475. [CrossRef]

27. Zhou, F.; Zhang, L.; Gu, L.; Zhang, Y.; Hou, C.; Bi, K.; Chen, X.; Zhang, H. Simultaneous quantification of 13 compounds in Guanxin Shutong Capsule by HPLC method. *J. Chromatogr. Sci.* **2016**, *54*, 971–976. [CrossRef]

28. Shukla, S.; Mehta, A.; Bajpai, V.K.; Shukla, S. In vitro antioxidant activity and total phenolic content of ethanolic leaf extract of *Stevia rebaudiana* Bert. *Food Chem. Toxicol.* **2009**, *47*, 2338–2343. [CrossRef]

29. Awe, F.B.; Fagbemi, T.N.; Ifesan, B.O.T.; Badejo, A.A. Antioxidant properties of cold and hot water extracts of cocoa, Hibiscus flower extract, and ginger beverage blends. *Food Res. Int.* **2013**, *52*, 490–495. [CrossRef]

30. Tohma, H.S.; Gulçin, I. Antioxidant and radical scavenging activity of aerial parts and roots of Turkish liquorice (*Glycyrrhiza glabra* L.). *Int. J. Food Prop.* **2010**, *13*, 657–671. [CrossRef]

31. Liu, J.; Mu, T.; Sun, H.; Fauconnier, M.L. Optimization of ultrasonic-microwave synergistic extraction of flavonoids from sweet potato leaves by response surface methodology. *J. Food Process Preserv.* **2019**, *43*, e13928. [CrossRef]

32. Dahmoune, F.; Spigno, G.; Moussi, K.; Remin, H.; Cherbal, A.; Madani, K. *Pistacia lentiscus* leaves as a source of phenolic compounds: Microwave-assisted extraction optimized and compared with ultrasound-assisted and conventional solvent extraction. *Ind. Crop. Prod.* **2014**, *61*, 31–40. [CrossRef]

33. Prommuak, C.; De-Eknamkul, W.; Shotipruk, A.; Shotipruk, A. Extraction of flavonoids and carotenoids from Thai silk waste and antioxidant activity of extracts. *Sep. Purif. Technol.* **2008**, *62*, 444–448. [CrossRef]

34. Guo, X.; Zou, X.; Sun, M. Optimization of extraction process by response surface methodology and preliminary characterization of polysaccharides from *Phellinus igniarius*. *Carbohydr. Polym.* **2010**, *80*, 344–349. [CrossRef]

35. Feng, S.; Luo, Z.; Tao, B.; Chen, C. Ultrasonic-assisted extraction and purification of phenolic compounds from sugarcane (*Saccharum officinarum* L.) rinds. *LWT Food Sci. Technol.* **2015**, *60*, 970–976. [CrossRef]

36. Chemat, F.; Rombaut, N.; Sicaire, A.G.; Meullemiestre, A.; Fabiano-Tixier, A.S.; Abert-Vian, M. Ultrasound assisted extraction of food and natural products. Mechanisms, techniques, combinations, protocols and applications. A review. *Ultrason. Sonochem.* **2017**, *34*, 540–560. [CrossRef]
37. Clarke, G.; Ting, K.; Wiart, C.; Fry, J. High correlation of 2, 2-diphenyl-1-picrylhydrazyl (DPPH) radical scavenging, ferric reducing activity potential and total phenolics content indicates redundancy in use of all three assays to screen for antioxidant activity of extracts of plants from the Malaysian rainforest. *Antioxidants* **2013**, *2*, 1–10.
38. Madsen, H.L.; Andersen, C.M.; Jørgensen, L.V.; Skibsted, L.H. Radical scavenging by dietary flavonoids. A kinetic study of antioxidant efficiencies. *Eur. Food Res. Technol.* **2000**, *211*, 240–246. [CrossRef]
39. Re, R.; Pellegrini, N.; Proteggente, A.; Pannala, A.; Yang, M.; Rice-Evans, C. Antioxidant activity applying an improved ABTS radical cation decolorization assay. *Free Radic. Biol. Med.* **1999**, *26*, 1231–1237. [CrossRef]
40. Pacher, P.; Beckman, J.S.; Liaudet, L. Nitric oxide and peroxynitrite in health and disease. *Physiol. Rev.* **2007**, *87*, 315–424. [CrossRef]
41. Vanacker, S.A.B.E.; Tromp, M.N.J.L.; Haenen, G.R.M.M.; Vandervijgh, W.J.F.; Bast, A. Flavonoids as scavengers of nitric oxide radical. *Biochem. Biophys. Res. Commun.* **1995**, *214*, 755–759. [CrossRef] [PubMed]
42. Gülçin, İ. Antioxidant activity of caffeic acid (3, 4-dihydroxycinnamic acid). *Toxicology* **2006**, *217*, 213–220. [CrossRef] [PubMed]

 *antioxidants*

Article

Compositional Features and Bioactive Properties of *Aloe vera* Leaf (Fillet, Mucilage, and Rind) and Flower

Mikel Añibarro-Ortega [1], José Pinela [1,*], Lillian Barros [1], Ana Ćirić [2], Soraia P. Silva [3], Elisabete Coelho [3], Andrei Mocan [4,5], Ricardo C. Calhelha [1], Marina Soković [2], Manuel A. Coimbra [3] and Isabel C. F. R. Ferreira [1,*]

[1] Centro de Investigação de Montanha (CIMO), Instituto Politécnico de Bragança, Campus de Santa Apolónia, 5300-253 Bragança, Portugal; mklanibarro@gmail.com (M.A.-O.); lillian@ipb.pt (L.B.); calhelha@ipb.pt (R.C.C.)

[2] Department of Plant Physiology, Institute for Biological Research "Siniša Stanković", University of Belgrade, Bulevar Despota Stefana 142, 11000 Belgrade, Serbia; rancic@ibiss.bg.ac.rs (A.Ć.); mris@ibiss.bg.ac.rs (M.S.)

[3] QOPNA & LAQV-REQUIMTE, Department of Chemistry, University of Aveiro, 3810-193 Aveiro, Portugal; soraiapiressilva@ua.pt (S.P.S.); ecoelho@ua.pt (E.C.); mac@ua.pt (M.A.C.)

[4] Department of Pharmaceutical Botany, "Iuliu Hațieganu" University of Medicine and Pharmacy, Gheorghe Marinescu Street 23, 400337 Cluj-Napoca, Romania; mocan.andrei@umfcluj.ro

[5] Laboratory of Chromatography, Institute of Advanced Horticulture Research of Transylvania, University of Agricultural Sciences and Veterinary Medicine, 400372 Cluj-Napoca, Romania

* Correspondence: jpinela@ipb.pt (J.P.); iferreira@ipb.pt (I.C.F.R.F); Tel.: +351-273-303-219 (I.C.F.R.F.)

Received: 8 September 2019; Accepted: 27 September 2019; Published: 1 October 2019

Abstract: This work aimed to characterize compositional and bioactive features of *Aloe vera* leaf (fillet, mucilage, and rind) and flower. The edible fillet was analysed for its nutritional value, and all samples were studied for phenolic composition and antioxidant, anti-inflammatory, antimicrobial, tyrosinase inhibition, and cytotoxic activities. Dietary fibre (mainly mannan) and available carbohydrates (mainly free glucose and fructose) were abundant macronutrients in fillet, which also contained high amounts of malic acid (5.75 g/100 g dw) and α-tocopherol (4.8 mg/100 g dw). The leaf samples presented similar phenolic profiles, with predominance of chromones and anthrones, and the highest contents were found in mucilage (131 mg/g) and rind (105 mg/g) extracts, which also revealed interesting antioxidant properties. On the other hand, the flower extract was rich in apigenin glycoside derivatives (4.48 mg/g), effective against *Pseudomonas aeruginosa* (MIC = 0.025 mg/mL and MBC = 0.05 mg/mL) and capable of inhibiting the tyrosinase activity (IC_{50} = 4.85 mg/mL). The fillet, rind, and flower extracts also showed a powerful antifungal activity against *Aspergillus flavus, A. niger, Penicillium funiculosum*, and *Candida albicans*, higher than that of ketoconazole. Thus, the studied *Aloe vera* samples displayed high potential to be exploited by the food or cosmetic industries, among others.

Keywords: *Aloe barbadensis* Mill.; nutritional composition; neutral sugars; phenolic compounds; organic acids; antioxidant capacity; antimicrobial activity; tyrosinase inhibitory activity; cytotoxicity

---

1. Introduction

Aloe vera (also known as *Aloe barbadensis* Mill.) is a flowering succulent plant of the family Asphodelaceae currently naturalized in many tropical and sub-tropical countries. In traditional medicine, it has been widely used for centuries to treat skin disorders and other ailments, as well as for its purgative effect [1]. Today, this species is used worldwide as a valuable ingredient for functional foods (such as healthy drinks and other beverages), cosmetics (including creams, lotions, soaps, and shampoos), and drugs (such as tablets and capsules) [2].

The dagger-shaped leaves are the most used part of the plant, in which two major fractions can be identified, namely the outer, photosynthetically active green cortex, usually known as rind, and the inner parenchyma, known as pulp or fillet. Furthermore, the leaf secretes two different exudates—the reddish-yellow latex produced by the pericyclic cells under the cutinized epidermis and the transparent, slippery mucilage or gel produced by the thin-walled tubular cells in the inner parenchyma [3,4]. The gel is approximately 98% moisture, and the non-aqueous remainder largely consists of acemannan (a bioactive acetylated glucomannan) and other polysaccharides, sugars, minerals, organic acids, and vitamins [1,4–6]. Traditionally, it is used topically to treat wounds, minor burns, and skin irritations and internally to treat constipation, coughs, ulcers, and diabetes, among other ailments [1,7]. The latex, on the other hand, contains hydroxyanthracene derivatives, including anthraquinone C- and O-glycosides, and is a natural drug well-known for its cathartic effect and also used as a bittering agent in alcoholic beverages [8,9].

Today, *Aloe vera* is produced on a large scale in many countries around the world to supply the still growing and economically important industry [1]. For production of *Aloe vera* gel, the leaves can be processed either by grinding the inner fillet after removing the rind (treated as bio-waste) and rinsing away the latex, or the whole leaf. However, in the second case, a subsequent filtration/purification step is required to remove unwanted constituents, especially those from latex [10,11]. The gel has been marketed fresh or in powdered concentrate and included in different formulations for food, health, medicinal, and cosmetic purposes [9,10,12].

Over the last few decades, the *Aloe vera* leaf has been the subject of several scientific studies that aimed to characterize its chemical and biological properties [4–6,13,14]. Still, there are compositional and bioactive parameters that deserve further study, and the flower remains an underexploited plant part. Moreover, a lack of information on the exact part of the plant analysed or even the species involved is common in many works. There are confusing descriptions, mostly about the inner part of the leaf, due to the different terms that have been used interchangeably, such as fillet, pulp, mucilage, gel, and parenchyma, among others. Technically, these terms do not refer to the same part, since fillet or pulp refer to the fleshy inner part of the leaf including the cell walls, while gel or mucilage refer to the viscous clear liquid contained within the parenchyma cells [4].

This comprehensive study was performed to evaluate and compare compositional and bioactive features of different parts of *Aloe vera*, namely leaf (which was divided into fillet, mucilage and rind) and flower. More specifically, it was intended to determine the nutritional and chemical composition of the edible fillet and the profiles in phenolic compounds of the four sample extracts, as well as their antioxidant, anti-inflammatory, antimicrobial, tyrosinase inhibition, and cytotoxic capacities. This will provide accurate and up-to-date research information on *Aloe vera*.

2. Materials and Methods

2.1. Chemicals

All standard compounds used for chromatographic quantifications (47885-U, 2-deoxyglucose, α, β, γ, and δ tocopherols, oxalic, quinic, malic, ascorbic, citric, and fumaric acids, from Sigma-Aldrich (St. Louis, MO, USA); tocol, from Matreya (Pleasant Gap, PA, USA); chlorogenic acid, *p*-coumaric acid, apigenin-6-glucoside, apigenin-7-glucoside, and luteolin-6-C-glucoside, from Extrasynthèse (Genay, France); and aloin, from Alfa Aesar (Ward Hill, MA, USA)) and bioactivity assays (trolox, kojic acid, dexamethasone, ellipticine, streptomycin, and ketoconazole, from Sigma-Aldrich) had a purity level of at least 95%. All other reagents were of analytical grade and purchased from common sources.

2.2. Plant Material

Freshly cut *Aloe vera* samples (leaves and flowers) of certified organic production were supplied by "aCoruela do Alentejo," a company located in the parish of São Brás e São Lourenço, municipality of Elvas, Portugal. Twelve three-year-old leaves with about 10 cm × 45 cm were weighed, and the green rind was separated from the inner fillet with a knife. Then, the transparent slippery exudate consisting mainly of gel was collected from the mucilage layer of the outer leaf pulp adjacent to the rind. The weight of the different leaf parts was determined to calculate the weight percentage (wt%). A batch of each plant part (fillet, mucilage, rind, and flower) was then lyophilized (FreeZone 4.5, Labconco, Kansas City, MO, USA), reduced to a fine powder (~20 mesh), and homogenized to obtain a representative sample that was kept at −20 °C until analysis.

2.3. Compositional Analysis of Fillet

2.3.1. Proximate Composition

The edible fillet was analysed for moisture, protein, fat, ash, and crude fibre contents following the AOAC procedures [15]. Briefly, the crude protein content (N × 6.25) was estimated by the macro-Kjeldahl method, using an automatic distillation and titration unit (Pro-Nitro-A, JP Selecta, Barcelona); the crude fat content was determined by Soxhlet extraction with petroleum ether; the ash content was determined by incineration in a muffle furnace at 550 ± 15 °C; and the crude fibre content was estimated by the Weende method through an acid/alkaline hydrolysis of insoluble residues. The available carbohydrate content was determined by the anthrone method [16]. The dietary fibre content was estimated by difference. The results were given as g per 100 g of plant material.

The energy value was calculated as follows: 4 × (g protein + g available carbohydrates) + 9 × (g fat) + 2 × (g dietary fibre) [17] and given as kcal per 100 g of plant material.

2.3.2. Fatty Acids, Tocopherols, and Organic Acids

Fatty acids were analysed in a DANI GC 1000 (DANI instruments, Contone, Switzerland) equipped with a split/splitless injector and a flame ionization detector (FID) at 260 °C. The fatty acids obtained by Soxhlet extraction were methylated with methanol:sulphuric acid:toluene 2:1:1 (v:v:v) during at least 12 h in a bath at 50 °C and 160 rpm. Then, deionised water was added to obtain phase separation, and the fatty acid methyl esters (FAME) were recovered with diethyl ether. The upper phase was dehydrated and filtered through 0.2 μm nylon filters for injection. Chromatographic separation was performed on a Zebron-Kame column (30 m × 0.25 mm i.d. × 0.20 μm film thickness, Phenomenex, Torrance, CA, USA). The oven temperature program was as follows: the initial temperature of the column was 100 °C, held for 2 min, then a 10 °C/min ramp to 140 °C, 3 °C/min ramp to 190 °C, 30 °C/min ramp to 260 °C, held for 2 min [18]. The carrier gas (hydrogen) flow rate was 1.1 mL/min, measured at 100 °C. Split injection (1:50) was carried out at 250 °C. The identification was made by chromatographic comparison of the retention times of the sample FAME peaks with those of commercial standards (standard 47885-U). The results were processed using the Clarity 4.0.1.7 Software (DataApex, Podohradska, Czech Republic) and given as relative percentage of each fatty acid.

Tocopherols were analysed in a high-performance liquid chromatography (HPLC) system (Knauer, Smartline system 1000, Berlin, Germany) coupled to a fluorescence detector (FP-2020, Jasco, Easton, MD, USA) programmed for excitation at 290 nm and emission at 330 nm, following an previously described analytical procedure [19]. The samples (500 mg) were spiked with a BHT solution (10 mg/mL) and tocol (internal standard, 50 μg/mL) and homogenized with methanol (4 mL) by shaking in vortex (1 min) and then with hexane (4 mL). After that, a saturated NaCl aqueous solution (2 mL) was added, the mixture was homogenized, centrifuged (5 min, 4000× *g*) and the clear upper layer was collected. The extraction was repeated twice with hexane. The obtained extracts were dried under a nitrogen stream, redissolved in 2 mL of *n*-hexane, dehydrated, and filtered through 0.22 μm disposable syringe filters for injection. Chromatographic separation was performed in normal phase on a Polyamide II column (5 μm particle size, 250 × 4.6 mm; YMC, Kyoto, Japan). Elution was performed with a mixture of *n*-hexane and ethyl acetate (70:30, v/v). The detected compounds were identified by chromatographic comparisons with authentic standards (α, β, γ, and δ isoforms) and quantified using the internal standard method. The results were processed using the Clarity 4.0.1.7 Software and given as μg per 100 g of plant material.

Organic acids were analysed in a ultra-fast liquid chromatography system (Shimadzu 20 A series UFLC, Shimadzu Corporation, Kyoto, Japan) coupled to a photodiode array detector (UFLC-PDA), as previously described [20]. The samples (1 g) were stirred with meta-phosphoric acid (25 mL) for 45 min and filtered, first through Whatman No. 4 paper and then through 0.2 μm nylon filters. Chromatographic separation was achieved in reverse phase on a C18 column (5 μm particle size, 250 × 4.6 mm; Phenomenex, Torrance, CA, USA). The elution was performed with sulphuric acid (3.6 mM). PDA detection was carried out at 215 and 245 nm (for ascorbic acid). The detected compounds were identified and quantified by chromatographic comparison of the peak area with calibration curves obtained from commercial standards (oxalic, quinic, malic, ascorbic, citric, and fumaric acids). The results were processed using the LabSolutions Multi LC-PDA software and expressed in mg per 100 g of plant material.

2.3.3. Sugars and Glycosidic-Linkage Composition

The fillet sample was dialysed with a membrane cut-off of 12–14 kDa to recover the high molecular weight (HMW) compounds. Neutral sugars of the initial and dialysed samples were analysed by gas chromatography-flame ionization detection (GC-FID) after conversion to their alditol acetates [21,22]. The quantification was carried out using 2-deoxyglucose as internal standard. Monosaccharides were released from polysaccharides with pre-hydrolysis of the samples using 0.2 mL of 72% (w/w) H_2SO_4 for 1 h at room temperature followed by 2.5 h hydrolysis in 1 M H_2SO_4 at 100 °C. After hydrolysis, the reduction ($NaBH_4$) and acetylation (acetic anhydride using methyl imidazole as catalyst) of the monosaccharides were performed. The alditol acetates were analysed using a DB-225 column (30 m, 0.25 mm i.d., 0.25 μm film thickness) and a GC-FID PerkinElmer-Clarus 400 [23,24]. Free sugars were also quantified on the initial sample (not dialysed) using the same method, by omitting the hydrolysis in the abovementioned steps. The oven temperature program was as follows: 220 °C, hold for 7 min, to 240 °C at a rate of 5 °C/min. The temperature of injector was 220 °C, and the detector was 240 °C. Hydrogen was used as the carrier gas.

Glycosidic-linkage composition of the dialysed sample (HMW) was determined by GC-qMS of the partially methylated alditol acetates as previously described [22]. The polysaccharides were methylated using CH_3I, hydrolysed (TFA 2M) and the resultant monosaccharides were reduced ($NaBD_4$) and acetylated. The partially methylated alditol acetates (PMAAs) obtained were analysed by gas chromatography mass spectrometry (GC-qMS) on a Shimadzu GCMS-QP2010 Ultra [25]. The GC was equipped with an SGE HT5 (Supelco, Bellefonte, PA, USA) fused silica capillary column (30 m length, 0.25 mm i.d., and 0.10 μm of film thickness).

2.4. Bioactive Analysis of Fillet, Mucilage, Rind, and Flower

2.4.1. Preparation of Hydroethanolic Extracts

The powdered fillet, mucilage, rind, and flower samples (~1 g) underwent solid-liquid extraction twice with a 80:20 (v/v) ethanol:water mixture (30 mL) for 1 h at room temperature, and the supernatants were then filtered through Whatman no. 4 filter paper [26]. Ethanol was separated from the filtrates in a rotary evaporator (Büchi R-210, Flawil, Switzerland), and the aqueous phase was lyophilized to obtain dried extracts.

2.4.2. Analysis of Phenolic Compounds

The dried extracts (10 mg) were dissolved in a 80:20 (v/v) ethanol:water mixture (2 mL) and filtered through 0.22-μm disposable syringe filters. The analysis was performed in a HPLC-DAD-ESI/MS[n] system (Dionex Ultimate 3000 UPLC, Thermo Scientific, San Jose, CA, USA), as previously described [27]. Chromatographic separation was made in a Waters Spherisorb S3 ODS-2 C18 column (3 μm, 4.6 mm × 150 mm; Waters, Milford, MA, USA). Double online detection was carried out with a diode array detector (DAD, using 280 and 370 nm as preferred wavelengths) and a Linear Ion

Trap (LTQ XL) mass spectrometer (MS, Thermo Finnigan, San Jose, CA, USA) equipped with an electrospray ionization (ESI) source. Phenolic compounds were identified by comparison of their retention times and UV-vis and mass spectra with those obtained from standard compounds, when available; otherwise, compounds were tentatively identified comparing the obtained information with available data reported in the literature. For quantitative analysis, a calibration curve for each available phenolic compound standard (aloin A (280 nm: $y = 3859.4x + 21{,}770$, $R^2 = 0.9996$; and 370 nm: $y = 7184.4x + 17{,}013$, $R^2 = 0.9996$); chlorogenic acid ($y = 168{,}823x - 161{,}172$, $R^2 = 0.9999$); p-coumaric acid ($y = 301{,}950x + 6966.7$, $R^2 = 0.9999$); apigenin-6-glucoside ($y = 107{,}025x + 61531$, $R^2 = 0.9989$); apigenin-7-glucoside ($y = 10{,}683x - 45{,}794$, $R^2 = 0.9906$); and luteolin-6-C-glucoside ($y = 4087.1x + 72{,}589$, $R^2 = 0.9988$)) was constructed based on the UV signal. Quantification of the phenolic compounds that are not commercially available as standards was performed by using the most similar available standard molecule. The results were expressed as mg per g of extract.

2.4.3. Evaluation of Bioactive Properties

Antioxidant Activity

The extracts antioxidant capacity was evaluated by the in vitro assays of oxidative haemolysis inhibition (OxHLIA), thiobarbituric acid reactive substances formation inhibition (TBARS), and β-carotene bleaching inhibition (β-CBI), following methodologies previously described [28,29]. Extract concentrations ranging from 5 to 0.0159 mg/mL were used. Trolox was the positive control.

OxHLIA assay—An erythrocyte solution (2.8%, v/v; 200 μL) was mixed with 400 μL of either extract solution in PBS, PBS solution (control), or water (for complete haemolysis). After pre-incubation at 37 °C for 10 min with shaking, AAPH (200 μL, 160 mM in PBS, from Sigma-Aldrich) was added, and the optical density (690 nm) was measured every 10 min in a microplate reader (Bio-Tek Instruments, ELX800) until complete haemolysis [28]. The results were expressed as IC_{50} values (μg/mL) at a Δt of 60 min, i.e., extract concentration required to keep 50% of the erythrocyte population intact for 60 min.

TBARS assay—A porcine brain cell solution (1:2, w/v; 0.1 mL) was incubated with the extract solutions (0.2 mL) plus $FeSO_4$ (10 μM; 0.1 mL) and ascorbic acid (0.1 mM; 0.1 mL) at 37 °C for 1 h. Then, trichloroacetic (28% w/v, 0.5 mL) and thiobarbituric (TBA, 2%, w/v, 0.38 mL) acids were added, and the mixture was heated at 80 °C for 20 min. After centrifugation at 3000× g for 10 min, the malondialdehyde (MDA)-TBA complexes formed in the supernatant were monitored at 532 nm (Specord 200 spectrophotometer, Analytik Jena, Jena, Germany) [29]. The results were expressed as EC_{50} values (μg/mL), i.e., extract concentration providing 50% of antioxidant activity.

β-CBI assay—A β-carotene-linoleic acid emulsion (4.8 mL) was mixed with the extract solutions (0.2 mL) and the absorbance was measured at 470 nm as soon as mixed ($A_{\beta T0}$) and after 2 h of incubation at 50 °C ($A_{\beta T0}$). The β-CBI capacity was calculated as follows: ($A_{\beta T2}/A_{\beta T0}$) × 100 [29]. The results were expressed as EC_{50} values (μg/mL).

Antimicrobial Activity

The extracts were tested against the Gram-negative bacteria *Staphylococcus aureus* (ATCC 11632), *Staphylococcus epidermidis* (clinical isolate Ibis 2999), *Staphylococcus lugdunensis* (clinical isolate Ibis 2996), *Micrococcus flavus* (ATCC 10240), and *Listeria monocytogenes* (NCTC 7973), and the Gram-positive bacteria *Escherichia coli* (ATCC 25922), *Pseudomonas aeruginosa* (ATCC 27853), and *Salmonella* Typhimurium (ATCC 13311), as described by Soković et al. [30]. The fungi *Aspergillus flavus* (ATCC 9643), *Aspergillus niger* (ATCC 6275), *Penicillium funiculosum* (ATCC 36839), *Candida albicans* (clinical isolate Ibis 475/15), *Trichophyton mentagrophytes* (clinical isolate Ibis 2979/18), *Trichophyton tonsurans* (clinical isolate Ibis16/17), *Microsporum gypseum* (clinical isolate Ibis 3277/18), and *Microsporum canis* (clinical isolate Ibis 2990/18) were tested as described by Soković and van Griensven [31]. The microorganisms were obtained from the Mycological laboratory, Department of Plant Physiology, Institute for Biological Research "Sinisa Stanković," University of Belgrade, Serbia. Streptomycin and ketoconazole were the positive controls used for the antibacterial and antifungal activities, respectively. The results were given as minimum inhibitory (MIC) and minimum bactericidal (MBC) or fungicidal (MFC) concentrations (mg/mL).

Anti-Inflammatory Activity

The extract capacity to inhibit the lipopolysaccharide (LPS)-induced nitric oxide (NO) production by a murine macrophage cell line (RAW 264.7) was determined as nitrite concentration in the culture medium [32]. Dexamethasone was used as a positive control, and negative controls were performed without LPS. NO production was determined using a Griess Reagent System kit containing sulfanilamide, N-1-naphthylethylenediamine dihydrochloride (NED), and nitrite solutions. The results were expressed as EC_{50} values (µg/mL), i.e., extract concentration providing 50% of NO production inhibition.

Tyrosinase Inhibitory Activity

The extracts capacity to inhibit the tyrosinase activity was measured according to the method of Chen et al. [33], using L-3,4-dihydroxyphenylalanine (L-DOPA) as substrate. For each extract, four wells of a 96-well plate were designated as A, B, C, and D, and each one contained a reaction mixture (200 µL) as follows: (A) 66 mM PBS (pH 6.8; 120 µL) and mushroom tyrosinase in PBS (46 U/mL; 40 µL); (B) PBS (160 µL); (C) PBS (80 µL), tyrosinase (40 µL), and aqueous extract solution with 5% DMSO (40 µL); and (D) PBS (120 µL) and aqueous extract solution with 5% DMSO (40 µL). After incubation of the reaction mixtures for 10 min at room temperature, 2.5 mM L-DOPA in PBS (40 µL) was added and the plate was incubated again for 20 min. The absorbance was measured at 475 nm (SPECTROstar Nano, BMG Labtech, Ortenberg, Germany). A kojic acid solution (0.10 mg/mL) was used as positive control. The inhibition percentage of the tyrosinase activity was calculated as follows:

$$I(\%) = \frac{(A - B) - (C - D)}{(A - B)} \times 100 \tag{1}$$

IC_{50} values were then calculated from the obtained inhibition percentage values.

Cytotoxic and Hepatotoxic Activities

The extracts cytotoxicity was evaluated by the sulforhodamine B (from Sigma-Aldrich) assay against five human tumour cell lines (acquired from Leibniz-Institut DSMZ): metastatic melanoma (MM127), breast adenocarcinoma (MCF-7), non-small cell lung carcinoma (NCI-H460), cervical carcinoma (HeLa), and hepatocellular carcinoma (HepG2), as previously described [34]. The same assay was also used to evaluate the extracts hepatotoxicity against a non-tumour cell line (porcine liver primary cells, PLP2) [35]. Ellipticine was used as positive control. The results were expressed in GI_{50} values (µg/mL), i.e., extract concentration providing 50% of cell growth inhibition.

2.5. Statistical Analysis

All analyses were performed in triplicate, and the results were presented as mean ± standard deviation. All statistical tests were performed at a 5% significance level using SPSS Statistics software (IBM SPSS Statistics for Windows, Version 22.0, IBM Corp., Armonk, NY, USA:). The fulfilment of the one-way analysis of variance (ANOVA) requirements, specifically the normal distribution of the residuals and the homogeneity of variance, was tested by means of the Shapiro Wilk's and Levene's tests, respectively. Depending on the homoscedasticity, the dependent variables were compared using Tukey's honestly significant difference (HSD; when homoscedastic, $p > 0.05$) or Tamhane's T2 multiple comparison (when heteroscedastic, $p < 0.05$) tests.

Differences between two samples were analysed by applying a two-tailed paired Student's t-test; significant differences were considered when the p-value was lower than 0.05.

A Pearson's correlation was run in SPSS to evaluate whether there is a relationship between the identified compounds and the measured bioactivities.

3. Results and Discussion

3.1. Nutritional Composition of Fillet

Aloe vera fillet has been used in the food industry to develop functional foods such as beverages, milk, yogurt, jam, jellies, ice cream, and food supplements, as well as in edible fruit coatings. It can also be used to improve the quality of meat products [36] and is often commercialized as concentrated dry powder. In this study, the fillet sample corresponded to 58 ± 4% of the total leaf weight, while 31 ± 2% consisted of green rind.

The nutritional composition of *Aloe vera* fillet is presented in Table 1. This inner part of the leaf consists of 98 ± 1 g/100 g of moisture, the same amount that was found in the mucilage. Similar moisture contents (98–99 g/100 g) were previously reported [5,9,37]. Lower values were found in the rind and flower samples (87 ± 1 and 84 ± 1, respectively; Table S1).

Table 1. Nutritional value and organic acids and tocopherols composition of *Aloe vera* fillet.

Nutritional Component	Fresh Fillet	Dry Powder
Moisture (g/100 g) [1]	98 ± 1	-
Protein (g/100 g)	0.044 ± 0.001	2.60 ± 0.05
Ash (g/100 g)	0.150 ± 0.003	9.0 ± 0.2
Fat (g/100 g)	0.0168 ± 0.0006	1.00 ± 0.04
Available carbohydrates (g/100 g)	0.630 ± 0.006	37.4 ± 0.3
Dietary fibre (g/100 g)	0.84 ± 0.02	50.1 ± 0.3
Crude fibre (g/100 g)	0.120 ± 0.003	7.1 ± 0.2
Energy (kcal/100 g)	4.54 ± 0.05	269 ± 3
Oxalic acid (mg/100 g)	2.39 ± 0.04	142 ± 2
Quinic acid (mg/100 g)	11.63 ± 0.07	689 ± 4
Malic acid (mg/100 g)	97 ± 1	5750 ± 66
Total organic acids (mg/100 g)	111 ± 1	6581 ± 73
α-Tocopherol (μg/100 g)	81 ± 2	4813 ± 104
β-Tocopherol (μg/100 g)	3.59 ± 0.06	213 ± 3
γ-Tocopherol (μg/100 g)	6.7 ± 0.1	396 ± 8
δ-Tocopherol (μg/100 g)	1.78 ± 0.02	106 ± 1
Total tocopherols (μg/100 g)	93 ± 2	5527 ± 98

[1] The results are presented as mean ± standard deviation.

Dietary fibre was a predominant macronutrient in *Aloe vera* fillet, with 50.1 ± 0.3 g/100 g dw, followed by available carbohydrates, which corresponded to 37.4 ± 0.3 g/100 g dw (Table 1). A slightly higher dietary fibre content (57.64 g/100 g dw) was described by Femenia et al. [5] in fillet samples of *Aloe vera* cultivated in Ibiza, Spain. In turn, the fraction corresponding to crude fibre was isolated through acid and alkaline digestion of the sample and may consist of cellulose and small amounts of hemicellulose and lignin [38], which do not dissolve in the used solutions of sulphuric acid and potassium hydroxide. This small amount of insoluble fibre may correspond to the cell walls of the parenchyma cells that contain the gel. A higher crude fibre content (12.95 g/100 g dw) was previously reported in *Aloe vera* samples from Coquimbo, Chile [39].

The fresh fillet revealed reduced levels of protein and fat and slightly higher amounts of ash (minerals) (Table 1). This sample was, therefore, characterized by a low energy value (269 ± 3 kcal/100 g dw). These values are lower than that previously reported for fat (4.21 g/100 g dw) and ash (15.37–17.64 g/100 g dw) but slightly higher for the protein content (3.72–7.26 g/100 g dw) [5,39]. Potassium and calcium were previously found as major minerals in fillet samples [5] and may contribute to the wound healing capacity of this medicinal plant. Such compositional variations can be justified by the different geographical and edaphoclimatic conditions where the *Aloe* samples were grown.

As shown in Table 1, oxalic, quinic, and malic acids were detected in the fillet. Malic acid was the most abundant, with a concentration of 97 ± 1 mg/100 g of fresh fillet and 5.75 ± 0.07 g/100 g of dried powder. This acid is a natural component of aloe gel and an excellent freshness indicator. It was also detected along with the other two acids in the mucilage sample (Table S1). In fact, it was more abundant in the mucilage collected from the vascularized layer of the leaf, but the amount of total organic acids found in fillet and mucilage did not differ significantly. Bozzi et al. [9] also detected malic acid in fresh *Aloe vera* gel and others, like citric, lactic, and succinic acids, in commercial gel powders. However, citric acid (which can be found in the rind; Table S1) is added to the concentrated powders as a natural preservative by adjusting the gel pH prior to its concentration and drying in order to improve flavour and prevent oxidation. In turn, lactic and succinic acids should be absent from these concentrates, since they are indicators of bacterial fermentation and enzymatic degradation [9]. As presented in Table S1, ascorbic acid was not detected in the fillet and mucilage samples, but it was found in the green rind. Fumaric acid, in turn, was detected in rind and flower.

Tocopherols are important fat-soluble chain-breaking antioxidants. As shown in Table 1, the four isoforms were detected in the fillet, and α-tocopherol (4.8 ± 0.1 mg/100 g dw) was the most abundant, followed by γ- and β-tocopherols. Therefore, a 100 g portion of dried fillet provides about 69%, 44%, and 32% of the recommended dietary allowances of vitamin E for children from 4–8 and 9–13 years old and individuals with 14 or more years old, respectively (values calculated based on the α-tocopherol content) [40]. Bashipour and Ghoreishi [41] obtained a lower amount (1.53 mg/100 g dw) of α-tocopherol from *Aloe vera* samples grown in Isfahan, Iran, when applying optimized supercritical CO_2 extraction conditions. Comparable α-tocopherol levels (4.70 mg/100 g dw) were reported by López-Cervantes et al. [42] in *Aloe vera* flowers harvested in south Sonora, México.

The fatty acids composition of *Aloe vera* fillet is presented in Table 2. Eighteen fatty acids were detected, with predominance of palmitic (C16:0, 32.1 ± 0.6%), stearic (C18:0, 16.4 ± 0.2%), linoleic (C18:2n6, 15.0 ± 0.2%), and oleic (C18:1n9, 12.9 ± 0.1%) acids. Thus, 67 ± 1% of the lipid fraction is constituted by saturated fatty acids (SFA) and 32.9 ± 0.4% corresponds to mono- and polyunsaturated (MUFA and PUFA) fatty acids. Essential PUFA such as C18:2n6 and linolenic acid (C18:3n3) play important biological functions and are involved in the modulation of inflammatory and chronic degenerative diseases [43]. Despite this, its concentration in the fillet is very low compared to other compounds identified with possible health-promoting effects. Odd-chain SFA, including pentadecanoic (C15:0) and heptadecanoic (C17:0) acids, were also found (Table 2). These two fatty acids have been gaining research interest because they are important as quantitative internal standards and biomarkers for assessing dietary food intake and the risk of coronary heart disease and type II diabetes mellitus [44]. A previous study also reported C18:2n6, C18:3n3, lauric acid (C12:0) and myristic acid (C14:0) in *Aloe vera* gel [45]. Regarding other *Aloe* species, the fatty acids C18:2n6, C16:0, lignoceric (C24:0), C18:0, tricosanoic (C23:0), behenic (C22:0), and C18:3n3, among others, were described in *Aloe ferox* gel [46]; and C18:3n3, C18:2n6, C16:0, C18:0, C18:1n9, and C14:0 were found in a whole leaf extract of *Aloe arborescence* [47].

Table 2. Fatty acids composition of *Aloe vera* fillet.

Fatty Acid	Relative Percentage (%) [1]
Caproic acid (C6:0)	0.51 ± 0.01
Caprylic acid (C8:0)	0.21 ± 0.01
Capric acid (C10:0)	0.70 ± 0.01
Lauric acid (C12:0)	6.83 ± 0.09
Myristic acid (C14:0)	3.57 ± 0.06
Pentadecanoic acid (C15:0)	0.41 ± 0.02
Palmitic acid (C16:0)	32.1 ± 0.6
Heptadecanoic acid (C17:0)	0.92 ± 0.02
Stearic acid (C18:0)	16.4 ± 0.2
Oleic acid (C18:1n9c)	12.9 ± 0.1
Linoleic acid (C18:2n6c)	15.0 ± 0.2
α-Linolenic acid (C18:3n3)	4.00 ± 0.07
Arachidic acid (C20:0)	0.689 ± 0.008
Heneicosanoic acid (C21:0)	0.212 ± 0.002
Behenic acid (C22:0)	1.14 ± 0.01
Erucic acid (C22:1)	0.95 ± 0.01
Tricosanoic acid (C23:0)	0.86 ± 0.04
Lignoceric acid (C24:0)	2.54 ± 0.01
Saturated fatty acids (SFA)	67 ± 1
Monounsaturated fatty acids (MUFA)	13.8 ± 0.04
Polyunsaturated fatty acids (PUFA)	19.0 ± 0.2

[1] The results are presented as mean ± standard deviation.

3.2. Sugars and Glycosidic-linkage Composition of Fillet

Table 3 shows that the fillet sample contained 64 g/100 g dw of sugars, mainly glucose, uronic acids, and mannose. In this sample, 35 g/100 g dw of the sugars were found in the free form, determined as glucose and mannose (Table 1). The identification of the alditol acetate corresponding to mannose is probably due to the presence of fructose, as the methodology used converts fructose in glucitol (57%) and mannitol (43%) [48]. These results show that fillet sample contained 8 g/100 g dw of fructose and 27 g/100 g dw of glucose. A previous study reported fructose and glucose contents ranging from 0.56 to 9.62 g/100 g dw and 4.57 to 28.27 g/100 g dw, respectively, in fresh gel and powdered gel concentrates of *Aloe vera* [9], thus comprising the value quantified in this study.

Table 3. Carbohydrates of the initial and dialysed *Aloe vera* fillet identified as alditol acetates.

Fillet Sample	Carbohydrate (mol%)						Total Carbohydrate	
	Ara	Xyl	Man	Gal	Glc	UA	(g/100 g)	RSD (%)
Initial sample								
Total sugars	1.0	2.0	21.4	3.5	50.1	22.0	64.0	3
Free sugars			9.7		90.3		34.9	12
HMW	1.4	1.5	65.2	3.2	15.9	12.7	76.5	6

HMW: high molecular weight (sample dialysed with a membrane cut-off of 12–14 kDa). Ara: arabinose; Xyl: xylose; Man: mannose; Gal: galactose; Glc: glucose; UA: uronic acids; RSD: relative standard deviation.

Upon dialysis (12–14 kDa cut off), the weight of the obtained high molecular weight (HMW) material accounted for 39.9%. The sample obtained after dialysis (HMW) was composed by 77% of sugars, mainly mannose (65%), although glucose was also present (16%). To disclose the polysaccharides composition of the HMW sample, a methylation analysis was performed (Table 4).

Table 4. Glycosidic-linkage composition of the dialysed *Aloe vera* fillet.

Glycosyl Linkage	HMW (> 14 kDa)	
	% mol	RSD (%)
t-Ara*f*	0.4	7
t-Ara*p*	0.5	24
5-Ara*f*	0.7	2
Total	**1.7**	**10**
t-Xyl*p*	0.2	9
4-Xyl*p*	1.7	21
Total	**1.9**	**20**
t-Man	1.0	16
4-Man	74.0	0
2,4-Man	1.6	36
3,4-Man	1.1	35
4,6-Man	2.6	12
Total	**80.2**	**2**
t-Gal	0.5	28
Total	**0.5**	**28**
t-Glc	0.6	3
4-Glc	14.0	7
4,6-Glc	1.1	17
Total	**15.6**	**8**
Total Man*p*/T-Man*p*	77.4	
% Branching	3.2	

HMW: high molecular weight (sample dialysed with a membrane cut-off of 12–14 kDa). RSD: relative standard deviation.

The glycosidic-linkage composition of dialysed *Aloe vera* fillet shows a mannan composed by a backbone of (β1→4)-mannose residues [4,6], as observed by the high amount of the (1→4)-linked mannose residues (74.0%). Although (β1→4)-glucose residues may be present as insertions of the mannan backbone [49], these can have also indicate the presence of a glucan, possibly cellulose. *Aloe vera* mannan is also reported to be acetylated at the C-2 and C-3 positions and containing some side chains, mainly of galactose attached to C-6 [4,6]. The analysis performed using alkali conditions does not allow to maintain the acetyl groups. Nevertheless, the presence of (1→2,4)- and (1→3,4)-linked mannose residues are probably resultant from resistant acetylation positions of the mannan (Table 4). The branching percentage (3.2%), which can be estimated by the ratio between the (1→4,6)- mannose and the total amount of mannose, is in accordance with the presence of terminally-linked galactose and arabinose residues, identified in both pyranose (0.5%) and furanose (0.4%) forms [6]. The ratio calculated by the relative amount of total mannose divided by the amount of terminally linked mannose shows that this polysaccharide had a higher molecular weight than those previously reported [6].

3.3. Phenolic Composition of Fillet, Mucilage, Rind, and Flower

Data related to the phenolic compounds identification in the obtained *Aloe vera* extracts are presented in Table 5, namely the retention time, λ_{max} in the UV-vis region, pseudomolecular ion, ions of major fragments in MS^2, and tentative identification (the obtained extraction yields and the phenolic contents that can be found in fresh and dried samples are shown in Table S1). The chromatographic profiles recorded at 280 and 370 nm are shown in Figures 1 and 2. Up to 17 phenolic compounds were identified in the leaf extracts and eight in the flower extract, which were classified into four groups—phenolic acids, flavonoids, chromones, and anthrones. Most of these compounds have already been previously reported in *Aloe vera* [50–53], so that their identities were attributed by interpreting data acquired from HPLC-DAD-ESI/MS^n with those of literature.

Table 5. Phenolic compounds identified in *Aloe vera* leaf and flower extracts. Retention time (Rt), wavelengths of maximum absorption in the UV-vis region (λmax), pseudomolecular and MS^2 fragment ions, and relative abundance in brackets.

Peak	Rt (min)	λ_{max} (nm)	$[M-H]^-$ (m/z)	MS^2 (m/z)	Tentative Identification [1]	Reference
				Part A: *Aloe vera* leaf (fillet, mucilage and rind)		
1	6.14	227, 245, 252, 298	393	375(3), 303(12), 273(100), 245(5), 203(3)	Aloesin (aloeresin B) isomer 1	[50]
2	7.13	213, 243, 252, 299	393	375(3), 303(14), 273(100), 245(4), 203(3)	Aloesin (aloeresin B) isomer 2	[50]
3	7.73	213, 244, 252, 299	455	473(5), 411(5), 391(7), 365(100), 341(3), 333(11), 275(3), 243(3)	Unknown	-
4	9.56	254, 271, 345	609	447(100), 357(5), 327(15)	Luteolin-6,8-C-diglucoside	[51]
5	11.53	306	337	191(100), 173(5), 163(10), 119(3)	*cis* 5-*p*-Coumaroylquinic acid	[51]
6	12.63	306	337	191(100), 173(5), 163(10), 119(3)	*trans* 5-*p*-Coumaroylquinic acid	[51]
7	13.41	270, 340	593	473(43), 431(100), 311(79)	Apigenin-6,8-C-diglucoside	[52]
8	14.71	254, 271, 346	447	357(52), 327(100)	Luteolin-6-C-glucoside	[52]
9	15.98	222, 272, 303, 355	433	343(3), 313(100), 271(5), 255(3)	10-Hydroxyaloin B	[50]
10	16.36	222, 273, 305, 355	433	343(3), 313(100), 271(5), 255(3)	10-Hydroxyaloin A	[50]
11	17.60	227, 272, 301, 350	433	343(5), 313(37), 271(100), 255(3)	5-Hydroxyaloin A	[52]
12	18.11	272, 340	431	413(8), 341(27), 311(100)	Apigenin-6-C-glucoside	[52]
13	18.88	219, 269, 301, 357	505	448(3), 343(100), 172(5)	6'-Malonylnataloin	[52]
14	23.97	225, 269, 298, 355	417	297(100), 255(3)	Aloin B (isobarbaloin)	[50]
15	25.71	225, 269, 298, 355	417	297(100), 255(3)	Aloin A (barbaloin)	[50]
16	26.56	225, 269, 298, 355	503	459(100), 417(5), 297(40)	Malonyl aloin B	-
17	27.62	229, 252, 300	553	407(100), 375(5), 347(11), 275(52), 259(5), 233(7), 191(4)	2'-*p*-Methoxycoumaroylaloresin	[52]
18	28.01	225, 269, 298, 355	503	459(100), 417(5), 297(48)	Malonyl aloin A	-
19	36.41	252, 284	585	567(5), 521(18), 495(100), 463(3), 373(3), 333(3)	Unknown	-
				Part B: *Aloe vera* flower		
20	7.13	298, 324	353	191(100), 179(10), 173(5), 135(3)	5-O-Caffeoylquinic acid	[52]
21	9.53	254, 271, 345	609	447(100), 357(5), 327(15)	Luteolin-6,8-C-diglucoside	[52]
22	13.34	270, 340	593	473(43), 431(100), 311(79)	Apigenin-6,8-C-diglucoside isomer 1	[52]
23	14.71	254, 271, 346	447	357(52), 327(100)	Luteolin-6-C-glucoside	[52]
24	15.80	270, 340	593	473(43), 431(100), 311(79)	Apigenin-6,8-C-diglucoside isomer 2	[52]
25	16.50	271, 340	563	443(7), 413(100), 323(5), 311(3), 293(27)	Apigenin-2''-O-pentoxide-C-hexoside	[53]
26	17.70	-	593	473(8), 443(100), 371(3), 353(3), 341(3), 323(28), 311(9), 285(3)	Methyl-luteolin-2''-O-pentoxide-C-hexoside	[53]
27	18.14	272, 340	431	413(8), 341(27), 311(100)	Apigenin-6-C-glucoside	[52]

[1] The compounds identity was attributed by interpreting data acquired from HPLC-DAD-ESI/MS^n with those of the literature.

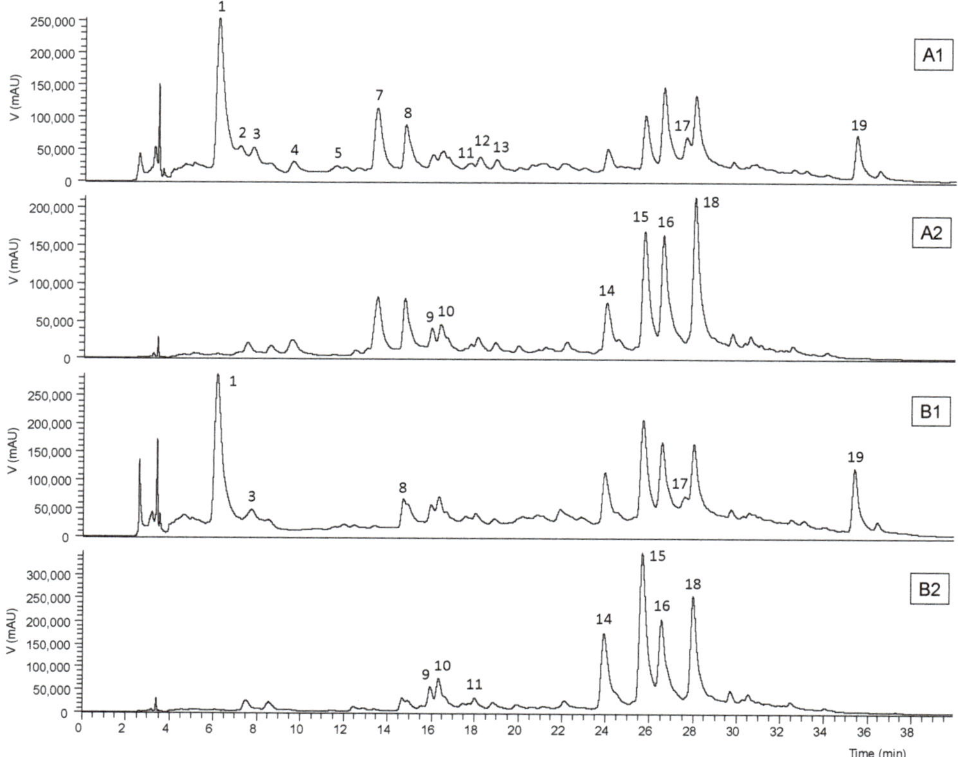

Figure 1. HPLC phenolic profiles of *Aloe vera* rind and mucilage extracts recorded at 280 nm (A1 and B1, respectively) and 370 nm (A2 and B2, respectively). See Table 5 for peak identification.

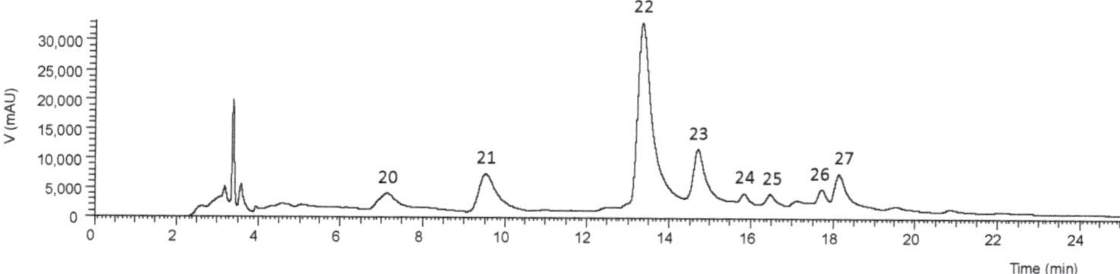

Figure 2. HPLC phenolic profile of *Aloe vera* flower extract recorded at 370 nm. See Table 5 for peak identification.

The chromones aloesin or aloeresin B (peak 1) and 2′-*p*-methoxycoumaroylaloresin (peak 17) and the anthrones 10-hydroxyaloin B (peak 9), 10-hydroxyaloin A (peak 10), aloin B or isobarbaloin (peak 14), aloin A or barbaloin (peak 15), malonyl aloin B (peak 16), and malonyl aloin A (peak 18) were detected in the three studied parts of the *Aloe vera* leaf (Table 6). The mucilage contained the highest content (131 ± 3 mg/g extract) of phenolic compounds, mostly anthrones (62.1%) and chromones (34.6%), followed by two luteolin glucosides (3.3%). This result is in accordance with the literature, which states that the vascularized layer that covers the inner fillet is rich in anthraquinone glycosides and anthrone derivatives [3]. The rind was ranked second, with 105 ± 3 mg/g extract of phenolic compounds, of which 44.9% anthrones and 43.8% chromones; it also contained luteolin and apigenin glucosides and the phenolic acid *p*-coumaroylquinic acid. However, the chromone levels found in the rind did not differ statistically from those of the mucilage (Table 6). Although the phenolic profiles of the fillet and mucilage were similar, a significantly lower concentration (11.2 ± 0.2 mg/g extract) of these secondary metabolites was found in the fillet. In addition, this leaf part had an equal ratio of anthrones and chromones (Table 6).

Table 6. Content of phenolic compounds in *Aloe vera* leaf (fillet, mucilage, and rind) and flower extracts. See Table 5 for peak identification.

Peak	Content (mg/g Extract) [1]				Statistics	
	Fillet	Mucilage	Rind	Flower	H [2]	*p*-Value [3]
1	5.4 ± 0.4 [c]	39 ± 3 [a]	34.4 ± 0.7 [b]	-	0.176	<0.001
2	-	-	4.8 ± 0.2	-	-	-
4	-	3.5 ± 0.2	tr	-	-	-
5	0.180 ± 0.008	-	0.31 ± 0.01	-	0.560	<0.001*
6	0.145 ± 0.002	-	-	-	-	-
7	-	-	6.67 ± 0.04	-	-	-
8	tr	0.79 ± 0.01 [b]	3.3 ± 0.1 [a]	-	0.095	<0.001
9	tr	3.2 ± 0.2 [a]	2.83 ± 0.06 [b]	-	0.120	<0.001
10	tr	5.1 ± 0.4 [a]	3.8 ± 0.2 [b]	-	0.149	<0.001
11	-	3.35 ± 0.06	2.4 ± 0.2	-	0.286	<0.001*
12	-	-	1.64 ± 0.05	-	-	-
13	-	-	3.23 ± 0.04	-	-	-
14	0.24 ± 0.03 [c]	13.3 ± 0.1 [a]	4.3 ± 0.3 [b]	-	0.240	<0.001
15	1.24 ± 0.06 [c]	22.2 ± 0.5 [a]	9.9 ± 0.4 [b]	-	0.292	<0.001
16	1.43 ± 0.08 [c]	16.6 ± 0.4 [a]	7.8 ± 0.4 [b]	-	0.350	<0.001
17	0.13 ± 0.04 [c]	6.08 ± 0.08 [b]	6.9 ± 0.4 [a]	-	0.139	<0.001
18	2.4 ± 0.1 [c]	17.6 ± 0.2 [a]	12.8 ± 0.8 [b]	-	0.157	<0.001
20	-	-	-	0.30 ± 0.01	-	-
21	-	-	-	tr	-	-
22	-	-	-	2.42 ± 0.04	-	-
23	-	-	-	tr	-	-
24	-	-	-	0.94 ± 0.01	-	-
25	-	-	-	0.96 ± 0.01	-	-
26	-	-	-	tr	-	-
27	-	-	-	0.152 ± 0.009	-	-
Σ Phenolic acids	0.33 ± 0.01 [a]	-	0.31 ± 0.01 [a,b]	0.30 ± 0.01 [b]	0.735	0.022
Σ Flavonoids	-	4.3 ± 0.2 [b]	11.6 ± 0.1 [a]	4.48 ± 0.05 [b]	0.298	<0.001
Σ Anthrones	5.3 ± 0.3 [c]	81 ± 1 [a]	47 ± 2 [b]	-	0.291	<0.001
Σ Chromones	5.5 ± 0.5 [b]	45 ± 3 [a]	46 ± 1 [a]	-	0.259	<0.001
Σ Phenolic compounds	11.2 ± 0.2 [c]	131 ± 3 [a]	105 ± 3 [b]	4.78 ± 0.05 [d]	0.134	<0.001

tr: traces. [1] The results are presented as mean ± standard deviation. [2] Homoscedasticity (H) was tested by the Levene's test: $p > 0.05$ indicates homoscedasticity and $p < 0.05$ indicates heteroscedasticity. [3] Statistically significant differences ($p < 0.05$) among two samples* were assessed by a Student's *t*-test and among more than two samples were assessed by a one-way ANOVA (and indicated by different letters), using Tukey's honestly significant difference (HSD) or Tamhane's T2 multiple comparison tests, when homoscedasticity was verified or not, respectively.

Variations in the phenolic profiles of *Aloe* species have been reported. According to Fan et al. [50], aloesin is more abundant in *A. barbadensis* and *A. ferox* than in *A. chinensis* and *A. arborescens*. In these species, aloin A predominated over aloin B (according to our results), and lower concentrations were also found in *A. chinensis*. In general, higher contents and more complex phenolic compounds were reported in *A. barbadensis*. Kanama et al. [54] found minimal qualitative variations in the phenolic profiles of *A. ferox* exudate samples obtained from different regions of South Africa. Despite this, aloin B content varied from 18.4 to 149.7 mg/g, aloin A ranged from 21.3 to 133.4 mg/g, and aloesin from 111.8 to 561.8 mg/g of dried exudate. This result corroborates the data presented in Table 6, since aloesin predominated over both aloins, despite lower levels have been quantified in our samples.

6′-Malonylnataloin (peak 13), a malonylated derivative of the rare anthrone nataloin, was detected in the rind extract (Table 6). This anthrone *C*-glycoside is considered of great importance in systematic discrimination of different *Aloe* species and has been reported in *A. vera*, *A. arborescens*, *A. ellenbeckii*, *A. eru*, *A. grandidentata*, *A. brevifolia*, and *A. ferox* [52,55].

The quantification of aloins (as hydroxyanthracene derivatives) is recommended in routine quality control analyses of *Aloe* samples. These compounds are highly valorised in the pharmaceutical industry, allowed in dietary supplements, and used in small quantities as a bittering agent in alcoholic beverages. However, because of their laxative properties, levels of aloin A and B in *Aloe* leaf preparations intended for oral consumption were limited by the International Aloe Science Council to 10 ppm (10 mg/kg) or less [56]. These levels can be controlled and limited by adding purification steps in the manufacturing process.

The phenolic profile of *Aloe vera* flower (Figure 2) was different from that of leaf (Figure 1), being constituted mainly by the flavonoids apigenin-6,8-C-diglucoside, apigenin-2″-O-pentoxide-C -hexoside, apigenin-6-C-glucoside, and traces of luteolin glucoside derivatives (accounting for 93.4% of the extract), and by the phenolic acid 5-O-caffeoylquinic acid (Table 5). As shown in Table 6, this extract had the lowest levels (4.78 ± 0.05 mg/g extract) of phenolic compounds. As far as we know, it is the first time that some of these compounds are described in *Aloe vera* flower. No anthraquinone glycosides were detected in this part of the plant as previously stated by Keyhanian and Stahl-Biskup [51].

3.4. Bioactive Properties of Fillet, Mucilage, Rind, and Flower

3.4.1. Antioxidant Capacity

The results of the antioxidant capacity of the *Aloe vera* extracts are presented in Table 7. For the OxHLIA assay, data were given as IC_{50} values, corresponding to the extract concentration capable of protecting 50% of the erythrocyte population from oxidative haemolysis for 60 min; whereas, for the TBARS and β-CBI assays, data were expressed as EC_{50} values, meaning the extract concentration capable of providing 50% of antioxidant activity. In both cases, the lower the EC_{50} or IC_{50} values, the higher the antioxidant capacity.

The mucilage extract (at 47 ± 2 μg/mL) was the most effective in inhibiting the formation of TBARS. This cell-based assay allowed for the measuring the extract capacity to inhibit the formation of malondialdehyde and other reactive substances, which are generated from the *ex vivo* decomposition of lipid peroxidation products resulting from the oxidation of the porcine brain cell membranes.

The rind extract provided the best protection against oxidative haemolysis (IC_{50} value, 56 ± 4 μg/mL) and β-carotene bleaching (EC_{50} value, 51 ± 4 μg/mL). In the OxHLIA assay, the erythrocytes were exposed to the haemolytic action of hydrophilic radicals that resulted from the thermal decomposition of the peroxyl radical initiator AAPH and, subsequently, to the action of lipophilic radicals generated through a lipid peroxidation phenomenon as a result of the initial attack [57]. In the β-CBI assay, β-carotene underwent discoloration in the absence of antioxidant extract, which results in a reduction in the absorbance of the test solution with increasing reaction time. The presence of antioxidants hindered the bleaching extension by neutralizing the linoleic hydroperoxyl radicals formed in the reaction emulsion [58]. These results are consistent with those of Lucini et al. [59], which concluded that the green rind is more antioxidant than the inner parenchyma.

Pearson's analysis indicated a strong correlation between the antihaemolytic and β-carotene bleaching inhibition capabilities and the flavonoids content ($r = -0.778$, $p = 0.003$, and $r = -0.865$, $p < 0.001$, respectively). In fact, the higher IC_{50} and EC_{50} values obtained with these two *in vitro* assays, respectively, were achieved with fillet extract (Table 7), in which no flavonoids were detected (Table 6). On the other hand, TBARS inhibition was strongly correlated with malic acid contents ($r = -0.946$, $p < 0.001$) and moderately with the levels of anthrones ($r = -0.667$, $p = 0.018$) and chromones ($r = -0.676$, $p = 0.016$). The results of a previous work [60] suggested that the antioxidant capacity of *Aloe vera* gel extract might be ascribed to a synergistic action of bioactive compounds. The same study also evidenced that the gel extract is able to protect the erythrocyte membrane from AAPH-induced oxidative injury and partially restore its normal protein profiles. This observation may support the fact that the best results of the antioxidant activity of the present study were achieved with the OxHLIA assay because the extracts had IC_{50} values closer to those of trolox.

Table 7. Antioxidant, anti-tyrosinase, and antimicrobial capacities of *Aloe vera* leaf (fillet, mucilage, and rind) and flower extracts and positive controls.

	Fillet		Mucilage		Rind		Flower		Positive Control	
	Antioxidant Activity [1]								Trolox	
OxHLIA (IC_{50}, µg/mL)	378 ± 18 [a]		105 ± 8 [b]		56 ± 4 [c]		80 ± 4 [b,c]		20.4 ± 0.4 [d]	
TBARS (EC_{50}, µg/mL)	87 ± 4 [b]		47 ± 2 [c]		97 ± 3 [b]		347 ± 14 [a]		5.4 ± 0.3 [d]	
β-CBI (EC_{50}, µg/mL)	78 ± 6 [a]		63 ± 4 [b]		51 ± 4 [c]		59 ± 4 [b,c]		0.20 ± 0.01 [d]	
	Anti-Tyrosinase Activity								Kojic Acid	
IC_{50} (mg/mL) or I(%) [2]	na		$30.38 \pm 0.01\%$		$27.2 \pm 0.7\%$		4.85 ± 0.07		0.078 ± 0.001	
	Antibacterial Activity								Streptomycin	
	MIC	MBC	MIC	MBC	MIC	MBC	MIC	MBC	MIC	MBC
Staphylococcus aureus	0.60	0.80	0.60	0.80	0.60	0.80	0.60	0.80	0.006	0.012
Staphylococcus epidermidis	0.60	0.80	0.40	0.80	0.40	0.80	0.60	0.80	0.003	0.006
Staphylococcus lugdunensis	0.40	0.80	0.20	0.40	0.40	0.80	0.40	0.80	0.025	0.050
Micrococcus flavus	0.80	1.60	0.60	0.80	0.80	1.20	0.40	0.80	0.20	0.30
Listeria monocytogenes	1.20	1.60	0.40	0.80	0.60	0.80	0.80	1.20	0.20	0.30
Escherichia coli	0.050	0.10	0.10	0.20	0.025	0.050	0.025	0.050	0.006	0.012
Pseudomonas aeruginosa	0.10	0.20	0.050	0.10	0.10	0.20	0.025	0.050	0.025	0.050
Salmonella Typhimurium	0.80	1.20	0.80	1.20	0.40	0.80	0.80	1.20	0.25	0.50
	Antifungal Activity								Ketoconazole	
	MIC	MFC	MIC	MFC	MIC	MFC	MIC	MFC	MIC	MFC
Aspergillus flavus	0.10	0.20	0.80	>1.60	0.10	0.20	0.10	0.20	0.25	0.50
Aspergillus niger	0.10	0.20	>1.60	>1.60	0.20	0.40	0.10	0.20	0.20	0.50
Penicillium funiculosum	0.050	0.10	>1.60	>1.60	0.050	0.10	0.10	0.20	0.20	0.50
Candida albicans	0.050	0.10	0.050	0.10	0.050	0.10	0.050	0.10	0.40	0.80
Trichophyton mentagrophytes	0.10	0.20	0.10	0.20	0.20	0.40	0.10	0.20	0.012	0.025
Trichophyton tonsurans	0.025	0.050	0.025	0.050	0.012	0.025	0.050	0.10	0.0015	0.003
Microsporum gypseum	0.025	0.050	0.20	0.40	0.050	0.10	0.012	0.025	0.006	0.012
Microsporum canis	0.025	0.050	0.050	0.10	0.050	0.10	0.025	0.050	0.003	0.006

na: no activity; MIC: minimum inhibitory concentration (mg/mL); MBC: minimum bactericidal concentration (mg/mL); MFC: minimum fungicidal concentration (mg/mL). [1] Statistics for antioxidant activity: ([a]) homoscedasticity was tested by the Levene's test: $p = 0.130$ for OxHLIA (homoscedastic); $p = 0.010$ for TBARS (heteroscedastic); and $p = 0.305$ for β-CBI (homoscedastic); and ([b]) Statistically significant differences ($p < 0.05$) were assessed by a one-way ANOVA (and indicated by different letters) using Tukey's honestly significant difference (HSD) or Tamhane's T2 multiple comparison tests, when homoscedasticity was verified or not, respectively: $p < 0.001$ in all cases. [2] Inhibition percentage of tyrosinase activity at 8 mg/mL.

3.4.2. Antimicrobial Activity

Since *Aloe vera* has been used in the development of topical cosmeceutical formulations, microorganisms of the skin flora were also tested. *Staphylococcus aureus*, *S. epidermidis*, and *S. lugdunensis* are commensal bacteria of the skin that can become opportunistic pathogens and cause a number of skin infections and also life-threatening diseases [61,62]. For these three bacteria, MIC and MBC values ≤ 0.6 and 0.8 mg/mL, respectively, were reached with the tested extracts (Table 7). The mucilage extract was the most effective against *S. lugdunensis* (with MIC and MBC values of 0.2 and 0.4 mg/mL, respectively), a Gram-positive bacterium that has been associated with a wide variety of infections, including skin and soft-tissue infections (furuncles, cellulitis, and abscesses), but also cardiovascular, central nervous infections, and urinary tract infections [63]. The mucilage extract was also effective in inhibiting and killing the *Micrococcus flavus* and also *Listeria monocytogenes* (Table 7), a food-borne pathogen capable of infecting both humans and animals. Occasionally, *L. monocytogenes* can cause localized skin infections, especially in people dealing with infected animals. The activity against this pathogenic bacterium was found strongly correlated with anthrones ($r = -0.845$ for MIC and $r = -0.816$ for MBC, $p = 0.001$) and chromones ($r = -0.795$ for MIC and $r = -0.861$ for MBC, $p \leq 0.002$) and moderately/strongly with flavonoids ($r = -0.611$, $p = 0.035$, for MIC and $r = -0.781$, $p \leq 0.003$, for MBC).

The rind extract had the best results against *Salmonella* Typhimurium (*S. enterica* serovar Typhimurium) (MIC of 0.4 mg/mL and MBC of 0.8 mg/mL) and rind and flower extracts against *Escherichia coli* (MIC of 0.025 mg/mL and MBC of 0.05 mg/mL). In the case of *Pseudomonas aeruginosa*, the flower extract was as effective as the antibiotic streptomycin in inhibiting (0.025 mg/mL) and killing (0.05 mg/mL) this multidrug resistant opportunistic pathogen. Interestingly, this carbapenem-resistant Gram-negative pathogen appears in the 2017 WHO list of threatening bacteria for which new antibiotics are urgently needed [64].

As presented in Table 7, three of the *Aloe vera* extracts (except for mucilage) had an antifungal activity (MIC and MFC values ≤ 0.1 and 0.4 mg/mL, respectively) against *Aspergillus flavus*, *A. niger*, and *Penicillium funiculosum* superior to that of the positive control ketoconazole (MIC and MFC values ≤ 0.25 and 0.5 mg/mL, respectively). All extracts were more effective than ketoconazole in inhibiting (MIC, 0.05 mg/mL) or killing (MFC, 0.1 mg/mL) the opportunistic yeast *Candida albicans*, which is the most prevalent fungal pathogen in humans, causing candidiasis. This fungal infection affects predominantly superficial skin and mucosa (oral and vaginal), but it can also lead to life-threatening systemic infections [65].

Filamentous fungi causing superficial infections in keratinized tissues, namely *Trichophyton mentagrophytes*, *T. tonsurans*, *Microsporum gypseum*, and *M. canis*, were also tested. As found for the other microorganisms, extract concentrations ≤ 0.2 and 0.4 mg/mL were sufficient to inhibit growth or kill, respectively, these dermatophytes (Table 7). The rind extract was the most effective against *T. tonsurans*, while the flower followed by the fillet extracts were the most effective against *M. gypseum* and *M. canis*.

3.4.3. Tyrosinase Inhibition Capacity

The extracts capacity to inhibit tyrosinase activity can be translated into their potential as skin whitening and anti-hyperpigmentation agents, since melanin production is impaired when this enzyme is inhibited, resulting in a less pigmented skin [66]. In this study, the flower extract was the most active (with an IC_{50} of 4.85 ± 0.07 mg/mL), but still not as effective as kojic acid (IC_{50} of 0.078 ± 0.001 mg/mL), the depigmenting agent used as a positive control (Table 7). The high percentage of flavonoids present in this extract (Table 6) may justify the observed inhibitory effect. Previous studies reported that plant extracts rich in flavonoids have a strong suppressive impact on tyrosinase, which enables its use in skin lightening cosmeceuticals [66]. For rind and mucilage extracts, it was not possible to calculate IC_{50} values, so the results were given as percentage of inhibition. The fillet did not provide any value in terms of tyrosinase inhibition. Despite this, chromones isolated from *Aloe vera*, including aloesin that was found mainly in rind and mucilage, have been reported to have tyrosinase inhibitory activity [14].

3.4.4. Anti-Inflammatory and Cytotoxic Activities

Although it is claimed that *Aloe vera* gel has important therapeutic properties, the performed bioassays indicated that none of the four extracts has anti-inflammatory activity or cytotoxicity against metastatic melanoma (MM127), breast adenocarcinoma (MCF-7), non-small cell lung carcinoma (NCI-H460), cervical carcinoma (HeLa), and hepatocellular carcinoma (HepG2) at the tested concentrations (EC_{50} and GI_{50} values >400 μg/mL). Hepatotoxicity was also not observed against the non-tumour PLP2 cell line, whereas a GI_{50} value of 8.6 ± 0.1 μg/mL was obtained for ellipticine, the anticancer agent used as positive control. These results are somewhat supported by the study of du Plessis and Hamman [67], in which the cytotoxic activity of *Aloe vera*, *Aloe marlothii*, *Aloe speciosa*, and *Aloe ferox* gels was investigated against HepG2, HeLa, and human neuroblastoma (SH-SY5Y) cells.

A decrease in cell viability was reported only at concentrations >10 mg/mL, and the half-maximal cytotoxic concentration (CC$_{50}$) values were above 1000 mg/mL, except for *Aloe vera* gel in HepG2 cells (CC$_{50}$ value = 269.3 mg/mL). Hussain et al. [68] reported that *Aloe vera* gel (extracted from dried leaves and used directly as a drug solution) displays cytotoxic effects against the MCF-7 and HeLa cell lines by inducing apoptosis and modulating the expression of effector molecules. In addition, no significant cytotoxicity toward normal lymphocytes was recorded for 24 h.

Most of the cytotoxicity studies available in the literature report the use of isolated compounds rather than crude extracts, which consist of a mixture of phytochemicals and other constituents with or without bioactive properties. El-Shemy et al. [69] tested aloesin, aloe-emodin, and barbaloin extracted from *Aloe vera* leaves against Ehrlich ascite carcinoma and found a significant inhibition as follows: barbaloin >aloe-emodin >aloesin. The authors also described that aloe-emodin was active against the human colon cancer cell lines DLD-1 and HT2. The antimetastatic potential of aloe-emodin on highly metastatic B16-F10 melanoma murine cells has also been described [70].

4. Conclusions

This study showed that 58% of the total *Aloe vera* leaf weight corresponds to inner fillet, but the green rind also accounted for a considerable percentage. The fillet consisted mainly of moisture, followed by dietary fibre and available carbohydrates (mainly glucose and fructose). Malic acid, which is an excellent freshness indicator, and α-tocopherol, a powerful fat-soluble chain-breaking antioxidant, were detected in high amounts in this jelly-like parenchyma. Based on the glycosidic-linkage composition, it was concluded that the fillet sample is composed mainly by mannans, possibly acemannan.

The three leaf samples revealed similar phenolic profiles, with predominance of chromones (aloesin and 2′-*p*-methoxycoumaroylaloresin) and anthrones (aloin A and B, malonyl aloin A and B, and 10-hydroxyaloin A and B). The highest contents of phenolic compounds were found in the mucilage and rind extracts, which also revealed interesting antioxidant properties. On the other hand, the flower extract was rich in apigenin glycoside derivatives, effective against the multidrug-resistant *P. aeruginosa*, and capable of inhibiting the activity of the enzyme tyrosinase. The fillet, rind, and flower extracts also showed a powerful antifungal activity against *A. flavus*, *A. niger*, *P. funiculosum*, and *C. albicans*, higher than that of ketoconazole. Therefore, all the studied *Aloe vera* samples present high potential to be exploited by food or cosmeceutical industries.

Supplementary Materials: The following are available online at http://www.mdpi.com/2076-3921/8/10/444/s1, Table S1. Moisture, organic acids and phenolic compounds composition of *Aloe vera* leaf (fillet, mucilage, and rind) and flower.

Author Contributions: Investigation, M.A.-O., J.P., L.B., S.P.S., E.C., M.S., and M.A.C.; Methodology, J.P., A.Ć., A.M., and R.C.C.; Supervision, J.P. and I.C.F.R.F.; Writing—original draft, M.A.-O., J.P., and L.B.; Writing—review & editing, M.S., M.A.C., and I.C.F.R.F.

Funding: The authors are grateful to the Foundation for Science and Technology (FCT, Portugal) and FEDER (European Regional Development Fund) under Programme PT2020 for financial support to CIMO (UID/AGR/00690/2019) and the research contracts of J. Pinela, R.C. Calhelha, and L. Barros (national funding by FCT, through the institutional scientific employment program-contract); to the Project AllNat—POCI-01-0145-FEDER-030463 (PTDC/EQU-EPQ/30463/2017), funded by FEDER funds through COMPETE2020—Programa Operacional Competitividade e Internacionalização (POCI)—and by national funds through FCT/MCTES; and to FEDER-Interreg España-Portugal programme for financial support through the project 0377_Iberphenol_6_E.

Acknowledgments: The authors are grateful to the company "aCourela do Alentejo" for having supplied the plant material.

Conflicts of Interest: The authors declare no conflict of interest.

References

1. Sánchez-Machado, D.I.; López-Cervantes, J.; Sendón, R.; Sanches-Silva, A. *Aloe vera*: Ancient knowledge with new frontiers. *Trends Food Sci. Technol.* **2017**, *61*, 94–102. [CrossRef]
2. Eshun, K.; He, Q. Aloe vera: A valuable ingredient for the food, pharmaceutical and cosmetic industries: A review. *Crit. Rev. Food Sci. Nutr.* **2004**, *44*, 91–96. [CrossRef] [PubMed]
3. Baruah, A.; Bordoloi, M.; Deka Baruah, H.P. *Aloe vera*: A multipurpose industrial crop. *Ind. Crops Prod.* **2016**, *94*, 951–963. [CrossRef]
4. Hamman, J.H. Composition and applications of *Aloe vera* leaf gel. *Molecules* **2008**, *13*, 1599–1616. [CrossRef]
5. Femenia, A.; Sánchez, E.S.; Simal, S.; Rosselló, C. Compositional features of polysaccharides from *Aloe vera* (*Aloe barbadensis* Miller) plant tissues. *Carbohydr. Polym.* **1999**, *39*, 109–117. [CrossRef]
6. Simões, J.; Nunes, F.M.; Domingues, P.; Coimbra, M.A.; Domingues, M.R. Mass spectrometry characterization of an *Aloe vera* mannan presenting immunostimulatory activity. *Carbohydr. Polym.* **2012**, *90*, 229–236. [CrossRef]
7. Maan, A.A.; Nazir, A.; Khan, M.K.I.; Ahmad, T.; Zia, R.; Murid, M.; Abrar, M. The therapeutic properties and applications of *Aloe vera*: A review. *J. Herb. Med.* **2018**, *12*, 1–10. [CrossRef]
8. Saccù, D.; Bogoni, P.; Procida, G. *Aloe* exudate: Characterization by reversed phase HPLC and headspace GC-MS. *J. Agric. Food Chemestry* **2001**, *49*, 4526–4530. [CrossRef]
9. Bozzi, A.; Perrin, C.; Austin, S.; Arce Vera, F. Quality and authenticity of commercial aloe vera gel powders. *Food Chem.* **2007**, *103*, 22–30. [CrossRef]
10. The International Aloe Science Council. Available online: https://www.iasc.org/Home.aspx (accessed on 22 July 2019).
11. Sehgal, I.; Winters, W.D.; Scott, M.; David, A.; Gillis, G.; Stoufflet, T.; Nair, A.; Kousoulas, K. Toxicologic assessment of a commercial decolorized whole leaf *Aloe vera* juice, lily of the desert filtered whole leaf juice with aloesorb. *J. Toxicol.* **2013**, *2013*, 802453. [CrossRef]
12. Kim, K.H.; Lee, J.G.; Gyuun, D.; Kim, M.K.; Park, J.H.; Shin, Y.G.; Lee, S.K.; Jo, T.H.; Oh, S.T. The development of a new method to detect the adulteration of commercial aloe gel powders. *Arch. Pharm. Res.* **1998**, *21*, 514–520. [CrossRef] [PubMed]
13. Cesar, V.; Jozić, I.; Begović, L.; Vuković, T.; Mlinarić, S.; Lepeduš, H.; Borović Šunjić, S.; Žarković, N. Cell-type-specific modulation of hydrogen peroxide cytotoxicity and 4-hydroxynonenal binding to human cellular proteins in vitro by antioxidant *Aloe vera* extract. *Antioxidants* **2018**, *7*, 125. [CrossRef] [PubMed]
14. Wu, X.; Yin, S.; Zhong, J.; Ding, W.; Wan, J.; Xie, Z. Mushroom tyrosinase inhibitors from *Aloe barbadensis* Miller. *Fitoterapia* **2012**, *83*, 1706–1711. [CrossRef] [PubMed]
15. *AOAC Official Methods of Analysis of AOAC International*, 20th ed.; AOAC International: Rockville, MD, USA, 2016.
16. Osborne, D.R.; Voogt, P.; Barrado, A.M. *Análisis de los Nutrientes de los Alimentos*; Editorial Acribia: Zaragoza, Spain, 1985; ISBN 8420005711.
17. Regulation (EU) No 1169/2011 of the European Parliament and of the Council of 25 Obtober 2011 on the provision of food information to consumers. *Official Journal of the European Union* **2011**, *L 304*, 16–63.
18. Iyda, J.H.; Fernandes, Â.; Ferreira, F.D.; Alves, M.J.; Pires, T.C.S.P.; Barros, L.; Amaral, J.S.; Ferreira, I.C.F.R. Chemical composition and bioactive properties of the wild edible plant *Raphanus raphanistrum* L. *Food Res. Int.* **2019**, *121*, 714–722. [CrossRef] [PubMed]
19. Pinela, J.; Barreira, J.C.M.; Barros, L.; Cabo Verde, S.; Antonio, A.L.; Carvalho, A.M.; Oliveira, M.B.P.P.; Ferreira, I.C.F.R. Suitability of gamma irradiation for preserving fresh-cut watercress quality during cold storage. *Food Chem.* **2016**, *206*, 50–58. [CrossRef] [PubMed]
20. Pereira, C.; Barros, L.; Carvalho, A.M.; Ferreira, I.C.F.R. Use of UFLC-PDA for the analysis of organic acids in thirty-five species of food and medicinal plants. *Food Anal. Methods* **2013**, *6*, 1337–1344. [CrossRef]
21. Nunes, F.M.; Coimbra, M.A. Chemical characterization of the high molecular weight material extracted with hot water from green and roasted Arabica coffee. *J. Agric. Food Chem.* **2001**, *49*, 1773–1782. [CrossRef]

22. Coimbra, M.A.; Delgadillo, I.; Waldron, K.W.; Selvendran, R.R. Isolation and Analysis of Cell Wall Polymers from Olive Pulp. In *Molecular Methods of Plant Analysis*; Linskens, H.F., Jackson, J.F., Eds.; Springer: Berlin/Heidelberg, Germany, 1996; pp. 19–44.

23. Blakeney, A.B.; Harris, P.J.; Henry, R.J.; Stone, B.A. A simple and rapid preparation of alditol acetates for monosaccharide analysis. *Carbohydr. Res.* **1983**, *113*, 291–299. [CrossRef]

24. Selvendran, R.R.; March, J.F.; Ring, S.G. Determination of aldoses and uronic acid content of vegetable fiber. *Anal. Biochem.* **1979**, *96*, 282–292. [CrossRef]

25. Nobre, C.; Sousa, S.C.; Silva, S.P.; Pinheiro, A.C.; Coelho, E.; Vicente, A.A.; Gomes, A.M.P.; Coimbra, M.A.; Teixeira, J.A.; Rodrigues, L.R. *In vitro* digestibility and fermentability of fructo-oligosaccharides produced by *Aspergillus ibericus*. *J. Funct. Foods* **2018**, *46*, 278–287. [CrossRef]

26. Pinela, J.; Barros, L.; Dueñas, M.; Carvalho, A.M.; Santos-Buelga, C.; Ferreira, I.C.F.R. Antioxidant activity, ascorbic acid, phenolic compounds and sugars of wild and commercial *Tuberaria lignosa* samples: Effects of drying and oral preparation methods. *Food Chem.* **2012**, *135*, 1028–1035. [CrossRef] [PubMed]

27. Bessada, S.M.F.; Barreira, J.C.M.; Barros, L.; Ferreira, I.C.F.R.; Oliveira, M.B.P.P. Phenolic profile and antioxidant activity of *Coleostephus myconis* (L.) Rchb.f.: An underexploited and highly disseminated species. *Ind. Crops Prod.* **2016**, *89*, 45–51. [CrossRef]

28. Lockowandt, L.; Pinela, J.; Roriz, C.L.; Pereira, C.; Abreu, R.M.V.; Calhelha, R.C.; Alves, M.J.; Barros, L.; Bredol, M.; Ferreira, I.C.F.R. Chemical features and bioactivities of cornflower (*Centaurea cyanus* L.) capitula: The blue flowers and the unexplored non-edible part. *Ind. Crops Prod.* **2019**, *128*, 496–503. [CrossRef]

29. Pinela, J.; Barros, L.; Carvalho, A.M.; Ferreira, I.C.F.R. Nutritional composition and antioxidant activity of four tomato (*Lycopersicon esculentum* L.) farmer' varieties in Northeastern Portugal homegardens. *Food Chem. Toxicol.* **2012**, *50*, 829–834. [CrossRef]

30. Soković, M.; Glamočlija, J.; Marin, P.D.; Brkić, D.; Griensven, L.J.L.D. van Antibacterial effects of the essential oils of commonly consumed medicinal herbs using an *in vitro* model. *Molecules* **2010**, *15*, 7532–7546. [CrossRef]

31. Soković, M.; van Griensven, L.J.L.D. Antimicrobial activity of essential oils and their components against the three major pathogens of the cultivated button mushroom, *Agaricus bisporus*. *Eur. J. Plant Pathol.* **2006**, *116*, 211–224. [CrossRef]

32. Sobral, F.; Sampaio, A.; Falcão, S.; Queiroz, M.J.R.P.; Calhelha, R.C.; Vilas-Boas, M.; Ferreira, I.C.F.R. Chemical characterization, antioxidant, anti-inflammatory and cytotoxic properties of bee venom collected in Northeast Portugal. *Food Chem. Toxicol.* **2016**, *94*, 172–177. [CrossRef]

33. Chen, C.-H.; Chan, H.-C.; Chu, Y.-T.; Ho, H.-Y.; Chen, P.-Y.; Lee, T.-H.; Lee, C.-K. Antioxidant activity of some plant extracts towards xanthine oxidase, lipoxygenase and tyrosinase. *Molecules* **2009**, *14*, 2947–2958. [CrossRef]

34. Vaz, J.A.; Heleno, S.A.; Martins, A.; Almeida, G.M.; Vasconcelos, M.H.; Ferreira, I.C.F.R. Wild mushrooms *Clitocybe alexandri* and *Lepista inversa*: *In vitro* antioxidant activity and growth inhibition of human tumour cell lines. *Food Chem. Toxicol.* **2010**, *48*, 2881–2884. [CrossRef]

35. Pereira, C.; Calhelha, R.C.; Barros, L.; Ferreira, I.C.F.R. Antioxidant properties, anti-hepatocellular carcinoma activity and hepatotoxicity of artichoke, milk thistle and borututu. *Ind. Crops Prod.* **2013**, *49*, 61–65. [CrossRef]

36. Shahrezaee, M.; Soleimanian-Zad, S.; Soltanizadeh, N.; Akbari-Alavijeh, S. Use of *Aloe vera* gel powder to enhance the shelf life of chicken nugget during refrigeration storage. *LWT* **2018**, *95*, 380–386. [CrossRef]

37. Rodríguez-González, V.M.; Femenia, A.; González-Laredo, R.F.; Rocha-Guzmán, N.E.; Gallegos-Infante, J.A.; Candelas-Cadillo, M.G.; Ramírez-Baca, P.; Simal, S.; Rosselló, C. Effects of pasteurization on bioactive polysaccharide acemannan and cell wall polymers from *Aloe barbadensis* Miller. *Carbohydr. Polym.* **2011**, *86*, 1675–1683. [CrossRef]

38. Rosa Santos, J. *Determinação do Teor de Fibra Alimentar em Produtos Hortofrutícolas*; Universidade de Lisboa: Lisbon, Portugal, 2013.

39. Miranda, M.; Maureira, H.; Rodríguez, K.; Vega-Gálvez, A. Influence of temperature on the drying kinetics, physicochemical properties, and antioxidant capacity of *Aloe vera* (*Aloe Barbadensis* Miller) gel. *J. Food Eng.* **2009**, *91*, 297–304. [CrossRef]

40. Otten, J.J.; Hellwig, J.P.; Meyers, L.D. *Dietary Reference Intakes: The Essential Guide to Nutrient Requirements*; Otten, J.J., Hellwig, J.P., Meyers, L.D., Eds.; The National Academies Press: Washington, DC, USA, 2006; ISBN 030965646X.

41. Bashipour, F.; Ghoreishi, S.M. Response surface optimization of supercritical CO_2 extraction of α-tocopherol from gel and skin of *Aloe vera* and almond leaves. *J. Supercrit. Fluids* **2014**, *95*, 348–354. [CrossRef]

42. López-Cervantes, J.; Sánchez-Machado, D.I.; Cruz-Flores, P.; Mariscal-Domínguez, M.F.; Servín de la Mora-López, G.; Campas-Baypoli, O.N. Antioxidant capacity, proximate composition, and lipid constituents of *Aloe vera* flowers. *J. Appl. Res. Med. Aromat. Plants* **2018**, *10*, 93–98. [CrossRef]

43. Yates, C.M.; Calder, P.C.; Ed Rainger, G. Pharmacology and therapeutics of omega-3 polyunsaturated fatty acids in chronic inflammatory disease. *Pharmacol. Ther.* **2014**, *141*, 272–282. [CrossRef]

44. Jenkins, B.; West, J.; Koulman, A. A review of odd-chain fatty acid metabolism and the role of pentadecanoic acid (C15:0) and heptadecanoic acid (C17:0) in health and disease. *Molecules* **2015**, *20*, 2425–2444. [CrossRef]

45. Nejatzadeh-Barandozi, F. Antibacterial activities and antioxidant capacity of *Aloe vera*. *Org. Med. Chem. Lett.* **2013**, *3*, 5. [CrossRef]

46. Loots, D.T.; van der Westhuizen, F.H.; Botes, L. *Aloe ferox* leaf gel phytochemical content, antioxidant capacity, and possible health benefits. *J. Agric. Food Chem.* **2007**, *55*, 6891–6896. [CrossRef]

47. Hirata, T.; Suga, T. Biologically active constituents of leaves and roots of Aloe arborescens var. natalensis. *Z. Naturforschung C* **1977**, *32*, 731–734. [CrossRef]

48. Silva, S.P.; Moreira, A.S.P.; Maria do Rosário, M.D.; Evtuguin, D.V.; Coelho, E.; Coimbra, M.A. Contribution of non-enzymatic transglycosylation reactions to the honey oligosaccharides origin and diversity. *Pure Appl. Chem.* **2019**, *91*, 1231–1242. [CrossRef]

49. Talmadge, J.; Chavez, J.; Jacobs, L.; Munger, C.; Chinnah, T.; Chow, J.T.; Williamson, D.; Yates, K. Fractionation of *Aloe vera* L. inner gel, purification and molecular profiling of activity. *Int. Immunopharmacol.* **2004**, *4*, 1757–1773. [CrossRef] [PubMed]

50. Fan, J.J.; Li, C.H.; Hu, Y.J.; Chen, H.; Yang, F.Q. Comparative assessment of *in vitro* thrombolytic and fibrinolysis activity of four aloe species and analysis of their phenolic compounds by LC-MS. *S. Afr. J. Bot.* **2018**, *119*, 325–334. [CrossRef]

51. Keyhanian, S.; Stahl-Biskup, E. Phenolic constituents in dried flowers of *Aloe vera* (*Aloe barbadensis*) and their *in vitro* antioxidative capacity. *Planta Med.* **2007**, *73*, 599–602. [CrossRef]

52. El Sayed, A.M.; Ezzat, S.M.; El Naggar, M.M.; El Hawary, S.S.; El Sayed, A.M.; Ezzat, S.M.; El Naggar, M.M.; El Hawary, S.S. *In vivo* diabetic wound healing effect and HPLC-DAD-ESI-MS/MS profiling of the methanol extracts of eight *Aloe species*. *Rev. Bras. Farmacogn.* **2016**, *26*, 352–362. [CrossRef]

53. Ferreres, F.; Gil-Izquierdo, A.; Andrade, P.B.; Valentão, P.; Tomás-Barberán, F.A. Characterization of C-glycosyl flavones O-glycosylated by liquid chromatography-tandem mass spectrometry. *J. Chromatogr. A* **2007**, *1161*, 214–223. [CrossRef]

54. Kanama, S.K.; Viljoen, A.M.; Kamatou, G.P.P.; Chen, W.; Sandasi, M.; Adhami, H.-R.; Van Wyk, B.-E. Simultaneous quantification of anthrones and chromones in *Aloe ferox* ("Cape aloes") using UHPLC–MS. *Phytochem. Lett.* **2015**, *13*, 85–90. [CrossRef]

55. Grace, O.M.; Kokubun, T.; Veitch, N.C.; Simmonds, M.S.J. Characterisation of a nataloin derivative from *Aloe ellenbeckii*, a maculate species from east Africa. *S. Afr. J. Bot.* **2008**, *74*, 761–763. [CrossRef]

56. Younes, M.; Aggett, P.; Aguilar, F.; Crebelli, R.; Filipič, M.; Frutos, M.J.; Galtier, P.; Gott, D.; Gundert-Remy, U.; Kuhnle, G.G.; et al. Safety of hydroxyanthracene derivatives for use in food. *EFSA J.* **2018**, *16*, e05090.

57. Prieto, M.A.; Vázquez, J.A. A time-dose model to quantify the antioxidant responses of the oxidative hemolysis inhibition assay (OxHLIA) and its extension to evaluate other hemolytic effectors. *BioMed Res. Int.* **2014**, *2014*, 15. [CrossRef] [PubMed]

58. Kulisic, T.; Radonic, A.; Katalinic, V.; Milos, M. Use of different methods for testing antioxidative activity of oregano essential oil. *Food Chem.* **2004**, *85*, 633–640. [CrossRef]

59. Lucini, L.; Pellizzoni, M.; Pellegrino, R.; Molinari, G.P.; Colla, G. Phytochemical constituents and *in vitro* radical scavenging activity of different *Aloe species*. *Food Chem.* **2015**, *170*, 501–507. [CrossRef] [PubMed]

60. Mazzulla, S.; Sesti, S.; Schella, A.; Perrotta, I.; Anile, A.; Drogo, S. Protective effect of *Aloe vera* (*Aloe barbadensis* Miller) on erythrocytes anion transporter and oxidative change. *Food Nutr. Sci.* **2012**, *3*, 1697–1702.

61. Argemi, X.; Hansmann, Y.; Prola, K.; Prévost, G.; Argemi, X.; Hansmann, Y.; Prola, K.; Prévost, G. Coagulase-negative Staphylococci pathogenomics. *Int. J. Mol. Sci.* **2019**, *20*, 1215. [CrossRef]

62. Chan, M.M.-Y. Antimicrobial effect of resveratrol on dermatophytes and bacterial pathogens of the skin. *Biochem. Pharmacol.* **2002**, *63*, 99–104. [CrossRef]

63. Babu, E.; Oropello, J. *Staphylococcus lugdunensis*: The coagulase-negative staphylococcus you don't want to ignore. *Expert Rev. Anti-Infect. Ther.* **2011**, *9*, 901–907. [CrossRef]

64. WHO. WHO Publishes List of Bacteria for Which New Antibiotics Are Urgently Needed. Available online: https://www.who.int/news-room/detail/27-02-2017-who-publishes-list-of-bacteria-for-which-new-antibiotics-are-urgently-needed (accessed on 24 June 2019).

65. Espino, M.; Solari, M.; de los Ángeles Fernández, M.; Boiteux, J.; Gómez, M.R.; Silva, M.F. NADES-mediated folk plant extracts as novel antifungal agents against *Candida albicans*. *J. Pharm. Biomed. Anal.* **2019**, *167*, 15–20. [CrossRef]

66. Ephrem, E.; Elaissari, H.; Greige-Gerges, H. Improvement of skin whitening agents efficiency through encapsulation: Current state of knowledge. *Int. J. Pharm.* **2017**, *526*, 50–68. [CrossRef]

67. du Plessis, L.H.; Hamman, J.H. *In vitro* evaluation of the cytotoxic and apoptogenic properties of aloe whole leaf and gel materials. *Drug Chem. Toxicol.* **2014**, *37*, 169–177. [CrossRef]

68. Hussain, A.; Sharma, C.; Khan, S.; Shah, K.; Haque, S. *Aloe vera* inhibits proliferation of human breast and cervical cancer cells and acts synergistically with cisplatin. *Asian Pac. J. Cancer Prev.* **2015**, *16*, 2939–2946. [CrossRef] [PubMed]

69. El-Shemy, H.; Aboul-Soud, M.; Nassr-Allah, A.; Aboul-Enein, K.; Kabash, A.; Yagi, A. Antitumor properties and modulation of antioxidant enzymes' activity by *Aloe vera* leaf active principles isolated via supercritical carbon dioxide extraction. *Curr. Med. Chem.* **2010**, *17*, 129–138. [CrossRef] [PubMed]

70. Tabolacci, C.; Lentini, A.; Mattioli, P.; Provenzano, B.; Oliverio, S.; Carlomosti, F.; Beninati, S. Antitumor properties of aloe-emodin and induction of transglutaminase 2 activity in B16–F10 melanoma cells. *Life Sci.* **2010**, *87*, 316–324. [CrossRef] [PubMed]

Article

Investigation of In Vitro Antioxidant and Antibacterial Potential of Silver Nanoparticles Obtained by Biosynthesis Using Beech Bark Extract

Corneliu Tanase [1,*,†], Lavinia Berta [2,†], Năstaca Alina Coman [1,*], Ioana Roșca [1], Adrian Man [3], Felicia Toma [3], Andrei Mocan [4,5], László Jakab-Farkas [6], Domokos Biró [6] and Anca Mare [3]

1   Department of Pharmaceutical Botany, "George Emil Palade" University of Medicine, Pharmacy, Sciences and Technology of Târgu Mureș, 38 Gheorghe Marinescu Street, Târgu Mureș, 540139 Mureș, Romania; rosca_ioana01@yahoo.com

2   Department of General and Inorganic Chemistry, "George Emil Palade" University of Medicine, Pharmacy, Sciences and Technology of Târgu Mureș, 38 Gheorghe Marinescu Street, Târgu Mureș, 540139 Mureș, Romania; lavinia.berta@umfst.ro

3   Department of Microbiology, "George Emil Palade" University of Medicine, Pharmacy, Sciences and Technology of Târgu Mureș, 38 Gheorghe Marinescu Street, Târgu Mureș, 540139 Mureș, Romania; anca.mare@umfst.ro (A.M.); felicia.toma@umfst.ro (F.T.); adrian.man@umfst.ro (A.M.)

4   Department of Pharmaceutical Botany, "Iuliu Hațieganu" University of Medicine and Pharmacy, 23 Gheorghe Marinescu Street, 400337 Cluj-Napoca, Romania; mocan.andrei@umfcluj.ro

5   Laboratory of Chromatography, Institute of Advanced Horticulture Research of Transylvania, University of Agricultural Sciences and Veterinary Medicine, 400372 Cluj-Napoca, Romania

6   Faculty of Technical and Human Sciences, Sapientia Hungarian University of Transylvania, 540485 Târgu Mureș, Romania; jflaci@ms.sapientia.ro (L.J.-F.); dbiro@ms.sapientia.ro (D.B.)

*   Correspondence: corneliu.tanase@umfst.ro (C.T.); alinacoman2194@yahoo.com (N.A.C.); Tel.: +40-7442-15543 (C.T.)

†   These authors share the first authorship.

Received: 10 September 2019; Accepted: 2 October 2019; Published: 8 October 2019

Abstract: Green synthesis is one of the rapid and best ways for silver nanoparticles (AgNP) synthesis. In the present study, synthesis and bioactivity of AgNPs has been demonstrated using water beech (*Fagus sylvatica* L.) bark extract. The physical and chemical factors such as time, metal ion solution, and pH, which play a vital role in the AgNPs synthesis, were assessed. The AgNPs were characterized by ultraviolet-visible (UV-Vis) spectrometry, Fourier transform infrared spectroscopy (FT-IR), and transmission electron microscopy (TEM). Antioxidant and antimicrobial activity of the obtained AgNPs was evaluated. AgNPs were characterized by color change pattern, and the broad peak obtained at 420–475 nm with UV-Vis confirmed the synthesis of AgNPs. FT-IR results confirmed that phenols and proteins of beech bark extract are mainly responsible for capping and stabilization of synthesized AgNPs. TEM micrographs showed spherical or rarely polygonal and triangular particles with an average size of 32 nm at pH = 9, and 62 nm at pH = 4. Furthermore, synthesized AgNPs were found to exhibit antioxidant activity and have antibacterial effect against *Staphylococcus aureus*, methicillin-resistant *Staphylococcus aureus* (MRSA), *Escherichia coli*, and *Pseudomonas aeruginosa*. These results indicate that bark extract of *F. sylvatica* L. is suitable for synthesizing stable AgNPs, which act as an excellent antimicrobial agent.

Keywords: antioxidant; antibacterial; beech bark; silver nanoparticles; polyphenols

---

1. Introduction

Lately, nanotechnology focuses on synthesis and characterization of metallic and non-metallic nanoparticles having different compositions, sizes, and shapes. Molecular aggregates with dimensions between 1 and 100 nm are called nanoparticles [1]. Due to their optical, magnetic, chemical, and mechanical properties, nanoparticles are used in many areas of advanced technology such as the electronic and optoelectronic industry; in the medical sector for diagnostics, antimicrobial properties, and transport of the drug to a specified location in order to fulfill its purpose; and for environmental protection and energy conversion [2]. The biosynthesis of nanomaterials becomes a large scientific area, because nanoparticles have a low obtaining cost and a wide range of uses: catalytic, biological and biomedical applications, physics, environmental remediation fields [3]. From all the types of nanoparticles, silver ones are the most commonly used. However, silver is known to be a good antimicrobial being used as an antiseptic agent [4].

The use of plant extracts in nanoparticle biosynthesis is advantageous because most plant extracts have a high content of bioactive compounds such as: flavonoids, terpenoids, tannins, and alkaloids responsible for reducing nanoparticles [5,6]. Silver nanoparticles (AgNPs) obtained by biosynthesis are characterized by different shapes and sizes, depending on the nature and concentration of the extract, pH, temperature, reaction time, and silver solution [7]. For example, flavonoids can chelate and actively reduce metal ions in nanoparticles due to the numerous hydroxyl and carbonyl groups [8]. In a study using leaf extracts from *Magnolia kobus* and *Diospyros kaki*, it has been observed that the presence of terpenoids and reducing sugars has led to the synthesis of nanoparticles, whose size and shape may vary by changing the reaction conditions [9]. Fruit extracts of *Solanum virginianum* have a high content of steroidal alkaloids: solanocarpine, carpesterol, and solanocarpidin [10]. Among the high quantity flavonoids are apigenin and quercetin which are considered as anti-capping agents and stabilizers of the formed nanoparticles [10]. In order to elucidate the mechanism of nanoparticle biosynthesis, Mittal et al. used extracts of *Syzygium cumini* fruit, observing that the presence of flavonic compounds led to the reduction of the metal ions through a redox reaction. Furthermore, the biomolecules present in the extract were responsible for AgNP aggregation and stabilization [11].

The shape and size of the biosynthesized nanoparticles depend on physical and chemical factors [12]. On the other hand, the extract used is also dependent on certain factors: the type of plant, the extraction solvent, and the extraction temperature [13,14]. In a study of turmeric extracts, Sathishkumar et al. confirmed that pH plays an important role in controlling the size and shape of the obtained nanoparticles. An alkaline plant extract has several functional groups that will facilitate the reduction process and, thus, a larger number of nanoparticles will be formed [15]. However, the optimum pH varies depending on the nature of the used ion solution.

In an experimental study using aqueous extracts of *Myrtus communis* as a reducer in nanoparticle biosynthesis, it was shown that the obtained nanoparticles have the ability to neutralize free radicals [15]. Furthermore, the ability to reduce Fe^{3+} compared to vitamin C was tested and it was observed that the reducing power increases with the increase of phenolic content [16]. However, using night jasmine extracts (*Cestrum nocturnum*), the obtained AgNPs have an antioxidant capacity almost 5% higher than that of vitamin C [17].

The increasingly common antibiotics resistance has led to the research of alternative treatment methods. Thus, in order to demonstrate the antibacterial activity, different bacteria were inoculated on culture media and it was observed that AgNPs can inhibit their growth. Inhibition was observed in comparison with the control and the strongest inhibitory activity was demonstrated on *Citrobacter* spp., *Salmonella typhi*, *Vibrio cholerae* [17]. In a recent study [18], AgNPs were synthesized from *Lantana camara*. The results showed significant antimicrobial activity against *Staphylococcus aureus*, *Pseudomonas aeruginosa*, and *Escherichia coli*. These results are comparable with the used standard (ciprofloxacin), but also with petroleum ether extract and essential oil from *L. camara* leaves, showing that nanoparticles have dose-dependent membrane permeation with respect to rate [18].

Taking into account the literature data, the main objectives of the current study were: (1) The biosynthesis of AgNPs using beech bark polyphenolic extract as a new bioresource; (2) the determination of the influence of the main factors on the biosynthesis process (type of reducing agent, time, and pH);

(3) the characterization of the biosynthesized AgNPs by specific analyses (UV–Vis, FT-IR, and TEM);
(4) the evaluation of the antioxidant and antibacterial activity of the biosynthesized AgNPs.

2. Materials and Methods

2.1. Materials

The beech bark used for experimental research comes as waste from a wood processing company from Vatra Dornei, Romania. Before the extraction process, the bark was dried under normal temperature and aeration conditions (room temperature), and ground using a grinding mill type VEB NOSSENER MASCHINENBAU 1980. After grinding, the plant product was sieved through sieve 5.

Silver nitrate ($AgNO_3$, M—169.87 g/mol, d—4.35 g/cm^3), silver acetate ($AgC_2H_3O_2$, M—166.91 g/mol, d—3.26 g/cm^3), Folin–Ciocalteu reagent, potassium bromide (KBr), and methanol were purchased from Merck Company (Darmstadt, Germany), 2,2'-azino-bis (3-ethylbenzothiazoline-6-sulphonic acid)-(ABTS) was from Sigma-Aldrich (Steinheim, Germany), 2,2-diphenyl-1-picrylhydrazyl (DPPH) and 6-hydroxy-2,5,7,8-tetramethylchroman-2-carboxylic acid (Trolox) were obtained from Alfa-Aesar (Karlsruhe, Germany).

The following bacterial reference strains, stored at −70 °C in the Microbiology Laboratory from UMFST Târgu Mureș, were used: *S. aureus* ATCC (American Type Culture Collection) 25923, methicillin-resistant *S. aureus* (MRSA) ATCC 43300, *E. coli* ATCC 25922, *Klebsiella pneumoniae* ATCC 700603, *P. aeruginosa* ATCC 27853.

2.2. Extraction Method

The ultrasound-assisted extraction was made according to the method described previously [19] with some modifications. Briefly, 10 g of beech bark was placed in an Erlenmeyer flask. Then, 100 mL of distilled water was added. The Erlenmeyer flask was then inserted into the ultrasonic water bath (Professional Ultrasonic Cleaner MRC: AC 150 H, 150 W, 40 KHz, heating power 300 W) preheated to 70 °C, for an extraction time of 30 min. At the end of the extraction, the solution obtained was filtered through the Buchner cone, transferring the filtrate to a 100 mL volumetric flask and filled up to the mark with distilled water. In order to obtain a clear solution and to avoid the interference of any bark residues in the biosynthesis process, the extract was centrifuged in centrifuge tubes, for 5 min at 5000 rpm.

2.3. Characterization of Extract

The determination of the polyphenol content was performed spectrophotometrically, using the Folin–Ciocalteu method [20]. The total polyphenol content (TPC), expressed in mg gallic acid (GAE)/g plant material, was calculated based on the absorbance read with the spectrophotometer at 765 nm, taking into account the calibration curve (concentration range between 0 and 200 mg/L) of the standard gallic acid solution.

2.4. Synthesis of Silver Nanoparticles

For the biosynthesis, 100 mL of silver nitrate/acetate solution (1 mM) was prepared in advance. For adjusting the pH, a 50 mL HNO_3/ NaOH solution (1 mM) was prepared. Biosynthesis consists of mixing 10 mL of beech bark extract with 90 mL of silver nitrate/acetate in an Erlenmeyer glass. From the obtained solution, the pH was measured, and the values were in the range of 5.56–5.63, depending on the experimental version. Using the pH meter with an agitator at 800 rpm the solutions were brought to pH = 4 using about 4 mL HNO_3, and to pH = 9 using about 5 mL NaOH, depending on the initial pH of the solution. Biosynthesis took place in the ultrasonic bath at 60 °C, for 3 h until the color transformation indicated finalization of nanoparticles biosynthesis. At the time of introducing the samples into the ultrasonic water bath, samples were taken for each experimental version in order to record the spectrum of the initial solution. Tested solutions (TS) were: TS1—beech bark extract, pH = 4, silver nitrate; TS2—beech bark extract, pH = 9, silver nitrate; TS3—beech bark extract, pH = 4, silver acetate; TS4—beech bark extract, pH = 9, silver acetate.

2.5. Characterization of Silver Nanoparticles

2.5.1. UV-Vis Analysis

The samples of synthesized nanoparticles were measured at different time intervals (T0 = 0, T1 = 1 h, T2 = 1 h 30′, T3 = 2 h, T4 = 3 h, T5 = 24 h) based on previous results (in publication process). The results were recorded with the Analytik Jena Specord 210 Plus190 (UV/Vis)-1100 nm Spectrophotometer using synthetic quartz spectrophotometer cells (190–2500 nm, path length 10 mm, volume 1.75 mL). The measurements were performed in the following parameters: 1 nm resolution, the wavelength range between 350 and 700 nm.

2.5.2. FT-IR Analysis

For the FT-IR analysis, the AgNP dry matter was used. FT-IR analysis involves mixing the obtained powder with KBr in the agate mortar in a ratio of 1:100. Using the FT-IR Thermo Nicolet 380 Spectrophotometer (Thermo Scientific, Waltham, USA), the measuring range 4000–400 cm^{-1}, spectra were recorded for each TS.

2.5.3. TEM Analysis

The nanoparticles in the solution were characterized morphologically and dimensionally using conventional electron transmission microscopy, using a JEOL 100 U TEM microscope (Japan Electron Optics Laboratory, Tokyo, Japan) at 100 kV. The nitrocellulose substrates coated with amorphous carbon layers, prepared on a 300 mesh copper microgrid, served as the substrate used for nanoparticle investigation. The thickness of the thermally evaporated carbon layer was about 4 nm. The SERS (Surface-Enhanced Raman Scattering) spectra were recorded in solution with a portable spectrometer (Raman Systems R3000 CN, Edinburgh, United Kingdom) equipped with a 785 nm diode coupled to a 100 km optical fiber. The laser power was 200 mV and the integration time was 30 s.

The obtained images were processed with the ImageJ software to determine the AgNP size and the histograms were made for each experimental version with Excel.

2.6. In Vitro Antioxidant Activities

2.6.1. Free Radical-Scavenging Activity Using 2,2-Diphenyl-1-picrylhydrazyl (DPPH)

The antioxidant activity was monitored according to the DPPH method described by Martins et al. [21] with modifications. This method was performed by using a SPECTRO star Nano microplate reader (BMG Labtech, Offenburg, Germany). The reaction mixture in each of the 96-wells consisted of 30 µL of sample solution (in an appropriated dilution, according to the reference range of the calibration curve) and a 0.004% methanolic solution of DPPH. The mixture was further incubated for 30 min in the dark, and the reduction of the DPPH radical was determined by measuring the absorption of the sample at 515 nm. Trolox was used as a standard reference and the results were expressed as Trolox equivalents (TE) per g of dry weight (mg TE/g dw TS).

2.6.2. Trolox Equivalents Antioxidant Capacity (TEAC) Assay

The radical scavenging activity of the tested solutions against the stable synthetic 2,2′-azino-bis 3-ethylbenzothiazoline-6-sulphonic acid (ABTS) radical cation was measured using the method previously described by Mocan et al. [22]. Briefly, 20 µL of sample is mixed with 200 µL of radical solution and incubated for 6 min. The absorbance of the final solution is measured at 760 nm after the incubation period. A Trolox calibration curve was plotted as a function of the percentage of ABTS radical scavenging activity. The final results were expressed as milligrams of TE per gram of dry weight tested solutions (mg TE/g dw TS).

2.7. Antibacterial Activity

2.7.1. Minimal Inhibitory Concentration (MIC) of Silver Nanoparticles

In order to determine the MICs of the obtained AgNPs against tested bacterial strains, we used the microdilution method, as previously described [20]. To obtain binary dilutions from the TS, 200 μL TS was added in the first column of the microplate. One-hundred microliters of sterile water were added in all the other wells of the microplate. One-hundred microliters from the first column were transferred with a multichannel pipette into the second column of the microplate. These steps were repeated until the last column, from which, the last 100 μL was discarded. Ten microliters of 0.5 McFarland bacterial suspension were mixed with 9990 μL of Mueller-Hinton broth medium 2X and 100 μL from this bacterial inoculum were dispensed in each well of the microplate. Additionally, wells with culture medium only, culture medium with TS (negative controls), and culture medium with bacterial inoculum (growth control) were prepared. The microplates were incubated at 37 °C for 24 h. MIC was considered in the last well in which no bacterial growth was noted. For TS with a high degree of turbidity, resazurin was used as an indicator of bacterial growth [23]. After the plates were incubated for 24 h at 37 °C, 3 μL 0.015% resazurin was added to each well, and the plates were further incubated for 2–4 h. A color change of resazurin, from purple to pink, indicated a bacterial growth. The last well in which the resazurin color did not change was considered MIC.

2.7.2. Minimal Bactericidal Concentration (MBC) of Silver Nanoparticles

From each well in which no bacterial growth was observed in the MIC method, 1 μL of suspension was spot-inoculated on blood agar with a calibrated bacteriological inoculation loop. MBC was considered the first TS concentration where no bacterial growth was observed on blood agar.

2.7.3. AgNP Effect on Bacterial Growth Rate (GR)

The following bacterial strains were used: *S. aureus* ATCC 25923 (SA); *E. coli* ATCC 25922; *P. aeruginosa* ATCC 27853. To determine the growth rate, the solution with the highest bactericidal activity was chosen (TS4). For the determination of GR, the steps described in a previous work were performed [20]. The content of the microplate well in which MIC was noted was reproduced in a 2 mL Eppendorf tube and it was incubated at 37 °C. At the initial moment and after 3 and 6 h of incubation, 50 μL of bacterial suspension were removed, serial diluted, and from these dilutions, 50 μL were evenly seeded on the surface of Mueller-Hinton agar with a bacteriological inoculation loop. The plates were incubated for 18–24 h and the colonies were automatically counted using "Flash & Go Automatic Colony Counter" instrument (IUL Instruments S.A., Bacelona, Spain) from the plate with the most countable colonies. As control for the bacterial growth, the same protocol was used, in the absence of TS. Mathematical adjustments were performed to compensate the dilution from where the colonies were counted and the inoculation volume.

2.8. Statistical Analysis

All analytical determinations were performed in triplicate, and the results are expressed as the mean ± standard deviation. The statistical significance was assessed by GraphPad InStat 3 software (GraphPad Software, San Diego, Canada), at a significance threshold value of $p < 0.05$.

3. Results

3.1. Characterization of Aqueous extracts

The aqueous extract was characterized in terms of total phenolic content (TPC), which is considered responsible for silver ion reduction and AgNP synthesis. Thus, the total phenolic content was found to be 20.59 mg GAE/g plant material.

3.2. Characterization of AgNPs

The first sign of nanoparticle formation is the color transformation of the solution. The influence of the extract and of the used pH has an important role and it is observed that the solutions change their color becoming brownish-red at pH = 9 and brown at pH = 4 (Figure 1). The appearance of the opaque aspect is achieved as the biosynthesis time passes.

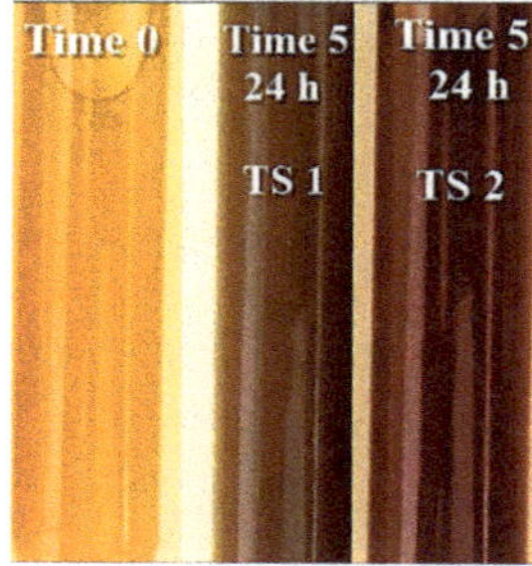

Figure 1. Color transformation of beech bark extract solution with $AgNO_3$ at pH = 4 (test solution 1 (TS1)) and pH = 9 (TS2).

3.2.1. UV-Visible Spectral Analysis of AgNPs

The optical and electronic properties of the obtained silver nanoparticles are observed by UV-Vis spectrophotometric analysis. Many factors that influence the absorption spectrum shift have been analyzed, observing that the wavelength at which the maximum absorbance is recorded is dependent on the time, pH, and silver ion solution. The biosynthesis time, visually determined, is 3 h for each experimental version, but the time of appearance of the nanoparticles is different and can be observed by UV-Vis spectra analysis (Figure 2). Use of beech bark extract (TS1) after 3 h of maintenance in the ultrasonic bath leads to the formation of AgNPs. The maximum absorbance was recorded at the wavelength of 475 nm (Figure 2a). The recording of the UV-Vis spectrum for TS2 shows a maximum absorbance corresponding to the wavelength 425 nm from time 0 (Figure 2b). For TS3, it can be observed that the time of appearance of AgNPs is 3 h and the maximum absorbance corresponds to the wavelength of 450 nm (Figure 2c). At basic pH (TS4), under the same experimental conditions it is observed (Figure 2d) that the appearance of the maximum absorbance corresponds to the wavelength 420 nm.

UV-Vis spectra recordings show that the wavelengths recorded for the TS are mainly pH dependent. Thus, at pH = 4, maximum absorbances are recorded at wavelengths in the range of 450–475 nm. When using the basic pH, the UV-Vis spectra shows a shift of the maximum absorbances at a wavelength range between 411 and 431 nm. These results are supported by the literature data in which the use of acidic pH records lower maximum absorbances [15]. The value of the acidic pH leads to the agglomeration of the nanoparticles. Thus, on the UV-Vis spectrum, there is a lower absorbance peak compared to the use of a basic pH which favors obtaining a larger number of nanoparticles and smaller dimensions. Thus, at basic pH, the reaction rate increases followed by crystallization into smaller particles, which involves the nucleation process and the growth of small nanoparticles [15,24]. In a study using extracts from oak bark, rich in polyphenolic compounds and tannins, it was shown that the synthesis rate of AgNPs increases with increasing pH. The pH with value 9 has the highest absorbance; above this value the synthesis rate decreases [25]. Analyzing the experimental versions comparatively, depending on the used silver solution, no significant differences were found.

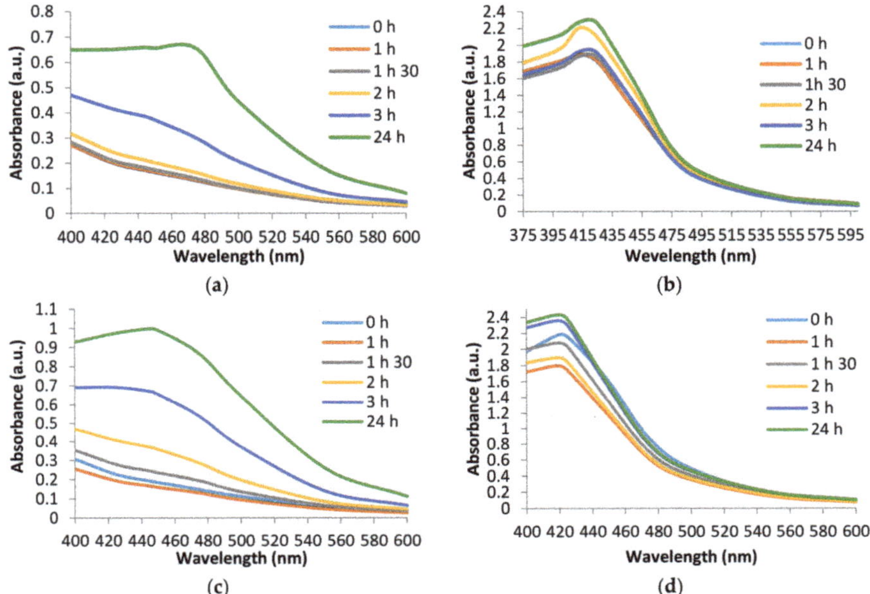

Figure 2. UV–visible absorption spectra of synthesized silver nanoparticles using tested solutions: (**a**) TS1—beech bark extract, pH = 4, $AgNO_3$; (**b**) TS2—beech bark extract, pH = 9, $AgNO_3$; (**c**) TS3—beech bark extract, pH = 4, $AgC_2H_3O_2$; (**d**) TS4—beech bark extract, pH = 9, $AgC_2H_3O_2$.

3.2.2. FT-IR Analysis of Biosynthesized AgNP

FT-IR analyses were conducted to identify the biomolecules that might be responsible for reducing the Ag+ ions in the beech bark extract, and also those that may be involved in the stabilization of silver nanoparticle synthesis. Figure 3 shows the spectrum of aqueous beech bark extract, where the bands corresponding to the –O–H bonds at 3414 and 1608 cm^{-1} specific to the carbonyl group can be observed.

The recording of the FT-IR spectra in the case of AgNPs obtained by reducing $AgNO_3$ (TS1–TS2) with the polyphenolic compounds in the aqueous beech extract shows the appearance of a band at the value of 1512 cm^{-1} and intensifies the band corresponding to the wavenumber 1384 cm^{-1} (Figure 4). These show that certain compounds in the aqueous extract modify their structure, thus, generating AgNPs. For TS3–TS4 solutions, there were no different results compared to TS1–TS2.

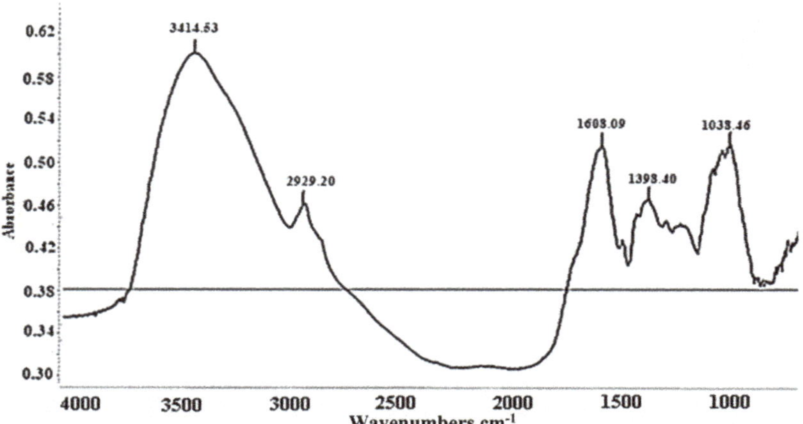

Figure 3. Fourier transform infrared spectra of aqueous bark extract.

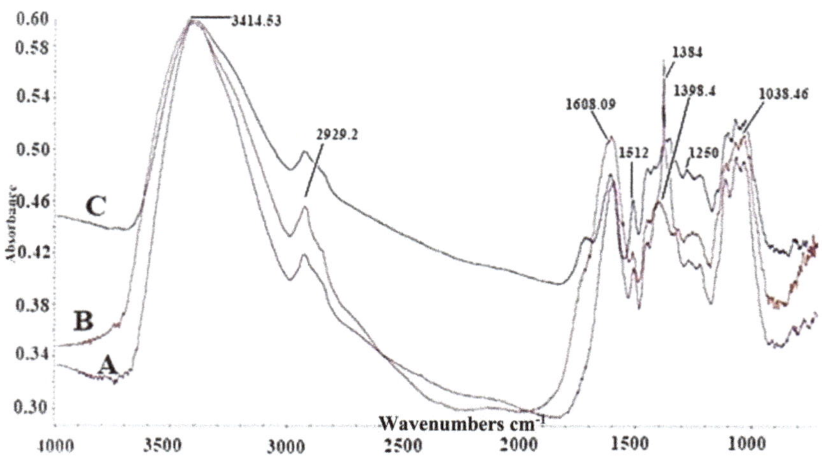

Figure 4. Fourier transform infrared spectra of gold nanoparticles (AgNPs): A—TS2 (AgNPs obtained with AgNO$_3$ at pH = 9), B—aqueous bark extract, C—TS1 (AgNPs obtained with AgNO$_3$ at pH = 4).

3.2.3. The Analysis by Transmission Electron Microscopy (TEM) of Biosynthesized AgNP

Transmission electron microscopy was used in order to characterize the obtained nanoparticles in terms of their morphology and size. Thus, the TEM photomicrographs obtained were analyzed using ImageJ program, obtaining information about the surface and diameter of the biosynthesized nanoparticles.

In Figure 5, it is observed that the AgNPs synthesized are nanometric and uniformly distributed. Regarding the morphology of the nanoparticles, it can be observed that it varies, encountering polygonal, spherical, and even triangular shapes at TS1 (Figure 5a). In contrast, at basic pH (TS2) only spherical shapes are observed (Figure 5c). The particles synthesized in TS1 have sizes between 10 and 420 nm (Figure 5b), with an average of 118.75 nm, and about 44% of these have a diameter below 100 nm. Regarding the size of the nanoparticles obtained at pH 9 (TS2) they fall in the range of 2–80 nm, with an average of 44.02 nm and 100% of the measured nanoparticles have dimensions smaller than 100 nm (Figure 5d).

Analyzing the synthesized AgNPs in the presence of the extract obtained from the beech bark and AgNO$_3$ (Figure 6), the results are similar to those presented previously. The morphology of the obtained AgNPs at pH = 4 (TS3) is varied, in this case observing polygonal, triangular, and spherical shapes (Figure 6a). In contrast, at basic pH (TS4) it can be observed that the AgNP morphology is constant, with only the spherical shape being met (Figure 6c). AgNPs obtained in the basic medium have dimensions in the range of 2–80 nm, with an average of 32.43 and 100% being smaller than 100 nm in diameter (Figure 6d).

The results obtained with the help of TEM confirm the results presented above, regarding the basic pH which favors the formation of a larger number of smaller nanoparticles. Thus, comparing the tested solution, it is found that at pH = 9 all AgNPs have dimensions below 100 nm, and at pH = 4, about half of AgNPs have dimensions below 100 nm. These findings, related to shape and size, are also confirmed by other researchers who analyzed synthesized nanoparticles in the presence of extracts obtained from the bark of woody plants [26,27]. The particles synthesized in TS3 have sizes between 10 and 200 nm (Figure 6b), with an average of 62.2 nm.

Thus, from literature data, AgNPs with a larger surface area, provide better contact with microorganisms [28]. These particles are able to penetrate into the cell membrane or attach to the bacterial surface based on their size. In addition, they have been reported to be highly toxic to bacterial strains and their antibacterial efficacy is increased by decreasing the particle size [29].

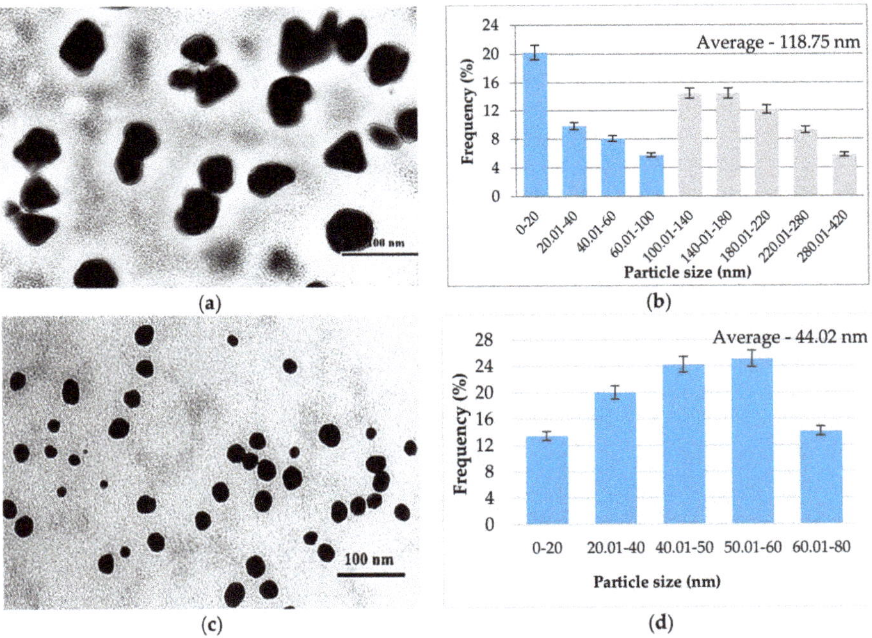

Figure 5. Graphical representation of AgNPs synthesized in the presence of the extract obtained from the beech bark and AgC$_2$H$_3$O$_2$: (**a**) TS3— TEM photomicrograph; (**b**) histogram of the distribution of AgNP size distribution in TS3; (**c**) TS4—TEM image; (**d**) histogram of the distribution of AgNP size distribution in TS4. (TS3—beech bark extract, pH = 4, AgC$_2$H$_3$O$_2$; TS4—beech bark extract, pH = 9, AgC$_2$H$_3$O$_2$).

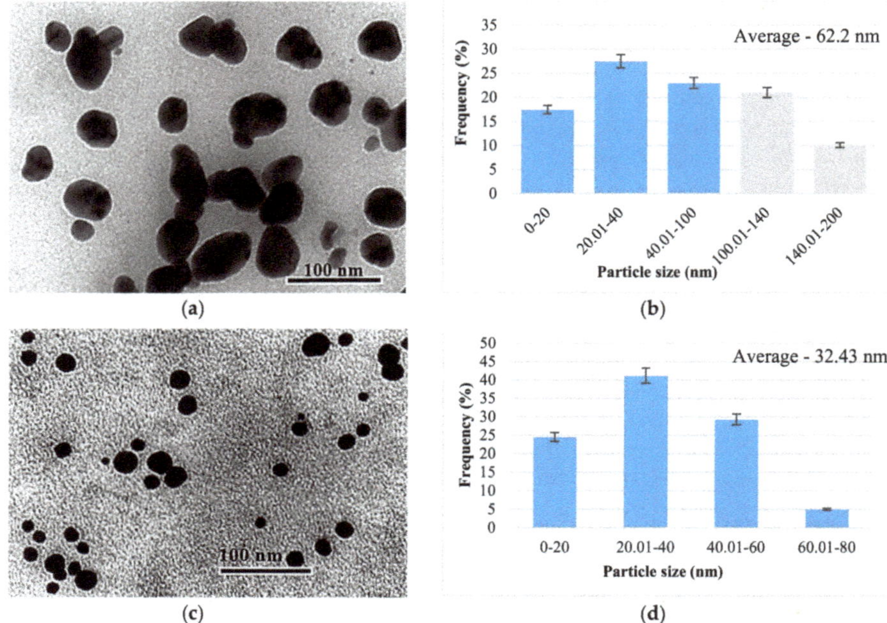

Figure 6. Graphical representation of AgNP synthesized in the presence of the extract obtained from the beech bark and $AgNO_3$: (**a**) TS1— TEM photomicrograph; (**b**) histogram of the distribution of the AgNP size distribution in the TS1; (**c**) TS2—TEM image; (**d**) histogram of the distribution of AgNP size distribution in TS2 (TS1—beech bark extract, pH = 4, $AgNO_3$; TS2—beech bark extract, pH = 9, $AgNO_3$).

4. Antioxidant Activity of AgNPs

The antioxidant activity of synthesized AgNPs was evaluated by DPPH and TEAC radical scavenging assays. The beech bark AgNPs exhibited free-radical-scavenging activity using both assays as shown in Table 1. The activity against DPPH and ABTS confirmed the antioxidant potential of the beech bark AgNPs. Significantly higher TEAC values were noted for TS1–TS2 as compared to TS3–TS4. Many AgNPs solutions have been evaluated for their antioxidant capacities, commonly associated with the content of phenolic compounds from plant extracts [17,30,31]. Beech bark contains a range of biologically active components, such as tannins and polyphenols, indicating that the antioxidant activity of the AgNPs has been enhanced by these bio-components.

Table 1. Antioxidant activity of AgNPs.

Sample	Beech Bark AgNP Characteristics	DPPH mg TE/g of Sample	TEAC mg TE/g of Sample
TS1	pH = 4, $AgNO_3$	11.68 ± 0.34	35.05 ± 0.69
TS2	pH = 9, $AgNO_3$	10.64 ± 0.35	37.23 ± 0.35
TS3	pH = 4, $AgC_2H_3O_2$	10.52 ± 0.21	28.61 ± 0.21
TS4	pH = 9, $AgC_2H_3O_2$	8.52 ± 0.15	20.39 ± 0.30

TE—Trolox equivalents; TS1—beech bark extract, pH = 4, $AgNO_3$; TS2—beech bark extract, pH = 9, $AgNO_3$; TS3—beech bark extract, pH = 4, $AgC_2H_3O_2$; TS4—beech bark extract, pH = 9, $AgC_2H_3O_2$. ± Standard deviation.

5. Antimicrobial Activity of AgNPs

To evaluate the antibacterial activity of the AgNPs against gram-negative bacteria (*E. coli* ATCC 25922, *K. pneumoniae* ATCC 700603, *P. aeruginosa* ATCC 27853) and gram-positive ones (*S. aureus* ATCC 25923 and MRSA ATCC 43300), MIC and MBC values were obtained. Afterwards, we studied how the TS influenced the bacterial growth rate (at the initial moment and after 3 and 6 h of incubation).

For therapeutic consideration, only antimicrobial activity obtained below 1 mM concentration was considered. The MIC and MBC of nanoparticles against pathogenic bacterial strains are as mentioned in Table 2. Thus, it is found that almost all tested solutions exhibit inhibitory activity on the growth of tested bacteria (except TS1, where the MIC was determined only for MRSA, at the maximum analyzed concentration). It can be seen that AgNPs obtained at pH = 9 (TS2, TS4) favor the inhibitory activity for the tested bacterial strains. They have a lower MIC compared to those obtained at pH = 4 (TS1, TS3). TS2 (for *E. coli*, *K. pneumoniae*, and *P. aeruginosa*) and TS4 (for *S. aureus*, *E. coli*, *K. pneumoniae*, and *P. aeruginosa*) have lower MIC and MBC compared to those obtained for the control beech bark extract. For the most part, the MBC was registered for the TS where the MIC was determined, the results being presented in Table 2. The lowest MBC was recorded for TS4 (MRSA, *E. coli*, and *P. aeruginosa*) and for TS2 (*E. coli*).

As can be seen from Figure 7, TS4 had a bactericidal effect on *E. coli* and inhibited the growth of *S. aureus* and *P. aeruginosa* after 6 h of incubation.

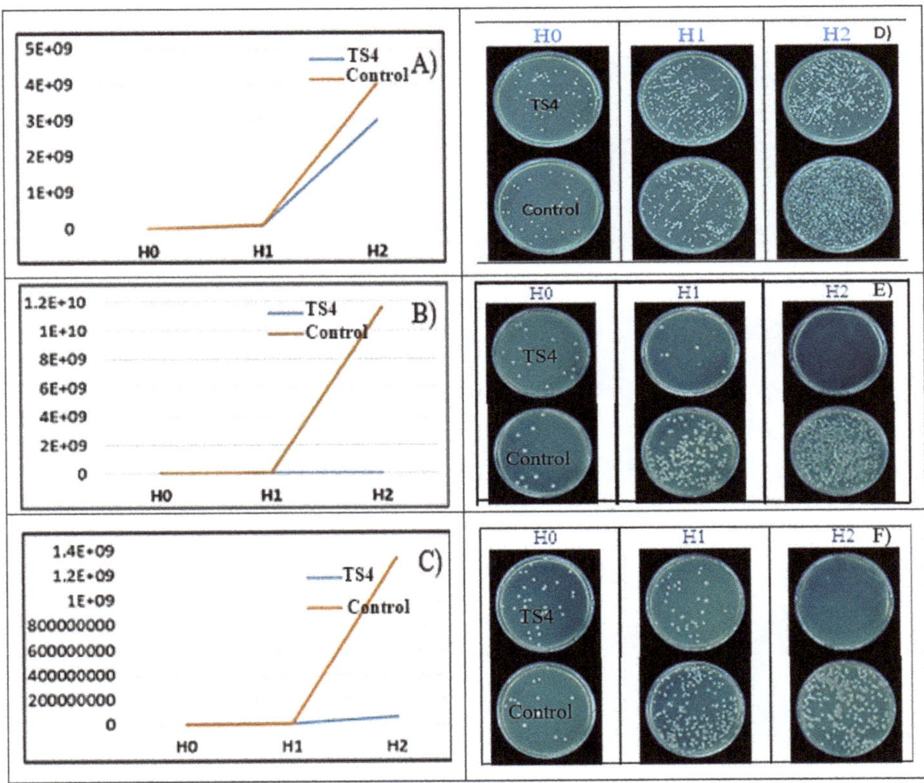

Figure 7. Graphical and visual representation of the growth rate for *S. aureus* (**A,D**), *E. coli* (**B,E**), *P. aeruginosa* (**C,F**) in the presence of TS4 (beech extract + $AgC_2H_3O_2$ pH = 9) and in the absence of TS4 (Control) at: initial time—H0, 3 h—H1, and 6 h—H2.

Table 2. Antimicrobial activity of AgNPs against pathogenic bacteria.

Pathogenic Bacteria	ATCC No.	AgNP Tested Solution	MIC	MBC
			mg/mL	mg/mL
Staphylococcus aureus	25923	TS1	>1.41	>1.41
		TS2	0.27	3.28
		TS3	1.21	>1.45
		TS4	0.09	2.37
		BBE	>2.5	>2.5
		$AgNO_3$	0.02	>0.15
		$AgC_2H_3O_2$	0.03	>0.15
MRSA	43300	TS1	1.41	1.41
		TS2	0.34	0.82
		TS3	0.42	1.45
		TS4	0.24	0.79
		BBE	>2.5	>2.5
		$AgNO_3$	0.02	0.05
		$AgC_2H_3O_2$	0.02	0.05
Escherichia coli	25922	TS1	>1.41	>1.41
		TS2	0.54	0.68
		TS3	1.45	1.45
		TS4	0.19	0,19
		BBE	>2.5	>2.5
		$AgNO_3$	0.02	0.02
		$AgC_2H_3O_2$	0.02	0.02
Klebsiella pneumoniae	700603	TS1	>1.41	>1.41
		TS2	2.74	3.28
		TS3	1.45	1.45
		TS4	0.99	1.38
		BBE	>2.5	>2.5
		$AgNO_3$	>0.15	>0.15
		$AgC_2H_3O_2$	>0.15	>0.15
Pseudomonas aeruginosa	27853	TS1	>1.41	>1.41
		TS2	0.41	0.82
		TS3	1.45	>1.45
		TS4	0.15	0.25
		BBE	>2.5	>2.5
		$AgNO_3$	0.02	0.02
		$AgC_2H_3O_2$	0.02	0.02

ATCC—American Type Culture Collection; MIC—minimal inhibitory concentration; MBC—minimum bactericidal concentration; MRSA—methicillin-resistant *S. aureus*, BBE—beech bark extract; $AgNO_3$—silver nitrate; $AgC_2H_3O_2$—silver acetate; TS1—beech bark extract, pH = 4, $AgNO_3$; TS2—beech bark extract, pH = 9, $AgNO_3$; TS3—beech bark extract, pH = 4, $AgC_2H_3O_2$; TS4—beech bark extract, pH = 9, $AgC_2H_3O_2$.

The antibacterial mechanism might beexplained by the interaction between silver nanoparticles and bacterial DNA. AgNPs can cause DNA damage, and ultimately bacterial cell death. This effect has been demonstrated on *S. aureus* and *E. coli* strains [32].

However, the antibacterial mechanism is different, depending on the type of bacteria to be inhibited. For example, silver nanoparticles have an inhibitory effect on *Helicobacter pylori*, stating two theories of action: small amounts of AgNPs enter the bacterium and inhibit the respiratory chain, while another theory considers that AgNPs inhibit urease from *H. pylori* [33]. Evidence of bacterial growth inhibition have also been demonstrated for strains of *S. aureus*, *Streptococcus pneumoniae*, *Proteus mirabilis*, and *E. coli* [34].

6. Conclusions

Following the performed analyses, it was found that the biosynthesis of silver nanoparticles is time and pH dependent. UV-Vis analysis shows that basic pH accelerates biosynthesis independently of the ion salt used. This is also supported by TEM analysis in which the appearance in large quantities and smaller size of AgNPs was demonstrated in all the versions where the basic pH was used.

FTIR analysis demonstrated that certain functional groups of the compounds present in the aqueous extracts contributed to the formation of AgNPs, a fact supported by the intense bands around 1384 cm^{-1}.

AgNPs tested solutions showed antioxidant activity as well as inhibitory and bactericidal activity. The lowest MBC was obtained for TS4 (0.19 mg/mL) against *E. coli* ATCC 25922. The small nanoparticles at pH = 9 have a lower MIC/MBC compared to those synthesized at pH = 4. This result could be explained by the fact that alkaline pH favors the penetration of AgNPs into the bacterial cell, having an inhibitory or bactericidal effect.

By determining the growth rate, it was observed that for *E. coli*, the tested AgNPs had a bactericidal effect; while for *S. aureus* and *P. aeruginosa* the growth was inhibited.

These results indicate that bark extract of *Fagus sylvatica* L. is suitable for synthesizing stable AgNPs which act as an antimicrobial agent.

Author Contributions: Conceptualization, C.T., L.B., and A.M. (Anca Mare); methodology, C.T., L.B., A.M. (Anca Mare), and A.M. (Andrei Mocan); software, N.A.C. and I.R.; validation, C.T., L.B., and A.M. (Anca Mare); formal analysis, N.A.C., I.R., A.M. (Anca Mare), and L.J.-F.; investigation, N.A.C., I.R., A.M. (Anca Mare), and A.M. (Adrian Man); resources, C.T., L.B., A.M. (Anca Mare), A.M. (Andrei Mocan), and D.B.; data curation, N.A.C., I.R., A.M. (Andrei Mocan), L.J.-F., D.B., F.T., and A.M. (Anca Mare); writing—original draft preparation, C.T. and L.B.; writing—review and editing, A.M. (Anca Mare); visualization, C.T. and A.M. (Adrian Man); project administration, C.T.; funding acquisition, C.T.

Funding: This work was supported by the University of Medicine and Pharmacy of Târgu Mureș Research grant number 15609/12/29.12.2017.

References

1. Chaturvedi, V.; Verma, P. Fabrication of silver nanoparticles from leaf extract of *Butea monosperma* (flame of forest) and their inhibitory effect on bloom-forming cyanobacteria. *Bioresour. Bioprocess.* **2015**, *2*, 18. [CrossRef]

2. Kouvaris, P.; Delimitis, A.; Zaspalis, V.; Papadopoulos, D.; Tsipas, S.A.; Michailidis, N. Green synthesis and characterization of silver nanoparticles produced using arbutus unedo leaf extract. *Mater. Lett.* **2012**, *76*, 18–20. [CrossRef]

3. Ahmad, N.; Sharma, S. Green synthesis of silver nanoparticles using extracts of *Ananas comosus*. *Green Sustain. Chem.* **2012**, *2*, 141–147. [CrossRef]

4. Prabhu, S.; Poulose, E.K. Silver nanoparticles: Mechanism of antimicrobial action, synthesis, medical applications, and toxicity effects. *Int. Nano Lett.* **2012**, *2*, 32. [CrossRef]

5. Parlinska-Wojtan, M.; Kus-Liskiewicz, M.; Depciuch, J.; Sadik, O. Green synthesis and antibacterial effects of aqueous colloidal solutions of silver nanoparticles using camomile terpenoids as a combined reducing and capping agent. *Bioprocess Biosyst. Eng.* **2016**, *39*, 1213–1223. [CrossRef] [PubMed]

6. Sarfraz, M.; Griffin, S.; Gabour Sad, T.; Alhasan, R.; Nasim, J.M.; Irfan Masood, M.; Schäfer, H.K.; Ejike, E.C.C.; Keck, M.C.; Jacob, C.; et al. Milling the Mistletoe: Nanotechnological conversion of african mistletoe (*Loranthus micranthus*) intoantimicrobial materials. *Antioxidants* **2018**, *7*, 60. [CrossRef] [PubMed]

7. Keat, C.L.; Aziz, A.; Eid, A.M.; Elmarzugi, N.A. Biosynthesis of nanoparticles and silver nanoparticles. *Bioresour. Bioprocess.* **2015**, *2*, 47. [CrossRef]

8. Makarov, V.V.; Love, A.J.; Sinitsyna, O.V.; Makarova, S.S.; Yaminsky, I.V.; Taliansky, M.E.; Kalinina, N.O. "Green" nanotechnologies: Synthesis of metal nanoparticles using plants. *Acta Nat.* **2014**, *6*, 35–44. [CrossRef]

9. Song, J.Y.; Jang, H.-K.; Kim, B.S. Biological synthesis of gold nanoparticles using magnolia kobus and diopyros kaki leaf extracts. *Process Biochem.* **2009**, *44*, 1133–1138. [CrossRef]

10. Amin, M.; Anwar, F.; Janjua, M.R.S.A.; Iqbal, M.A.; Rashid, U. Green Synthesis of Silver Nanoparticles through Reduction with Solanum Xanthocarpum, L. Berry extract: Characterization, antimicrobial and urease inhibitory activities against *Helicobacter pylori*. *Int. J. Mol. Sci.* **2012**, *13*, 9923–9941. [CrossRef]

11. Mittal, A.K.; Bhaumik, J.; Kumar, S.; Banerjee, U.C. Biosynthesis of silver nanoparticles: Elucidation of prospective mechanism and therapeutic potential. *J. Colloid Interface Sci.* **2014**, *415*, 39–47. [CrossRef] [PubMed]

12. Punjabi, K.; Choudhary, P.; Samant, L.; Mukherjee, S.; Vaidya, S.; Chowdhary, A. Biosynthesis of nanoparticles: A review. *Int. J. Pharm. Sci. Rev. Res.* **2015**, *24*, 219–226.

13. Mohamad, N.A.N.; Arham, N.A.; Jai, J.; Hadi, A. Plant extract as reducing agent in synthesis of metallic nanoparticles: A review. *Adv. Mater. Res.* **2014**, *832*, 350–355. [CrossRef]

14. Mohd Zainol, M.K.; Abdul-Hamid, A.; Abu Bakar, F.; Pak Dek, S. Effect of different drying methods on the degradation of selected flavonoids in *Centella asiatica*. *Int. Food Res. J.* **2009**, *16*, 531–537.

15. Sathishkumar, M.; Sneha, K.; Yun, Y.-S. Immobilization of silver nanoparticles synthesized using *Curcuma longa* tuber powder and extract on cotton cloth for bactericidal activity. *Bioresour. Technol.* **2010**, *101*, 7958–7965. [CrossRef]

16. Shahat, A.S.; Assar, N.H. Biochemical and antimicrobial studies of biosynthesised silver nanoparticles using aqueous extract of *Myrtus communis* L. *Ann. Biol. Res.* **2015**, *6*, 90–103.

17. Keshari, A.K.; Srivastava, R.; Singh, P.; Yadav, V.B.; Nath, G. Antioxidant and antibacterial activity of silver nanoparticles synthesized by *Cestrum nocturnum*. *J. Ayurveda Integr. Med.* **2018**. [CrossRef]

18. Shriniwas, P.; Subhash, K. Antioxidant, antibacterial and cytotoxic potential of silver nanoparticles synthesized using terpenes rich extract of *Lantana camara* L. leaves. *Biochem. Biophys. Rep.* **2017**, *10*, 76–81. [CrossRef]

19. Tanase, C.; Domokos, E.; Coşarcă, S.; Miklos, A.; Imre, S.; Domokos, J.; Dehelean, C.A. Study of the ultrasound-assisted extraction of polyphenols from beech (*Fagus sylvatica* L.) bark. *BioResources* **2018**, *13*, 2247–2267. [CrossRef]

20. Tanase, C.; Cosarca, S.; Toma, F.; Mare, A.; Man, A.; Miklos, A.; Imre, S.; Boz, I. Antibacterial activities of beech bark (*Fagus sylvatica* L.) polyphenolic extract. *Environ. Eng. Manag. J.* **2018**, *17*. [CrossRef]

21. Martins, N.; Barros, L.; Dueñas, M.; Santos-Buelga, C.; Ferreira, I.C.F.R. Characterization of phenolic compounds and antioxidant properties of *Glycyrrhiza glabra* L. rhizomes and roots. *R. Soc. Chem. Adv.* **2015**, *5*, 26991–26997. [CrossRef]

22. Mocan, A.; Schafberg, M.; Crişan, G.; Rohn, S. Determination of lignans and phenolic components of *Schisandra chinensis* (Turcz.) Baill. using HPLC-ESI-ToF-MS and HPLC-Online TEAC: Contribution of individual components to overall antioxidant activity and comparison with traditional antioxidant assays. *J. Funct. Foods* **2016**, *24*, 579–594. [CrossRef]

23. Elshikh, M.; Ahmed, S.; Funston, S.; Dunlop, P.; McGaw, M.; Marchant, R.; Banat, I.M. Resazurin-based 96-well plate microdilution method for the determination of minimum inhibitory concentration of biosurfactants. *Biotechnol. Lett.* **2016**, *38*, 1015–1019. [CrossRef] [PubMed]

24. Anigol, L.B.; Charantimath, J.S.; Gurubasavaraj, P.M. Effect of concentration and ph on the size of silver nanoparticles synthesized by green chemistry. *Org. Med. Chem. Int. J.* **2017**, *3*, 1–5. [CrossRef]

25. Heydari, R.; Rashidipour, M. Green Synthesis of silver nanoparticles using extract of oak fruit hull (jaft): Synthesis and in vitro cytotoxic effect on Mcf-7 cells. *Int. J. Breast Cancer* **2015**, *2015*, 846743. [CrossRef] [PubMed]
26. Mittal, A.K.; Chisti, Y.; Banerjee, U.C. Synthesis of metallic nanoparticles using plant extracts. *Biotechnol. Adv.* **2013**, *31*, 346–356. [CrossRef] [PubMed]
27. Kumar, D.; Kumar, G.; Agrawal, V. Green synthesis of silver nanoparticles using *Holarrhena antidysenterica* (L.) Wall. bark extract and their larvicidal activity against dengue and filariasis vectors. *Parasitol. Res.* **2018**, *117*, 377–389. [CrossRef]
28. Rai, M.; Yadav, A.; Gade, A. Silver nanoparticles as a new generation of antimicrobials. *Biotechnol. Adv.* **2009**, *27*, 76–83. [CrossRef]
29. Agnihotri, S.; Mukherji, S.; Mukherji, S. Size-controlled silver nanoparticles synthesized over the range 5–100 nm using the same protocol and their antibacterial efficacy. *R. Soc. Chem. Adv.* **2014**, *4*, 3974–3983. [CrossRef]
30. Abbasi, B.H.; Nazir, M.; Muhammad, W.; Hashmi, S.S.; Abbasi, R.; Rahman, L.; Hano, C. A comparative evaluation of the antiproliferative activity against Hepg2 liver carcinoma cells of plant-derived silver nanoparticles from basil extracts with contrasting anthocyanin contents. *Biomolecules* **2019**, *9*, 320. [CrossRef]
31. Otunola, G.A.; Afolayan, A.J. In Vitro Antibacterial, antioxidant and toxicity profile of silver nanoparticles green-synthesized and characterized from aqueous extract of a spice blend formulation. *Biotechnol. Biotechnol. Equip.* **2018**, *32*, 724–733. [CrossRef]
32. Mitiku, A.A.; Yilma, B. Antibacterial and antioxidant activity of silver nanoparticles synthesized using aqueous extract of *Moringa stenopetala* leaves. *Afr. J. Biotechnol.* **2017**, *16*, 1705–1716. [CrossRef]
33. Johnston, H.J.; Hutchison, G.; Christensen, F.M.; Peters, S.; Hankin, S.; Stone, V. A Review of the in vivo and in vitro toxicity of silver and gold particulates: Particle attributes and biological mechanisms responsible for the observed toxicity. *Crit. Rev. Toxicol.* **2010**, *40*, 328–346. [CrossRef] [PubMed]
34. Zhang, X.-F.; Liu, Z.-G.; Shen, W.; Gurunathan, S. Silver nanoparticles: Synthesis, characterization, properties, applications, and therapeutic approaches. *Int. J. Mol. Sci.* **2016**, *17*, 1534. [CrossRef] [PubMed]

Article

Enhanced Recovery of Antioxidant Compounds from Hazelnut (*Corylus avellana* L.) Involucre Based on Extraction Optimization: Phytochemical Profile and Biological Activities

Marius Emil Rusu [1,†], **Ionel Fizeşan** [2,†], **Anca Pop** [2,*], **Ana-Maria Gheldiu** [3,*], **Andrei Mocan** [3], **Gianina Crişan** [3], **Laurian Vlase** [1], **Felicia Loghin** [2], **Daniela-Saveta Popa** [2] and **Ioan Tomuta** [1]

[1] Department of Pharmaceutical Technology and Biopharmaceutics, Faculty of Pharmacy, "Iuliu Hatieganu" University of Medicine and Pharmacy, 8 Victor Babes, 400012 Cluj-Napoca, Romania; marius.e.rusu@gmail.com (M.E.R.); laurian.vlase@umfcluj.ro (L.V.); tomutaioan@umfcluj.ro (I.T.)

[2] Department of Toxicology, Faculty of Pharmacy, "Iuliu Hatieganu" University of Medicine and Pharmacy, 8 Victor Babes, 400012 Cluj-Napoca, Romania; ionel.fizesan@umfcluj.ro (I.F.); floghin@umfcluj.ro (F.L.); dpopa@umfcluj.ro (D.-S.P.)

[3] Department of Pharmaceutical Botany, Faculty of Pharmacy, "Iuliu Hatieganu" University of Medicine and Pharmacy, 8 Victor Babes, 400012 Cluj-Napoca, Romania; mocan.andrei@umfcluj.ro (A.M.); gcrisan@umfcluj.ro (G.C.)

[*] Correspondence: anca.pop@umfcluj.ro (A.P.); Gheldiu.Ana@umfcluj.ro (A.-M.G.); Tel.: +40-264-450-555 (A.P.); +40-264-595-770 (A.-M.G.)

[†] These authors contributed equally to this work.

Received: 13 August 2019; Accepted: 5 October 2019; Published: 8 October 2019

Abstract: Tree nut by-products could contain a wide range of phytochemicals, natural antioxidants, which might be used as a natural source for dietary supplements. The aim of the present study was to evaluate the phenolic and sterolic composition, as well as the antioxidant and other biological activities, of hazelnut involucre (HI) extracts. Experimental designs were developed in order to select the optimum extraction conditions (solvent, temperature, time) using turbo-extraction by Ultra-Turrax for obtaining extracts rich in bioactive compounds. Qualitative and quantitative analyses were performed by LC-MS and LC-MS/MS and they revealed important amounts of individual polyphenols and phytosterols, molecules with antioxidant potential. The richest polyphenolic HI extract with the highest antioxidant activity by TEAC assay was further evaluated by other in vitro antioxidant tests (DPPH, FRAP) and enzyme inhibitory assays. Additionally, the cytotoxic and antioxidant effects of this extract on two cancerous cell lines and on normal cells were tested. This is the first study to analyze the composition of both hydrophilic and lipophilic bioactive compounds in HI extracts. Our findings reveal that this plant by-product presents strong biological activities, justifying further research, and it could be considered an inexpensive source of natural antioxidants for food, pharmaceutical, or cosmetic industry.

Keywords: hazelnut involucre; antioxidants; polyphenols; phytosterols; biological activity; experimental design; LC-MS; turbo-extraction; Ultra-Turrax; enzymatic inhibition

---

1. Introduction

The interest in traditional bioactive herbal compounds with less detrimental effects on human body than their synthetic counterparts is growing. Many clinical trials and cohort studies showed that consumption of foods (vegetables, fruits, nuts) rich in biologically active molecules with demonstrated antioxidant capacity, through different mechanisms of actions, has the potential to protect against cardiometabolic diseases, neurodegenerative disorders, cancer, or age-related sarcopenia and frailty [1–5]. Increased intake of antioxidants from tree nuts and peanuts influences risk factors associated with aging and can extend health span and lifespan [6].

Hazelnut (*Corylus avellana* L.) is known for the nutritional and therapeutic properties of its fruits, rich sources of healthy fatty acids, vitamins, minerals, and polyphenols [7,8]. Polyphenols, common secondary plant metabolites characterized by several hydroxyl groups linked to a phenol ring, can act synergistically with other phytochemicals to lower the oxidation and inflammation processes which may trigger many pathological conditions or age-related chronic diseases [9–11]. These biologically active molecules donate electrons or hydrogen atoms to reactive free radicals preventing lipid oxidation or cellular damage and acting as natural antioxidants with many health benefits [12]. In addition, plant polyphenols act as signaling molecules and can participate in many enzymatic pathways [13].

Recently, special interest was given to tree nut by-products, waste plant matrices, with great potential as sources of biologically active compounds. As little is known about the biochemical profile of hazelnut involucre (HI) [14], we intended to extend the knowledge about this promising by-product. In a previous study, we successfully applied a D-optimal experimental design to optimize the bioactive compound extraction from walnut septum, another tree nut by-product [15]. In that experiment, acetone was more efficient than ethanol to extract both polyphenols and phytosterols from the plant matrix. The objectives of this study were the optimization of the extraction process in order to identify the optimal extraction conditions for attaining rich bioactive compound extracts from HI and the emphasizing of key biological activities for the optimally selected extract, which would justify further research on this plant matrix. The first aim of our study was the analysis of phenolic and phytosterol compounds from HI based on experimental designs. Turbo-extraction (TE) by Ultra-Turrax, an efficient and economical extraction method, with higher outcomes compared to other methods, was chosen [15]. Three extraction variables, stirring time, pH, and percentage of water in acetone, were selected to define the optimal extraction conditions. LC-MS analysis was chosen for the identification and quantification of several individual polyphenols and phytosterols from involucre. Based on the obtained results, optimal conditions were selected to achieve the HI extract with the highest content of polyphenolic compounds and the highest antioxidant activity (AA) determined by Trolox equivalent antioxidant capacity (TEAC) assay. The biological activities of this optimal extract were further examined by other in vitro antioxidant assays (DPPH and FRAP), for enzymatic inhibitory capacity (tyrosinase and α-glucosidase), and a potential cytotoxic effect on lung and breast cancer cell lines.

2. Materials and Methods

2.1. Chemicals

2,2-Diphenyl-1-2,4,6-trinitro-phenyl hydrazine (DPPH), 2,4,6-tris(2-pyridyl)-S-triazine (TPTZ) (≥99%), 3,4-dihydroxy-L-phenylalanine (L-DOPA) (≥98%), 6-hydroxy-2,5,7,8-tetramethylchromane-2-carboxylic acid (Trolox) (≥97%), 2,2′-azino-bis(3-ethylbenzothiazoline-6-sulfonic acid) diammonium salt (ABTS) (≥98%), dimethyl sulfoxide (DMSO) (≥99%), ferric chloride, hydrogen peroxide 30% solution, kojic acid, mushroom tyrosinase, *N*-acetyl-L-cysteine (≥99%), phosphate buffer, *p*-nitrophenyl-β-D-glucuronide (PNPG), sodium carbonate, vanillin (99%), resazurin, neutral red, 2,7 dichloro-fluorescein diacetate (DCFH-DA), acetic acid (≥99%) were acquired from Sigma Aldrich (Schnelldorf, Germany). Acetone, Folin–Ciocâlteu reagent, hydrochloric acid (HCl) (37%), sodium phosphate (≥99%) were acquired from Merck (Darmstadt, Germany). α-Glucosidase solution (*Saccharomyces cerevisiae*, EC 3.2.1.20) was obtained from Sigma (Darmstadt, Germany). Aluminum chloride (≥98%) was acquired from Carl Roth (Karlsruhe, Germany), Dulbecco's modified Eagle medium (DMEM) from Gibco (Paisley, UK), and fetal bovine serum (FBS) from Sigma Aldrich (Steinheim, Germany).

All solvents were of LC grade and all reagents were of analytical grade.

The standards used for both spectrophotometric and liquid chromatography–mass spectrometry (LC-MS) analyses were: apigenin (≥95%), caffeic acid (≥98%), brassicasterol (≥98%), caftaric acid (≥97%), (+)-catechin (≥96%), chlorogenic acid (≥95%), (−)-epicatechin (≥90%), fisetin (≥98%), gentisic acid

(≥98%), L-glutathione (≥98%), hyperoside (quercetin 3-D-galactoside) (≥97%), isoquercitrin (quercetin 3-β-D-glucoside) (≥98%), kaempferol (≥97%), luteolin (≥98%), myricetin (≥96%), patuletin (≥98%), *p*-coumaric acid (≥98%), protocatechuic acid (3,4-dihydroxybenzoic acid) (≥97%), quercetin (≥95%), quercitrin (quercetin 3-rhamonoside) (≥78%), rutoside (quercetin-3-O-rutinoside) (≥97%), syringic acid (≥95%), vanillic acid (≥97%), campesterol (~65%), ergosterol (≥95%), stigmasterol (~95%) acquired from Sigma-Aldrich, ferulic acid (≥99%), gallic acid (≥98%) acquired from Merck (Darmstadt, Germany), beta-sitosterol (≥80%), sinapic acid (≥98%) acquired from Carl Roth (Karlsruhe, Germany).

2.2. Plant Samples

Hazelnut (*Coryllus avelana* L.) involucre of high quality was provided by an organic orchard in Cluj County, Romania. In the autumn of 2017, hazelnuts were harvested and separated from the involucre. The by-product was dried in the shade for three days at 18–22 °C, and then it was delivered to the Faculty of Pharmacy, "Iuliu Hatieganu" University of Medicine and Pharmacy Cluj-Napoca, Romania.

Preparation of the Extracts

The involucre was ground in a coffee grinder (Argis, RC-21, Electroarges SA, Curtea de Arges, Romania) for 5 min and the powder was screened through a 200 μm Retsch sieve. HI was weighed (2 g) and mixed with the extraction solvent (20 mL) in Falcon tubes. TE was accomplished using an Ultra-Turrax homogenizer (T 18; IKA Labortechnik, Staufen, Germany) for 1 to 3 min (12,000 rpm) and a Vortex RX-3 (Velp Scientifica, Usmate, Italy) for 2 min. The homogenates were centrifuged (Hettich, Micro 22R, Andreas Hettich GmbH & Co., Tuttlingen, Germany) 15 min at 5000 rpm. The supernatant was carefully separated and, using a rotary evaporator (Hei-VAP, Heidolph Instruments GmbH & Co., Schwabach, Germany), the solvent was removed under vacuum at 45 °C. The dry residue was taken up in water, placed in amber glass vials, and lyophilized (Advantage 2.0, SP Scientific, Warminster, PA, USA). After lyophilization, the samples (weight between 6 and 520 mg) were stored at room temperature. For further analyses, the lyophilized extracts were dissolved in 70% EtOH (10 mg/mL), if not specified otherwise. All assays were performed in triplicate.

Two D-optimal experimental designs, implemented by Modde software, version 11.0 (Sartorius Stedim Data Analytics AB, Umeå, Sweden), were developed for the extraction process optimization. Three factors, stirring time, pH, and percentage of water in acetone, were the independent variables for both D-optimal experimental designs. The total phenolic content (TPC), total flavonoid content (TFC), condensed tannin content (CTC), and the AA-TEAC were the dependent variables in the first experimental design used in the screening step (Table 1). The extracts were prepared according to this first experimental design. The individual concentrations of the bioactive compounds quantified by LC/MS methods were the dependent variables in the second experimental design used in the optimization step (Table 2).

Table 1. Independent and dependent variables of the experimental design used in the screening step.

Variables	Level		
	−1	0	1
Independent variables (factors)			
Stirring time (min) (X_1)	1	2	3
pH (X_2)	3	5	7
Water in solvent (%, *v/v*) (X_3)	0	25	50
Dependent variables (responses)			
Total phenolic content (TPC, mg GAE/g dw [1]) (Y_1)			
Total flavonoid content (TFC, mg QE/g dw [2]) (Y_2)			
Condensed tannin content (CTC, mg CE/g dw [3]) (Y_3)			
Antioxidant activity (AA, mg TE/g dw [4]) (Y_4)			

[1]—mg GAE/g dw = gallic acid equivalents per dry weight of hazelnut involucre, [2]—mg QE/g dw = quercetin equivalents per dry weight of hazelnut involucre, [3]—mg CE/g dw = catechin equivalents per dry weight of hazelnut involucre, [4]—mg TE/g dw = Trolox equivalents per dry weight of hazelnut involucre.

Table 2. Independent and dependent variable of experimental design used in the optimization step.

Variables	Level		
	−1	**0**	**1**
Independent variables (factors)			
Stirring time (min) (X_1)	1	2	3
pH (X_2)	3	5	7
Water in solvent (%, *v/v*) (X_3)	0	25	50
Dependent variables (responses)			
Epicatechin (µg/g dw) (Y_1)			
Catechin (µg/g dw) (Y_2)			
Syringic acid (µg/g dw) (Y_3)			
Gallic acid (µg/g dw) (Y_4)			
Protocatechuic acid (µg/g dw) (Y_5)			
Vanillic acid (µg/g dw) (Y_6)			
p-Coumaric acid (µg/g dw) (Y_7)			
Ferulic acid (µg/g dw) (Y_8)			
Hyperoside (µg/g dw) (Y_9)			
Isoquercitrin (µg/g dw) (Y_{10})			
Quercitrin (µg/g dw) (Y_{11})			
Stigmasterol (µg/g dw) (Y_{12})			
Campesterol (µg/g dw) (Y_{13})			
Beta-sitosterol (µg/g dw) (Y_{14})			

dw—dry weight hazelnut involucre.

2.3. Determination of Total Bioactive Compounds

2.3.1. Total Phenolic Content

The TPC of the HI extracts was assessed by Folin–Ciocâlteu (FC) spectrophotometric method as previously reported [15,16]. A Synergy HT multi-detection microplate reader with 96-well plates (BioTek Instruments, Inc., Winooski, VT, USA) was used for high sample throughput. At first, 20 µL of each sample (HI extracts diluted five times) were mixed with 100 µL of FC reagent (diluted 1:10) and, after 3 min, 80 µL of sodium carbonate solution (7.5% *w/v*) was added. The mixtures were incubated in the dark at room temperature for 30 min. The absorbance was measured at 760 nm against a solvent blank. Gallic acid was used as a reference standard, and the TPC was expressed as gallic acid equivalents (GAE) per dry weight (dw) of involucre (mg GAE/g dw).

2.3.2. Total Flavonoid Content

The TFC of the HI extracts was assessed according to a method previously described [17]. In a 96-well plate, 100 µL of sample extract was added to 100 µL of 2% $AlCl_3$ aqueous solution. The plate was incubated in the dark at room temperature for 15 min and the absorbance was measured at 420 nm against a solvent blank. The TFC was expressed as quercetin equivalents (QE) per dw of involucre (mg QE/g dw).

2.3.3. Condensed Tannin Content

The CTC in HI extracts was assessed according to a modified version of the vanillin assay formerly described [18]. In a 96-well plate, 50 µL of HI sample was added to 250 µL 0.5% vanillin in 4% concentrated HCl in methanol. After the plate was incubated in the dark at 30 °C for 20 min, the absorbance was measured at 500 nm against a solvent blank. The CTC was expressed as catechin equivalents (CE) per dw of involucre (mg CE/g dw).

2.4. Determination of the Antioxidant Activity

2.4.1. TEAC Assay

The antiradical activity of HI extracts was assessed according to the TEAC assay previously described [17]. Briefly, 20 µL of the sample was mixed with 200 µL of ABTS radical solution, incubated for 6 min, and the absorbance of the mixture was measured at 760 nm. The scavenging activity against ABTS radical cation was calculated and used to plot the Trolox calibration curve. The AA according to this assay was expressed as Trolox equivalents (TE) per dw involucre (mg TE/g dw). This assay was used during the screening phase of the study for the evaluation of AA of all HI extracts obtained within the experimental plan in order to optimize the extraction method.

2.4.2. DPPH Radical Scavenging Activity

The antiradical activity of HI extracts against the free radical DPPH was measured using a method previously described [19]. In a 96-well plate, 30 µL of sample solution was mixed with a 0.004% methanol solution of DPPH, and then incubated in the dark for 30 min. The absorbance was measured at 517 nm against a solvent blank. Trolox was used as a reference standard and the results were expressed as TE per dw extract (mg TE/g dw extract). This assay was performed only on the richest polyphenolic HI extract.

2.4.3. FRAP Assay

The reduction capacity of the HI extract was assessed by ferric reducing antioxidant power (FRAP) assay (analyzes the reduction of Fe^{3+}-TPTZ to the blue-colored Fe^{2+}-TPTZ) using a slight modified previously described method [20]. Briefly, a quantity of 25 µL sample was incubated with 175 µL FRAP reagent (300 mM acetate buffer, pH 3.6: 10 mM TPTZ in 40 mM HCl: 20 mM $FeCl_3 \cdot 6H_2O$ in 40 mM HCl, 10:1:1, *v/v/v*) in the dark for 30 min. The absorbance was measured at 593 nm and the results were expressed as TE per dw extract (mg TE/g dw extract). This assay was done only on the richest polyphenolic HI extract.

2.5. Phytochemical Analysis by LC-MS

The phytochemical profiles of the lyophilized HI extracts were analyzed by LC-MS. For this assay, an Agilent 1100 HPLC Series system (Agilent, Santa Clara, CA, USA), equipped with degasser, binary gradient pump, column thermostat, autosampler, and UV detector, coupled with an Agilent Ion Trap 1100 SL mass spectrometer (LC/MSD Ion Trap VL) were used.

2.5.1. Identification and Quantification of Individual Polyphenolic Compounds

An LC-MS method previously described [21,22] was used for the identification of individual polyphenols in the HI extracts. The 18 external standards were: apigenin, caffeic acid, caftaric acid, chlorogenic acid, ferulic acid, fisetin, gentisic acid, hyperoside, isoquercitrin, kaempferol, luteolin, myricetin, patuletin, *p*-coumaric acid, quercetin, quercitrin, rutoside, and sinapic acid (Table S1). In brief, the chromatographic separation was performed on a reverse-phase analytical column (Zorbax SB-C18, 100 mm × 3.0 mm i.d., 3.5 µm particles) with a mixture of methanol/acetic acid 0.1% (*v/v*) as mobile phase and a binary gradient. The elution started with a linear gradient, beginning with 5% methanol and ending at 42% methanol at 35 min, isocratic elution followed with 42% methanol for the next 3 min, rebalancing with 5% methanol in the next 7 min. The flow rate was 1 mL/min, the injection volume was 5 µL, and the column temperature was 48 °C. The detection procedure was performed on both UV and MS mode. The UV detector was set at 330 nm until 17 min (for the detection of polyphenolic acids), then at 370 nm until the end of analysis time (for the detection of flavonoids and their aglycones). The MS system operated using an electrospray ion (ESI) source in negative mode (capillary 3000 V, nebulizer 60 psi (nitrogen), dry gas temperature 360 °C, and dry nitrogen gas at 12 L/min).

Another LC-MS method (LC-MS method II) previously described [23] was used to detect the other six polyphenols (epicatechin, catechin, syringic acid, gallic acid, protocatechuic acid, and vanillic acid) in HI extracts (Table S2). The chromatographic separation was accomplished on the same analytical column and in the same chromatographic conditions as mentioned before but with a slightly different binary gradient (start: 3% methanol; at 3 min: 8% methanol; at 8.5 min: 20% methanol; at 10 min: rebalance column with 3% methanol). The detection of the compounds was performed on MS mode. All identified polyphenols were measured both in the HI non-hydrolyzed and hydrolyzed extracts (equal amounts of extract and 4 M HCl kept 30 min on 100 °C water bath) on the basis of their peak areas and comparison with a calibration curve of their corresponding standards. The results were expressed as micrograms of phenolic compounds per dw of involucre (μg/g dw).

2.5.2. Identification and Quantification of Phytosterols

The phytosterols in the HI extracts were determined according to an LC-UV-MS/MS method previously described [24]. In brief, apparatus and chromatographic analytical column were the same, but elution of compounds was performed in an isocratic mode, mobile phase containing acetonitrile/methanol (90:10, *v/v*), with a flow rate of 1 mL/min at 45 °C and 5 μL injection volume. For the detection of the analytes, the same ion trap mass spectrometer was used, fitted with an atmospheric pressure chemical ionization (APCI) interface in a positive mode. Operating conditions (dry nitrogen gas temperature 325 °C at a flow rate of 7 L/min, nebulizer pressure 60 psi, capillary voltage -4000 V) were adjusted for achieving maximum sensitivity values.

The full identification of compounds was performed by comparing the retention times and mass spectra with five external standards (ergosterol, brassicasterol, stigmasterol, campesterol, beta-sitosterol) (Table S3). Under the ionization conditions used for their determination, all sterols lose a water molecule and their pseudo-molecular ions are in the form $M-H_2O+H^+$ [25]. The multiple reactions monitoring (MRM) analysis mode was used for detection in order to avoid or reduce the interference from the background. The quantification was performed based on the intensity of major daughter ions in the mass spectra [25]. The results were expressed as micrograms of phytosterols per dw involucre (μg/g dw).

The Agilent ChemStation (vA09.03) and DataAnalysis (v5.3) software were used for the acquisition and investigation of chromatographic data.

2.6. Selection and Biological Activities of the Optimal Hazelnut Involucre Extract

Based on the experimental data obtained during the screening and the optimization steps, the extract with the richest polyphenolic content and the best antioxidant activity by TEAC assay was further selected as the optimal HI extract in order to evaluate its biological effects. The antioxidant activities by DPPH and FRAP assays, the enzymatic inhibitory potential (tyrosinase and α-glucosidase), and the cytotoxicity and antioxidant effects on cellular cultures were further evaluated for this extract. All determinations were performed in triplicate on the lyophilized extract.

2.6.1. Enzyme Inhibitory Activities

Tyrosinase Inhibitory Activity

The tyrosinase inhibitory activity of HI extract was calculated using a slightly modified previously described method [26]. In a 96-well plate, four wells were designated (HI lyophilized extract dissolved in water containing 5% DMSO) as follows: (A) 46 U/mL (40 μL) mushroom tyrosinase (MT) in 66 mM phosphate buffer (PB), pH 6.6 (120 μL); (B) only PB (160 μL); (C) PB (80 μL), MT (40 μL) and the sample (40 μL); (D) PB (120 μL) and the sample (40 μL). After incubation at room temperature for 10 min, 2.5 mM L-DOPA prepared in PB (40 μL) was added in all wells. After room temperature incubation for 20 min, the absorbance was measured at 475 nm. The tyrosinase inhibitory activity was calculated using kojic acid as an external standard (0.01–0.10 mg/mL). The inhibition percentage of enzymatic activity was assessed by the following equation: $[(A - B) - (C - D)] \times 100/(A - B)$. The results were expressed as milligram kojic acid equivalents (KAE) per dw extract (mg KAE/g dw).

α-Glucosidase Inhibitory Assay

The *α*-glucosidase inhibitory activity was performed using the protocol previously described [27]. Briefly, 50 μL sample solution (2 mg HI lyophilized extract/mL), 50 μL glutathione (0.5 mg/mL), 50 μL of 10 mM PNPG (*p*-nitrophenyl-β-D-glucuronide) solution, and 50 μL α-glucosidase solution in PB (pH 6.8) were mixed in a 96-wells microplate and incubated at 37 °C for 15 min. Similarly, the blank was prepared by adding a sample solution to all reaction reagents without the enzyme (α-glucosidase) solution. The reaction was finally stopped by adding 50 μL of 0.2 M sodium carbonate. The absorbance for sample and blank were read at 400 nm and then the absorbance of blank was subtracted from the absorbance of the sample.

2.6.2. Biological Activities of HI Extract on Cell Lines

Cell Culture

The normal human gingival fibroblasts (HGF) (CLS Cell Lines Service, Eppelheim, Germany), and the cancerous cell lines A549 (human lung adenocarcinoma) and T47D-KBluc (human breast cancer) (ATCC, Manassas, United States of America) were maintained in Dulbecco's modified Eagle medium (DMEM) supplemented with 10% fetal bovine serum (FBS) on culture flasks at 37 °C in a humidified incubator with 5% CO_2 supplementation. The medium was changed every two to three days and the cells were subcultured once they reached 70%–80% confluence.

Preparation of Extract Solutions

A 100 mg/mL stock solution was prepared in dimethyl sulfoxide (DMSO) from lyophilized extract. The stock solution was further diluted in DMSO to obtain working solutions of 0.25, 6.25, 12.5, 18.75, 25, 37.25, 50, and 75 mg/mL. These working solutions were then used to obtain the desired concentrations, ranging from 400 to 1 μg/mL, in the cell culture medium.

Viability Assays

Cells were seeded in 96-well plates and left to attach for 24 h. Dead and unattached cells were washed with PBS while the remaining viable cells were further exposed for 24 h/48 h to the HI extract. Following the exposure, the cells were washed with PBS and viability was assessed by two complementary assays, Alamar Blue assay, and Neutral Red assay.

Alamar Blue (AB) assay was used to measure the metabolic ability of exposed cells to convert resazurin, a non-fluorescent compound, to resorufin, a fluorescent product. The cells were exposed to a resazurin solution of 200 μM for 3 h and the fluorescence was measured at $\lambda_{excitation} = 530/25$, $\lambda_{emission} = 590/35$, using Synergy 2 multi-mode microplate reader.

Neutral Red (NR) assay was used to measure the ability of the exposed cells to incorporate the supravital dye neutral red. The cellular uptake of this dye reveals the capacity of the cell to maintain pH gradients through the production of ATP. Briefly, 150 μL 40 μg/mL neutral red dye was added to each well. The cells were incubated with the dye at 37 °C for 2 h. Afterwards, the cells were washed twice with PBS and accumulated dye was extracted by adding a solution containing 50% ethanol, 49% water, and 1% glacial acetic acid to each well. Cell viability was determined by measuring the fluorescence at $\lambda_{excitation} = 530/25$, $\lambda_{emission} = 620/40$, using Synergy 2 multi-mode microplate reader.

The experiments were performed with three biological replicates, each one including six technical replicates. The results were expressed as relative values compared to the negative control (100%) (cells exposed to culture medium containing 0.2% DMSO).

IC$_{50}$ values were calculated, when possible, from the dose-response curves obtained for each condition using a four-parameter logistic curve, in order to allow a comparison between the conditions tested.

Dichloro-Fluorescein Diacetate (DCFH-DA) Assay

The ability of the HI extract to protect against the oxidative stress in T47D-KBluc, A549, and HGF cells was monitored using the reactive oxygen species (ROS) sensitive 2,7 dichloro-fluorescein diacetate (DCFH-DA) dye. After a 24 h treatment with non-toxic HI extract concentrations, the cells were washed with PBS and further loaded with 50 μM DCFH-DA in Hanks' balanced salt solution (HBSS) for 2 h. Following incubation, the excess of DCFH-DA was washed, and the cells were exposed to 250 μM H$_2$O$_2$ for 2 h. The conversion of DCFH-DA to the fluorescent compound dichlorofluorescein (DCF) was measured using Synergy 2 multi-mode microplate reader at $\lambda_{excitation}$ = 485/20, $\lambda_{emission}$ = 528/20. The potency of HI extracts, to mitigate the induction of oxidative stress after H$_2$O$_2$ treatment, was compared to N-Acetyl Cysteine (NAC) treatment (20 mM solution).

2.7. Statistical Analysis

All samples were analyzed in triplicate (n = 3) and the outcomes were reported as the mean ± standard deviation (SD). The data were statistically analyzed by Student's t-test using Microsoft Office Excel computer software and by one-way analysis of variance (ANOVA) with post hoc Dunnett's test using SigmaPlot 11.0 computer software. The difference showing a p level of 0.05 or lower was considered statistically significant.

3. Results and Discussion

3.1. Fitting the Experimental Data with the Models

The matrix of the experimental design containing the 17 formulations generated by the software and the results achieved after performing all the experimental runs are summarized in Table 3.

For data fitting, the partial least squares (PLS) method was employed and several statistical parameters were further used to assess the fitting results. The coefficient of correlation (R^2) represents the variation of the response explained by the selected model, or goodness of fit, while the Q^2 indicates the model capacity to be predictive. Furthermore, ANOVA test was employed to assess the experimental model validity, significance and lack of fit. The model proved good predictive ability, as shown by Q^2 > 0.7, and a good correlation between the predicted and observed values was found, as suggested by R^2 > 0.9 (Table 4). Differences of less than 0.2–0.3 between R^2 and Q^2 indicate a high predictive power of a good model. Models proved adequate validity (>0.4) corroborated with a reduced lack of fit for each evaluated response. The reproducibility values were >0.95, which means that the replicates generated similar responses by working under identical experimental conditions, thus making the experimental setup adequate for the purpose of the study. The results of ANOVA test showed a significant influence of the evaluated factors over TPC, TFC, CTC, and AA-TEAC, with p-value for the regression <0.001 (Table 4). Considering the results shown in Table 4, the fitting models were found to be appropriate to describe the experimental data, as the values for the lack of fit were not significant in extent with the pure error. The regression equation coefficients for the responses are presented in Table S4.

Table 3. Matrix of experimental design and results for total phenolic content, total flavonoid content, condensed tannin content, and antioxidant activity of hazelnut involucre extracts based on a factorial design during the screening step.

Sample Code	Run Order	Factorial Design with Coded Values			Determination (Experimental)			
		X_1	X_2	X_3	Y_1 (TPC)	Y_2 (TFC)	Y_3 (CTC)	Y_4 (AA-TEAC)
N1	12	1	3	0	2.62 ± 0.35	2.23 ± 0.23	0.04 ± 0.02	9.86 ± 1.13
N2	3	3	3	0	4.88 ± 0.66	2.88 ± 0.10	0.26 ± 0.14	18.89 ± 1.62
N3	2	1	7	0	4.33 ± 0.41	3.64 ± 0.18	1.48 ± 0.85	16.01 ± 1.07
N4	14	3	7	0	2.59 ± 0.19	1.80 ± 0.09	0.57 ± 0.08	8.65 ± 0.53
N5	10	1	3	50	320.83 ± 24.42	32.12 ± 1.14	226.74 ± 3.69	1049.75 ± 25.43
N6	7	3	3	50	377.43 ± 26.74	43.10 ± 1.59	280.69 ± 7.85	1296.51 ± 19.25
N7	1	1	7	50	332.78 ± 19.62	39.82 ± 0.92	274.27 ± 14.73	1207.62 ± 30.23
N8	13	3	7	50	334.68 ± 16.78	37.32 ± 0.70	242.31 ± 1.06	1137.48 ± 47.46
N9	4	1	5	25	292.98 ± 9.49	34.47 ± 1.39	244.06 ± 4.03	1065.73 ± 55.15
N10	8	3	5	25	210.67 ± 5.46	25.04 ± 2.77	174.81 ± 3.02	885.31 ± 17.82
N11	9	2	3	25	188.82 ± 0.60	22.01 ± 1.02	150.18 ± 1.19	718.15 ± 21.35
N12	5	2	7	25	257.22 ± 8.50	28.06 ± 1.04	228.27 ± 2.33	823.02 ± 46.84
N13	16	2	5	0	15.50 ± 0.39	10.56 ± 0.51	2.19 ± 0.15	109.79 ± 26.74
N14	6	2	5	50	313.21 ± 8.00	32.56 ± 0.45	225.72 ± 1.99	1261.77 ± 180.09
N15	17	2	5	25	197.39 ± 16.29	25.53 ± 0.21	180.87 ± 4.21	760.24 ± 114.77
N16	11	2	5	25	180.95 ± 10.42	25.38 ± 1.43	174.78 ± 13.29	749.62 ± 97.78
N17	15	2	5	25	196.32 ± 15.63	25.30 ± 0.58	169.95 ± 2.18	714.15 ± 93.65

X_1, stirring time; X_2, pH; X_3, water in solvent (%, v/v). TPC: Total phenolic content expressed as mg GAE/g dw = gallic acid equivalents per dry weight of hazelnut involucre; TFC: Total flavonoid content expressed as mg QE/g dw = quercetin equivalents per dry weight of hazelnut involucre; CTC: Condensed tannin content expressed as mg CE/g dw = catechin equivalents per dry weight of hazelnut involucre; AA-TEAC: Antioxidant activity by TEAC assay expressed as mg TE/g dw = Trolox equivalents per dry weight of hazelnut involucre. Data are shown as mean ± SD (standard deviation).

Table 4. Statistical parameters after data analysis and fit with factorial model.

Evaluated Response	Total Phenolic Content ($R^2 = 0.96, Q^2 = 0.91$)					Total Flavonoid Content ($R^2 = 0.92, Q^2 = 0.78$)					Condensed Tannin Content ($R^2 = 0.95, Q^2 = 0.88$)					Antioxidant Activity ($R^2 = 0.97, Q^2 = 0.92$)				
Reproducibility	0.95					0.98					0.97					0.99				
	SS	DF	MS	F-v	p-v	SS	DF	MS	F-v	p-v	SS	DF	MS	F-v	p-v	SS	DF	MS	F-v	p-v
Regression	286,363	5	57,272.6	61.46	0.001	2819.57	5	563.91	26.71	0.001	1728.27	5	345.65	46.74	0.001	5.26×10^5	6	878,126	69.9	0.001
Lack of fit	8678.93	9	964.32	1.22	0.527	227.15	9	25.23	9.99	0.094	75.94	9	8.43	3.12	0.266	122,720	8	15,340	10.58	0.089
Pure error	1570.49	2	785.24			5.04	2	2.52			5.39	2	2.69			2899.26	2	1449.63		

R^2: Coefficient of correlation/goodness of fit; Q^2: Goodness of prediction; SS: Sum of squares; DF: Degrees of freedom; MS: Mean square; F-v: F-value, Fischer's ratio; p-v: p-value, probability.

The influence of the extraction conditions (factors) on the quantified individual compound levels (responses) was studied using the second experimental design during the optimization step (Table 2). The factors were the same as in the screening step and the responses were the levels of the 14 bioactive compounds (polyphenols and sterols) (Table 5). Moreover, the samples were hydrolyzed and the results of the recovery for the main bioactive compounds were compared to those obtained from non-hydrolyzed samples (Table S5). The statistical parameters R^2, Q^2, regression, lack of fit and pure error were determined for fitting the experimental data with the experimental design. The selected model presented good quality, with R^2 values between 0.75 and 0.94 and Q^2 results in the range of 0.47–0.84. For all the evaluated responses, the results of ANOVA test for the model had statistical significance ($p < 0.05$) and lack of fit p-values in the range of 0.064–0.91 (Table S6). Considering the values obtained for these statistical parameters, the experimental data for the bioactive compounds were adequately described by the fitting models, with a quadratic model statistically significant and a low lack of fit. The evaluated responses were significantly influenced by the chosen factors. The regression equation coefficients for all the bioactive compounds determined in HI extracts are shown in Table S7.

Table 5. Matrix of experimental design and results for bioactive compound recovery from hazelnut involucre extracts.

Sample Code	Run Order	Factorial Design with Coded Values			Determination (Experimental Results)													
		X_1	X_2	X_3	Y_1	Y_2	Y_3	Y_4	Y_5	Y_6	Y_7	Y_8	Y_9	Y_{10}	Y_{11}	Y_{12}	Y_{13}	Y_{14}
N1	12	1	3	0	ND	ND	ND	ND	ND	ND	0.37	0.23	1.23	0.43	3.78	30.05	3.45	916.69
N2	3	3	3	0	0.16	13.62	0.54	5.29	21.73	3.10	0.94	0.54	2.69	1.13	8.36	61.06	5.75	2444.00
N3	2	1	7	0	0.11	10.28	0.41	3.78	14.55	2.44	0.70	0.44	3.09	1.01	7.32	30.05	2.32	1442.87
N4	14	3	7	0	ND	ND	ND	ND	ND	ND	0.81	0.49	2.04	0.80	6.51	36.02	5.88	1408.88
N5	10	1	3	50	1.48	155.09	2.65	55.39	103.98	15.78	ND	ND	31.61	9.21	76.21	ND	ND	ND
N6	7	3	3	50	3.61	201.95	2.58	63.59	131.15	16.61	ND	ND	51.72	17.74	114.26	ND	ND	ND
N7	1	1	7	50	3.17	158.06	2.77	53.14	124.58	21.10	ND	3.23	43.90	10.25	97.15	ND	ND	ND
N8	13	3	7	50	2.05	161.14	ND	42.87	103.52	12.29	ND	ND	42.26	13.29	94.52	ND	ND	ND
N9	4	1	5	25	3.48	216.97	3.33	69.07	140.91	18.71	6.58	3.97	50.99	15.71	112.05	ND	25.35	3145.78
N10	8	3	5	25	1.62	108.01	2.01	37.15	70.58	9.83	3.19	ND	30.93	9.10	71.32	195.28	45.04	5166.14
N11	9	2	3	25	1.65	150.77	2.63	53.23	97.26	14.51	2.74	2.49	31.04	8.64	71.34	185.63	31.38	3792.66
N12	5	2	7	25	3.73	243.03	5.53	91.93	227.37	25.41	4.41	3.30	36.23	10.31	85.39	145.24	24.79	3480.22
N13	16	2	5	0	0.27	19.98	0.87	11.83	33.32	4.71	2.14	1.51	5.89	ND	17.99	77.68	12.66	2843.16
N14	6	2	5	50	ND	78.05	2.41	44.25	100.40	13.61	ND	ND	29.52	8.60	67.21	ND	ND	ND
N15	17	2	5	25	2.41	172.79	3.39	54.21	119.09	14.81	3.14	2.15	2.44	9.57	85.62	141.49	28.96	3821.43
N16	11	2	5	25	1.98	159.99	2.65	61.39	103.95	12.48	3.91	2.89	36.71	10.82	80.39	197.31	31.49	5305.01
N17	15	2	5	25	1.89	186.77	3.78	57.64	116.71	15.31	3.61	2.65	39.23	11.98	88.73	178.96	22.76	3213.27

X_1, stirring time (min); X_2, pH; X_3, water in solvent (%, v/v). Y_1: Epicatechin; Y_2: Catechin; Y_3: Syringic acid; Y_4: Gallic acid; Y_5: Protocatechuic acid; Y_6: Vanillic acid; Y_7: p-Coumaric acid; Y_8: Ferulic acid; Y_9: Hyperoside; Y_{10}: Isoquercitrin; Y_{11}: Quercitrin; Y_{12}: Stigmasterol; Y_{13}: Campesterol; Y_{14}: Beta-sitosterol. All responses are expressed as µg bioactive compound per gram of dry weight hazelnut involucre. ND—not determined.

3.2. The Influence of Experimental Conditions on TPC, TFC, CTC, and AA-TEAC

As it can be noticed from the results for TPC, TFC, CTC, and AA-TEAC (Table 3), the responses were influenced by the factors used in the experimental design.

The stirring time was in the range of 1–3 min, the pH varied from acidic to neutral (3–5–7), and the two solvents used were water and acetone mixed in various proportions (0%–25%–50% water in acetone). The best extraction yields were obtained when using binary-solvent systems, while mono-solvent systems displayed a lower extraction power. These findings confirm the rational choice of the solvent mixture based on the previously reported determination of bioactive compounds in a study that aimed to evaluate the recovery efficiency from walnut septum extracts [15]. For a higher extraction yield, the best solvent mixture is between a polar, protic solvent (water) with a polar, relatively acidic, aprotic solvent (acetone). The best results for TPC, TFC, CTC, and AA-TEAC were obtained for pH 3 and 50% water in acetone, proving a positive relationship between the content of bioactive compounds and the antioxidant activity. The influences of working conditions on TPC, TFC, CTC, and AA-TEAC as scaled and centered coefficient plots are depicted in Figure 1, while the response surfaces for predicting the extraction efficiency for the aforementioned parameters are shown in Figure 2. From these figures, it can be observed that the factor with the highest positive impact on the recovery of bioactive compounds is the amount of water in the mixture solvent, a higher amount of water being associated with an increased extraction yield. In addition, the water amount proved to have a statistically significant influence upon TPC, TFC, CTC, AA-TEAC, and bioactive compound recovery (see Tables S4 and S7). The optimal extraction conditions for TPC, TFC, CTC, and AA-TEAC, generated by the software used, are given in Table 6.

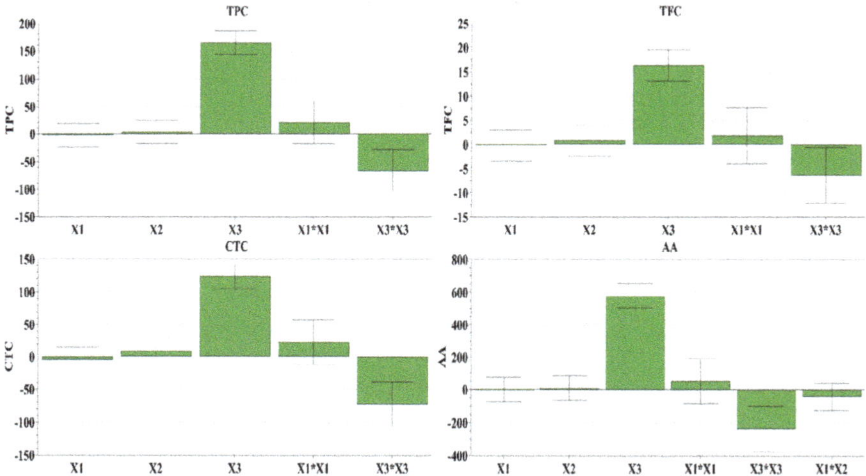

Figure 1. Influence of working conditions on total phenolic content (**TPC**), total flavonoid content (**TFC**), condensed tannin content (**CTC**), and antioxidant activity (**AA**) by TEAC assay of hazelnut involucre extracts, presented as scaled and centered coefficient plots. X_1, stirring time (min); X_2, pH; X_3, water in solvent (% *v/v*).

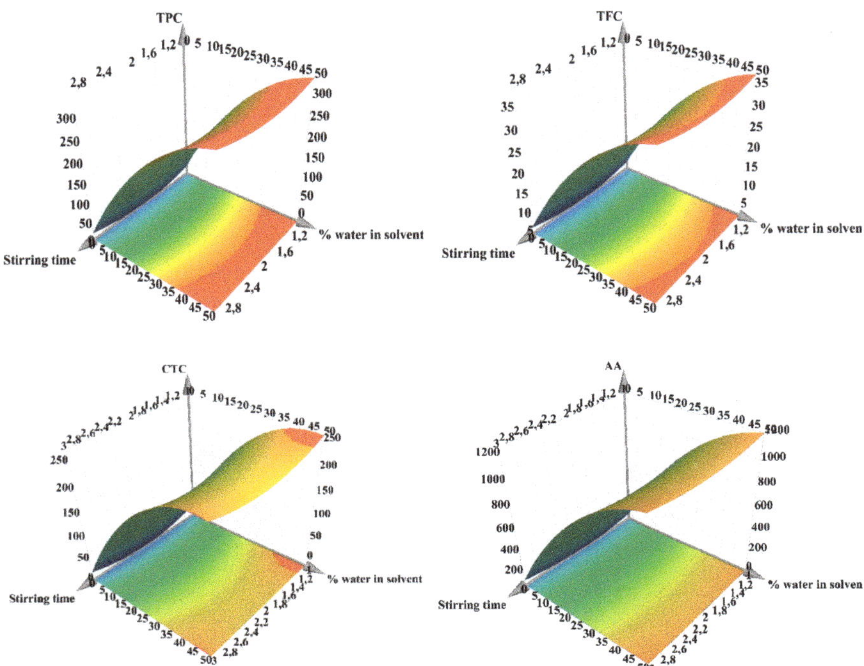

Figure 2. Response surface for predicting the recovery yield for total phenolic content (**TPC**), total flavonoid content (**TFC**), condensed tannin content (**CTC**,) and antioxidant activity (**AA**) by TEAC assay for hazelnut involucre extracts with regard to: X_1, stirring time (min); X_3, water in solvent (% *v/v*); X_2, pH = 5. The regions in red represent the domains of working conditions assuring the maximum extraction yield for the evaluated bioactive compounds.

Table 6. Optimum experimental conditions for improved recovery of TPC, TFC, CTC, AA-TEAC from hazelnut involucre extracts.

Parameters	TPC	TFC	CTC	AA-TEAC
Stirring time (min)	3	3	3	3
pH	3	3	3	3
Water in solvent (%)	50	50	50	50
Predicted	345.96	38.41	269.67	1253.09
Determined	370.42	41.97	279.30	1291.22
Bias (%)	7.07	9.26	3.57	3.04

TPC: Total phenolic content expressed as mg GAE/g dw = gallic acid equivalents per dry weight of hazelnut involucre; TFC: Total flavonoid content expressed as mg QE/g dw = quercetin equivalents per dry weight of hazelnut involucre; CTC: Condensed tannin content expressed as mg CE/g dw = catechin equivalents per dry weight of hazelnut involucre; AA-TEAC: Antioxidant activity by TEAC assay expressed as mg TE/g dw = Trolox equivalents per dry weight of hazelnut involucre.

3.3. Quantitative Determinations of Total Bioactive Compounds

As expected from previous studies [28], better extraction results of phenolic compounds found in HI, and responsible for the antioxidant activity and human health benefits, were obtained using a binary-solvent mixture of water and acetone.

3.3.1. Total Phenolic Content

A significant difference in TPC values was noticed when an equal mixture of water and acetone was used compared to pure acetone. The value for the richest phenolic compound extract was 377.43 mg GAE/g dw HI when aqueous acetone solvent was used (N6, run order 7, Table 3). Because no similar data were found in the literature, where TPC was expressed using another phenolic compound reference (CE vs. GAE), we could not effectively compare the results. Thus, in the study of Alasalvar et al. [29], the TPC in hazelnut green leafy cover extracted with an aqueous 80% acetone or 80% ethanol were 201 or 156 mg catechin equivalent (CE)/g extract, respectively. The TPC in hazelnut green leafy cover obtained by Shahidi et al. [14], using a 80:20 (*v/v*) ethanol/water mixture as extraction solvent, was 127.3 mg CE/g extract, while the TPC for hazelnut skin, hard shell, tree leaf, and kernel were 577.7, 214.1, 134.7, and 13.7 mg CE/g extract, respectively. As seen in another study [30], TPC in tree nut by-products is several times higher compared to nut kernels. The quantification of TPC in different hazelnut kernel extracts exposed values in the range of 11.17–14.77 mg GAE/g [31] and 10.21–23.28 mg GAE/g [32].

Esposito et al. [33] detected quantities of 193.8 mg GAE/g methanol hazelnut shell extract for TPC, while Masullo et al. [34], for the same by-product, noticed that TPC was 340.44 μg GAE/mg extract, corresponding to 340.44 mg GAE/g.

3.3.2. Total Flavonoid Content

Flavonoids exhibit both in vitro and in vivo biological activities and may elicit health benefits including cardiovascular protection [35]. In this study, the highest value for the TFC was 43.10 ± 1.59 mg QE/g dw involucre (N6, run order 7, Table 3). As in the case of phenolics, the best values for total flavonoids were obtained at equal volumes of water and acetone in the solvent mixture.

To the best of our knowledge, there is no research regarding the flavonoid content in HI, therefore TFC values for nuts and other by-products were used as a comparison. The highest flavonoid content in walnut septum was 9.76 ± 0.23 mg QE/g [15], while hazelnuts presented TFC values between 0.09 ± 0.01 and 0.36 ± 0.07 mg rutin equivalent/g extract [36].

3.3.3. Condensed Tannin Content

Condensed tannins or proanthocyanidins had revealed positive effects on neurogenesis, cognitive improvement, and prevention of neuron death in neurodegenerative diseases such as Alzheimer's disease [37].

In our study, the best value for CTC was 280.69 ± 7.85 mg CE/g dw HI, using as solvent 50% aqueous acetone solution (N6, run order 7, Table 3). Kim et al. [38] noticed values of 4.91 mg proanthocyanidins per gram of hazelnuts, while Lainas et al. [39] detected higher amounts of condensed tannins, 31.30 mg CE/g natural hazelnuts and 36.70 mg CE/g roasted hazelnut skin. The quantities for CTC obtained by Alasalvar et al. [29] were 385 and 542 mg CE/g extract from green leafy cover extracted with 80% ethanol in water and 80% acetone in water, respectively. The variations observed between the CTC results are due to different extraction conditions performed on various herbal matrices. The results of CTC found in the current study are in line with those obtained by Alasalvar et al. [29] on the same matrix.

3.4. Determination of Antioxidant Activity

The antioxidant capacity of phenolic compounds depends basically on the number and position of the substituents on the aromatic ring [12]. However, because of their reduced bioavailability, due to extensive phase 2 metabolism, polyphenols might function more as up-regulators of the antioxidant activity than as direct antioxidants [40].

3.4.1. TEAC Assay

The antioxidant activity against the stable synthetic ABTS radical cation of different HI extracts is depicted in Table 3. This analysis is based on electron transfer reactions in order to assess the radical scavenging activity of several compounds. The highest values for AA were for 50% aqueous acetone extracts, with the strongest activity of 1296.51 ± 19.25 mg TE/g dw involucre (5.18 ± 0.08 mmol TE/g dw involucre). The AA for green leafy cover noticed by Shahidi et al. [14] was 117 μmol TE/g extract after extraction with 80% ethanol, while the values for hazelnut kernel, skin, hard shell, and tree leaf were between 29 and 148 μmol TE/g extract. Alasalvar et al. [29] reported AA values of 1.29 and 1.14 mmol TE/g extract obtained from green leafy cover using 80% acetone and 80% ethanol, respectively. In another source, the antioxidant activities were in the range of 3063–3573 μmol TE/100 g dw natural hazelnuts [41]. The AA of hazelnut by-product extracts was 4–5-fold greater than that of kernel extracts, meaning that hazelnut by-product extracts compared to hazelnut kernel extract would have a stronger antioxidant activity.

3.4.2. DPPH Radical Scavenging Activity

The DPPH radical scavenging assay was applied to measure the capability of HI to deactivate and scavenge this stable free radical. Antioxidant molecules can reduce DPPH free radicals, through hydrogen atom-donating capacity, and the absorbance will decrease faster if the antioxidant potential of the extract is more powerful [42].

In our study, the in vitro DPPH radical scavenging activity was 292.23 mg TE/g involucre extract obtained using 50% aqueous acetone solvent, at pH 3, and 3 min stirring time. The reported DPPH values on related matrices were between 84.9% and 93.6%, and 8.1% and 41.1% for natural hazelnuts and roasted hazelnuts, respectively [32]. In the study of Oliveira et al. [43], the EC_{50} values for hazelnut leaves obtained for DPPH radical scavenging were <0.3 mg/mL, while Masullo et al. [34] found the highest DPPH scavenging activity value (EC_{50} = 54.3 μg/mL) for hazelnut shells corresponding to the highest TPC value. Similarly, Yuan et al. [44] found a value of around 11 mg GAE/g for the richest TPC hazelnut shell extract. As there were no data regarding the DPPH assay for HI, a comparison with the results of other studies was not possible.

3.4.3. FRAP Assay

The FRAP assay, a fast and sensitive way for measuring the antioxidant capacity in samples via the reduction of ferric iron (Fe^{3+}) to ferrous iron (Fe^{2+}) by the antioxidants present in the samples, is a significant indication of the antioxidant activity.

In our analysis, the reducing power for the richest polyphenolic HI extract achieved using water/acetone (1:1) at pH 3 and stirring time 3 min was 350.52 mg TE/g involucre extract. FRAP data for related matrices were 12.0 mg GAE/g hazelnut shell [44] and 400.97 mg TE/g walnut septum extract [15].

Our findings are in agreement with previous studies, which show a positive correlation between phenolic compounds and the antioxidant activity [45,46].

3.5. The Influence of Experimental Conditions on Individual Bioactive Compounds

The individual polyphenols and phytosterols that were identified and quantified by LC-MS and LC-MS/MS in HI extracts are presented in Table 5. For epicatechin (Y_1), catechin (Y_2), syringic acid (Y_3), gallic acid (Y_4), protocatechuic acid (Y_5), and vanillic acid (Y_6) (Figure S1), the highest extraction yield was obtained for 2 min of stirring, at pH of 7, in a mixture of acetone/water (3:1). For *p*-coumaric acid (Y_7) and ferulic acid (Y_8) (Figure S2), the best recovery was attained for 1 min stirring time, pH 5, solvent mixture of acetone/water (3:1), while for hyperoside (Y_9), isoquercitrin (Y_{10}), and quercitrin (Y_{11}) (Figure S3), the highest extraction power was observed for 3 min of stirring, at pH 3, in a mixture of solvents acetone/water in equal proportions (1:1). As far as stigmasterol and beta-sitosterol are concerned, their best recovery was obtained after 2 min of stirring the HI extracts in a mixture of solvents acetone/water (3:1) at pH 5, while the extraction of campesterol was the highest when stirring the sample for 3 min, at pH 5, in a mixture of acetone/water (3:1) (Table 5) (Figure S4).

From the 11 depicted polyphenols, catechin and protocatechuic acid were found in the highest quantity, whereas from the three revealed phytosterols, beta-sitosterol presented the highest amount. The influence of the working conditions on the evaluated bioactive compound recovery from HI extracts is depicted in Figure 3. Furthermore, for predicting the recovery of bioactive compounds considering the working conditions, the response surfaces were generated and presented in Figure 4. By analyzing the two figures, it can be concluded that the amount of water in the solvent mixture exhibited the highest impact on the recovery of bioactive compounds, followed by the stirring time. The pH displayed a minimum impact on the extraction yield.

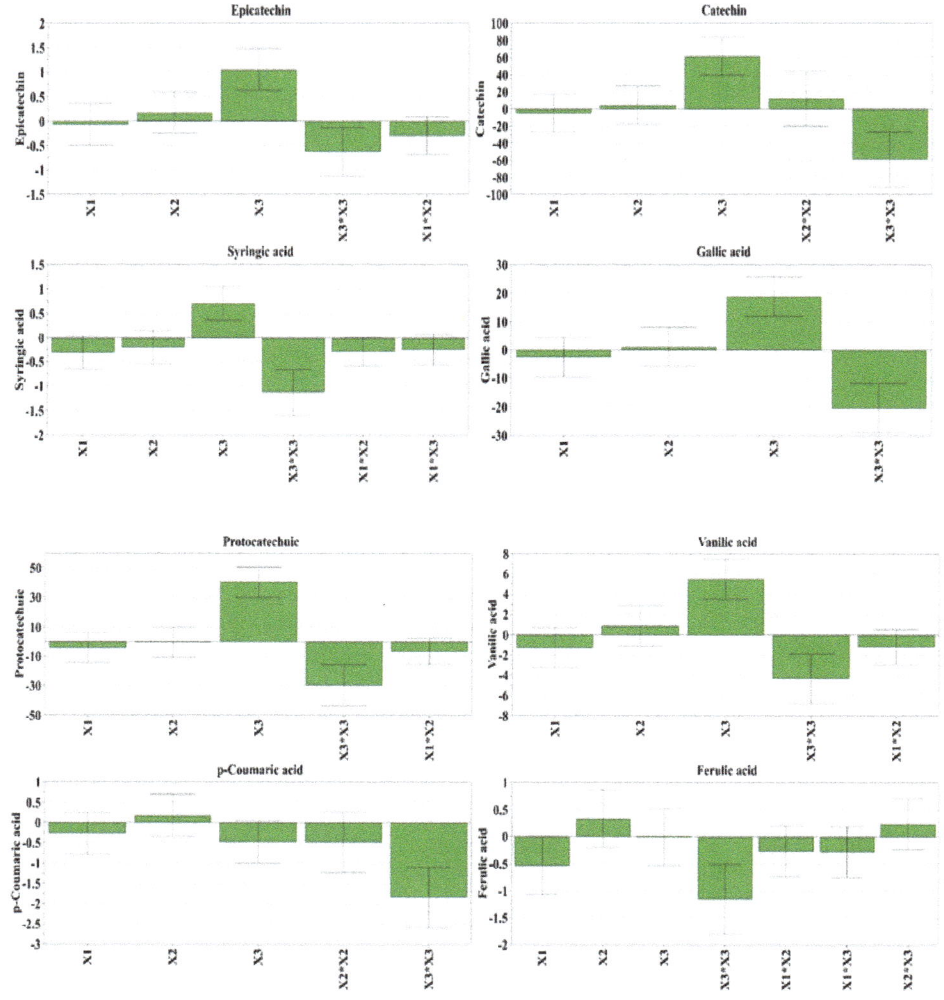

Figure 3. *Cont.*

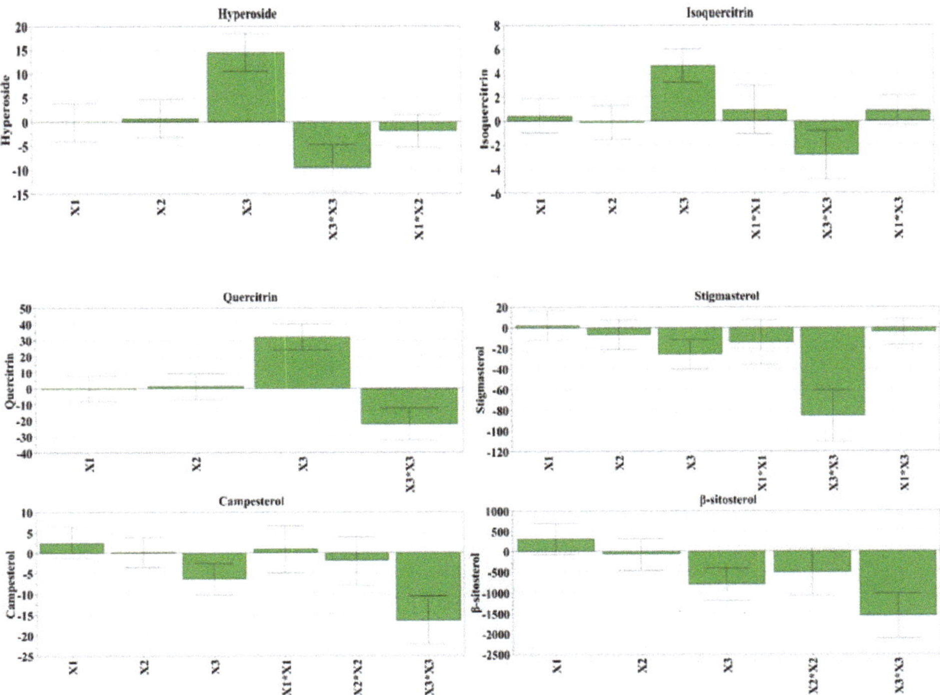

Figure 3. Influence of working conditions on the bioactive compound recovery from hazelnut involucre extracts, depicted as scaled and centered coefficient plots. X_1, stirring time (min); X_2, pH; X_3, water in solvent (% *v/v*).

We consider this information especially valuable because it presents the optimal experimental conditions for obtaining the maximum extraction yield.

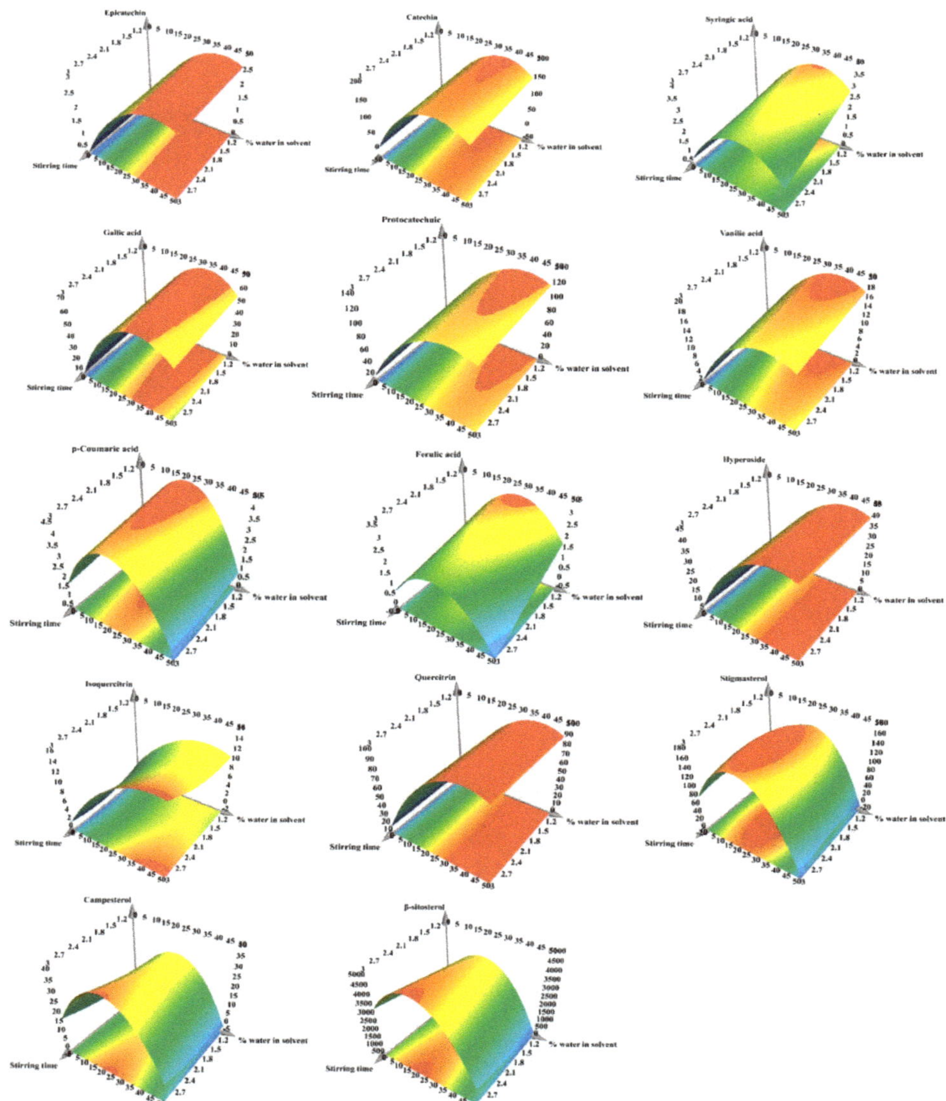

Figure 4. Response surface for prediction of bioactive compound recovery from hazelnut involucre extracts with respect to: X_1, stirring time (min); X_2, pH; X_3, water in solvent (% *v/v*). The regions in red represent the domains of working conditions assuring the maximum extraction yield for the evaluated bioactive compounds.

3.5.1. Quantification of Individual Polyphenols

From the 18 phenolic compounds analyzed by the validated LC-MS method, only five polyphenols (*p*-coumaric acid, ferulic acid, hyperoside, isoquercitrin, quercitrin) were quantified in the HI extracts. Using the LC-MS method II, all six polyphenols (epicatechin, catechin, syringic acid, gallic acid, protocatechuic acid, and vanillic acid) were quantified in the HI extracts. For this method, the coefficient of linearity (R^2) was in the range of 0.9922–0.9997 and the accuracy bias was ≤15% [15].

The highest values for epicatechin and catechin were 3.73 and 243.02 µg/g dw HI, respectively (Table 5). Montella et al. [47] attained higher amounts of epicatechin and catechin, 342 and 2500 µg/g hazelnut skin extracts, a by-product richer in polyphenols.

The maximum amounts for syringic acid, gallic acid, protocatechuic acid, and vanillic acid were 5.53, 91.93, 227.37, and 25.41 µg/g dw HI, respectively. In other matrices, Jakopic et al. [8] found quantities of 0.52 µg gallic acid and 2.92 µg protocatechuic acid per gram of hazelnut kernels, while Montella et al. [47], for the same hydroxybenzoic acids, gallic and protocatechuic acids, obtained concentration of 62.1 and 21.1 µg/g hazelnut skin extracts, respectively.

The best quantities for two cinnamic acid derivatives, *p*-coumaric, and ferulic acids were 6.58 and 3.97 µg/g dw HI, respectively (Table 5). For the same two acids, Shahidi et al. [14] obtained values of 1662 and 327 µg/g green leafy cover extracts, respectively. Again, it is worth stating that many factors, including cultivar type, location, agricultural practices or growing conditions, degree of ripeness, storage conditions, and industrial processing, can affect the chemical composition of tree nut kernels and their by-products [11].

The highest quantities for hyperoside, quercitrin, and isoquercitrin were 51.72, 17.74, and 114.26 µg/g dw HI, respectively.

The 17 hazelnut involucre extracts were hydrolyzed (as previously mentioned) and further analyzed by LC-MS for quantification of the main polyphenolic compounds (epicatechin, catechin, syringic acid, gallic acid, protocatechuic acid, vanillic acid). When comparing the non-hydrolyzed to the hydrolyzed samples (Table S5), the evaluated bio compounds registered a downward trend. This could be explained by the degradation of these polyphenols during the hydrolysis process, which was more significant than their release from the biological matrix or polymeric compounds.

3.5.2. Quantification of Phytosterols

The three phytosterols (stigmasterol, campesterol, beta-sitosterol) found in HI extracts were quantified on the basis of their peak areas and evaluation with a calibration curve of the corresponding standards [25]. The richest extract in campesterol was 45.04 µg/g dw HI, while the highest quantities of stigmasterol and beta-sitosterol were 197.30 and 5305.01 µg/g dw HI extracts, respectively (Table 5).

Phytosterols are known for the decreasing effect against total cholesterol and LDL-cholesterol, important risk factors in cardiovascular diseases [48]. Moreover, the consumption of plant sterols has been shown to increase the antioxidant activity [49]. Our study found quantities of phytosterols in HI which are in line with those in nuts [50] and proved that this by-product can be a source of natural sterols with several health benefits.

3.6. Enzyme Inhibitory Activities

3.6.1. Selection of the Optimal HI Extract

Polyphenols are bioactive compounds with potent antioxidant activity exercised by several mechanisms. In brief, they can act as direct and/or indirect antioxidants modulating many important cellular signaling pathways, such as: The nuclear factor erythroid 2-related factor 2/electrophile-responsive elements (Nrf2/EpRE) (involved in cell protection and detoxification) and the nuclear factor kappa B (NF-κB) (involved in the induction of pro-inflammatory processes). Polyphenols protect cells against free radicals and ROS, inhibiting mutagenesis and carcinogenesis by promoting apoptosis and autophagy in tumor cells. Moreover, they are epigenetic modulators, having real anticancer potential demonstrated in many in vitro and in vivo assays [5]. At the molecular level, extracts rich in polyphenols obtained from tree nut skin or by-products [15] showed inhibitory activity on key enzymes related to pathological conditions including type 2 diabetes mellitus (T2DM), obesity or skin hyperpigmentation [5].

Based on the experimental findings from the screening and optimization steps, the HI extract with the highest content in total polyphenols also displayed the best antioxidant capacity by TEAC assay

(N6, run order 7, Table 3). Besides, this extract is the richest in hyperoside, isoquercitrin, and quercitrin (Table 5), flavonoids which are powerful antioxidants [15]. The extract was further evaluated for other biological activities: The antioxidant activities by two other assays, DPPH and FRAP (results presented above), the tyrosinase and α-glucosidase inhibitory activities, and the cytotoxic and antioxidant activities on tumor and normal cell lines.

3.6.2. Tyrosinase Inhibitory Activity

Tyrosinase is a key enzyme, which catalyzes melanin production, primarily responsible for the pigmentation of human skin, hair, and eyes. It also catalyzes the synthesis of neuromelanin in the human brain, and overproduction of neuromelanin is linked with neuronal damage and neurodegeneration in Parkinson's disease and Huntington's diseases. Tyrosinase was also associated with vegetable and fruit browning, which can lead to rapid postharvest degradation [51]. Tyrosinase inhibition may prevent skin hyperpigmentation and neurodegeneration. Therefore, tyrosinase downregulation, through natural inhibitors of this crucial enzyme, specifically targets melanogenesis in the cell with no side effects.

In our study, the tyrosinase inhibitory activity of 50% aqueous acetone HI extract was 165.17 ± 1.88 mg KAE/g. To the best of our knowledge, there were no previous studies of tyrosinase inhibitory activity of HI, thus no direct comparison could be made. Other plant matrices offered lower results, 129.98 ± 3.03 mg KAE per gram of walnut septum [15] and 2.28 ± 0.01 mg KAE/g of hazelnut extract [36]. Thus, HI may be an alternative natural source as a tyrosinase inhibitor in preventing hyperpigmentation, economical and convenient for the cosmetic and food industry.

3.6.3. α-Glucosidase Inhibitory Activity

Delayed glucose absorption and reducing postprandial blood glucose levels are antidiabetic strategies. The inhibition of the activity of α-glucosidase, an enzyme located on the epithelium of the small intestine, and the slowing of starch metabolism can be part of dietary therapy in the treatment of diabetes. Several plant extracts, rich in phenolic compounds and having structural features which may block the active sites of the enzyme, were shown to inhibit intestinal α-glucosidase [52]. Yeast α-glucosidase assay can be a fast and inexpensive method to screen for potential α-glucosidase inhibitors [27].

In our study, the IC_{50} for the 50% aqueous acetone HI extract was 0.10 mg/mL, which was much lower compared to acarbose, at 0.80 mg/mL, revealing HI extract as a more potent α-glucosidase inhibitor than acarbose, the anti-diabetic drug used in the treatment of T2DM. The enzyme inhibitory value for hazelnut was 7.57 ± 0.01 mmol acarbose equivalents per gram of extract [36].

It is possible that the enzymatic inducing effect of HI extract on both tyrosinase and α-glucosidase may be induced not only by the content of polyphenols but also by other compounds present in the extract. However, further research is needed to complete the phytochemical profile of this by-product and for the elucidation of the mechanisms of action.

3.7. Biological Activities of HI Extract on Cell Lines

3.7.1. Viability Assay

Exposure of T47D-KBluc and A549 cell lines to the polyphenolic richest HI extract resulted in a dose-dependent decrease in cellular viability, with a comparable susceptibility to the extract observed for both cancerous cell lines in both assays. The calculated IC_{50}, defined here as the concentration of the HI extract that induces the cellular death of 50% of the total viable cells, at 24 h and 48 h after exposure of the cell lines, based on the data from Alamar Blue and Neutral Red assays are presented in Table 7. In comparison with the T47D-KBluc cell line, the A549 cell line appeared to be more sensitive to the toxic effects of the HI extracts. In addition, when the A549 cells were exposed for 24 h to the HI extract, a hormetic effect was observed at intermediate concentrations in the case of Alamar Blue assay (Figure 5). The toxicity of the extract was more prominent after 48 h exposure for all the cell lines tested, with IC$_{50}$ reaching almost half of the values observed at 24 h post-exposure in the case of A549 and T47D-KBluc (Table 7). Contrary to the results observed when using cancerous cell lines, exposure of the normal HGF cells to the HI extract resulted in a mild cytotoxic effect, observed only at the highest tested doses.

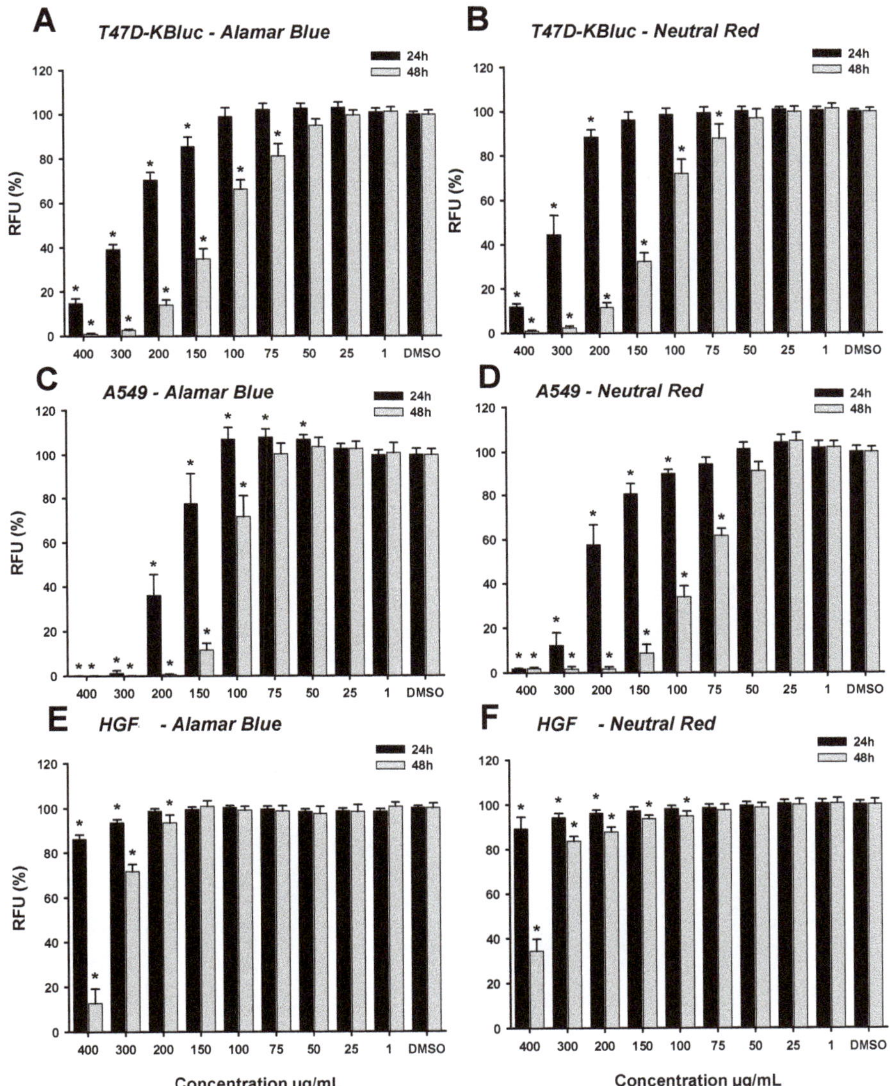

Figure 5. Cytotoxic effect of the HI extract observed using Alamar Blue assay on T47D-KBluc (**A**), A549 (**C**), and HGF (**E**) and using Neutral Red assay on T47D-KBluc (**B**), A549 (**D**), and HGF (**F**). The results are expressed as relative means ± standard deviations (six technical replicates for each of the three biological replicates) where the negative control (DMSO 0.2%) is 100%. (*) indicates significant differences compared to negative control (ANOVA + Dunnett's; $p < 0.05$).

Table 7. IC_{50} values (µg extract/mL) obtained by Alamar Blue and Neutral Red assays for T47D-KBluc, A549, and HGF at 24 h and 48 h post-exposure.

Cell Lines	24 h		48 h	
	Alamar Blue	Neutral Red	Alamar Blue	Neutral Red
T47D-KBluc	281.41 ± 22.7	294.56 ± 4.03	123.62 ± 1.97	125.042 ± 1.15
A549	180.28 ± 4.6	222.10 ± 15.05	112.53 ± 1.44	83.06 ± 1.33
HGF	>400	>400	>300	>300

T47D-KBluc: Human breast cancer; A549: Human lung adenocarcinoma; HGF: Human gingival fibroblasts.

Similarly to the results reported by Gallego et al. [53], the present results corroborate to the idea that extracts from different parts of hazelnut plant can exert a cytotoxic effect on cancerous cell lines. A direct comparison between our study and the previously mentioned study is not possible as only a limited number of higher concentrations were tested in the latter study, while the extracts were obtained from stems and leaves of hazel trees [53]. In agreement with other studies in the literature, due to the presence of active compounds such as polyphenols, flavonoids, or other plant-specific compounds such as taxanes present in *Corylus avellana* L., plant-derived extracts can inhibit the growth of breast, gastric, and pulmonary tumor-derived cell lines [54–57].

3.7.2. DCFH-DA Assay

In order to evaluate a potential antioxidant effect of the HI extract, three concentrations (25, 50, 75 µg/mL) that did not affect the cellular viability were selected and further tested on the two cancerous cell lines and on the normal HGF cells (Figure 6). The treatment with the HI extract alone modified the oxidative status of all three cell lines, with a statistical decrease of ROS being observed at the highest tested doses. Similar to the HI extract, NAC treatment alone led to a decrease in the basal oxidative status compared to the negative control. Exposure to H_2O_2 alone resulted in an approximately threefold increase of the signal when compared to the negative control, while pre-exposure to the antioxidant NAC decreased the ROS generation by approximately 50%. Pre-incubation with HI extract decreased the ROS in a dose-dependent manner, reaching statistical significance at the two highest doses of 50 and 75 µg/mL. In the case of T47D-KBluc cells, at the highest dose, the reduction of oxidative stress was almost equal to the one exerted by NAC.

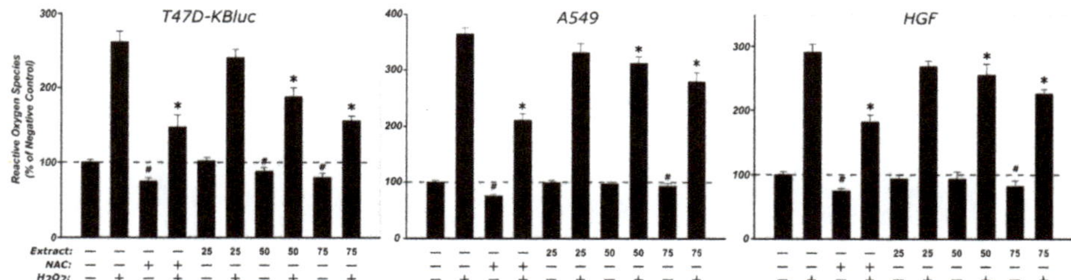

Figure 6. Antioxidant effect of the HI extract evaluated using DCFH-DA assay on T47D-KBluc, A549, and HGF. The cellular model was pre-exposed to the extract (25, 50, and 75 µg/mL) or NAC (20 mM) for 24 h, and further incubated with 50 µM DCFH-DA. The antioxidant effect of the HI extract was assessed after 2 h from oxidative stress induced by exposure to 250 µM H_2O_2. The results are expressed as relative means ± standard deviations (six technical replicates for each of the three biological replicates) where the negative control (DMSO 0.2%) is 100%. (*) indicates significant differences compared to H_2O_2 exposure alone; (#) indicates significant differences compared to negative control (ANOVA + Dunnett's; $p < 0.05$).

The obtained results are in agreement with the previously detailed antioxidant activity observed in the non-cellular assays, but due to insufficient data present in literature, no direct comparison was possible.

4. Conclusions

This study characterized the phytochemical profile of hazelnut (*Corylus avellana* L.) involucre and described the optimal experimental conditions for maximizing the extraction efficiency of biologically active compounds from this tree nut by-product. We aimed to obtain hazelnut involucre extracts with high contents of phenolics, strong antioxidant capacity, and biological potential based on experimental designs. Besides, the phytochemical profile of the extracts using LC-MS methods was assayed. In order to determine the ideal extraction conditions, several parameters (stirring time, pH, solvent) were coupled with chemical analysis and statistical tools. From the eleven polyphenols (epicatechin, catechin, syringic acid, gallic acid, protocatechuic acid, vanillic acid, *p*-coumaric acid, ferulic acid, hyperoside, isoquercitrin, quercitrin) and three phytosterols (stigmasterol, campesterol, beta-sitosterol) determined and quantified, catechin (up to 0.243 mg/g dw HI), protocatechuic acid (up to 0.227 mg/g dw HI), and beta-sitosterol (up to 5.305 mg/g dw HI) were found in the highest amount. From the extraction conditions, the amount of water in the solvent proved to have a statistically significant influence upon all evaluated responses.

The best extraction conditions to attain the richest extract in total phenols were accomplished using turbo-extraction by Ultra-Turrax at stirring time 3 min, pH 3, and 50% acetone in water (*v/v*) as extraction solvent, while the best extraction conditions for sterols were 75% acetone in water (*v/v*), pH 5, and stirring time 2 or 3 min. The richest extract in polyphenols presented strong in vitro antioxidant activity in classical tests (TEAC, DPPH, FRAP) as well as in two cancer cell line assays (T47D-KBluc and A549). Cytotoxic and antioxidant effects of HI extract were more intense on these cancer cells than on a normal cell line (HGF). These results are promising and justify further research to characterize HI as a valuable source of bioactive compounds and to assay its anticancer potential. Moreover, the richest polyphenol HI extract demonstrated good enzyme inhibitory potential on tyrosinase (165.17 ± 1.88 mg KAE/g) and α-glucosidase (eight times stronger inhibition than acarbose), two key enzymes involved in age-related diseases. Phytosterols, compounds with proven cardioprotective effects and antioxidant capacity, are other valuable molecules found in HI.

The results of our study successfully accomplished the proposed objectives and justify future scientific investigations, including a comprehensive identification of individual polyphenols. Moreover, the content in bioactive compounds, correlated with good results for antioxidant and enzyme inhibitory activities, warrants more research in order to understand the bioavailability of specific molecules and to reveal their potential mechanisms of action for future use of hazelnut involucre in the cosmetic industry, in addition to food and pharmaceutical production.

Supplementary Materials: The following are available online at http://www.mdpi.com/2076-3921/8/10/460/s1, Figure S1: The chromatograms for the individual polyphenolic compounds quantified by the LC-MS method II according to the retention time and m/z values in Supplementary Table S2, Figure S2: The chromatograms for the individual polyphenolic compounds quantified by an LC-MS method according to the retention time and m/z values in Supplementary Table S1, Figure S3: The chromatograms for the individual polyphenolic compounds quantified by an LC-MS method according to the retention time and m/z values in Supplementary Table S1, Figure S4: The chromatograms for phytosterols quantified by an LC-MS/MS method according to the retention time and m/z values in Supplementary Table S3; Table S1: Retention times and values of m/z for the individual polyphenolic compounds, Table S2: Retention times and values of m/z for the individual polyphenolic compounds quantified by LC-MS method II, Table S3: Retention times (RT) and specific ions of the phytosterols, Table S4: Regression equation coefficients for total bioactive compounds in hazelnut involucre extracts, Table S5: Quantitative evaluation of the recovery of main bioactive compounds in non-hydrolyzed and hydrolyzed samples of hazelnut involucre extracts, Table S6: Statistical parameters after data analyze and fit with factorial model for bioactive compounds in hazelnut involucre extracts, Table S7: Regression equation coefficients for individual bioactive compounds in hazelnut involucre extracts.

Author Contributions: Conceptualization, M.E.R., A.M., L.V., D.-S.P., and I.T.; methodology, M.E.R., A.M., D.-S.P., and I.T.; software, M.E.R., A.-M.G., I.F., A.P., and I.T.; formal analysis, M.E.R., A.-M.G., I.F., A.P., A.M., L.V., and D.-S.P.; investigation, M.E.R., I.F., A.P., A.M., L.V., and I.T.; supervision, L.V., G.C., F.L., D.-S.P., and I.T.; writing—original draft preparation, M.E.R., A.-M.G., I.F., A.P., and D.-S.P.; writing—review and editing, M.E.R., A.-M.G., I.F., A.P., A.M., G.C., L.V., F.L., D.-S.P., and I.T.

Funding: This research was funded in part by "Iuliu Hatieganu" University of Medicine and Pharmacy in Cluj-Napoca, Romania through a Ph.D. grant (PCD No. 1529/60/18.01.2019).

Conflicts of Interest: The authors declare no conflict of interest.

References

1. Chen, G.; Zhang, R.; Martínez-González, M.; Zhang, Z.; Bonaccio, M.; van Dam, R.; Qin, L. Nut consumption in relation to all-cause and cause-specific mortality: A meta-analysis 18 prospective studies. *Food Funct.* **2017**, *18*, 3893–3905. [CrossRef] [PubMed]

2. Carughi, A.; Feeney, M.J.; Kris-Etherton, P.; Fulgoni, V., III; Kendall, C.W.C.; Bulló, M.; Webb, D. Pairing nuts and dried fruit for cardiometabolic health. *Nutr. J.* **2016**, *15*, 23. [CrossRef] [PubMed]

3. Poulose, S.M.; Miller, M.G.; Shukitt-Hale, B. Role of walnuts in maintaining brain health with age. *J. Nutr.* **2014**, *144*, 561S–566S. [CrossRef] [PubMed]

4. Arias-Fernández, L.; Machado-Fragua, M.; Graciani, A.; Guallar-Castillón, P.; Banegas, J.; Rodríguez-Artalejo, F.; Lana, A.; Lopez-Garcia, E. Prospective Association Between Nut Consumption and Physical Function in Older Men and Women. *J. Gerontol. A Biol. Sci. Med. Sci.* **2018**, *74*, 1091–1097.

5. Rusu, M.E.; Simedrea, R.; Gheldiu, A.-M.; Mocan, A.; Vlase, L.; Popa, D.-S.; Ferreira, I.C.F.R. Benefits of tree nut consumption on aging and age-related diseases: Mechanisms of actions. *Trends Food Sci. Technol.* **2019**, *88*, 104–120. [CrossRef]

6. Rusu, M.E.; Mocan, A.; Ferreira, I.C.F.R.; Popa, D.-S. Health Benefits of Nut Consumption in Middle-Aged and Elderly Population. *Antioxidants* **2019**, *8*, 302. [CrossRef] [PubMed]

7. Alasalvar, C.; Shahidi, F.; Liyanapathirana, C.M.; Ohshima, T. Turkish Tombul Hazelnut (*Corylus avellana* L.). 1. Compositional Characteristics. *J. Agric. Food Chem.* **2003**, *51*, 3790–3796. [CrossRef]

8. Jakopic, J.; Petkovsek, M.M.; Likozar, A.; Solar, A.; Stampar, F.; Veberic, R. HPLC-MS identification of phenols in hazelnut (*Corylus avellana* L.) kernels. *Food Chem.* **2011**, *124*, 1100–1106. [CrossRef]

9. Alasalvar, C.; Shahidi, F. Natural antioxidants in tree nuts. *Eur. J. Lipid Sci. Technol.* **2009**, *111*, 1056–1062. [CrossRef]

10. Hever, J.; Cronise, R.J. Plant-based nutrition for healthcare professionals: Implementing diet as a primary modality in the prevention and treatment of chronic disease. *J. Geriatr. Cardiol.* **2017**, *14*, 355–368.

11. Rusu, M.E.; Gheldiu, A.-M.; Mocan, A.; Vlase, L.; Popa, D.-S. Anti-aging potential of tree nuts with a focus on phytochemical composition, molecular mechanisms and thermal stability of major bioactive compounds. *Food Funct.* **2018**, *9*, 2554–2575. [CrossRef] [PubMed]

12. Shahidi, F.; Ambigaipalan, P. Phenolics and polyphenolics in foods, beverages and spices: Antioxidant activity and health effects — A review. *J. Funct. Foods* **2015**, *18*, 820–897. [CrossRef]

13. Sánchez-González, C.; Ciudad, C.J.; Noé, V.; Izquierdo-Pulido, M. Health benefits of walnut polyphenols: An exploration beyond their lipid profile. *Crit. Rev. Food Sci. Nutr.* **2017**, *57*, 3373–3383. [CrossRef] [PubMed]

14. Shahidi, F.; Alasalvar, C.; Liyana-Pathirana, C. Antioxidant phytochemicals in hazelnut kernel (*Corylus avellana* L.) and hazelnut byproducts. *J. Agric. Food Chem.* **2007**, *55*, 1212–1220. [CrossRef] [PubMed]

15. Rusu, M.E.; Gheldiu, A.-M.; Mocan, A.; Moldovan, C.; Popa, D.-S.; Tomuta, I.; Vlase, L. Process Optimization for Improved Phenolic Compounds Recovery from Walnut (*Juglans regia* L.) Septum: Phytochemical Profile and Biological Activities. *Molecules* **2018**, *23*, 2814. [CrossRef] [PubMed]

16. Mocan, A.; Zengin, G.; Simirgiotis, M.; Schafberg, M.; Mollica, A.; Vodnar, D.C.; Crişan, G.; Rohn, S. Functional constituents of wild and cultivated Goji (*L. barbarum* L.) leaves: Phytochemical characterization, biological profile, and computational studies. *J. Enzym. Inhib. Med. Chem.* **2017**, *32*, 153–168. [CrossRef] [PubMed]

17. Mocan, A.; Schafberg, M.; Crisan, G.; Rohn, S. Determination of lignans and phenolic components of Schisandra chinensis (Turcz.) Baill. using HPLC-ESI-ToF-MS and HPLC-online TEAC: Contribution of individual components to overall antioxidant activity and comparison with traditional antioxidant assays. *J. Funct. Foods* **2016**, *24*, 579–594. [CrossRef]

18. Price, M.L.; Van Scoyoc, S.; Butler, L.G. A Critical Evaluation of the Vanillin Reaction as an Assay for Tannin in Sorghum Grain. *J. Agric. Food Chem.* **1978**, *26*, 1214–1218. [CrossRef]

19. Mocan, A.; Fernandes, A.; Barros, L.; Crişan, G.; Smiljkovic, M.; Sokovic, M.; Ferreira, I. Chemical composition and bioactive properties of the wild mushroom Polyporus squamosus (Huds.) Fr: A study with samples from Romania. *Food Funct.* **2018**, *9*, 160–170. [CrossRef] [PubMed]

20. Damiano, S.; Forino, M.; De, A.; Vitali, L.A.; Lupidi, G.; Taglialatela-Scafati, O. Antioxidant and antibiofilm activities of secondary metabolites from Ziziphus jujuba leaves used for infusion preparation. *Food Chem.* **2017**, *230*, 24–29. [CrossRef] [PubMed]

21. Mocan, A.; Vlase, L.; Raita, O.; Hanganu, D.; Paltinean, R.; Dezsi, S.; Gheldiu, A.M.; Oprean, R.; Crisan, G. Comparative studies on antioxidant activity and polyphenolic content of *Lycium barbarum* L. and Lycium chinense Mill. leaves. *Pak. J. Pharm. Sci.* **2015**, *28*, 1511–1515. [PubMed]

22. Pop, C.E.; Pârvu, M.; Arsene, A.L.; Pârvu, A.E.; Vodnar, D.C.; Tarcea, M.; Toiu, A.M.; Vlase, L. Investigation of antioxidant and antimicrobial potential of some extracts from *Hedera helix* L. *Farmacia* **2017**, *65*, 624–629.

23. Babotă, M.; Mocan, A.; Vlase, L.; Crisan, O.; Ielciu, I.; Gheldiu, A.M.; Vodnar, D.C.; Crişan, G.; Păltinean, R. Phytochemical analysis, antioxidant and antimicrobial activities of Helichrysum arenarium (L.) Moench. and Antennaria dioica (L.) Gaertn. Flowers. *Molecules* **2018**, *23*, 409. [CrossRef] [PubMed]

24. Toiu, A.; Mocan, A.; Vlase, L.; Pârvu, A.E.; Vodnar, D.C.; Gheldiu, A.M.; Moldovan, C.; Oniga, I. Phytochemical composition, antioxidant, antimicrobial and in vivo anti-inflammatory activity of traditionally used Romanian Ajuga laxmannii (Murray) Benth. ("Nobleman's beard" - barba împăratului). *Front. Pharmacol.* **2018**, *9*, 7. [CrossRef] [PubMed]

25. Vlase, L.; Parvu, M.; Parvu, E.A.; Toiu, A. Chemical constituents of three Allium species from Romania. *Molecules* **2013**, *18*, 114–127. [CrossRef] [PubMed]

26. Masuda, T.; Fujita, N.; Odaka, Y.; Takeda, Y.; Yonemori, S.; Nakamoto, K.; Kuninaga, H. Tyrosinase inhibitory activity of ethanol extracts from medicinal and edible plants cultivated in okinawa and identification of a water-soluble inhibitor from the leaves of Nandina domestica. *Biosci. Biotechnol. Biochem.* **2007**, *71*, 2316–2320. [CrossRef] [PubMed]

27. Mocan, A.; Diuzheva, A.; Carradori, S.; Andruch, V.; Massafra, C.; Moldovan, C.; Sisea, C.; Petzer, J.P.; Petzer, A.; Zara, S.; et al. Development of novel techniques to extract phenolic compounds from Romanian cultivars of Prunus domestica L. and their biological properties. *Food Chem. Toxicol.* **2018**, *119*, 189–198. [CrossRef] [PubMed]

28. Albuquerque, B.R.; Prieto, M.A.; Vazquez, J.A.; Barreiro, M.F.; Barros, L.; Ferreira, I.C.F.R. Recovery of bioactive compounds from Arbutus unedo L. fruits: Comparative optimization study of maceration/microwave/ultrasound extraction techniques. *Food Res. Int.* **2018**, *109*, 455–471. [CrossRef] [PubMed]

29. Alasalvar, C.; Karamać, M.; Amarowicz, R.; Shahidi, F. Antioxidant and antiradical activities in extracts of hazelnut kernel (*Corylus avellana* L.) and hazelnut green leafy cover. *J. Agric. Food Chem.* **2006**, *54*, 4826–4832. [CrossRef] [PubMed]

30. Siriwardhana, S.S.; Shahidi, F. Antiradical Activity of Extracts of Almond and Its By-products. *JAOCS* **2002**, *79*, 903–908. [CrossRef]

31. Oliveira, I.; Sousa, A.; Morais, J.S.; Ferreira, I.C.F.R.; Bento, A.; Estevinho, L.; Pereira, J.A. Chemical composition, and antioxidant and antimicrobial activities of three hazelnut (*Corylus avellana* L.) cultivars. *Food Chem. Toxicol.* **2008**, *46*, 1801–1807. [CrossRef] [PubMed]

32. Locatelli, M.; Coïsson, J.D.; Travaglia, F.; Bordiga, M.; Arlorio, M. Impact of Roasting on Identification of Hazelnut (*Corylus avellana* L.) Origin: A Chemometric Approach. *J. Agric. Food Chem.* **2015**, *63*, 7294–7303. [CrossRef] [PubMed]

33. Esposito, T.; Sansone, F.; Franceschelli, S.; Del Gaudio, P.; Picerno, P.; Aquino, R.P.; Mencherini, T. Hazelnut (*Corylus avellana* L.) Shells Extract: Phenolic Composition, Antioxidant Effect and Cytotoxic Activity on Human Cancer Cell Lines. *Int. J. Mol. Sci.* **2017**, *18*, 392. [CrossRef] [PubMed]

34. Masullo, M.; Cerulli, A.; Mari, A.; de Souza Santos, C.C.; Pizza, C.; Piacente, S. LC-MS profiling highlights hazelnut (Nocciola di Giffoni PGI) shells as a byproduct rich in antioxidant phenolics. *Food Res. Int.* **2017**, *101*, 180–187. [CrossRef] [PubMed]

35. Faggio, C.; Sureda, A.; Morabito, S.; Sanches-Silva, A.; Mocan, A.; Nabavi, S.F.; Nabavi, S.M. Flavonoids and platelet aggregation: A brief review. *Eur. J. Pharmacol.* **2017**, *807*, 91–101. [CrossRef] [PubMed]

36. Mollica, A.; Zengin, G.; Stefanucci, A.; Ferrante, C.; Menghini, L.; Orlando, G.; Brunetti, L.; Locatelli, M.; Dimmito, M.P.; Novellino, E.; et al. Nutraceutical potential of Corylus avellana daily supplements for obesity and related dysmetabolism. *J. Funct. Foods* **2018**, *47*, 562–574. [CrossRef]

37. Gorji, N.; Moeini, R.; Memariani, Z. Almond, hazelnut and walnut, three nuts for neuroprotection in Alzheimer's disease: A neuropharmacological review of their bioactive constituents. *Pharmacol. Res.* **2018**, *129*, 115–127. [CrossRef]

38. Kim, Y.; Keogh, J.; Clifton, P. Benefits of nut consumption on insulin resistance and cardiovascular risk factors: Multiple potential mechanisms of actions. *Nutrients* **2017**, *9*, 1271. [CrossRef]

39. Lainas, K.; Alasalvar, C.; Bolling, B.W. Effects of roasting on proanthocyanidin contents of Turkish Tombul hazelnut and its skin. *J. Funct. Foods* **2016**, *23*, 647–653. [CrossRef]

40. Bolling, B.W. Almond Polyphenols: Methods of Analysis, Contribution to Food Quality, and Health Promotion. *Compr. Rev. Food Sci. Food Saf.* **2017**, *16*, 346–368. [CrossRef]

41. Arcan, I.; Yemeniciog, A. Antioxidant activity and phenolic content of fresh and dry nuts with or without the seed coat. *J. Food Compost. Anal.* **2009**, *22*, 184–188. [CrossRef]

42. Delgado, T.; Malheiro, R.; Pereira, J.A.; Ramalhosa, E. Hazelnut (*Corylus avellana* L.) kernels as a source of antioxidants and their potential in relation to other nuts. *Ind. Crop. Prod.* **2010**, *32*, 621–626. [CrossRef]

43. Oliveira, I.; Sousa, A.; Valentão, P.; Andrade, P.B.; Ferreira, I.C.F.R.; Ferreres, F.; Bento, A.; Seabra, R.; Estevinho, L.; Pereira, J.A. Hazel (*Corylus avellana* L.) leaves as source of antimicrobial and antioxidative compounds. *Food Chem.* **2007**, *105*, 1018–1025. [CrossRef]

44. Yuan, B.; Lu, M.; Eskridge, K.M.; Isom, L.D.; Hanna, M.A. Extraction, identification, and quantification of antioxidant phenolics from hazelnut (*Corylus avellana* L.) shells. *Food Chem.* **2018**, *244*, 7–15. [CrossRef] [PubMed]

45. Bolling, B.W.; McKay, D.L.; Blumberg, J.B. The phytochemical composition and antioxidant actions of tree nuts. *Asia Pac. J. Clin. Nutr.* **2010**, *19*, 117–123. [PubMed]

46. Alasalvar, C.; Bolling, B. Review of nut phytochemicals, fat-soluble bioactives, antioxidant components and health effects. *Br. J. Nutr.* **2015**, *113*, S68–S78. [CrossRef] [PubMed]

47. Montella, R.; Coisson, J.D.; Travaglia, F.; Locatelli, M.; Malfa, P.; Martelli, A.; Arlorio, M. Bioactive compounds from hazelnut skin (*Corylus avellana* L.): Effects on Lactobacillus plantarum P17630 and Lactobacillus crispatus P17631. *J. Funct. Foods* **2013**, *5*, 306–315. [CrossRef]

48. Ras, R.T.; Geleijnse, J.M.; Trautwein, E.A. LDL-cholesterol-lowering effect of plant sterols and stanols across different dose ranges: A meta-analysis of randomised controlled studies. *Br. J. Nutr.* **2014**, *112*, 214–219. [CrossRef] [PubMed]

49. Hsu, C.C.; Kuo, H.C.; Huang, K.E. The Effects of Phytosterols Extracted from Diascorea alata on the Antioxidant Activity, Plasma Lipids, and Hematological Profiles in Taiwanese Menopausal Women. *Nutrients* **2017**, *9*, 1320. [CrossRef]

50. Phillips, K.; Ruggio, D.; Ashraf-Khorassani, M. Phytosterol Composition of Nuts and Seeds Commonly Consumed in the United States. *J. Agric. Food Chem.* **2005**, *53*, 9436–9445. [CrossRef]

51. Pillaiyar, T.; Manickam, M.; Namasivayam, V. Skin whitening agents: Medicinal chemistry perspective of tyrosinase inhibitors. *J. Enzym. Inhib. Med. Chem.* **2017**, *32*, 403–425. [CrossRef] [PubMed]

52. Mocan, A.; Moldovan, C.; Zengin, G.; Bender, O.; Locatelli, M.; Simirgiotis, M.; Atalay, A.; Vodnar, D.C.; Rohn, S.; Crişan, G. UHPLC-QTOF-MS analysis of bioactive constituents from two Romanian Goji (*Lycium barbarum* L.) berries cultivars and their antioxidant, enzyme inhibitory, and real-time cytotoxicological evaluation. *Food Chem. Toxicol.* **2018**, *115*, 414–424. [CrossRef] [PubMed]

53. Gallego, A.; Metón, I.; Baanante, I.V.; Ouazzani, J.; Adelin, E.; Palazon, J.; Bonfill, M.; Moyano, E. Viability-reducing activity of Coryllus avellana L. extracts against human cancer cell lines. *Biomed. Pharmacother.* **2017**, *89*, 565–572. [CrossRef] [PubMed]

54. Bestoso, F.; Ottaggio, L.; Armirotti, A.; Balbi, A.; Damonte, G.; Degan, P.; Mazzei, M.; Cavalli, F.; Ledda, B.; Miele, M. In vitro cell cultures obtained from different explants of Corylus avellana produce Taxol and taxanes. *BMC Biotechnol.* **2006**, *6*, 45. [CrossRef] [PubMed]

55. Danihelová, M.; Veverka, M.; Šturdík, E.; Jantová, S. Antioxidant action and cytotoxicity on HeLa and NIH-3T3 cells of new quercetin derivatives. *Interdiscip. Toxicol.* **2013**, *6*, 209–216.

56. Jamshidi, M.; Ghanati, F. Taxanes content and cytotoxicity of hazel cells extract after elicitation with silver nanoparticles. *Plant Physiol. Biochem.* **2017**, *110*, 178–184. [CrossRef]
57. Glei, M.; Fischer, S.; Lamberty, J.; Ludwig, D.; Lorkowski, S.; Schlörmann, W. Chemopreventive Potential of In Vitro Fermented Raw and Roasted Hazelnuts in LT97 Colon Adenoma Cells. *Anticancer Res.* **2018**, *38*, 83–93.

 *antioxidants*

Article

In Vitro Toxicity Assessment of Stilbene Extract for Its Potential Use as Antioxidant in the Wine Industry

Concepción Medrano-Padial [1], María Puerto [1], F. Javier Moreno [2], Tristan Richard [3], Emma Cantos-Villar [4] and Silvia Pichardo [1,*]

[1] Area of Toxicology, Faculty of Pharmacy, C/Profesor García González n°2, Universidad de Sevilla, 41012 Seville, Spain; cmpadial@us.es (C.M.-P.); mariapuerto@us.es (M.P.)
[2] Area of Cellular Biology, Faculty of Biology, Avda. Reina Mercedes s/n, Universidad de Sevilla, 41012 Seville, Spain; onorato@us.es
[3] Faculté des Sciences Pharmaceutiques, Unité de Recherche OEnologie EA 4577, USC 1366 INRA, Equipe Molécules d'Intérêt Biologique (Gesvab) - Institut des Sciences de la Vigne et du Vin, Université de Bordeaux, CS 50008 - 210, Chemin de Leysotte, 33882 Villenave d'Ornon, France; tristan.richard@u-bordeaux.fr
[4] Instituto de Investigación y Formación Agraria y Pesquera (IFAPA), Centro Rancho de la Merced, Consejería de Agricultura, Ganadería, Pesca y Desarrollo Sostenible (Junta de Andalucía). Cañada de la Loba, 11471 Jerez de la Frontera, Spain; emma.cantos@juntadeandalucia.es
* Correspondence: spichardo@us.es; Tel.: +34-954-556762

Received: 19 September 2019; Accepted: 7 October 2019; Published: 9 October 2019

Abstract: The reduction of sulfur dioxide in wine is a consumer's demand, considering the allergic effects that may occur in people who are sensitive to it. Stilbenes are candidates of great interest for this purpose because of their antioxidant/antimicrobial activities and health properties, and also because they are naturally found in the grapevine. In the present study, the in vitro toxicity of an extract from grapevine shoots (with a stilbene richness of 45.4%) was assessed in two human cell lines. Significant damage was observed from 30 μg/mL after 24 h, and 40 μg/mL after 48 h of exposure. Similarly, the ultrastructural study revealed a significant impairment of cell growing. The extract was able to protect cells against an induced oxidative stress at all concentrations studied. In view of the promising results, a more exhaustive toxicological assessment of the extract is needed to confirm the safety of its further use as additive in wine.

Keywords: cytotoxicity; stilbene; wine; antioxidant; Caco-2; Hep-G2

---

1. Introduction

The most widely used preservative in the wine industry is sulfur dioxide (SO_2) so far. Nevertheless, the exposure to SO_2 may have health side effects such as dermatitis, urticarial, angioedema, diarrhea, abdominal pain, bronchoconstriction, and anaphylaxis [1]. Therefore, alternatives to the use of SO_2 in wines are required. Moreover, the International Organization of Vine and Wine (OIV), in agreement with previous European Commission regulations (Ruling n° 606/2009) [2], recommended that the total SO_2 content should not exceed 150 mg/L in red wines and 200 mg/L in white and rosé organic/conventional wines [3]. In this regard, the wine industry is developing strategies to reduce and/or replace SO_2. The most promising tools for the replacement of SO_2 in wine are the use of physical methods (ultrasounds, ultraviolet, pulsed electric fields, high hydrostatic pressure, etc.) and the addition of different compounds (dimethyl dicarbonate, bacteriocin, phenolic compounds, enzymes, colloidal silver complex, etc.) [1]. An alternative must be accomplished with SO_2 action on antioxidant and antimicrobial capacity, wine oenological parameters, and organoleptic characters. However, despite the antimicrobial and antioxidant properties presented by these methods, these technologies require complex and expensive equipment. Most importantly, none of them have proven to be as effective as

SO_2 by itself so far [1,3]. For the scientific community and the wine industry, it is a challenge developing new alternatives to completely or partly eliminate the use of SO_2 in the winemaking process to produce healthier wines but maintaining the quality requirements of consumers [4]. Consumers' demand for natural food additives can lead more food manufacturers to substitute synthetic antioxidants for natural antioxidant compounds [5]. In this sense, the use of phenolic compounds should be highlighted due to the favorable results recently obtained. Natural extracts rich in stilbenes have been assayed for this purpose. Grape stems extracts are especially rich in flavonoids and stilbenes, with high concentrations of *trans*-resveratrol and ε-viniferin, showing high antioxidant activity and good antimicrobial properties [6–9]. Thus, wines treated with extracts obtained from grapevine shoots have reported excellent enological parameters, higher color intensity, and purity than wines treated with SO_2, and satisfactory organoleptic features analysis [10,11]. Therefore, grapevine shoots have a promising development as a source of naturally available additives.

Despite natural additives are perceived as posing no health risk to consumers, the safety of these compounds needs to be assured before its commercialization [5]. In this concern, the toxicological studies required by the European Food Safety Authority (EFSA) comprise toxicokinetics, genotoxicity, and in vivo toxicity studies [12]. The first approach to the toxicity effect of any compound should be the in vitro cytotoxicity tests to define basal cytotoxicity, directly related to cell death induction. These studies are very useful to set the concentration range to perform further in vitro testing (genotoxicity studies) and confirm its safety to be used in the food industry [13].

In this regard, the cytotoxicity of different stilbenes have been already assayed in different cell lines. The viability of cultured macrophages, tumor-derived human T cells, and human epidermoid carcinoma cells have been reported to be inhibited by resveratrol and piceatannol in a concentration-dependent manner [14]. Vineatrol® (an extract of grapevine-shoot containing resveratrol and other stilbenes) has exhibited a higher antiproliferative effect than resveratrol *per se* in various cancer cells assayed in vitro [15–17]. However, resveratrol and Vineatrol®30 exerted a concentration-dependent cytotoxic effect on V79 cells, the effect being much more pronounced with resveratrol than with Vineatrol® [8]. Therefore, although the effect of some of the major stilbenes contained in the grapevine shoot extract is known, the concomitant presence of different stilbenes and other substances may modulate the individual response. As instance, Vineatrol® has a different antiproliferative activity than their main compounds, with a possible synergistic effect of both resveratrol and ε-viniferin [17]. For this reason, the cytotoxicity study of the extract is mandatory considering that it is not possible to infer the summative effect of stilbenes.

In addition, it is known that stilbenes have antibacterial, antifungal, cardioprotective, neuroprotective, and pharmacological properties including antiaging effects [18,19]. While its antioxidant ability can be highlighted, stilbenes are capable of scavenging or activating cellular-enzymatic antioxidant defenses decreasing the production of intracellular reactive oxygen species (ROS) [20]. However, phenolic compounds usually exhibit both antioxidant and prooxidant activities at different doses [21,22]. Consequently, the scavenging activity of the grapevine shoot extract is studied in the present work together with the oxidative stress status and the protective and reversion properties of this extract against an oxidant agent in Caco-2 and HepG2 cell lines.

Considering the promising use of this stilbene-rich extract in the wine industry, the present work studies the cytotoxicity of the extract, including Caco-2 (colorectal adenocarcinoma cells) and HepG2 (epithelial liver cancer cells). Furthermore, an exhaustive morphological assay was carried out in order to evidence ultrastructural cellular injures that would clarify the mechanism of action of the extract. In addition, the alteration in the oxidative status and glutathione (GSH) content as well as the protective/reversion effect were investigated in both cell lines after short term exposure to this extract.

2. Materials and Methods

2.1. Supplies, Chemicals and Model Systems

Culture medium, fetal bovine serum, and cell culture reagents were provided by Gibco (Biomol, Sevilla, Spain). The rest of the chemicals were purchased from Sigma-Aldrich (Madrid, Spain) and VWR International Eurolab (Barcelona, Spain).

The human cell lines Caco-2 and HepG2 were maintained as described in Gutierrez-Praena et al. (2012) [23].

2.2. Stilbene-Enriched Extract

The method used for the preparation of stilbene extract was reported in a previous work [24]. Grapevine shoots were harvested in 2015 in Bordeaux region (France) and were composed of a mixture of Merlot and Cabernet Sauvignon varieties of *Vitis vinifera*. Before extraction, shoots were dried in open air for at least two months. Finely ground grapevine shoots were extracted with two times 20 L of acetone-water (6:4, *v/v*) at room temperature under agitation, twice for 12 h. After filtration, acetone was removed by evaporation under reduced pressure and the aqueous phase was lyophilized. Finally, the extract was deposited on an Amberlite XAD-7 column and washed with water. The column was then eluted with acetone. The efficiency of this process is 5.5% giving 55 g of stilbene extract per kilogram of stems.

Furthermore, the extract was fractionated by centrifugal partition chromatography (CPC) using the method of Biais et al. (2017) [7]. Briefly, the extract was eluted using the biphasic Arizona solvent system K in descending mode with the organic phase acting as stationary phase. For each run, 10 g of the extract were injected. Peak detection was monitored at 254, 280, 306, and 320 nm leading to six fractions. Only the fractions containing stilbenes were collected. The nine main stilbenes were quantified by HPLC-DAD, indicating that the stilbene-enriched extract contained at least 45.38% ± 5% of total stilbenes (*w/w*) [25]. Compounds were identified by UV spectrum and retention time from standards. *Trans*-resveratrol was quantified at 306 nm as *trans*-resveratrol; piceatannol was quantified at 320 nm as *trans*-piceid; ε-viniferin, r2-viniferin and ω-viniferin were quantified at 320 nm as ε-viniferin; hopeaphenol, isohopeaphenol, pallidol, miyabenol, and ampelopsin A were quantified at 280 nm as ampelopsin A [25].

2.3. Test Solutions

The range of the extract concentrations for the cytotoxicity tests was selected considering the concentration to be incorporated in wine (100 µg/mL). Serial test solutions (0–100 µg/mL) were prepared from stock solution (1000 µg/mL) in dimethyl sulfoxide (DMSO), being the final concentration in DMSO below 0.5%.

Concentrations of the extract used for both oxidative stress assays were calculated based on the cytotoxicity study previously performed. The mean effective concentration (EC_{50}) value obtained (31.18 µg/mL and 20.56 µg/mL in HepG2, and 55.77 µg/mL and 39.02 µg/mL in Caco-2 cells for 24 h and 48 h, respectively) was chosen as the highest exposure concentration, along with the fractions $EC_{50/2}$ and $EC_{50/4}$.

To measure protection and reversion abilities, test solutions of the extract to cell viability greater than 75% were selected ($EC_{50/2}$ and $EC_{50/4}$).

2.4. Cytotoxicity Assays

The exposure concentrations were prepared in medium, with the highest concentration being 100 µg/mL. For the control group culture medium without the extract was used. Moreover, a control of solvent (0.5% of DMSO) has also been included. The exposure concentrations were added to both cells and the cytotoxicity assays were performed after 24 and 48 h of exposure.

Neutral red uptake (NR) was evaluated according to Borenfreund and Puerner (1984) [26] with modifications [27]. MTS (3-(4,5-dimethylthiazol-2-yl)-5-(3-carboxymethoxyphenyl)- 2-(4-sulfophenyl)-2H- tetrazolium salt) reduction was measured according to the procedure of Baltrop et al. (1991) [28]. Moreover, total protein content (TP) was performed according to the procedure described by Bradford (1976) [29].

2.5. Morphological Study under Light and Transmission Electron Microscope

Light and electron microscope observations were performed according to Gutiérrez-Praena et al. (2019) [30]. Cultured cells were exposed to three different concentrations of the extract, the EC_{50} value, and their fractions ($EC_{50}/2$, $EC_{50}/4$). HepG2 were exposed to 7.79, 15.59 and 31.18 µg/mL and Caco-2 cells were exposed to 13.90, 27.88, and 55.77 µg/mL.

2.6. Oxidative Stress and Antioxidant Ability Assays

The oxidative stress endpoints measured, ROS content and GSH levels, were carried out following the methods described in Gutiérrez-Praena et al. (2012) [23]. The results of both assays were expressed as relative light units (RLU).

For the estimation of the protection and reversion abilities of extract, H_2O_2 100 µM was administered to induced changes in the cell membranes and antioxidant system in HepG2 and Caco-2 cells [31].

For the protection assay, after discarding the previous medium, exposure solutions ($EC_{50/2}$ and $EC_{50/4}$) of the extract were first added to the cells and incubated at 37 °C for 24 h or 48 h. After the treatment time, the medium was discarded and then exposed to 100 µM H_2O_2 for 2 h. Similarly, Caco-2 and HepG2 were pre-treated with H_2O_2 for 2 h for the reversion assay and a later exposure of the extract for 24 or 48 h. Unexposed cells were included in the figures in order to compare the results with basal levels of ROS and GSH. A control of DMSO was also incorporated in all plates.

Then, both abilities were evaluated by measuring the ROS levels and GSH content as previously described.

2.7. Calculations and Statistical Analysis

All experiments were performed three times per assay. The data for all experiments were given as the arithmetic mean percentage ± standard deviation in relation to control. Statistical analysis was performed using analysis of variance (ANOVA) followed by Dunnett's multiple comparison tests. Differences were considered significant in respect to the control group at $p < 0.05$ (*), $p < 0.01$ (**) and at $p < 0.001$ (***).

3. Results

3.1. Presence of Stilbenes in the Extract

The main compounds were ε-viniferin (16.34%, *w/w*), *trans*-resveratrol (8.07%, *w/w*), isohopeaphenol (4.11%, *w/w*), ampelopsin A (3.21%, *w/w*), pallidol (3.12%, *w/w*), ω-viniferin (2.77 %, *w/w*), miyabenol C (2.75%, *w/w*), r2-viniferin (2.24%, *w/w*), hopeaphenol (2.01%, *w/w*), and piceatannol (0.76%, *w/w*). The stilbene-enriched extract contained 45.38% (*w/w*) of total stilbenes.

3.2. Cytotoxicity Studies

Both cell types exposed to the extract underwent a concentration and time-dependent decrease in all endpoints. In HepG2 cells, after 24 h of exposure, the MTS assay indicated a significant reduction in the cellular viability at concentrations of 40 µg/mL and above, showing greater alteration than the TP and NR assays (Figure 1A). Similarly, the most sensitive endpoints after 48 h were MTS reduction and NR uptake (Figure 1B). Considering the EC_{50} values, toxic effects were more evident in HepG2 cells in the longer exposure time, being 31.2 ± 2.4 µg/mL for 24 h, and 20.6 ±2.7 µg/mL for 48 h of exposure.

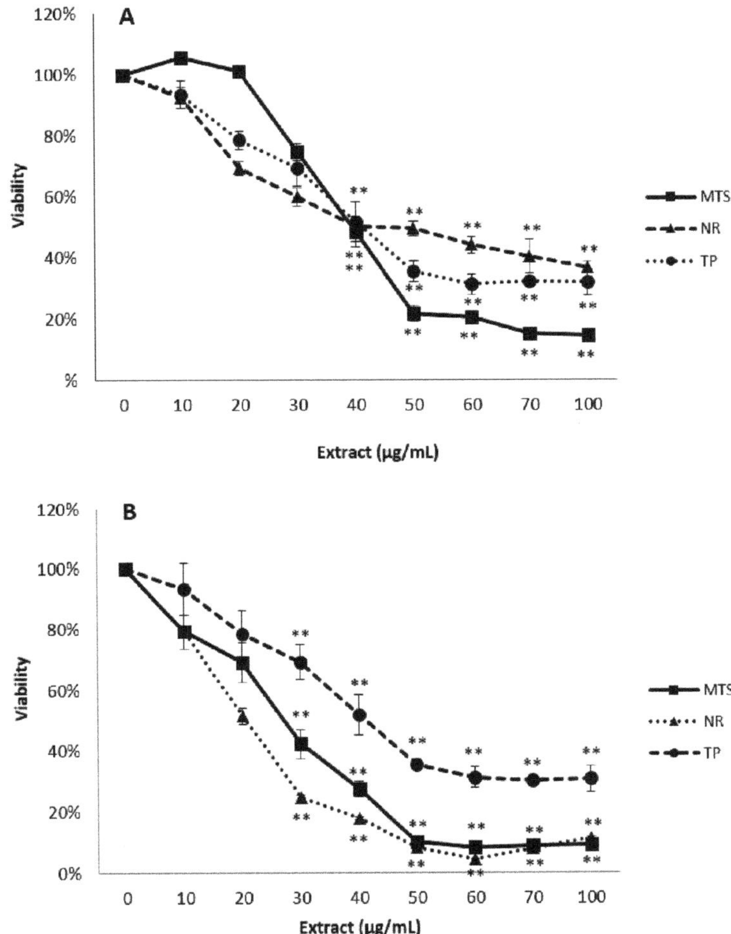

Figure 1. Reduction of tetrazolium salt (MTS), neutral rep uptake (NR), and total protein content (TP) of HepG2 cells exposed for 24 h (**A**) and 48 h (**B**) to 0–100 µg/mL of the stilbene extract (45%). All values expressed as mean ± SD. Significant differences in respect to the control from $p < 0.01$ (**).

After the exposure of Caco-2 cells to the extract, all assays showed similar concentration-dependent reduction in cell viability. MTS assay indicated a marked reduction in the cellular viability, being the EC_{50} value in this endpoint of 55.8 ± 4.0 µg/mL and 39.0 ± 2.7 µg/mL after 24 h and 48 h, respectively. These decreases were significantly different from the control group at the concentration 40 µg/mL and above at 24h, but when cells were exposed to the extract for two days, the cell viability was significantly reduced from 30 µg/mL (Figure 2).

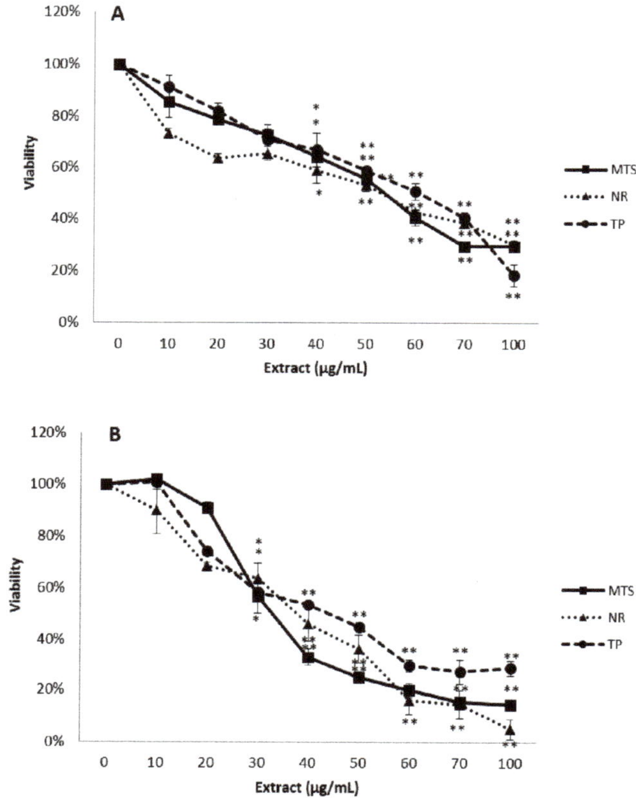

Figure 2. Reduction of tetrazolium salt (MTS), neutral rep uptake (NR), and total protein content (TP) of Caco-2 cells exposed for 24 h (**A**) and 48 h (**B**) to 0–100 µg/mL of the stilbene extract (45%). All values expressed as mean ± SD. Significant differences in respect to the control from $p < 0.05$ (*) and $p < 0.01$ (**).

3.3. Light and Electron Microscopic Observation in HepG2 and Caco-2 Cells

HepG2 and Caco-2 cells exposed to different concentrations of the extract experienced a significant impairment of cell growing which compromised the survival of the cells at high concentrations.

Under light microscope, unexposed HepG2 cells underwent normal mitotic processes (Figure 3A). However, when cells were treated with 15.6 and 31.2 µg/mL of the extract they revealed an intense lipid degeneration in the cytoplasm with vacuoles that tends to confluency (Figure 3B,C). Moreover, aberrant mitotic figures were also detected, which suggest that the extract was able to stop the cell growing at any step, including mitosis (Figure 3B,C). These morphological features were also observed under electron microscopy with cells showing big lipid drops (Figure 3D–F).

The damage observed in Caco-2 cells exposed to the extract was less profuse in comparison to HepG2 cells, although a marked vacuolization was shown as well as cell death. In fact, the presence of apoptotic nuclei was more frequent in the case of Caco-2 cells. Whereas cells underwent normal mitotic process in the control group (Figure 4A), aberrant mitosis can be observed from the lowest concentration assayed (13.9 µg/mL) (Figure 4B) showing the first stages in the process of apoptosis characterized by a continuous ring of condensed chromatin at the interior surface of the nuclear envelope (Figure 4C). Caco-2 cells were analyzed by transmission electron microscopy showing a nucleus with an irregular surface, decondensed chromatin, and very developed nucleoli with well-known fibrillar center. Autophagosomal vacuoles were observed in the cytoplasm (Figure 4D). Similar findings were observed in the ultrastructural study, with the lowest concentration showing lipid degeneration (Figure 4E,F). In addition, Caco-2 cells showed big citoplasmatic inclusions as a result of autophagic processes that lead to cell death (Figure 4E,F).

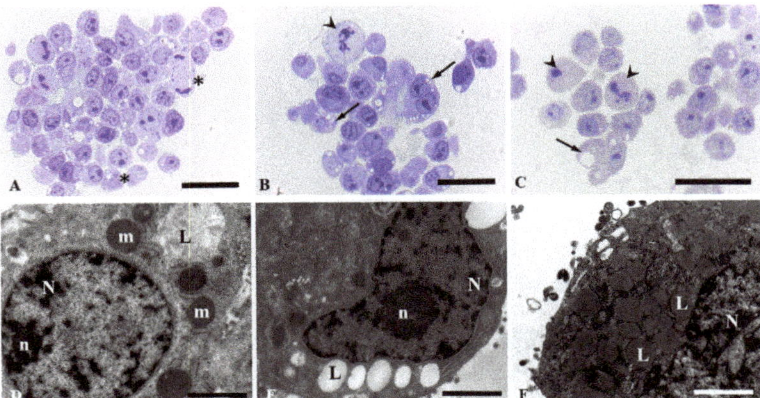

Figure 3. Morphology of HepG2 cells after 24 h of exposure to the extract observed by light microscopy (**A–C**, bars = 25 μm) and electron microscopy (**D–F**, bars = 2 μm). Unexposed control cultures (**A**) and HepG2 cells exposed to 7.79 μg/mL of the stilbene extract (45%) (**D**), 15.59 μg/mL of the extract (**B,E**) and 31.18 μg/mL of the stilbene extract (45%) (**C,F**). (**A**) Unexposed cells undergoing mitotic processes (arrow heads). (**B**) and (**C**) Lipid degeneration with confluent lipid drops (arrow) and aberrant mitotic figures (arrow head). (**D**) Cells showing nucleus (N) and nucleolus (n); mitochondria (m); lipid drops (L). (**E**) Big lipid drops (L); nucleus (N) and nucleolus (n) are also observed. (**F**) High amount of lipid drops that tends to confluency (L).

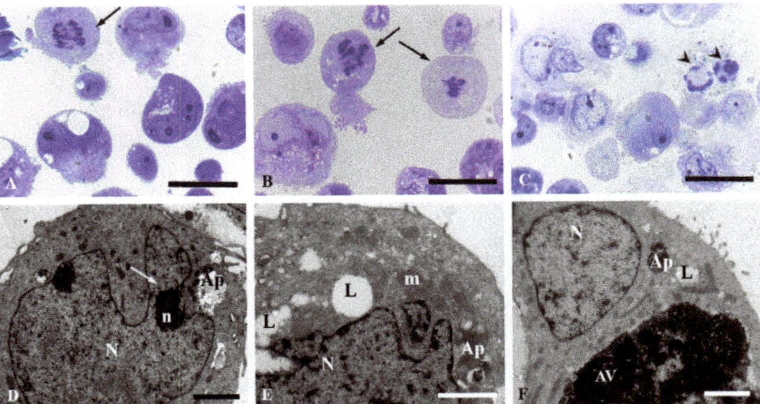

Figure 4. Morphology of Caco-2cells after 24 h of exposure to the stilbene extract (45%) observed by light microscopy (**A–C**, bars = 25 μm) and electron microscopy (**D–F**, bars = 2 μm). Unexposed control cultures (**A,D**), and Caco-2 cells exposed to 13.9 μg/mL of the extract (**B,E**) and 27.88 μg/mL of the extract (**C,F**). (**A**) Unexposed cells undergoing mitotic processes (black arrow). (**B**) Cells undergoing aberrant mitosis (black arrow). (**C**) Cells showing apoptotic nuclei condensation (arrow head). (**D**) Unexposed cells showing normal nucleus with an irregular surface (N), nucleolus (n) with large fibrillar center (white arrow), and autophagosomes (Ap) in the cytoplasm. (**E**) Damaged nucleus and mitochondria are observed as well as big lipid drops. Autophagosomes (Ap) evidenced degenerative process. (**F**) Lipid drops (L), autophagosomes (Ap), and autophagic vacuoles (AV) are shown.

3.4. Oxidative Stress Assays

The control of solvent evidenced no significant changes when cells were exposed to 0.3% of DMSO. However, HepG2 cells experienced a significant decrease in ROS levels when they were exposed to all the concentrations of the extract tested after 24 and 48 h, showing a greater decrease after the exposure to 20.6 µg/mL (Figure 5A). Moreover, GSH content underwent concentration-dependent enhancements in comparison to the negative control group after both times of exposure (Figure 5B).

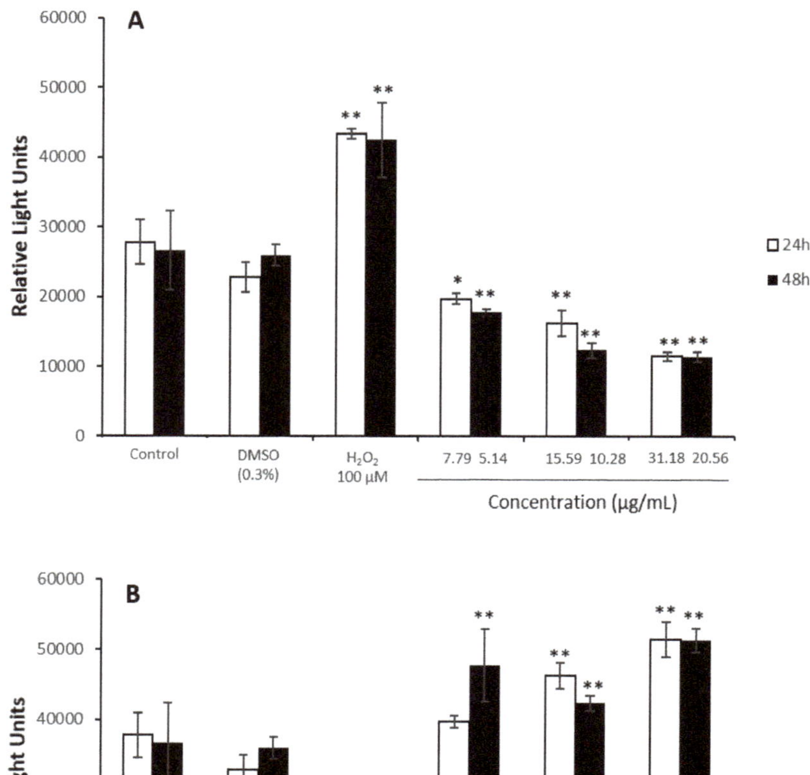

Figure 5. ROS content (**A**) and GSH content (**B**) in HepG2 cells exposed to 0–31.18 µg/mL or 0–20.56 µg/mL stilbene extract (45%) during 24 h or 48 h, respectively. All values are expressed as mean ± SD. Differences were considered significant compared to the control group from $p < 0.05$ (*), $p < 0.01$ (**) and $p < 0.001$ (***).

When Caco-2 cells were exposed to increasing concentrations of the extract, ROS content was significantly reduced (Figure 6A). GSH activity was increased significantly at the highest concentrations assayed, 55.77 µg/mL and 39.02 µg/mL for 24 and 48 h, respectively, with enhancements of 1.6 folds compared to the control negative group (Figure 6B).

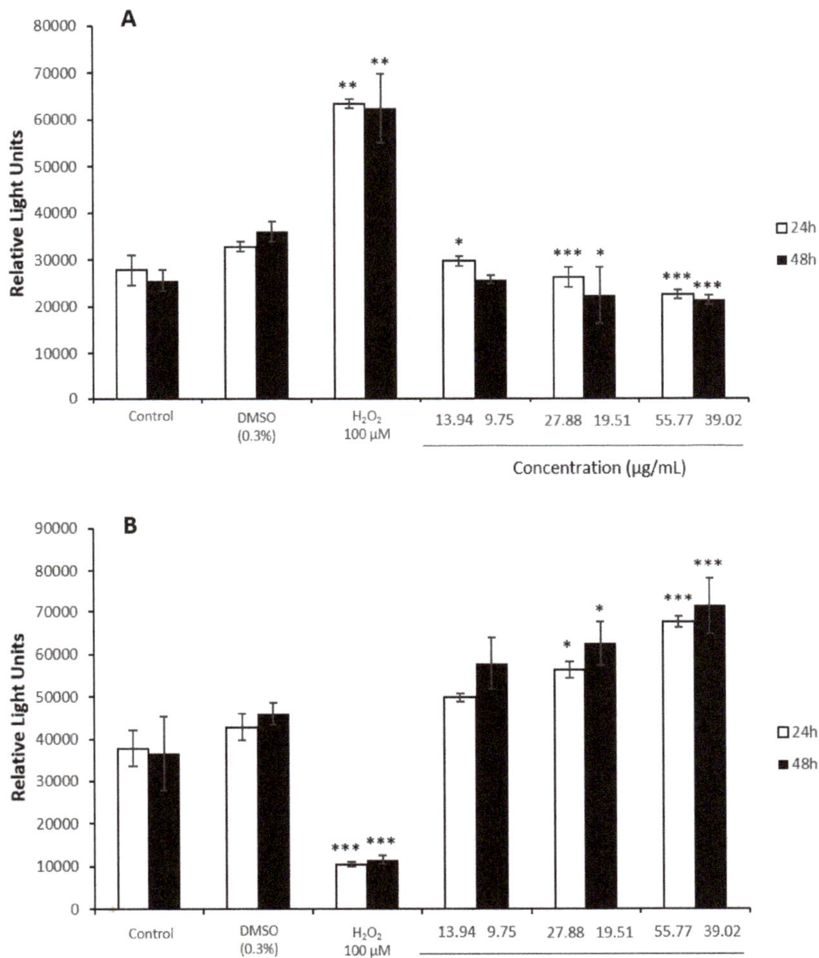

Figure 6. ROS content (**A**) and GSH content (**B**) in Caco-2 cells exposed to 0–55.77 µg/mL or 0–39.02 µg/mL stilbene extract (45%) during 24 h or 48 h, respectively. All values are expressed as mean ± SD. Differences were considered significant compared to the control group from $p < 0.05$ (*), $p < 0.01$ (**) and $p < 0.001$ (***).

3.5. Antioxidant Assays

The antioxidant ability of the extract was evaluated taking into account their capacity to protect the cells against a further exposure of H_2O_2 or their capacity to revert the damage induced by this substance after a previous exposure by measuring both ROS and GSH levels. No significant changes were recorded when cells were exposed to 0.3% of DMSO (data not shown).

The results showed that the extract was able to protect HepG2 cells against an induced oxidative stress at all concentrations studied, showing a marked decrease of ROS content at both treatment times (Figure 7A). Similarly, after the pre-treatment with H_2O_2 for 2 h, the extract presented a greater reversion role in a concentration and time-dependent manner. This reduced ROS content even lower than basal levels at the highest concentrations tested for 24 h and after 48 h of exposure of all studied concentrations (Figure 7B). By contrast, in both protection and reversion assays, GSH levels of the hepatic cells were not affected by the administration of 7.8 µg/mL during 24 h and 5.1 µg/mL for 48 h of the extract, while they experienced a significant increase when they were exposed to the highest concentration (Figure 7C,D). After the pre-treatment with 15.6 µg/mL and 10.3 µg/mL during 24 h and 48 h respectively and a later exposure of H_2O_2, the results showed higher GSH levels than basal content.

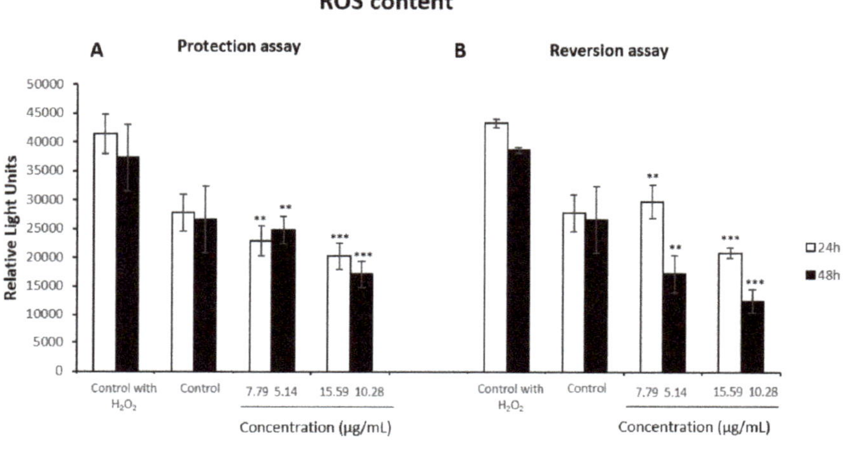

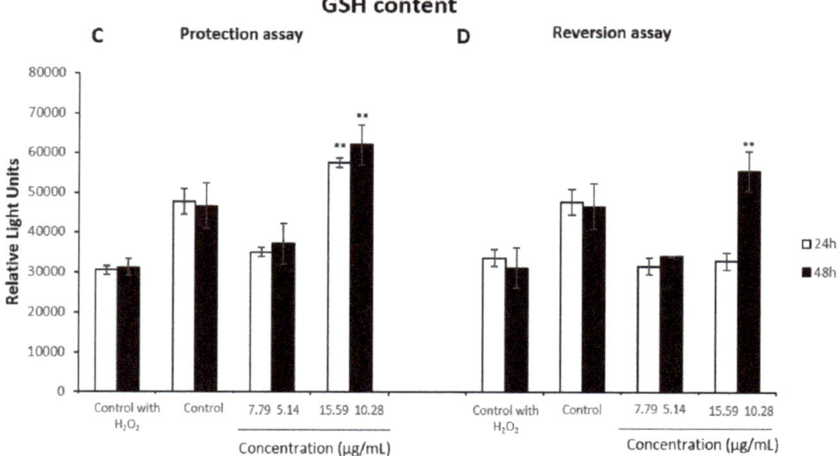

Figure 7. ROS (**A**) and GSH content (**C**) in HepG2 cells first pretreated with 0–15.59 µg/mL or with 0–10.28 µg/mL of the stilbene extract (45%) for 24 h or 48 h, respectively, and a later exposure to H_2O_2 100 µM for 2 h. ROS (**B**) and GSH (**D**) content in HepG2 cells exposed to H_2O_2 100 µM first and a later 24 h or 48 h-treatment with 0–15.59 µg/mL or 0–10.28 µg/mL, respectively. Cells exposed to H_2O_2 100 µM 2h were used as control. All values are expressed as mean ± SD. Differences were considered significant compared to the control group from $p < 0.01$ (******) and $p < 0.001$ (*******).

The cell line Caco-2 presented lower protection and reversion capacity when compared to the effects observed in HepG2 cells. In the protection assay, the extract was able to significantly reduced ROS content with respect to the control group treated with H_2O_2 at all concentrations assayed after both pre-treatment times (Figure 8A). Similar results were obtained when Caco-2 cells were exposed to H_2O_2 prior the extract (Figure 8B). At the highest concentrations assayed ROS content was reduced down to basal levels. With respect to GSH levels, in both reversion and protection assays, GSH content were higher than basal levels after both times of treatment at the highest concentration assayed. The results showed a significant increase between control group with H_2O_2 treatment and those exposed to the highest concentration tested of the extract for 24 h and 48 h prior H_2O_2 (Figure 8C). Moreover, after the exposure to H_2O_2 for 2 h, only Caco-2 cells treated with 22.9 µg/mL during 24h or with 19.5 µg/mL for 48 h presented a significant increase of GSH levels (Figure 8D).

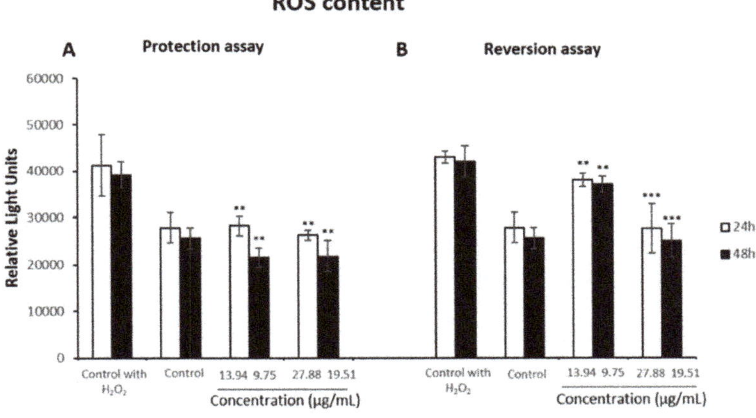

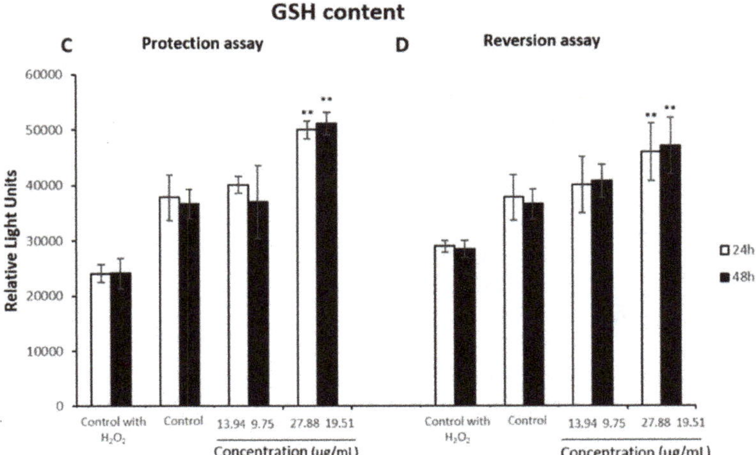

Figure 8. ROS (**A**) and GSH content (**C**) in Caco-2 cells first pretreated with 0–27.88 µg/mL or with 0–19.51 µg/mL of the extract for 24 h or 48 h, respectively, and a later exposure to H_2O_2 100 µM for 2 h. ROS (**B**) and GSH (**D**) content in HepG2 cells exposed to H_2O_2 100 µM first and a later 24 h or 48 h-treatment with 0–27.88 µg/mL or 0–19.51 µg/mL respectively. Cells exposed to H_2O_2 100 µM 2 h were used as control. All values are expressed as mean ± SD. Differences were considered significant compared to the control group from $p < 0.01$ (**) and $p < 0.001$ (***).

4. Discussion

Previous reports have revealed the interesting properties of stilbene-rich extracts form grapevine-shoot to be used in the vine and wine industry [3,10,23,32]. Therefore, the cytotoxicity study in human cell lines (Caco-2 and HepG2) is a key step in order to evaluate its safety. The cytotoxicity study performed in the present work evidenced that our extract (45% of stilbenes) reduced cell viability from 40 µg/mL after 24 h of exposure and from 30 µg/mL at 48 h in both cell lines. Similarly, Vineatrol® (29% of stilbenes) inhibits cell proliferation in another human colon carcinoma cell line (SW480 cells), showing a concentration and time-dependent reduction in the MTT assay [16]. The latter extract also affected cell cycle progression of different human cancer cell lines (SW480, SW620, and HTC 116), making them more prone to suffer cytotoxic effects of toxicants [33]. Moreover, these authors have previously reported that Vineatrol® disrupts cells proliferation more efficiently than resveratrol on HepG2 cells [17].

Studies using single stilbenes are more numerous in comparison to those using stilbenes extracts. In this sense, a concentration-dependent cytotoxicity of resveratrol and piceatannol have been reported in cultured macrophages, tumor-derived human T cells, and human epidermoid carcinoma cells [14]. These results showed cell inhibition at high concentration of resveratrol (50 µmol/L) in all cell types, while at low concentration (5 µmol/L) stimulation of macrophages was detected. Moreover, Colin et al. (2009) reported that the mixture of *trans*-resveratrol and ε-viniferin (Vineatrol®) as well as resveratrol itself affect cell cycle progression of human colon cancer cell lines [33]. Similarly, both resveratrol and Vineatrol® exerted a concentration-dependent cytotoxic effect on V79 cells, the effect being much more pronounced with resveratrol than with Vineatrol® [8]. In addition, Mizuno et al., 2017 reported that different modifications of stilbenes have different cytotoxic effect in CHO-K1 and HepG2 cells [34]. The latter authors hypothesized that the cell viability decrease may be reduced because of the appearance of necrosis or late stages of apoptosis when the metabolic activity is harshly reduced. This finding has been observed in the morphological study carried out in the present work, showing the first stages of apoptosis such as condensed chromatin in the nuclear envelope in Caco-2 cells exposed to the stilbene extract. Similarly, nuclear staining of the human colorectal cells SW480, evidenced that resveratrol-analogs and Vineatrol® preparation caused a nuclear redistribution of cyclin A, could be the first step for apoptosis considering that this finding was observed in both normal and apoptotic cells [33]. SW480 cells experienced apoptosis in a dose and time-dependent way. Moreover, the resveratrol tetramer, r-viniferin, provoked a cell cycle arrest followed by apoptosis in the prostate cancer cell line LNCaP, being this effect higher than in the exposure to resveratrol where apoptosis was not observed [35]. Also, different cultures of leukemic cells from chronic B cell malignancies and normal peripheral blood-derived mononuclear cells were exposed to Vineatrol®. Whereas impairment of cell proliferation and apoptotic processes were observed in exposed leukemic cells, the survival of normal peripheral blood mononuclear cells was little affected in the presence of these polyphenolic compounds and higher concentrations were required in order to elicit cell death [15]. This finding is particularly interesting since normal cells are usually more sensitive than cancer cell lines, highlighting the anti-cancer effect of stilbenes [36,37]. Ji et al. (2018) reported that cancer cells observed under microscope became flattened and elongated, which were morphological changes of cellular senescence [38]. In addition, resveratrol arrested cancer cell proliferation in a concentration-dependent way. However, no effect was recorded in cell viability for non-tumorigenic human breast epithelial cells MCF-10A. In this regards, Elshaer et al. (2018) reviewed the anti-cancer mechanisms of resveratrol by inducing autophagy and apoptosis [39]. In addition to death cell and autophagy, other morphological changes were observed in cells exposed to stilbenes. Scanning electron microscopy revealed that resveratrol altered the morphology of T cells and skin cells at concentrations ≥50 µmol/L, especially affecting the cell membrane [14]. Besides cell death, previous degradation steps have been evidenced in the present work. Hence, one of the most remarkable morphological features were presence of big lipid drops evidencing lipid degeneration, probably due to the degenerative processes in the organelles [40].

The defense against oxidative damage is one of the most commonly described characteristics of stilbenes [21]. However, phenolic compounds frequently exhibit prooxidant or antioxidant activities depending on the dose, alkali pH, high concentrations of transition metal ions, and the presence of oxygen molecules [21,41,42]. In this regard, Cotoras et al., (2014) assessed in vitro and in vivo antioxidant and prooxidant properties of phenolic compounds obtained from *Vitis vinifera* pomace, evidencing a prooxidant effect by accumulation of ROS [43]. By contrast, our results suggest that the extract modulates important functions related to the maintenance of Caco-2 and HepG2 redox environment, preventing the increase in ROS levels induced by H_2O_2, and even reverting them down to basal levels, but no prooxidant effect was recorded. Goutzourelas et al. (2015) also found that two extracts from stems of Greek grape varieties significantly decreased the ROS levels by 21.8 ± 2.0% and 16.5 ± 3.7%, respectively, compared to the control group, in C2C12 cells after 24 h of exposure [44]. This antioxidant activity can be explained by the ability of polyphenols to enhance antioxidant mechanisms, such as GSH synthesis, which acts as a scavenger to directly remove hydroxyl radical and singlet oxygen and participates in the elimination of hydrogen peroxide. In addition, Müller et al. (2009) found that the grapevine-shoot extract Vineatrol®30, acted as a free radical scavenger and potent antioxidant enhancing the glutathione peroxidase activity at non-cytotoxic concentrations in V79 Chinese hamster lung fibroblast cells [8]. Similarly, an increase of GSH levels in endothelial cells was observed after 24 h of exposure of polyphenolic extracts derived from the stems of three Greek grape varieties [45].

5. Conclusions

Besides all beneficial abilities of stilbenes, such as antioxidant activity, they may also have potential toxic effects. In this regard, the present work entail the first approach to the toxicological studies required before being used in wines. The cytotoxicity study evidenced that the stilbene extract reduced cell viability in both cell lines in the range of concentrations from 40–100 µg/mL after 24 h of exposure and from 30–100 µg/mL at 48 h of exposure. In addition, ultrastructural changes were observed from 15.6 µg/mL in HepG2 cells and 27.9 µg/mL in Caco-2 cells, which highlighted the higher sensitivity of the morphological study in comparison to the cytotoxicity endpoints. Moreover, this work also presents a wide assessment of the prooxidant and antioxidant profile of our extract in human cells, which had not been previously studied. In general, at the exposed concentrations, the extract presented a protective and reductive role against an induced oxidative stress. Taking into account all these findings as a starting approach, more research is needed to establish effective and safe concentrations of this stilbene extract intended to be used in wines as a preservative.

Author Contributions: Formal analysis, C.M.-P., T.R. and E.C.-V.; Funding acquisition, S.P. and E.C.-V.; Investigation, C.M.-P., M.P., F.J.M. and T.R.; Methodology, C.M.-P., M.P., F.J.M. and T.R.;; Project administration, E.C.-V. and S.P.; Resources, T.R. and E.C.-V.; Supervision, M.P. and S.P.; Writing—original draft, C.M.-P., F.J.M. and S.P.; Writing—review & editing, C.M.-P., T.R., E.C.-V. and S.P.

Funding: This work was supported by the Ministerio de Economía, Industria y Competitividad and INIA for the financial support for this project (RTA2015-00005-C02-02). Moreover, it was also supported by the Bordeaux Metabolome Facility and MetaboHUB (ANR-11-INBS-0010 project).

Acknowledgments: The authors thank the CITIUS Biology Service (University of Seville) for the technical assistance offered. Moreover, we would like to thank M.M. Merchán-Gragero, M-L. Iglesias and A. Palos-Pinto for their technical assistance.

Conflicts of Interest: The authors declare no conflict of interest. The funders had no role in the design of the study; in the collection, analyses, or interpretation of data; in the writing of the manuscript, or in the decision to publish the results.

References

1. Guerrero, R.F.; Cantos-Villar, E. Demonstrating the efficiency of sulphur dioxide replacements in wine: A parameter review. *Food. Sci. Technol.* **2015**, *42*, 27–43. [CrossRef]

2. EC, European Parliament and of the Council. Directive 2006/52/EC of the European Parliament and of the Council of 5 July 2006 amending directive 95/2/EC on food additives other than colours and sweeteners and Directive 94/35/EC on sweeteners for use in foodstuffs. *Off. J. Eur. Union* **2006**, *204*, 10–22.

3. Cruz, S.; Raposo, R.; Ruiz-Moreno, M.J.; Garde-Cerdán, T.; Puertas, B.; Gonzalo-Diago, A.; Moreno-Rojas, J.M.; Cantos-Villar, E. Grapevine-shoot stilbene extract as a preservative in white wine. *Food Packag. Shelf Life* **2018**, *18*, 164–172. [CrossRef]

4. Santos, M.C.; Numes, C.; Saraiva, J.A.; Coimbra, M.A. Chemical and physical methodologies for the replacement/reduction of sulfur dioxide use during winemaking: Review of their potentialities and limitations. *Eur. Food Res. Technol.* **2012**, *234*, 1–12. [CrossRef]

5. Shahidi, F.; Ambigaipalan, P. Phenolics and polyphenolics in foods, beverages and spices: Antioxidant activity and health effects—A review. *J. Funct. Foods* **2015**, *18*, 820–897. [CrossRef]

6. Anastasiadi, M.; Pratsinis, H.; Kletsas, D.; Skaltsounis, A.L.; Haroutounian, S.A. Grape stem extracts Polyphenolic content and assessment of their in vitro antioxidant properties. *LWT-Food Sci. Technol.* **2012**, *48*, 316–322. [CrossRef]

7. Biais, B.; Krisa, S.; Cluzet, S.; Da Costa, G.; Waffo-Teguo, P.; Mérillon, J.M.; Richard, T. Antioxidant and Cytoprotective Activities of Grapevine Stilbenes. *J. Agric. Food Chem.* **2017**, *65*, 4952–4960. [CrossRef]

8. Müller, C.; Ullmann, K.; Wilkens, A.; Winterhalter, P.; Toyokuni, S.; Steinberg, P. Potent Antioxidative Activity of Vineatrol®30 Grapevine-shoot Extract. *Biosci. Biotechnol. Biochem.* **2009**, *73*, 1831–1836. [CrossRef] [PubMed]

9. Ruiz-Moreno, M.J.; Raposo, R.; Cayuela, J.M.; Zafrilla, P.; Piñeiro, Z.; Moreno-Rojas, J.M.; Mulero, J.; Puertas, B.; Giron, F.; Guerrero, R.F.; et al. Valorization of grape stems. *Ind. Crop. Prod.* **2015**, *63*, 152–157. [CrossRef]

10. Raposo, R.; Ruiz-Moreno, M.J.; Garde-Cerdán, T.; Puertas, B.; Moreno-Rojas, J.M.; Gonzalo-Diago, A.; Guerrero, R.F.; Ortíz, V.; Cantos-Villar, E. Grapevine-shoot stilbene extract as a preservative in red wine. *Food Chem.* **2016**, *197*, 1102–1111. [CrossRef]

11. Raposo, R.; Chinnici, F.; Ruiz-Moreno, M.J.; Puertas, B.; Cuevas, F.J.; Carbú, M.; Guerrero, R.F.; Ortíz-Somovilla, V.; Moreno-Rojas, J.M.; Cantos-Villar, E. Sulfur free red wines through the use of grapevine shoots: Impact on the wine quality. *Food Chem.* **2018**, *243*, 453–460. [CrossRef] [PubMed]

12. EFSA (European Food Safety Authority). Guidance for submission for food additive evaluations. *EFSA J.* **2012**, *10*, 2760.

13. Maisanaba, S.; Llana-Ruiz-Cabello, M.; Gutiérrez-Praena, D.; Pichardo, S.; Puerto, M.; Prieto, A.I.; Jos, A.M.; Cameán, A.M. New advances in active packaging incorporated with essential oils or their main components for food preservation. *Food Rev. Int.* **2017**, *33*, 447–515. [CrossRef]

14. Radkar, V.; Hardej, D.; Lau-Cam, C.; Billack, B. Evaluation of resveratrol and piceatannol cytotoxicity in macrophages, T cells, and skin cells. *Arhiv za Higijenu Rada i Toksikologiju* **2007**, *58*, 293–304. [CrossRef] [PubMed]

15. Billard, C.; Izard, J.C.; Roman, V.; Kern, C.; Mathiot, C.; Mentz, F.; Kolb, J.P. Comparative Antiproliferative and Apoptotic Effects of resveratrol, ε-viniferin and vine-shots Derived Polyphenols (Vineatrols) on Chronic B Lymphocytic Leukemia Cells and Normal Human Lymphocytes. *Leuk. Lymphoma* **2002**, *43*, 1991–2002. [CrossRef]

16. Marel, A.K.; Lizard, G.; Izard, J.C.; Latruffe, N.; Delmas, D. Inhibitory effects of *trans*-resveratrol analogs molecules on the proliferation and the cell cycle progression of human colon tumoral cells. *Mol. Nutr. Food Res.* **2008**, *52*, 538–548. [CrossRef]

17. Colin, D.; Lancon, A.; Delmas, D.; Lizard, G.; Abrossinow, J.; Kahn, E.; Jannin, B.; Latruffe, N. Antiproliferative activities of resveratrol and related compounds in human hepatocyte derived HepG2 cells are associated with biochemical cell disturbance revealed by fluorescence analyses. *Biochimie* **2008**, *90*, 1674–1684. [CrossRef]

18. Guerrero, F.R.; García-Parrilla, M.C.; Puertas, B.; Cantos-Villar, E. Wine, resveratrol and health: A review. *Nat. Prod. Commun.* **2009**, *4*, 635–658. [CrossRef]

19. Richard, T.; Poupard, P.; Nassra, M.; Papastamoulis, Y.; Iglésias, M.L.; Krisa, S.; Waffo-Teguo, P.; Mérillon, J.M.; Monti, J.P. Protective effect of ε-viniferin on b-amyloid peptide aggregation investigated by electrospray ionization mass spectrometry. *Bioorg. Med. Chem.* **2011**, *19*, 3152–3155. [CrossRef]

20. Frombaum, M.; Le Clanche, S.; Bonnefont-Rousselot, D.; Borderie, D. Antioxidant effects of resveratrol and other stilbene derivatives on oxidative stress and NO bioavailability: Potential benefits to cardiovascular diseases. *Biochimie* **2012**, *94*, 269–276. [CrossRef]

21. Ferguson, L.R. Role of plant polyphenols in genomic stability. *Mutat. Res.* **2001**, *475*, 89–111. [CrossRef]

22. Ozkan, A.; Erdogan, A. A comparative evaluation of antioxidant and anticancer activity of essential oil from *Origanum onites* (Lamiaceae) and its two major phenolic components. *Turk. J. Biol.* **2011**, *35*, 735–742. [CrossRef]

23. Gutiérrez-Praena, D.; Pichardo, S.; Jos, A.; Moreno, F.J.; Cameán, A.M. Biochemical and pathological toxic effects induced by the cyanotoxin Cylindrospermopsin on the human cell line Caco-2. *Water Res.* **2012**, *49*, 1566–1575. [CrossRef] [PubMed]

24. Gabaston, J.; El Khawand, T.; Waffo-Teguo, P.; Decendit, A.; Richard, T.; Mérillon, J.M.; Roman, P. Stilbenes from grapevine root: A promising natural insecticide against *Leptinotarsa decemlineata*. *J. Pest Sci.* **2018**, *91*, 897–906. [CrossRef]

25. Guerrero, R.F.; Biais, B.; Richard, T.; Puertas, B.; Waffo-Teguo, P.; Merillon, J.M.; Cantos-Villar, E. Grapevine cane's waste is a source of bioactive stilbenes. *Ind. Crop. Prod.* **2016**, *94*, 884–892. [CrossRef]

26. Borenfreud, E.; Puerner, J.A. A simple quantitative procedure using monolayer culture for cytotoxicity assays. *J. Tissue Cult. Methods* **1984**, *9*, 7–9. [CrossRef]

27. Zhang, S.Z.; Lipsky, M.M.; Trump, B.F.; Hsu, I.C. Neutral red (NR) assay for cell viability and xenobiotic-induced cytotoxicity in primary cultures of human and rat hepatocytes. *Cell Biol. Toxicol.* **1990**, *6*, 219–234. [CrossRef]

28. Baltrop, J.A.; Owen, T.C.; Cory, A.H. 5-((3-carboxyphenyl)-3-(4,5 dimethilthiazolyl)-3-(4- sulfophenyl) tetrazolium, inner salt (MTS) and related analogs of 2- (4,5- dimethylthiazolyl)- 2,5- diphenylterazolium bromide (MTT) reducing to purple water soluble formazan as cell viability indicators. *Bioorg. Med. Chem. Lett.* **1991**, *1*, 611. [CrossRef]

29. Bradford, M.M. A Rapid and Sensitive Method for the Quantitation of Microgram Quantities of Protein Utilizing the Principle of Protein-Dye Binding. *Anal. Biochem.* **1976**, *72*, 248–254. [CrossRef]

30. Gutiérrez-Praena, D.; Guzmán-Guillén, R.; Pichardo, S.; Moreno, F.J.; Vasconcelos, V.; Jos, A.; Cameán, A.M. Cytotoxic and morphological effects of microcystin-LR, cylindrospermopsin, and their combinations on the human hepatic cell line HepG2. *Environ. Toxicol.* **2019**, *34*, 240–251. [CrossRef]

31. Wijeratne, S.S.K.; Cuppett, S.L.; Schlegel, V. Hydrogen Peroxide Induced Oxidative Stress Damage and Antioxidant Enzyme Response in Caco-2 Human Colon Cells. *J. Agric. Food Chem.* **2005**, *53*, 8768–8774. [CrossRef] [PubMed]

32. Schnee, S.; Queiroz, E.F.; Voinesco, F.; Marcourt, L.; Dubuis, P.H.; Wolfender, J.L.; Gindro, K. *Vitis vinifera* canes, a new source of antifungal compounds against *Plasmopara viticola*, *Erysiphe necator*, and *Botrytis cinerea*. *J. Agric. Food Chem.* **2013**, *61*, 5459–5467. [CrossRef] [PubMed]

33. Colin, D.; Gimazane, A.; Lizard, G.; Izard, J.C.; Solary, E.; Latruffe, N.; Delmas, D. Effects of resveratrol analogs on cell cycle progression, cell cycle associated proteins and 5fluoro-uracil sensitivity in human derived colon cancer cells. *Int. J. Cancer* **2009**, *124*, 2780–2788. [CrossRef] [PubMed]

34. Mizuno, C.S.; Ampomaah, W.; Mendonça, F.R.; Andrade, G.C.; Nazaré da Silva, A.M.; Goulart, M.O.; Alves dos Santos, R. Cytotoxicity and genotoxicity of stilbene derivatives in CHO-K1 and HepG2 cell lines. *Genet. Mol. Biol.* **2017**, *40*, 656–664. [CrossRef] [PubMed]

35. Empl, M.T.; Albers, M.; Wang, S.; Steinberg, P. The Resveratrol Tetramer r-Viniferin Induces a Cell Cycle Arrest Followed by Apoptosis in the Prostate Cancer Cell Line LNCaP. *Phytother. Res.* **2015**, *29*, 1640–1645. [CrossRef]

36. Yilmazer, A. Cancer cell lines involving cancer stem cell populations respond to oxidative stress. *Biotechnol. Rep.* **2018**, *17*, 24–30. [CrossRef] [PubMed]

37. De Filippis, B.; Ammazzalorso, A.; Fantacuzzi, M.; Giampietro, L.; Maccallini, C.; Amoroso, R. Anticancer Activity of Stilbene-Based Derivatives. *Chem. Med. Chem.* **2017**, *12*, 1–14. [CrossRef] [PubMed]

38. Ji, S.; Zheng, Z.; Liu, S.; Ren, G.; Gao, J.; Zhang, Y.; Li, G. Resveratrol promotes oxidative stress to drive DLC1 mediated cellular senescence in cancer cells. *Exp. Cell Res.* **2018**, *370*, 292–302. [CrossRef]

39. Elshaer, M.; Chen, Y.; Wang, X.J.; Tang, X. Resveratrol: An overview of its anti-cancer mechanisms. *Life Sci.* **2018**, *207*, 340–349. [CrossRef]

40. De Bortoli, M.; Taverna, E.; Maffioli, E.; Casalini, P.; Crisafi, F.; Kumar, V.; Caccia, C.; Polli, D.; Tedeschi, G.; Bongarzone, I. Lipid accumulation in human breast cancer cells injured by iron depletors. *J. Exp. Clin. Cancer Res.* **2018**, *37*, 75. [CrossRef] [PubMed]

41. Blokhina, O.; Virolainen, E.; Fagerstedt, K.V. Antioxidants, oxidative damage and oxygen deprivation stress: A review. *Ann. Bot.* **2003**, *91*, 179–194. [CrossRef] [PubMed]

42. Llana-Ruiz-Cabello, M.; Gutiérrez-Praena, D.; Puerto, M.; Pichardo, S.; Jos, A.; Cameán, A.M. In vitro pro-oxidant/antioxidant role of carvacrol, thymol and their mixture in the intestinal Caco-2 cell line. *Toxicol. In Vitro* **2015**, *29*, 647–656. [CrossRef] [PubMed]

43. Cotoras, M.; Vivanco, H.; Melo, R.; Aguirre, M.; Silva, E.; Mendoza, L. In vitro and in vivo evaluation of the antioxidant and prooxidant activity of phenolic compounds obtained from grape (*Vitis vinifera*) pomace. *Molecules* **2014**, *19*, 21154–21167. [CrossRef] [PubMed]

44. Goutzourelas, N.; Stagos, D.; Spanidis, Y.; Liosi, M.; Apostolou, A.; Priftis, A.; Haroutounian, S.; Spandidos, D.A.; Tsatsakis, A.M.; Kouretas, D. Polyphenolic composition of grape stem extracts affects antioxidant activity in endothelial and muscle cells. *Mol. Med. Rep.* **2015**, *12*, 5846–5856. [CrossRef] [PubMed]

45. Goutzourelas, N.; Stagos, D.; Housmekeridou, A.; Karapouliou, C.; Kerasioti, E.; Aligiannis, N.; Skaltsounis, A.L.; Spandidos, D.A.; Tsatsakis, A.M.; Kouretas, D. Grape pomace extract exerts antioxidant effects through an increase in GCS levels and GST activity in muscle and endothelial cells. *Int. J. Mol. Med.* **2015**, *36*, 433–441. [CrossRef] [PubMed]

 *antioxidants*

Article

Methyl Jasmonate Reduces Inflammation and Oxidative Stress in the Brain of Arthritic Rats

Heloisa V. Pereira-Maróstica [1], Lorena S. Castro [1], Geferson A. Gonçalves [1], Francielli M.S. Silva [2], Lívia Bracht [1], Ciomar A. Bersani-Amado [2], Rosane M. Peralta [1], Jurandir F. Comar [1], Adelar Bracht [1] and Anacharis B. Sá-Nakanishi [1,*]

[1]   Department of Biochemistry, University of Maringá, 7020900 Maringá, Brazil; heloisavialle@gmail.com (H.V.P.-M.); losacastro@hotmail.com (L.S.C.); gefersonag2@gmail.com (G.A.G.); lbracht@uem.br (L.B.); rmperalta@uem.br (R.M.P.); jfcomar@uem.br (J.F.C.); abracht@uem.br (A.B.)

[2]   Department of Pharmacology and Therapeutics, University of Maringá, 87020900 Maringá, Brazil; franciellimss@gmail.com (F.M.S.S.); cabamado@uem.br (C.A.B.-A.)

*   Correspondence: absnakanishi@uem.br

Received: 10 July 2019; Accepted: 4 October 2019; Published: 15 October 2019

Abstract: Methyl jasmonate (MeJA), common in the plant kingdom, is capable of reducing articular and hepatic inflammation and oxidative stress in adjuvant-induced arthritic rats. This study investigated the actions of orally administered MeJA (75–300 mg/kg) on inflammation, oxidative stress and selected enzyme activities in the brain of Holtzman rats with adjuvant-induced arthritis. MeJA prevented the arthritis-induced increased levels of nitrites, nitrates, lipid peroxides, protein carbonyls and reactive oxygen species (ROS). It also prevented the enhanced activities of myeloperoxidase and xanthine oxidase. Conversely, the diminished catalase and superoxide dismutase activities and glutathione (GSH) levels caused by arthritis were totally or partially prevented. Furthermore, MeJA increased the activity of the mitochondrial isocitrate dehydrogenase, which helps to supply NADPH for the mitochondrial glutathione cycle, possibly contributing to the partial recovery of the GSH/oxidized glutathione (GSSG) ratio. These positive actions on the antioxidant defenses may counterbalance the effects of MeJA as enhancer of ROS production in the mitochondrial respiratory chain. A negative effect of MeJA is the detachment of hexokinase from the mitochondria, which can potentially impair glucose phosphorylation and metabolism. In overall terms, however, it can be concluded that MeJA attenuates to a considerable extent the negative effects caused by arthritis in terms of inflammation and oxidative stress.

Keywords: adjuvant-induced arthritis; rheumatoid arthritis; methyl jasmonate; brain mitochondria; hexokinase; oxidative stress

---

1. Introduction

Methyl jasmonate (MeJA) is a cyclopentanone widely spread over the plant kingdom where it functions as signaling molecule associated with biotic and abiotic stress [1]. It was first isolated from the flowers of jasmine (*Jasminum grandiflorum*), whose infusions have been traditionally used to relieve stress, depression, irritability and memory deficit [2]. MeJA has been reported to present anti-tumor activity without affecting healthy cells [3]. In fact, MeJA was revealed to be cytotoxic against various murine and human cancer cell lines [1,3]. In the cancer cells the compound mainly affects the mitochondria, where it detaches the outer membrane-bound hexokinase, stimulates the production of reactive oxygen species (ROS), depletes ATP and induces apoptosis [4,5].

MeJA shares structural similarity with anti-inflammatory prostaglandins and, therefore, has been lately investigated as a promising anti-inflammatory agent [3,6]. The compound itself and its synthetic derivatives inhibit the synthesis of tumor necrosis factor alpha (TNFα), interleukin 1 beta (IL-1β), interleukin 6 (IL-6), prostaglandin E and nitric oxide (NO) in murine macrophages (RAW264.7) stimulated by lipopolysaccharide (LPS) [6–8]. A recent study also showed that orally administered MeJA improves the systemic and articular inflammation in rats with adjuvant-induced arthritis, an experimental model of human severe rheumatoid arthritis [9]. Thus, MeJA emerges also as a promising agent for treatment of rheumatoid arthritis and other systemic inflammatory disorders.

Rheumatoid arthritis is a chronic and autoimmune inflammatory disease that can lead to progressive joint destruction and affects approximately 1.0% of the adult population worldwide [10,11]. An accentuated hyperplasia of the synovial membrane and cartilage and intense production of pro-inflammatory cytokines (TNFα, IL-1β and IL-6) are normally associated with rheumatoid arthritis [12]. The overproduction of reactive species and metalloproteinases are additionally stimulated by cytokines and mediate the tissue injury [13,14]. Rheumatoid arthritis is a systemic disease and in addition to affecting the joints, it evokes inflammatory and oxidative alterations in other organs, such as lungs, liver and heart [12]. In fact, the oxidative stress biomarkers are modified in the serum of patients with rheumatoid arthritis and in the extra-articular tissues of animal models of rheumatoid arthritis [14–19]. Metabolic alterations also occur in rheumatoid arthritis, as for example, the condition of muscle wasting known as rheumatoid cachexia [20]. The hepatic metabolism is also significantly modified in rats with adjuvant-induced arthritis, a phenomenon associated with pronounced oxidative stress in the organ [17,21–23].

Inflammation in rheumatoid arthritis also affects the brain where it causes fatigue and reduced cognitive function [12]. Cerebral atrophy and other structural modifications have been reported in patients with severe rheumatoid arthritis [24,25]. The levels of ROS, NO, lipoperoxides and protein carbonyl groups are increased in the brain of rats with adjuvant-induced arthritis, particularly in the mitochondria, where the transmembrane potential is also increased [26]. These alterations are accompanied by decreased levels of reduced glutathione (GSH) and diminished activities of antioxidant enzymes. Additionally, the activity of the pro-oxidant enzyme xanthine oxidase (XO) and the pro-inflammatory enzyme iNOS are increased in the brain of arthritic rats [26].

An earlier study has shown that intraperitoneally administered MeJA attenuates the memory dysfunction, decreases the levels of prostaglandin E, TNFα and IL-1 and suppresses the expression of COX-2 and iNOS in the brain of mice with LPS-induced neuroinflammation [27]. In addition, MeJA reverses in mice the memory impairment caused by scopolamine and unpredictable chronic mild stress (UCMS), phenomena that were associated with the improvement in the brain's levels of oxidative stress biomarkers, specifically lipoperoxides and GSH [28,29]. Considering these actions of MeJA on the cerebral tissue of mice, it is reasonable to hypothesize that MeJA will be capable of attenuating the inflammation and oxidative stress in the brain of arthritic rats. The present work was thus planned to investigate the effects of orally administered MeJA on inflammation and on the oxidative status of the brain of rats using the model of adjuvant-induced arthritis. Because orally administered MeJA stimulated ROS production in isolated hepatic mitochondria [9], this work has also evaluated the production of ROS in isolated brain mitochondria from healthy and arthritic rats as well as the activities of several mitochondrial dehydrogenases. Furthermore, the activity of hexokinase was also evaluated, as the liver of arthritic rats presents a considerably enhanced glucose phosphorylation capacity [30]. The data obtained in the current study should allow to infer about the possible actions of MeJA on the brain of patients with severe rheumatoid arthritis.

2. Materials and Methods

2.1. Chemicals

Methyl jasmonate, *o*-phthalaldehyde (OPT), 2,4-dinitrophenylhydrazine (DNPH), 1,1',3,3'-tetraethoxypropane, horseradish peroxidase, nitrate reductase, β-nicotinamide adenine dinucleotide (NADH), 2'-7'-dichlorofluorescein-diacetate (DCFH-DA), oxidized dichlorofluorescein

(DCF), reduced glutathione (GSH) and oxidized glutathione (GSSG) were purchased from Sigma Chemical Co (St. Louis, MO, USA). All other chemicals were of analytical grade.

2.2. Animals and Induction of Arthritis

Male Holtzman rats weighing approximately 180 g were obtained from the Center of Animal Breeding of the University of Maringá (UEM, Maringá, Brazil) and kept in the Animal Care Unit of our laboratory under standard temperature (24 ± 3 °C) in a 12 h light/dark cycle. The animals were maintained in steel cages (three rats/cage) and fed ad libitum with laboratory diet (Nuvilab®, Colombo, Brazil). For arthritis induction, animals were subcutaneously injected in the left hind paw with 0.1 mL (500 µg) of Freund's adjuvant (heat inactivated *Mycobacterium tuberculosis*, derived from the human strain H37Rv), suspended in Nujol® (Sigma Chemical Co-St. Louis, MO, USA) [31]. Rats of similar weights and age were used as controls. The procedures followed the guidelines of the Brazilian Council for the Control of Animal Experimentation (CONCEA) and were previously approved by the Ethics Committee for Animal Experimentation of the State University of Maringá (Protocol number CEUA 4475171016).

2.3. Experimental Design

Forty-two animals were randomly distributed into seven groups: controls (C), which received corn oil; controls (C300) treated with MeJA at the dose of 300 mg/kg; arthritic rats (A), which received corn oil; arthritic rats (A75, A150 and A300) treated with MeJA, respectively, at the doses of 75, 150 and 300 mg/kg; and arthritic rats (A_{IBU}) treated with ibuprofen (30 mg/kg). This procedure was repeated three times (42 × 3 = 126 rats in total) throughout the study. Animals were orally treated (by gavage) once a day with MeJA, corn oil or ibuprofen during the 18 days after arthritis induction. The doses of MeJA were defined according to previous studies showing that they are biologically active and well tolerated when the compound is administered to rodents [9,32].

2.4. Brain Homogenate Preparation and Mitochondria Isolation

Fasted (12 h) rats were anesthetized with an overdose of sodium thiopental (100 mg/kg) and the brain immediately removed, freeze-clamped and stored in liquid nitrogen. The tissue was homogenized in a van Potter–Elvehjem homogenizer with nine volumes of ice-cold 0.1 M potassium phosphate buffer (pH 7.4) and an aliquot was separated as total homogenate. The remaining homogenate was centrifuged (11,000g/15 min) and the supernatant separated as homogenate supernatant. Brain mitochondria were isolated by differential centrifugation as previously described [33] with some modifications. The fresh brain was placed in ice-cold solution containing 200 mM mannitol, 76 mM sucrose, 1 mM ethylene glycol-bis(2-aminoethylether)-N,N,N′,N′-tetraacetic acid (EGTA), 0.1 mM PMSF (phenylmethylsulfonyl fluoride) and 10 mM 2-amino-2-hydroxymethyl- propane-1,3-diol (TRIS; pH 7.4). The brain was cut into small pieces and homogenized in a van Potter–Elvehjem homogenizer. The homogenate was then first centrifuged at 650g for 10 min and the mitochondria in the supernatant were sedimented by centrifuging at 8000g (10 min). The resulting mitochondria were washed and resuspended in the same buffer. An aliquot of this preparation was used to prepare disrupted mitochondria by a repeated freeze-thawing procedure in liquid nitrogen.

2.5. Brain Oxidative Stress and Inflammatory Parameters

Protein carbonyl groups were measured spectrophotometrically in the brain homogenate supernatant using 2,4-dinitrophenylhydrazine (DNPH) as previously described [34]. The molar extinction coefficient (ε) of 2.20×10^4 M^{-1} cm^{-1} was used in the calculations.

The levels of lipoperoxides were quantified by means of the thiobarbituric acid reactive substances (TBARS) assay [35]. The content in TBARS was calculated from the standard curve prepared with 1,1′,3,3′-tetraethoxypropane.

The content of ROS was quantified via the 2′-7′-dichlorofluorescein-diacetate (DCFH-DA) assay [36], which quantifies the oxidation of DCFH to the fluorescent 2′,7′-dichlorofluorescein (DCF) in the presence of ROS. The formation of DCF was measured by spectrofluorimetry (Shimadzu RF-5301; 504 nm for excitation and 529 nm for emission). Although ROS have short half-lives, decomposition in frozen tissue is not very pronounced and the fractional loss and inactivations are independent of their concentrations, so that the proportions between the various conditions are preserved when they are assayed at the same time [9,17].

The rate of mitochondrial ROS production was estimated by measuring the linear fluorescence increase (504 nm for excitation and 529 nm for emission) due to DCF formation from DCFH via oxidation by H_2O_2 in the presence of horseradish peroxidase [37]. Briefly, intact mitochondria (0.5 mg) were suspended in 2 mL of a mixture containing 250 mM mannitol, 1.36 µM DCFA-DA, 10 mM HEPES buffer (pH 7.2), 10 µM rotenone and 10 mM succinate as respiratory substrate. Fluorescence was recorded during 10 min under agitation. The results were expressed as nmol min^{-1} (mg protein)$^{-1}$.

The production of NO was indirectly quantified by measuring the levels of nitrite plus nitrate in the brain homogenate supernatant. Nitrate was first converted into nitrite by adding nitrate reductase and the total nitrite was quantified by the Griess reagent [38].

Reduced (GSH) and oxidized (GSSG) glutathione were measured in the brain homogenate by spectrofluorimetry (excitation at 350 nm and emission at 420 nm) with the *o*-phthalaldehyde (OPT) assay [39]. The activities of catalase, superoxide dismutase (SOD), xanthine oxidase (XO) and myeloperoxidase (MPO) were assayed by spectrophotometry in the supernatant of the brain homogenate. The catalase activity was estimated spectrophotometrically at 240 nm using H_2O_2 as substrate [40]. The activity of SOD was estimated by the pyrogallol autoxidation method [41]. The activity (MPO) was measured by spectrophotometry with *o*-dianisidine [42]. The XO activity was estimated as the increase in absorbance at 295 nm due to uric acid formation [43].

2.6. Mitochondrial Transmembrane Potential ($\Delta\Psi_m$) and Enzyme Activities

Mitochondrial membrane energization (transmembrane potential; $\Delta\Psi_m$) was measured by spectrofluorimetry using safranin as fluorescent probe [44]. The wavelengths for excitation and emission were 520 and 580 nm, respectively. Energization was achieved by addition of 50 µM succinate plus 2 µM rotenone and the full de-energization was achieved by the addition of carbonyl cyanide-4-(trifluoro-methoxy) phenylhydrazone (FCCP; 10 µM).

The activity of NADH dehydrogenase in disrupted mitochondria was estimated by spectrophotometry at 420 nm using potassium ferricyanide as the electron acceptor and the results were calculated using the molar extinction coefficient (ε) of 1.04×10^3 M^{-1} cm^{-1} [45].

The ATPase activity was measured in intact (coupled and uncoupled) and disrupted mitochondria as previously described [46]. Briefly, intact mitochondria were incubated in a medium (0.5 mL) containing 50 mM KCl, 0.2 M sucrose, and 10 mM Tris–HCl (pH 7.4) plus 0.2 mM EGTA and 5.0 mM ATP for 20 min, at 37 °C. The reaction was initiated by the addition of ATP and stopped with 5% trichloroacetic acid. The ATPase activity was quantified by measuring inorganic phosphate release from ATP.

The activity of succinate dehydrogenase was measured by spectrophotometry (500 nm) in a medium containing 100 mM triethanolamine (pH 8.3), 0.5 mM EDTA, 2 mM KCN, 6.5 µM phenazine methosulfate, 0.6 mM iodonitrotetrazolium, and aliquots of disrupted mitochondria [40]. The reaction was initiated with the addition of 10 mM succinate. Values were calculated using the molar extinction coefficient (ε) of reduced iodonitrotetrazolium (1.93×10^4 M^{-1} cm^{-1}).

The activity of malate dehydrogenase was assayed by spectrophotometry at 340 nm in a medium (1.5 mL) containing 120 mM phosphate buffer (pH 7.8), 0.25 mM NADH, and aliquots of the supernatant obtained after centrifuging disrupted mitochondria at 10,000*g*. The reaction was started by the addition of 0.1 mM oxaloacetate [47]. Values were calculated using the molar extinction coefficient (ε) of NADH (6.22×10^3 M^{-1} cm^{-1}).

The activity of the NADP$^+$-dependent isocitrate dehydrogenase was assayed in a reaction medium (1 mL) containing 0.1 M TRIS (pH 7.4), 2 mM MgCl$_2$, 2 mM NADP$^+$, and aliquots of the supernatant [48].

The reaction was started by the addition of 1.25 mM isocitrate and the increase in absorbance was monitored at 340 nm ($\varepsilon = 6.22 \times 10^3$ M^{-1} cm^{-1}).

The activity of L-glutamate dehydrogenase was measured in a reaction medium (1 mL) containing 50 mM triethanolamine (pH 8.0), 0.1 M ammonium sulfate, 95 μM NADH, 2.5 mM EDTA, 1 mM ADP, and aliquots of the supernatant described above [49]. The reaction was started by addition of α-ketoglutarate (8.0 mM) and the decrease in absorbance was monitored at 340 nm.

The activity of α-ketoglutarate dehydrogenase was measured in a medium containing 100 mM phosphate buffer (pH 7.4), 2 mM NAD$^+$, 0.2 mM thiamine pyrophosphate, 1 mM MgCl$_2$, 0.1% Triton X-100, 0.3 mM dithiothreitol, 10 mM α-ketoglutarate and aliquots of disrupted mitochondria suspensions [40]. The reaction was initiated by the addition of coenzyme A (0.2 mM) and monitored spectrophotometrically as the reduction of NAD$^+$ at 340 nm ($\varepsilon = 6.22 \times 10^3$ M^{-1} cm^{-1}).

2.7. Brain Hexokinase (HK) Activity

Fresh brains were homogenized in medium containing 200 mM mannitol, 76 mM sucrose, 1 mM EGTA, 0.1 mM PMSF and 10 mM TRIS (pH 7.4). The homogenate was centrifuged at 650g for 8 min. The activity of the hexokinase (HK) in the cytosol was measured using the liquid fraction obtained by an additional centrifugation of the brain homogenate supernatant (40 min at 29,000g) [50]. This procedure precipitates the mitochondria and other cell components. The incubation medium contained in a final volume of 1 mL: 100 mM Tris-HCl (pH 7.2), 20 mM glucose, 2 mM ATP, 10 mM MgCl$_2$, 1.5 mM NAD$^+$, 5 units of glucose 6-phosphate dehydrogenase from *Leuconostoc mesenteroides* and 50 μL of high speed centrifugation supernatant. The reaction was monitored spectrophotometrically at 340 nm ($\varepsilon = 6.22 \times 10^3$ M^{-1} cm^{-1}).

2.8. Statistical Analysis

Statistical analysis was done using the GraphPad Prism Software (version 5.0). Statistical significance was inferred from ANOVA one-way with Newman–Keuls post-hoc testing. The 5% level of significance was adopted ($p < 0.05$).

3. Results

3.1. Indicators of Inflammation and Oxidative Stress

The results of the measurements of indicators of inflammation and oxidative stress in the brain of the various groups of experimental animals are shown in Figure 1. Figure 1a,b illustrates the effects of arthritis and the MeJA and ibuprofen treatments on the myeloperoxidase (MPO) activity and the nitrite plus nitrate levels. The latter are indicators for the NO production. The MeJA treatment did not affect the MPO activity in healthy rats. In arthritic rats, however, in which the activity of MPO was increased 2.5-fold, the MeJA treatment caused progressive diminutions as the doses were increased. At the highest dose (300 mg/kg) the MPO activity was close to the activity in the control animals and in the rats treated with ibuprofen. The levels of nitrite plus nitrate were increased approximately 1.4-fold in the brain of arthritic rats. Here again the MeJA treatment did not modify these levels in control rats. In arthritic rats both the MeJA and the ibuprofen treatments diminished the elevated nitrite plus nitrate levels. For the MeJA dose of 300 mg/kg the nitrite plus nitrate levels were very close to those observed in control rats or in ibuprofen-treated rats.

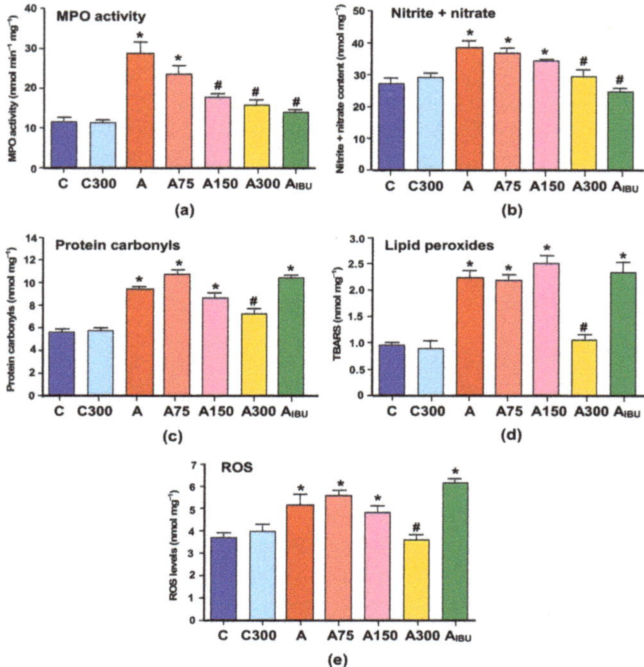

Figure 1. Effects of the methyl jasmonate (MeJA) treatment on parameters of inflammation and oxidative stress in the brain. The activity of (**a**) myeloperoxidase (MPO); and (**b**) the levels of nitrite plus nitrate; (**c**) protein carbonyl groups; (**d**) thiobarbituric acid reactive substances (TBARS); and (**e**) reactive oxygen species (ROS) were measured as described in Materials and Methods. C, controls treated with corn oil; C300, controls treated with 300 mg/kg MeJA; A, arthritic rats treated with corn oil; A75, A150 and A300, arthritic rats treated with 75, 150 and 300 mg/kg MeJA, respectively; A_{IBU}, arthritic rats treated with 30 mg/kg ibuprofen. Data are the means ± standard errors of the mean of five animals for each experimental condition. Statistical analysis: ANOVA one-way with Newman–Keuls post-hoc testing. * $p < 0.05$, different from the controls (C); # $p < 0.05$, different from non-treated arthritic rats (A).

Figure 1c–e shows the levels of protein carbonyls, lipid peroxides (TBARS) and reactive oxygen species (ROS) that were found in the brain of arthritic and control rats and the effects of MeJA and ibuprofen treatments. The levels of carbonyl groups and TBARS were increased 1.7- and 2.2-fold, respectively, in the brain of arthritic rats. No modifications were found when MeJA was given to healthy rats. In arthritic rats the MeJA doses of 75 and 150 mg/kg were also ineffective in modifying the protein carbonyls and TBARS levels. Actually, there was even a small tendency toward higher values, though lacking statistical significance. Treatment with 300 mg/kg MeJA, however, produced clear decreases in the levels of carbonyl groups and TBARS to values close to the control ones. The levels of ROS were increased 1.4-fold in the brain of arthritic rats. The MeJA treatment of control rats was without effect. Here again the lower MeJA doses were ineffective and only the 300 mg/kg dose produced a significant and pronounced decrease to a value close to that of the control condition. Treatment of arthritic animals with ibuprofen did not improve any of the oxidative stress indicators.

3.2. GSH Levels

Figure 2 shows the levels of reduced glutathione (GSH) and the GSH/GSSG ratio. The levels of GSH and the GSH/GSSG ratio were 35% and 70% lower, respectively, in the brain of arthritic animals. The level of GSSG was not modified by arthritis, but it was slightly increased in rats treated with 75 and 150 mg/kg MeJA and ibuprofen (not shown). Treatment of arthritic rats with 300 mg/kg MeJA, but not with ibuprofen, increased the level of GSH to a value close to the control one. In consequence, an almost similar increase in the GSH/GSSG ratio was also found at the 300 mg/kg dose, although a decreasing tendency was apparent at lower doses.

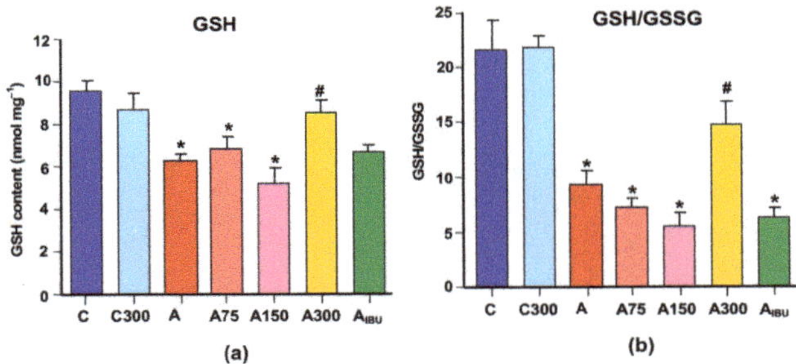

Figure 2. Effects of the MeJA treatment on (**a**) glutathione (GSH) levels and (**b**) GSH/ oxidized glutathione (GSSG) ratios. Symbols are those defined in Figure 1. Data are the means ± standard errors of the means of five animals for each experimental condition. Statistical analysis: ANOVA one-way with Newman–Keuls post-hoc testing. *$p < 0.05$, different from the controls (C); #$p < 0.05$, different from non-treated arthritic rats (A).

3.3. Mitochondrial ROS Generation and Membrane Potential

In isolated respiring rat liver mitochondria, it has been found that MeJA increases net ROS generation [9]. For this reason, experiments were done in which the action of MeJA treatment on ROS generation by brain mitochondria was measured. Figure 3a shows the results obtained when brain mitochondria isolated from rats under various conditions were incubated in a medium containing succinate as the electron donor. ROS generation was measured as the fluorescence increase due to the formation of the indicator DCF. In mitochondria from arthritic rats, ROS generation was clearly increased by a factor of 1.36. Treatment with MeJA did not reverse this modification although there was some tendency in this direction. There was also a small tendency toward increasing ROS generation in mitochondria from healthy rats treated with MeJA. ROS generation is a dynamic variable which could be dependent on the continuous presence of MeJA. Since the latter is no longer present in mitochondria from treated rats, attempts were made at investigating the short term effects of the compound. The results shown in Figure 3a represent the possible reversible and short-term effects of MeJA on ROS generation in isolated mitochondria from healthy and arthritic rats in the range of up to 10 mM. They show that the compound increased ROS generation in brain mitochondria from both healthy and arthritic rats. The increments relative to the starting conditions were similar for both healthy and arthritic conditions in the range up to 1.25 mM. After this concentration there was a declining tendency, more pronounced in mitochondria from healthy rats.

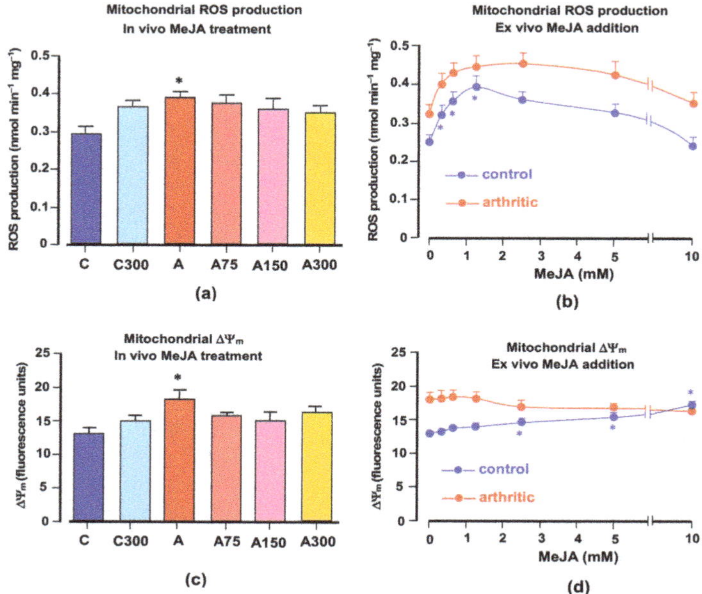

Figure 3. Effects of MeJA on ROS production and membrane potential in brain mitochondria from healthy and arthritic rats. (**a**) ROS production in mitochondria isolated from healthy and arthritic rats, treated or not with different MeJA doses (C, controls; C300, controls treated with 300 mg/kg MeJA; A, arthritic rats; A75, A150 and A300, arthritic rats treated with 75, 150 and 300 mg/kg MeJA, respectively). (**b**) ROS production by brain mitochondria isolated from healthy and arthritic rats and incubated with varying concentrations of MeJA. (**c**) Membrane potential ($\Delta\Psi_m$) of brain mitochondria isolated from healthy and arthritic rats, treated or not with different MeJA doses. (**d**) Membrane potential ($\Delta\Psi_m$) of brain mitochondria isolated from healthy and arthritic rats and incubated with varying concentrations of MeJA. Data are the means ± standard errors of the mean of five animals for each experimental condition. Statistical analysis: ANOVA one-way with Newman–Keuls post-hoc testing. *$p < 0.05$, different from the corresponding controls.

It is generally accepted that ROS generation in mitochondria shows a positive correlation with the membrane potential [51]. The measurements illustrated by Figure 3c,d were done in order the explore this relationship in terms of the effects of arthritis and MeJa. As described in the Materials and Methods section, energization was achieved by the addition of succinate, the same substrate used in the experiments in which ROS generation was measured. Figure 3c reveals a clearly increased membrane energization in mitochondria from arthritic rats. This correlates nicely with the increased rates of ROS generation shown in Figure 3a. The effects of MeJA treatment, however, are unclear, as no statistically significant modifications were detected even though there is a general tendency toward a lower energization. When mitochondria isolated from control and arthritic rats were incubated with varying MeJA concentrations in the range of up to 10 mM in order to explore possible reversible actions (Figure 3d), only minimal modifications were observed. These were restricted to the control rats, in which there was a small increment at high MeJA concentrations.

3.4. Antioxidant and Prooxidant Enzyme Activities

Figure 4a,b shows the influence of arthritis and the subsequent MeJA treatment on the activities of the antioxidant enzymes catalase (CAT) and superoxide dismutase (SOD). The CAT and SOD activities were 25% and 40% lower, respectively, in the brain of arthritic rats. Treatment with 150 and 300 mg/kg MeJA or ibuprofen increased the catalase activity to values close to the control ones. Treatment of arthritic rats with 75, 150 and 300 mg/kg MeJA, but not with ibuprofen, increased the SOD activity to values close to the control ones.

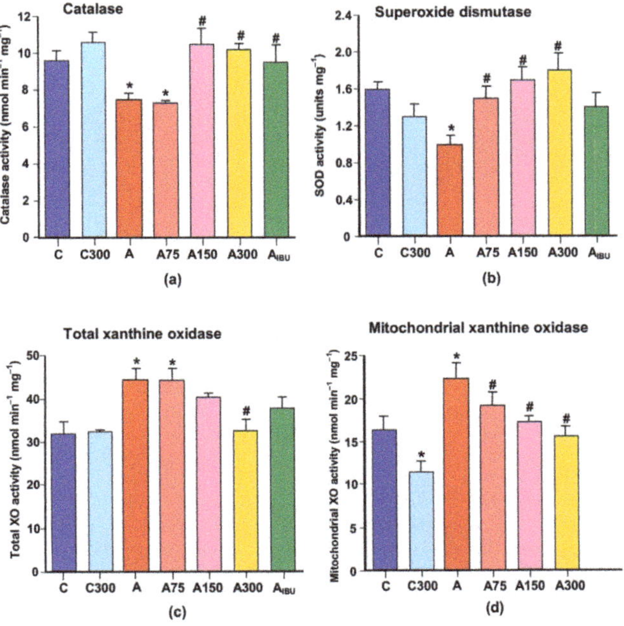

Figure 4. Effects of MeJA treatment on enzymes linked to the oxidative homeostasis. (**a**) Catalase; (**b**) superoxide dismutase; (**c**) total xanthine oxidase; (**d**) mitochondrial xanthine oxidase. Abbreviations: C, controls; C300, controls treated with 300 mg/kg MeJA; A, arthritic rats; A75, A150 and A300, arthritic rats treated with 75, 150 and 300 mg/kg MeJA, respectively; A$_{IBU}$, arthritic rats treated with 30 mg/kg ibuprofen. Data are the means ± standard errors of the mean of five animals for each experimental condition. Statistical analysis: ANOVA one-way with Newman–Keuls post-hoc testing. *$p < 0.05$, different from the controls (C); # $p < 0.05$, different from non-treated arthritic rats (A).

Figure 4c,d illustrates the actions of arthritis and MeJA on the activity of xanthine oxidase (XO), which can be considered a prooxidant enzyme whose activity is usually recognized as an important source of reactive oxygen species in the brain [26]. In the brain homogenate from arthritic rats the activity of XO was increased 1.4-fold. The treatment with 300 mg/kg MeJA almost abolished this increase. The activity of XO was also increased 1.4-fold in isolated brain mitochondria of arthritic rats. In this case the treatment of arthritic rats with MeJA showed a well-defined dose-dependent diminution to values close to the control ones.

3.5. Mitochondrial Enzymes Linked to Oxidative Metabolism

Several mitochondrial enzymes linked to oxidative metabolism were assayed with the purpose of searching for possible modifications in energy metabolism and in reactions leading to the production of NADPH for feeding the glutathione cycle with reducing equivalents. To the latter category belong L-glutamate dehydrogenase (which also operates with NADP$^+$) and the NADP$^+$-dependent isocitrate dehydrogenase. The activity of the first one was not modified by arthritis or by the subsequent MeJA treatment (not shown). The NADP$^+$-dependent isocitrate dehydrogenase, however, was diminished in arthritic rats by 23%, as shown in Figure 5a. The MeJA treatment increased this enzyme in both healthy and arthritic rats. In the latter, this effect led to a complete prevention of the diminution caused by arthritis.

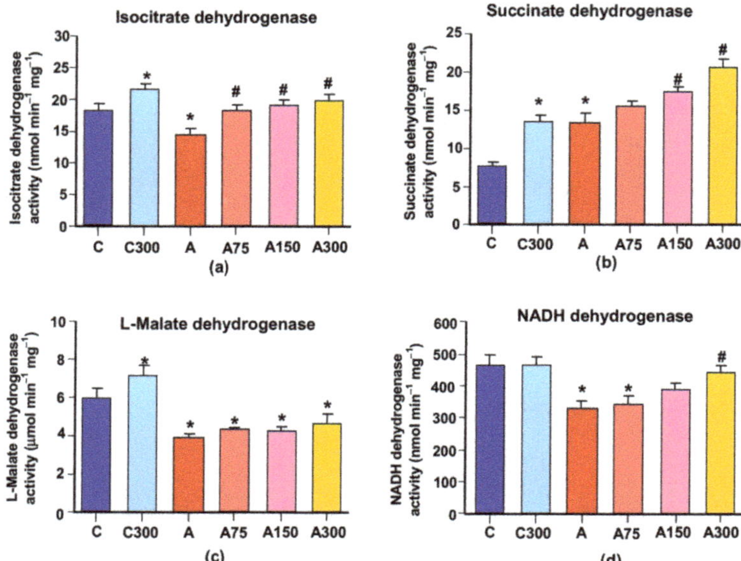

Figure 5. Effects of MeJA treatment on selected mitochondrial enzyme activities. (**a**) Succinate dehydrogenase; (**b**) isocitrate dehydrogenase; (**c**) L-malate dehydrogenase; (**d**) NADH dehydrogenase. Abbreviations: C, controls; C300, controls treated with 300 mg/kg MeJA; A, arthritic rats; A75, A150 and A300, arthritic rats treated with 75, 150 and 300 mg/kg MeJA, respectively; A_{IBU}, arthritic rats treated with 30 mg/kg ibuprofen. Data are the means ± standard errors of the mean of five animals for each experimental condition. Statistical analysis: ANOVA one-way with Newman–Keuls post-hoc testing. * $p < 0.05$, different from the controls (C); # $p < 0.05$, different from non-treated arthritic rats (A).

The other enzymatic activities linked to oxidative metabolism that were measured were ATPase, α-ketoglutarate dehydrogenase, L-malate dehydrogenase, succinate dehydrogenase and NADH-dehydrogenase. No modifications by arthritis or MeJA treatment were found for the first two. The other three, however, suffered modifications. Succinate dehydrogenase was increased by a factor of 1.8 in arthritic rats; almost the same increment was caused by the MeJA treatment of healthy rats. The latter effect was apparently maintained in arthritic rats in which the already higher activity was further increased by the MeJA treatment. For the 300 mg/kg dose this led to an increase of 2.76-fold when compared to the healthy controls. The activities of the other two enzymes were diminished by arthritis. The L-malate dehydrogenase activity was diminished by 30%. Although the MeJA treatment of healthy rats produced an increase in the L-malate dehydrogenase activity, this effect was not prominent enough in arthritic rats so as to conduct to a significant recovery. The diminution of NADH dehydrogenase caused by arthritis amounted to 31%. The treatment of healthy rats with 300 mg/kg MeJA did not produce modifications. However, the same treatment of arthritic rats almost completely prevented the diminution of the NADH-dehydrogenase activity.

3.6. Hexokinase Activity

The liver of arthritic rats presents a considerably increased glucose phosphorylation capacity [30], which is diminished when the rats are treated with MeJA [9]. For these reasons and because glucose oxidation may be related to the oxidative state of the brain tissue, experiments were done in which the influence of arthritis and MeJA on the hexokinase activity was investigated. Figure 6 illustrates the results that were obtained. Figure 6a shows the measurements of the hexokinase activity in the cytosolic fraction of brains from healthy and arthritic rats, treated or not with the current different MeJA doses. Arthritis had no significant effect on the hexokinase activity present in the cytosolic fraction. The MeJA treatment, on the other hand, increased the hexokinase activity in the cytosolic fraction obtained from healthy rats. A similar effect was found in arthritic rats treated with MeJA, with a maximal effect at the dose of 300 mg/kg.

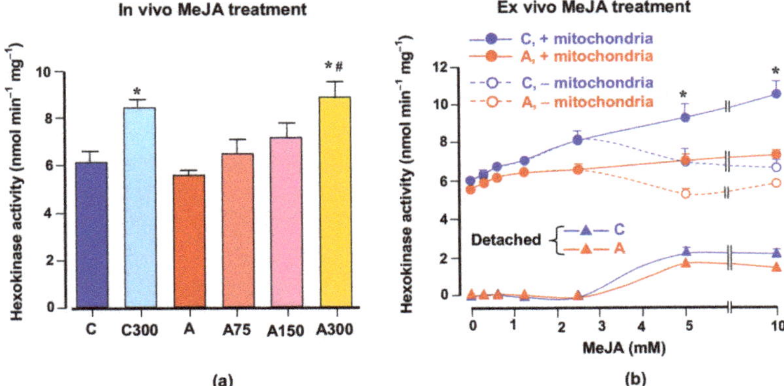

Figure 6. Effects of MeJA treatment on hexokinase activity. (**a**) Hexokinase activities in the cytosolic fraction of brains from healthy and arthritic rats, treated or not with different MeJA doses (C, controls; C300, controls treated with 300 mg/kg MeJA; A, arthritic rats; A75, A150 and A300, arthritic rats treated with 75, 150 and 300 mg/kg MeJA, respectively). (**b**) Hexokinase activities in the cytosolic fraction of brains from healthy and arthritic rats incubated with various MeJA concentrations with or without previous centrifugation to eliminate mitochondria and other cell components. Data are the means ± standard errors of the mean of five animals for each experimental condition. Statistical analysis: ANOVA one-way with Newman–Keuls post-hoc testing; *$p < 0.05$, different from the corresponding controls (C); #$p < 0.05$, different from non-treated arthritic rats (A).

In brain cells the hexokinase is attached to the mitochondrial membrane and it is known that MeJA is able to detach the enzyme [5]. Since mitochondria were absent in the cytosolic fraction used for the hexokinase assay, the possibility exists that the higher hexokinase activities found in the brain of MeJA-treated animals can actually represent detached enzyme due to the continuous presence of MeJA in the tissue. To investigate this possibility the series of experiments shown in Figure 6b were done. MeJA was added at varying concentrations to the low-speed centrifugation supernatant of the brain homogenate either before or after the high-speed centrifugation for precipitating the mitochondria and other cell components. The time of exposure to MeJA was in all cases 30 min. The hexokinase activity in the cytosolic fractions was represented against the MeJA concentrations. Up to the MeJA concentration of 2.5 mM, precipitation of the mitochondria did not modify the activity of hexokinase found in the cytosolic fraction, although a small increasing tendency was found for both types of incubations. Above this concentration, however, the hexokinase activity continued to increase with the MeJA concentration in those preparations in which the mitochondria were still present when the compound was added. In the preparations in which the mitochondria had been precipitated previous to the addition of MeJA, the hexokinase activity ceased to increase. The enhancement of the activity when the incubation containing mitochondria was used probably reflects the detachment of the hexokinase bound to the organelles caused by MeJA. The detached hexokinase, shown in Figure 6b, calculated as the difference between incubations with and without mitochondria, was similar for the healthy and arthritic condition.

4. Discussion

The actions of MeJA in the brain of arthritic rats clearly point in the direction of a diminution of both inflammation and oxidative stress. The modifications that MeJA caused, however, were multivariate and complex, and frequently not restricted to the arthritic rat. The scheme in Figure 7 offers an overview of the actions of MeJA in the brain cells in opposition to the modifications induced by arthritis, which should be a helpful guide in the discussion that follows. The scheme in Figure 7 assumes that MeJA easily crosses biological membranes including the blood brain barrier, as demonstrated experimentally and indeed expected from a lipid soluble substance [3,52]. The black arrows refer mainly to the modifications caused by arthritis whereas the red dotted arrows indicate the main sites of action of MeJA. In the brain, the increased oxidative stress induced by arthritis has been attributed to both an increased production of ROS and an impaired ROS scavenging system [26]. In fact, previous reports and the results of this study show that the activity of XO and mitochondrial ROS production are increased in the arthritic condition, while the GSH levels and the activity of superoxide dismutase and catalase are reduced in the brain [26,53]. This imbalance between pro- and antioxidant systems results in higher levels of ROS, protein oxidation and lipoperoxides in the arthritic brain.

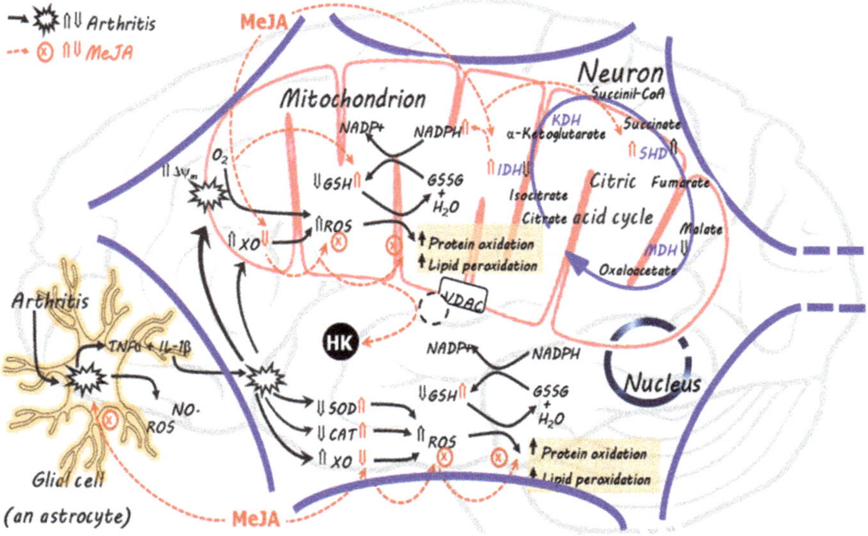

Figure 7. Events modifying inflammation and the oxidative state in the brain of rats with adjuvant-induced arthritis and the actions of methyl jasmonate (MeJA). The scheme is based mainly on the results of the present study and literature data. The symbol ⇑ means up-regulation and ⇓ means down-regulation. Black arrows indicate events in the absence of MeJA and red arrows indicate the effects of MeJA. Abbreviations: TNF-α, tumor necrosis factor alpha; IL-1β, interleukin 1 β; GSH, reduced glutathione; GSSG, oxidized glutathione; ROS, reactive oxygen species; NO, nitric oxide; HK, hexokinase; SOD, superoxide dismutase; CAT, catalase; IDH, isocitrate dehydrogenase; MDH, malate dehydrogenase; KDH, alpha-ketoglutarate dehydrogenase; SDH, succinate dehydrogenase; XO, xanthine oxidase; $\Delta\psi_m$, mitochondrial membrane potential; VDAC: voltage-dependent anion channel.

In the present study the indicators of inflammation in the brain of arthritic rats that were measured are the activity of MPO and the level of nitrite plus nitrate as an indicator of the NO concentration. The MPO activity has been considered as one of the best inflammatory and oxidative stress markers for several inflammatory diseases [54]. The nitrite (NO_2^-) plus nitrate (NO_3^-) levels, on the other hand, are an index of NO production and indirectly also of the inflammatory cytokines. Inflammatory cytokines stimulate glial cells to produce nitric oxide (NO) by the inducible NO synthase (iNOS) in the brain [55,56]. Both the activity of MPO and the nitrite plus nitrate levels, appeared substantially increased in arthritic rats and both were reduced to the control levels by the MeJA treatment at the doses of 150 and 300 mg/kg. These effects were similar to ibuprofen (30 mg/kg), an anti-inflammatory agent currently used to attenuate the symptoms of rheumatoid arthritis. This is consistent with a previous study showing that orally administered MeJA improves the systemic and articular inflammation in rats with adjuvant-induced arthritis [9]. The mechanism by which MeJA decreases inflammation is not yet sufficiently known, but it has been reported that it influences the activation of nuclear factor kappa B (NF-κB). Experiments have shown that MeJA attenuates the LPS-induced activation of NF-κB in RAW267.4 cells and reduces the expression of NF-κB in the brain of mice with LPS-induced neuroinflammation [7,27]. In the latter, intraperitoneally administered MeJA additionally inhibits the production of IL-1β, IL-6, TNFα and prostaglandin E, possibly by glial cells, as illustrated by Figure 7. It also downregulates the expression of iNOS and COX-2 in the brain. These results were recently corroborated by the observation that MeJA reduces the levels of TNF-α and IL-6 and nitrite in the brain of mice with neuroinflammation induced by rotenone, a phenomenon that is also likely to be related to the downregulation of the NF-κB expression [57]. In addition, in mice, MeJA improves the memory dysfunction associated with the neuroinflammation [27]. The effective doses of MeJA in the present study, i.e., 150 and 300 mg/kg, may be relatively high for clinical purposes, but it must be stressed that they have been associated to a notable absence of toxicity [3,9,32].

Comparison of the dose dependences of the decreases in the inflammatory indicators (MPO and nitrite plus nitrate), on one side, with the dose dependences of the decreases in the oxidative stress indicators (protein carbonyls, TBARS and ROS), on the other side, reveals that there was a dose shift toward higher ones in the latter actions (see Figure 1). Diminution of oxidative stress occurred only at the highest dose. At the lowest doses there was even an increasing tendency of the oxidative stress indicators. This may bear relation to the fact that MeJA actually exerts a dual role, i.e., although it certainly acts as an antioxidant agent, it can also be regarded as a prooxidant agent in spite of its anti-inflammatory action. The prooxidant action is exerted in the respiratory chain. The latter is an important source of reactive oxygen species [58] and our experimental data show that the compound increases ROS output in respiring brain mitochondria in accordance to previous observations in rat liver mitochondria [9]. This effect depends on the presence of MeJA as it was not preserved when the mitochondria were isolated from the brain tissue of treated rats. Furthermore, this action was not different in healthy and arthritic rats. It is unlikely that the phenomenon bears relation to the membrane potential, as MeJA did not modify this parameter. Its mechanism remains to be clarified.

The net diminution in oxidative stress caused by MeJA derives, thus, from its effects on ROS generation systems other than the mitochondrial electron transport chain or from its role as an enhancer of the antioxidant defenses. In the first category one may include ROS-producing enzymes whose activity is increased by the inflammatory mediators of arthritis. One such enzyme, which is generally considered an important source of ROS in the brain, namely xanthine oxidase, was analyzed in the present work (see Figure 7). Xanthine oxidase, under specific circumstances, is able to generate H_2O_2, superoxide anion and nitric oxide [26,59,60]. Its activity, which was substantially increased by arthritis, was gradatively diminished to control values by MeJA, especially in the mitochondria. This may be an important factor that helps to reduce the production of ROS in the mitochondria under in vivo conditions. With respect to the antioxidant defenses, MeJA clearly restored the activity of both catalase and superoxide dismutase. These actions can be attributed at least in part to an inhibition of the production of proinflammatory cytokines (see Figure 7). MeJA also caused a recovery in the GSH levels and in the GSH/GSSG ratio, corroborating the conclusion that the compound enhances the antioxidant defenses. In this respect the increase in the activity of isocitrate dehydrogenase induced by MeJA may play a significant role as the production of NADPH can occur directly via the action of this enzyme. Consequently, preservation of the activity of the mitochondrial isocitrate dehydrogenase by the MeJA treatment in arthritic rats is expected to contribute to the preservation of a high $NADPH/NADP^+$ ratio. This, in turn, will contribute to an increased regeneration of GSH and reduction of the oxidative stress via the glutathione system. Similar MeJA actions were also found in the liver of arthritic rats [9], which were attributed to microRNA 101-mediated upregulation of the nuclear factor erythroid 2-related 2 (Nfr2), a redox-sensible transcription factor that upregulates antioxidant defense genes, including catalase and enzymes required for GSH synthesis and sources of NADPH like isocitrate dehydrogenase [61–63].

In addition to isocitrate dehydrogenase, adjuvant-induced arthritis also modified the activities of other respiratory enzymes, namely NADH dehydrogenase, L-malate dehydrogenase and succinate dehydrogenase. This is in line with previous reports that inflammatory mediators and reactive oxygen or nitrogen species may affect mitochondrial functions in brain cells and in human chondrocytes [64–67]. The MeJA treatment did not modify the diminished L-malate dehydrogenase activity, but it prevented the diminished NADH dehydrogenase activity. Surprisingly the MeJA treatment further increased the mitochondrial succinate dehydrogenase, which was possibly increased in arthritic rats by the action of TNFα [63]. It is difficult to infer from the available data the consequences and significance of these actions of MeJA for the mitochondrial functions.

In the brain, arthritis did not modify the hexokinase activity, contrary to what happens in the liver where the activity of the analogous glucose phosphorylating enzyme (glucokinase) is substantially increased, leading to augmented hepatic glycolysis [9]. In the brain the treatment with MeJA resulted in an apparent increase in the hexokinase activity, a phenomenon that has already been described for cancer and brain cells [5]. This apparent increase when the hexokinase is assayed in the cytosolic fraction results from the detachment of the enzyme from the mitochondria. Although the phenomenon was only detected at high concentrations in the present work after 30 min incubation, it is worth mentioning that the phenomenon is time-dependent [5], which makes it a likely occurrence in animals exposed to low doses during several days. The enzyme isoforms normally attached to the mitochondrial membrane are types I and II. Type IV hexokinase (glucokinase), the isoform normally found in the liver cells, is not attached to the mitochondrial membrane [5]. Our results suggest that detachment of the hexokinase from the mitochondria in healthy and arthritic animals by MeJA was similar. Consequently, the phenomenon should in principle not be considered as a mechanism capable of having contributed to the antioxidant or prooxidant activities of MeJA. In cancer cells the detachment of the hexokinase from the mitochondria is considered one of the main reasons for its cytotoxicity because these cells depend heavily on the oxidation of glucose for survival and ATP for its phosphorylation is more readily available at the surface of the mitochondria [68,69]. Detachment of the hexokinase from the brain mitochondria by MeJA has not been considered neurotoxic however, probably because the brain cells are less dependent on glycolysis for survival [5], even though a limited impairment of glycolysis in the brain can be expected to occur.

The behavioral and cognitive changes induced by MeJA and reported by several authors require a final set of comments. MeJA improves the cognitive dysfunction associated with neuro-inflammation. It was reported that MeJA attenuates LPS-induced memory dysfunction via mechanisms involving inhibition of pro-inflammatory mediators and beta-amyloid generation in mice [27]. In other investigations the effects of MeJA were assessed in mice with memory impairment caused by scopolamine and unpredictable chronic mild stress (UCMS) [28,29]. In these studies, MeJA affected oxidative stress in the brain in a way that was similar to that found in the present study. In these studies, MeJA was intraperitoneally administered at doses of up to 100 mg/kg and the oxidative status improvement was associated with neuroprotection and reduced memory dysfunctions. This allows to hypothesize that the antioxidant and anti-inflammatory properties of MeJA may also improve the brain fatigue and cognitive function in patients with severe rheumatoid arthritis [12,24].

5. Conclusions

In conclusion, MeJA treatment decreased inflammation and oxidative stress in the brain of rats with adjuvant-induced arthritis. The improvement of oxidative stress occurred in consequence of reduced inflammation associated with increased antioxidant defenses. On the other hand, MeJA induced the hexokinase (HK) detachment from the brain mitochondria of both healthy and arthritic rats a phenomenon that can diminish glucose phosphorylation and metabolism in the brain. The effective doses of MeJA were relatively high, but the results of the present study make this compound a potential precursor for developing anti-rheumatic and anti-inflammatory drugs. Finally, future investigations may take into account the role of Nfr2 in the actions of MeJA and the perspective of modifications in the metabolism of glucose in the brain tissue.

Author Contributions: Conceptualization, J.F.C. and A.B.S.-N.; investigation, H.V.P.-M., J.F.C., L.S.C., L.B., F.M.S.S. and G.A.G.; resources, J.F.C., A.B., R.M.P.; data curation, C.A.B.-A., A.B.S.-N. and J.F.C.; writing—original draft preparation, A.B.S.-N. and J.F.C.; writing—review and editing, A.B. and R.M.P..; visualization, A.B.; funding acquisition, J.F.C. and A.B.S.-N.

Funding: This work was funded by the Conselho Nacional de Desenvolvimento Científico e Tecnológico (CNPq) [Grant number: 40974720165].

Acknowledgments: Authors wish to thank the financial support of the Conselho Nacional de Desenvolvimento Científico e Tecnológico (CNPq) and Coordenação de Aperfeiçoamento Pessoal de Nível Superior (CAPES).

Conflicts of Interest: The authors declare no conflict of interest. The funders had no role in the design of the study; in the collection, analyses, or interpretation of data; in the writing of the manuscript, or in the decision to publish the results.

References

1. Raviv, Z.; Cohen, S.; Reischer-Pelech, D. The anti-cancer activities of jasmonates. *Cancer Chemother. Pharmacol.* **2013**, *71*, 275–285. [CrossRef] [PubMed]

2. Kuroda, K.; Inoue, N.; Ito, Y.; Kubota, K.; Sugimoto, A.; Kakuda, T.; Fushiki, T. Sedative effects of the jasmine tea odor and (R)-(−)-linalool, one of its major odor components on autonomic nerve activity and mood states. *Eur. J. Appl. Physiol.* **2005**, *95*, 107–114. [CrossRef] [PubMed]

3. Cesari, I.M.; Carvalho, E.; Rodrigues, M.F.; Mendonça, B.S.; Amôedo, N.D.; Rumjanek, F.D. Methyl jasmonate: Putative mechanisms of action on cancer cells cycle, metabolism and apoptosis. *Int. J. Cell Biol.* **2014**, *2014*, ID572097. [CrossRef] [PubMed]

4. Zhang, M.; Zhang, M.W.; Zhang, L.; Zhang, L. Methyl jasmonate and its potential in cancer therapy. *Plant Signal Behav.* **2015**, *10*, e106219. [CrossRef] [PubMed]

5. Goldin, N.; Arzoine, L.; Heyfets, A.; Israelson, A.; Zaslavsky, Z.; Bravman, T.; Bronner, V.; Notcovich, A.; Shoshan-Barmatz, V.; Flescher, E. Methyl jasmonate binds to and detaches mitochondria bound hexokinase. *Oncogene* **2008**, *27*, 4636–4643. [CrossRef] [PubMed]

6. Dang, H.T.; Lee, H.J.; Yoo, E.S.; Hong, J.; Bao, B.; Choi, J.S.; Jung, J.H. New jasmonate analogues as potential anti-inflammatory agents. *Bioorg. Med. Chem.* **2008**, *16*, 10228–10235. [CrossRef]

7. Kim, M.J.; Kim, S.S.; Park, K.J.; An, H.J.; Choi, Y.H.; Lee, N.H.; Hyun, C.G. Methyl jasmonate inhibits lipopolysaccharide-induced inflammatory cytokine production via mitogen-activated protein kinase and nuclear factor-κB pathways in RAW 264.7 cells. *Die Pharm.* **2016**, *71*, 540–543. [CrossRef]

8. Lee, H.J.; Maeng, K.; Dang, H.T.; Kang, G.J.; Ryou, C.; Jung, J.H.; Kang, H.K.; Prchal, J.T.; Yoo, E.S.; Yoon, D. Anti-inflammatory effect of methyl dehydrojasmonate (J2) is mediated by the NF-kappaB pathway. *J. Mol. Med.* **2011**, *89*, 83–90. [CrossRef]

9. Sá-Nakanishi, A.B.; Soni-Neto, J.; Moreira, L.S.; Gonçalves, G.A.; Silva-Comar, F.M.S.; Bracht, L.; Bersani-Amado, C.A.; Peralta, R.M.; Bracht, A.; Comar, J.F. Anti-inflammatory and antioxidant actions of methyl jasmonate are associated with metabolic modifications in the liver of arthritic rats. *Oxid. Med. Cell. Longev.* **2018**, *2018*, 2056250. [CrossRef]

10. Uhlig, T.; Moe, R.H.; Kvien, T.K. The burden of disease in rheumatoid arthritis. *Pharmacoeconomics* **2014**, *32*, 841–851. [CrossRef]

11. Kitas, G.D.; Gabriel, S.E. Cardiovascular disease in rheumatoid arthritis: State of the art and future perspectives. *Ann. Rheum. Dis.* **2011**, *70*, 8–14. [CrossRef] [PubMed]

12. McInnes, I.B.; Schett, G. The pathogenesis of rheumatoid arthritis. *N. Engl. J. Med.* **2011**, *365*, 2205–2209. [CrossRef] [PubMed]

13. Misko, T.P.; Radabaugh, M.R.; Highkin, M.; Abrams, M.; Friese, O.; Gallavan, R.; Bramson, C.; Le Graverand, M.P.H.; Lohmander, L.S.; Roman, D. Characterization of nitrotyrosine as a biomarker for arthritis and joint injury. *Osteoarthr. Cart.* **2013**, *21*, 151–156. [CrossRef] [PubMed]

14. Stamp, L.K.; Khalilova, I.; Tarr, J.M.; Senthilmhan, R.; Turner, R.; Haigh, R.C.; Winyard, P.G.; Kettle, A.J. Myeloperoxidase and oxidative stress in rheumatoid arthritis. *Rheumatology* **2012**, *51*, 1796–1803. [CrossRef] [PubMed]

15. Lemarechal, H.; Allanore, Y.; Chenevier-Gobeaux, C.; Kahan, A.; Ekindjian, O.G.; Borderie, D. Serum protein oxidation in patients with rheumatoid arthritis and effects of infliximab therapy. *Clin. Chim. Acta* **2006**, *372*, 147–153. [CrossRef]

16. Haruna, Y.; Morita, Y.; Yada, T.; Satoh, M.; Fox, D.A.; Kashihara, N. Fluvastatin reverses endothelial dysfunction and increased vascular oxidative stress in rat adjuvant-induced arthritis. *Arthritis Rheum.* **2007**, *56*, 1827–1835. [CrossRef]

17. Comar, J.F.; Sá-Nakanishi, A.B.; Oliveira, A.L.; Wendt, M.M.N.; Bersani-Amado, C.A.; Ishii-Iwamoto, E.L.; Peralta, R.M.; Bracht, A. Oxidative state of the liver of rats with adjuvant-induced arthritis. *Free Rad. Biol. Med.* **2013**, *58*, 144–153. [CrossRef]

18. Schubert, A.C.; Wendt, M.M.N.; Sá-Nakanishi, A.B.; Bersani-Amado, C.A.; Peralta, R.M.; Comar, J.F.; Bracht., A. Oxidative status and oxidative metabolism of the heart from rats with adjuvant-induced arthritis. *Exp. Mol. Pathol.* **2016**, *100*, 393–401. [CrossRef]

19. Bracht, A.; Silveira, S.S.; Castro-Ghizoni, C.V.; Sá-Nakanishi, A.B.; Oliveira, M.R.N.; Bersani-Amado, C.A.; Peralta, R.M.; Comar, J.F. Oxidative changes in the blood and serum albumin differentiate rats with monoarthritis and polyarthritis. *Springer Plus* **2016**, *5*, 36–50. [CrossRef]

20. Roubenoff, R. Rheumatoid cachexia: A complication of rheumatoid arthritis moves into the 21st century. *Arthritis Res. Ther.* **2009**, *11*, 108–109. [CrossRef]

21. Wendt, M.N.M.; de Oliveira, M.C.; Castro, L.S.; Franco-Salla, G.B.; Parizotto, A.V.; Silva, F.M.S.; Natali, M.R.M.; Bersani-Amado, C.A.; Bracht, A.; Comar, J.F. Fatty acids uptake and oxidation are increased in the liver of rats with adjuvant-induced arthritis. *Biochim. Biophys. Acta (BBA) Mol. Basis. Dis.* **2019**, *1865*, 696–707. [CrossRef] [PubMed]

22. Ames-Sibin, A.P.; Barizão, C.L.; Castro-Ghizoni, C.V.; Silva, F.M.S.; Sá-Nakanishi, A.B.; Bracht, L.; Bersani-Amado, C.A.; Natali, M.R.M.; Bracht, A.; Comar, J.F. β-Caryophyllene, the major constituent of copaiba oil, reduces systemic inflammation and oxidative stress in arthritic rats. *J. Cell. Biochem.* **2018**, *119*, 10262–10277. [CrossRef] [PubMed]

23. Castro-Ghizoni, C.V.; Ames, A.P.A.; Lameira, O.A.; Bersani-Amado, C.A.; Sá-Nakanishi, A.B.; Bracht, L.; Natali, M.R.M.; Peralta, R.M.; Bracht, A.; Comar, J.F. Anti-inflammatory and antioxidant actions of copaiba oil (*Copaifera reticulata*) are associated with histological modifications in the liver of arthritic rats. *J. Cell. Biochem.* **2017**, *118*, 3409–3423. [CrossRef] [PubMed]

24. Wartolowska, K.; Hough, M.G.; Jenkinson, M.; Andersson, J.; Wordsworth, B.P.; Tracey, I. Structural changes of the brain in rheumatoid arthritis. *Arthritis Rheum.* **2012**, *64*, 371–379. [CrossRef] [PubMed]

25. Bekkelund, S.I.; Pierre-Jerome, C.; Husby, G.; Mellgren, S.I. Quantitative cerebral MR in rheumatoid arthritis. *Am. J. Neuroradiol.* **1995**, *16*, 767–772. [PubMed]

26. Wendt, M.M.N.; Sá-Nakanishi, A.B.; Ghizoni, C.V.C.; Bersani-Amado, C.A.; Peralta, R.M.; Bracht., A.; Comar, J.F. Oxidative state and oxidative metabolism in the brain of rats with adjuvant-induced arthritis. *Exp. Mol. Pathol.* **2015**, *98*, 549–557. [CrossRef]

27. Solomon, U.; Taghogho, E.A. Methyl jasmonate attenuates memory dysfunction and decreases brain levels of biomarkers of neuroinflammation induced by lipopolysaccharide in mice. *Brain Res. Bull.* **2017**, *131*, 133–141. [CrossRef]

28. Eduviere, A.T.; Umukoro, S.; Aderibigbe, A.O.; Ajayi, A.M.; Adewole, F.A. Methyl jasmonate enhances memory performance through inhibition of oxidative stress and acetylcholinesterase activity in mice. *Life Sci.* **2015**, *132*, 20–26. [CrossRef]

29. Umukoro, S.; Aluko, O.M.; Eduviere, A.T.; Owoeye, O. Evaluation of adaptogenic-like property of methyl jasmonate in mice exposed to unpredictable chronic mild stress. *Brain Res. Bull.* **2016**, *121*, 105–114. [CrossRef]

30. Fedatto, Z., Jr.; Ishii-Iwamoto, E.L.; Bersani-Amado, C.; Maciel, E.R.M.; Bracht, A.; Kelmer-Bracht, A.M. Glucose phosphorylation capacity and glycolysis in the liver of arthrityic rats. *Inflamm. J.* **2000**, *49*, 128–132.

31. Pearson, C.M.; Wood, F.D. Studies of arthritis and other lesions induced in rats by the injection of mycobacterial adjuvant. *Am. J. Pathol. Phila.* **1963**, *42*, 93–95.

32. Umukoro, S.; Olugbemide, A.S. Antinociceptive effects of methyl jasmonate in experimental animals. *J. Nat. Med.* **2011**, *65*, 466–470. [CrossRef] [PubMed]

33. Katyare, S.S.; Rajan., R.R. Influence of thyroid hormone treatment on the respiratory activity of cerebral mitochondria from hypothyroid rats. A critical re-assessment. *Exp. Neurol.* **2005**, *195*, 416–422. [CrossRef] [PubMed]

34. Levine, R.L.; Garland, D.; Oliver, C.N.; Amici, A.; Climent, I.; Lenz, A.G.; Ahn, B.W.; Shaltiel, S.; Stadtman, E.R. Determination of carbonyl content in oxidatively modified proteins. *Meth. Enzymol.* **1990**, *186*, 464–478. [CrossRef]

35. Buege, J.A.; Aust, S.D. Microsomal lipid peroxidation. *Meth. Enzymol.* **1978**, *52*, 302–310. [CrossRef]

36. Siqueira, I.R.; Fochesatto, C.; Torres, I.L.S.; Dalmaz, C.; Netto, C.A. Aging affects oxidative state in hippocampus, hypothalamus and adrenal glands of Wistar rats. *Life Sci.* **2005**, *78*, 271–278. [CrossRef]

37. Biazon, A.C.B.; Wendt, M.M.N.; Moreira, J.R.; Castro-Ghizoni, C.V.; Soares, A.A.; Silveira, S.S.; Sá-Nakanishi, A.B.; Bersani-Amado, C.A.; Peralta, R.M.; Bracht, A.; et al. The in vitro antioxidant capacities of hydroalcoholic extracts from roots and leaves of *Smallanthus sonchifolius* (yacon) do not correlate with their in vivo antioxidant action in diabetic rats. *J. Biosci. Med.* **2016**, *4*, 15–27. [CrossRef]

38. Bryan, N.S.; Grisham, M.B. Methods to detect nitric oxide and its metabolites in biological samples. *Free Rad. Biol. Med.* **2007**, *43*, 645–657. [CrossRef]

39. Hissin, P.J.; Hilf, R. A fluorometric method for determination of oxidized and reduced glutathione in tissues. *Anal. Biochem.* **1976**, *74*, 214–226. [CrossRef]

40. Bergmeyer, H.U. *Methods of Enzymatic Analysis*; Verlag Chemie/Academic Press: London, UK; Weinheim, Germany, 1974; ISBN 352725370X.

41. Marklund, S.; Marklund, G. Involvement of the superoxide anion radical in the oxidation of pyrogallol and a convenient assay for superoxide dismutase. *Eur. J. Biochem.* **1974**, *47*, 469–474. [CrossRef]

42. Bradley, P.P.; Christensen, R.D.; Rothstein, G. Cellular and extracellular myeloperoxidase in pyogenic inflammation. *Blood* **1982**, *60*, 618–622. [CrossRef] [PubMed]

43. Galilea, J.; Canela, E.I.; Bozal, J. The course analysis of guanine and hypoxanthine transformation to uric acid by bovine liver guanine aminohydrolase and xanthine oxidase. *J. Mol. Catal.* **1981**, *12*, 27–36. [CrossRef]

44. Maciel, E.N.; Vercesi, A.E.; Castilho, R.F. Oxidative stress in (Ca^{2+})-induced membrane permeability transition in brain mitochondria. *J. Neurochem.* **2001**, *79*, 1237–1245. [CrossRef] [PubMed]

45. Sá-Nakanishi, A.B.; Soares, A.A.; de Oliveira, A.L.; Comar, J.F.; Peralta, R.M.; Bracht, A. Effects of treating old rats with an aqueous *Agaricus blazei* extract on oxidative and functional parameters of the brain tissue and brain mitochondria. *Oxid. Med. Cell. Longev.* **2014**, *2014*, 563179. [CrossRef] [PubMed]

46. Saling, S.C.; Comar, J.F.; Mito, M.S.; Peralta, R.M.; Bracht, A. Actions of juglone on energy metabolism in the rat liver. *Toxicol. Appl. Pharmacol.* **2011**, *257*, 319–327. [CrossRef]

47. Acharya, M.M.; Katyare, S.S. Structural and functional alterations in mitochondrial membrane in picrotoxin-induced epileptic rat brain. *Exp. Neurol.* **2005**, *192*, 79–88. [CrossRef]

48. Maeng, O.; Kim, Y.C.; Shin, H.J.; Li, J.O.; Huh, T.L.; Kang, K.I.; Kim, Y.S.; Paik, S.G.; Lee, H. Cytosolic $NADP^{+}$-dependent isocitrate dehydrogenase protects macrophages from LPS-induced nitric oxide and reactive oxygen species. *Biochem. Biophys. Res. Comm.* **2004**, *317*, 558–564. [CrossRef]

49. Hussain, M.M.; Zannis, V.I.; Plaitakis, A. Characterization of glutamate dehydrogenase isoproteins purified from the cerebellum of normal subjects and patients with degenerative neurological disorders, and from human neoplastic cell lines. *J. Biol. Chem.* **1989**, *264*, 20730–20735.

50. Vilela, V.R.; Oliveira., A.L.; Comar, J.F.; Peralta, R.M.; Bracht., A. Effects of tadalafil on the cAMP stimulated glucose output in the rat liver. *Chem. Biol. Interact.* **2014**, *220*, 1–11. [CrossRef]

51. Nicholls, D.G. Mitochondrial membrane potential and aging. *Aging Cell* **2004**, *3*, 35–40. [CrossRef]

52. Hossain, S.J.; Aoshima, H.; Koda, H.; Kiso, Y. Fragrances in oolong tea that enhance the response of GABAA receptors. *Biosci. Biotechnol. Biochem.* **2004**, *68*, 1842–1848. [CrossRef] [PubMed]

53. Gonçalves, G.A.; Sá-Nakanishi, A.B.; Wendt, M.M.N.; Comar, J.F.; Bersani-Amado, C.A.; Bracht, A.; Peralta, R.M. Green tea extract improves the oxidative state of the liver and brain in rats with adjuvant-induced arthritis. *Food Funct.* **2015**, *6*, 2701–2711. [CrossRef] [PubMed]

54. Khan, A.; Alsahli, M.; Rahmani, A. Myeloperoxidase as an active disease biomarker: Recent biochemical and pathological perspectives. *Med. Sci.* **2018**, *6*, 33. [CrossRef] [PubMed]

55. Murakami, A.; Ohigashi, H. Targeting NOX, INOS and COX-2 in inflammatory cells: Chemoprevention using food phytochemicals. *Int. J. Cancer* **2007**, *121*, 2357–2363. [CrossRef] [PubMed]

56. Wakita, T.; Shintani, F.; Yagi, G.; Asai, M.; Nozfawa, S. Combination of inflammatory cytokines increases nitrite and nitrate levels in the paraventricular nucleus of conscious rats. *Brain Res.* **2001**, *905*, 12–20. [CrossRef]

57. Alabi, A.O.; Ajayi, A.M.; Ben-Azu, B.; Bakre, A.G.; Umukoro, S. Methyl jasmonate abrogates rotenone-induced parkinsonian-like symptoms through inhibition of oxidative stress, release of pro-inflammatory cytokines, and down-regulation of immnopositive cells of NF-κB and α-synuclein expressions in mice. *Neurotoxicology* **2019**, *74*, 172–183. [CrossRef]

58. Murphy, M.P. How mitochondria produce reactive oxygen species. *Biochem. J.* **2005**, *417*, 1–13. [CrossRef]

59. Halliwell, B.; Gutteridge, J.M.C. *Free Radicals in Biology and Medicine*; Oxford University Press: London, UK, 2007.

60. Harrison, R. Structure and function of xanthine oxidoreductase: Where are we now? *Free Rad. Biol. Med.* **2002**, *33*, 774. [CrossRef]

61. Dinkova-Kostova, A.T.; Abramov, A.Y. The emerging role of Nrf2 in mitochondrial function. *Free Rad. Biol. Med.* **2015**, *88*, 179–188. [CrossRef]

62. Peng, Z.; Zhang, Y. Methyl jasmonate induces the apoptosis of human colorectal cancer cells via downregulation of EZH2 expression by microRNA-101. *Mol. Med. Rep.* **2016**, *15*, 957–962. [CrossRef]

63. Wang, Y.; Xiang, W.; Wang, M.; Huang, T.; Xiau, X.; Wang, L.; Tao, D.; Dong, L.; Zheng, F.; Jiang, G. Methyl jasmonate sensitizes human bladder cancer cells to gambogic acid-induced apoptosis through down-regulation of EZH2 expression by miR-101. *Br. J. Pharmacol.* **2014**, *171*, 618–631. [CrossRef] [PubMed]

64. Kim, J.; Xu, M.; Xo, R.; Mates, A.; Wilson, G.; Pearsall, A.T.; Grishko, V. Mitochondrial DNA damage is involved in apoptosis caused by pro-inflammatory cytokines in human OA chondrocytes. *Osteoarthr. Cart.* **2010**, *18*, 424–432. [CrossRef] [PubMed]

65. López-Armada, M.; Caramés, B.; Martín, M.; Cillero-Pastor, B.; Lires-Dean, M.; Fuentes-Boquete, I.; Arenas, J.; Blanco, F. Mitochondrial activity is modulated by TNFalpha and IL-1beta in normal human chondrocyte cells. *Osteoarthr. Cart.* **2006**, *14*, 1011–1022. [CrossRef]

66. Tretter, L.; Adam-Vizi, V. Inhibition of Krebs cycle enzymes by hydrogen peroxide: A key role of α-ketoglutarate dehydrogenase in limiting NADH production under oxidative stress. *J. Neurosci.* **2000**, *20*, 8972–8979. [CrossRef] [PubMed]

67. Levrat, C.; Larrick, J.W.; Wright, S.C. Tumor necrosis factor induces activation of mitochondrial succinate dehydrogenase. *Life Sci.* **1991**, *49*, 1731–1737. [CrossRef]

68. Li, J.; Chen, K.; Wang, F.; Dai, W.; Li, S.; Feng, J.; Wu, L.; Liu, T.; Xu, S.; Xia, Y.; et al. Methyl jasmonate leads to necrosis and apoptosis in hepatocelular carcinoma via inhibition of glycolysis and represses tumor growth in mice. *Oncotarget* **2017**, *8*, 45965–45980. [CrossRef]

69. Chen, Z.; Zhang, H.; Lu, W.; Huang, P. Role of mitochondria-associated hexokinase II in cancer cell death induced by 3-bromopyruvate. *Biochim. Biophys. Acta (BBA) Bioenerg.* **2009**, *1787*, 553–570. [CrossRef]

Article

Sideritis Perfoliata (Subsp. *Perfoliata*) Nutritive Value and Its Potential Medicinal Properties

Namrita Lall [1,2,3,*], Antonios Chrysargyris [4], Isa Lambrechts [1], Bianca Fibrich [1],
Analike Blom Van Staden [1], Danielle Twilley [1], Marco Nuno de Canha [1],
Carel Basson Oosthuizen [1], Dikonketso Bodiba [1] and Nikolaos Tzortzakis [4,*]

[1] Department of Plant and Soil Sciences, University of Pretoria, Pretoria 0002, South Africa;
ihlambrechts@gmail.com (I.L.); bianca.fibrich@gmail.com (B.F.); analikeblom@gmail.com (A.B.V.S.);
berrington.danielle@gmail.com (D.T.); marcodecanhasa@gmail.com (M.N.d.C.);
u04405765@tuks.co.za (C.B.O.); dikonketso.bodiba@gmail.com (D.B.)
[2] School of Natural Resources, University of Missouri, Columbia, MO 65211, USA
[3] College of Pharmacy, JSS Academy of Higher Education and Research, Mysuru, Karnataka 570015, India
[4] Department of Agricultural Sciences, Biotechnology and Food Science, Cyprus University of Technology,
3036 Lemesos, Cyprus; a.chrysargyris@cut.ac.cy
* Correspondence: namrita.lall@up.ac.za (N.L.); nikolaos.tzortzakis@cut.ac.cy (N.T.);
Tel.: +27-124206670 (N.L.); +357-25002280 (N.T.)

Received: 11 September 2019; Accepted: 25 October 2019; Published: 30 October 2019

Abstract: *Sideritis perfoliata* L. subsp. *perfoliata* is an endemic species of the Eastern Mediterranean region with several uses in traditional medicine. The present study aims to explore the unknown properties of *S. perfoliata* investigating the nutritional content as well as the antioxidant, anticancer, antituberculosis, antiwrinkle, anti-acne, hyper/hypo-pigmentation and antibacterial activities. Mineral content, nutritional value, the composition and antioxidant properties of the essential oil, the antityrosinase, the antibacterial activity and anti-elastase potential of the extract, were evaluated. The antiproliferative activity of *S. perfoliata* against cervical cancer (HeLa), human melanoma (UCT-Mel-1), human hepatocellular carcinoma (HepG2) and human epidermoid carcinoma (A431) was investigated. Cytotoxic effects on normal human keratinocyte (HaCat) and kidney epithelial (Vero) cell lines were also determined. *Sideritis perfoliata* exhibited high nutritional value of proteins and minerals (K, P, Mg, Fe, Zn, Cu). The most abundant components of the essential oil were found to be α-pinene, β-phelladrene, valeranone, β-pinene and sabinene. The ethanolic extract of *S. perfoliata* displayed moderate antioxidant potential and antibacterial activity against *Prevotella intermedia*. Noteworthy elastase and moderate anticancer potential against the human liver cancer cell line (HepG2) was observed with IC_{50} values of 57.18 ± 3.22 µg/mL and 64.27 ± 2.04 µg/mL respectively. The noteworthy *in vitro* activity of *S. perfoliata* could be due to the presence of flavonoids and phenols in the leaves, having high nutritional value. *Sideritis perfoliata* could potentially be useful to reduce the appearance of wrinkles and for the treatment of liver cancer. The moderate antibacterial, antioxidant and elastase activity of the plant can be linked to the traditional use of *S. perfoliata* for the treatment of wounds and inflammation.

Keywords: *Sideritis perfoliata*; wrinkles; *Cutibacterium acnes*; antimycobacterial; melanin inhibition; antioxidant capacity

---

1. Introduction

The genus *Sideritis*, belonging to the Lamiaceae (Labiatae) family, is comprised of more than 150 species. The genus is further divided in 2 subgenera and 7 sections [1]. The subgenus *Sideritis* divided into four sections includes approximately 125 species, the majority of which are distributed in the

Mediterranean [2]. The taxonomy of *Sideritis* species is highly intricate due to frequent hybridization between species, the variation of ecotypes and the degree of polymorphism [3]. It is crucial to distinguish the different *Sideritis* species and it is for this reason that secondary metabolites, in particular diterpenoids and flavonoids, are used for the chemotaxonomic identification of species within the genus [1]. However, modern methods such as high resolution melting analysis (HMR) coupled with polymerase chain reaction (PCR) offer rapid detection of these species [3].

Sideritis perfoliata L. (subsp. *perfoliata*) belongs to section Empedoclea, and is distributed mainly in Greece, Cyprus, Turkey and in other Balkan and East Mediterranean countries, where it grows in the wild or is cultivated [4]. This perennial species has a woody base, opposite and entire leaves (dentate or serrate) and is widely distributed in the Mediterranean Basin which is characterized by a rich tradition in phytotherapy, dating back to ancient times [5]. The usage of *Sideritis* species in folklore medicine has been known since the Dioskourides era (1st century) when the plant was used to treat wounds from iron weapons like swords or knives. The name *Sideritis* comes from the word "sidero" which means iron in Greek [1].

Plants of the *Sideritis* genus, commonly known in Cyprus, Greece and other Balkan countries as "mountain tea", are used as teas, for feeding and as flavoring agents. *Sideritis* species are used extensively in European and Mediterranean countries as traditional medicines with anti-inflammatory, antimicrobial, vulnerary, antioxidant, antispasmodic, analgesic, stomachic and carminative activities [1, 6,7]. Decoctions and infusions of *S. perfoliata* are widely used for the treatment of common cold, flu, cough, gastrointestinal disorders such as dyspepsia and as a calmative agent [6,8]. Charami et al. [9] reported the use of *S. perfoliata* for its antirheumatic, anti-ulcerative, digestive and vaso-protective properties. Recently, botanical and pharmacological studies focusing on *Sideritis*, is likely due to the above-mentioned properties which have been proven and have been attributed to the diverse chemical profile of the genus, that is rich in flavonoids, terpenoids, coumarins, sterols and iridoids [1,7]. These properties, the advances in technology and the current popularity and the use of natural products for self-medication, and as dietary supplements, generates a growing interest in the pharmaceutical industry to exploit plant-derived products to serve as discovery leads to new prototypes [10].

Medicinal and aromatic plants are enriched with antioxidants such as polyphenols, which act as scavengers of free radicals by adsorbing and neutralizing reactive oxygen species (ROS), quenching singlet and triplet oxygen or decomposing peroxides [9]. The Mediterranean Basin is characterized by a rich tradition of phytotherapy dating back to ancient times, in addition to being home to the majority of the *Sideritis* species. The aim of the current study was to explore the unknown properties of *S. perfoliata* investigating nutritional, antioxidant, anticancer, antituberculosis, antiwrinkle, anti-acne, hyper/hypo-pigmentation and antibacterial properties.

2. Materials and Methods

2.1. Plant Material and Growing Conditions

Sideritis perfoliata seedlings with 3–4 established leaves were purchased from the Cypriot National Agricultural Department and grown using an organic farming cultivation protocol. Plants were grown in a silt loam soil (6.76% clay, 39.2% sand and 54% silt) with organic matter of 3.02%, pH of 8.42, Electrical Conductivity (EC) of 0.85 mS/cm, and $CaCO_3$ of 21.23%, allowing the seedlings to grow for 2.5 months. The plantation was fully irrigated with no pesticides or chemicals being applied during the growing period. Weeds were removed manually or mechanically. Above ground (~5 cm) plant material (~10 kg) including leaves, stems and flowers, was collected during spring at the early flowering stage and raw material processed in the lab.

2.2. Mineral, Chlorophyll and Nutritional Content

The mineral content of the leaves was determined using six replicates (two plants pooled/replicate) according to Chrysargyris et al. [11]. Dried plant material was ashed in a furnace (Carbolite, AAF 1100,

GERO, Neuhausen, Germany) at 500 °C for 6 h and then acid digested using 2 M HCl. Determination of Potassium (K), Calcium (Ca), Magnesium (Mg), Sodium (Na), Iron (Fe), Copper (Cu) and Zinc (Zn) content was performed using atomic absorption spectrophotometry (PG Instruments AA500FG, Leicestershire, UK) and Nitrogen (N) content was determined using the Kjeldahl method (BUCHI, Digest automat K-439 and Distillation Kjelflex K-360). A colorimetric method using molybdate/vanadate was used for the determination of Phosphorus (P) content [12].

Leaf chlorophylls were extracted with 10 mL dimethyl sulfoxide (DMSO) at 65 °C for 30 min. The photosynthetic leaf pigments, Chlorophyll a (Chl a), Chlorophyll b (Chl b) and total Chlorophyll (t-Chl) content were then calculated [13].

The nutritional value of *S. perfoliata* was assessed by determining moisture, protein, fat, carbohydrate and ash composition, following AOAC procedures [14]. Briefly, nutrient content was determined by using the macro-Kjeldahl (N × 6.25), petroleum ether Soxhlet extraction, and incineration (600 °C) methods for protein, fat and moisture, respectively. The carbohydrate content was determined by difference in dry weight and the energetic value was estimated by applying the formula:

$$\text{Energy} \left(\frac{\text{kcal}}{100 \text{ g}} \text{ d.w.} \right) = 4 \times \left(\frac{\text{g protein}}{100 \text{ g}} \text{ d.w.} + \frac{\text{g carbohydrate}}{100 \text{ g}} \text{ d.w.} \right) + 9 \times \left(\frac{\text{g fat}}{100 \text{ g}} \text{ d.w.} \right)$$

The results were expressed in g/100 g d. w.

2.3. Ethanolic Extract Preparation

Aerial parts (leaves and stems/shoots) of *S. perfoliata* were air dried at 42 °C in an oven and then milled to a fine powder. The plant material was then macerated with pure ethanol (1:2.5 v/v) in a shaking incubator at 160 rpm, in glass bottles. After 72 h of shaking, the material was filtered using filter paper (Whatman No. 1, Merck, Darmstadt, Germany). The remaining ethanol was removed using a rotary evaporation unit, until complete dryness.

2.4. Phytochemical Analysis of the Ethanolic Extract

The phytochemicals present in the ethanolic extract of *S. perfoliata* were determined using the methods described by Mushtaq et al. [15] with minor changes. The stock solutions (1 mg/mL in dH$_2$O) of the ethanolic extract of *S. perfoliata* were used to determine the presence of tannins, saponins and terpenes. The formation of a yellow brown precipitate after the addition of 2 mL of 5% FeCl$_3$ in dH$_2$O, indicated the presence of tannins. The presence of saponins was determined by vigorously shaking the stock solution and observing a stable persistent froth. The development of a reddish-brown interface after the addition of 5 mL CHCl$_3$, 2 mL acetic anhydride, and a few drops of concentrated H$_2$SO$_4$ to 2 mL of the stock solution, indicated the presence of terpenes.

Stock solutions of the ethanolic extract of *S. perfoliata* (1 mg/mL in methanol) were prepared to determine the presence of alkaloids, cardiac glycosides, flavonoids and phenolics. The presence of alkaloids was determined by first adding 1.5 mL of 1% HCl to 2 mL of the stock solution and heating the mixture in a water bath set at 90 °C. Following the incubation, 6 drops of Dragendroff's reagent were added and a formation of an orange precipitate was indicative of the presence of alkaloids. The presence of cardiac glycosides was determined by adding 1 mL glacial acetic acid and 1–2 drops of FeCl$_3$, followed by the addition of 1 mL of concentrated H$_2$SO$_4$ to 2 mL of the ethanolic stock solution. The formation of a brown ring at the interface indicated the presence of a deoxysugar, characteristic of cardenolides. The presence of flavonoids was determined by observing the presence of a pink color after adding a few drops of concentrated HCl followed by small pieces of magnesium turnings. The presence of phenolics, on the other hand was indicated by the presence of a blue or green color, after the addition of 1 mL of 1% ferric chloride solution to the stock solution.

2.5. Antioxidant Activity of the Ethanolic Extract

The nitric oxide (NO) scavenging activity of the extract was measured according to the method described by Mayur et al. [16]. In the wells of a 96-well microtiter plate, the ethanolic extract of *S. perfoliata* was allowed to react with an equal volume of sodium nitroprusside (10 mM) for 90 min at room temperature. To the wells of the test sample, 100 µL of Griess reagent was added and to the blank color control wells, dH_2O was added. The final concentration of the extract ranged from 15.6 to 2000 µg/mL and. L-ascorbic acid was used as the positive control. The nitrite content was measured after 5 min at 546 nm and the percentage scavenging activity was determined.

To determine the DPPH scavenging activity 20 µL of extract, L-ascorbic acid (positive control) and 100% ethanol (negative control) were allowed to react with 90 µL of 90 µM DPPH (2,2-Diphenyl-1-picrylhydrazyl) methanolic solution in a 96-well microtiter plate. To the blank extract color control wells, DPPH was substituted with dH_2O. The final concentrations of the sample and L-ascorbic acid ranged from 7.8 to 500 µg/mL and 1.6 to 100 µg/mL respectively. The plate was incubated for 30 min in the dark and the absorbance measured at 515 nm using a BIO-TEK Power Wave Multiwell plate reader (Analytical and Diagnostic Products, Weltevreden Park, South Africa).

The total antioxidant capacity was assayed by the phosphomolybdenum (PM) method [17].The extract and gallic acid were prepared to stock solutions of 5 mg/mL. In test tubes, solutions of 0.3 mL of the samples were added to 1 mL of 0.6 M sulphuric acid (H_2SO_4), 1 mL of 4 mM ammonium molybdate and 1 mL of 28 mM sodium phosphate (total of 3 mL reagent). The blank was prepared by adding 0.3 mL of methanol instead of the extract. The test tubes were incubated at 95 °C for 90 min. Absorbance was measured at 695 nm. The fifty percent inhibitory concentration (IC_{50}) concentration was calculated for each of the antioxidant assays.

2.6. Antiwrinkle, Hyper/Hypo-Pigmentation, Anti-Acne, Antimycobacterial Activity of the Ethanolic Extract

2.6.1. Elastase Inhibition

The antiwrinkle potential of the *S. perfoliata* was tested using the elastase inhibition assay as described by Aumeeruddy et al. [18]. The reaction mixture contained 100 mM Tris buffer (pH 8.0), 0.5 M HCl and the ethanolic extract of *S. perfoliata*. The test concentration ranged between 3.13 and 250 µg/mL for the extract and the positive control (ursolic acid). Porcine pancreatic elastase (PPE) (5 mM, 0.02 mL) was added to the reaction mixture and incubated for 15 min at 37 °C followed by the addition of the substrate, 4 mM N-succinyl-(Ala)3-p-nitroanilide. The change in the absorbance of the reaction mixture was measured kinetically at 405 nm for 15 min at 37 °C using KC Junior software and a BIO-TEK Power-Wave XS multiwell plate reader (A.D.P, Weltevreden Park, South Africa). One unit of elastolytic activity is defined as the release of 1 µM of p-nitroaniline/min from N-succinyl-(Ala)3-p-nitroanilide. The IC_{50} value of *S. perfoliata* was calculated.

2.6.2. Tyrosinase Inhibition

The tyrosinase assay was conducted according to the method described by Curto et al. [19] with slight modifications. The enzymatic rate of mushroom tyrosinase (Sigma-Aldrich Co. LLC, St. Louis, MO, USA) was evaluated spectrophotometrically. In a 96-well plate, 70 µL of the stock solution (600 µg/mL in 0.38 mM DMSO and pH 6.5 potassium phosphate buffer) was combined with 30 µL of tyrosinase (333 Units/mL in potassium phosphate buffer) and incubated for 5 min at room temperature. Following incubation, 110 µL substrate (2 mM L-tyrosine) was added to each well. Final concentrations of the test sample and the positive control, kojic acid (Sigma-Aldrich Co. LLC, St. Louis, MO, USA), ranged between 1.5 and 200 µg/mL. The negative control consisted of 0.01% DMSO in potassium phosphate buffer. The rate of tyrosinase was determined for the sample (Abs_{sample}), positive control ($Abs_{positive}$) and the negative control wells ($Abs_{control}$), by measuring the absorbance over a period of 30 min at 492 nm using the BIO-TEK Power-Wave XS multi-well plate reader (A.D.P., Weltevreden

Park, RSA). The percentage tyrosinase inhibition for the sample was calculated as follows and the IC_{50} value was then calculated.

$$\text{Percentage inhibition (\%)} = 100\% - \left[\left(\frac{Abs_{sample/positive}}{Abs_{control}}\right) \times 100\right]$$

2.6.3. Melanin Inhibition

The effect of the ethanolic extract of *S. perfoliata* on melanin production in UCT-MEL-1 cells was determined as described by Matsuda et al. [20]. Following culturing, as described in Section 2.7.1, the cells were plated in 96-well plates and incubated for 24 h at 37 °C in the CO_2 incubator. Stock solutions of the ethanolic extract of *S. perfoliata* and kojic acid (400 µg/mL), were prepared in 1.25% DMSO in phosphate buffered saline (PBS). The stock solutions were serial diluted in Dulbecco's Modified Eagle's Medium (DMEM) and added to the 96-well plates, such that the final concentrations in the plates ranged between 6.25 and 200 µg/mL and the final concentration of DMSO present was 0.6%. The treated cells were incubated for 24 h at 37 °C in the CO_2 incubator. Colour controls, without cells, were prepared in the same manner. An untreated cell control was also included where media was used instead of the sample. Following incubation, the supernatant was removed, combined with HEPES buffer (0.4 M, pH 6.8) at a 1:1 ratio and assayed for extracellular melanin at 475 nm. The attached cells were washed with 100 mL of PBS and lysed following the addition of 40 µL of 1 M NaOH and 10 µL trypsin, for 16 h at room temperature. The optical density of the intracellular melanin was measured at 475 nm.

2.6.4. Antimycobacterial Activity

The antimycobacterial activity of the extract was assessed using the microtiter Alamar blue assay [21]. Briefly, *Mycobacterium smegmatis* (MC^2 155) was cultured and maintained on 7H11 agar plates. A single colony was transferred into fresh 7H9 media supplemented with 2% glycerol and 0.5% Tween 80 and allowed to grow for 24 h at 37 °C. The bacterial inoculum was prepared by adjusting the 24 h logarithmic culture to a concentration of 1.5×10^6 CFU/mL. The extract was tested at concentrations ranging from 15.6 to 1000 µg/mL. Ciprofloxacin (0.08–5 µg/mL) was used as the positive control and a solvent (2.5% DMSO), untreated bacterial and negative control were included in the assay. The final assay volume was 200 µL. The plates were covered and incubated at 37 °C for 18–24 h, followed by the addition of 20 µL of Alamar blue solution. The plates were left to incubate for an additional 30 min and the Minimum Inhibitory Concentration (MIC) was determined as the lowest concentration where no color change could be observed from blue to pink.

2.6.5. Antibacterial Activity against *Cutibacterium Acnes*

Susceptibility of the *Cutibacterium acnes* bacteria (ATCC 6919) was tested using the microdilution broth assay as described by Lall et al. [21]. Briefly, 0.1 mL of extract (2 mg/mL) and tetracycline (0.2 mg/mL) were serially diluted in 100 µL of Brain Heart Infusion (BHI) broth in a 96-well plate. Concentrations of plant extract ranged from 3.91 to 500 µg/mL, while the positive control, tetracycline was tested from 0.39 to 50 µg/mL. Cultures of *C. acnes* grown for 72 h at 37 °C on BHI agar supplemented with 1% glucose were inoculated in BHI broth at a concentration of 1.5×10^8 CFU/mL ($OD_{600} = 0.132$). One hundred microlitres of the bacterial suspension was then added to the relevant test wells. The 96-well plates were then incubated at 37 °C for 72 h in an Anaerocult® jar with Anaerocult® A (Merck, Darmstadt, Germany) for the generation of CO_2. Control plates were included and consisted of *C. acnes* without treatment, an extract vehicle DMSO treatment (2.5% v/v) and a BHI media control. After 72 h, 0.02 mL of PrestoBlue® (LTC Tech (Pty) Ltd., Johannesburg, South Africa) reagent was added. The MIC was then determined after 1 h of incubation.

2.6.6. Antibacterial Activity against *P. Intermedia* and *S. Mutans*

The antibacterial activity was determined using the microtiter plate assay, where the MIC was calculated; against *Streptococcus mutans* (ATCC 25175) and *Prevotella intermedia* (ATCC 25611). In short, a 96-well plate was supplemented with 0.1 mL of Brain Heart Infusion broth for *S. mutans* and Tryptone Soy Broth for *P. intermedia*. To that, 100 μL of plant extract dissolved in 2.5% DMSO (50 mg/mL) and 100 μL of Chlorhexidine (CHX) dissolved in water (1.25 mg/mL) were serially diluted down to give concentrations ranging from 0.048 to 12.5 mg/mL and 4.8×10^{-3} to 0.625 mg/mL, for the plant extract and positive control respectively. After which 24 h old *S. mutans* and *P. interdia* inocula (3×10^8 CFU/mL), grown at 37 °C under anaerobic conditions in an anaerobic jar containing Anaerocult®A, were added. The plates were then incubated for a further 24 h, followed by the addition of 20 μL of PrestoBlue® viability reagent. In addition to CHX, DMSO (2.5%) and the respective media were used as negative controls [21,22].

2.7. Cytotoxicity

2.7.1. Cell Culture

The human keratinocyte (HaCat), normal African green monkey kidney (Vero), human liver carcinoma (HepG2) and cervical cancer (HeLa) cell lines were purchased from Separation Scientific SA (Pty) Ltd. (Johannesburg, South Africa). The human epidermoid carcinoma (A431) cell line was purchased from Sigma-Aldrich (Pty) Ltd. (Johannesburg, South Africa). The pigmented human malignant melanoma cell line (UCT-MEL-1) was originally isolated from the metastatic lymph nodes of a patient from Groote Schuur Hospital, Cape Town, South Africa and was donated by Prof. Lester Davids (University of Cape Town). The human keratinocyte (HaCat), human epidermoid carcinoma (A431: ATCC CRL-1555) cell lines were maintained in culture flasks containing DMEM, while the normal African green monkey kidney (Vero: ATCC CCL-81), human liver carcinoma (HepG2: ATCC HB-8065), and human cervical cancer (HeLa: ATCC CCL-2) cell lines were maintained in Eagle's Minimal Essential Medium (EMEM) as previously described [23]. Both media were supplemented with 1% antibiotics (100 U/mL penicillin, 100 μg/mL streptomycin and 250 μg/mL fungizone) and 10% heat-inactivated fetal bovine serum (FBS). The cells were cultured to 80% confluency in a humidified incubator set at 5% CO_2 and 37 °C and sub cultured. Detachment was achieved through treatment with trypsin-EDTA (0.25% trypsin containing 0.01% EDTA) for 10 min followed by the addition of supplemented media to inhibit the reaction.

2.7.2. Cell Viability

Cell viability was measured using the method as described by Berrington and Lall [24] using the 2,3-Bis-(2-methoxy-4-nitro-5-sulfophenyl)-2H-tetrazolium-5-carboxanilide salt (XTT) Cell Proliferation Kit II (Merck Millipore, USA). The HepG2 cells were seeded at a concentration of 2000 cells/mL, whereas the remaining cell lines were seeded at a concentration of 1.0×10^6 cells/mL in 96-well plates (100 μL) and allowed to adhere for 24 h. The *S. perfoliata* extract was prepared at stock concentration of 20 mg/mL, serially diluted and added to the 96-well plate at final concentrations ranging from 1.56 to 200 μg/mL. Controls included a 2% DMSO vehicle control, cells grown in medium only and Actinomycin D at final concentration ranging from 3.9×10^{-4} to 0.05 μg/mL. Cells were incubated for a further 72 h with the respective samples and controls. Thereafter, 0.05 mL XTT (0.3 mg/mL) was added to the cells and incubated for 2 h where after the absorbance was measured at 490 nm (reference wavelength of 690 nm) using a BIO-TEK powerwave XS plate reader. Blank plates were included, which were prepared in the same manner as mentioned above, without the additional of cells, to allow for color compensation of the samples. The percentage viability was calculated as follows and the IC_{50} value was then calculated.

$$\text{Percentage viability } (\%) = \left[\left(\frac{\text{Abs}_{\text{sample/positive}}}{\text{Abs}_{\text{control}}}\right) \times 100\right]$$

2.8. Essential Oil Yield and Composition

Aerial parts of *S. perfoliata* plants were harvested and three biological samples (pooled of three individual plants/sample) were air-dried at 42 °C in an oven, chopped and were hydro-distilled for 3 h, using a Clevenger apparatus for essential oil (EO) extraction. The EOs were analyzed by Gas Chromatography-Mass Spectrometry (GC/MS) (Shimadzu GC/MSQP-2010 Plus, Tokyo, Japan) and their constituents were determined as described previously [25].

2.9. Total Phenolics and Antioxidant Activity of the Essential Oil

Total phenolics in the EOs was determined using the Folin-Ciocalteu method, as described previously [26]. Results were expressed as milligrams of gallic acid equivalents per gram of EO.

The DPPH assay was used to evaluate the scavenging activity of the EO as described by Oke et al. [27]. A volume of 400 µL of serial EO dilutions (0–100 µg/mL) or of the methanol control was mixed with 100 µL of 0.2 mM DPPH solution (in methanol). The mixture was incubated in the dark for 30 min and the absorbance was measured at 517 nm. Butylatedhydroxytoluene (BHT) was used as a positive control. Results were expressed as IC_{50} values.

The 2,2'-azino-bis 3-ethylbenzothiazoline-6-sulphonic acid (ABTS) method was also assayed [28]. The ABTS radical was prepared by mixing 7 mM ABTS and 2.45 mM potassium persulfate and incubated for 16 h in the dark. The solution was then diluted with 80% methanol until the absorbance at 734 nm was 0.700 ± 0.02. Therafter, 1 mL of the diluted ABTS solution was mixed with 100 µL of the EO dilutions (0–100 µg/mL) and dilutions of the positive control, ascorbic acid (0–250 µg/mL). The absorbance was recorded after 5 min at 743 nm. The ABTS radical scavenging activity was expressed as IC_{50} values.

The ability of the *S. perfoliata* oil to reduce Fe^{3+} was performed according to Bettaieb Rebey et al. [29]. Dilutions of essential oils (0–100 µg/mL) and ascorbic acid (0–250 µg/mL) as positive control, were mixed with 500 µL of 200 mM sodium phosphate buffer (pH 6.6) and 500 µL of 1% potassium ferricyanide ($K_3Fe(CN)_6$). The mixture was incubated at 50 °C for 20 min. Then 500 µL of 10% trichloroacetic acid was added and the reaction was centrifuged at 650× g for 10 min. A volume 500 µL of the supernatant was then mixed with 500 µL of distilled water and 100 µL of 0.1% ferric chloride. The absorbance was then measured at 700 nm. The EC_{50} (effective concentration) was defined as the concentration at which the absorbance was 0.5 for the reducing power.

The EO was tested for the bleaching of β-carotene in linoleic acid system, by measuring the coupled autoxidation of β-carotene and linoleic acid [30]. Briefly, 5 mg of β-carotene were dissolved in 50 mL of chloroform. A 3 mL volume of the mixture was added to 40 mg of linoleic acid and 400 mg of Tween 20. Chloroform was removed using nitrogen gas, after which 100 mL of distilled water was added to the emulsion and mixed well. Thereafter, 1.5 mL of the emulsion was added to a test tube containing 20 µL of the essential oil or BHT. Absorbance was measured immediately (t = 0) at 470 nm, after an incubation time of 1 h in a water bath at 50 °C (t = 60). The antioxidant activity was expressed as percentage inhibition compared with the negative control (0.02 mL of methanol instead of EO) using the following equation:

$$\text{Antioxidant activity } (\%) = 100 \times \frac{(\text{DR}_\text{C} - \text{DR}_\text{S})}{\text{DR}_\text{C}}$$

where DRc is the degradation rate of the control, DRs is the degradation rate of the sample. DR_C and DR_S were calculated using the following equation:

$$\text{Degradation rate (DR)} = \left[\frac{(\ln a/b)}{60} \right]$$

where a = absorbance at t = 0 and b = absorbance at t = 60 min.

2.10. Statistical Analysis

Experiments were performed in triplicate in three independent experiments, unless otherwise stated. Data were statistically analyzed using IBM SPSS v.21 for analysis of variance (ANOVA) and presented as means ± standard deviation (SD). Duncan's multiple range test (DMRT) was used for mean comparisons when significant differences were detected (p-value < 0.05). The absolute 50% inhibitory concentration (IC_{50}) was calculated using the GraphPad Prism version 4.00 for Windows, GraphPad Software (La Jolla, CA, USA).

3. Results and Discussion

3.1. Mineral, Chlorophyll and Nutritional Content

The results of the nutritional analysis of the aerial parts of *S. perfoliata* are demonstrated in Table 1. This species has a high nutritional value, with 14.64% of protein which is higher than numerous edible species which also form part of the Lamiaceae family. Like the majority of *Sideritis* species, it is rich in micronutrients as previously described when compared to the potassium, phosphorus and magnesium content of *S. scardica* or *S. raeseri* [31,32]. Mineral content of *S. perfoliata* might vary after application of different cultivation practices and multiple harvest at a commercial farm [32].

Table 1. Mineral analysis, nutritional value and chlorophyll content of aerial plant tissue. Data are presented as the mean of six replicates ± SD.

Mineral Content		Nutritional Value		Chlorophll Content	
N (mg/kg d.w.)	23.34 ± 0.39	Dry Matter (%)	19.69 ± 0.50	Total Chlorophyll (mg/g f.w.)	0.968 ± 0.063
K (mg/kg d.w.)	27.96 ± 0.77	Moisture (%)	80.31 ± 0.50	Chlorophyll a (mg/g f.w.)	0.779 ± 0.054
P (mg/kg d.w.)	2.06 ± 0.09	Ash (%)	9.91 ± 0.51	Chlorophyll b (mg/g f.w.)	0.189 ± 0.009
Mg (mg/kg d.w.)	3.24 ± 0.42	Protein (%)	14.64 ± 0.24		
Ca (mg/kg d.w.)	6.95 ± 1.78	Total Fats (%)	1.76 ± 0.05		
Na (mg/kg d.w.)	0.34 ± 0.03	Carbohydrates (%)	73.52 ± 0.19		
Cu (mg/kg d.w.)	62.06 ± 13.97	Energy (kcal/100 g d.w.)	368.40 ± 0.77		
Zn (mg/kg d.w.)	62.22 ± 2.25				
Fe (mg/kg d.w.)	149.08 ± 31.02				

The total Chlorophyll content in the aerial parts of *S. perfoliata* was 0.986 ± 0.06 mg/g of fresh tissue (0.779 and 0.189 mg/g for Chlorophyll a and b respectively). The Chlorophyll content of *S. perfoliata* compared well with the Chlorophyll content in other Lamiaceae species. In a study by Chrysargyris et al. reported the total Chlorophyll content of sage and lavender which were 1.18 and 1.08 mg/g fresh weight (fw) respectively [13].

3.2. Phytochemical Analysis

The phytochemical characteristics from the ethanolic extract of *S. perfoliata* were analyzed. Alkaloids, cardiac glycosides, flavonoids, phenolics, saponins, tannins and terpenes were found to be present within the extract. These phytochemical constituents are known for their medicinal potential. Phenols are a group of compounds that have been proven to have anti-ageing, antioxidant and anti-inflammatory potential. This is confirmed by the noteworthy elastase inhibitory activity and free radical scavenging potential of *S. perfoliata* ethanolic extract seen in the present study (Table 2). Tannins are compounds that readily bind to proteins and interferes with protein synthesis. Flavonoids are compounds known to have antimicrobial, anticancer and antioxidant activity. Their antimicrobial activity is due to their ability to interact with the bacterial cell wall. The moderate anticancer potential of the ethanolic extract of *S. perfoliata* against the human liver cancer cell line (HepG2) could be due to the presence of flavonoids in the leaves. The characteristics of saponins, identified to be present in the ethanolic extract, include the coagulation of red blood cells and anti-inflammatory activity. Alkaloids have been known for centuries for their medicinal potential. These compounds are cytotoxic, antibacterial, analgesic and antispasmodic [33]. This could explain the moderate cytotoxic effect of *S. perfoliata* observed in the present study (Table 2). The phytochemicals identified in this study proves the medicinal potential of *S. perfoliata* and could attract the interest of pharmaceutical industries to isolate its major components. Recently, six iridoids, three flavonoids, two phenylethanoid glucosides and one phenolic acid have been isolated from the plant, providing new inputs on plant metabolism related to cultivation practices, with possible industrial applications [32].

Table 2. Biological activity of *Sideritis perfoliata* ethanolic extract. Data presented is the average of three replicates ± standard deviation (SD).

Assay		*S. perfoliata* EtOH Extract MIC [a]/IC$_{50}$[b] ± SD (µg/mL)	Positive Control MIC/IC$_{50}$ ± SD (µg/mL)
Antibacterial	*M. smegmatis*	NI[l]	0.62[c]
	P. intermedia	3.1×10^3	0.48[e]
	C. acnes	500	0.78[d]
	S. mutans	6.2×10^3	0.48[e]
Cytotoxicity	Vero	201.5 ± 3.32	$0.02 \pm 8 \times 10^{-3}$[f]
	HaCat	134.3 ± 10.1	$0.01 \pm 1.4 \times 10^{-3}$[f]
Anticancer	HeLa	102.5 ± 0.99	$1 \times 10^{-3} \pm 8 \times 10^{-3}$[f]
	A431	133.25 ± 10.45	$0.04 \pm 7 \times 10^{-3}$[f]
	UCT-MEL-1	103.15 ± 0.92	$0.04 \pm 2 \times 10^{-3}$[f]
	HepG2	64.27 ± 2.04	$1 \times 10^{-3} \pm 7 \times 10^{-4}$[f]
Antioxidant	DPPH	23.9 ± 0.85	1.90 ± 0.05[h]
	Nitric oxide	266.0 ± 7.1	56.2 ± 41.0[h]
	TAC	2.004 ± 0.28 (PM) 1.596 ± 0.40 (FRC)	0.889 ± 0.26[i] (PM) 2.538 ± 0.38[i] (FRC)
Pigmentation	Tyrosinase	NI[m]	2.84 ± 0.23[j]
	Melanin	NI[m]	1.56 ± 0.18[j]
Wrinkles	Elastase	57.18 ± 3.22	18.45 ± 2.23[k]

[a] Minimum Inhibitory Concentration, [b] Inhibitory concentration where 50% activity/viability is inhibited, [c] Ciprofloxacin, [d] Tetracycline, [e] Chlorhexidine, [f] Actinomycin-D, [h] L-Ascorbic acid, [I] Gallic acid, [j] Kojic acid, [k] Ursolic acid, NI: No inhibition at the highest concentration tested [l] (1000 µg/mL), [m] (200 µg/mL), PM—phosphomolybdenum method, FRC—ferric reducing capacity.

3.3. Biological Activities

3.3.1. Antioxidant Activity

The results of the antioxidant activities of the extract are presented in Table 2. The human body is constantly exposed to external factors that contribute to the development of reactive oxygen species (ROS), resulting in oxidative stress. Studies have confirmed the contribution of oxidative stress to the onset of various diseases such as cancer, inflammatory processes and skin diseases such as acne vulgaris. Currently, there is a trend towards using antioxidants from a natural origin due to the accumulation of synthetic antioxidants in the body [34,35]. Nitric oxide plays a central role in the regulation of several physiological processes in the human body. However, nitric oxide that is a product of sodium nitroprusside has a strong nitrosonium (NO^+) character that can change and interfere with the normal function of several cellular processes [16]. Pro-inflammatory cytokines and bacterial cell wall components have been linked with the overproduction and release of increased amounts of nitric oxide. The increased nitric oxide free radicals are involved in the onset of inflammation associated with diseases such as acne vulgaris and post-inflammatory hyperpigmentation [36].

Sideritis perfoliata displayed a moderate nitric oxide scavenging activity with an IC_{50} of 266.0 ± 7.1 µg/mL (Table 2). The noteworthy free radical scavenging activity for *S. perfoliata* can be attributed to the high phenolic content of *Sideritis* spp. [37]. Previous studies done by Zengin et al. [37] speculated the significant nitric oxide scavenging activity of *S. galatica* to be due to the presence of tocopherols, carotenoids and ascorbic acid in the plant extract.

Phongpaichit et al. [38] reported an extract with an IC_{50} between 10 and 50 µg/mL to have a significant antioxidant capacity. In the present study, *S. perfoliata* displayed noteworthy antioxidant activity with an IC_{50} of 23.9 ± 0.85 µg/mL. Studies done by Güvenç et al. [39] on the DPPH antioxidant potential of various *Sideritis* spp, confirmed the free radical scavenging potential of flavonoids together with phenylethanoid glycosides present in these species.

The estimation of the total antioxidant capacity of an extract uses a combination of different antioxidant parameters. These include determining the reducing capacity of the extract against a specific free radical agent such as ammonium molybdate as well as calculating the phenolic content of the extract. High phenolic content is associated with high antioxidant capacity [17,40]. The *S. perfoliata* extract had significantly lower phenol content than that of the positive control gallic acid, while showing a significantly higher capacity to scavenge the ammonium molybdate free radical when compared with gallic acid. This means that the *S. perfoliata* extract has an overall moderate antioxidant capacity.

3.3.2. Elastase Inhibition

The connective tissue of the skin has the main component of elastic fibres known as elastin. Elastic fibres and collagenous fibres in the skin create a network and they develop in the epidermis. Elastin is degraded by elastase; therefore, inhibition of the elastase enzyme can retain the elasticity of the skin [41]. *Sideritis perfoliata* showed significant inhibition of the enzyme elastase, with an IC_{50} value of 57.18 ± 3.22 µg/mL (Table 2). Ursolic acid was used as the positive control, the IC_{50} value was 18.45 ± 2.23 µg/mL. No elastase inhibition has been reported previously for *S. perfoliata* and species in the same genus. The IC_{50} value of 57.18 ± 3.22 µg/mL indicates good enzyme inhibition, meaning that the plant has the potential to maintain skin elasticity. Lee et al. [41] studied the elastase inhibition of *Areca catechu*, and reported noteworthy activity, with an IC_{50} value of 42.4 µg/mL. In 2013, Ndlovu et al. [42] investigated the elastase inhibition of *Peltophorum africanum*, they reported the plant to be a potential enzyme inhibitor with an IC_{50} of 55.83 µg/mL.

According to the Gas Chromatography-Mass Spectrometry (GC-MS) results, the essential oil of *S. perfoliata* is characterized by the presence of monoterpenes, which are known to inhibit the activity of the elastase enzyme, according to Kacem and Meraihi [43]. They investigated the elastase inhibition of the essential oil extracted from seeds of *Nigella sativa* (L.) and reported that the main components of the essential oil were p-cymene, thymoquinone, carvone, thymol and carvacrol, which are monoterpenes. These monoterpenes had IC_{50} values between 12 and 104 µM.

3.3.3. Tyrosinase and Melanin Inhibition

The effect of the *S. perfoliata* on melanogenesis has not yet been determined. It was reported that the acetone and methanol extract of *S. stricta*, being in the same genus, exhibited 15.66 ± 0.11% and 23.29 ± 0.56% inhibition of tyrosinase, respectively, at 200 µg/mL [44]. Similar inhibition of tyrosinase was observed for the ethanol extract of *S. perfoliata* at concentrations higher than 25 µg/mL (Figure 1). However, no inhibition was observed at concentrations lower than 12.5 µg/mL. Additionally, no significant inhibition of melanin production in UCT-MEL-1 cells was observed at the highest concentration tested (200 µg/mL). The inhibitors of elastase and tyrosinase enzymes can potentially have applications as skin hyperpigmentation, anti-ageing and antiwrinkle agents as well as in the treatment of other dermatological disorders [45].

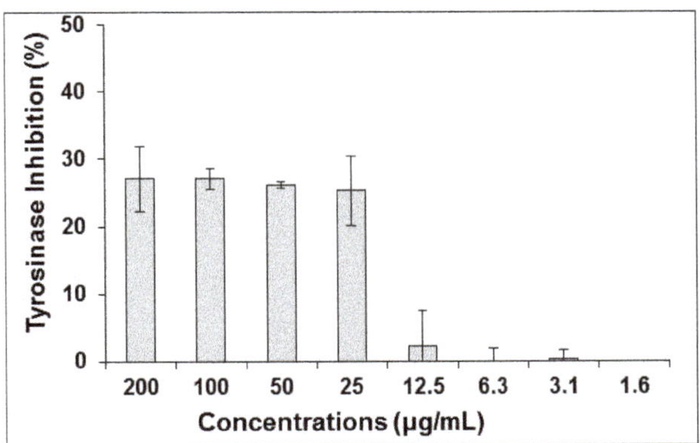

Figure 1. The inhibitory effect of *S. perfoliata* on mushroom tyrosinase at different treatment concentrations. Data is represented as are means ± SD (n = 3).

Compounds previously isolated from *S. perfoliata* that have been reported for their tyrosinase and/or melanin production inhibitory activity were acteoside, ajugoside, caffeic acid, leucoceptoside A, lavandulifolioside and martynoside [9,46–51]. A study conducted by Song and Sim. in 2009 concluded that acteoside inhibited both melanogenesis and the tyrosinase protein activity in B16F10 melanoma cells, but the effects were dose-dependent [49]. The same effects of the inhibition of tyrosinase and melanogenesis were confirmed by Son et al. [50]. Furthermore, acteoside lowered cyclic AMP levels in cells that were stimulated by α-melanocyte stimulating hormone [49]. Acteoside inhibited melanogenesis on a post-translational level by reducing the levels of microphthalmia associated transcription factor (MITF) proteins, tyrosinase-related protein-1 (TRP-1) and tyrosinase [50]. It was reported that acteoside resulted in a 25.78% reduction in tyrosinase activity at a concentration of 53 µM [48], however a more recent study reported a 50% reduction in enzyme activity at a concentration of 20.67 µM [51]. Other compounds potentialy responsible for the tyrosinase inhibitory activity, exhibited by the ethanol extract of *S. perfoliata*, were caffeic acid, leucosceptoside A, lavandulifolioside and martynoside exhibiting 27.00%, 21.65%, 12.00% and 20.55% enzyme inhibition at 90 µM, 51 µM, 44 µM and 52 µM, respectively [47,48].

3.3.4. Antimycobacterial Activity

In a study conducted by Basile et al. [52], the acetone extracts of *S. italica* leaves and flowers exhibited antibacterial activity against various strains of *Pseudomonas aeruginosa*, *Proteus mirabilis*, *Salmonella typhi* and *Proteus vulgaris*. Moreover, Askun et al. [53] reported the antibacterial and antimycobacterial activity of different *Sideritis* spp. The authors found that the methanolic extract of *S. leptoclada* exhibited antibacterial activity, with an MIC of 640 µg/mL on *Enterobacter aerogenes* and *Salmonella typhimurium*. A low mycobacterial activity was observed by the extracts of *S. albiflora* and *S. leptoclada* with an MIC of 1568 µg/mL. A similar result was observed in the present study with the *S. perfoliata* extract exhibiting an MIC greater than 1000 µg/mL.

3.3.5. Antibacterial Activity against *Cutibacterium Acnes*

The major causative microorganism involved in the progression of acne is *Cutibacterium acnes*. This study provides the first report of the extract of *S. perfoliata* against *C. acnes* (ATCC 6919). The ethanolic extract of this species exhibited an MIC of 500 µg/mL, while the positive control, tetracycline had an MIC of 0.78 µg/mL (Table 2). Extracts with an MIC less than 1000 µg/mL are regarded as active [54], which is reflected in the findings of the present study. Additionally, antibacterial activity of *S. perfoliata* has been reported from its essential oil, which inhibited the growth of *Staphylococcus epidermidis* with an MIC of 250 µg/mL, another microorganism involved in progression of acne [55,56].

Kirimer et al. [55] reported the major constituent of the *S. perfoliata* essential oil as limonene (37.70%) followed by sabinene (18.8%), while Ezer et al. [57] also identified limonene as the major constituent making up 22.40% of the EO composition. The essential oils of two Korean citrus varieties containing high quantities of limonene (81.63% and 83.38%), inhibited the growth of *C. acnes* (ATCC 6919) at concentrations of 0.31 μL/mL, as reported by Kim et al. [58]. However, the essential oil of *S. perfoliata* in the present study did not show high concentrations of limonene present as reported by Kirimer et al. [55] and this could be due to different localities, method used for oil preparation, growth stage of the plant and cultivation management.

3.3.6. Antibacterial Activity against *S. mutans* and *P. intermedia*

Streptococcus mutans and *Prevotella intermedia* are the main causative agents of periodontal diseases and are known for their resistance [59–61]. According to Eloff [22] and van Vuuren [62], an extract with an MIC lower than 8 mg/mL is considered to have some antimicrobial activity, while extracts having an MIC below 1 mg/mL are considered worth exploring further. The *S. perfoliata* extract in the present study, showed low antibacterial activity for *S. mutans* and moderate activity for *P. intermedia* with MICs of 6.25 mg/mL and 3.125 mg/mL respectively.

The essential oils of some members of the *Sideritis spp.* namely; *S. curvidens* and *S. lanata* were reported to have high antimicrobial activity against *S. mutans*, using the disc diffusion method. The zones of inhibition measured at 22 mm for *S. curvidens* and 23 mm for *S. lanata* using 25 μL of essential oil [63]. There were no previous reports of the activity of *S. perfoliata* on either of the microorganisms tested in the present study, providing new knowledge on the antibacterial activity of the examined species.

3.3.7. Cytotoxicity

Secondary metabolites from various plant species have become highly researched as alternative therapeutic agents for various diseases, including cancer. Interestingly, in the United States more than 60% of anticancer drugs approved from between 1983 and 1994 were of natural origin [64,65]. While their efficacy towards certain cancer types is encouraging, it is crucial that their toxicity towards normal healthy cells remain low to obtain the highest level of efficacy and specificity towards cancer cells. The toxicity of the ethanolic extract of *S. perfoliata* was investigated against cancerous and non-cancerous cells (Table 2). According to Kuete and Efferth [66], a plant extract is considered significantly toxic if it shows an IC_{50} value lower than 100 μg/mL when tested on normal cell lines in vitro, while IC_{50}'s between 100 and 300 μg/mL constitute moderate toxicity, those between 300 and 1000 μg/mL exhibit low toxicity, and those above 1000 μg/mL exhibit no in vitro toxicity. When investigating cancerous cell lines, plant extracts exhibiting IC_{50}'s lower than 20 μg/mL are considered to have significant anticancer activity, while moderate toxicity constitutes those values ranging between 20 and 50 μg/mL, low toxicity between 50 and 200 μg/mL, and no toxicity towards cancerous cells for values higher than 200 μg/mL. According to these thresholds, the ethanolic extract of *S. perfoliata* showed moderate toxicity towards the non-cancerous Vero and HaCat cells with IC_{50} values of 201.5 ± 3.32 and 134.3 ± 10.1 μg/mL, respectively. Low toxicity towards all the cancerous cell lines was noted with IC_{50} values of 103.15 ± 0.92, 133.25 ± 10.45, 64.27 ± 2.04 and 102.5 ± 0.99 μg/mL against UCT-MEL-1, A431, HepG2 and HeLa cells respectively (Table 2). Although the activity of the ethanolic extract was found to have low toxicity against HepG2 cells, the toxicity against this cell line was much higher than on the other cancerous cell lines. In a study by Loizzo et al. [67], the activity of the essential oil of *S. perfoliata* against amelanotic melanoma (C32) and renal cell adenocarcinoma (ACHN) was found to exhibit moderate toxicity with IC_{50} values of 95.58 and 100.90 μg/mL respectively, further supporting the cytotoxic effects of the *S. perfoliata*. The essential oil was, however, not toxic to human breast cancer (MCF-7) and human prostate cancer (LNCap) cells with IC_{50} values above 400 μg/mL.

3.4. Essential Oil Yield, Composition and Antioxidant Activity

The essential oil yield in terms of dry weight basis, in general, for all *Sideritis* genus species is considerably low, in contrast with other Lamiaceae family genera [1]. There is a reported yield of *S. perfoliata* from Turkey as low as 0.08% [68]. Kirimer et al. [69] reported the yield of many *Sideritis* species varying from traces to 0.85%, with *S. perfoliata* ranging from 0.12% to 0.36%. When *S. perfoliata* subjected to drought stress (deficit irrigation) which is considered as an abiotic stress factor, EO yield increased up to 0.75% and 0.83% at a conventional and organic cultivation [32]. In the present study the yield was 0.30% ± 0.08%, in accordance with the species records (Table 3). The lower EO yield could be related that the collection of plant material in our study was from an organic plantation for seed/seedling production (''mother'' plants) and not a commercial plantation as in previous studies [32]. They have also reported the connection between the yield and the percentage of total monoterpenes or sesquiterpenes in the oils.

Table 3. Composition (%) of essential oils of *S. perfoliata* aerial parts, after GC/MS analysis and compound identification. Values are mean percentage (%) of three replicates ±SD.

RI	Compound	Mean ± SD	RI	Compound	Mean ± SD
926	α-Thujene	0.71 ± 0.09	1100	*trans*-Sabinene hydrade	0.47 ± 0.03
933	α-Pinene	27.92 ± 1.47	1178	Terpinen-4-ol	0.22 ± 0.04
948	Camphene	0.08 ± 0.01	1191	α-Terpineol	0.10 ± 0.02
973	Sabinene	4.59 ± 0.51	1204	γ-Terpineol	0.17 ± 0.00
977	β-Pinene	7.12 ± 0.64	1244	Carvone	0.37 ± 0.34
989	α-Myrcene	2.45 ± 0.29	1271	Geranial	0.06 ± 0.06
1003	3-Octanol	0.05 ± 0.00	1425	β-caryophyllene	2.89 ± 0.29
1005	α-Phellandrene	0.88 ± 0.09	1462	α-caryophyllene	0.09 ± 0.01
1013	3-Carene	2.29 ± 0.04	1479	Caryophyllene-9-epi	0.97 ± 0.17
1017	α-Terpinene	0.18 ± 0.01	1495	Germacrene D	1.75 ± 0.21
1024	o-Cymene	0.15 ± 0.01	1581	*trans*-Sesquisabinene hydrate	1.02 ± 0.14
1029	β-Phellandrene	26.59 ± 1.60	1587	Caryophyllene oxide	0.13 ± 0.01
1041	Benzene acetaldehyde	0.06 ± 0.01	1617	Cubenol-1-epi	1.83 ± 0.27
1046	*trans*-β-Ocimene	0.06 ± 0.01	1673	Valeranone	11.21 ± 2.05
1058	γ-Terpinene	0.38 ± 0.03	1704	δ-Dodecalactone	0.34 ± 0.07
1067	*cis*-Sabinene hydrate	0.09 ± 0.00	1737	Mint Sulfide	0.09 ± 0.00
1089	Terpinolene	2.48 ± 0.09	1990	Isokaurene	0.29 ± 0.10

Summary of identified EO components	
Parameter	**Percentage (%)**
Total Identified	98.10
Not Identified	1.74
<0.05%	0.16
Total of which are Monoterpenes	75.88
Total of which are Oxygenated Monoterpenes	1.48
Total of which are Sesquiterpenes	5.71
Total of which are Oxygenated Sesquiterpenes	14.19

Not many reports are available in literature characterizing the components of the essential oil of *S. perfoliata* subs *perfoliata*, particularly from plants collected in Cyprus. The essential oil analysis revealed 34 identified compounds, representing the 98.10% of the total chromatogram of the oil profile. The most abundant components were α-pinene (27.92%), β-phelladrene (26.59%), valeranone (11.21%), β-pinene (7.12%), sabinene (4.59%), and α-myrcene, terpinolene, careen, β-caryophyllene with percentages 2–3%. Plants from the Eastern Mediterranean Region of Turkey also revealed the similar oil profile, with high percentage of α and β-pinene and β-phelladrenne [70]. Loizzo et al. [67] also found high percentages of β-phelladrene (32.85%), sabinene (12.76%), α and β-pinene (8.66% and 8.90%) when essential oils from *S. perfoliata* sampled in Lebanon were analyzed, in accordance with the analysis presented here. However, there are studies on *Sideritis* reporting differences between the components and their participation in the oil chromatograph [55,68].

The antioxidant activity of the *S. perfoliata* oils is reported in this study for the first time. The EO exhibited high antioxidant activity, with a two-fold increase of ABTS scavenging activity and Reducing Power whereas a four-fold increase for β-carotene activity was observed, when compared to their respective positive controls (Table 4). A study comparing the antioxidant activity of *S. italica* prepared from the leaves and flower heads reported higher in the EO extracted from the leaves [52].

Table 4. Total phenolics, flavonoids, and antioxidants in the leaf essential oil extracted from *Sideritis perforata*.

	Essential Oil	Positive Control
Essential oil yield (%)	0.30 ± 0.08	
Total phenolics (mg GA/g of oil)	0.53 ± 0.01	
DPPH IC$_{50}$ (μg/mL)	17.61 ± 0.40	17.93 ± 2.56 [a]
ABTS IC$_{50}$ (μg/mL)	10.10 ± 1.50	26.25 ± 0.41 [b]
Reducing Power (μg/mL)	22.62 ± 0.61	38.62 ± 0.21 [b]
β-carotene (AA%)	4.06 ± 0.50	18.10 ± 1.23 [a]

[a] Butylatedhydroxytoluene, [b] Ascorbic acid.

4. Conclusions

Sideritis perfoliata has a high protein and antioxidant content due to the presence of flavonoids and phenolics. Moreover, we revealed the wrinkle reducing potential of *S. perfoliata* as it showed noteworthy elastase inhibition. Traditionally, *S. perfoliata* is used for the treatment of wounds which could be attributed to the antibacterial and antioxidant activity reported in this study. The ethanol extract showed moderate toxicity towards the non-cancerous cell lines, whereas low toxicity was noted against the cancerous cell lines. However, the most promising anticancer activity was noted against human liver cancer cells.

Author Contributions: A.C. performed chemical analysis, A.C. and I.L. composition and wrote part of the original draft; I.L. performed chemical analysis, data curation, methodology and coordinated the South African team manuscript draft preparation; B.F., A.B.V.S., D.T., M.N.d.C., C.B.O., D.B. performed chemical analysis and analysis of results; N.L. administered and supervised the project, reviewed and edited the final manuscript; N.T. conceived and designed the research, administered and supervised the project, obtained funding, reviewed and edited the final manuscript.

Funding: The research performed at the University of Pretoria was funded by the National Research Foundation (NRF) and the University of Pretoria. Funding for publication was granted by the Cyprus University of Technology Open Access Author Fund.

Conflicts of Interest: The authors declare no conflict of interest.

References

1. González-Burgos, E.; Carretero, M.E.; Gómez-Serranillos, M.P. Sideritis spp.: Uses, chemical composition and pharmacological activities—A review. *J. Ethnopharmacol.* **2011**, *135*, 209–225. [CrossRef] [PubMed]

2. Stanoeva, J.P.; Stefova, M.; Stefkov, G.; Kulevanova, S.; Alipieva, K.; Bankova, V.; Aneva, I.; Evstatieva, L.N. Chemotaxonomic contribution to the Sideritis species dilemma on the Balkans. *Biochem. Syst. Ecol.* **2015**, *61*, 477–487. [CrossRef]

3. Kalivas, A.; Ganopoulos, I.; Xanthopoulou, A.; Chatzopoulou, P.; Tsaftaris, A.; Madesis, P. DNA barcode ITS2 coupled with high resolution melting (HRM) analysis for taxonomic identification of Sideritis species growing in Greece. *Mol. Biol. Rep.* **2014**, *41*, 5147–5155. [CrossRef] [PubMed]

4. Dimopoulos, P.; Raus, T.; Bergmeier, E.; Constantinidis, T.; Iatrou, G.; Kokkini, S.; Strid, A.; Tzanoudakis, D. Vascular plants of Greece: An annotated checklist. Supplement. *Willdenowia* **2016**, *46*, 301–347. [CrossRef]

5. Strid, A.; Tan, K. *Mountain Flora of Greece*; Vol. II; Edinburgh University Press: Edinburgh, UK, 1991.

6. Kirimer, N.; Tabanca, N.; Tümen, G.; Duman, H.; Başer, K.H.C. Composition of the essential oils of four endemic Sideritis species from Turkey. *Flavour Fragr. J.* **1999**, *14*, 421–425. [CrossRef]

7. Todorova, M.; Trendafilova, A. Sideritis scardica Griseb., an endemic species of Balkan peninsula: Traditional uses, cultivation, chemical composition, biological activity. *J. Ethnopharmacol.* **2014**, *152*, 256–265. [CrossRef] [PubMed]

8. Uysal, I.; Gücel, S.; Tütenocakli, T.; Öztürk, M. Studies on the medicinal plants of Ayvacik-Çanakkale in Turkey. *Pakistan J. Bot.* **2012**, *44*, 239–244.

9. Charami, M.T.; Lazari, D.; Karioti, A.; Skaltsa, H.; Hadjipavlou-Litina, D.; Souleles, C. Antioxidants and anti-inflammatory activities of Sideritis perfoliata subsp. Perfoliata (Lamiaceae). *Phytother. Res.* **2008**, *22*, 450–454. [CrossRef]

10. Levine, R.; Walsh, C.; Schwarts-Bloom, R. *Pharmacology: Drug Actions and Reactions*, 6th ed.; Parthenon Publishing Group: New York, NY, USA, 2000.

11. Chrysargyris, A.; Xylia, P.; Botsaris, G.; Tzortzakis, N. Antioxidant and antibacterial activities, mineral and essential oil composition of spearmint (*Mentha spicata* L.) affected by the potassium levels. *Ind. Crops Prod.* **2017**, *103*, 202–212. [CrossRef]

12. Chrysargyris, A.; Antoniou, O.; Athinodorou, F.; Vassiliou, R.; Papadaki, A.; Tzortzakis, N. Deployment of olive-stone waste as a substitute growing medium component for Brassica seedling production in nurseries. *Environ. Sci. Pollut. Res.* **2019**. [CrossRef]

13. Chrysargyris, A.; Laoutari, S.; Litskas, V.D.; Stavrinides, M.C.; Tzortzakis, N. Effects of water stress on lavender and sage biomass production, essential oil composition and biocidal properties against Tetranychus urticae (Koch). *Sci. Hortic. (Amsterdam).* **2016**, *213*, 96–103. [CrossRef]

14. Latimer, G.W. *Official Methods of Analysis of AOAC International*, 20th ed.; AOAC Int.: Gaithersburg, MD, USA, 2016.

15. Mushtaq, A.; Akbar, S.; Zargar, M.A.; Wali, A.F.; Malik, A.H.; Dar, M.Y.; Hamid, R.; Ganai, B.A. Phytochemical screening, physicochemical properties, acute toxicity testing and screening of hypoglycaemic activity of extracts of eremurus himalaicus baker in normoglycaemic wistar strain albino rats. *Biomed. Res. Int.* **2014**, *2014*. [CrossRef] [PubMed]

16. Mayur, B.; Sandesh, S.; Shruti, S.; Sung-yum, S. Analyses of the methanolic extract of the leaves of Rhinacanthus nasutus. *J. Med. Plants Res.* **2010**, *4*, 1554–1560.

17. Phatak, R.S.; Hendre, A.S. Total antioxidant capacity (TAC) of fresh leaves of Kalanchoe pinnata. *J. Pharmacogn. Phytochem. JPP* **2014**, *2*, 32–35.

18. Aumeeruddy, M.Z.; Aumeeruddy-Elalfi, Z.; Neetoo, H.; Zengin, G.; Blom van Staden, A.; Fibrich, B.; Lambrechts, I.A.; Rademan, S.; Szuman, K.M.; Lall, N.; et al. Pharmacological activities, chemical profile, and physicochemical properties of raw and commercial honey. *Biocatal. Agric. Biotechnol.* **2019**, *18*, 101005. [CrossRef]

19. Curto, E.V.; Kwong, C.; Hermersdörfer, H.; Glatt, H.; Santis, C.; Virador, V.; Hearing, V.J., Jr.; Dooley, T.P. Inhibitors of mammalian melanocyte tyrosinase: In vitro comparisons of alkyl esters of gentisic acid with other putative inhibitors. *Biochem. Pharm.* **1999**, *57*, 663–672. [CrossRef]

20. Matsuda, H.; Kawaguchi, Y.; Yamazaki, M.; Hirata, N.; Naruto, S.; Asanuma, Y.; Kaihatsu, T.; Kubo, M. Melanogenesis stimulation in murine B16 melanoma cells by *Piper nigrum* leaf extract and its lignan constituents. *Biol. Pharm. Bull.* **2004**, *27*, 1611–1616. [CrossRef]

21. Lall, N.; Henley-Smith, C.J.; De Canha, M.N.; Oosthuizen, C.B.; Berrington, D. Viability reagent, prestoblue, in comparison with other available reagents, utilized in cytotoxicity and antimicrobial assays. *Int. J. Microbiol.* **2013**, *2013*. [CrossRef]

22. Eloff, J. Which extractant should be used for the screening and isolation of antimicrobial components from plants? *J. Ethnopharmacol.* **1998**, *60*, 1–8. [CrossRef]

23. Rademan, S.; Anantharaju, P.G.; Madhunapantula, S.V.; Lall, N. the Anti-Proliferative and Antioxidant Activity of Four Indigenous South African Plants. *Afr. J. Tradit. Complement. Altern. Med.* **2019**, *16*, 13–23. [CrossRef]

24. Berrington, D.; Lall, N. Anticancer activity of certain herbs and spices on the cervical epithelial carcinoma (HeLa) cell line. *Evid.-Based Complement. Altern. Med.* **2012**, *2012*. [CrossRef] [PubMed]

25. Chrysargyris, A.; Panayiotou, C.; Tzortzakis, N. Nitrogen and phosphorus levels affected plant growth, essential oil composition and antioxidant status of lavender plant (*Lavandula angustifolia* Mill.). *Ind. Crops Prod.* **2016**, *83*, 577–586. [CrossRef]

26. Kavoosi, G.; Rowshan, V. Chemical composition, antioxidant and antimicrobial activities of essential oil obtained from *Ferula assa-foetida* oleo-gum-resin: Effect of collection time. *Food Chem.* **2013**, *138*, 2180–2187. [CrossRef]

27. Oke, F.; Aslim, B.; Ozturk, S.; Altundag, S. Essential oil composition, antimicrobial and antioxidant activities of *Satureja cuneifolia* Ten. *Food Chem.* **2009**, *112*, 874–879. [CrossRef]

28. Xylia, P.; Chrysargyris, A.; Botsaris, G.; Tzortzakis, N. Potential application of spearmint and lavender essential oils for assuring endive quality and safety. *Crop Prot.* **2017**, *102*, 94–103. [CrossRef]

29. Bettaieb Rebey, I.; Jabri-Karoui, I.; Hamrouni-Sellami, I.; Bourgou, S.; Limam, F.; Marzouk, B. Effect of drought on the biochemical composition and antioxidant activities of cumin (*Cuminum cyminum* L.) seeds. *Ind. Crops Prod.* **2012**, *36*, 238–245. [CrossRef]

30. Zhang, H.; Chen, F.; Wang, X.; Yao, H.Y. Evaluation of antioxidant activity of parsley (*Petroselinum crispum*) essential oil and identification of its antioxidant constituents. *Food Res. Int.* **2006**, *39*, 833–839. [CrossRef]

31. Karapandzova, M.; Qazimi, B.; Stefkov, G.; Bačeva, K.; Stafilov, T.; Panovska, T.K.; Kulevanova, S. Chemical characterization, mineral content and radical scavenging activity of Sideritis scardica and S. raeseri from R. Macedonia and R. Albania. *Nat. Prod. Commun.* **2013**, *8*, 639–644. [CrossRef]

32. Chrysargyris, A.; Kloukina, C.; Vassiliou, R.; Tomou, E.-M.; Skaltsa, H.; Tzortzakis, N. Cultivation strategy to improve chemical profile and anti-oxidant activity of Sideritis perfoliata L. subsp. perfoliata. *Ind. Crops Prod.* **2019**, *140*, 111694. [CrossRef]

33. Yadav, R.N.S.; Agarwala, M. Phytochemical analysis of some medicinal plants. *J. Phytol.* **2011**, *3*, 10–14.

34. Athwal, G.; Hui, A.M.; Dabagh, B.A.; Hui, A.M.; Dabagh, B.A. Biomarine Actives. In *Cosmeceuticals and Active Cosmetics*; CRC Press: Boca Raton, FL, USA, 2015; pp. 403–410.

35. Lambrechts, I.A.; de Canha, M.N.; Lall, N. Exploiting medicinal plants as possible treatments for acne vulgaris. In *Medicinal Plants for Holistic Health and Well-Being*; Academic Press: San Diego, CA, USA, 2018; pp. 117–143.

36. Bojovic, D.; Jankovic, S.; Potpara, Z.; Tadic, V. Of the phytochemical research performed to date on sideritis species. *Serbian J. Exp. Clin. Res.* **2011**, *12*, 109–122. [CrossRef]

37. Zengin, G.; Sarikurkcu, C.; Aktumsek, A.; Ceylan, R. Sideritis galatica Bornm.: A source of multifunctional agents for the management of oxidative damage, Alzheimer's and diabetes mellitus. *J. Funct. Foods* **2014**, *11*, 538–547. [CrossRef]

38. Phongpaichit, S.; Nikom, J.; Rungjindamai, N.; Sakayaroj, J.; Hutadilok-Towatana, N.; Rukachaisirikul, V.; Kirtikara, K. Biological activities of extracts from endophytic fungi isolated from Garcinia plants. *FEMS Immunol. Med. Microbiol.* **2007**, *51*, 517–525. [CrossRef] [PubMed]

39. Güvenç, A.; Houghton, P.J.; Duman, H.; Coşkun, M.; Şahin, P. Antioxidant activity studies on selected Sideritis species native to Turkey. *Pharm. Biol.* **2005**, *43*, 173–177. [CrossRef]

40. Blainski, A.; Lopes, G.C.; De Mello, J.C.P. Application and analysis of the folin ciocalteu method for the determination of the total phenolic content from *Limonium brasiliense* L. *Molecules* **2013**, *18*, 6852–6865. [CrossRef] [PubMed]

41. Lee, K.K.; Kim, J.H.; Cho, J.J.; Choi, J.D. Inhibitory effects of 150 plant extracts on elastase activity, and their anti-inflammatory effects. *Int. J. Cosmet. Sci.* **1999**, *21*, 71–82. [CrossRef]

42. Ndlovu, G.; Fouche, G.; Tselanyane, M.; Cordier, W.; Steenkamp, V. In vitro determination of the anti-aging potential of four southern African medicinal plants. *BMC Complement. Altern. Med.* **2013**, *13*, 304. [CrossRef]

43. Kacem, R.; Meraihi, Z. Effects of essential oil extracted from *Nigella sativa* (L.) seeds and its main components on human neutrophil elastase activity. *Yakugaku Zasshi* **2006**, *126*, 301–305. [CrossRef]

44. Deveci, E.; Tel-çayan, G.; Duru, M.E. Phenolic profile, antioxidant, anticholinesterase, and anti-tyrosinase activities of the various extracts of ferula elaeochytris and sideritis stricta. *Int. J. Food Prop.* **2018**, *21*, 771–783. [CrossRef]

45. Chiocchio, I.; Mandrone, M.; Sanna, C.; Maxia, A.; Tacchini, M.; Poli, F. Screening of a hundred plant extracts as tyrosinase and elastase inhibitors, two enzymatic targets of cosmetic interest. *Ind. Crops Prod.* **2018**, *122*, 498–505. [CrossRef]

46. Kermasha, S.; Bisakowski, B.; Ramaswamy, H.; Van de Voort, F.R. Thermal and microwave inactivation of soybean lipoxygenase. *LWT - Food Sci. Technol.* **1993**, *26*, 215–219. [CrossRef]

47. Iwai, K.; Kishimoto, N.; Kakino, Y.; Mochida, K.; Fujita, T. In vitro antioxidative effects and tyrosinase inhibitory activities of seven hydroxycinnamoyl derivatives in green coffee beans. *J. Agric. Food Chem.* **2004**, *52*, 4893–4898. [CrossRef] [PubMed]

48. Karioti, A.; Protopappa, A.; Megoulas, N.; Skaltsa, H. Identification of tyrosinase inhibitors from *Marrubium velutinum* and *Marrubium cylleneum*. *Bioorganic Med. Chem.* **2007**, *15*, 2708–2714. [CrossRef] [PubMed]

49. Song, H.S.; Sim, S.S. Acetoside inhibits α-MSH-induced melanin production in B16 melanoma cells by inactivation of adenyl cyclase. *J. Pharm. Pharmacol.* **2009**, *61*, 1347–1351. [CrossRef]

50. Son, Y.O.; Lee, S.A.; Kim, S.S.; Jang, Y.S.; Chun, J.C.; Lee, J.C. Acteoside inhibits melanogenesis in B16F10 cells through ERK activation and tyrosinase down-regulation. *J. Pharm. Pharmacol.* **2011**, *63*, 1309–1319. [CrossRef]

51. Tundis, R.; Bonesi, M.; Pugliese, A.; Nadjafi, F.; Menichini, F.; Loizzo, M.R. Tyrosinase, acetyl- and butyryl-cholinesterase inhibitory activity of *Stachys lavandulifolia* Vahl (Lamiaceae) and Its major constituents. *Rec. Nat. Prod.* **2015**, *9*, 81–93.

52. Basile, A.; Senatore, F.; Gargano, R.; Sorbo, S.; Del Pezzo, M.; Lavitola, A.; Ritieni, A.; Bruno, M.; Spatuzzi, D.; Rigano, D.; et al. Antibacterial and antioxidant activities in *Sideritis italica* (Miller) Greuter et Burdet essential oils. *J. Ethnopharmacol.* **2006**, *107*, 240–248. [CrossRef]

53. Askun, T.; Tumen, G.; Satil, F.; Ates, M. Characterization of the phenolic composition and antimicrobial activities of Turkish medicinal plants. *Pharm. Biol.* **2009**, *47*, 563–571. [CrossRef]

54. Da Silva, A.P.S.A.; Nascimento Da Silva, L.C.; Martins Da Fonseca, C.S.; de Araújo, J.M.; dos Santos Correia, M.T.; da Silva Cavalcanti, M.; de Menezes Lima, V.L. Antimicrobial activity and phytochemical analysis of organic extracts from *Cleome spinosa* Jaqc. *Front. Microbiol.* **2016**, *7*, 1–10. [CrossRef]

55. Kirimer, N.; Demirci, B.; Iscan, G.; Baser, K.H.C.; Duman, H. Composition of the essential oils of two Sideritis species from Turkey and antimicrobial activity. *Chem. Nat. Compd.* **2008**, *44*, 121–123. [CrossRef]

56. Kumar, B.; Pathak, R.; Mary, P.B.; Jha, D.; Sardana, K.; Gautam, H.K. New insights into acne pathogenesis: Exploring the role of acne-associated microbial populations. *Dermatol. Sin.* **2016**, *34*, 67–73. [CrossRef]

57. Ezer, N.; Vila, R.; Cañigueral, S.; Adzet, T. Essential oil composition of four Turkish species of Sideritis. *Phytochemistry* **1996**, *41*, 203–205. [CrossRef]

58. Kim, S.S.; Baik, J.S.; Oh, T.H.; Yoon, W.J.; Lee, N.H.; Hyun, C.G. Biological activities of Korean *Citrus obovoides* and *Citrus natsudaidai* essential oils against acne-inducing bacteria. *Biosci. Biotechnol. Biochem.* **2008**, *72*, 2507–2513. [CrossRef]

59. Samaranayake, L.P. *Essential Microbiology for Dentistry*, 5th ed.; Elsevier: Amsterdam, The Netherlands, 2002.

60. Falsetta, M.L.; Klein, M.I.; Colonne, P.M.; Scott-Anne, K.; Gregoire, S.; Pai, C.H.; Gonzalez-Begne, M.; Watson, G.; Krysan, D.J.; Bowen, W.H.; et al. Symbiotic relationship between *Streptococcus mutans* and *Candida albicans* synergizes virulence of plaque biofilms in vivo. *Infect. Immun.* **2014**, *82*, 1968–1981. [CrossRef]

61. Popova, C.; Dosseva-Panova, V.; Panov, V. Microbiology of periodontal diseases. A review. *Biotechnol. Biotechnol. Equip.* **2013**, *27*, 3754–3759. [CrossRef]

62. van Vuuren, S.F. Antimicrobial activity of South African medicinal plants. *J. Ethnopharmacol.* **2008**, *119*, 462–472. [CrossRef]

63. Uğur, A.; Varol, Ö.; Ceylan, Ö. Antibacterial activity of *Sideritis curvidens* and *Sideritis lanata* from Turkey. *Pharm. Biol.* **2005**, *43*, 47–52. [CrossRef]

64. Stevigny, C.; Bailly, C.; Quetin-Leclercq, J. Cytotoxic and antitumor potentialities of aporphinoid alkaloids. *Curr. Med. Chem. Anti-Cancer Agents* **2005**, *5*, 173–182. [CrossRef]

65. Newman, D.J.; Cragg, G.M. Natural products as sources of new drugs over the last 25 years. *J. Nat. Prod.* **2007**, *70*, 461–477. [CrossRef]

66. Kuete, V.; Efferth, T. African flora has the potential to fight multidrug resistance of cancer. *Biomed. Res. Int.* **2015**, *2015*, 1–24. [CrossRef] [PubMed]

67. Loizzo, M.R.; Tundis, R.; Menichini, F.; Saab, A.M.; Statti, G.A.; Menichini, F. Cytotoxic activity of essential oils from Labiatae and Lauraceae families against in vitro human tumor models. *Anticancer Res.* **2007**, *27*, 3293–3299. [PubMed]

68. Özkan, G.; Krüger, H.; Schulz, H.; Özcan, M. Essential oil composition of three sideritis species used as herbal teas in Turkey. *J. Essent. Oil-Bear. Plants* **2005**, *8*, 173–177. [CrossRef]

69. Kirimer, N.; Baser, K.H.C.; Dermici, B.; Duman, H. Essential oils of Sideritis species of Turkey belonging to the section Empedoclia. *Chem. Nat. Compd.* **2004**, *40*, 19–23. [CrossRef]
70. Karaborklu, S. Chemical characterization of *Sideritis perfoliata* L. essential oil and its fumigant toxicity against two pest insects. *J. Food Agric. Environ.* **2014**, *12*, 434–437.

 *antioxidants*

Article

Polyphenols from *Lycium barbarum* (Goji) Fruit European Cultivars at Different Maturation Steps: Extraction, HPLC-DAD Analyses, and Biological Evaluation

Andrei Mocan [1,2], Francesco Cairone [3], Marcello Locatelli [4], Francesco Cacciagrano [4], Simone Carradori [4], Dan C. Vodnar [5], Gianina Crișan [1], Giovanna Simonetti [6] and Stefania Cesa [3,*]

[1] Department of Pharmaceutical Botany, "Iuliu Hațieganu" University of Medicine and Pharmacy, 23 Gheorghe Marinescu Street, 400337 Cluj-Napoca, Romania; mocan.andrei@umfcluj.ro (A.M.); gcrisan@umfcluj.ro (G.C.)

[2] Laboratory of Chromatography, Institute of Advanced Horticulture Research of Transylvania, University of Agricultural Sciences and Veterinary Medicine, 400372 Cluj-Napoca, Romania

[3] Dipartimento di Chimica e Tecnologie del Farmaco, Università degli Studi di Roma "La Sapienza", P.le Aldo Moro 5, 00185 Rome, Italy; francesco.cairone@uniroma1.it

[4] Department of Pharmacy, "G. d'Annunzio" University of Chieti–Pescara, Via dei Vestini 31, 66100 Chieti, Italy; marcello.locatelli@unich.it (M.L.); francesco.cacciagrano@studenti.unich.it (F.C.); simone.carradori@unich.it (S.C.)

[5] Department of Food Science, University of Agricultural Sciences and Veterinary Medicine, 400372 Cluj-Napoca, Romania; dan.vodnar@usamvcluj.ro

[6] Dipartimento di Sanità Pubblica e Malattie Infettive, Università degli Studi di Roma "La Sapienza" P.le Aldo Moro 5, 00185 Rome, Italy; giovanna.simonetti@uniroma1.it

[*] Correspondence: stefania.cesa@uniroma1.it; Tel.: +39-06-49913198; Fax: +39-06-49913133

Received: 11 October 2019; Accepted: 7 November 2019; Published: 16 November 2019

Abstract: Goji berries are undoubtedly a source of potentially bioactive compounds but their phytochemical profile can vary depending on their geographical origin, cultivar, and/or industrial processing. A rapid and cheap extraction of the polyphenolic fraction from *Lycium barbarum* cultivars, applied after homogenization treatments, was combined with high-performance liquid chromatography (HPLC) analyses based on two different methods. The obtained hydroalcoholic extracts, containing interesting secondary metabolites (gallic acid, chlorogenic acid, catechin, sinapinic acid, rutin, and carvacrol), were also submitted to a wide biological screening. The total phenolic and flavonoid contents, the antioxidant capacity using three antioxidant assays, tyrosinase inhibition, and anti-*Candida* activity were evaluated in order to correlate the impact of the homogenization treatment, geographical origin, and cultivar type on the polyphenolic and flavonoid amount, and consequently the bioactivity. The rutin amount, considered as a quality marker for goji berries according to European Pharmacopeia, varied from ≈200 to ≈400 µg/g among the tested samples, showing important differences observed in relation to the influence of the evaluated parameters.

Keywords: goji berry; *Lycium barbarum*; HPLC-DAD; antioxidant capacity; TPC; TFC; anti-tyrosinase activity; anti-*Candida* activity

1. Introduction

Lycium barbarum L. is a traditional Chinese medicinal plant, specifically a shrub belonging to Solanaceae family, whose fruits, well known as goji berries, have acquired increasing popularity in Europe and North America in recent years. Goji berries and their derived products are considered a relevant source of (micro)nutrients, especially natural antioxidants, which contribute to the extraordinary nutritional quality of this matrix [1]. In fact, a healthy functional role is recognized as belonging to its fruits and their derived extracts and infusions, containing polysaccharides, carotenoids, and flavonoids, as well as salts, vitamins, and other micronutrients. The original habitat of goji is not certainly established, although more than 70 different species of *Lycium* exist [2]; *L. barbarum* represents the most relevant species from a biological point of view [3,4]. Overall, the chemical diversity of this fruit could provide many opportunities for food supplement development given the important growth forecasts of the polyphenol market worldwide.

Actually, the most researched components from goji fruits are the water-soluble arabinogalactans (*Lycium barbarum* polysaccharides or LBP), which are estimated to comprise up to 5–8% of the dried fruits [5] and include in their composition six types of monosaccharides (arabinose, rhamnose, xylose, mannose, galactose, and glucose), galacturonic acid, and almost all the proteinogenic amino acids [6]. LBP have been shown to control blood glucose, modulate glucose metabolism leading to the improvement of oxidative stress markers, alleviate insulin resistance, and stimulate the increase of glucose transporter type 4 (GLUT4) expression [7,8].

Dried fruits are composed of a carotenoid fraction of 0.03–0.5% [9], in which 11 free carotenoids and 7 carotenoid esters were detected from unsaponified and saponified *L. barbarum* extracts [6]. Zeaxanthin dipalmitate, which can vary from 30% to 80% of total carotenoids, represents the main molecule [10], followed by β-carotene, neoxanthin, and cryptoxanthin at lower concentrations. The carotenoid content of goji berry was recently the object of great interest for researchers for its beneficial effects on retinopathy. The ability to protect the visual function and the overall antioxidant properties [11] make this matrix an interesting ally for the prevention of the onset and the care of age-related macular degeneration (AMD), which represents one of the main causes of vision loss, which is estimated as cause of 5% of all cases of blindness [12]. Among carotenoids, lutein and zeaxanthin, the most represented pigments in the macular area of the human retina, are particularly effective in the macula protection from oxidative damage by scavenging harmful reactive oxygen species. Therefore, lutein and zeaxanthin may play a pivotal role in preventing AMD if further studies can clarify the explicit effects that can be expected in terms of their function regarding the development and progression of AMD [13].

Concerning phenylpropanoids, flavonoids, and isoflavonoids, the goji berry's protective and antioxidant role was correlated with the presence of caffeic and chlorogenic acid, with a high content of quercetin-3-*O*-rutinoside and kaempferol-3-*O*-rutinoside, and finally with representative coumarins and lignans [14,15]. Furthermore, phenolics are the most abundant secondary metabolites of plants [16]. Quercetin, kaempferol, and relative derivatives, among which rutin is the most frequent and representative, are well-known as radical scavengers and anti-oxidants capable of preventing cancer, cardiovascular disease, and other chronic disease onset [3,17]. In recent years, polyphenols were shown to exert a protective role against neurodegeneration, an ability related to their antioxidant and scavenging abilities. They could interact positively with vitamins E and C in lipid bilayer membranes via non-covalent associations, enabling the regeneration of these endogenous antioxidants [18,19].

Very recently, goji extracts have also been considered as potential candidates for designing innovative functional products (cosmetics and cosmeceuticals, among others) from natural matrices in the treatment of pigmentation disorders, and *L. barbarum* has also shown the capacity to inhibit the oxidation of L-DOPA, which is catalyzed by tyrosinase [20–22]. Tyrosinase is a ubiquitous and multifunctional enzyme that can catalyze hydroxylation and successive oxidation of phenolic compounds to quinones, regulating the melanogenesis process in humans.

Moreover, according to a preliminary *in vitro* study, *L. shawii* extracts have been reported to show antimicrobial effects on different species of bacteria [23]. Indeed, the anti-*Candida* activity of plants rich in polyphenols and polymeric flavan-3-ols has been studied [24–26]. Among the monoterpene phenols, carvacrol was investigated for its antifungal and antibacterial effects [27], and different studies demonstrated that this molecule not only is able to explicate a potent action against *C. albicans*, but also impairs the growth of different morphological forms, such as yeast, hyphae, and even the most resistant forms of biofilm [28].

Considering the multifactorial parameters, such as origin, cultivar, harvest dates, climatic factors, and applied technologies that could severely impact on the phytochemical profile and consequently on the associated biological activity, the objective of the present work is to determine the content and composition of the main functional constituents of four goji berry cultivars harvested in Italy and Romania at different dates with a particular emphasis on the polyphenolic derivatives. Their antioxidant potential, inhibitory activity on tyrosinase, and anti-fungal activity after two different treatments of homogenization (Ultraturrax® IKA, Staufen, Germany, or domestic mixer Girmi, Omegna, Italy) were evaluated. Besides the total phenolic and flavonoid content, the antioxidant capacity was evaluated using three assays—2,2-diphenyl-1-picryl-hydrazyl (DPPH), 2,2′-azino-bis-(3-ethylbenzothiazoline-6-sulfonate (ABTS), and ferric reducing antioxidant potential (FRAP)—to detect the radical scavenging activity of the investigated samples.

2. Materials and Methods

2.1. Materials

All HPLC-grade solvents were purchased from Carlo Erba (Milan, Italy) and Fluka (Milan, Italy), whereas chemical standards were supplied by Sigma-Aldrich (Milan, Italy).

Goji berries were generously gifted by Azienda Natural Goji® (Fondi, Italy), and by two local providers from Northwest Romania. Samples given by Azienda Natural Goji®, belonging to "Poland" and "Wild" varieties, P and W, were harvested at different commercial harvesting periods (2015: 6 July, 23 July, 3 August; 2016: 26 July; 4 August, P1–P5 and W1–W5) in Fondi (Italy) based on their stage maturity as determined by the producer. Samples given by the two local providers from Romania belonged to the Biglifeberry and Erma (B, E) varieties, were from only one collection (B1, E1), and were harvested at maturity in the summer of 2014 from two organic cultures from Ciuperceni (E variety, SM county) and Ploscoș (B variety, Cluj county). No specific permissions were requested for the plant sample collection, which did not involve endargered or protected species.

All samples were quickly frozen at −80 °C and stored at −18 °C until the analyses were performed, with no direct sunlight exposure.

2.2. Sample Preparation

Sample preparation was made according to the method previously described [10]. In brief, defrosted, washed, and wiped goji berries were homogenized at room temperature for 2 min by a T18 Ultraturrax® homogenizer (IKA®, Staufen, Germany) at 10,000 rpm (U samples) or by a domestic mixer at 16,000 rpm (TR01, Girmi, Omegna Italy) (D samples), and the resulting purees were subjected to the extraction procedure. Aliquots of fresh whole and homogenized samples were dehydrated to constant weight with the only aim to evaluate their water content.

2.3. Extraction of Hydroalcoholic Fractions

Extractions were performed with a hydroalcoholic mixture (ethanol:water = 70:30, water acidified with 0.5% acetic acid) and a small amount of cyclohexane for 3 h. The resulting suspension was centrifuged at 12,000× g for 10 min at 4 °C, and the upper organic phase was discarded. The hydroalcoholic phase, whose separation from the lower solid phase was completed using paper filtration, was transferred to a rotary evaporator at a reduced pressure and 40 °C. The dryness was completed using lyophilization. The obtained residues, protected from light and air exposure, were either immediately analyzed or stored at −18 °C until further analyses.

2.4. HPLC Analysis

Chromatographic analyses were carried out following a validated method [29] for the detection and quantification of the main secondary metabolites.

Moreover, rutin was quantified at 360 nm according to a slightly modified method based on the method reported by Imbaraj et al. [30] and described in a previous work [10]. Rutin was identified via comparison of the retention time and of the UV spectra of a pure external standard and quantified using a calibration curve ($y = 62x - 0.02$, $R^2 = 0.9972$, in the range 0.5–50 µg/mL). Quercetin and kaempferol derivatives were approximately identified using the evaluation of the UV spectra and indicatively quantified as rutin equivalents.

2.5. Total Phenolic Content (TPC)

The TPC was assessed using the Folin-Ciocâlteu method [21] with a SPECTROstar Nano Multi-Detection Microplate Reader in 96-well plates (BMG Labtech, Ortenberg, Germany). Briefly, a mixture solution consisting of 20 µL of extract, 100 µL of Folin-Ciocâlteu reagent, and 80 µL of sodium carbonate (Na_2CO_3, 7.5% *w/v*) was homogenized and incubated at room temperature in the dark (30 min). The TPC was expressed as gallic acid equivalents (GAE) in milligrams per gram dry weight (dw) of plant material (mg GAE/g dw plant material) after reading the absorbance of the samples at 760 nm.

2.6. Total Flavonoid Content (TFC)

The total flavonoid content (TFC) was determined using the $AlCl_3$ method, with the absorbance of the samples being measured at 420 nm. The results of the TFC were expressed as quercetin equivalents (QE) in milligrams per gram dry weight (dw) of plant material (mg QE/g dw plant material) [31].

2.7. Antioxidant Activity

2.7.1. DPPH Radical Scavenging Assay

The DPPH antioxidant capacity was determined according to the method described in Mocan et al. [32], with some modifications, by using a SPECTROstar Nano microplate reader (BMG Labtech, Offenburg, Germany). Each of the 96 wells consisted of 30 µL of sample solution (in an appropriated dilution) and a 0.004% methanolic solution of DPPH. After 30 min of incubation in the dark, the absorbance of the sample was read at 515 nm. Results were expressed as Trolox equivalents per gram of dry weight herbal extract (mg TE/g dw herbal extract).

2.7.2. Trolox Equivalent Antioxidant Capacity (TEAC) Assay

The TEAC of the samples was measured using the method previously described [21] and the results were expressed as milligrams of Trolox equivalents (TE) per gram of dry herbal extract (mg TE/g dw extract).

2.7.3. FRAP Assay

The FRAP assay was assessed using the method described in Damiano et al. [33] with slight modifications. The FRAP reagent was prepared by mixing ten volumes of acetate buffer (300 mM, pH 3.6), one volume of 2,4,6-tris(2-pyridyl)-*s*-triazine (TPTZ) solution (10 mM TPTZ in 40 mM HCl), and one volume of $FeCl_3$ solution (20 mM $FeCl_3·6H_2O$ in 40 mM HCl). After mixing the samples with the reagent (25 µL sample and 175 µL FRAP reagent), the samples were incubated in the dark for 30 min at room temperature and the absorbance was measured at 593 nm in 96-well plates. The final results were expressed as milligrams of Trolox equivalents (TE) per milligram of extract (mg TE/g extract).

2.8. Tyrosinase Inhibitory Activity

The tyrosinase inhibitory activity of each sample was determined via a method previously described [34] using a SPECTROstar Nano Multi-Detection Microplate Reader with 96-well plates (BMG Labtech, Ortenberg, Germany). Samples were disolved in water containing 5% dimethylsulfoxide (DMSO). For each sample, four wells were designated as A, B, C, and D, where each one contained a reaction mixture (200 µL) as follows: (A) 140 µL of 66 mM phosphate buffer solution (pH = 6.6) (PBS), 40 µL of mushroom tyrosinase in PBS (23 U/mL) (Tyr); (B) 160 µL PBS; (C) 80 µL PBS, 40 µL Tyr, 40 µL sample, and 80 µL PBS; and (D) 120 µL PBS and 40 µL sample. The plate was then incubated at room temperature for 10 min; after incubation, 40 µL of 2.5 mM L-DOPA in PBS solution were added in each well and the mixtures were incubated again at room temperature for 20 min. The absorbance of each well was measured at 475 nm and the inhibition percentage of the tyrosinase activity was calculated using the following Equation (1):

$$\%I = \frac{(A-B)-(C-D)}{(A-B)} \times 100 \tag{1}$$

The results were expressed as mg kojic acid equivalents per gram of dry weight of extract (mg KAE/g extract) using a calibration curve between 0.01–0.10 mg kojic acid/mL of solution.

2.9. Antifungal Susceptibility Testing

The *in vitro* antifungal susceptibility was evaluated using extracts, rutin, and carvacrol (Sigma Aldrich, St. Louis, MO, USA). To evaluate the minimal inhibitory concentration (MIC) of the extracts and compounds, a broth microdilution method was performed according to a standardized protocol for yeasts [35].

The assay was carried out with *C. albicans* ATCC 24433, *C. albicans* ATCC 10261, and *C. albicans* ATCC 90028 coming from the American Type Culture Collection (ATCC, Rockville, MD, USA). *Candida* strains were grown on Sabouraud dextrose agar at 37 °C for 24 h. Then, cell suspensions of the strains were prepared in a RPMI 1640 medium (Sigma-Aldrich, Milan, Italy) buffered to pH 7.0 with 0.165 mmol L^{-1} 3-(*N*-morpholino)propanesulfonic acid (MOPS). The final concentration of the inoculum was 1×10^3–5×10^3 cells mL^{-1}. The extracts were dissolved in DMSO and diluted 100 times in RPMI 1640 broth. Ten concentrations ranging from 1000 to 1.9 µg mL^{-1} for the extracts and from 64 to 0.125 µg mL^{-1} for the compounds were tested against *Candida albicans* strains in 96-well round-bottom microtitration plates. The MIC$_{50}$, MIC$_{90}$, and MIC$_{100}$—the lowest concentrations of extracts that caused growth inhibitions ≥50%, ≥90%, and 100%, respectively—were evaluated. Data was reported as the median of the MIC.

3. Results and Discussion

3.1. Polyphenols Extraction and HPLC Analysis

The selected freshly defrozen berry samples were submitted to a mild and quick, one-step double-phase extraction (graphical abstract), which allowed us to obtain the hydroalcoholic fraction separated by the solid residue and purified by the lipid fraction as already described in a previous work [10]. The only centrifugation step, followed by a preliminary evaporation of ethanol and the subsequent freeze-drying, allowed for obtaining a residue that was directly ready for the HPLC analysis and for the evaluation using the biological tests.

Fresh samples as such, or after homogenization, were also submitted to a dehydration step to evaluate their water content (about 88%), and the dried samples were extracted in the same conditions that were applied to the fresh samples. Results, not shown, demonstrate that the dehydration process did not influence the extraction yield nor the chemical composition of the hydroalcoholic fraction.

Extraction yields (% *w/w*) of the hydroalcoholic fraction from goji berries performed with the above-described conditions from the selected samples accounted for 35–60% in dry weight with mean values around 43–46%. No statistically relevant differences were observed comparing cultivars type, harvesting date, seasons, or adopted homogenization technique. Indeed, the homogenization process, as shown later, had an influence on the quali-quantitative distribution of each analyte. A lower yield (medium 378 vs 523 mg/g dry weight) was shown for the harvesting dates of season 2016 with respect to those of 2015, but only for sample P. This could depend on the different maturation stages of the selected samples, but it seems not relevant for an overall rating.

3.2. HPLC Analysis at 278 nm

From the analysis of the data reported in Table 1 regarding the phenolic profiling of goji berries, we can first note that, among the 21 phenolic standards, vanillic acid, 4-hydroxy benzoic acid, 3-hydroxy-4-methoxy benzaldehyde, *p*-coumaric acid, *t*-ferulic acid, naringin, 2,3-dimethoxy benzoic acid, benzoic acid, *o*-coumaric acid, quercetin, harpagoside, and *t*-cinnamic acid were not detected or were present in trace amounts (below Limit Of Quantification or Limit Of Detection, data not shown). The inter- and intra-cultivar differences were huge and did not follow a specific trend. Generally speaking, Italian cultivars were richer in terms of phenolic compounds with respect to Romanian ones. Among Italian cultivars (especially for Wild), the last two harvesting dates gave berries richer in phenolic compounds regardless of the mixing treatment. Chlorogenic acid, carvacrol, and at lower extent, sinapinic acid were the most-detected metabolites. Conversely, gallic acid, catechin, and syringic acid were only present in some cultivars and their amounts differed a lot depending on the applied treatment. Lastly, a few samples were also characterized by low amounts of naringenin (P2D) and epicatechin (P4U).

Table 1. Phenolic pattern of goji berries.

Compound	Concentration (µg/mg ± SD)																					
	P1D	P2D	P3D	P4D	P5D	P1U	P2U	P3U	P4U	P5U	W1D	W2D	W3D	W4D	W5D	W1U	W1U	W3U	W4U	W5U	E1D	B1D
Gallic acid		0.77 ± 0.07				0.03 ± 0.01		tr														
Catechin						0.23 ± 0.01			0.23 ± 0.01		tr				0.37 ± 0.01		0.03 ± 0.01		0.31 ± 0.01	0.25 ± 0.01		
Chlorogenic acid		0.19 ± 0.01	0.16 ± 0.01	0.55 ± 0.02	0.54 ± 0.02	0.56 ± 0.01	0.11 ± 0.01	0.13 ± 0.01	0.57 ± 0.02	0.37 ± 0.02	0.47 ± 0.02	0.10 ± 0.01		0.77 ± 0.03	0.43 ± 0.02	0.22 ± 0.01	0.12 ± 0.01	0.28 ± 0.01	0.43 ± 0.02	0.45 ± 0.02		
Epicatechin									0.07 ± 0.01													
Syringic acid										0.06 ± 0.01			0.14 ± 0.01	0.09 ± 0.01		0.04 ± 0.01	0.21 ± 0.01	0.27 ± 0.01				
3-Hydroxy benzoic acid															0.08 ± 0.01							
Sinapinic acid			tr	0.03 ± 0.01	0.03 ± 0.01	0.03 ± 0.01		tr	0.03 ± 0.01	0.03 ± 0.01	tr			0.03 ± 0.01	0.03 ± 0.01	tr			0.03 ± 0.01	0.03 ± 0.01		
Naringenin		0.06 ± 0.01																				
Carvacrol	0.03 ± 0.01	0.05 ± 0.01	0.05 ± 0.01	0.05 ± 0.01	0.05 ± 0.01	0.08 ± 0.01	0.05 ± 0.01	0.04 ± 0.01	0.05 ± 0.01	0.04 ± 0.01	0.04 ± 0.01	0.06 ± 0.01	0.05 ± 0.01	0.04 ± 0.01	0.07 ± 0.01	0.04 ± 0.01	0.05 ± 0.01	0.05 ± 0.01	0.04 ± 0.01	0.05 ± 0.01		0.23 ± 0.02
Total (µg/mg)	0.03	1.07	0.21	0.63	0.62	0.93	0.16	0.17	0.95	0.50	0.51	0.16	0.18	0.93	0.98	0.27	0.24	0.54	1.09	0.78		0.23

tr: traces; P: Polonia cultivar; W: Wild cultivar; E: Erma cultivar; B: Biglifeberry cultivar; D: domestic mixer; U: Ultraturrax®.

The chromatograms of the flavonoid fraction, recorded at 360 nm, showed the presence of about 10–15 peaks, according to the different analyzed cultivars, among which the most representative was rutin. In Table 2, rutin and the total flavonoid contents are reported, calculated as rutin equivalents, after considering all the peaks whose areas were at least 10% of rutin area and taking into account the standard calibration curve of this molecule. Values are expressed in relation to dry weight (dw) for a comparison with the literature data and in µg/mg extract for a better evaluation of other data on bioactivity presented in this work and obtained by the extracts evaluation. Rutin accounted for about 185–400 µg/g dw of goji berries. In the literature, reported values of 281 [30], 296 [36], and 326 µg/g dw [37] are found, perfectly overlapping with our results. Wider ranges between 159–629 µg/g dw were reported in Zhang et al. [38] and higher values (730–1380 µg/g dw) were detected in Dong et al. [39]. Many different results were available in the literature regarding the other representative flavonoids recognized mainly as quercetin hexosides, and secondly as kaempferol and isorhamnetin derivatives. Zhang et al. [38] reported the quercetin-rhamno-di-hexoside content as 435–1065 µg/g, Donno et al. [14] reported a hyperoside content of 116 mg/100 g fresh weight, whereas Inbaraj et al. [30] studied different quercetin-rhamno-hexoside derivatives (in the range of 70–440 µg/g). In our chromatograms, many different peaks were detected in the flavonoid component, although their identification was not possible. These were reported as a sum and expressed as rutin equivalents on the basis of its calibration curve. According to this evaluation, rutin accounted for about 20–65% of the total flavonoid components. The selected samples did not show relevant differences among cultivars (rutin content mean values: P = 276 ± 8.3, W = 291 ± 8.7, E = 288 ± 8.6, and B = 255 ± 7.6 µg/g), nor among applied homogenization processes (D = 287 ± 8.6 vs. U = 279 ± 8.4 µg/g), whereas differences could be revealed in the rutin content along the maturation stage (i.e., rutin increased by 70%, from P1 to P3, collected a month apart in July–August 2015 and decreased by 45% from P4 to P5, collected fifteen days apart, in July–August 2016). The total flavonoid content ranged from 380–1600 µg/g with a slight difference between cultivars P and W (770 ± 38 vs. 910 ± 45 µg/g) and no differences after homogenization treatments D and U (815 ± 41 vs. 862 ± 43 µg/g).

Table 2. Rutin and flavonoid content of analyzed goji berries hydroalcoholic (HA) extracts. Flavonoids are expressed as rutin equivalent using the HPLC areas collected at 360 nm.

Sample	HA Extract (mg/g dw)	Rutin (± 3%) (µg/g dw)	Flavonoids (± 5%) (µg/g dw)	Rutin (± 3%) (µg/mg Extract)	Flavonoids (± 5%) (µg/mg Extract)
P1D	467	210	790	0.45	1.69
P2D	548	263	661	0.48	1.28
P3D	528	360	658	0.68	1.25
P4D	410	330	950	0.80	2.31
P5D	340	205	564	0.60	1.66
W1D	422	371	995	0.88	2.36
W2D	472	274	699	0.58	1.48
W3D	493	256	874	0.52	1.77
W4D	430	268	1033	0.62	2.40
W5D	448	336	925	0.75	2.06
P1U	450	184	591	0.41	1.31
P2U	558	335	735	0.60	1.32
P3U	589	316	705	0.54	1.20
P4U	414	317	1598	0.77	3.86
P5U	345	238	428	0.69	1.24
W1U	456	405	1134	0.89	2.49
W2U	506	309	642	0.61	1.27
W3U	494	242	1237	0.49	2.50
W4U	464	198	1183	0.43	2.55
W5U	413	250	374	0.60	0.90
E1D	528	288	1605	0.54	3.04
B1D	420	255	752	0.61	1.79

dw: dry weight.

As shown by the data, the rutin content increase was usually correlated with a decrease in the amount of other flavonoid components and vice versa, but this trend was not confirmed by the P4–P5 series. The comparison between the D and U series demonstrated differences between P4D and P4U samples only in terms of an increase of other flavonoids, whereas W5D and W5U samples were characterized by discrepancies in terms of rutin and flavonoid amount. Overall, the different ratios between rutin and other flavonoids in the goji phytocomplex depended on the cultivar type, but cultivars also seemed highly influenced by the maturation step, as is clearly shown by the sample P.

3.3. Total Phenolic and Flavonoid Content

The total phenolic (TPC)/flavonoid (TFC) content assays are rapid and low-cost methods used for the screening of herbal or food samples subjected to different processing or treatment procedures [39,40]. In this study, the highest value for the TPC was found in the W4U sample, followed by P4D, whereas the highest TFC was obtained from the P4U and W4U samples, as seen in Table 3. Interestingly, the lowest values for the TPC were presented by the samples collected from Romania, while Italian samples presented overall higher values for TPC. The results obtained in this study for the Italian samples are comparable with the ones obtained for samples collected in the Umbria region (central Italy) [41], while the results obtained for the Romanian samples are more comparable with the findings concerning samples collected from Northern Italy [14]. Furthermore, these results support the evidence that TPC in our tested goji berries could vary among different genotypes and was strictly dependent on the pre-harvest practices, cultivar/clone, environmental conditions, and maturity stage at harvest. All these experimental variables could modulate the accumulation of specific phenolics [42], which are majorly responsible of the antioxidant and biological activity.

Table 3. Total phenolic (TPC) and flavonoid (TFC) content, 2,2-diphenyl-1-picryl-hydrazyl (DPPH), 2,2′-azino-bis-(3-ethylbenzothiazoline-6-sulfonate (ABTS) scavenging capacity, ferric reducing antioxidant power (FRAP), and tyrosinase inhibition of the extracts of *Lycium* cultivars (values expressed are means ± SD of three measurements).

Sample	TPC (mg GAE/g dw Extract)	TFC (mg QE/g dw Extract)	FRAP (mg TE/g dw Extract)	DPPH Scavenging (mg TE/g dw Extract)	ABTS Scavenging (mg TE/g dw Extract)	Tyrosinase Inhibition (mg KAE/g dw Extract)
P1D	19.22 ± 0.52	4.05 ± 0.61	23.84 ± 0.31	26.29 ± 1.65	40.45 ± 0.26	2.12 ± 0.25
P2D	17.09 ± 1.59	3.58 ± 0.11	19.47 ± 0.99	24.62 ± 0.24	34.79 ± 1.13	2.50 ± 0.86
P3D	18.09 ± 0.13	3.87 ± 0.76	20.80 ± 0.74	16.77 ± 4.38	36.85 ± 0.39	2.39 ± 0.53
P4D	22.64 ± 0.97	2.89 ± 0.13	36.63 ± 0.84	36.47 ± 1.42	44.97 ± 0.48	9.69 ± 0.19
P5D	19.50 ± 0.67	2.64 ± 0.17	33.28 ± 2.03	35.58 ± 2.14	44.91 ± 0.61	6.54 ± 0.78
W1D	12.28 ± 1.87	3.18 ± 0.21	16.76 ± 1.03	18.66 ± 2.22	31.88 ± 0.58	2.59 ± 0.35
W2D	20.29 ± 0.88	4.26 ± 0.13	21.24 ± 1.46	16.77 ± 2.83	39.10 ± 1.68	3.17 ± 0.83
W3D	15.33 ± 0.16	3.45 ± 0.13	20.96 ± 0.75	17.99 ± 3.98	37.71 ± 0.33	3.33 ± 0.70
W4D	15.27 ± 0.69	2.39 ± 0.06	29.90 ± 1.57	32.97 ± 2.60	37.84 ± 0.74	8.03 ± 0.85
W5D	19.54 ± 0.70	3.56 ± 0.24	32.14 ± 3.49	40.32 ± 4.01	42.03 ± 0.20	8.42 ± 0.75
P1U	18.86 ± 0.45	3.42 ± 0.10	28.54 ± 0.59	32.58 ± 2.70	40.67 ± 1.16	5.05 ± 0.58
P2U	18.39 ± 1.57	3.87 ± 0.26	25.93 ± 1.79	21.78 ± 3.53	39.10 ± 0.46	3.50 ± 0.29
P3U	19.59 ± 0.73	3.61 ± 0.14	23.86 ± 0.59	21.11 ± 3.94	38.02 ± 0.63	3.72 ± 0.63
P4U	7.69 ± 1.78	11.03 ± 1.24	23.89 ± 2.09	10.75 ± 0.06	45.24 ± 4.69	57.60 ± 0.01
P5U	17.71 ± 0.93	2.77 ± 0.17	30.86 ± 0.86	26.23 ± 2.15	39.43 ± 0.46	6.48 ± 0.16
W1U	17.43 ± 0.30	4.02 ± 0.02	20.90 ± 0.49	19.89 ± 1.65	37.57 ± 1.13	4.33 ± 0.72
W2U	11.83 ± 0.67	3.13 ± 0.24	15.71 ± 0.05	12.20 ± 2.53	27.81 ± 0.46	4.66 ± 0.99
W3U	17.00 ± 0.65	3.24 ± 0.12	20.49 ± 0.62	17.99 ± 0.77	35.96 ± 0.38	1.56 ± 0.74
W4U	23.49 ± 3.36	10.54 ± 0.12	51.93 ± 2.97	14.98 ± 4.40	63.58 ± 13.07	62.79 ± 0.66
W5U	15.10 ± 1.41	2.58 ± 0.09	28.59 ± 2.55	29.13 ± 0.94	35.76 ± 0.30	9.30 ± 0.44
E1D	5.76 ± 0.24	10.47 ± 1.83	23.00 ± 3.66	2.18 ± 0.84	51.14 ± 4.27	59.49 ± 1.33
B1D	3.31 ± 0.66	9.29 ± 0.60	21.56 ± 2.28	na	76.78 ± 7.91	49.35 ± 4.33

na: not active.

3.4. Antioxidant Activity

The antioxidant capacity of plant extracts and food matrices can be evaluated using several *in vitro* assays. Since the results of the antioxidant effect generated by an assay are influenced by the used protocol, a combination panel of assays is being required. The antioxidant capacity of the several goji berries cultivars was tested using three different assays to underline the different antioxidant facets of the investigated samples (Table 3).

3.4.1. Trolox Equivalent Antioxidant Capacity (TEAC) Assay

The TEAC assay measured the ability of goji berries' phenolic extracts to reduce the *in vitro* formed radicals. In this study, among Italian samples, the highest TEAC value was presented by sample W4U (63.58 ± 13.07 mg TE/g extract), which also presented high values for TPC and TFC. However, the highest value for the TEAC assay was obtained from the Romanian B1D sample (76.78 ± 7.91 mg TE/g extract). Zhang et al. [38] reported TEAC values ranging from 47.8 ± 6.6 to 78.2 ± 4.8 μM TE/g fresh weight for eight different Chinese goji berries genotypes, while Abdennacer et al. [43] reported a TEAC value of 0.08 mg TE/mg dw for *Lycium intricatum* Boiss. fruit methanolic extracts. Nonetheless, Mocan et al. [15] obtained values of 24.86 ± 2.15 and 25.12 ± 2.11 mg TE/g extract for 70% methanolic extracts obtained using sonication of the same Romanian cultivars of goji berries.

3.4.2. DPPH Radical Scavenging Assay

The DPPH radical scavenging assay is among the most frequently used antioxidant assays and offers a first approach for evaluating the antioxidant capacity of a complex mixture [44]. In the current research, the DPPH activity of the tested goji berries ranged from na (not active, B1D sample) to the highest value of 40.32 ± 4.01 mg TE/g extract (for the Italian W5D sample). Zhang et al. [38] obtained values ranging between 35.88 ± 2.2 to 85.46 ± 1.9 μM TE/g fresh weight for the DPPH antioxidant capacity assay from eight Chinese goji genotypes, while Abdennacer et al. [43] reported a value of 0.05 mg TE/mg dry weight for the methanolic extracts of fruits of Tunisian *Lycium intricatum*.

3.4.3. FRAP Assay

The FRAP assay is an electron transfer-based method that measures the reduction of the ferric ion (Fe^{3+})-ligand complex to the intensely blue colored ferrous (Fe^{2+}) complex via antioxidants in an acidic environment [44]. In this study, the highest reducing power was presented by the Italian W4U sample with a value of 51.93 ± 2.97 mg TE/g extract, followed by the P4D sample (36.63 ± 0.84 mg TE/g extract), while the lowest value was obtained for Italian W2U sample (15.71 ± 0.05 mg TE/g extract).

Overall, the different trends and results obtained in the antioxidant capacity assays show that the contents of total phenolics and flavonoids could not sufficiently explain the observed antioxidant activity of fruit and plant phenolic extracts, which are mixtures of different compounds with various activities in the tested samples [42].

3.5. Enzyme Inhibitory Activity

Tyrosinase inhibitors can be attractive for cosmetic and pharmaceutical industries as depigmentation agents, as well as in food and agriculture industries as anti-browning compounds. Goji berry extracts have been previously described as tyrosinase inhibitors [15] and are currently used as natural ingredients in several market-available cosmetic or dermato-cosmetic preparations. In this study, the tyrosinase inhibitory activity of the tested goji extracts ranged between 1.56 ± 0.74 mg KAE/g extract for the W3U sample to 62.79 ± 0.66 for the W4U sample. The second-highest value in terms of tyrosinase inhibition was obtained for the Romanian E1D sample (59.49 ± 1.33 mg KAE/g extract), while the two samples from Romania generally exhibited a higher tyrosinase inhibitory potential than samples from Italy, except sample W4U.

3.6. Anti-Candida Activity of Extracts

Antifungal activity was demonstrated via a broth microdilution method, using the standard drug fluconazole as a positive control. Among the different goji extracts tested against three *Candida albicans* strains, P samples displayed the best inhibitory activity. In particular, P1D, P1U, W1D, W1U, E1D, and B1D showed MIC_{50} values of 138 µg mL^{-1}, 186 µg mL^{-1}, 238 µg mL^{-1}, 250 µg mL^{-1}, >1000 µg mL^{-1}, and >1000 µg mL^{-1}, respectively, against all *Candida albicans* strains. Samples harvested at different commercial harvesting periods showed different MIC values. Interestingly, the P1D sample, harvested on 6 July 2015, was endowed with the best antifungal activity compared to the other harvesting periods (P2D, P3D, P4D and P5D) with a MIC_{100} of 476 µg mL^{-1}, 552 µg mL^{-1}, 609 µg mL^{-1}, >1000 µg mL^{-1}, and >1000 µg mL^{-1}, respectively against all the considered strains (Table 4).

Table 4. Antifungal activity of goji berry extracts and reference compounds expressed as an MIC (µg/mL).

Sample	MIC (µg/mL)* at 24 h								
	Candida albicans ATCC 24433			*Candida albicans* ATCC 10261			*Candida albicans* ATCC 90028		
	50	90	100	50	90	100	50	90	100
P1D	125	250	500	281.25	312.5	625	93.75	250	500
P2D	250	250	500	375	375	750	125	250	500
P3D	250	250	500	375	500	1000	187.5	250	500
P4D	125	1000	>1000	500	1000	>1000	500	1000	>1000
P5D	250	1000	>1000	500	1000	>1000	500	1000	>1000
P1U	125	187.5	375	375	500	1000	250	375	750
P2U	250	250	500	375	500	1000	250	250	500
P3U	250	250	500	375	500	1000	250	250	500
P4U	125	250	1000	250	500	1000	250	500	1000
P5U	250	500	1000	250	500	1000	500	1000	1000
W1D	250	250	500	500	500	1000	187.5	250	500
W2D	250	250	500	500	500	1000	187.5	250	500
W3D	250	250	500	500	500	1000	250	250	500
W4D	125	1000	>1000	500	1000	>1000	500	1000	>1000
W5D	125	1000	>1000	500	1000	>1000	500	1000	>1000
W1U	187.5	250	500	500	500	1000	250	375	500
W2U	250	375	500	500	500	1000	375	375	750
W3U	250	250	500	500	500	1000	125	250	500
W4U	250	250	>1000	500	1000	>1000	500	1000	>1000
W5U	250	1000	>1000	500	1000	>1000	500	1000	>1000
E1D	>1000	>1000	>1000	>1000	>1000	>1000	>1000	>1000	>1000
B1D	>1000	>1000	>1000	>1000	>1000	>1000	>1000	>1000	>1000
Rutin	64	>64	>64	64	>64	>64	64	>64	>64
Carvacrol	0.25	1	2	0.25	1	2	0.125	0.25	0.5
Fluconazole	0.5	2	8	1	4	>64	0.5	2	8

* The data are expressed as a median.

Due to the presence in these extracts of considerable amounts of well-recognized antimicrobial metabolites, we further tested rutin and carvacrol against the same fungal strains [45]. Rutin was almost inactive with MIC values ≥64 µg/mL. Conversely, carvacrol displayed a strong inhibitory activity against *C. albicans* with MIC_{50} ranging from 0.125 µg/mL to 0.25 µg/mL. Despite the high amount of carvacrol present in B1D (Table 1), it displayed a weak anti-fungal activity thus suggesting that the concurrent presence of other secondary metabolites can enhance or decrease the bioactivity of these goji extracts.

Furthermore, considering the quantified components and the explicated anti-*Candida* effects, W4U represented the sample characterized by a higher amount of detected secondary metabolites and a positive upward trend in the biological activity. Many components, other than those quantified using the HPLC analyses, could contribute to the TPC and TFC of the W4U sample, whose activity was evidently correlated not only with the single analytes, even though they represented the most representative molecules in weight terms, but also in the complexity of the phytocomplex, evidently characterized by many different minor components. However, this clearly shows how important the work-up is in the preservation of many components (especially for flavonoid compounds, considering the effect of U on P) represented in extremely low concentrations. The high flavonoid component seemed to correlate, in particular, with a marked activity on tyrosinase.

4. Conclusions

Despite the long tradition in nutrition and medicine, the goji berry's composition is strictly related to the geographical origin (soil type, climate), cultivar type, harvesting time, and industrial/domestic processing. The changes of the phytochemical profile have a deep influence on the bioactivity of this herbal medicinal/edible product. Our investigation aimed at the identification of the main bioactive components that are responsible for selected biological activities. The HPLC analytical fingerprint was characterized by a variable amount of phenolic compounds, mainly depending on the geographical origin, cultivar type, homogenization treatment, and above all, harvesting period. Moreover, we highlighted, by means of an array of different assays, this valuable natural matrix in terms of the antioxidant activity, anti-fungal activity, and inhibition of tyrosinase, aiming at the correlation between the phytochemical profile and the reported bioactivity. These data could be of high interest in the formulation of new goji-based food supplements and for further pharmaceutical development of this valuable plant matrix.

Author Contributions: Conceptualization and data curation, S.C. (Stefania Cesa) and S.C. (Simone Carradori); methodology, A.M. and D.C.V.; HPLC validation, M.L. and F.C. (Francesco Cacciagrano); formal analysis, F.C. (Francesco Cairone) and G.S.; writing—review and editing, S.C. (Stefania Cesa) and G.C.; supervision, D.C.V. and S.C. (Simone Carradori); funding acquisition, S.C. (Stefania Cesa) and A.M.

Funding: This work was financially supported by funding from the "Sapienza" University of Rome, Scientific Research Programs 2016–2017. RP116154E7012EA9. Andrei Mocan acknowledges the support by a grant from the Romanian Ministry of Research and Innovation, CNCS - UEFISCDI, project number PN-III-P1-1.1-PD-2016- 1900, within PNCDI III.

Conflicts of Interest: The authors declare no conflict of interest.

References

1. Yao, R.; Heinrich, M.; Weckerle, C.S. The genus *Lycium* as food and medicine: A botanical, ethnobotanical and historical review. *J. Ethnopharmacol.* **2018**, *212*, 50–66. [CrossRef]
2. Kafkaletou, M.; Christopoulos, M.V.; Tsaniklidis, G.; Papadakis, I.; Ioannou, D.; Tzoutzoukou, C.; Tsantili, E. Nutritional value and consumer-perceived quality of fresh goji berries (*Lycium barbarum* L. and *L. chinense* L.) from plants cultivated in Southern Europe. *Fruits* **2018**, *73*, 5–12. [CrossRef]
3. Dong, J.Z.; Wang, S.H.; Zhu, L.; Wang, Y. Analysis on the main active components of *Lycium barbarum* fruits and related environmental factors. *J. Med. Plants Res.* **2012**, *6*, 2276–2283.
4. Potterat, O. Goji (*Lycium barbarum* and *L. chinense*): Phytochemistry, pharmacology and safety in the perspective of traditional uses and recent popularity. *Planta Med.* **2010**, *76*, 7–19. [CrossRef]
5. Wang, C.C.; Chang, S.C.; Inbaraj, B.S.; Chen, B.H. Isolation of carotenoids, flavonoids and polysaccharides from *Lycium barbarum* L. and evaluation of antioxidant activity. *Food Chem.* **2010**, *120*, 184–192. [CrossRef]
6. Amagase, H.; Farnsworth, N.R. A review of botanical characteristics, phytochemistry, clinical relevance in efficacy and safety of *Lycium barbarum* fruit (Goji). *Food Res. Int.* **2011**, *44*, 1702–1717. [CrossRef]
7. Zhao, R.; Li, Q.; Xiao, B. Effect of *Lycium barbarum* polysaccharide on the improvement of insulin resistance in NIDDM rats. *Yakugaku Zasshi* **2005**, *125*, 981–988. [CrossRef]
8. Masci, A.; Carradori, S.; Casadei, M.A.; Paolicelli, P.; Petralito, S.; Ragno, R.; Cesa, S. *Lycium barbarum* polysaccharides: Extraction, purification, structural characterisation and evidence about hypoglycaemic and hypolipidaemic effects. A review. *Food Chem.* **2018**, *254*, 377–389. [CrossRef] [PubMed]
9. Peng, Y.; Ma, C.; Li, Y.; Leung, K.S.; Jiang, Z.H.; Zhao, Z. Quantification of zeaxanthin dipalmitate and total carotenoids in *Lycium* fruits (*Fructus Lycii*). *Plant Foods Hum. Nutr.* **2005**, *60*, 161–164. [CrossRef] [PubMed]

10. Patsilinakos, A.; Ragno, R.; Carradori, S.; Petralito, S.; Cesa, S. Carotenoid content of Goji berries: CIELAB, HPLC-DAD analyses and quantitative correlation. *Food Chem.* **2018**, *2698*, 49–56. [CrossRef] [PubMed]

11. Benzie, I.F.; Chung, W.Y.; Wang, J.; Richelle, M.; Bucheli, P. Enhanced bioavailability of zeaxanthin in a milk-based formulation of wolfberry (Gou Qi Zi; Fructus barbarum L.). *Br. J. Nutr.* **2006**, *96*, 154–160. [CrossRef] [PubMed]

12. Pascolini, D.; Mariotti, S.P. Global estimates of visual impairment: 2010. *Br. J. Ophthalmol.* **2012**, *96*, 614–618. [CrossRef] [PubMed]

13. Kim, E.; Kim, H.; Kwon, O.; Chang, N. Associations between fruits, vegetables, vitamin A, β-carotene and flavonol dietary intake, and age-related macular degeneration in elderly women in Korea: The Fifth Korea National Health and Nutrition Examination Survey. *Eur. J. Clin. Nutr.* **2018**, *72*, 161–167. [CrossRef] [PubMed]

14. Donno, D.; Beccaro, G.L.; Mellano, M.G.; Cerutti, A.K.; Bounous, G. Goji berry fruit (*Lycium* spp.): Antioxidant compound fingerprint and bioactivity evaluation. *J. Funct. Foods* **2015**, *18*, 1070–1085. [CrossRef]

15. Mocan, A.; Moldovan, C.; Zengin, G.; Bender, O.; Locatelli, M.; Simirgiotis, M.; Atalay, A.; Vodnar, D.C.; Rohn, S.; Crişan, G. UHPLC-QTOF-MS analysis of bioactive constituents from two Romanian Goji (*Lycium barbarum* L.) berries cultivars and their antioxidant, enzyme inhibitory, and real-time cytotoxicological evaluation. *Food Chem. Toxicol.* **2018**, *115*, 414–424. [CrossRef]

16. Ignat, I.; Volf, I.; Popa, V.I. A critical review of methods for characterisation of polyphenolic compounds in fruits and vegetables. *Food Chem.* **2011**, *126*, 1821–1835. [CrossRef]

17. Mikulic-Petkovsek, M.; Schmitzer, V.; Slatnar, A.; Stampar, F.; Veberic, R. Composition of sugars, organic acids, and total phenolics in 25 wild or cultivated berry species. *J. Food Sci.* **2012**, *77*, C1064–C1070. [CrossRef]

18. Sharma, S.; Narang, J.K.; Ali, J.; Baboota, S. Synergistic antioxidant action of vitamin E and rutin SNEDDS in ameliorating oxidative stress in a Parkinson's disease model. *Nanotechnol* **2016**, *27*, 375101. [CrossRef]

19. Fabre, G.; Bayach, I.; Berka, K.; Paloncýová, M.; Starok, M.; Rossi, C.; Duroux, J.L.; Otyepka, M.; Trouillas, P. Synergism of antioxidant action of vitamins E, C and quercetin is related to formation of molecular associations in biomembranes. *Chem. Commun.* **2015**, *51*, 7713–7716. [CrossRef]

20. García-Borrón, J.C.; Solano, F. Molecular anatomy of tyrosinase and its related proteins: Beyond the histidine-bound metal catalytic center. *Pigment Cell Res.* **2002**, *15*, 162–173. [CrossRef]

21. Mocan, A.; Schafberg, M.; Crisan, G.; Rohn, S. Determination of lignans and phenolic components of *Schisandra chinensis* (Turcz.) Baill. using HPLC-ESI-ToF-MS and HPLC-online TEAC: Contribution of individual components to overall antioxidant activity and comparison with traditional antioxidant assays. *J. Funct. Food* **2016**, *24*, 579–594. [CrossRef]

22. Zolghadri, S.; Bahrami, A.; Hassan Khan, M.T.; Munoz-Munoz, J.; Garcia-Molina, F.; Garcia-Canovas, F.; Saboury, A.A. A comprehensive review on tyrosinase inhibitors. *J. Enzyme Inhib. Med. Chem.* **2019**, *34*, 279–309. [CrossRef] [PubMed]

23. Dahech, I.; Farah, W.; Trigui, M.; Ben Hssouna, A.; Belghith, H.; Belghith, K.S.; Ben Abdallah, F. Antioxidant and antimicrobial activities of *Lycium shawii* fruits extract. *Int. J. Biol. Macromol.* **2013**, *60*, 328–333. [CrossRef] [PubMed]

24. Simonetti, G.; D'Auria, F.D.; Mulinacci, N.; Milella, R.A.; Antonacci, D.; Innocenti, M.; Pasqua, G. Phenolic content and in vitro antifungal activity of unripe grape extracts from agro-industrial wastes. *Nat. Prod. Res.* **2019**, *33*, 1–5. [CrossRef] [PubMed]

25. Simonetti, G.; Santamaria, A.R.; D'Auria, F.D.; Mulinacci, N.; Innocenti, M.; Cecchini, F.; Pericolini, E.; Gabrielli, E.; Panella, S.; Antonacci, D.; et al. Evaluation of anti-*Candida* activity of *Vitis vinifera* L. seed extracts obtained from wine and table cultivars. *BioMed Res. Int.* **2014**, *2014*, 127021. [CrossRef]

26. Simonetti, G.; Valletta, A.; Kolesova, O.; Pasqua, G. Plant products with antifungal activity: From field to biotechnology strategies. In *Natural Products as Source of Molecules with Therapeutic Potential*; Springer: Cham, Switzerland, 2018; pp. 35–71.

27. Can Baser, K.H. Biological and pharmacological activities of carvacrol and carvacrol bearing essential oils. *Curr. Pharm. Des.* **2008**, *14*, 3106–3119. [CrossRef]

28. Chaillot, J.; Tebbji, F.; Remmal, A.; Boone, C.; Brown, G.W.; Bellaoui, M.; Sellam, A. The monoterpene carvacrol generates endoplasmic reticulum stress in the pathogenic fungus *Candida albicans*. *Antimicrob. Agents Chemother.* **2015**, *59*, 4584–4592. [CrossRef]

29. Locatelli, M.; Ferrante, C.; Carradori, S.; Secci, D.; Leporini, L.; Chiavaroli, A.; Leone, S.; Recinella, L.; Orlando, G.; Martinotti, S.; et al. Optimization of aqueous extraction and biological activity of *Harpagophytum procumbens* root on *ex vivo* rat colon inflammatory model. *Phytother. Res.* **2017**, *31*, 937–944. [CrossRef]
30. Inbaraj, B.S.; Lu, H.; Kao, T.H.; Chen, B.H. Simultaneous determination of phenolic acids and flavonoids in *Lycium barbarum* Linnaeus by HPLC–DAD–ESI-MS. *J. Pharm. Biomed. Anal.* **2010**, *51*, 549–556. [CrossRef]
31. Mocan, A.; Crişan, G.; Vlase, L.; Crişan, O.; Vodnar, D.C.; Raita, O.; Gheldiu, A.M.; Toiu, A.; Oprean, R.; Tilea, I. Comparative studies on polyphenolic composition, antioxidant and antimicrobial activities of *Schisandra chinensis* leaves and fruits. *Molecules* **2014**, *19*, 15162–15179. [CrossRef]
32. Martins, N.; Barros, L.; Dueñas, M.; Santos-Buelga, C.; Ferreira, I.C.F.R. Characterization of phenolic compounds and antioxidant properties of *Glycyrrhiza glabra* L. rhizomes and roots. *RSC Adv.* **2015**, *5*, 26991–26997. [CrossRef]
33. Damiano, S.; Forino, M.; De, A.; Vitali, L.A.; Lupidi, G.; Taglialatela-Scafati, O. Antioxidant and antibiofilm activities of secondary metabolites from *Ziziphus jujuba* leaves used for infusion preparation. *Food Chem.* **2017**, *230*, 24–29. [CrossRef] [PubMed]
34. Masuda, T.; Yamashita, D.; Takeda, Y.; Yonemori, S. Screening for tyrosinase inhibitors among extracts of seashore plants and identification of potent inhibitors from *Garcinia subelliptica*. *Biosci. Biotechnol. Biochem.* **2005**, *69*, 197–201. [CrossRef] [PubMed]
35. Clinical and Laboratory Standards Institute. *Reference Method for Broth Dilution Antifungal Susceptibility Testing of Yeasts*, 4th ed.; CLSI document M27; CLSI: Wayne, PA, USA, 2017.
36. Le, K.; Chiu, F.; Ng, K. Identification and quantification of antioxidants in *Fructus lycii*. *Food Chem.* **2007**, *105*, 353–363. [CrossRef]
37. De Moura, C.; dos Reis, A.S.; da Silva, L.D.; de Lima, V.A.; Oldoni, T.L.C.; Pereira, C.; Carpes, S.T. Optimization of phenolic compounds extraction with antioxidant activity from açaí, blueberry and goji berry using response surface methodology. *Emir. J. Food Agr.* **2018**, *30*, 180–189.
38. Zhang, Q.; Chen, W.; Zhao, J.; Xi, W. Functional constituents and antioxidant activities of eight Chinese native goji genotypes. *Food Chem.* **2016**, *200*, 230–236. [CrossRef]
39. Dong, J.Z.; Lu, D.Y.; Wang, Y. Analysis of flavonoids from leaves of cultivated *Lycium barbarum* L. *Plant Foods Hum. Nutr.* **2009**, *64*, 199–204. [CrossRef]
40. Granato, D.; Shahidi, F.; Wrolstad, R.; Kilmartin, P.; Melton, L.D.; Hidalgo, F.J.; Miyashita, K.; Camp, J.V.; Alasalvar, C.; Ismail, A.B.; et al. Antioxidant activity, total phenolics and flavonoids contents: Should we ban in vitro screening methods? *Food Chem.* **2018**, *264*, 471–475. [CrossRef]
41. Ceccarini, M.R.; Vannini, S.; Cataldi, S.; Moretti, M.; Villarini, M.; Fioretti, B.; Albi, E.; Beccari, T.; Codini, M. *In vitro* protective effects of *Lycium barbarum* berries cultivated in Umbria (Italy) on human hepatocellular carcinoma cells. *Biomed Res. Int.* **2016**, *2016*, 7529521. [CrossRef]
42. Abeywickrama, G.; Debnath, S.C.; Ambigaipalan, P.; Shahidi, F. Phenolics of selected cranberry genotypes (*Vaccinium macrocarpon* Ait.) and their antioxidant efficacy. *J. Agric. Food Chem.* **2016**, *64*, 9342–9351. [CrossRef]
43. Abdennacer, B.; Karim, M.; Yassine, M.; Nesrine, R.; Mouna, D.; Mohamed, B. Determination of phytochemicals and antioxidant activity of methanol extracts obtained from the fruit and leaves of Tunisian *Lycium intricatum* Boiss. *Food Chem.* **2015**, *174*, 577–584. [CrossRef] [PubMed]
44. Shahidi, F.; Zhong, Y. Measurement of antioxidant activity. *J. Funct. Food* **2015**, *18*, 757–781. [CrossRef]
45. Orhan, D.D.; Ozçelik, B.; Ozgen, S.; Ergun, F. Antibacterial, antifungal, and antiviral activities of some flavonoids. *Microbiol. Res.* **2010**, *165*, 496–504. [CrossRef] [PubMed]

 *antioxidants*

Article

Exploring Target Genes Involved in the Effect of Quercetin on the Response to Oxidative Stress in *Caenorhabditis elegans*

Begoña Ayuda-Durán [1], Susana González-Manzano [1], Antonio Miranda-Vizuete [2], Eva Sánchez-Hernández [1], Marta R. Romero [3], Montserrat Dueñas [1], Celestino Santos-Buelga [1,*] and Ana M. González-Paramás [1]

[1] Grupo de Investigación en Polifenoles, Universidad de Salamanca, Campus Miguel de Unamuno, 37007 Salamanca, Spain; bego_ayuda@usal.es (B.A.-D.); susanagm@usal.es (S.G.-M.); evasanher@usal.es (E.S.-H.); mduenas@usal.es (M.D.); paramas@usal.es (A.M.G.-P.)
[2] Instituto de Biomedicina de Sevilla, Hospital Universitario Virgen del Rocío/CSIC/Universidad de Sevilla, 41013 Sevilla, Spain; amiranda-ibis@us.es
[3] Center for the Study of Liver and Gastrointestinal Diseases (CIBERehd), Experimental Hepatology and Drug Targeting (HEVEFARM), Universidad de Salamanca, Institute for Biomedical Research of Salamanca (IBSAL), 37007 Salamanca, Spain; marta.rodriguez@usal.es
* Correspondence: csb@usal.es

Received: 15 November 2019; Accepted: 20 November 2019; Published: 25 November 2019

Abstract: Quercetin is one the most abundant flavonoids in the human diet. Although it is well known that quercetin exhibits a range of biological activities, the mechanisms behind these activities remain unresolved. The aim of this work is to progress in the knowledge of the molecular mechanisms involved in the biological effects of quercetin using *Caenorhabditis elegans* as a model organism. With this aim, the nematode has been used to explore the ability of this flavonoid to modulate the insulin/insulin-like growth factor 1(IGF-1) signaling pathway (IIS) and the expression of some genes related to stress response. Different methodological approaches have been used, i.e., assays in knockout mutant worms, gene expression assessment by RT-qPCR, and *C. elegans* transgenic strains expressing green fluorescent protein (GFP) reporters. The results showed that the improvement of the oxidative stress resistance of *C. elegans* induced by quercetin could be explained, at least in part, by the modulation of the insulin signaling pathway, involving genes *age-1, akt-1, akt-2, daf-18, sgk-1, daf-2,* and *skn-1*. However, this effect could be independent of the transcription factors DAF-16 and HSF-1 that regulate this pathway. Moreover, quercetin was also able to increase expression of *hsp-16.2* in aged worms. This observation could be of particular interest to explain the effects of enhanced lifespan and greater resistance to stress induced by quercetin in *C. elegans*, since the expression of many heat shock proteins diminishes in aging worms.

Keywords: quercetin; IIS pathway; *C. elegans*; oxidative stress

---

1. Introduction

Quercetin (Q) is the most abundant flavonol in the human diet, being present in a wide variety of plant-derived foods, such as nuts, grapes, onions, broccoli, apples, or black tea. Dietary quercetin consumption has been associated to different health benefits, including antioxidant and anti-inflammatory effects and protection against aging-related diseases, such as cardiovascular pathologies, cancer, and neurodegenerative disorders [1–3]. Classically, these actions have been explained, at least in part, by its antioxidant and free-radical scavenging properties, as demonstrated in in vitro studies [4,5], although the actual mechanisms through which quercetin and other flavonoids exert their in vivo effects remain unresolved. Understanding the molecular mechanisms by which Q can exert its biological activity is important in order to develop strategies to modulate the physiological changes associated with aging that lead to chronic diseases. Although studies with quercetin and other flavonoids using in vivo models have increased in recent years, most of the knowledge on their biological activity still derives from in vitro or ex vivo findings, while assays that consider complex interactions of various processes, such as absorption, metabolism, and interaction with organs and tissues are more limited [6].

Some important molecular pathways in complex organisms can be explored using the model organism *Caenorhabditis elegans*. There is a high degree of homology between *C. elegans* and human genes involved in different processes, such as aging, apoptosis, cell signaling, cell polarity, metabolism, or cell cycle, which are conserved between mammals and the nematode [7]. In fact, *C. elegans* offers promising possibilities to study mechanisms of action and effects of secondary compounds of foods and plants, due to its simplicity of handling, conservation of metabolic pathways, and the possibility of manipulating cell signaling routes by biotechnological methods. In addition, it is not pathogenic, and no ethical boundaries exist to its experimental usage.

Different studies have evaluated the biological effects of quercetin and related compounds in *C. elegans*. In a previous work, we have shown that growing *C. elegans* in the presence of 200 μM of Q or its 3′- and 4′-*O*-methylated metabolites significantly prolonged the lifespan and increased the resistance against thermal- and juglone-induced oxidative stress [8]. Further support was obtained by checking the oxidation status of proteins in the nematode, which was greatly reduced [8]. Similar observations were also made by other authors [9–13]. Treatment with Q was also seen to lead to a reduction in the intracellular levels of reactive oxygen species (ROS), either in nematodes subjected to or not subjected to stress [11,13–15]. In general, these studies show that Q possesses a relevant biological activity in an in vivo system, which results in greater protection against oxidative processes. Regarding worm lifespan, an extension has been found in worms grown in the presence of concentrations of Q between 70 and 200 μM, while no lifespan extension was observed for Q concentrations outside that range. On the contrary, a hormetic response of the worm to this phytochemical has been demonstrated, so that the increase in the concentration of quercetin above certain levels results in decreased survival [16]. Nevertheless, the molecular mechanisms through which these effects are produced are still unclear, and different explanations, sometimes apparently contradictory, have been offered by distinct authors [9–14].

The response to oxidative stress in *C. elegans* is regulated through several pathways, including those of insulin/IGF-1 (IIS), c-Jun N-terminal kinases (JNK), and the signaling p38 MAPK (mitogen-activated protein kinases) pathway [17–19]. The IIS pathway (Figure 1) controls many important biological processes, including development, reproduction, metabolism, somatic maintenance, and stress resistance [20]. Insulin-like peptides binding to DAF-2, the orthologue of the insulin/IGF-1 receptor in *C. elegans* [21], activate its tyrosine kinase activity. This activation triggers a cascade of phosphorylation events through different kinases (AGE-1/PI3k, PDK-1, AKT-1/2, and SGK-1) that promote the phosphorylation-dependent cytoplasmic sequestration of the factors DAF-16/FoxO, HSF-1, and SKN-1/Nrf, preventing their transcriptional activity [22]. On the other hand, loss of insulin signaling in *C. elegans* results in several cytoprotective phenotypes resistant to both thermal and oxidative stress and also increases pathogen resistance and lifespan [23]. Flavonoids could influence cellular systems changing the expression of different genes through the modulation of distinct transcription factors, acting simultaneously on various signaling pathways, including the IIS pathway [24].

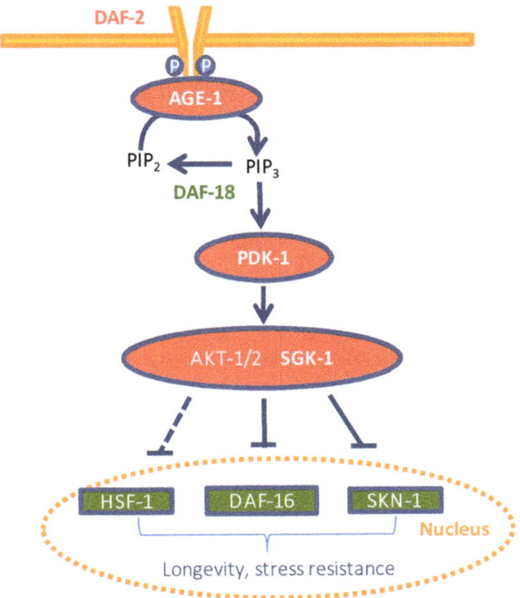

Figure 1. Scheme of the insulin/IGF-1 signaling pathway (IIS) in *C. elegans*. The components of the pathway that promote IIS are colored red and those that either antagonize IIS or are antagonized by IIS are colored green.

In order to contribute to elucidate the mechanisms involved in the effects of flavonoids, in this work, the ability of quercetin to modulate the insulin/IGF-1 signaling pathway (IIS) has been explored. Specifically, the influence on the resistance to thermally-induced oxidative stress has been assessed using *C. elegans* strains with loss-of-function mutations in genes of the IIS pathway (i.e., *daf-2*, *age-1*, *daf-16*, *akt-1*, *akt-2*; *sgk-1*, *hsf-1*, *skn-1*, and *daf-18*). Additionally, the ability of Q to modify the expression of some genes related to stress, namely *daf-16*, *hsf-1*, *skn-1*, *hsp-16.2*, *hsp-70*, *sod-3*, and *gst-4*, has also been determined by RT-qPCR or using the GFP fluorescent reporter in *C. elegans* transgenic strains.

2. Materials and Methods

2.1. Standards and Reagents

Quercetin (Q), ampicillin sodium salt, nistatine, agar, yeast extract, fluorodeoxyuridine (FUdR), phosphate-buffered saline (PBS), cholesterol, and 2-mercaptoethanol, were purchased from Sigma-Aldrich (Madrid, Spain). Dimethyl sulfoxide (DMSO) was obtained from Panreac (Barcelona, Spain). SYBR® SelectMaster Mix and high-capacity cDNA reverse transcription kit were from Applied Biosystems (Carlsbad, CA, USA), and the Illustra™ RNAspin mini isolation kit was from GE Healthcare (Buckinghamshare, UK).

2.2. Strains and Maintenance Conditions

The wild-type strain N2 and the mutant strains VC475, *hsp-16.2(gk249)* V; CB1270, *daf-2 (e1370)* III; TJ1052, *age-1(hx546)* II; CF1038, *daf-16(mu86)* I; CB1375, *daf-18(e1375)* IV; BQ1, *akt-1(mg306)* V; KQ1323, *akt-2(tm812) sgk-1(ft15)* X; PS3551, *hsf-1(sy441)* I; EU1, *skn-1(zu67)* IV/nT1(unc-?(n754)let-?) (IV;V); CF1553, *muIs84 ((Psod-3::gfp))*; TJ356, *zIs356 (Pdaf-16::daf-16::gfp; rol-6 (su1006))* IV; CL2166, *dvIs19 ((Pgst-4::gfp::NLS; rol-6 (su1006))* III; AM446, *rmIs223 (Phsp70::gfp; rol-6(su1006))*; CL2070, *dvIs70 (Phsp-16.2::gfp; rol-6 (su1006))*, as well as the *E. coli* OP50 bacterial strain, were obtained from the *Caenorhabditis* Genetics Center (CGC) at the University of Minnesota (Minneapolis, MN, USA). Worms were routinely propagated at 20 °C on nematode growth medium (NGM) plates with *E. coli* OP50 as a food source.

Synchronization of worm cultures was achieved by treating gravid hermaphrodites with bleach: 5N NaOH (2:1). Eggs are resistant whereas worms are dissolved in the bleach solution. The suspension was shaken with a vortex mixer during 1 min and kept for a further minute on rest, this process was repeated five times. The suspension was centrifuged (2 min, 9500 g). The pellet containing the eggs was washed six times with an equal volume of buffer M9 (3 g KH_2PO_4, 6 g Na_2HPO_4, 5 g NaCl, 1 mL 1M $MgSO_4$, H_2O to 1 L). Quercetin solution (200 mM) in DMSO was added to the nematode growth medium during its preparation to get a 200 µM final concentration on the plates. Control plates were also prepared without the flavonoid but containing the same volume of DMSO (0.1% DMSO, *v/v*). Around 100 to 300 µL of the M9 with eggs (depending on eggs Øconcentration) were transferred and incubated on NGM agar plates with or without Q. When the worms reached the L4 stage, they were transferred to new plates with or without Q but also containing FUdR at a concentration of 150 µM to prevent reproduction and progeny overgrowth. The worms were transferred every 2 days to fresh plates with FUdR for the different treatments (with or without Q) until they reached the day of the assay.

2.3. Stress Assays

Oxidative stress in worms was induced by subjecting the animals to 35 °C heat-shock treatment. Worms were incubated at 20 °C on NGM-*E. coli* OP50 plates with or without Q until days 2 and 9 of adulthood. Then they were transferred with a platinum wire to agar plates Ø 35 mm, 20 worms per plate) and switched to 35 °C for 4, 6, or 8 h. The time was decided depending on the thermotolerance of the specific worm strain used in the assay, which was previously checked. After that time, dead and alive nematodes were counted. In the studies involving the use of worm mutants, in addition to the mutant control, a parallel control using N2 wild-type (WT) worms was also included. For the assays, ten plates were used per treatment containing 20 worms per plate, resulting in a total of 200 worms, although a small percentage of worms was usually lost in the score. Only in the case of the *skn-1* mutant (EU1) were 300 worms (20 worms per plate/15 plates) used. According to the information supplied by the CGC, the work with this mutant required special considerations, as only the homozygote worms with WT appearance are considered for the assay, while the uncoordinated heterozygote worms are only employed to maintain the lineage. To have reproducible results in the heat shock assays, that is, to have the smallest possible temperature oscillations, some precautions were followed, in agreement with the guidelines stated in the reference paper [25]. During assays, the door of the incubator was only opened when the survival rate had to be measured. The plates were not stacked inside the incubator, they were placed in a row on the same shelf of the incubator, leaving enough space between them for the air to circulate properly. The temperature was controlled with the incubator's own thermostat and with an external thermostat with sensors inside the incubator to monitor possible temperature oscillations.

2.4. RT-qPCR Assays

Adult worms of the N2 *C. elegans* strain were treated with or without 200 µM of Q for 4 days. The worms were collected with M9 buffer, centrifuged at 10,000 g for 1 min, and the pellet was dissolved in 300 µL of M9, to which 3.5 µL of 2-mercaptoethanol was added. Total RNA was extracted using the RNAspin Mini RNA Isolation Kit (GE Healthcare). In order to maximize cell breakage, in the first stage of the extraction, 10 stainless-steel beads (2 mm) were added. The mixture was vortex shaken vigorously and further homogenized in a Thermo Savant FastPrep 120 Cell Disrupter System, with a speed of 5.5 m/s and run time duration of 10 s, five times. cDNA was produced with high-capacity cDNA reverse transcription kit (Applied Biosystems) using 2 µg of total RNA per reaction. The expression of mRNA was assessed by quantitative real-time PCR, using SYBR green as the detection method. The gene expression data were analyzed using the comparative $2^{-\Delta\Delta Ct}$ method, with *act-1* as the normalizer [26]. Nine independent experiments were performed. *Act-1* was used as a normalizer both in the assays carried out in non-stressed and stressed worms. The information related to gene-specific primers used in this work can be found in the Supplementary Table S1.

2.5. Fluorescence Quantification and Visualization

Synchronized L1 larvae expressing an inducible green fluorescent protein (GFP) reporter for *gst-4*, *hsp-16.2*, *hsp-70*, *sod-3*, and *daf-16* genes were grown on NMG plates in the presence or absence of Q until the day of the assay, when they were subjected to or not subjected to thermally-induced oxidative stress (35 °C, 1 h). The precise day of assay was defined when a higher intensity of the fluorescence was observed after carrying out a screening with the different strains throughout the life of the worm, namely, day 3 in *gst-4* and day 5 in *hsp-70*. For the remaining strains, as no clear increase in the fluorescence was observed, the assessment was made in young (day 2 of adulthood) and older adult worms (day 9 of adulthood). In the cases of *hsp-16.2* and *hsp-70* reporter strains, after thermal stress, the worms were allowed to recover at 20 °C for 2 or 3 h respectively, before pictures were taken. The expression of *gst-4*, *hsp-16.2*, *hsp-70*, and *sod-3* was measured by quantifying the fluorescence of the GFP reporter. To analyze the subcellular localization of the DAF-16::GFP reporter, worms were classified as diffuse cytoplasmic, intermediate cytoplasmic/nuclear, and strong nuclear translocation. Approximately 35 randomly selected worms for each experiment were mounted in a 5 µL drop of 10 mM levamisole (except for DAF-16::GFP in 2% sodium azide) on a 3% agarose pad covered with a coverslip. All fluorescence determinations were done in an Olympus BX61 fluorescence microscope equipped with a filter set (excitation 470 ± 20 mn, emission 500 ± 20 nm) and a DP72 digital camera coupled to CellSens Software for image acquisition and analysis. ImageJ software was used to quantify fluorescence intensity. Three independent experiments were performed per assay and reporter strain.

2.6. Statistical Analysis

The statistical analyses were performed using the PC software package SPSS (version 23.0; SPSS Inc., Chicago, IL, USA). Analysis of variance (ANOVA) was applied for multiple comparisons of values to determine possible significant differences between treated and control groups. To analyze survival to thermal stress, contingency tables were prepared, and statistical significance was calculated using the Chi Square Test.

3. Results and Discussion

3.1. Assays in Wild Type and Mutant Worms

An enhancement in the survival was observed in N2 wild-type worms treated with Q 200 µM after being subjected to thermal stress compared with non-treated controls. Specifically, in the assays carried out on the second day of adulthood, the average proportion of living worms after stress was 64.78% in the control group and 81.11% in Q-treated worms ($p = 0.000$), while on the ninth day of adulthood, the survival rate was 35.1% in untreated worms and 47.2% in treated animals ($p = 0.001$) (Figure 2).

The molecular mechanisms involved in this enhancement of thermotolerance were explored, checking the ability of Q to modulate the stress resistance in *C. elegans* mutant strains. Initially, the effect of Q on thermal stress resistance was evaluated in worms carrying loss of function mutations in genes of the IIS pathway, namely *age-1(hx546)*, *akt-1(mg306)*, *daf-2(e1370)*, and the double mutant *akt-2(tm812); sgk-1(ft15)*, all of them long-lived and with greater resistance to stress than the wild-type strain. Similar to the wild-type, mutant worms were subjected to a thermal shock (35 °C, 6–8 h) on days 2 and 9 of adulthood. The obtained results showed that the stress resistance was not increased in any of the mutant worms treated with quercetin at either day 2 or 9 (Figure 3). These results suggested that *age-1*, *akt-1*, *akt-2*, *sgk-1*, and *daf-2* were necessary to mediate stress resistance induced by Q in *C. elegans*.

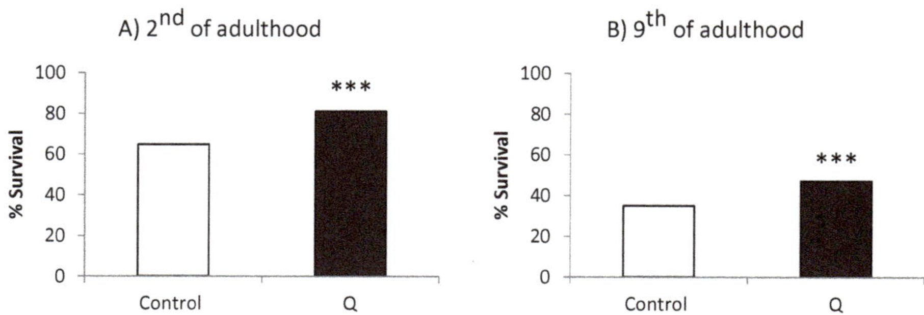

Figure 2. Percentages of survival following thermal stress (35 °C, 8 h) applied at days (**A**) 2 and (**B**) 9 of adulthood in N2 wild-type *C. elegans* strain not treated (controls) and treated with Q (200 µM). Statistical significance was calculated using the Chi Square Test (200–300 individuals per assay in both controls and treated worms). The asterisks (***) indicate significant differences at $p < 0.001$.

Pietsch et al. [11] also studied the effect of Q in some of these mutants. As in our case, they did not find an improvement in the survival after thermal stress in *daf-2* and *age-1* mutants treated with 200 µM quercetin, concluding that that those genes were required to mediate the stress-protective effects of the flavonol. However, they found that the treatment with Q unexpectedly produced an improvement in the resistance to stress in the *akt-2* mutant, a gene located downstream of DAF-2 and AGE-1 in the IIS pathway. They proposed that this could be explained because SGK-1 was more important for resistance to stress than the AKT kinases [11]. In the present study, no increase in the survival after thermal stress was observed in the double-mutant *akt-2;sgk-1* treated with Q (Figure 3), suggesting that *sgk-1* could actually be necessary to improve the resistance to stress by quercetin. Similarly, no changes in the survival after stress were produced in Q-treated *akt-1* mutants, a gene that encodes an ortholog of serine/threonine kinase AKT/PKB and interacts with the IIS pathway. This indicates that the kinase *akt-1* was also necessary for the improvement in the resistance to stress in *C. elegans*.

The resistance to thermal stress in response to Q was also studied in *daf-16, hsf-1, skn-1, daf-18*, and *hsp-16.2* mutant worms. The transcription factors DAF-16, HSF-1, and SKN-1 produce changes in the expression of several genes in response to a reduced IIS pathway. The genes regulated by these transcription factors are functionally relevant, including stress response genes, such as catalases, glutathione-S-transferases, metallothioneins, and genes that encode antimicrobial peptides, chaperones like *hsp-16.2*, apolipoproteins, and lipases [27]. As it can observed in Figure 4, the treatment with Q did not improve the survival in *daf-18, skn-1*, and *hsp-16.2* mutants exposed to thermal stress, suggesting that these genes would be necessary to explain the effect of Q in worm resistance against stress. However, the treatment with Q continued to produce an improvement in the resistance to thermal stress in *daf-16* and *hsf-1* mutants, indicating that the effect of Q in *C. elegans* was independent of these genes. Similar results on the influence of Q were obtained in young and older adults in the studied mutants (Figure 4).

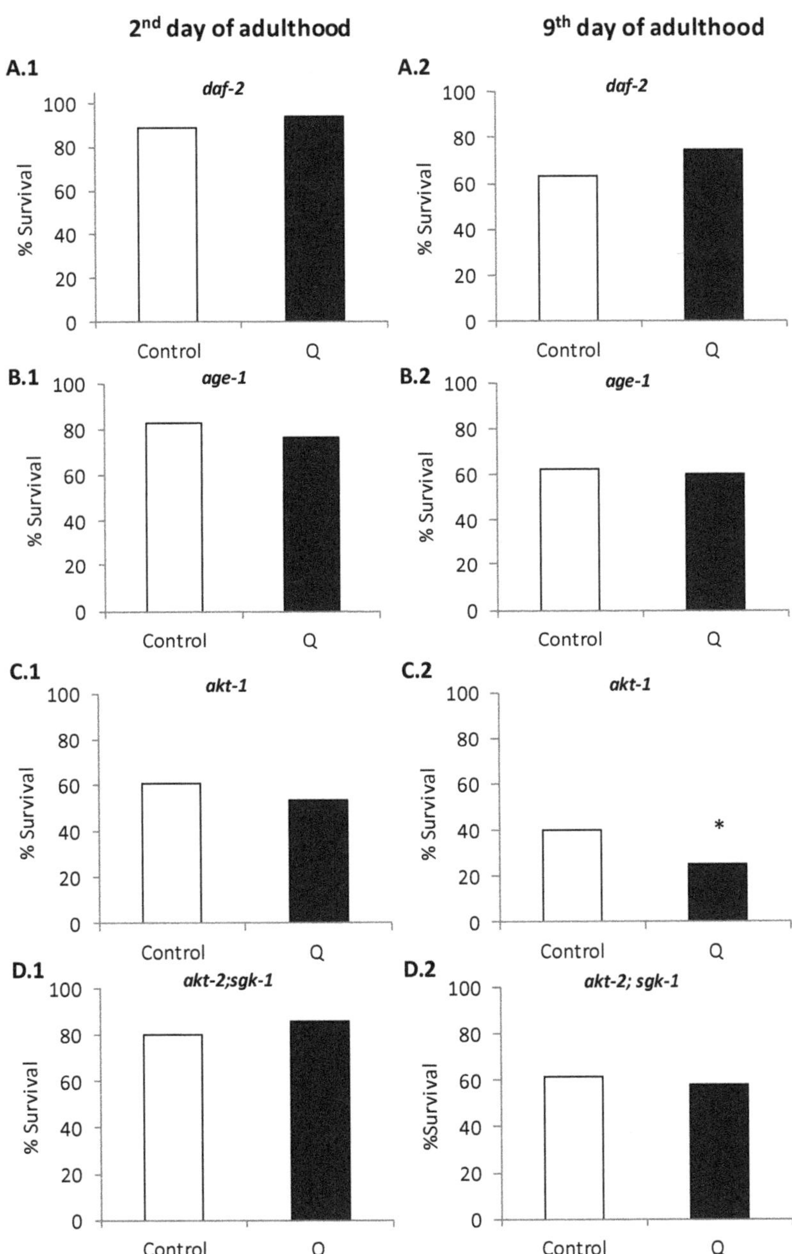

Figure 3. Percentages of survival following thermal stress (35 °C, 8 h) applied at days 2 (**A–D.1**) and 9 (**A–D.2**) of adulthood in different long-lived *C. elegans* mutants from the IIS pathway: *daf-2(e1370)* (**A**), *age-1(hx546)* (**B**), *akt-1(mg306)* and *akt-2(tm812)* (**C**); *sgk-1(ft15)* (**D**) cultivated in the absence (controls) and presence of Q (200 μM). Statistical significance was calculated using the Chi Square Test. In both controls and treated worms, ten plates were used per assay containing 20 worms per plate (i.e., 200 worms in total). The asterisk (*) indicates significant differences at $p < 0.05$.

2nd day of adulthood   9th day of adulthood

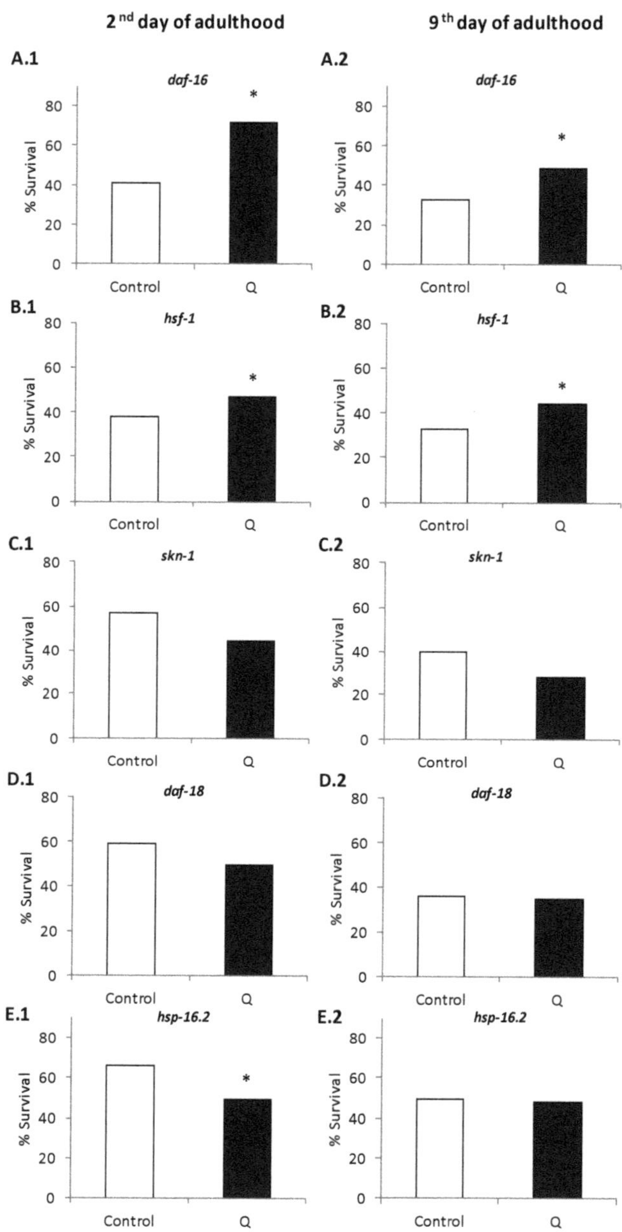

Figure 4. Percentages of survival following thermal stress (35 °C, 6 h; except for *hsp-16.2* (35 °C, 4 h)) applied at days 2 (**A–E.1**) and 9 (**A–E.2**) of adulthood in *daf-16 (mu86)* (**A**), *hsf-1(sy441)* (**B**), *skn-1(zu67)* (**C**), *daf-18(e1375)* (**D**) *and hsp-16.2(gk249)* (**E**) mutants cultivated in the absence (controls) and presence of Q (200 µM). Statistical significance was calculated using the Chi Square Test. In both controls and treated worms, ten plates were used per assay containing 20 worms per plate (i.e., 200 worms in total), but in the case of the *skn-1* mutant, 300 worms (20 worms per plate/15 plates) were used. The asterisk (*) indicates significant differences at $p < 0.05$ on comparing with control.

3.2. q-RT-PCR Analyses

The influence of quercetin on some transcription factors and target genes of the IIS pathway was also explored by quantification of the expression of *daf-16, hsf-1, skn-1, daf-18,* and *hsf-16.2* by RT-qPCR in wild type worms, both submitted and not submitted to thermal stress (5 h, 35 °C) after growing 4 days in the presence of the flavonol. As it can be seen in Figure 5, the treatment with Q did not modify the expression of *daf-18* either after subjecting or not subjecting the worms to thermal stress. However, the results previously obtained with the mutants (Figure 4) indicated that *daf-18* could be involved in the protective effects of quercetin against stress. A possible explanation to this apparent contradiction could be that the result obtained in the *daf-18* mutant is rather reflecting the involvement of the protein kinase AGE-1, as the DAF-18/phosphatase and tensin homolog (PTEN)protein is responsible for dephosphorylating and inhibiting AGE-1/PI3K, counteracting its activity [21]. Another possibility is that DAF-18 was regulated by quercetin at the post-transcriptional or activity level.

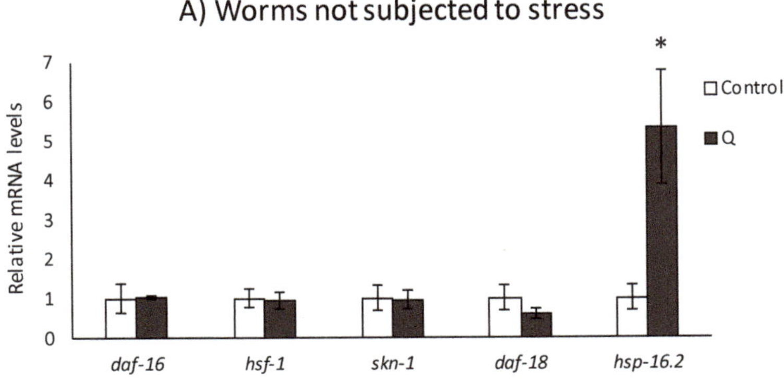

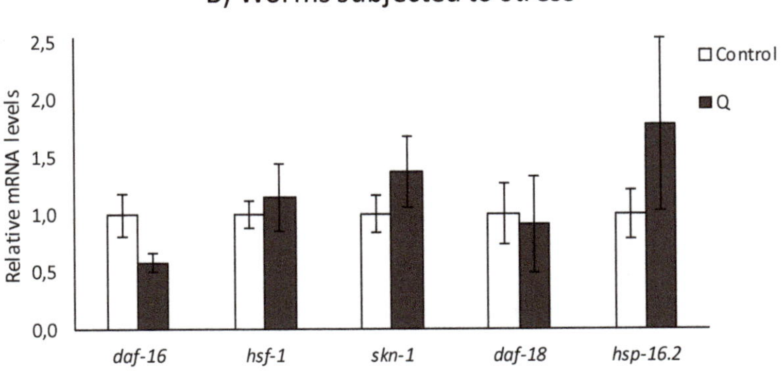

Figure 5. Effect of Q on the expression of *daf-16, hsf-1, skn-1, daf-18,* and *hsp-16.2* genes in N2 *C. elegans* cultivated in the absence (controls) and presence of Q (200 µM) (**A**) under non-stressed conditions or (**B**) after subjecting them to thermal stress. The expression level was determined by RT-qPCR. *act-1* was used as housekeeping control. Nine independent experiments were performed. The results are presented as the mean values ± Standard Error of the Mean (SEM). Statistical significance was calculated using by one-way analysis of variance (ANOVA). The asterisk (*) indicates significant differences at $p < 0.05$.

Similarly, the expression of *hsf-1* was not modified by the treatment with Q, neither in normal growing conditions nor after application of thermal stress (35 °C, 5 h) (Figure 5). This result would confirm that the heat transcription factor *hsf-1* is not involved in the effects of Q on stress resistance, supporting the above-described observations on *hsf-1* mutants (Figure 4). In a previous study on *hsf-1* loss of function mutants, Fitzenberger et al. [28] also found that this gene was not required to explain the ability of Q to prevent the glucose-induced reduction of survival in *C. elegans*.

The treatment with Q did not produce changes in the expression of *daf-16* (Figure 5). Together with the findings obtained in the assays with mutants, these results seem to confirm that the effect of Q on worm stress resistance is independent of *daf-16*. Other authors had already studied the involvement of *daf-16* in the effects of Q using knockout worms [10,11,13], finding that Q continued improving longevity and resistance to thermal and hydrogen peroxide-induced oxidative stress in *daf-16* mutants, which suggest that this gene was not essential for the effects of quercetin, observations that are consistent with the results obtained herein, either in worms subjected or not subjected to stress.

SKN-1 is the homologue of Nrf-2 transcription factor, which regulates oxidative stress response and lifespan, mobilizing the conserved phase 2 detoxification response [29]. No differences were found in the expression of *skn-1* between worms treated or not treated with Q and not subjected to stress (Figure 5A), which seems in agreement with the observations of Pietsch et al. [11], who reported that quercetin induced an increase in the lifespan of *skn-1* mutants under normal growth conditions, indicating that the effects of this flavonoid in the absence of stress are independent of that gene. However, the results obtained with the *skn-1* mutants (Figure 4) indicated that this gene is a mediator in the protective effects of Q against stress. Tullet et al. [29,30] suggested that SKN-1 is a transcriptional co-regulator of DAF-16 regarding resistance to oxidative stress and the expression of detoxification genes in response to a reduced IIS signal, but it extends worm half-life independently of DAF-16. This dual function of *skn-1* might be coherent with the results observed herein, where Q did not change the *skn-1* expression in absence of stress, although this gene seems to be necessary in the improvement of resistance to thermal stress mediated by Q.

The influence of Q on thermal shock proteins HSP-16.2, whose expression is influenced by the IIS, was also studied. As shown in Figure 5, the treatment with Q induced an increase in the expression of *hsp-16.2*, either without or with stress. However, the differences were more noticeable in the absence of stress, which could possibly be explained by the already strong induction of HSP-16.2 caused by the thermal stress, which would make the effect induced by Q treatment less relevant. Together with the results in the *hsp-16.2* mutants, where a decrease in the resistance to thermal stress was observed in quercetin-treated worms (Figure 4), it appears that thermal shock proteins can be involved in the protective effects of the flavonol against stress.

All in all, the obtained results showed that, at least in part, the IIS pathway would be involved in the improvement to thermotolerance induced by Q, entailing the genes *age-1, akt-1, akt-2, sgk-1, daf-2, daf-18, skn-1*, and *hsp-16.2*, but independent of *daf-16* and *hsf-1*. In addition, the involvement of these genes was not modified by the age of the worm. Previous studies by our group with epicatechin also showed that the enhanced stress resistance induced by this flavan-3-ol in *C. elegans* was also mediated by the IIS pathway, although it did not necessarily involve the same genes, as in that case the expression levels of the main transcription factors of the pathway (*daf-16, skn-1*, and *hsf-1*) were modified by the compound [31]. Actually, even compounds belonging to the same flavonoid class seem to act through different mechanisms. Thus, the lifespan extension produced by quercetin-3-*O*-diglucoside in *C. elegans* was explained by upregulation of the genes *daf-2, old-1, osr-1*, and *sek-1*, whereas no modification was produced in the expression of *daf-16, age-1*, and *sir-2.1* [12]. However, assays with a flavonol-rich extract obtained from *Baccharis trimera* concluded that the improvement in the stress resistance was independent of several stress-related signaling pathways (p38, JNK, and ERK) and transcription factors SKN-1 and DAF-16 [32].

3.3. Assays with Fluorescent Reporters

The insulin signaling pathway transmits signals in response to the environmental conditions that could change the expression of different genes related to stress and longevity, thus regulating important processes such as aging, metabolism, or dauer formation. The expression of some such genes, namely those encoding heat shock proteins (*hsp-16.2* and *hsp-70*) and antioxidant enzymes (*sod-3* and *gst-4*), was assessed in order to gain further insight into the mechanisms of action underlying the effects of Q on the modulation of lifespan and stress in *C. elegans*. With this aim, transgenic strains that express GFP under the control of *gst-4*, *sod-3*, *hsp-16.2*, and *hsp-70* promoters were employed. GFP expression levels were analyzed in animals grown in the presence or absence of quercetin under non-stress conditions for *Pgst-4::gfp* and *Psod-3::gfp* reporters, whereas for *Phsp-70::gfp* and *Phsp-16.2::gfp* reporters, worms were previously submitted to a heat shock (35 °C, 1 h) and further allowed to recover for 3 h (*hsp-70*) or 2 h (*hsp-16.2*) at 20 °C.

As it can be seen in Figure 6, the treatment with Q did not produce an increase in the expression of any of the studied genes (*gst-4*, *sod-3*, *hsp-16.2*, and *hsp-70*) in young worms of reproductive age. Indeed, there was even a decrease in the expression of *hsp-70* in the worms treated with Q ($p = 0.018$). The expression of *hsp-16.2* and *sod-3* was also studied using reporter strains in older worms (day 9), in order to establish if the mechanism of action underlying the effects of Q was dependent on the age of the worm. Similar results were obtained regarding the expression of *sod-3* (Figure 7) as compared to younger worms (Figure 6), finding no differences in the expression by the treatment with Q. However, the expression of *hsp-16.2* was increased by the treatment with Q at day 9 ($p = 0.008$) (Figure 7), which was not observed at day 2 (Figure 6).

Some of these genes have already been studied by other authors using fluorescent reporters to explore worm response to Q, specifically *gst-4* and *sod-3*. Opposed results were reported regarding the effects of Q on *sod-3*. Whereas, Kampkötter et al. [9] observed a decrease in the expression of *sod-3* in worms exposed to 100 µM Q, Grünz et al. [13] found that growing the worms in the presence of 100 µM of Q produced a significant increase in *sod-3* expression. As for the present study, no significant changes were detected in the expression of *sod-3* after treatment with 200 µM of Q in any of the two days studied. On the other hand, Kampkötter et al. [14] observed that the expression of *gst-4* was not modified by Q 100 µM under normal growth conditions, although it decreased the expression of *gst-4* when worms were subjected to oxidative stress with juglone 20 µM [14]. Those results would be in agreement with the observations made herein, where no increase in the expression of *gst-4* was found under normal growth conditions (200 µM Q).

The results for *Phsp-16::gfp* obtained at different ages of the worms could help to understand the effects of increased lifespan and improvement of resistance to stress induced by Q in *C. elegans*, since many heat shock proteins are regulated positively at the beginning of adult life to further decrease throughout life [33]. The obtained result might indicate that this decrease could be reversed by the treatment with Q. It is pertinent to indicate that in previous studies on longevity, greater survival started to be observed from approximately day 8 onwards in worms treated with quercetin in relation to non-treated worms [8], suggesting changes in *C. elegans* metabolism favored by prolonged exposure to the flavonol, among which, the upregulation of some heat shock proteins could be involved. Actually, the increased lifespan and thermotolerance observed in certain mutants, such as the long-lived *age-1*, have been explained by an increase in the regulation and accumulation of HSP-16 [34]. The involvement of *hsp-16.2* was also supported by the increase in the expression of *hsp-16.2* found in the RT-qPCR studies (Figure 5) and the loss of the improvement in resistance to thermal stress in the *hsp-16.2* mutant (Figure 4). On the other hand, the results showed a decrease in the reporter of *hsp-70* in worms treated with Q (Figure 6). Differences in the expression of genes encoding distinct heat shock proteins, were also found by Pietsch et al. [15], observing that Q produced an increase in the expression of *hsp-3*, *hsp-12.6*, *hsp-16.1*, and *hsp-16.41*, but a decrease of *hsp-70* and *hsp-17*.

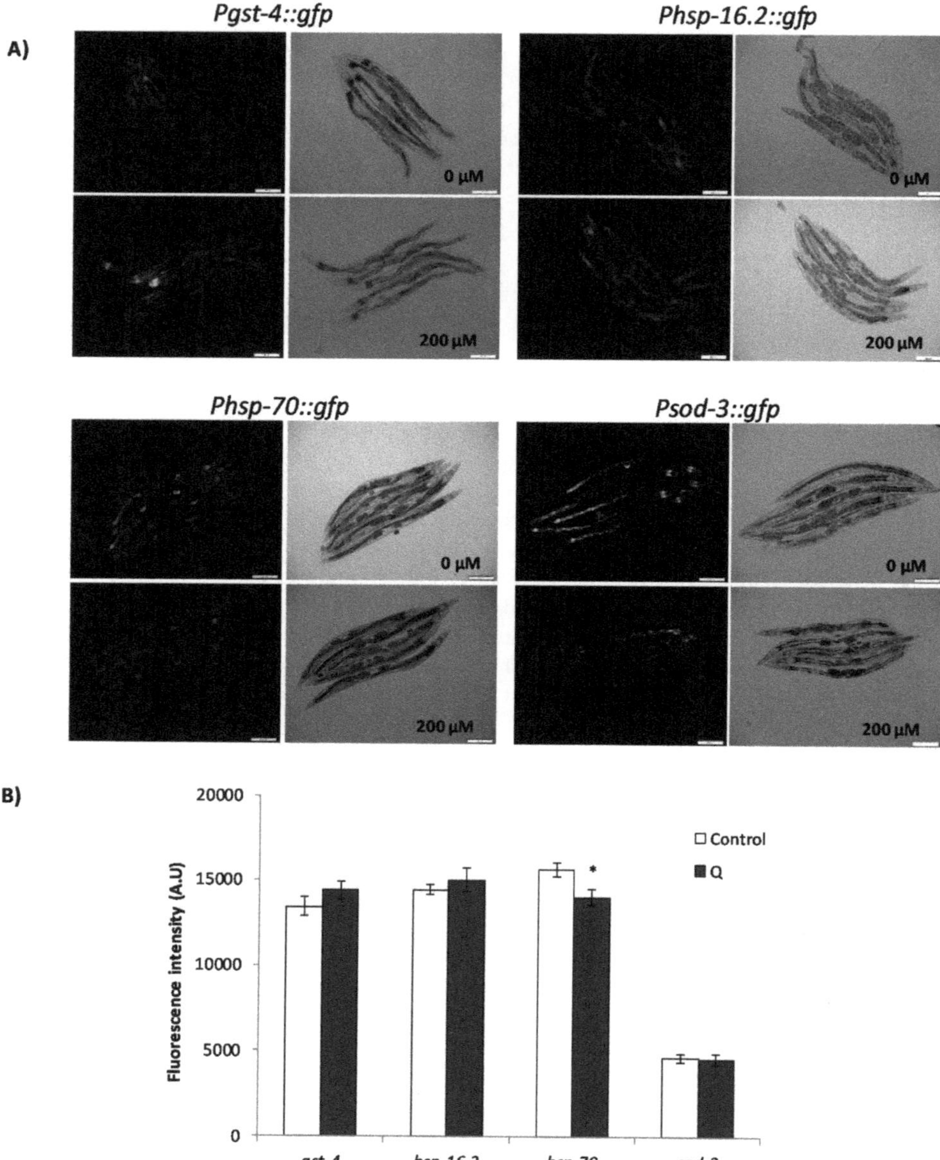

Figure 6. Effect of Q on the expression of *gst-4*, *hsp-16.2*, *hsp-70*, and *sod-3* after cultivation of *C. elegans* in the absence and presence of Q (200 μM). (**A**) Representative fluorescence images of control and Q-treated worm strains. (**B**) Quantification of the relative fluorescence intensities of transgenic worms. Total green fluorescent protein (GFP) fluorescence of each whole worm was quantified using Image J software. Three independent experiments were performed. The results are presented as the mean values ± SEM. Approximately 35 randomly selected worms from each set of experiments were examined. Differences compared with the control (0 μM Q, 0.1% dimethyl sulfoxide (DMSO)) were considered statistically significant at * ($p < 0.05$) by one-way ANOVA.

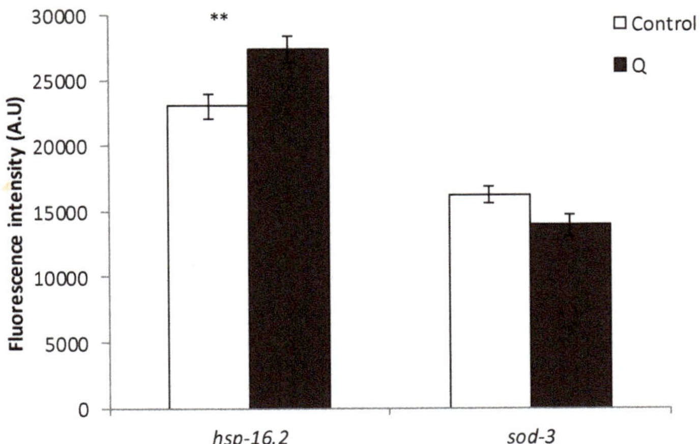

Figure 7. Effect of Q on the expression of *hsp-16.2* and *sod-3* in old worms (day 9 of adulthood) after cultivation of *C. elegans* in the absence and presence of Q (200 µM). Total GFP fluorescence of each whole worm was quantified using Image J software. Three independent experiments were performed. The results are presented as the mean values ± SEM. Approximately 35 randomly selected worms from each set of experiments were examined. Differences compared with the control (0 µM Q, 0.1% DMSO) were considered statistically significant at ** ($p < 0.01$).

The modulation of the subcellular localization of the DAF-16 forkhead transcription factor was examined using a transgenic strain expressing a fusion protein DAF-16::GFP. The treatment with Q did not affect the translocation of DAF-16 to the nucleus with respect to the control worms, neither under normal growth conditions nor after thermal stress (Figure 8). The localization of DAF-16 was also studied in older worms (day 9), obtaining similar results that at day 2 regarding its subcellular localization.

As discussed above, the treatment with Q did not produce an increase in the expression of *daf-16* in any of the assayed conditions, while it led to an increase in the resistance to thermal stress on *daf-16* mutant strains. All these results indicate that the effect of this flavonol in the improvement of worm resistance to stress is independent of *daf-16*. A similar conclusion has been obtained by some authors [10,11], although others observed greater translocation of DAF-16 from the cytosol to the nucleus following Q treatment [9,13,14]. In this respect, Saul et al. [10] suggested that the translocation of DAF-16 to the nucleus in response to quercetin could be more of a circumstantial effect than proof of an underlying longevity mechanism. In fact, although DAF-16 is a key factor in the control of stress response and longevity [17,35,36], it has also been pointed out that its translocation to the nucleus does not guarantee a longer lifespan, suggesting that DAF-16 would not necessarily act only in the regulation of longevity [37,38].

The finding that the effects of Q can be independent of *daf-16* but dependent on *daf-2* and other components of the IIS, such as *akt-1*, *age-1*, and *akt-2/sgk-1*, could appear surprising. Nevertheless, Hekimi et al. [37] and Lin et al. [39] pointed out that when the IIS pathway is altered in response to different environmental signals, besides the stress resistance genes, other unidentified signals are regulated by *daf-2*, which are not dependent on *daf-16*, but that are also essential for the extended longevity in *daf-2* mutants.

In the end, the obtained results indicated that two key transcription factors that could be involved in the expression of heat shock proteins, DAF-16 and HSF-1 [40], were not related with the effects exerted by Q on *C. elegans* (Figures 4 and 5). Actually, it is described that HSF-1 is not a key factor for all HSPs [41]. Mertenskötter et al. [42] showed that in the expression of chaperone genes, protein biosynthesis and protein degradation was positively influenced by the MAPK pathway and established the importance of this pathway in heat stress responses, possibly by a PMK-1-mediated activation of the transcription factor SKN-1 in *C. elegans*. Other authors have also found that components of related pathways, such as UNC-43, SEK-1, and OSR-1 are involved in the molecular mechanisms of the response to quercetin and other polyphenols [11,43,44]. Thus, the MAPK pathways could also be implicated in the effects of quercetin, which might contribute to explain the role of SKN-1 and the activation of certain HSPs observed in the present study.

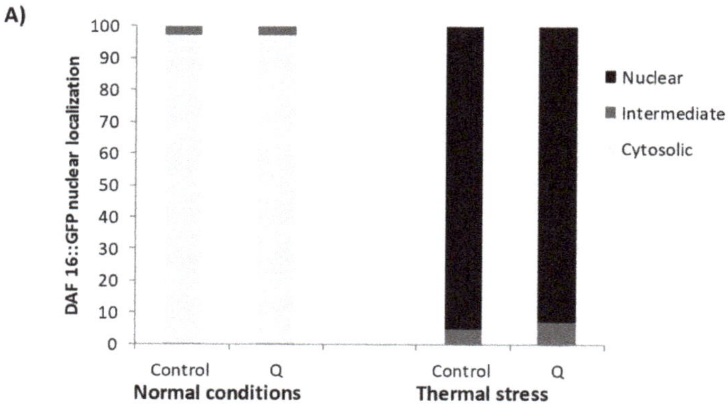

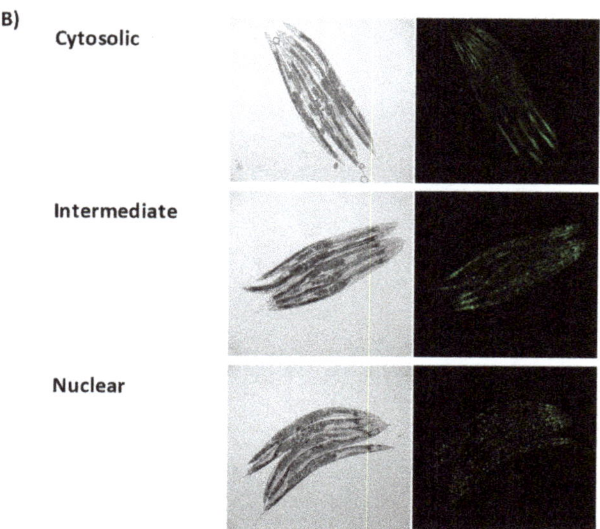

Figure 8. (**A**) Effect of Q on DAF-16::GFP subcellular distribution and (**B**) representative pictures of the subcellular location of DAF-16::GFP, i.e., cytosolic, intermediate and nuclear. Transgenic worms expressing the fusion protein DAF-16::GFP were cultivated in the absence (controls) and presence of Q (200 µM) after subjecting or not subjecting the worms to thermal stress and evaluated at day 2 of adulthood.

4. Conclusions

The enhancement of the oxidative stress resistance of *C. elegans* induced by Q could be explained, at least in part, by the modulation of the insulin signaling pathway, namely the genes *age-1, akt-1, akt-2, daf-18, sgk-1, daf-2, skn-1,* and *hsp-16.2.* However, this effect would be independent of *daf-16* and *hsf-1.* The implication of *daf-2* and not of *daf-16* in the observed effects seems to reinforce the idea that there are signals regulated by *daf-2* independent of *daf-16* that are also essential for the extension of longevity and resistance to stress.

The expression of *skn-1* was not changed by the treatment with Q when worms were submitted to stress, but the results obtained in the *skn-1* mutants pointed out that it was necessary to mediate the resistance to thermal stress. It has been reported that SKN-1 promotes longevity by a different mechanism to the protection against oxidative damage [29,30], and also that the increase of lifespan by Q was maintained in *skn-1* mutants [11]. This could explain why quercetin does not alter *skn-1* expression under normal conditions, while this gene seems involved in the increased resistance to stress induced by Q.

The studies with transgenic strains showed that Q did not produce an increase in the expression of the antioxidant enzymes GST-4 and SOD-3, nor of the heat shock proteins HSP-16.2 and HSP-70 in young worms, whereas an increase was produced in the expression of HSP-16.2 in aged worms. This observation could be important to explain, at least in part, the effects of enhanced lifespan and greater resistance to stress produced by Q in *C. elegans*, since the expression of many heat shock proteins diminishes throughout worm life, a process that might be counteracted by this flavonol.

In summary, the network of signaling pathways that could be modulated by quercetin and other flavonoids is diverse and complex, and the molecular mechanisms of action can vary depending on the compound. For instance, while DAF-16 and HSF-1 transcription factors were previously demonstrated to be involved in the effects of epicatechin [31], they do not seem to be required in the stress resistance effects of Q. In this work, just the insulin signaling pathway has been explored, but other routes are surely involved, which should be considered in future studies.

Supplementary Materials: The following are available online at http://www.mdpi.com/2076-3921/8/12/585/s1, Table S1: Oligonucleotide sequence of primers used to determine the expression levels of *C. elegans* genes by RT-QPCR.

Author Contributions: Conceptualization, C.S.-B. and A.M.G.-P.; Methodology B.A.-D., S.G.-M. and M.R.R.; Formal Analysis, B.A.-D., E.S.-H., M.D. and S.G.-M.; Resources, C.S.-B., A.M.-V, M.R.R. and A.M.G.-P.; Writing—Original Draft Preparation, B.A.-D. and C.S.-B.; Writing—Review and Editing, B.A.-D., C.S.-B., A.M.G.-P. and A.M.-V.; Supervision, C.S.-B., A.M.G.-P. and A.M.-V.; Project Administration, C.S.-B., A.M.G.-P. and A.M.-V.; Funding Acquisition, C.S.-B., A.M.G.-P. and A.M.-V.

Funding: This research was funded by the Spanish Ministerio de Economía y Competitividad (MINECO Projects AGL2015-64522-C2 and BFU2015-64408-P) and Fondo Europeo de Desarrollo Regional (FEDER)-Interreg España-Portugal Programme (Project ref. 0377_IBERPHENOL_6_E).

Acknowledgments: The authors are thankful to Francisco J. Martín-Vallejo for statistical advice.

Conflicts of Interest: The authors declare no conflict of interest.

References

1. Kaliora, A.C.; Dedoussis, G.V. Natural antioxidant compounds in risk factors for CVD. *Pharm. Res.* **2007**, *56*, 99–109. [CrossRef] [PubMed]

2. Boots, A.W.; Haenen, G.R.; Bast, A. Health effects of quercetin: From antioxidant to nutraceutical. *Eur. J. Pharm.* **2008**, *585*, 325–337. [CrossRef] [PubMed]

3. Darband, S.G.; Kaviani, M.; Yousefi, B.; Sadighparvar, S.; Pakdel, F.G.; Attari, J.A.; Mohebbi, I.; Naderi, S.; Majidinia, M. Quercetin: A functional dietary flavonoid with potential chemo-preventive properties in colorectal cancer. *J. Cell. Physiol.* **2018**, *233*, 6544–6560. [CrossRef] [PubMed]

4. Cos, P.; Ying, L.; Calomme, M.; Hu, J.P.; Cimanga, K.; Van Poel, B.; Pieters, L.; Vlietinck, A.J.; Vanden Berghe, D. Structure-activity relationship and classification of flavonoids as inhibitors of xanthine oxidase and superoxide scavengers. *J. Nat. Prod.* **1998**, *61*, 71–76. [CrossRef] [PubMed]

5. Middleton, E.; Kandaswami, C.; Theoharides, T.C. The effects of plant flavonoids on mammalian cells: Implications for inflammation, heart disease, and cancer. *Pharm. Rev.* **2000**, *52*, 673–751. [PubMed]

6. González-Paramás, A.M.; Ayuda-Durán, B.; Martínez, S.; González-Manzano, S.; Santos-Buelga, D. The Mechanisms Behind the Biological Activity of Flavonoids. *Curr. Med. Chem.* **2018**, *26*, 1–14. [CrossRef]

7. Kyriakakis, E.; Markaki, M.; Tavernarakis, N. *Caenorhabditis elegans* as a model for cancer research. *Mol. Cell. Oncol.* **2015**, *2*, e975027. [CrossRef]

8. Surco-Laos, F.; Cabello, J.; Gómez-Orte, E.; González-Manzano, S.; González-Paramás, A.M.; Santos-Buelga, C.; Dueñas, M. Effects of *O*-methylated metabolites of quercetin on oxidative stress, thermotolerance, lifespan and bioavailability on *Caenorhabditis elegans*. *Food Funct.* **2011**, *2*, 445–456. [CrossRef]

9. Kampkötter, A.; Timpel, C.; Zurawski, R.F.; Ruhl, S.; Chovolou, Y.; Proksch, P.; Wätjen, W. Increase of stress resistance and lifespan of *Caenorhabditis elegans* by quercetin. Comparative Biochemistry and Physiology. *Comp. Biochem. Part B* **2008**, *149*, 314–323. [CrossRef]

10. Saul, N.; Pietsch, K.; Menzel, R.; Steinberg, C.E. Quercetin-mediated longevity in *C. elegans*: Is DAF-16 involved? *Mech. Ageing Dev.* **2008**, *129*, 611–613. [CrossRef]

11. Pietsch, K.; Saul, N.; Menzel, R.; Stürzenbaum, S.R.; Steinberg, C.E. Quercetin mediated lifespan extension in *Caenorhabditis elegans* is modulated by age-1, daf-2, sek-1 and unc-43. *Biogerontology* **2009**, *10*, 565–578. [CrossRef] [PubMed]

12. Xue, Y.L.; Ahiko, T.; Miyakawa, T.; Amino, H.; Hu, F.; Furihata, K.; Kita, K.; Shirasawa, T.; Sawano, Y.; Tanokura, M. Isolation and *Caenorhabditis elegans* lifespan assay of flavonoids from onion. *J. Agric. Food Chem.* **2011**, *59*, 5927–5934. [CrossRef]

13. Grünz, G.; Haas, K.; Soukup, S.; Klingenspor, M.; Kulling, S.E.; Daniel, H.; Spanier, B. Structural features and bioavailability of four flavonoids and their implications for lifespan-extending and antioxidant actions in *C. elegans*. *Mech. Ageing Dev.* **2012**, *133*, 1–10. [CrossRef] [PubMed]

14. Kampkötter, A.; Nkwonkam, C.G.; Zurawski, R.F.; Timpel, C.; Chovolou, Y.; Wätjen, W.; Kahl, R. Investigations of protective effects of the flavonoids quercetin and rutin on stress resistance in the model organism *Caenorhabditis elegans*. *Toxicology* **2007**, *234*, 113–123. [CrossRef] [PubMed]

15. Pietsch, K.; Saul, N.; Chakrabarti, S.; Stürzenbaum, S.R.; Menzel, R.; Steinberg, C.E. Hormetins, antioxidants and prooxidants: Defining quercetin-, caffeic acid- and rosmarinic acid-mediated life extension in *C. elegans*. *Biogerontology* **2011**, *12*, 329–347. [CrossRef]

16. Dueñas, M.; Surco-Laos, F.; González-Manzano, S.; González-Paramás, A.M.; Gómez-Orte, E.; Cabello, J.; Santos-Buelga, C. Deglycosylation is a key step in biotransformation and lifespan effects of quercetin-3-O-glucoside in *Caenorhabditis elegans*. *Pharmacol. Res.* **2013**, *76*, 41–48. [CrossRef]

17. Oh, S.W.; Mukhopadhyay, A.; Svrzikapa, N.; Jiang, F.; Davis, R.J.; Tissenbaum, H.A. JNK regulates lifespan in *Caenorhabditis elegans* by modulating nuclear translocation of forkhead transcription factor/DAF-16. *Proc. Natl. Acad. Sci. USA* **2005**, *102*, 4494–4499. [CrossRef]

18. Troemel, E.R.; Chu, S.W.; Reinke, V.; Lee, S.S.; Ausubel, F.M.; Kim, D.H. p38 MAPK regulates expression of immune response genes and contributes to longevity in *C. elegans*. *PLoS Genet.* **2006**, *2*, e183. [CrossRef]

19. Altintas, O.; Park, S.; Lee, S.J.V. The role of insulin/IGF-1 signaling in the longevity of model in vertebrates, *C. Elegans D. Melanogaster*. *BMB Rep.* **2016**, *49*, 81–92. [CrossRef]

20. Lapierre, L.R.; Hansen, M. Lessons from *C. elegans*: Signaling pathways for longevity. *Trends Endocrin. Met.* **2012**, *23*, 637–644. [CrossRef]

21. Kimura, K.D.; Tissenbaum, H.A.; Liu, Y.; Ruvkun, G. Daf-2, an insulin receptor-like gene that regulates longevity and diapause in *Caenorhabditis elegans*. *Science* **1997**, *277*, 942–946. [CrossRef] [PubMed]

22. Ogg, S.; Paradis, S.; Gottlieb, S.; Patterson, G.I.; Lee, L.; Tissenbaum, H.A.; Ruvkun, G. The fork head transcription factor DAF-16 transduces insulin-like metabolic and longevity signals in *C. elegans*. *Nature* **1997**, *389*, 994–999. [CrossRef] [PubMed]

23. Mohri-Shiomi, A.; Garsin, D.A. Insulin signaling and the heat shock response modulate protein homeostasis in the *Caenorhabditis elegans* intestine during infection. *J. Biol. Chem.* **2008**, *283*, 194–201. [CrossRef] [PubMed]

24. Mansuri, M.L.; Parihar, P.; Solanki, I.; Parihar, M.S. Flavonoids in modulation of cell survival signaling pathways. *Genes Nutr.* **2014**, *9*, 400. [CrossRef]

25. Zevian, S.C.; Yanowitz, J.L. Methodological considerations for heat shock of the nematode *Caenorhabditis elegans*. *Methods* **2014**, *68*, 450–457. [CrossRef]

26. Livak, K.J.; Schmittgen, T.D. Analysis of relative gene expression data using real-time quantitative PCR and the $2^{-\Delta\Delta CT}$ method. *Methods* **2001**, *25*, 402–408. [CrossRef]

27. Kenyon, C.J. The genetics of ageing. *Nature* **2010**, *464*, 504–512. [CrossRef]
28. Fitzenberger, E.; Deusing, D.J.; Marx, C.; Boll, M.; Lüersen, K.; Wenzel, U. The polyphenol quercetin protects the *mev-1* mutant of *Caenorhabditis elegans* from glucose-induced reduction of survival under heat-stress depending on SIR-2.1, DAF-12, and proteasomal activity. *Mol. Nutr. Food Res.* **2014**, *58*, 984–994. [CrossRef]
29. Tullet, J.M.; Hertweck, M.; An, J.H.; Baker, J.; Hwang, J.Y.; Liu, S.; Oliveira, R.P.; Baumeister, R.; Blackwell, T.K. Direct inhibition of the longevity-promoting factor SKN-1 by insulin-like signaling in *C. elegans*. *Cell* **2008**, *132*, 1025–1038. [CrossRef]
30. Tullet, J.M.A.; Green, J.W.; Au, C.; Benedetto, A.; Thompson, M.A.; Clark, E.; Gilliat, A.F.; Young, A.; Schmeisser, K.; Gems, D. The SKN-1/Nrf2 transcription factor can protect against oxidative stress and increase lifespan in *C. elegans* by distinct mechanisms. *Aging Cell* **2017**, *16*, 1191–1194. [CrossRef]
31. Ayuda-Durán, B.; González-Manzano, S.; Miranda-Vizuete, A.; Dueñas, M.; Santos-Buelga, C.; González-Paramás, A.M. Epicatechin modulates stress-resistance in *C. elegans* via insulin/IGF-1 signaling pathway. *PLoS ONE* **2019**, *14*, e0199483. [CrossRef] [PubMed]
32. Paiva, F.A.; Bonomo, L.F.; Boasquivis, P.F.; de Paula, I.T.; Guerra, J.F.; Leal, W.M.; Silva, M.E.; Pedrosa, M.L.; Oliveira, R.P. Carqueja (*Baccharis trimera*) Protects against Oxidative Stress and β-Amyloid-Induced Toxicity in *Caenorhabditis elegans*. *Oxid. Med. Cell. Longev.* **2015**, 740162. [CrossRef]
33. Lund, J.; Tedesco, P.; Duke, K.; Wang, J.; Kim, S.K.; Johnson, T.E. Transcriptional profile of aging in *C. elegans*. *Curr. Biol.* **2002**, *12*, 1566–1573. [CrossRef]
34. Walker, G.A.; White, T.M.; McColl, G.; Jenkins, N.L.; Babich, S.; Candido, E.P.; Johnson, T.E.; Lithgow, G.J. Heat shock protein accumulation is upregulated in a long-lived mutant of *Caenorhabditis elegans*. *J. Gerontol. Ser. A* **2001**, *56*, B281–B287. [CrossRef] [PubMed]
35. Tullet, J.M. DAF-16 target identification in *C. elegans*: Past, present and future. *Biogerontology* **2015**, *16*, 221–234. [CrossRef] [PubMed]
36. Kenyon, C.; Chang, J.; Gensch, E.; Rudner, A.; Tabtiang, R.A. *C. elegans* mutant that lives twice as long as wild type. *Nature* **1993**, *366*, 461–464. [CrossRef]
37. Hekimi, S.; Burgess, J.; Bussière, F.; Meng, Y.; Bénard, C. Genetics of lifespan in *C. elegans*: Molecular diversity, physiological complexity, mechanistic simplicity. *Trends Genet.* **2001**, *17*, 712–718. [CrossRef]
38. Uno, M.; Nishida, E. Lifespan-regulating genes in *C. elegans*. *NPJ Aging Mech. Dis.* **2016**, *2*, 16010. [CrossRef]
39. Lin, K.; Hsin, H.; Libina, N.; Kenyon, C. Regulation of the *Caenorhabditis elegans* longevity protein DAF-16 by insulin/IGF-1 and germline signaling. *Nat. Genet.* **2001**, *28*, 139–145. [CrossRef]
40. Hsu, A.; Coleen, T.; Kenyon, C. Regulation of aging and age-related disease by DAF-16 and heat-shock factor. *Science* **2003**, *300*, 1142–1145. [CrossRef]
41. Tang, S.; Chen, H.; Cheng, Y.; Nasir, M.A.; Kemper, N.; Bao, E. The interactive association between heat shock factor 1 and heat shock proteins in primary myocardial cells subjected to heat stress. *Int. J. Mol. Med.* **2016**, *37*, 56–62. [CrossRef] [PubMed]
42. Mertenskötter, A.; Keshet, A.; Gerke, P.; Paul, R.J. The p38 MAPK PMK-1 shows heat-induced nuclear translocation, supports chaperone expression, and affects the heat tolerance of *Caenorhabditis elegans*. *Cell Stress Chaperones.* **2013**, *18*, 293–306. [CrossRef] [PubMed]
43. Pietsch, K.; Saul, N.; Swain, S.C.; Menzel, R.; Steinberg, C.E.; Stürzenbaum, S.R. Meta-Analysis of Global Transcriptomics Suggests that Conserved Genetic Pathways are Responsible for Quercetin and Tannic Acid Mediated Longevity in *C. elegans*. *Front. Genet.* **2012**, *3*, 48. [CrossRef] [PubMed]
44. Bonomo, L.F.; Silva, D.N.; Boasquivis, P.F.; Paiva, F.A.; Guerra, J.F.; Martins, T.A.; de Jesus-Torres, Á.G.; de Paula, I.T.; Caneschi, W.L.; Jacolot, P.; et al. Açaí (*Euterpe oleracea* Mart.) modulates oxidative stress resistance in *Caenorhabditis elegans* by direct and indirect mechanisms. *PLoS ONE* **2014**, *9*, e89933. [CrossRef] [PubMed]

 *antioxidants*

Article

3,3′-Diindolylmethane Promotes BDNF and Antioxidant Enzyme Formation via TrkB/Akt Pathway Activation for Neuroprotection against Oxidative Stress-Induced Apoptosis in Hippocampal Neuronal Cells

Bo Dam Lee [1,†], Jae-Myung Yoo [2,3,†], Seong Yeon Baek [1], Fu Yi Li [1], Dai-Eun Sok [4] and Mee Ree Kim [1,*]

[1] Department of Food and Nutrition, Chungnam National University, Daejeon 34134, Korea; bodam_lee@naver.com (B.D.L.); qor7683@naver.com (S.Y.B.); kaihuadouer@naver.com (F.Y.L.)
[2] Korean Medicine-Application Center, Korea Institute of Oriental Medicine, Daegu 41062, Korea; jmyoo@cnu.ac.kr
[3] Korean Medicine R&D Team 1, National Institute for Korean Medicine Development, Gyeongsan 38540, Korea
[4] College of Pharmacy, Chungnam National University, Daejeon 34134, Korea; daesok@cnu.ac.kr
* Correspondence: mrkim@cnu.ac.kr; Tel.: +82-42-821-6837; Fax: +82-42-821-8887
† These authors contributed equally to this work.

Received: 15 November 2019; Accepted: 16 December 2019; Published: 18 December 2019

Abstract: 3,3′-Diindolylmethane (DIM), a metabolite of indole-3-carbinol present in Brassicaceae vegetables, possesses various health-promoting effects. Nonetheless, the effect of DIM on neurodegenerative diseases has not been elucidated clearly. In this study, we hypothesized DIM may protect neuronal cells against oxidative stress-induced apoptosis by promoting the formation of brain-derived neurotrophic factor (BDNF) and antioxidant enzymes through stabilizing the activation of the tropomyosin-related kinase receptor B (TrkB) cascade and we investigated the effect of DIM on oxidative stress-mediated neurodegenerative models. DIM protected neuronal cells against oxidative stress-induced apoptosis by regulating the expression of apoptosis-related proteins in glutamate-treated HT-22 cells. Additionally, DIM improved the expression of BDNF and antioxidant enzymes, such as heme oxygenase-1, glutamate-cysteine ligase catalytic subunit, and NAD(P)H quinine oxidoreductase-1, by promoting the activation of the TrkB/protein kinase B (Akt) pathway in the cells. Consistent with in vitro studies, DIM attenuated memory impairment by protecting hippocampal neuronal cells against oxidative damage in scopolamine-treated mice. Conclusionally, DIM exerted neuroprotective and antioxidant actions through the activation of both BDNF production and antioxidant enzyme formation in accordance with the TrkB/Akt pathway in neuronal cells. Such an effect of DIM may provide information for the application of DIM in the prevention of and therapy for neurodegenerative diseases.

Keywords: 3,3′-diindolylmethane; antioxidant enzymes; brain-derived neurotrophic factor; hippocampal neuronal cells; neurodegenerative disease

---

1. Introduction

3,3′-Diindolylmethane (DIM) is a metabolite of indole-3-carbinol, which is present in Brassicaceae vegetables containing glucosinolates (206–3895 mg/kg) [1]. Hydrolysis products of glucosinolates, such as isothiocyanates and indol-3-carbinol, are well known for exerting health-promoting effects, including for neurodegenerative diseases [2]. DIM, a dimer of indole-3-carbinol that is converted in the acidic conditions of stomach after intake [3], is known to possess beneficial effects especially in antioxidant [4], anti-cancer [5], and hepatoprotective actions [6]. Recently, it was reported that DIM could protect neuronal cells against inflammation [7,8] and ischemia [7,8] in brain tissue. Moreover, DIM at the doses used did not show any serious side effects in healthy volunteers [9]. Nonetheless, the effect of DIM on neurodegenerative diseases has not been elucidated clearly.

Neurodegenerative diseases, which are increasing along with the extension of life spans and the occurrence of metabolic diseases, are known to be associated with the generation of reactive oxygen species (ROS) [10], which are responsible for the apoptosis of neuronal cells [11,12]. Therefore, the regulation of ROS generation is suggested as a key target for the treatment or prevention of neurodegenerative diseases. Brain-derived neurotrophic factor (BDNF), which is mainly produced from the central nervous system [13–15] and is concerned with the preservation of neuronal cells in in vivo systems, is known to play a crucial role [16]; it participates in the proliferation and differentiation of neuronal cells through the activation of tropomyosin-related kinase receptor B (TrkB) [17]. In addition, BDNF protects neuronal cells against oxidative stress [16,18]. Recently, it was observed that some specific antioxidant compounds ameliorate scopolamine-induced memory impairment by promoting BDNF production and suppressing oxidative damage in scopolamine-exposed mice [11,12,19]. Therefore, the regulation of BDNF formation may be a key point for the treatment or prevention of neurodegenerative diseases such as Alzheimer's disease.

Currently, inhibitors of acetylcholinesterase (AChE) have been used for treatment of Alzheimer's disease [20]. However, these inhibitors only delay the progression of Alzheimer's disease and also produce side effects [20]. In this respect, phytochemicals derived from vegetables may be preferable alternatives for the prevention or treatment of neurodegenerative diseases such as Alzheimer's disease.

In this study, we investigated whether DIM was able to protect hippocampal neuronal cells against oxidative stress-induced apoptosis, and then we examined how DIM protects neuronal cells against oxidative stress-induced apoptosis in both in vitro and in vivo models. Herein, we report that the neuroprotective and antioxidant properties of DIM are closely associated with the production of BDNF and antioxidant enzymes by activation of the pathways involving TrkB/protein kinase B (Akt) in hippocampal neuronal cells. Such an effect of DIM may help to address the neuroprotective action of indole-3-carbinol and some vegetables of the brassicaceae family and may provide information for the application of DIM as a therapeutic or preventive supplement for neurodegenerative diseases.

2. Materials and Methods

2.1. Materials

Antibiotics, FBS, 0.25% trypsin-EDTA, and 2,7-dichlorofluorescein diacetate (DCFDA) were procured from Invitrogen (Carlsbad, CA, USA). Dulbecco's modified eagle medium (DMEM), 1× PBS, and 1× Tris buffered saline (TBS) were purchased from Welgene, Inc. (Gyeongsan, Gyeongbuk, Korea). EZ-Cytox cell viability assay kit was obtained from DAEIL Lab (Seoul, Korea). DIM, K252a (a TrkB inhibitor), and MK-2206 (a specific Akt inhibitor) were procured from Cayman Chemical Company (Ann Arbor, MI, USA). Specific antibodies against cytochrome c (ab90529), glutamate-cysteine ligase catalytic subunit (GCLC, ab53179), and NAD(P)H quinone oxidoreductase-1 (NQO-1, ab34173) were purchased from Abcam, Inc. (Cambridge, UK). A specific antibody against BDNF (sc-65514) was obtained from Santa Cruz Biotechnology, Inc. (Dallas, TX, USA). Specific antibodies against apoptosis-inducing factor (AIF, #4642), B-cell lymphoma 2 (Bcl-2, #3498), Bcl-2-associated X protein (Bax, #2772), cleaved caspase-3 (#9661), heme oxygenase-1 (HO-1, #70081), NF-E2-related factor-2 (Nrf2, #12721), phospho-Akt (#9271), phospho-cAMP response element-binding protein (CREB, #9198), phospho-TrkB (#4619), and β-actin (#4970) as well as a horseradish peroxidase-conjugated IgG secondary antibody were procured from Cell Signaling Technology (Beverly, MA, USA). Muse Annexin V and Dead Cell Assay Kits were purchased from Merck Millipore, Inc. (Darmstadt, Germany). All other chemicals which were used in this study were analytical grade.

2.2. Animals

ICR mice [21] were obtained from Nara Biotech Co. (Pyeongteak, Korea). Their gender, age, and body weight were male, 6 weeks, and 25–30 g. The mice were housed in cages (five mice/cage) under specific pathogen-free conditions (21–24 °C and 40–60% relative humidity) with a 12 h light/dark cycle. In addition, they were given free access to standard rodent food (Orientbio Inc., Seongnam, Korea) and water. All animal experiments were approved by the Committee of Animal Care and Experiment of Chungnam National University (Daejeon, Korea) with a reference number (CNU-00973) and carried out following the guidelines of the Animal Care and Use Committee at Chungnam National University.

2.3. Behavior Tests

2.3.1. Morris Water Maze Test

The cognitive ability of mice was evaluated by Morris water maze test, which was reported previously [12]. One hour before the Morris water maze test, mice were administered with DIM (0, 10, or 20 mg/kg, orally). After three consecutive days, the mice were administered with scopolamine (2 mg/kg, i.p.) after the DIM treatment.

2.3.2. Passive Avoidance Test

The cognitive ability of the mice was evaluated by the passive avoidance test, which was reported previously [12]. One hour before the acquisition trial, the mice were orally administered with DIM before the scopolamine challenge.

2.4. Analysis of Antioxidant or Neurobiological Biomarkers in Brain Tissues

2.4.1. Determination of Lipid Peroxidation

Lipid peroxidation was determined following a method reported previously [22]. Briefly, brains isolated from sacrificed mice were homogenized in ice-cold 1× PBS, and then the homogenates were centrifuged (17,000× g at 4 °C) for 10 min. The supernatant (100 µL) was mixed with 50 µL of 20% aqueous trichloroacetic acid and 100 µL of 0.67% aqueous thiobarbituric acid, boiled for 10 min, and then centrifuged (60× g, room temperature (RT) for 30 min. The absorbance of the supernatant was measured at 532 nm using a microplate reader (DU650, Beckman Coulter, Brea, CA, USA).

2.4.2. Measurement of Glutathione

Glutathione (GSH) level in brain tissues or cell pellets was analyzed using a Quanticrom Glutathione Assay Kit obtained from BioAssay Systems (Hayward, CA, USA) according to the manufacturer's protocol.

2.4.3. Activities of Glutathione Reductase and Glutathione Peroxidase

The activities of glutathione reductase (GR) and glutathione peroxidase (GPx) in brain tissues were determined as described previously [19].

2.4.4. Activities of Acetylcholinesterase and Choline Acetyltransferase

The activities of acetylcholinesterase (AChE) and choline acetyltransferase (ChAT) in brain tissues were determined as described previously [19].

2.5. Hematoxylin and Eosin Staining Assay

Histological analysis was conducted by following a modification of a method reported previously [23]. Briefly, deparaffinized brain tissue slices were stained with hematoxylin-eosin.

The stained tissue slices were embedded with mounting solution. Histological changes of the brain tissues were observed under a light microscope with 200× magnification.

2.6. Cell Culture

The HT-22 cell line was procured from Medifron (Seoul, Korea), and HT-22 cells were maintained in DMEM medium with 10% FBS, 100 units/mL of penicillin, and 100 μg/mL of streptomycin at 37 °C in a humidified atmosphere of 5% CO_2. Passages three to 11 of the cells were used in this study. All the in vitro studies included a vehicle control group (0.1% dimethyl sulfoxide).

2.7. Cell Viability Assay

Cell viability was assessed according to a previous method [24]. HT-22 cells were preincubated with or without DIM (0–80 μM) for 30 min before glutamate challenge. After 12 h, cell viability was determined using the EZ-Cytox cell viability assay kit following the manufacturer's instructions.

2.8. Measurement of Intracellular ROS Level

The intracellular ROS level was assessed according to a previous method [24]. After glutamate treatment, HT-22 cells were stained with 10 μM DCFDA in Hank's balanced salt solution for 30 min in the darkness, and the fluorescence (an excitation wavelength of 485 nm and an emission wavelength of 525 nm) was measured by a microplate reader (Beckman Coulter DTX 880 Multimode Detector, Brea, CA, USA).

2.9. Extraction of Nuclear and Cytosolic Protein

Cytosolic and nuclear proteins were fractionated using a Nuclear Extraction Kit (Cayman Chemical, Ann Arbor, MI, USA) following the manufacturer instructions.

2.10. Immunoblot Analysis

Immunoblot analysis was carried out according to a previous method [12]. Briefly, blotted proteins on PVDF membrane were visualized by a WEST One western blot detection system (iNtRON Biotechnology, Inc., Gyeonggi-do, Korea). All the target protein levels were compared with loading control levels (β-actin or Lamin B), and the results were expressed as a density ratio of each protein, identified by a protein standard size marker (BIOFACT, Daejeon, Korea). The relative density of the protein expression was quantitated by Matrox Inspector software (version 2.1 for Windows; Matrox Electronic Systems Ltd., Dorval, Quebec, Canada).

2.11. Statistical Analysis

The experimental results were expressed as the mean ± SD for the in vitro data and as the mean ± SEM for the in vivo data. One-way analysis of variance (ANOVA) was used for multiple comparisons (GraphPad Prism version 5.03 for Windows, San Diego, CA, USA). The Dunnett test and the Tukey's test were applied for significant variations between treated groups. Differences at the * $p < 0.05$ and ** $p < 0.01$ levels were considered statistically significant.

3. Results

3.1. Neuroprotective Effect of 3,3′-Diindolylmethane on Glutamate-Treated HT-22 Cells

DIM possesses some beneficial properties, such as antioxidant action [4] and neuroprotective action [7,8]. Considering this, we investigated the effect of DIM on glutamate-induced cytotoxicity in HT-22 cells. When HT-22 cells were incubated with DIM prior to glutamate exposure, DIM at the concentrations used (10–80 μM) enhanced cell viability up to a concentration of 80 μM (Figure 1A). In addition, DIM dose-dependently reduced the ROS level and also restored the GSH level in the cells

(Figure 1B,C). A DIM concentration of 40 µM was sufficient to restore the level of ROS or GSH to that of the control group. These results suggest that DIM exerts neuroprotective activity by suppressing oxidative stress in hippocampal neuronal cells.

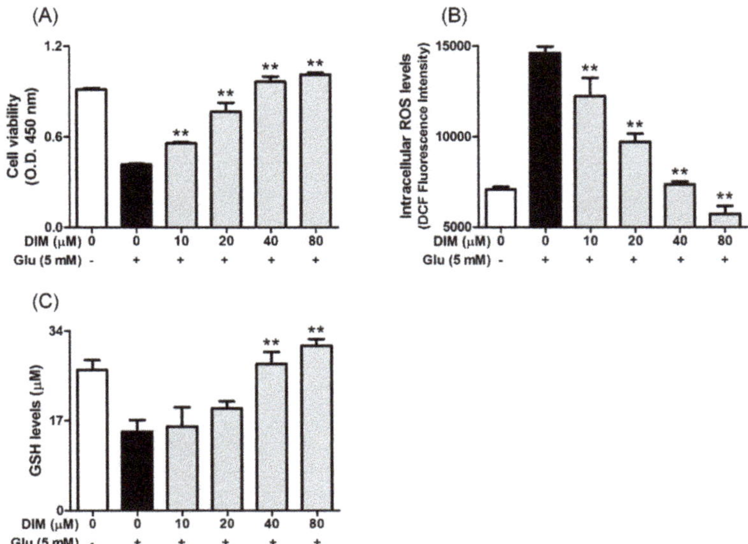

Figure 1. Effect of 3,3′-diindolylmethane (DIM) on glutamate-induced cytotoxicity, and reduction of ROS and glutathione levels in HT-22 cells. HT-22 cells, seeded on a 96 well-plates and incubated for 24 h, were incubated with or without DIM (0–80 µM) for 30 min before glutamate challenge (5 mM). After 12 h, cell viability, ROS level, and GSH level were measured as described in Materials and Methods. (**A**) Cell viability, (**B**) ROS level, and (**C**) GSH level. Data are the mean ± SD values of triple determinations. ** $p < 0.01$ versus glutamate-treated group. − is absence, + is presence.

3.2. Inhibitory Effect of 3,3′-Diindolylmethane on Oxidative Stress-Induced Apoptosis

After we found that DIM protects hippocampal neuronal cells against oxidative stress, we investigated the effect of DIM on the expression of apoptosis-related proteins in oxidative stress-exposed neuronal cells. In the previous report, we found that the elevation of ROS formation activated apoptosis signaling pathway in neuronal cells [12]. DIM downregulated the expression of pro-apoptotic factors such as Bax, cytochrome c, cleaved caspase-3, and AIF, whereas it upregulated that of Bcl-2, an anti-apoptotic factor (Figure 2). These findings suggest that DIM protects hippocampal neuronal cells against oxidative stress-induced apoptosis by regulating the expression of apoptosis-related proteins.

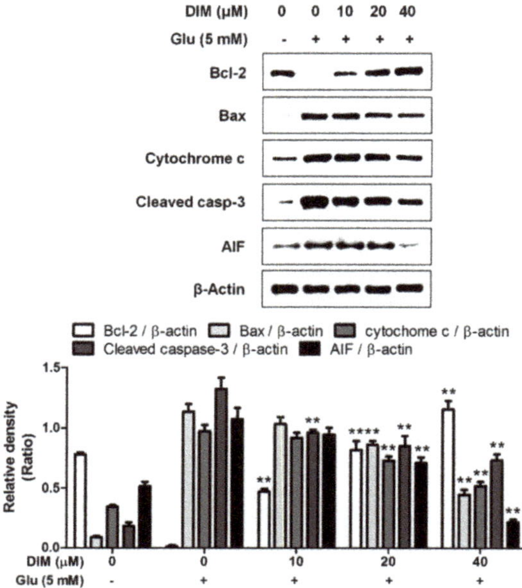

Figure 2. Suppressive effect of 3,3′-diindolylmethane on glutamate-induced apoptosis in HT-22 cells. HT-22 cells were seeded on a 60 mm dish, and then incubated for 24 h. The cells were challenged with glutamate after preincubation with or without DIM (0–40 μM) for 30 min. After 12 h, the expression of Bcl-2, Bax, cytochrome c, cleaved caspase-3, AIF, or β-actin was examined as described in Materials and Methods. The data were contained from three independent experiments. ** $p < 0.01$ versus glutamate-treated group. – is absence, + is presence.

3.3. Activatory Effect of 3,3′-Diindolylmethane on Both TrkB/CREB/BDNF Pathway and Akt/Nrf2/ARE Pathway

It was reported that activation of the TrkB/CREB/BDNF pathway and the Akt/Nrf2/antioxidant response element (ARE) pathway contributed to the neuroprotective action of *N*-acetyl serotonin [12]; therefore, we supposed that both pathways could be activated by DIM, which was found to protect hippocampal neuronal cells against oxidative stress-induced apoptosis. DIM dose-dependently improved the expression of p-TrkB, p-CREB, BDNF, or p-Akt and antioxidant enzymes such as HO-1, GCLC, and NQO-1 in glutamate-treated HT-22 cells (Figure 3). In addition, DIM also promoted the nuclear translocation of Nrf2 in glutamate-treated HT-22 cells. Especially, DIM at 40 μM almost fully restored the expression of all the proteins. Based on these results, we suppose that DIM is able to promote the activation of both the TrkB/CREB/BDNF pathway and the Akt/Nrf2/ARE pathway in oxidative stress-exposed HT-22 cells. Therefore, the neuroprotective action of DIM may be closely associated with the activation of both signaling pathways.

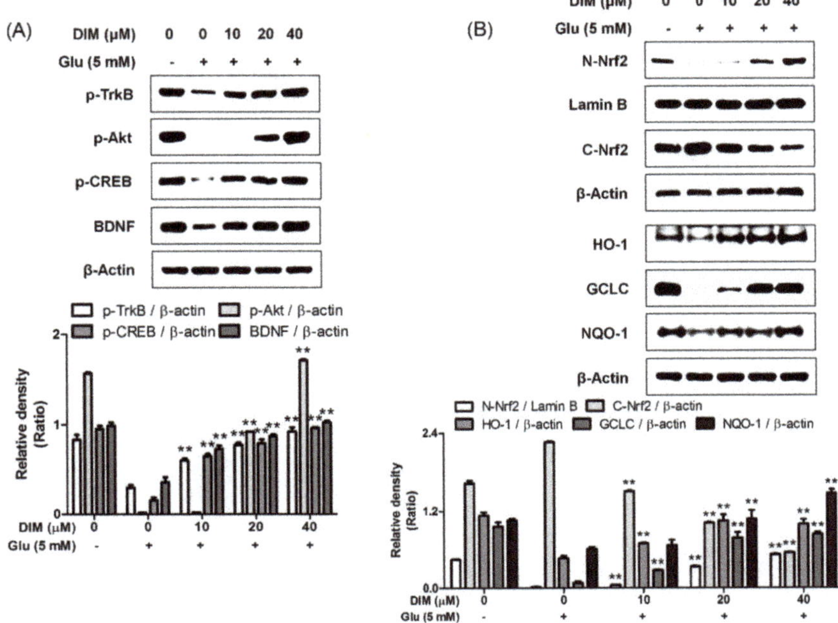

Figure 3. Effect of 3,3′-diindolylmethane on phosphorylation of TrkB, Akt, or CREB, and expression of BDNF, Nrf2, or antioxidant enzymes. The experiments were performed as described in the Figure 2 legend. The data were obtained from three independent experiments. (**A**) p-TrkB, p-Akt, p-CREB, and BDNF and (**B**) nuclear and cytosolic Nrf2, HO-1, GCLC, and NQO-1. ** $p < 0.01$ versus glutamate-treated group. − is absence, + is presence.

3.4. Suppressive Effects of K252a and MK-2206 on Neuroprotective Action of 3,3′-Diindolylmethane

Subsequently, we attempted to clarify the mechanism by which DIM promoted the activation of the TrkB/CREB/BDNF pathway and the Akt/Nrf2/ARE pathway in oxidative stress-exposed hippocampal neuronal cells. For this, we investigated the suppressive effects of K252a, a TrkB inhibitor, and MK-2206, a selective Akt inhibitor, on the neuroprotective action of DIM in oxidative stress-exposed hippocampal neuronal cells. When DIM was preincubated with HT-22 cells in combination with K252a or MK-2206 prior to glutamate treatment, the inclusion of either K252a or MK-2206 significantly attenuated the preventive action of DIM against oxidative stress-induced cell death (Figure 4A,B). In addition, the inhibitors reversed the inhibitory effect of DIM on the generation of ROS in glutamate-treated HT-22 cells (Figure 4C,D). Overall, these results support the notion that DIM protects hippocampal neuronal cells against oxidative stress by promoting the expression of both BDNF and antioxidant enzymes via activation of both the TrkB/CREB/BDNF pathway and the Akt/Nrf2/ARE pathway.

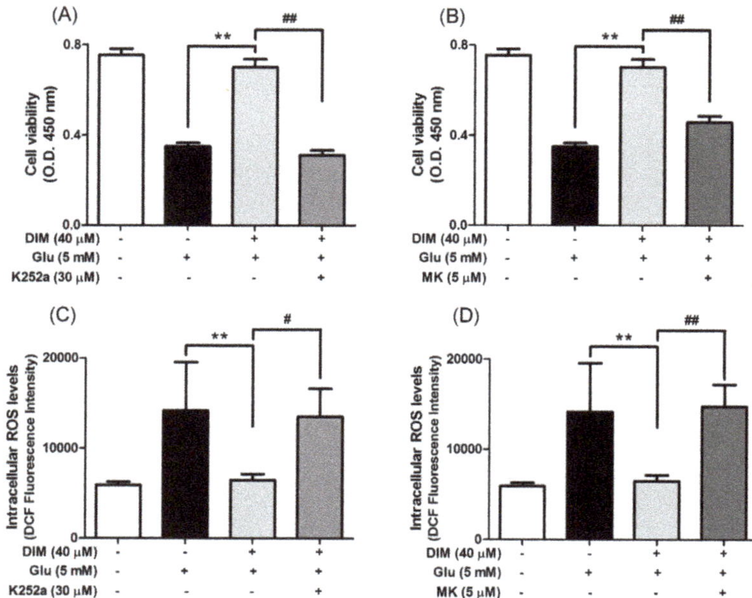

Figure 4. Inhibitory effect of K252a or MK-2206 on neuroprotective action of DIM. HT-22 cells were preincubated with or without DIM in combination with K252a or MK-2206 for 30 min before glutamate challenge. After 12 h, cell viability and ROS level were measured as described in Materials and Methods. (**A**), (**B**) Cell viability and (**C**), (**D**) ROS level. Data are the mean ± SD values of quintuple determinations. ** $p < 0.01$ versus glutamate-treated group; # $p < 0.05$ and ## $p < 0.01$ versus DIM with glutamate-treated group.

3.5. Improving Effect of 3,3′-Diindolylmethane on Scopolamine-Induced Memory Impairment in Mice

Finally, to confirm the neuroprotective action of DIM in an in vivo system, we further investigated the neuroprotective action of DIM in scopolamine-treated mice. When mice were orally administrated with DIM (10 or 20 mg/kg) before scopolamine challenge, DIM at 20 mg/kg decreased the time of escape latency and increased the number of platform area crossings and latency time in scopolamine-treated mice (Figure 5). However, DIM at 10 mg/kg did not improve the latency time in the mice, which suggests that the administration of DIM at 10 mg/kg failed to reach the effective concentration in brain because of the rapid metabolism of DIM [25]. Consistent with its antioxidant action in an in vitro system, DIM reduced lipid peroxidation and also improved the level of GSH and the activities of GR and GPx in brain tissue (Figure 6A–D). Additionally, DIM reduced AChE activity and enhanced ChAT activity compared with the scopolamine group (Figure 6E,F), further supporting the antioxidant action of DIM. Also, DIM prevented scopolamine-induced oxidative damage of neuronal cells in the hippocampal CA1 and CA3 regions of mice brains (Figure 7). These findings suggest that DIM improves scopolamine-induced memory impairment by protecting neuronal cells against oxidative damage in the hippocampus of mice. Based on this, we suggest that the neuroprotective action of DIM may be largely ascribed to the activation of both the TrkB/CREB/BDNF pathway and the Akt/Nrf2/ARE pathway.

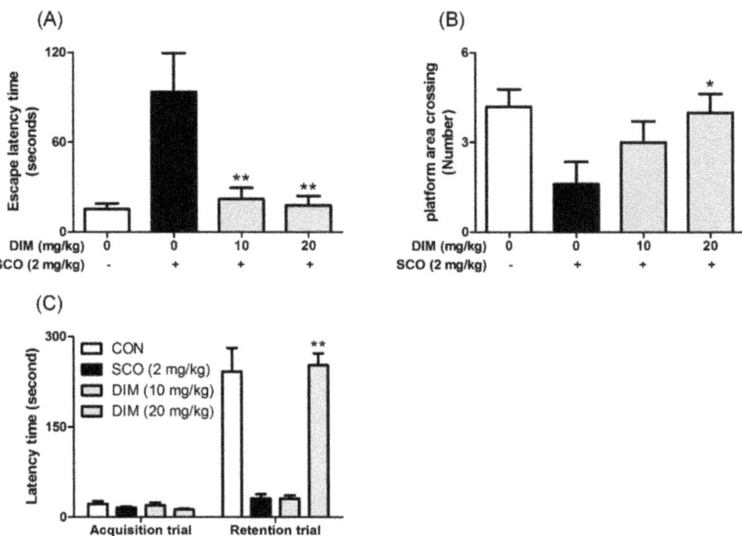

Figure 5. Inhibitory effect of 3,3′-diindolylmethane on memory impairment induced by scopolamine in mice. Mice were orally administrated with DIM (10 or 20 mg/kg) before scopolamine treatment (2 mg/kg, i.p.). After 1 h, the mice were tested for Morris water maze or passive avoidance (Acquisition trial). (**A**) Escape latency time, (**B**) number of platform area crossings, and (**C**) latency time. Data are the mean ± SEM values of sextuple determinations. * $p < 0.05$ and ** $p < 0.01$ versus scopolamine-treated group.

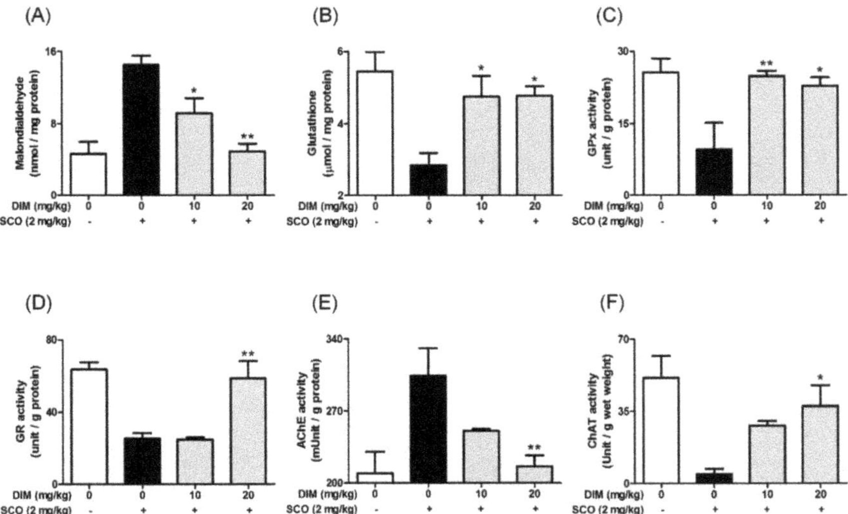

Figure 6. Effect of 3,3′-diindolylmethane on antioxidant biomarkers and cholinergic enzymes in brain tissue of mice treated with scopolamine. Experiments were performed as described in the Figure 5 legend. Lipid peroxidation, GSH, and the activities of GPx, GR, AChE, or ChAT in brain tissues were determined as described in Materials and Methods. (**A**) Lipid peroxidation, (**B**) GSH, (**C**) GPx activity, (**D**) GR activity, (**E**) AChE activity, and (**F**) ChAT activity. Data are the mean ± SEM values of sextuple determinations. * $p < 0.05$ and ** $p < 0.01$ versus scopolamine-treated group.

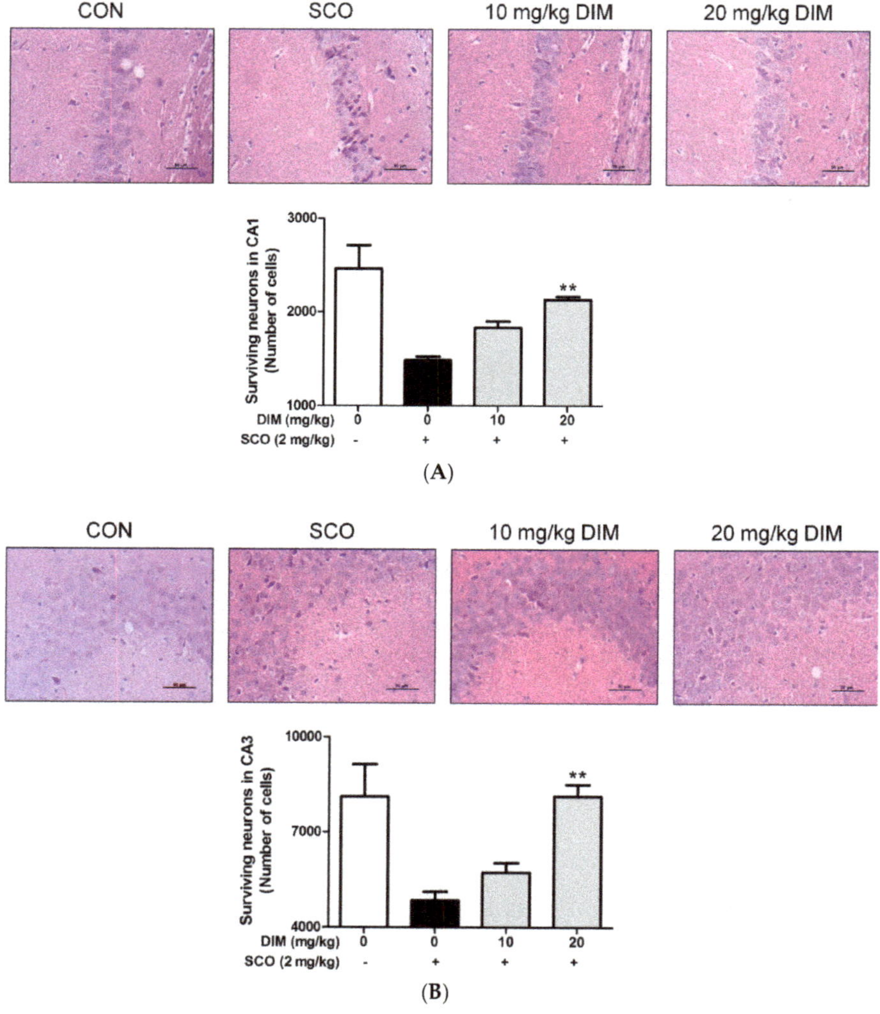

Figure 7. Neuroprotective effect of 3,3′-diindolylmethane on hippocampal CA1 and CA3 regions of mice treated with scopolamine. Experiments were performed as described in the Figure 5 legend. Neuronal cell staining of brain tissues was carried out as described in Materials and Methods. The results were obtained from three independent experiments. (**A**) CA1 region and (**B**) CA3 region. ** $p < 0.01$ versus scopolamine-treated group.

4. Discussion

DIM, a dimer of indole-3-carbinol, is known to have some beneficial effects, such as antioxidant action [4], anti-cancer activity [5], hepatoprotection [6], and neuroprotection [7,8]. Recently, it was reported that DIM at 10 μM completely protected neuronal cells against hypoxia through reducing the expression of the aryl hydrocarbon receptor modulator and aryl hydrocarbon receptor induced by hypoxia [7]. In addition, the activation of the aryl hydrocarbon receptor by α-naphthoflavone promoted apoptosis through upregulating ROS generation in neuronal cells [26]. However, DIM could not protect neuronal cells against hypoxia in neuronal cells treated with siRNA targeted to the aryl hydrocarbon receptor modulator or hydrocarbon receptor [7]. Therefore, it remains unknown how DIM protects neuronal cells against oxidative stress-induced apoptosis.

Recently, we reported that *N*-acetyl serotonin protected neuronal cells against oxidative stress-induced apoptosis through activating both the TrkB/CREB/BDNF and Akt/Nrf2/antioxidant enzyme pathways [12], and *N*-palmitoyl serotonin protected neuronal cells against oxidative stress-induced apoptosis by stabilizing the activation of the BDNF autocrine loop, although it could not directly activate the phosphorylation of TrkB [11]. Based on these findings, we surmised that DIM may protect hippocampal neuronal cells against oxidative stress-induced apoptosis by maintaining the activation of both the TrkB/CREB/BDNF pathway and the Akt/Nrf2/ARE pathway. Therefore, we investigated the effect of DIM on oxidative stress-induced apoptosis in hippocampal neuronal cells using an in vitro study. From the results, we found that DIM protected hippocampal neuronal cells against oxidative stress-induced apoptosis by promoting the expression of both BDNF and antioxidant enzymes, such as HO-1, NQO-1, and GCLC. Moreover, in an in vivo study, DIM attenuated scopolamine-induced memory impairment by protecting hippocampal neuronal cells against oxidative damage in mice. Furthermore, such a neuroprotective effect of DIM was possibly associated with the stabilized activation of an antioxidant enzyme-generating system such as the Akt/Nrf2/ARE pathway in hippocampal neuronal cells. These results lead to the assumption that TrkB activation stabilized by DIM may be related to the activation of the Akt/Nrf2/ARE pathway in neuronal cells.

One possible mechanism for the neuroprotective action of DIM may be related to the upregulation of BDNF expression through activation of the TrkB/Akt/CREB/BDNF pathway [18]. BDNF, a ligand of TrkB with a stronger neuroprotective action than that of other neurotrophic factors, is mainly produced from astrocytes [13], microglia [14], and neuronal cells [15]. Here, the stimulation of TrkB by BDNF is known to promote proliferation and differentiation of neural stem cells [17]. In the signaling pathway, the activation of TrkB by BDNF induces the phosphorylation of Akt [27], and then the activated Akt is able to produce BDNF through the activation of CREB in neuronal cells [28]. Thus, maintaining the production of BDNF in neuronal cells may be an important strategy for the prevention or treatment of neurodegenerative diseases, such as Alzheimer's disease. Consistent with this, our present data indicates that DIM protects hippocampal neuronal cells against oxidative stress-induced apoptosis by elevating the phosphorylation of TrkB, Akt, and CREB. In support of this, the inclusion of K252a, an inhibitor of TrkB, neutralized the neuroprotective effect of DIM on oxidative stress-induced cell death of hippocampal neuronal cells. A similar result was observed previously when HT-22 cells were incubated with glutamate in the presence of *N*-acetyl serotonin [12]. Therefore, we suggest that DIM is able to protect neuronal cells by promoting the formation of BDNF through the activation of the TrkB/Akt/CREB/BDNF pathway in oxidative stress-exposed neuronal cells.

Another possible mechanism for the neuroprotective action of DIM is associated with the formation of antioxidant enzymes such as HO-1, NQO-1, and GCLC through the activation of the Akt/Nrf2/ARE pathway. In support of this, DIM augmented the expression of antioxidant enzymes in accordance with the Nrf2/ARE pathway [29]. Moreover, the action of DIM may be extended to activation of the TrkB/Akt/Nrf2/ARE pathway; DIM activates TrkB to produce p-TrkB as mentioned earlier. In turn, activated Akt liberates Nrf2 from the keap1-Nrf2 complex [12], and then Nrf2 is combined into the ARE region on DNA after it translocates into the nucleus [29]. As a result, Nrf2 promotes the formation of antioxidant enzymes, such as HO-1, NQO-1, and GCLC [29]. In support of this, our present study showed that DIM upregulated the expression of p-TrkB and p-Akt and also improved nuclear translocation of Nrf2. Consequently, DIM restored the level of antioxidant enzymes, such as HO-1, NQO-1, and GCLC, to the control level in oxidative stress-exposed hippocampal neuronal cells. In further support of the TrkB/Akt/Nrf2/ARE pathway, the inclusion of K252a, a TrkB inhibitor, or MK-2206, a selective Akt inhibitor, nullified the suppressive effect of DIM on both ROS generation and cell death in oxidative stress-exposed neuronal cells. Probably consistent with the in vitro results, DIM at low doses (10–20 mg/kg) improved scopolamine-induced memory impairment by protecting neuronal cells against oxidative damage in the hippocampal CA1 and CA3 regions of the brain in mice. Additionally, the cholinolytic effect of scopolamine, mainly due to an increase of AChE activity and a decrease of ChAT activity, was remarkably abrogated by DIM. A similar result was observed previously with *N*-palmitoyl serotonin [19]. A reason for these results is that DIM is capable of crossing to the gastrointestinal tract as well as the blood–brain barrier [25]. In addition, the concentrations of DIM are equal to 0.8 to 1.6 mg/kg (48 to 96 mg/60 kg) in humans, based on body surface area [30].

Taking these results together, it is proposed that the activation of TrkB stabilized by DIM, followed by Akt activation, promotes the expression of both BDNF and antioxidant enzymes in neuronal cells. Therefore, the activation of the TrkB/Akt/Nrf2/ARE pathway, accompanied by stabilization of the TrkB/CREB/BDNF autocrine loop, may be important for the prevention and treatment of neurodegenerative diseases. In addition, the present study shows that DIM may be a candidate compound for the prevention and treatment of neurodegenerative diseases such as Alzheimer's disease.

5. Conclusions

The present study demonstrates that DIM exerts a neuroprotective action by producing BDNF and antioxidant enzymes through activation of the TrkB/Akt signal pathway in oxidative stress-exposed hippocampal neuronal cells. In addition, DIM ameliorates cognitive ability by maintaining the cholinergic system in scopolamine-exposed mice. These findings reveal a feature of the mechanism for the neuroprotective action of DIM against oxidative stress-induced apoptosis in neuronal cells. Such effects of DIM may provide further information for the application of DIM as a neuroprotective agent for the prevention and treatment of neurodegenerative diseases. In addition, this study may be helpful to establish new diagnostic and theranostic paradigms of neurodegenerative diseases, to explore a new biomarker of the diagnostic criteria of neurodegenerative diseases, and to develop innovative pharmacological protocols for the patients of neurodegenerative diseases. Further animal model studies may be required to confirm the feasibility of the use of DIM as a neuroprotective agent for specific neurodegenerative diseases.

Author Contributions: Conceptualization, M.R.K., B.D.L., and J.-M.Y.; methodology, M.R.K.; validation, S.Y.B. and F.Y.L.; formal analysis, B.D.L. and J.-M.Y.; investigation, B.D.L., S.Y.B., and F.Y.L.; data curation, M.R.K.; writing—original draft preparation, J.-M.Y.; writing—review and editing, D.-E.S. and M.R.K.; visualization, J.-M.Y.; supervision, M.R.K.; project administration, M.R.K.; funding acquisition, M.R.K. All authors have read and agreed to the published version of the manuscript.

Funding: This research was supported by the Basic Science Research Program through the National Research Foundation of Korea (NRF) funded by the Ministry of Education (2017R1D1A3B03027867).

Conflicts of Interest: The authors declare no conflict of interest.

Abbreviations

AChE, acetylcholinesterase; AIF, apoptosis-inducing factor; Akt, protein kinase B; Bax, Bcl-2-associated X protein; Bcl-2, B-cell lymphoma 2; BDNF, brain-derived neurotrophic factor; ChAT, choline acetyltransferase; CREB, cAMP response element-binding protein; DIM, 3,3′-diindolylmethane; GCLC, glutamate-cysteine ligase catalytic subunit; GPx, glutathione peroxidase; GR, glutathione reductase; GSH, glutathione; HO-1, heme oxygenase-1; NQO-1, NAD(P)H quinine oxidoreductase-1; Nrf2, NF-E2-related factor-2; ROS, reactive oxygen species; TrkB, Tropomyosin-related kinase receptor B.

References

1. McNaughton, S.A.; Marks, G.C. Development of a food composition database for the estimation of dietary intakes of glucosinolates, the biologically active constituents of cruciferous vegetables. *Br. J. Nutr.* **2003**, *90*, 687–697. [CrossRef]

2. Higdon, J.V.; Delage, B.; Williams, D.E.; Dashwood, R.H. Cruciferous vegetables and human cancer risk: Epidemiologic evidence and mechanistic basis. *Pharmacol. Res.* **2007**, *55*, 224–236. [CrossRef]

3. Zhang, W.W.; Feng, Z.; Narod, S.A. Multiple therapeutic and preventive effects of 3,3′-diindolylmethane on cancers including prostate cancer and high grade prostatic intraepithelial neoplasia. *J. Biomed. Res.* **2014**, *28*, 339–348.

4. Fuentes, F.; Paredes-Gonzalez, X.; Kong, A.N. Dietary glucosinolates sulforaphane, phenethyl isothiocyanate, indole-3-carbinol/3,3'-diindolylmethane: Anti-oxidative stress/inflammation, nrf2, epigenetics/epigenomics and in vivo cancer chemopreventive efficacy. *Curr. Pharmacol. Rep.* **2015**, *1*, 179–196. [CrossRef]

5. Li, Y.; Kong, D.; Ahmad, A.; Bao, B.; Sarkar, F.H. Antioxidant function of isoflavone and 3,3'-diindolylmethane: Are they important for cancer prevention and therapy? *Antioxid. Redox Signal.* **2013**, *19*, 139–150. [CrossRef]

6. Wang, S.Q.; Cheng, L.S.; Liu, Y.; Wang, J.Y.; Jiang, W. Indole-3-carbinol (i3c) and its major derivatives: Their pharmacokinetics and important roles in hepatic protection. *Curr. Drug Metab.* **2016**, *17*, 401–409. [CrossRef]

7. Rzemieniec, J.; Litwa, E.; Wnuk, A.; Lason, W.; Krzeptowski, W.; Kajta, M. Selective aryl hydrocarbon receptor modulator 3,3'-diindolylmethane impairs ahr and arnt signaling and protects mouse neuronal cells against hypoxia. *Mol. Neurobiol.* **2016**, *53*, 5591–5606. [CrossRef] [PubMed]

8. Kim, H.W.; Kim, J.; Kim, J.; Lee, S.; Choi, B.R.; Han, J.S.; Lee, K.W.; Lee, H.J. 3,3'-diindolylmethane inhibits lipopolysaccharide-induced microglial hyperactivation and attenuates brain inflammation. *Toxicol. Sci. Off. J. Society Toxicol.* **2014**, *137*, 158–167. [CrossRef] [PubMed]

9. Reed, G.A.; Sunega, J.M.; Sullivan, D.K.; Gray, J.C.; Mayo, M.S.; Crowell, J.A.; Hurwitz, A. Single-dose pharmacokinetics and tolerability of absorption-enhanced 3,3'-diindolylmethane in healthy subjects. *Cancer Epidemiol. Biomark. Prev. Publ. Am. Assoc. Cancer Res. Cosponsored Am. Soc. Prev. Oncol.* **2008**, *17*, 2619–2624. [CrossRef] [PubMed]

10. Tarafdar, A.; Pula, G. The role of nadph oxidases and oxidative stress in neurodegenerative disorders. *Int. J. Mol. Sci.* **2018**, *19*, 3824. [CrossRef] [PubMed]

11. Yoo, J.-M.; Lee, B.D.; Lee, S.J.; Ma, J.Y.; Kim, M.R. Anti-apoptotic effect of n-palmitoyl serotonin on glutamate-mediated apoptosis through secretion of bdnf and activation of trkb/creb pathway in ht-22 cells. *Eur. J. Lipid Sci. Technol.* **2018**, *120*, 1700397. [CrossRef]

12. Yoo, J.M.; Lee, B.D.; Sok, D.E.; Ma, J.Y.; Kim, M.R. Neuroprotective action of n-acetyl serotonin in oxidative stress-induced apoptosis through the activation of both trkb/creb/bdnf pathway and akt/nrf2/antioxidant enzyme in neuronal cells. *Redox Biol.* **2017**, *11*, 592–599. [CrossRef] [PubMed]

13. Zafra, F.; Lindholm, D.; Castren, E.; Hartikka, J.; Thoenen, H. Regulation of brain-derived neurotrophic factor and nerve growth factor mrna in primary cultures of hippocampal neurons and astrocytes. *J. Neurosci. Off. J. Soc. Neurosci.* **1992**, *12*, 4793–4799. [CrossRef]

14. Miwa, T.; Furukawa, S.; Nakajima, K.; Furukawa, Y.; Kohsaka, S. Lipopolysaccharide enhances synthesis of brain-derived neurotrophic factor in cultured rat microglia. *J. Neurosci. Res.* **1997**, *50*, 1023–1029. [CrossRef]

15. Conner, J.M.; Lauterborn, J.C.; Yan, Q.; Gall, C.M.; Varon, S. Distribution of brain-derived neurotrophic factor (bdnf) protein and mrna in the normal adult rat cns: Evidence for anterograde axonal transport. *J. Neurosci. Off. J. Soc. Neurosci.* **1997**, *17*, 2295–2313. [CrossRef]

16. Rossler, O.G.; Giehl, K.M.; Thiel, G. Neuroprotection of immortalized hippocampal neurones by brain-derived neurotrophic factor and raf-1 protein kinase: Role of extracellular signal-regulated protein kinase and phosphatidylinositol 3-kinase. *J. Neurochem.* **2004**, *88*, 1240–1252. [CrossRef]

17. Vilar, M.; Mira, H. Regulation of neurogenesis by neurotrophins during adulthood: Expected and unexpected roles. *Front. Neurosci.* **2016**, *10*, 26. [CrossRef]

18. Almeida, R.D.; Manadas, B.J.; Melo, C.V.; Gomes, J.R.; Mendes, C.S.; Graos, M.M.; Carvalho, R.F.; Carvalho, A.P.; Duarte, C.B. Neuroprotection by bdnf against glutamate-induced apoptotic cell death is mediated by erk and pi3-kinase pathways. *Cell Death Differ.* **2005**, *12*, 1329–1343. [CrossRef]

19. Min, A.Y.; Doo, C.N.; Son, E.J.; Sung, N.Y.; Lee, K.J.; Sok, D.E.; Kim, M.R. N-palmitoyl serotonin alleviates scopolamine-induced memory impairment via regulation of cholinergic and antioxidant systems, and expression of bdnf and p-creb in mice. *Chem.-Biol. Interact.* **2015**, *242*, 153–162. [CrossRef]

20. Habtemariam, S. Natural products in alzheimer's disease therapy: Would old therapeutic approaches fix the broken promise of modern medicines? *Molecules* **2019**, *24*, 1519. [CrossRef]

21. Schwartz, J.N.; Daniels, C.A.; Shivers, J.C.; Klintworth, G.K. Experimental cytomegalovirus ophthalmitis. *Am. J. Pathol.* **1974**, *77*, 477–492. [PubMed]

22. Jeon, H.L.; Yoo, J.M.; Lee, B.D.; Lee, S.J.; Sohn, E.J.; Kim, M.R. Anti-inflammatory and antioxidant actions of n-arachidonoyl serotonin in raw264.7 cells. *Pharmacology* **2016**, *97*, 195–206. [CrossRef] [PubMed]

23. Yoo, J.M.; Park, K.I.; Ma, J.Y. Anticolitic effect of viscum coloratum through suppression of mast cell activation. *Am. J. Chin. Med.* **2019**, *47*, 203–221. [CrossRef] [PubMed]

24. Jin, M.C.; Yoo, J.M.; Sok, D.E.; Kim, M.R. Neuroprotective effect of n-acyl 5-hydroxytryptamines on glutamate-induced cytotoxicity in ht-22 cells. *Neurochem. Res.* **2014**, *39*, 2440–2451. [CrossRef]
25. Anderton, M.J.; Manson, M.M.; Verschoyle, R.; Gescher, A.; Steward, W.P.; Williams, M.L.; Mager, D.E. Physiological modeling of formulated and crystalline 3,3′-diindolylmethane pharmacokinetics following oral administration in mice. *Drug Metab. Dispos. Biol. Fate Chem.* **2004**, *32*, 632–638. [CrossRef]
26. Yu, A.R.; Jeong, Y.J.; Hwang, C.Y.; Yoon, K.S.; Choe, W.; Ha, J.; Kim, S.S.; Pak, Y.K.; Yeo, E.J.; Kang, I. Alpha-naphthoflavone induces apoptosis through endoplasmic reticulum stress via c-src-, ros-, mapks-, and arylhydrocarbon receptor-dependent pathways in ht22 hippocampal neuronal cells. *Neurotoxicology* **2019**, *71*, 39–51. [CrossRef]
27. Guo, W.; Ji, Y.; Wang, S.; Sun, Y.; Lu, B. Neuronal activity alters bdnf-trkb signaling kinetics and downstream functions. *J. Cell Sci.* **2014**, *127*, 2249–2260. [CrossRef]
28. Cunha, C.; Brambilla, R.; Thomas, K.L. A simple role for bdnf in learning and memory? *Front. Mol. Neurosci.* **2010**, *3*, 1. [CrossRef]
29. Saw, C.L.; Cintron, M.; Wu, T.Y.; Guo, Y.; Huang, Y.; Jeong, W.S.; Kong, A.N. Pharmacodynamics of dietary phytochemical indoles i3c and dim: Induction of nrf2-mediated phase ii drug metabolizing and antioxidant genes and synergism with isothiocyanates. *Biopharm. Drug Dispos.* **2011**, *32*, 289–300. [CrossRef]
30. Nair, A.B.; Jacob, S. A simple practice guide for dose conversion between animals and human. *J. Basic Clin. Pharm.* **2016**, *7*, 27–31. [CrossRef]

 *antioxidants*

Article

Optimization of the Extraction of Bioactive Compounds from Walnut (*Juglans major 209 x Juglans regia*) Leaves: Antioxidant Capacity and Phenolic Profile

Adela Fernández-Agulló, Aída Castro-Iglesias, María Sonia Freire and Julia González-Álvarez *

Department of Chemical Engineering, School of Engineering, Universidade de Santiago de Compostela, 15782 Santiago de Compostela, Spain; adela.fernandez@usc.es (A.F.-A.); aida_9790@hotmail.com (A.C.-I.); mariasonia.freire@usc.es (M.S.F.)

* Correspondence: julia.gonzalez@usc.es; Tel.: +34-881816761

Received: 27 November 2019; Accepted: 20 December 2019; Published: 24 December 2019

Abstract: This work studies the extraction of phenolic compounds from walnut leaves of the hybrid *Juglans major 209 x Juglans regia* based on extract antioxidant capacity. Once the solid/liquid ratio was selected (1/10 g/mL), by means of a Box-Benkhen experimental design, the influence of temperature (25–75 °C), time (30–120 min), and aqueous ethanol concentration (10–90%) on extraction yield and ferric reducing antioxidant power (FRAP), 2,2-diphenyl-1-picrylhydrazyl (DPPH), and 2,2'-azinobis-3-ethylbenzothiazoline-6-sulfonic acid (ABTS) antioxidant activities were analyzed. In all cases, the quadratic effect of % EtOH was the most significant, followed by the linear effect of temperature and, for most of the responses, the effect of time was almost negligible. Response surface analysis allowed to select the optimal extraction conditions: 75 °C, 120 min and 50% ethanol, which led to the following extract properties: extraction yield, 30.17%; FRAP, 1468 nmol ascorbic acid equivalents (AAE)/mg extract d.b.; DPPH, 1.318 mmol Trolox equivalents (TRE)/g extract d.b.; DPPH EC_{50}, 0.11 mg/mL; ABTS, 1.256 mmol TRE/g extract (on dry basis) and ABTS EC_{50}, 0.985 mg/mL. Quercetin 3-β-D-glucoside, neochlorogenic acid, and chlorogenic acid, in this order, were the main compounds identified in this extract by ultra-performance liquid chromatography coupled with electrospray ionization and time-of-flight mass spectrometry (UPLC/ESI-QTOF-MS), with various potential applications that support this valorization alternative for walnut leaves.

Keywords: walnut leaves; *Juglans major 209 x Juglans regia*; maceration; phenolic compounds; antioxidant activity; UPLC/ESI-QTOF-MS; response surface methodology

---

1. Introduction

Juglans major 209 x Juglans regia is a walnut hybrid species intended to produce good-quality wood. Although wood is the main product, the use of other fractions that can be considered as wastes, such as the leaves, would contribute to more profitable production and to a more sustainable plantation management.

Walnut leaves have been intensively used in traditional medicine, and various studies have demonstrated the antimicrobial and antioxidant properties of the extracts from the leaves of several walnut (*Juglans regia*) cultivars [1–4]. Carvalho et al. [3] also demonstrated the antihemolytic and human renal cancer cell antiproliferative activities of walnut leaf methanolic extracts. Quantitative and qualitative determinations of the phenolic compounds present in walnut (*Juglans regia*) leaves have also been carried out [1,4–6] demonstrating quite significant variations in the extract composition. Amaral et al. [5] identified seven phenolic compounds in methanol and acidified water extracts of walnut leaves, being quercetin 3-*O*-galactoside the major compound while 4-*p*-*coumaroylquinic* acid was the minor one.

3-*O*-caffeoylquinic acids and quercetin *O*-pentoside were the main phenolic compounds among the 25 phenolic compounds identified in methanol and decoction extracts by Santos et al. [4]. Pereira et al. [1] identified 10 compounds in aqueous extracts: 3- and 5-caffeoylquinic acids, 3- and 4-*p*-coumaroylquinic acids, *p*-coumaric acid, quercetin 3-*O*-galactoside, quercetin 3-*O*-pentoside derivative, quercetin 3-*O*-arabinoside, quercetin 3-*O*-xyloside and quercetin 3-*O*-rhamnoside. In addition to phenolic acids and flavonoids, Nour et al. [6] also reported the presence of juglone (5-hydroxy-1,4-napthoquinone) in walnut leaves, ellagic acid as the dominating phenolic acid and myricetin, catechin hydrate and rutin as the main flavonoids.

Different solvents have been used for the extraction of phenolic compounds from walnut leaves: Aqueous ethanol [2,7], methanol [4,5], methanol with 1% butylated hydroxytoluene (BHT) [6], water [1,4], acidified water [5], and chloroform [5]. In this work, looking for a green extraction process, extraction was performed by maceration with aqueous ethanol, a bio-solvent produced from biomass, completely biodegradable [8], and generally recognized as safe (GRAS) solvent. In addition to the type of solvent, other factors that can affect the efficiency of the extraction process that were examined in this work are the extraction temperature, the extraction time, and the solid to solvent ratio [9].

The aim of this work was to provide a means for the disposal and valorization of walnut leaves based on extract antioxidant activity, giving the walnut plantation an additional value to that provided by wood. As far as we know, this is the first work on the extraction and characterization of phenolic compounds of walnut leaves of the hybrid *Juglans major 209 x Juglans regia*. An experimental design combined with response surface methodology was used to optimize the extraction of phenolic compounds from walnut leaves based on extract antioxidant activity. The phytochemical characterization of the extracts was carried out and the results compared with those previously obtained for leaves from the *Juglans regia* species.

2. Materials and Methods

2.1. Reagents and Standards

Acetic acid, sodium acetate, $FeCl_3 \cdot 6H_2O$, HCl, l-ascorbic acid, methanol, ethanol, and potassium persulfate were purchased from Panreac (Barcelona, Spain). 2,4,6-Tripyridyl-s-triazine (TPTZ), Trolox (6-hydroxy-2,5,7,8-tetramethylchroman-2-carboxylic acid) and DPPH (2,2-diphenyl-1-picrylhydrazyl) were purchased from Fluka (Steinheim, Germany). ABTS (2,2′-azinobis-3-ethylbenzothiazoline-6-sulfonic acid) and HPLC standards: (−)-epigallocatechin, (−)-gallocatechin, catechin hydrate, chlorogenic acid, ellagic acid, epicatechin, ferulic acid, gallic acid, isorharmnetin, kaempferol, neochlorogenic acid, *p*-coumaric acid, procyanidin B2, quercetin, quercetin 3-*β*-D glucoside, and taxifolin, were obtained from Sigma (Steinheim, Germany).

2.2. Materials

Walnut (*Juglans major 209 x Juglans regia*) leaves were collected in July 2018 in a plantation located in A Coruña, Spain, dried at room temperature, milled, sieved, and the fraction of particle size between 0.1 and 1 mm was selected and stored protected from light till analysis.

2.3. Extraction

Walnut leaves were extracted with aqueous ethanol in an orbital shaker (UNITRONIC-OR, Selecta (Barcelona, Spain) with temperature control at a shaking speed of 90 rpm. The solid/liquid ratio, temperature, time, and EtOH concentration were fixed for each experiment according to the established experimental planning (Tables 1 and 2). The extract was separated by vacuum filtration, concentrated in a Büchi R-210 rotavapor and finally dried under vacuum to obtain a dry powder. Extraction yield was determined as the weight loss percentage of the initial walnut leaves.

The choice of the independent variables to be analyzed and their respective variation intervals was based on previous investigations on the extraction of phenolic compounds from various lignocellulosic materials using aqueous alcohols [10–12]. In the first stage, the influence of the solid/liquid ratio (1/5, 1/7.5 and 1/10 g/mL) was analyzed for fixed values of the other variables: Temperature, 50 °C, time, 60 min, and ethanol concentration, 50% (Table 1). Extraction yield and extract ferric reducing antioxidant power (FRAP) antioxidant activity were determined. All the assays were replicated, and the results expressed as mean value and standard deviation. The existence of significant differences among the results depending on the solid/liquid ratio used was analyzed by applying one-way ANOVA together with the Tukey's test at a confidence level of 95% using the IBM SPSS Statistics 24 software (New York, NY, USA).

Once S/L ratio was fixed at 1/10, a Box-Behnken experimental design was applied to analyze the influence of temperature (x_1; 25, 50, and 75 °C), time (x_2; 30, 75, and 120 min) and aqueous ethanol concentration (x_3; 10%, 50%, and 90%) on extraction yield (Y_1, g extract/100 g leaves on dry basis (d.b.)) and extract antioxidant activity determined according to the FRAP (Y_2, nmol AAE/mg extract d.b.), DPPH (Y_3, mmol TRE/g extract d.b.), and ABTS (Y_4, mmol TRE/g extract d.b.) assays. The phenolic profile of the extract selected as the optimum was analyzed by ultra-performance liquid chromatography coupled with electrospray ionization and time-of-flight mass spectrometry (UPLC/ESI-QTOF-MS).

Table 1. Influence of the solid/liquid ratio on extraction yield and ferric reducing antioxidant power (FRAP) antioxidant activity of walnut leaf extracts (50 °C, 60 min, and 50% aqueous ethanol).

Exp.	S/L Ratio (g/mL)	Extraction Yield (%)	FRAP (nmol AAE/mg Extract)
A	1/5	25.82 ± 0.25 [a]	1246 ± 48 [a]
B	1/7.5	27.04 ± 0.42 [b]	1350 ± 53 [b]
C	1/10	27.86 ± 0.04 [b]	1512 ± 61 [c]

Values are presented as mean ± standard deviation. [a–c] In each column, values with different letters are significantly different ($p < 0.05$).

Table 2. Box-Benkhen experimental design with the experimental and predicted values of the responses.

Exp	x_1^*	x_2^*	x_3^*	$Y_{1\,exp}$	$Y_{1\,pred}$	$Y_{2\,exp}$	$Y_{2\,pred}$	$Y_{3\,exp}$	$Y_{3\,pred}$	$Y_{4\,exp}$	$Y_{4\,pred}$
1	−1	−1	0	21.03 ± 0.89	22.35	1181 ± 39	1262	1.114 ± 0.026	1.050	0.951 ± 0.026	0.964
2	−1	0	−1	20.35 ± 0.18	19.86	842 ± 46	908	0.663 ± 0.044	0.666	0.671 ± 0.027	0.774
3	−1	0	1	19.15 ± 2.13	18.71	930 ± 47	908	0.873 ± 0.013	0.852	0.807 ± 0.017	0.774
4	−1	1	0	23.89 ± 0.28	24.16	1250 ± 58	1262	1.008 ± 0.159	1.050	1.053 ± 0.049	0.964
5	0	−1	−1	23.13 ± 0.27	22.85	1160 ± 37	1041	0.913 ± 0.024	0.932	0.893 ± 0.041	0.875
6	0	−1	1	20.03 ± 1.16	19.96	1138 ± 17	1041	0.919 ± 0.039	0.932	0.866 ± 0.050	0.875
7	0	1	−1	24.52 ± 0.11	24.66	943 ± 65	1041	0.873 ± 0.037	0.932	0.869 ± 0.030	0.875
8	0	1	1	21.93 ± 1.76	21.77	876 ± 18	1041	1.003 ± 0.036	0.932	0.875 ± 0.028	0.875
9	1	−1	0	29.08 ± 1.43	28.40	1364 ± 42	1529	1.493 ± 0.106	1.396	1.166 ± 0.047	1.250
10	1	0	−1	27.01 ± 0.30	27.64	1199 ± 50	1174	1.193 ± 0.021	1.198	1.089 ± 0.035	1.060
11	1	0	1	22.35 ± 0.88	23.02	1238 ± 32	1174	1.031 ± 0.023	1.012	1.095 ± 0.032	1.060
12	1	1	0	30.17 ± 0.09	30.21	1468 ± 20	1529	1.318 ± 0.049	1.396	1.256 ± 0.017	1.250
13	0	0	0	25.85 ± 0.05	26.28	1389 ± 78	1395	1.239 ± 0.043	1.223	1.299 ± 0.049	1.315
14	0	0	0	27.19 ± 0.51	26.28	1560 ± 49	1395	1.190 ± 0.031	1.223	1.330 ± 0.032	1.315

Independent Variables		Levels	
	−1	0	+1
x_1, Temperature (°C)	25	50	75
x_2, Time (min)	30	75	120
x_3, % EtOH	10	50	90

Experimental values are presented as mean ± standard deviation. Y_1, extraction yield (g extract/100 g leaves d.b.); antioxidant activity: Y_2, FRAP (nmol AAE/mg extract d.b.); Y_3, DPPH (mmol TRE/g extract d.b.); Y_4, ABTS (mmol TRE/g extract d.b.); x_1^*, codified temperature; x_2^*, codified time; x_3^*, codified ethanol concentration.

2.4. Box-Behnken Experimental Design

The Box-Behnken experimental design applied consisted of 12 replicated experiments and 2 replicates in the central point (Table 2). The experiments were randomized to avoid unpredictable effects on the responses. Experimental results were analyzed using the IBM SPSS Statistics 24 software and fitted to polynomials of the form:

$$Y = a_0 + \sum_{i=0}^{3} a_i\, x_i^* + \sum_{i=1}^{2} \sum_{\substack{j=2 \\ j > i}}^{3} a_{ij} x_i^* x_j^* + \sum_{i=1}^{3} a_{ii}\, x_i^{*2} \tag{1}$$

where Y is the dependent variable or response, a_0 is a scaling constant, a_i represents the linear coefficients, a_{ij} the interaction coefficients, a_{ii} the quadratic coefficients, and x_i^* the independent variables coded at three levels: -1 (lower limit), 0 (central point), and $+1$ (upper limit) (Table 2). Analysis of variance (ANOVA) was applied to determine the validity of the quadratic model as well as the statistical significance of the regression coefficients at a 95% confidence level. Moreover, to confirm the model's accuracy, predicted values for each dependent variable were calculated and compared with the experimental ones (Table 2). The equations obtained for each dependent variable were visualized as response surface plots.

2.5. Analytical Techniques

In order to have a more complete characterization of the antioxidant activity of the extracts, various methods were used, namely, the DPPH and ABTS assays based on the capacity to scavenge free radicals and the FRAP assay that measures the capacity to reduce a metal ion.

The DPPH radical scavenging ability of the extracts was determined following the method proposed by Barreira et al. [13] modified as described in Vázquez et al. [11]. The results were expressed as mmol Trolox equivalent (TRE) per g extract d.b. and as the EC_{50} value, or extract concentration necessary to achieve a 50% DPPH radical inhibition. ABTS scavenging activity was determined according to the method of Re et al. [14], and the results expressed as mmol Trolox equivalent (TRE) per g extract d.b. and as the EC_{50} value. The FRAP assay was done according to Szöllösi and Szöllösi-Varga [15]. The results were expressed as nmol ascorbic acid equivalent (AAE) per mg extract d.b.

Phenolic compounds in the extract selected as the optimum were determined by UPLC/ESI-QTOF-MS using a Bruker Elute UHPLC (Billerica, MA, USA) and a Bruker TimsTOF (Billerica, MA, USA). Separations were performed using a Bruker Intensity Solo C18 2 μm (2.1 mm × 100 mm) column (Billerica, MA, USA) and a binary gradient of 0.1% aqueous formic acid for mobile phase A and 0.1% formic acid in methanol for mobile phase B at a flow rate of 0.25 mL/min. The LC gradient was 5% B from 0 to 0.4 min, from 5% to 35% B from 0.4 to 0.5 min, from 35% to 100% B from 0.5 to 7 min, 100% B from 7 to 12 min, from 100% to 5% B from 12 to 12.1 min and 5% B from 12.1 to 15 min. Table 3 shows the regression equation for each standard compound analyzed. As observed, all the compounds showed good linearity in a relatively wide concentration range.

Table 3. Calibration curves for the standard compounds analyzed by ultra-performance liquid chromatography coupled with electrospray ionization and time-of-flight mass spectrometry (UPLC/ESI-QTOF-MS).

Compound	Linear Range (mg/L)	Calibration Curve	R^2
(–)-Gallocatechin	1–200	$y = 8068x + 17{,}932$	0.9941
Catechin hydrate	1–200	$y = 15{,}461x + 52{,}943$	0.9958
Chlorogenic acid	1–1000	$y = 8181x + 22{,}563$	0.9936
Ellagic acid	1–1000	$y = 9647x + 38{,}987$	0.9959
Epicatechin	1–1000	$y = 9780x + 17{,}102$	0.9976
Ferulic acid	1–200	$y = 3455x + 22{,}085$	0.9936
Gallic acid	1–1000	$y = 3814x + 4250$	0.9977
Isorharmnetin	1–200	$y = 61{,}453x + 78{,}552$	0.9915
Kaempferol	1–200	$y = 61{,}712x + 75{,}935$	0.9923
Neochlorogenic acid	1–200	$y = 10{,}675x + 15{,}125$	0.9989
p-Coumaric acid	1–200	$y = 4442x + 4535$	0.9972
Procyanidin B2	1–200	$y = 4014x + 7252$	0.9925
Quercetin	1–1000	$y = 45{,}006x + 111{,}541$	0.9922
Quercetin 3-β-D-glucoside	1–1000	$y = 13{,}239x + 42{,}498$	0.9836
Taxifolin	1–200	$y = 21{,}398x + 28{,}956$	0.9903

3. Results and Discussion

3.1. Effect of the Solid/Liquid Ratio on Extraction Yield and Extract Antioxidant Activity

The influence of the solvent to walnut leaves ratio on the extraction yield and extract FRAP antioxidant activity is shown in Table 1. The extraction yield was positively influenced by decreasing S/L from 1/5 to 1/7.5. However, a further decrease to 1/10 didn't improve the extraction yield significantly. On the other hand, the extract FRAP antioxidant activity increased with decreasing the S/L ratio in the whole range analyzed. These results were in agreement with previous research on the extraction of phenolic compounds from various plant materials [7,16,17] and can be explained by analyzing the concentration gradient of phenolics between walnut leaves and the solvent. The lower the S/L ratio, the greater the concentration gradient and consequently the extraction rate. Therefore, a S/L ratio of 1/10 was selected for subsequent experiments on analyzing the impact of temperature, time, and ethanol concentration on the extraction of phenolic compounds from walnut leaves. A greater decrease in the S/L ratio was not considered since it would mean not only a higher solvent consumption but also a higher energy cost in the extraction and solvent recovery stages.

3.2. Relationship between Extract Antioxidant Properties

Antioxidants may respond in a different manner to different radical or oxidant sources [18]. For this reason, three methods based on different reaction mechanisms were used to determine the antioxidant activity of walnut leaf extracts and the relationship between them was analyzed. A regression analysis was performed between the values of the FRAP, DPPH, and ABTS antioxidant activities for the experiments in Table 2. As shown in Figure 1, quite good positive linear relationships were found between the three methods used to measure the antioxidant activity, which suggests that only one method could be used in practice to provide reliable information on the antioxidant properties of walnut leaf extracts.

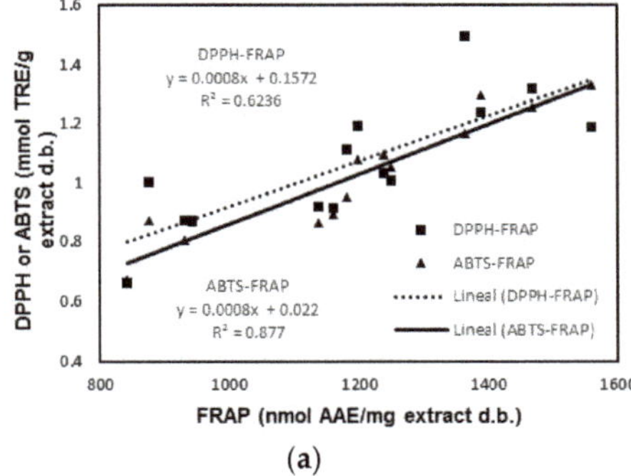

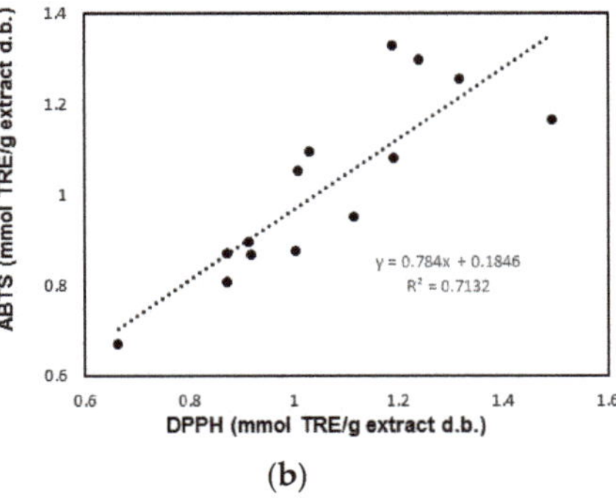

Figure 1. Linear relationships between FRAP and DPPH (2,2-diphenyl-1-picrylhydrazyl) and ABTS (2,2-azinobis-3-ethylbenzothiazoline-6-sulfonic acid) antioxidant activities (**a**) and between DPPH and ABTS antioxidant activities (**b**).

3.3. Optimization of the Extraction Conditions by a Box-Behnken Design

In order to determine the best extraction conditions for the extraction of phenolic compounds from walnut leaves, a Box-Behnken experimental design was applied. The independent variables selected were temperature, time, and ethanol concentration. The extraction conditions together with the experimental results obtained for the dependent variables: Extraction yield (Y_1), FRAP (Y_2), DPPH (Y_3), and ABTS (Y_4) antioxidant activities are shown in Table 2, that also shows the good agreement between experimental and predicted values. The fitting model (Equation (1)) was found appropriate to represent all the responses analyzed ($p < 0.05$). Table 4 shows the significant regression coefficients together with the statistical parameters used to evaluate the fitting results. For all the responses, the quadratic effect of %EtOH (in negative mode) was the most significant, followed by the linear effect of temperature (in positive mode), and the effects on which extraction time was involved were the least significant or not significant at all. Equations (2)–(5) show the dependence of the responses Y_1 to Y_4 on the significant independent variables, double interactions, and quadratic effects:

$$Y_1\ (\%) = 26.28 + 3.024\ x_1^* + 0.905\ x_2^* - 1.444\ x_3^* - 0.865\ x_1^* x_3^* - 3.973\ x_3^{*2} \tag{2}$$

$$Y_2\ (\text{nmol AAE/mg extract d.b.}) = 1395.25 + 13.25\ x_1^* - 354.50\ x_3^{*2} \tag{3}$$

$$Y_3\ (\text{mmol TRE/g extract d.b.}) = 1.223 + 0.173\ x_1^* - 0.093\ x_1^* x_3^* - 0.291\ x_3^{*2} \tag{4}$$

$$Y_4\ (\text{mmol TRE/g extract d.b.}) = 1.315 + 0.143\ x_1^* - 0.083\ x_1^{*2} - 0.125\ x_2^{*2} - 0.315\ x_3^{*2} \tag{5}$$

Figure 2a–c show the response surfaces for the extraction yield (Y_1) in the function of temperature and ethanol concentration for extraction times of 30 (a), 75 (b), and 120 min (c). Extraction yield varied between 17.80% and 30.55% depending on the extraction conditions. Regarding the influence of the linear effects, temperature and time were significant in the positive mode, whereas %EtOH in the negative mode. Only the quadratic effect of %EtOH and the double interaction temperature x %EtOH had significant influences on the response, both in the negative mode. The highest extraction yield in the ranges essayed of 30.55% was attained at the highest temperature and extraction time essayed, 75 °C and 120 min, and at an intermediate ethanol concentration, 38%.

Figures 3–5 show, respectively, the response surfaces for FRAP (Y_2), DPPH (Y_3), and ABTS (Y_4) antioxidant activities in function of temperature and ethanol concentration. All show a similar trend with respect to the influence of temperature and %EtOH, showing the presence of a maximum located in the vicinity of the design space for FRAP and DPPH and in the case of ABTS within it. Additionally, the responses for FRAP and DPPH were independent of time. FRAP ranged between 908 and 1529 nmol AAE/mg extract d.b. (Figure 3) and depended only on the negative quadratic effect of % EtOH and the linear effect of temperature in positive mode, reaching the highest value in the range assayed, at 75 °C and 50% ethanol. With respect to DPPH, it depended additionally on the interaction temperature x %EtOH in the negative mode, and a value of 1.40 mmol TRE/g extract d.b. was reached at 75 °C and 44% ethanol (Figure 4). ABTS depended on the quadratic effects of %EtOH, time and temperature, in this order of importance, all in the negative mode, and on the positive linear effect of temperature. For ABTS, a maximum located in the region studied, 1.38 mmol TRE/g extract d.b., was reached at 71.3 °C, 75 min, and 50% ethanol (Figure 5).

In summary, solvent concentration, temperature, and extraction time, in this order, are important factors influencing the extraction of phenolic compounds from walnut leaves. Both extraction yield and extract antioxidant activity depended significantly on the % ethanol used and aqueous solutions in the range of 38% to 50% ethanol were found to be superior to the more concentrated ones in any of the solvents, behavior that might be explained by the "likes dissolves like" principle [19]. High ethanol concentrations dissolve more lipophilic compounds, whereas higher proportions of hydrophilic compounds are extracted at low ethanol concentrations [20]. Otherwise, extraction yield increases when decreasing %EtOH up to 38%, whereas extract antioxidant capacity diminishes, which can be explained due to the solubilization of other types of compounds such as proteins and polysaccharides that impair selectivity [21]. According to these results, the ethanol concentration selected as the optimum was 50%. Concerning extraction temperature, a temperature rise of up to 75 °C increased the extraction yield as the extraction process was favored by temperature as the solubility of phenolic compounds and the mass transfer rate were enhanced [22]. Moreover, FRAP, DPPH, and ABTS antioxidant activity of the extracts also increased with increasing extraction temperature to 75 °C. However, greater extraction temperatures were not recommended as oxidation, epimerization, and degradation of phenolic compounds was promoted [17] with the consequent decrease of the extract antioxidant activity. Therefore, 75 °C was selected as the optimum temperature. Finally, 120 min was the selected time to favor the extraction yield, as its influence on antioxidant properties was almost negligible. Longer extraction times were not proposed since the effects were the same as those previously indicated for high temperatures [17].

In brief, the optimal conditions for the extraction of phenolic compounds from *Juglans major 209 x Juglans regia* leaves selected by analyzing together the response surfaces for all the dependent variables (Figures 2–5) were those corresponding to Experiment 12 (Table 2): 75 °C, 120 min, and 50% ethanol. Under these conditions, the predicted responses were: $Y_{1,\,pred} = 30.21\%$, $Y_{2,\,pred} = 1529$ nmol AAE/mg extract d.b., $Y_{3,\,pred} = 1.396$ mmol TRE/g extract d.b. and $Y_{4,\,pred} = 1.25$ mmol TRE/g extract d.b. These values were equal or very close to those corresponding to the model optimum for each response that were $Y_{1,pred} = 30.55\%$ at 75 °C, 120 min, and 38% EtOH, $Y_{2,pred} = 1529$ nmol AAE/mg extract d.b. at 75 °C and 50% EtOH, $Y_{3,pred} = 1.40$ mmol TRE/g extract d.b. at 75 °C and 44% EtOH and $Y_{4,pred} = 1.38$ mmol TRE/g extract d.b. at 71.3 °C, 75 min and 50% EtOH. Moreover, these extraction conditions were close to those found by Vieira et al. [7] as the global optimum conditions for the extraction of some of the main phenolic compounds found in this work, namely, 3-*O*-caffeoylquinic acid and quercetin 3-*O*-glucoside, from walnut (*Juglans regia*) leaves by maceration with aqueous ethanol (61.3 °C, 112.5 min, and 50.4% ethanol). 50% ethanol also led to higher antioxidant activities for walnut (*Juglans regia*) green husk extracts than pure water and ethanol, although the highest extraction yield was obtained with water [23].

Table 4. Coefficients of the models (Equation (1)) and statistical parameters.

	Y_1			Y_2			Y_3			Y_4		
	Coeff.	SE	p	Coeff	SE	p	Coeff.	SE	p	Coeff.	SE	p
a_0	26.281	0.274	0.000	1395.250	40.672	0.000	1.223	0.019	0.000	1.315	0.028	0.000
a_1	3.024	0.274	0.000	13.25	40.672	0.006	0.173	0.019	0.000	0.143	0.019	0.000
a_2	0.905	0.274	0.008	-	-	NS	-	-	NS	-	-	NS
a_3	−1.444	0.274	0.000	-	-	NS	-	-	NS	-	-	NS
a_{12}	-	-	NS	-	-	NS	-	-	NS	-	-	NS
a_{13}	−0.865	0.387	0.049	-	-	NS	−0.093	0.027	0.005	-	-	NS
a_{23}	-	-	NS	-	-	NS	-	-	NS	-	-	NS
a_{11}	-	-	NS	-	-	NS	-	-	NS	−0.083	0.028	0.012
a_{22}	-	-	NS	-	-	NS	-	-	NS	−0.125	0.028	0.001
a_{33}	−3.973	0.387	0.000	−354.500	57.519	0.000	−0.291	0.027	0.000	−0.315	0.028	0.000
R^2	0.964			0.789			0.947			0.951		
R^2corr.	0.947			0.757			0.934			0.933		
SE	0.774			115.038			0.053			0.055		
p	0.000			0.000			0.000			0.000		

NS: Non-significant for a 95% confidence level; SE: Standard error; p: Probability; R^2: Regression coefficient; R^2_c: Corrected regression coefficient; a_0: Scaling constant; a_i: Linear coefficients; a_{ij}: Interaction coefficients; a_{ii}: Quadratic coefficients; Y_1, extraction yield (g extract/100 g leaves d.b.); antioxidant activity: Y_2, FRAP (nmol AAE/mg extract d.b.); Y_3, DPPH (mmol TRE/g extract d.b.); Y_4, ABTS (mmol TRE/g extract d.b.).

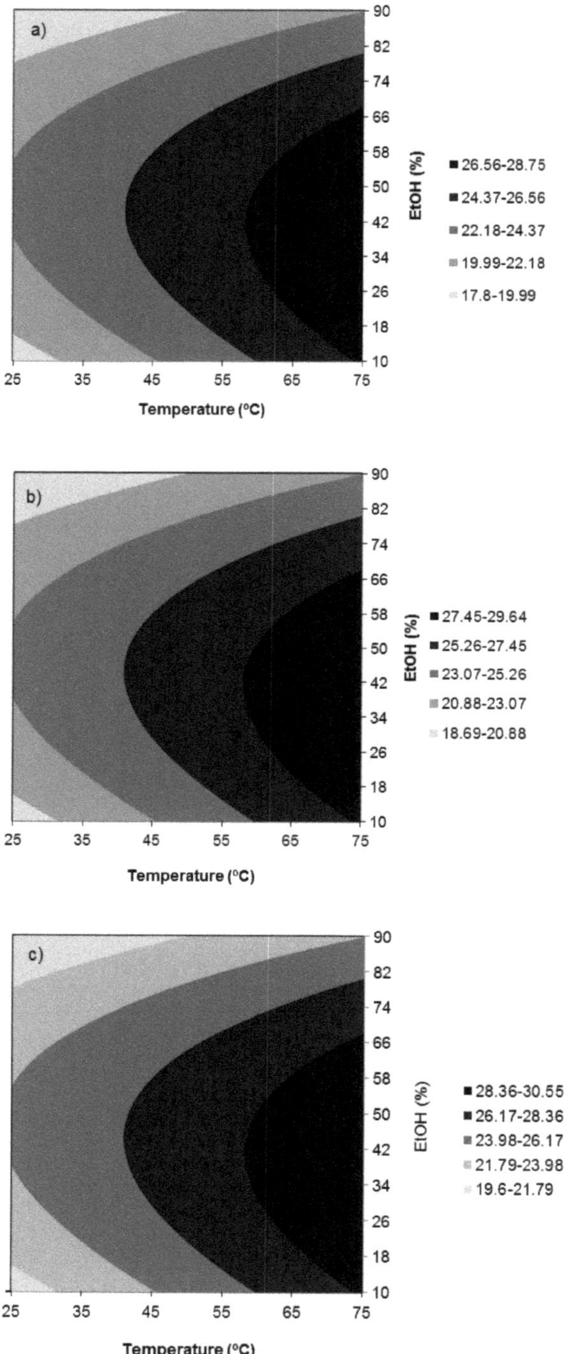

Figure 2. Response surfaces for extraction yield (Y_1, %) in function of temperature and ethanol concentration for extraction times of (**a**) 30, (**b**) 75, and (**c**) 120 min.

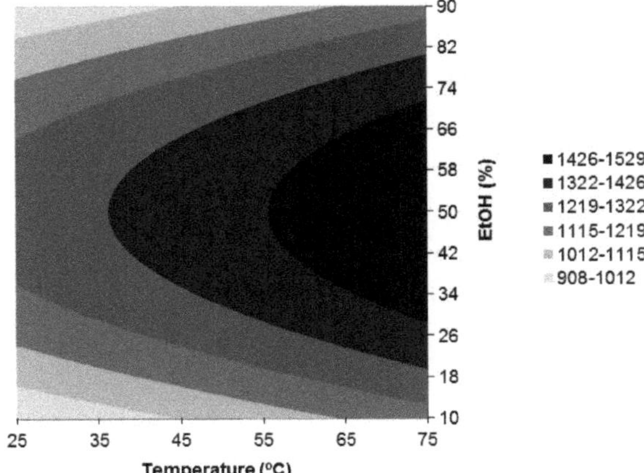

Figure 3. Response surfaces for FRAP antioxidant activity (Y_2, nmol AAE/mg extract d.b.) in the function of temperature and ethanol concentration.

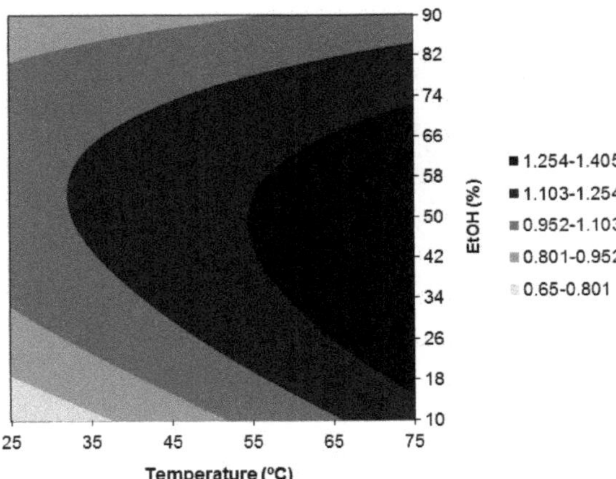

Figure 4. Response surfaces for DPPH antioxidant activity (Y_3, mmol TRE/g extract d.b.) in the function of temperature and ethanol concentration.

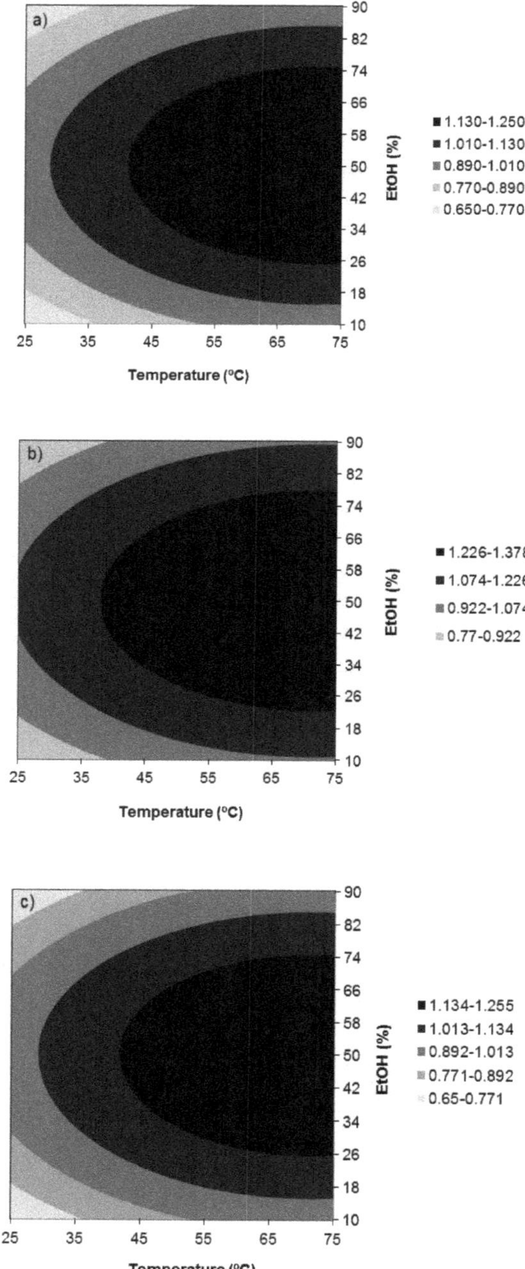

Figure 5. Response surfaces for ABTS antioxidant activity (Y_4, mmol TRE/g extract d.b.) in the function of temperature and ethanol concentration for extraction times of (**a**) 30, (**b**) 75, and (**c**) 120 min.

3.4. Characterization of the Optimum Walnut Leaf Extract

Both the experimental values and those predicted by the models (Equations (2)–(5), Table 4) for the extraction yield and antioxidant properties of the walnut leaf extract selected as the optimum are presented in Table 2 (Experiment 12). Additionally, Figure 6a,b show the ABTS and DPPH radical inhibition percentages, respectively, versus extract concentration for this walnut extract and for Trolox, the synthetic antioxidant used as a reference. The corresponding EC_{50} values were 0.985 and 0.11 mg/mL, for ABTS and DDPH, respectively. The DPPH EC_{50} value obtained in this work was lower (higher antioxidant activity) than those found for extracts obtained from leaves of different *Juglans regia* cultivars using other solvent systems: Aqueous extracts (0.151–0.202 mg/mL) [1], methanolic extracts (0.199 mg/mL), and petroleum ether extract (2.921 mg/mL) [3]. However, even lower EC_{50} values were obtained for methanolic (65.91 µg/mL) and decoction (78.97 µg/mL) extracts from certain walnut cultivars [4]. The DPPH EC_{50} value obtained in this work for the optimum walnut leaf extract was also significantly lower than that for *Juglans regia* green husk extracts, 0.33 mg/mL for 50% aqueous ethanol extracts [23], and 0.41 for methanolic extracts [3], and also slightly lower than that for the methanolic extract from walnut seeds (0.14 mg/mL) [3]. With respect to extraction yield it was slightly higher than that reported by Carvalho et al. [3] for the methanolic extract, 27.7%, and significantly higher than that of the petroleum ether ones, 1.1%.

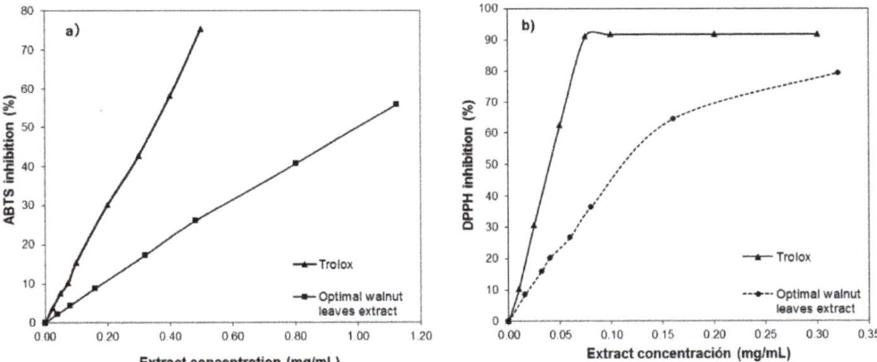

Figure 6. Percentage of ABTS (**a**) and DPPH (**b**) radical inhibition against extract concentration for the walnut leaf extract selected as the optimum with Trolox as reference.

Finally, to identify and quantify the compounds responsible for the antioxidant activity of the optimum *Juglans major 209 x Juglans regia* leaf extract, it was analyzed by UPLC/ESI-QTOF-MS. Table 5 and Figure 7 show the phenolic profile determined. Fifteen phenolic compounds were identified and quantified, with the flavonoid quercetin 3-β-D-glucoside being the major compound, representing 49.35% of the total phenolic composition, followed by chlorogenic acid (29.16%), and neochlorogenic acid (18.34%). Both quercetin 3-β-D-glucoside and chlorogenic acid were among the main phenolic compounds found in *Juglans regia* leaf extracts from some Spanish and Portuguese cultivars [1,4,7] that also presented significant quantities of other flavonoids such as quercetin 3-*O*-galactoside, quercetin *O*-pentoside, or quercetin *O*-xyloside [1,4,5,7] not identified in this work. However, the outstanding presence of neochlorogenic acid in the *Juglans major 209 x Juglans regia* leaf extracts, also reported by Pereira et al. [1], makes the main difference with the *Juglans regia* ones.

Concerning the interest of the main phenolic compounds identified, quercetin 3-β-D glucoside has been attracting increasing research interest. It is widely distributed in fruits, vegetables, and cereals and exhibits a broad number of chemoprotective effects both in vitro and in vivo, against oxidative stress, cancer, cardiovascular disorders, diabetes, and allergic reactions [24]. Chlorogenic acids are also present in many vegetables and fruits and some health benefits have been associated with them such as the prevention of cardiovascular diseases and types 2 diabetes [25] and also antioxidants, anticancer, anti-inflammatory, and immunomodulatory properties have been shown [26]. The importance of these phenolic compounds for various industrial applications demonstrate the interest of the valorization of *Juglans major 209 x Juglans regia* leaves.

Table 5. Identified compounds in the optimized walnut leaf extract by UPLC/ESI-QTOF-MS (T = 75 °C, t = 120 min, EtOH concentration = 50%, S/L ratio = 1/10 g/mL).

Compound	Retention Time (min)	*m/z*	ng/mg Extract d.b.	mg/100 g Leaves d.b.
(–)-Gallocatechin	2.4	305.06	14.17	0.45
Catechin hydrate	2.6	289.07	55.83	1.78
Chlorogenic acid	2.7	353.09	5737.50	183.14
Ellagic Acid	4	300.99	68.92	2.20
Epicatechin	2.8	289.07	12.08	0.39
Ferulic acid	3.6	193.04	158.58	5.06
Gallic acid	2.3	169.01	432.58	13.81
Isorharmnetin	5.4	315.05	3.17	0.10
Kaempferol	5.3	285.04	14.25	0.45
Neochlorogenic acid	2.5	353.08	9125.00	291.27
p-Coumaric ccid	3.6	163.04	10.83	0.35
Procyanidin B2	2.6	577.14	45.50	1.45
Quercetin	4.8	301.04	163.50	5.22
Quercetin 3-β-D-glucoside	3.8	463.09	15,441.67	492.90
Taxifolin	3.4	303.05	4.50	0.14

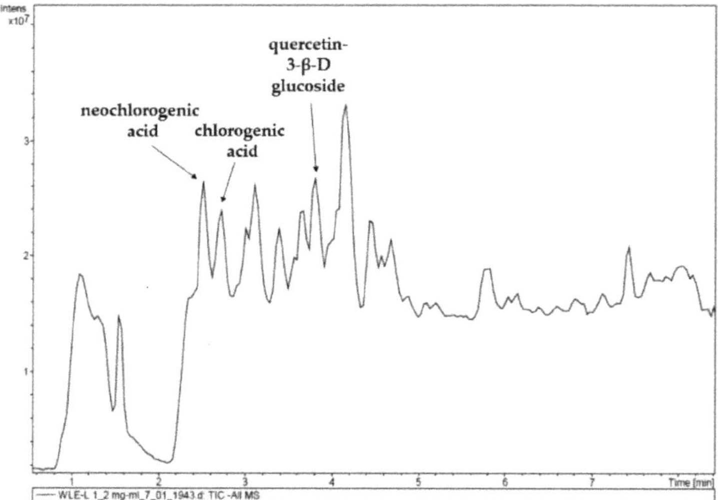

Figure 7. UPLC chromatogram of the walnut leaf extract selected as the optimum (T = 75 °C, t = 120 min, EtOH concentration = 50%, S/L ratio = 1/10 (g/mL).

4. Conclusions

Phenolic compounds such as quercetin 3-β-D-glucoside (4.92 mg/g), neochlorogenic acid (2.91 mg/g), and chlorogenic acid (1.83 mg/g) have been recovered from walnut (*Juglans major 209 x Juglans regia*) leaves by maceration in aqueous ethanol solutions under optimized conditions, 75 °C, 120 min, and 50% aqueous ethanol, selected based on extraction yield and extract antioxidant properties. The experimental values for the optimal extract properties were: Extraction yield, 30.17%; FRAP, 1468 nmol AAE/mg extract d.b.; DPPH, 1.318 mmol TRE/g extract d.b.; DPPH EC$_{50}$, 0.11 mg/mL; ABTS, 1.256 mmol TRE/g extract d.b. and ABTS EC$_{50}$, 0.985 mg/mL. Therefore, the present work proposes a valorization way of walnut leaves as a source of valuable compounds for various industrial applications based on their antioxidant properties.

Author Contributions: Conceptualization, M.S.F. and J.G.-Á.; formal analysis, M.S.F. and J.G.-Á.; investigation, A.F.-A., A.C.-I., M.S.F., and J.G.-Á.; methodology, A.F.-A. and A.C.-I.; resources, J.G.-Á.; software, A.F.-A. and A.C.-I.; supervision, M.S.F. and J.G.-Á.; validation, M.S.F. and J.G.-Á.; writing—original draft, J.G.-Á.; writing—review and editing, A.F.-A., A.C.-I., M.S.F. and J.G.-Á. All authors have read and agreed to the published version of the manuscript.

Funding: This research received no external funding.

Acknowledgments: The authors want to thank the use of RIAIDT-USC analytical facilities and Bosques Naturales S.A. for supplying walnut leaves.

Conflicts of Interest: The authors declare no conflict of interest.

References

1. Pereira, J.A.; Oliveira, I.; Sousa, A.; Valentao, P.; Andrade, P.B.; Ferreira, I.C.F.R.; Ferreres, F.; Bento, A.; Seabra, R.; Estevinho, L. Walnut (*Juglans regia* L.) leaves: Phenolic compounds, antibacterial activity and antioxidant potential of different cultivars. *Food Chem. Toxicol.* **2007**, *45*, 2287–2295. [CrossRef] [PubMed]
2. Almeida, I.F.; Fernandes, E.; Lima, J.L.F.C.; Costa, P.C.; Bahia, M.F. Walnut (*Juglans regia*) leaf extracts are strong scavengers of pro-oxidant reactive species. *Food Chem.* **2008**, *106*, 1014–1020. [CrossRef]
3. Carvalho, M.; Ferreira, P.J.; Mendes, V.S.; Silva, R.; Pereira, J.A.; Jerónimo, C.; Silva, B.M. Human cancer cell antiproliferative and antioxidant activities of *Juglans regia* L. *Food Chem. Toxicol.* **2010**, *48*, 441–447. [CrossRef] [PubMed]
4. Santos, A.; Barros, L.; Calhelha, R.C.; Dueñas, M.; Carvalho, A.M.; Santos-Buelga, C.; Ferreira, I.C.F.R. Leaves and decoction of *Juglans regia* L.: Different performances regarding bioactive compounds and in vitro antioxidant andantitumor effects. *Ind. Crops Prod.* **2013**, *51*, 430–436. [CrossRef]
5. Amaral, J.S.; Seabra, R.M.; Andrade, P.B.; Valentão, P.; Pereira, J.A.; Ferreres, F. Phenolic profile in the quality control of walnut (*Juglans regia* L.) leaves. *Food Chem.* **2004**, *88*, 373–379. [CrossRef]
6. Nour, V.; Trandafir, I.; Cosmulescu, S. HPLC determination of phenolic acids, flavonoids and juglone in walnut leaves. *J. Chromatogr. Sci.* **2013**, *51*, 883–890. [CrossRef]
7. Vieira, V.; Prieto, M.A.; Barros, L.; Coutinho, J.A.P.; Ferreira, O.; Ferreira, I.C.F.R. Optimization and comparison of maceration and microwave extraction systems for the production of phenolic compounds from *Juglans regia* L. for the valorization of walnut leaves. *Ind. Crops Prod.* **2017**, *107*, 341–352. [CrossRef]
8. Chemat, F.; Vian, M.A.; Cravotto, G. Green extraction of natural products: Concept and principles. *Int. J. Mol. Sci.* **2012**, *13*, 8615–8627. [CrossRef]
9. Moure, A.; Cruz, J.M.; Franco, D.; Domínguez, J.M.; Sineiro, J.; Domínguez, H. Natural antioxidants from residual sources. *Food Chem.* **2001**, *72*, 145–171. [CrossRef]
10. Piwowarska, N.; González-Alvarez, J. Extraction of antioxidants from forestry biomass: Kinetics and optimization of extraction conditions. *Biomass Bioenerg.* **2012**, *43*, 42–51. [CrossRef]
11. Vázquez, G.; Fernández-Agulló, A.; Gómez-Castro, C.; Freire, M.S.; Antorrena, G.; González-Álvarez, J. Response surface optimization of antioxidants extraction from chestnut (*Castanea sativa*) bur. *Ind. Crops Prod.* **2012**, *35*, 126–134. [CrossRef]
12. Xavier, L.; Freire, M.S.; González-Álvarez, J. Modeling and optimizing the solid–liquid extraction of phenolic compounds from lignocellulosic subproduct. *Biomass Conv. Bioref.* **2019**, *9*, 737–747. [CrossRef]
13. Barreira, J.C.M.; Ferreira, I.C.F.R.; Oliveira, M.B.P.P.; Pereira, J.A. Antioxidant activities of the extracts from chestnut flower, leaf, skins and fruit. *Food Chem.* **2008**, *107*, 1106–1113. [CrossRef]
14. Re, R.; Pellegrini, N.; Proteggente, A.; Pannala, A.; Yang, M.; Rice-Evans, C. Antioxidant activity applying an improved ABTS radical cation decolorization assay. *Free Radic. Biol. Med.* **1999**, *26*, 1231–1237. [CrossRef]
15. Szöllösi, R.; Szöllösi-Varga, I. Total antioxidant power in some species of Labiatae (adaptation of FRAP method). *Acta Biol. Szeged.* **2002**, *46*, 125–127.
16. Vuong, Q.V.; Hirun, S.; Roach, P.D.; Bowyer, M.C.; Phillips, P.A.; Scarlett, C.J. Effect of extraction conditions on total phenolic compounds and antioxidant activities of *Carica* papaya leaf aqueous extracts. *J. Herb. Med.* **2013**, *3*, 104–111. [CrossRef]
17. Tchabo, W.; Ma, Y.; Kwaw, E.; Xiao, L.; Wu, M.; Apaliya, M.T. Impact of extraction parameters and their optimization on the nutraceuticals and antioxidant properties of aqueous extract mulberry leaf. *Int. J. Food Prop.* **2018**, *21*, 717–732. [CrossRef]

18. Prior, R.L.; Wu, X.; Schaich, K. Standardized methods for the determination of antioxidant capacity and phenolics in foods and dietary supplements. *J. Agric. Food Chem.* **2005**, *53*, 4290–4302. [CrossRef]
19. Zhang, Z.-S.; Li, D.; Wang, L.-J.; Ozkan, N.; Chen, X.D.; Mao, Z.-H.; Yang, H.-Z. Optimization of ethanol–water extraction of lignans from flaxseed. *Sep. Purif. Technol.* **2007**, *57*, 17–24. [CrossRef]
20. Thoo, Y.Y.; Ho, S.K.; Liang, J.Y.; Ho, C.W.; Tan, C.P. Effects of binary solvent extraction system, extraction time and extraction temperature on phenolic antioxidants and antioxidant capacity from mengkudu (*Morinda citrifolia*). *Food Chem.* **2010**, *120*, 290–295. [CrossRef]
21. Jokic, S.; Velic, D.; Bilic, M.; Bucic-Kojic, A.; Planinic, M.; Tomas, S. Modelling of the process of solid- liquid extraction of total polyphenols from soybeans. *Czech J. Food Sci.* **2010**, *28*, 206–212. [CrossRef]
22. Dai, J.; Mumper, R.J. Plant phenolics: Extraction, analysis and their antioxidant and anticancer properties. *Molecules* **2010**, *15*, 7313–7352. [CrossRef]
23. Fernández-Agulló, A.; Pereira, E.; Freire, M.S.; Valentao, P.; Andrade, P.B.; González-Álvarez, J.; Pereira, J.A. Influence of solvent on the antioxidant and antimicrobial properties of walnut (*Juglans regia* L.) green husk extracts. *Ind. Crops Prod.* **2013**, *42*, 126–132.
24. Valentová, K.; Vrba, J.; Bancír̆ová, M.; Ulrichová, J.; Kr̆en, V. Isoquercitrin: Pharmacology, toxicology, and metabolism. *Food Chem. Toxicol.* **2014**, *68*, 267–282. [CrossRef]
25. Gutiérrez Ortiz, A.L.; Berti, F.; Navarini, L.; Monteiro, A.; Resmini, M.; Forzato, C. Synthesis of p-coumaroylquinic acids and analysis of their interconversion. *Tetrahedron Asymmetry* **2017**, *28*, 419–427. [CrossRef]
26. Jin, U.H.; Lee, J.Y.; Kang, S.K.; Kim, J.K.; Park, W.H.; Kim, J.G.; Moon, S.K.; Kim, C.H. A phenolic compound, 5-caffeoylquinic acid (chlorogenic acid), is a new type and strong matrix metalloproteinase-9 inhibitor: Isolation and identification from methanol extract of Euonymus. *Life Sci.* **2005**, *77*, 2760–2769. [CrossRef]

 *antioxidants*

Article

The Metabolite Urolithin-A Ameliorates Oxidative Stress in Neuro-2a Cells, Becoming a Potential Neuroprotective Agent

Guillermo Cásedas [1], Francisco Les [1,2], Carmen Choya-Foces [3], Martín Hugo [3] and Víctor López [1,2,*]

[1] Facultad de Ciencias de la Salud, Universidad San Jorge, 50830 Villanueva de Gállego (Zaragoza), Spain; gcasedas@usj.es (G.C.); fles@usj.es (F.L.)
[2] Instituto Agroalimentario de Aragón-IA2 (CITA-Universidad de Zaragoza), 50059 Zaragoza, Spain
[3] Unidad de Investigación, Hospital Universitario Santa Cristina, Instituto de Investigación Sanitaria Princesa (IIS-IP), E-28009 Madrid, Spain; carmenchoyaf@gmail.com (C.C.-F.); martin.hugo@salud.madrid.org (M.H.)
* Correspondence: ilopez@usj.es

Received: 7 February 2020; Accepted: 17 February 2020; Published: 21 February 2020

Abstract: Urolithin A is a metabolite generated from ellagic acid and ellagitannins by the intestinal microbiota after consumption of fruits such as pomegranates or strawberries. The objective of this study was to determine the cytoprotective capacity of this polyphenol in Neuro-2a cells subjected to oxidative stress, as well as its direct radical scavenging activity and properties as an inhibitor of oxidases. Cells treated with this compound and H_2O_2 showed a greater response to oxidative stress than cells only treated with H_2O_2, as mitochondrial activity (MTT assay), redox state (ROS formation, lipid peroxidation), and the activity of antioxidant enzymes (CAT: catalase, SOD: superoxide dismutase, GR: glutathione reductase, GPx: glutathione peroxidase) were significantly ameliorated; additionally, urolithin A enhanced the expression of cytoprotective peroxiredoxins 1 and 3. Urolithin A also acted as a direct radical scavenger, showing values of 13.2 μM Trolox Equivalents for Oxygen Radical Absorbance Capacity (ORAC) and 5.01 μM and 152.66 μM IC_{50} values for superoxide and 2,2-diphenyss1-picrylhydrazyl (DPPH) radicals, respectively. Finally, inhibition of oxidizing enzymes, such as monoamine oxidase A and tyrosinase, was also detected in a dose-dependent manner. The cytoprotective effects of urolithin A could be attributed to the improvement of the cellular antioxidant battery, but also to its role as a direct radical scavenger and enzyme inhibitor of oxidases.

Keywords: urolithin A; pomegranate; ellagic acid; dietary polyphenols; neuroprotection; peroxiredoxins

---

1. Introduction

Urolithins are natural polyphenolic compounds obtained from ellagic acids and ellagitannins by the gut microbiota. Ellagic acid is a phenolic antioxidant compound present in numerous fruits, such as pomegranates, strawberries, or nuts. Ellagitannins are a class of polymeric ellagic acid derivatives also present in pomegranates (punicalagins and punicalins) and other sources, such as oak or chestnut wood (castalagin, castalin, roburin A, grandinin) or *Melaleuca quinquenervia* leaves (castalin and grandinin) [1–3].

The beneficial properties of these plants and foods seem to be in relation with these polyphenolic metabolites; however, the metabolism of polyphenols from food seems to be insufficient to achieve adequate levels of urolithins in the body. In addition, it has been proven that apparently beneficial foods such as pomegranates have had a lot of interindividual variability due to the different urolithin metabotypes present in the population [4]. In fact, only 1 in 3 people have the right microbiota to perform this metabolism with maximum efficiency [5].

Therefore, it is very important to evaluate the activity of isolated urolithins as potential therapeutic agents. In addition, the use if urolithin A in humans and the safety profile of this compound have been widely evaluated, with no adverse effects on health observed [6].

Although urolithins are a group of metabolites, urolithin A (UA), also known as 3,8-dihydroxyurolithin, is one of the most representative compounds. There are numerous studies that demonstrate an important role of this compound in metabolic syndrome, improving cardiovascular function, decreasing the formation of triglycerides, inhibiting enzymes such as lipase or glucosidase, or relieving insulin resistance [7–9]. It has also been observed that UA may have an important role in the prevention of certain cancers, such as colorectal or prostate cancers [10,11]. UA also has an important role at the mitochondrial level, being able to activate mitophagy and prolonging lifespan in *Caenorhabditis elegans* worms, as well as beneficial mitochondrial effects in the skeletal muscle [12,13].

The set of all these beneficial properties for health may be due to the antioxidant capacity of polyphenols. However, there are few studies that link the antioxidant properties of this metabolite with a potential therapeutic activity in neurodegenerative diseases, where the redox status is essential. Therefore, the objective of this study was to evaluate whether urolithin A has antioxidant and neuroprotective effects using Neuro-2a cells and other in vitro models involving the use of central nervous system (CNS) enzymes or free radicals.

2. Materials and Methods

2.1. Reagents and Chemicals

Urolithin A (3,8-dihydroxyurolithin) (Figure 1) was purchased from Toronto Research Chemicals (TRC, Toronto, ON, Canada). Neuro-2a (N2a) cell line was provided from the American Type Culture Collection (ATCC, Manassas, VA, USA), while Monoamine oxidase A (MAO-A), 5,5′-dithiobis-(2-nitrobenzoic acid) (DTNB), Tris, galantamine, levodopa (L-DOPA), tyramine, horseradish peroxidase, 2,2′-azobis(2-methyl-propionamidine)-di-hydrochloride (AAPH), hydrogen peroxide (30% *w/w*), 2,4,6-Tris(2-pyridyl)-1,3,5-triazine (TPTZ), 2′,7′-dichloro-dihy-drofluorescein diacetate (DCFH-DA), 3-(4,5-Dimethyl-2-thiazolyl)-2,5- diphenyltetrazolium bromide (MTT), 6-hydroxy-2,5,7,8-tetra-methylchromane-2-carboxylic acid (Trolox), bovine serum albumin (BSA), vanillic acid, 4-aminoantipyrine, penicillin, streptomycin, tyrosinase, thiobarbituric acid (TBA), trichloroacetic acid (TCA), bicinchoninic acid (BCA), and pyrogallol were obtained from Sigma-Aldrich (Madrid, Spain). Clorgyline and α-kojic were from Cymit química (Barcelona, Spain). Na$_2$CO$_3$, HCl, NaCl, dimethyl sulfoxide (DMSO), MeOH, and potassium phosphate were supplied from Panreac (Barcelona, Spain). Dulbecco's modified Eagle's medium (DMEM), phosphate-buffered saline (PBS), and fetal bovine serum (FBS) were acquired from Gibco (Invitrogen, Paisley, UK).

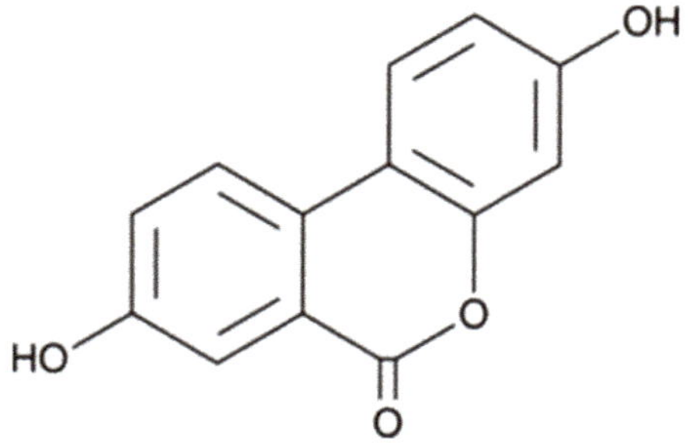

Figure 1. Structure of Urolithin A. Synonyms: 3,8-Dihydroxyurolithin; 3,8-Hydroxydibenzo-α-pyrone; 2′,7-Dihydroxy-3,4-benzocoumarin; δ-Lactone 2′,4,4′-Trihydroxy-2-biphenylcarboxylic acid.

2.2. Cytoprotective Properties of Urolithin A in Neuro-2a Cells

2.2.1. Neuro-2a Cell Culture and Treatments with Urolithin A and Hydrogen Peroxide

Neuro-2a cells (ATCC® CCL-131™) were thawed and cultured in 10% FBS-supplemented DMEM and 1% penicillin-streptomycin, seeded in a T175 flask, and placed into the incubator (5% CO_2, 37 °C). Once they reached the state of confluence, they were sub-cultured in a 96-well plate at a density of 1×10^4 cells/well in DMEM (10%) and incubated (37 °C, 5% CO_2) for 24 h before the treatment.

Stock solutions of urolithin A (438 mM) were prepared in sterilized PBS containing 1% DMSO (final concentration in the cells less than 0.1%). The sample was vortexed and filtered with a 0.22-µm syringe filter. Five dilution series were done from the stock solution. Stock solutions of hydrogen peroxide (1000 µM) were prepared and incorporated into the cells at a final concentration of 250 µM (45 min) to induce oxidative stress in the cells.

2.2.2. Mitochondrial Activity in Neuro-2a Cells Subjected to Oxidative Stress after Urolithin A Treatment

Cells were cultured in 96-well plates at a concentration of 1×10^4 cells per well. The cytotoxicity of urolithin A (0.5 µM–20 µM) was measured after 24 h by adding MTT (3-(4,5-dimethylthiazol -2-yl)-2,5-diphenyltetrazolium bromide) to the cell culture. Additionally, this assay can be performed to assess the potential cytoprotective effect of urolithin A after inducing a neuronal injury with 250 µM hydrogen peroxide for 45 min. After treatments (24 h exposition to urolithin, 45 min to hydrogen peroxide), cells were incubated for an additional 24 h; DMEM was then removed from every well and replaced with MTT solution (2 mg/mL in DMEM) and incubated at 37 °C for 3 h. Finally, MTT solution was removed and 100 µL of DMSO was added in each well to dissolve formazan crystals. Via a Synergy H1 Hybrid Multi-Mode Reader (Biotek, Bad Friedrichshall, Germany), the absorbance was read at 550 nm. Experiments were carried out in different weeks and diverse passages and expressed as percentage of control.

2.2.3. ROS Production in Neuro-2a Cells Subjected to Oxidative Stress after Exposition to Urolithin A

Neuro-2a cells were seeded in a 96-well plate. After 24 h, the medium was replaced with PBS supplemented with glucose and DCFH-DA (2,7-di-chloro-dihydrofluorescein diacetate, 0.01 M) for 30 min at 37 °C. After this time, PBS was removed and cells were washed twice with new PBS and treated with different concentrations of urolithin A (0.5–4 µM), as well as hydrogen peroxide (250 µM). The absorbance was checked at 480 nm ($\lambda_{excitation}$) and 520 nm ($\lambda_{emission}$) wavelengths [14]. The kinetic was performed over 90 min in a Synergy H1 Hybrid Multi-Mode Reader (Winooski, VT, USA). Results were represented as a percentage of intracellular ROS production (100% of control).

2.2.4. Lipid Peroxidation in Neuro-2a Cells Subjected to Oxidative Stress after Exposition to Urolithin A (TBARS assay)

Lipid peroxidation was assessed using the protocol by Mitsuru Uchiyama [15]. First, 100 µL of thawed pellets were placed into 200 µL of TBA–TCA–HCl cocktail and subsequently boiled at 100 °C for 10 min to accelerate the reaction. The reaction was broken by placing the samples on ice and stunning them in a vortex three times for 20 min and centrifuged at 4 °C (3000 rpm, 10 min). The obtained supernatant was placed in three different wells of a 96-well plate and absorbance was read at 530 nm using a Synergy H1 Hybrid Multi-Mode Reader.

2.2.5. Activity of Antioxidant Enzymes in Neuro-2a Cells Subjected to Oxidative Stress after Exposition to Urolithin A

Before analyzing the activity of antioxidant enzymes, the protein content of Neuro-2a cells was calculated by colorimetric bicinchoninic acid (BCA) assay and normalized lysis buffer for 20 min (Ethylenediaminetetraacetic acid (EDTA) 1 mM, Tris 25 mM, NaCl 150 mM and 0.1% Triton; pH = 7.4).

Moreover, leupeptin, pepstatin, and phenylmethylsulfonyl fluoride (PMSF) proteinase inhibitors (20, 10, and 35 µL/mL, respectively) were aggregated into the buffer. Finally, supernatant was kept for the experiments and the precipitate was discard.

The activity of catalase was measured as follows: hydrogen peroxide (1970 µL, 15 mM) was prepared in sodium phosphate buffer (pH = 7.5) supplemented with 30 µL of supernatant [16]; the mixture was prepared in a quartz cuvette, measuring the absorbance for 30 secs at 240 nm, using a Shimadzu Spectrophotometer UV-1800 (Duisburg, Germany). The activity of the enzyme was expressed following the next equation: Catalase activity = $((\Delta Abs/min) \times 2 \times F)/(0.0436 \times Vs \times C)$

43.6 mL $nmol^{-1}$ cm^{-1}: molar extinction coefficient of H_2O_2

F: dilution factor

C: Protein concentration (mg/mL)

Vs: Sample volume (mL)

$\Delta Abs/min$: activity of the kinetic

For superoxide dismutase (SOD), a mixture of 1555 µL of Tris–DTPA buffer (50 mM, pH 8.2), 20 µL of pyrogallol (23.78 mM) diluted in HCl (10 mM), and 25 µL total cell extracts was placed in a quartz cuvette to measure the oxidation of pyrogallol at 420 nm for 1 min [17], using a Shimadzu Spectrophotometer UV-1800 (Duisburg, Germany). The activity of SOD was expressed with the following equation: % Inhibition = $(\Delta Abs$ control $- \Delta Abs$ sample$)/(\Delta Abs$ control$) \times 100$

SOD activity = $(\%$ Inhibition $\times 2 \times F)/(50 \times Vs \times C)$

Glutathione reductase activity (GR) was quantified with the protocol developed by Staal et al. [18]: 1180 µL of 50 mM phosphate buffer—EDTA (6.3 mM, pH 7.4), 50 µL of total cell extracts, 35 µL of glutathione disulfide (GSSG) 5 mM, and 35 µL of nicotinamide adenine dinucleotide phosphate (NADPH; 2.4 mM), which were mixed in a quartz cuvette. Glutathione peroxidase activity (GPx) was detected by mixing 1220 µL of 50 mM phosphate buffer—EDTA 6.3 mM, 20 µL of total cell extracts, and 20 µL of glutathione (GSH) 10 mM. After 5 min in dark conditions, 20 µL of NADPH (2.4 mM) and 20 µL H_2O_2 63.5 mM were added to the cuvette [19,20]. GR activity was read for three minutes with 60 sec of delay at 340 nm using a Shimadzu Spectrophotometer UV-1800 (Duisburg, Germany). GPx activity was measured for 3 min at 340 nm and expressed in UI/mg protein. GR activity = $((\Delta Abs/min) \times 1.3 \times F)/(0.00622 \times Vs \times C)$ GPx activity = $((\Delta Abs/min) \times 1.3 \times F)/(0.00622 \times Vs \times C)$

0.00622 mL $nmol^{-1}$ cm^{-1}: molar extinction coefficient of NADPH

F: dilution factor

C: Protein concentration (mg/mL)

Vs: Sample volume (mL)

$\Delta Abs/min$: activity of the kinetic

2.2.6. Peroxiredoxin Expression in Neuro-2a Cells by Immunoblotting

In order to evaluate the effect of urolithin A on the expression of Prx1 and Prx3, Neuro-2a were grown in 6-well culture plates and treated with 0.5–4 µM urolithin A for 24h. Twenty-four hours later, cells were washed with PBS and scraped in lysis buffer (EDTA 1 mM, Tris 25 mM, NaCl 150 mM, 0.1% Triton; PMSF, leupeptin, and pepstatin; pH = 7.4) for 20 min. Supernatants were collected for protein determination with the bicinconinc acid method and dilutions were prepared to obtain the concentrations of proteins. Then, 10 µg of protein extract per sample was mixed with Laemmli buffer with β-mercaptoethanol, loaded onto 12% sodium dodecyl sulfate-polyacrylamide gel electrophoresis (SDS-PAGE), and subsequently transferred to nitrocellulose membranes. Immediately after this, protein transfer and loading were controlled by Ponceau red staining, followed by membrane blocking with bovine serum albumin (BSA). The following antibodies were used: anti-Peroxiredoxin 1 (ab41906; Abcam, Cambridge, UK) and anti-Peroxiredoxin 3 (AV52341; Sigma-Aldrich, St. Louis, MO, USA). Antibody binding was detected by chemiluminescence with species-specific secondary antibodies labeled with horseradish peroxidase (HRP) and visualized on a digital luminescent imager analyzer

(Fujifilm LAS-4000, Cambridge, MA, USA). Images were quantified using ImageQuant TL software (Global Life Sciences Solutions, Pittsburgh, PA, USA).

2.3. Urolithin A and Its Role as an Inhibitor of CNS Enzymatic Targets

2.3.1. Tyrosinase (TYR) Inhibition

Tyrosinase inhibitory activity of urolithin A was evaluated using 96-well plates and L-DOPA as the substrate [21]. The α-Kojic acid was used as a reference inhibitor. The reaction mixture contained 10 µL of urolithin A at different concentrations in DMSO, 80 µL phosphate buffer (pH = 6.8), 40 µL of L-DOPA, and 40 µL of tyrosinase (200 U/mL) in each well. Controls wells contained 10 µL of DMSO in place of the sample. The absorbance was measured at 475 nm (endpoint) using a Synergy H1 Hybrid Multi-Mode Reader (Biotek, Bad Friedrichshall, Germany). Inhibitory activity was determined as: % Inhibition = (Absorbance Control − Absorbance Urolitihin A)/(Absorbance Control) × 100

2.3.2. Acetylcholinesterase (AChE) Inhibition

The inhibition of AChE was carried out in 96-well microplates, measuring the absorbance at 450 nm 11 times using a Synergy H1 Hybrid Multi-Mode Reader [22]. Each well contained 25 µL of ATCI (15 mM) in MilliQ water, 50 µL of buffer B (50 mM Tris-HCl, pH = 8, 0.1% bovine serum), 125 µL of DTNB (3 mM) in buffer C (50 mM Tris-HCl, pH = 8, 100 mM NaCl, 20 mM $MgCl_2$), and 25 µL of urolithin A at different concentrations in buffer A (50 mM Tris-HCl, pH = 8). Finally, 25 µL of the enzyme (0.22 U/L) was added to the control and samples to complete the reaction. Blanks contained 25 µL of buffer A instead of the enzyme. Galantamine (Sigma-Aldrich, St. Louis, MO, USA), a drug used for Alzheimer's disease, was assayed for comparative purposes as a reference inhibitor. Percentages of AChE inhibition were calculated with the following formula: % Inhibition = [1 − ((Inhibitory Slope)/(Control Slope))] × 100

2.3.3. Monoamine Oxidase A (MAO-A) Inhibition

The MAO-A inhibition assay was performed following a protocol described by Olsen et al. [23]. Here, 50 µL of urolithin A at different concentrations in DMSO, 50 µL chromogenic solution (417 mM 4-aminoantipyrine, 800 µM vanillic acid, 4 U/mL horseradish peroxidase in potassium phosphate buffer pH = 7.6.), 100 µL of tyramine (300 µM), and 50 µL of MAO-A (8 U/mL) was supplemented into the well. As a standard reference inhibitor, clorgyline was selected. Control wells contained 50 µL of DMSO instead of urolithin A. The absorbance was read at 490 nm every 5 min over 30 min in a Synergy H1 Hybrid Multi-Mode Reader (Biotek, Bad Friedrichshall, Germany). Percentages of MAO-A inhibitions were calculated with the following formula: % Inhibition = [1 − ((Inhibitory Slope)/(Control Slope))] × 100

2.4. Urolithin A and Its Role as a Direct Free Radical Scavenger

2.4.1. Oxygen Radical Antioxidant Capacity ORAC Assay

The capacity or urolithin A to scavenge peroxyl radicals was measured by the oxygen radical antioxidant capacity (ORAC). Trolox was used as a reference standard for this assay. Therefore, data were represented as µmol Trolox equivalents (TE)/mg sample. Different concentrations of urolithin A (4.4 µM–4.4 mM) were dissolved in PBS and methanol (50:50) and placed into the wells. Urolithin A and Trolox were incubated with fluorescein (70 mM) in 96-well plates at 37 °C. Finally, AAPH (12 mM) was added and fluorescence was measured every 70 s for 1 h and 33 min at 485 nm (excitation) and 520 nm (emission), in a Synergy H1 Hybrid Multi-Mode Reader (Biotek, Bad Friedrichshall, Germany) [24].

2.4.2. Superoxide Radicals Generated by Xanthine/Xanthine Oxidase (X/XO) System

Another antioxidant test was performed in order to evaluate the antioxidant activity of urolithin A using a more physiological system of superoxide radical generation [25]. Here, 22.8 μM nitroblue tetrazolium (NBT), 90 μM xanthine, and 16 mM Na_2CO_3 were mixed in phosphate buffer (pH = 6.9). Then, 240 μL of this cocktail was added to the well. Next, 30 μL of urolithin A and 30 μL of xanthine oxidase (168 U/L) were added to start the reaction. Before measurement, the plate was incubated for 2 min at 37 °C. The superoxide radicals' ($O_2{}^-$) scavenging activity was assessed spectrophotometrically at 560 nm. The inhibitory activity of XO was also assayed spectrophotometrically at 295 nm. Gallic acid was used as a reference antioxidant compound.

2.4.3. DPPH Radical Assay

Here, 2,2-diphenyl-1-picrylhydrazyl (DPPH) is a purple free radical. Antioxidant reducing compounds are able to scavenge these free radicals, inducing a change in the color of this compound. A methanolic stock solution of DPPH (0.11 mM) was prepared and 150 μL was added together with 150 μL of different urolithin A concentrations dissolved in MeOH. Ascorbic acid and gallic acid were also measured as antioxidant standards. Control wells contained 150 μL of MeOH instead of urolithin A. The plate was incubated for 30 min under dark conditions and measured at 517 nm [26]. Radical scavenging capacity was calculated by the formula: % RSC = ((Absorbance Control−Absorbance Urolithin A))/(Absorbance Control) × 100

2.5. Statistical Analysis

Each experiment was performed at least three times on different days and results were expressed as the mean ± standard error (SE) of different assays. GraphPad Prism v.6 (GraphPad Software, San Diego, CA, USA) was required to perform data analyses, nonlinear regressions, and statistics (such as two-way ANOVA or one-way ANOVA). Statistical differences were detected by comparing IC_{50} values of urolithin A and reference compounds using Student's *t*-test.

3. Results

3.1. Cytoprotective Properties of Urolithin A in Neuro-2a Cells

3.1.1. Urolithin A Improves Mitochondrial Activity in Neuro-2a Cells Subjected to Oxidative Stress (MTT Assay)

The viability of Neuro-2a cells was evaluated by the MTT assay. In this case, different physiological concentrations of urolithin (0.5–50 μM) were tested in neurons for 24 h (Figure 2A). Results were as expected because this range of concentration was non-toxic, as mitochondrial activity was not significantly reduced.

The next purpose was to evaluate the protective effects of urolithin A on Neuro-2a cells using hydrogen peroxide as a neurotoxic insult. Different conditions (100 μM to 1000 μM of H_2O_2) and exposure times (15, 30, 45, 60 min) determined that incubation of hydrogen peroxide for 45 min at 250 μM was the most appropriate time period for inducing oxidative stress in N2a cells. Figure 2B shows how urolithin A improves mitochondrial activity against hydrogen peroxide (250 μM) in this cell line.

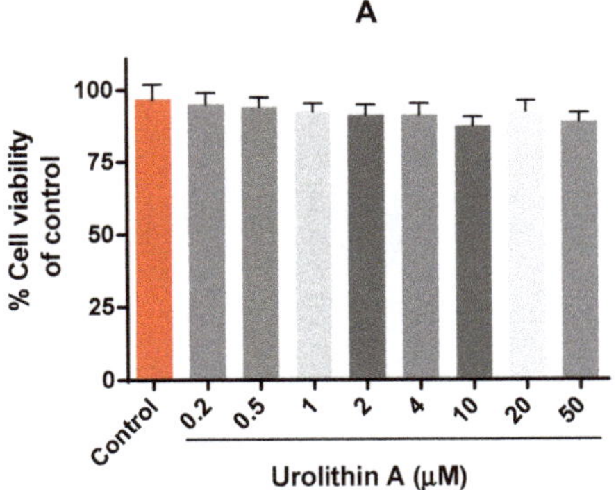

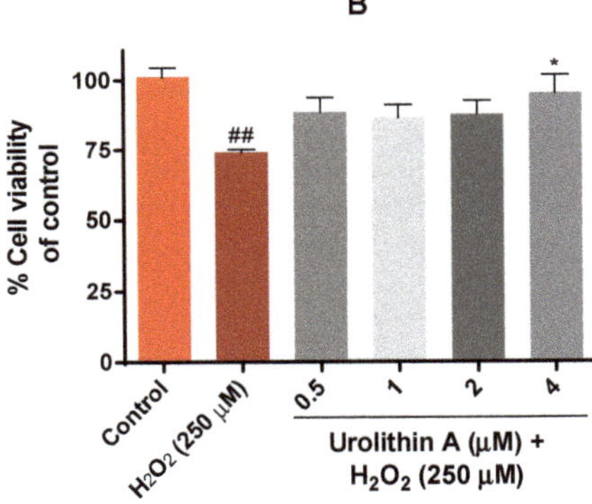

Figure 2. Mitochondrial activity in Neuro-2a cells culture (MTT assay). (**A**) Cytotoxicity of Neuro-2a cells after exposure to different concentrations of urolithin A. (**B**) Cytoprotective effects of urolithin A versus hydrogen peroxide (250 μM). Note: $* p < 0.05$ versus H_2O_2; $\#\# p < 0.01$ versus control.

3.1.2. Urolithin A Decreases Intracellular ROS Production in Neuro-2a Cells Subjected to Oxidative Stress (DCFHA-DA Assay)

Figure 3 shows the intracellular ROS production for 90 min. After 40 min of exposure, intracellular ROS reached its highest formation (165%) for cells treated with hydrogen peroxide, whereas control cells (non-treated) maintained a regular level close to 100%; cells stressed with hydrogen peroxide and treated with urolithin A at different concentrations showed lower values for ROS production. Next, the transition (50 min) showed a slight decrease in ROS formation for every treatment and maintained a similar level for the rest of the experiment.

3.1.3. Urolithin A Decreases Lipid Peroxidation in Neuro-2a Cells Subjected to Oxidative Stress (Thiobarbituric Acid Reactive Species, TBARS)

Figure 4 shows how thiobarbituric acid reactive species (TBARS) are generated. The results are exhibited as a percentage over the control (100%). As expected, lipid peroxidation generated by hydrogen peroxide was higher (147%) compared to control cells. Urolithin A response to lipid peroxidation was more effective at lower concentrations (0.5 and 1 μM).

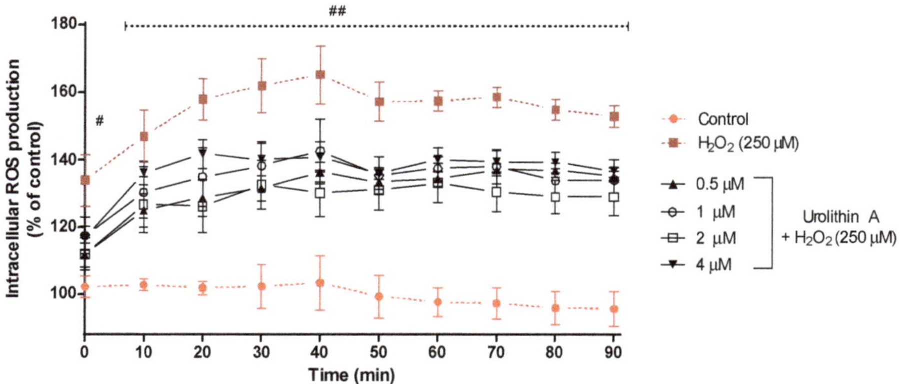

Figure 3. ROS production in Neuro-2a cells subjected to oxidative stress by hydrogen peroxide (250 µM) and treatments with urolithin A (0.5–4 µM). Data are expressed as percentage over control cells and the assay was carried out for 90 min in order to measure intracellular ROS production. Note: [#] $p < 0.005$ versus control; [##] $p < 0.001$ versus control. Significant differences appeared at the starting point for H_2O_2-N2a cells over control cells ($p < 0.001$). However, pre-treatments with 0.5 and 2 µM urolithin A at 0 and 10 min were associated with significant differences ($p < 0.01$). After 20 min, significant differences were reached at 1 µM ($p < 0.01$). Finally, 4 µM of the antioxidant exhibited significant differences ($p < 0.01$) between 40 and 60 min. Urolithin A (2 µM) displayed a greater mitochondrial response than any other co-treatment.

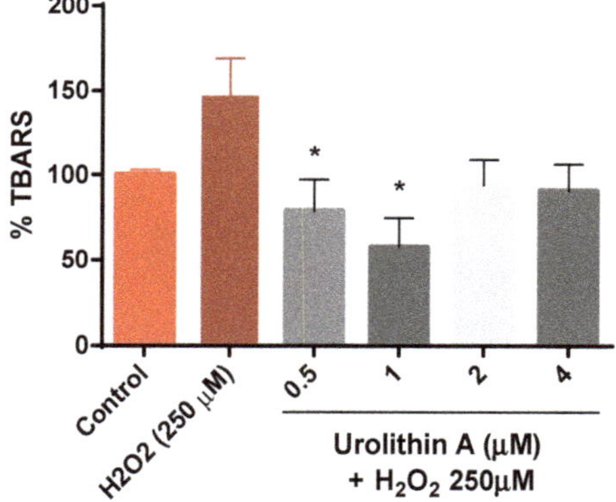

Figure 4. Thiobarbituric acid reactive species (TBARS) formation in Neuro-2a cells. Note: $* p < 0.05$ versus H_2O_2.

3.1.4. Urolithin A Enhanced the Activity of Antioxidant Enzymes in Neuro-2a Cells Subjected to Oxidative Stress (CAT, SOD, GR, GPx)

Results obtained after three replications in three different lysates showed a dose-dependent tendency of catalase for the treatments with urolithin A (Figure 5A). Surprisingly, the activity of the enzyme for the lowest concentration was even higher than the control. Therefore, the in vitro antioxidant effect observed for this metabolite may be more effective at lower doses. When the cells were treated only with hydrogen peroxide, the response of catalase decreased compared to urolithin A treatments, detecting significant differences against basal cells ($p < 0.005$) and the metabolite ($p < 0.05$; 0.01; 0.001).

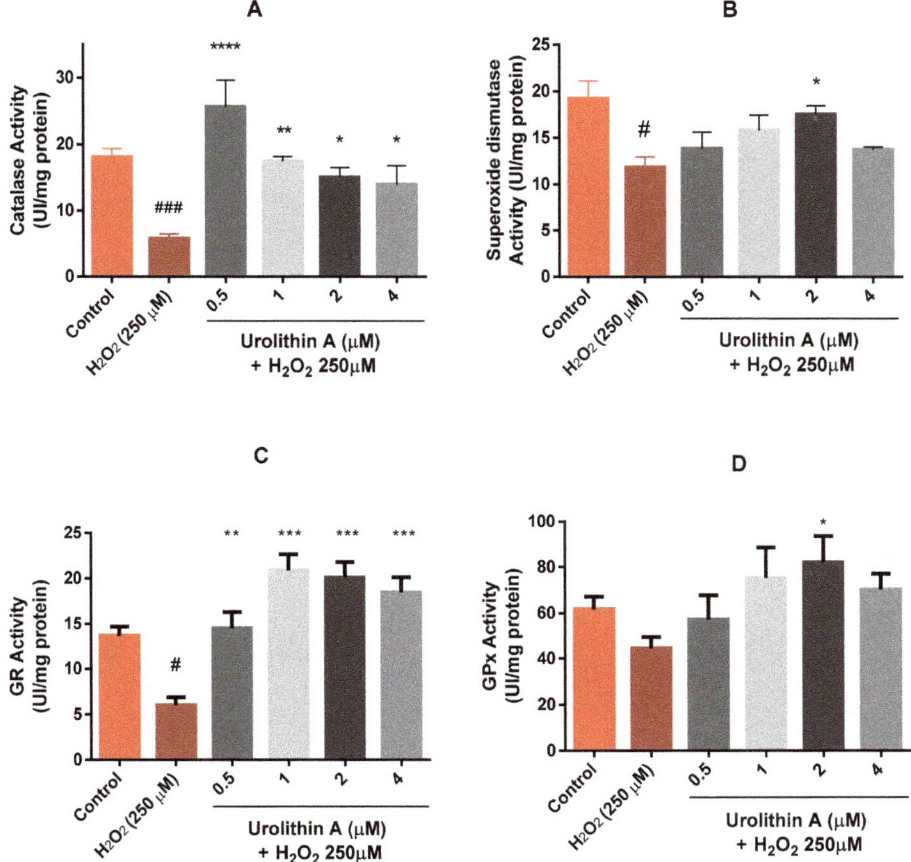

Figure 5. Neuro-2a cell culture redox status. (**A**) Catalase activity. (**B**) Superoxide dismutase activity. (**C**) Glutathione reductase activity. (**D**) Glutathione peroxidase activity. Note: * $p < 0.05$; ** $p < 0.01$; *** $p < 0.005$; **** $p < 0.001$ versus H_2O_2; # $p < 0.05$ versus Control; ### $p < 0.005$ versus Control.

Different outcomes were found for the superoxide dismutase (SOD) enzyme (Figure 5B); a dose-dependent tendency appeared up to 2 μM for SOD activity. Similar activity was observed at the highest concentration as the lowest. Significant differences appeared at 2 μM ($p < 0.05$).

Comparable effects were obtained for GR and GPx activities (Figure 5C,D). Activities for 0.5 μM urolithin A looked similar to control activities for both enzymes. Each concentration of the antioxidant compound exerted significant differences for GR activity ($p < 0.01$; $p < 0.005$). However, significant differences in GPx activity were only found at 2 μM urolithin A ($p < 0.05$).

3.1.5. Peroxiredoxins Expression

Figure 6 represents the expression of peroxiredoxins, a ubiquitous family of thiol-dependent peroxidases involved in peroxide detoxification and redox signaling [27,28]. Prx1 was increased in a dose-dependent manner over control cells (Figure 6A,B). Interestingly, mitochondrial Prx3 reached higher levels at the lower urolithin concentration tested (Figure 6C,D). Urolithin A (1 and 2 μM) treatments displayed significant differences in Prx1 expression in N2a cells over control cells ($p < 0.05$; $p < 0.01$).

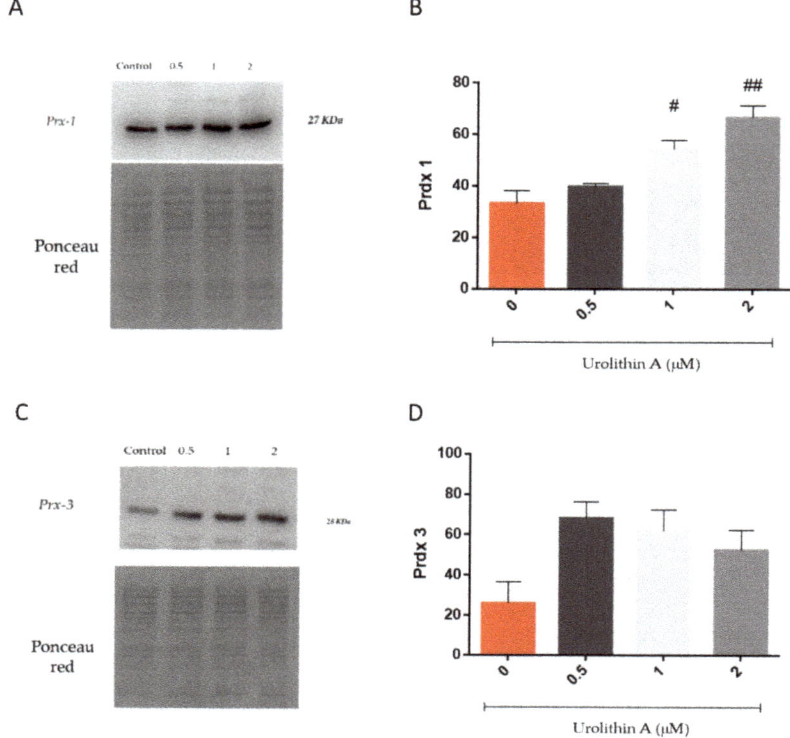

Figure 6. Urolithin A-induced cytoprotection of Neuro-2a cells is mediated by induction of peroxiredoxin 1 and 3 (Prx1, Prx3). Prx 1 (**A,B**) and Prx 3 (**C,D**) expressions were determined by Western blot in 10 µg of protein extract and expressed as Prx densitometry. Ponceau red staining of membranes prior to blotting was performed to further check protein transfer and loading. Note: # $p < 0.05$ versus Control; ## $p < 0.01$ versus Control

3.2. Urolithin A Inhibits Oxidases (Monoamine Oxidase A and Tyrosinase)

As described above, the potential neuroprotective activity of urolithin A was tested on different enzymes present in the central nervous system (tyrosinase, monoamine oxidase A, and acetylcholinesterase). IC_{50} values were calculated by non-linear regression for kojic acid, clorgyline, and galantamine reference inhibitors, respectively.

Figure 7 shows the profile of urolithin A as an enzyme inhibitor; in particular, Figure 7A compares the reference inhibitor and urolithin against tyrosinase (IC_{50} values were 24.39 ± 5.97 µM and 71.44 ± 10.07 µM for kojic acid and urolithin A, respectively). Figure 7B shows how clorgyline and urolithin A are able to inhibit MAO-A. In the case of the reference inhibitor, its IC_{50} value is relatively lower than urolithin A (0.09 ± 0.02 µM and 29.41 ± 9.01 µM, respectively). Again, it is seen how the metabolite draws a dose-dependent curve.

Finally, urolithin A was not considered as an inhibitor of AChE because very high non-physiological concentrations (876 µM) were used to reach 50% of inhibition.

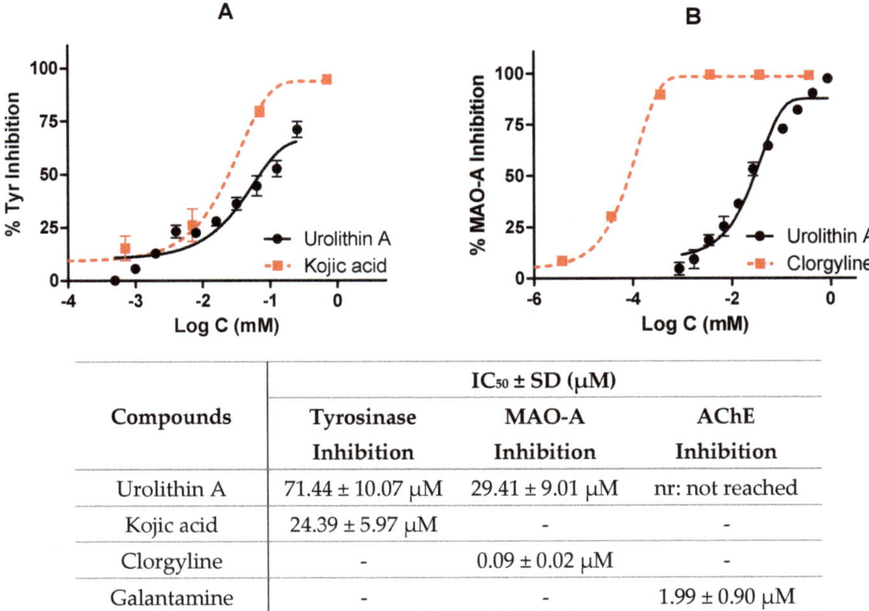

Compounds	IC$_{50}$ ± SD (µM)		
	Tyrosinase Inhibition	MAO-A Inhibition	AChE Inhibition
Urolithin A	71.44 ± 10.07 µM	29.41 ± 9.01 µM	nr: not reached
Kojic acid	24.39 ± 5.97 µM	-	-
Clorgyline	-	0.09 ± 0.02 µM	-
Galantamine	-	-	1.99 ± 0.90 µM

Figure 7. Enzymatic inhibition of urolithin A. IC$_{50}$ values were calculated by non-linear regression. (**A**) Tyrosinase inhibition profiles of urolithin A and kojic acid. (**B**) Monoamine oxidase A (MAO-A) inhibition profiles of urolithin A and clorgyline.

3.3. The Role of Urolithin A as a Direct Free Radical Scavenger

The direct antioxidant and reducing activities of urolithin A were measured by different free radical scavenging methods. To verify this fact, reference antioxidants such as gallic acid and ascorbic (vitamin C) acid were also tested and IC$_{50}$ values were calculated (Figure 8).

The antioxidant activity measured by the ORAC assay was 13.1 µmol TE/mg for urolithin A; this result reflects a great antioxidant capacity to neutralize peroxyl radicals compared to other high-content urolithin A extracts, such as pomegranate, the activity of which was 0.49 µmol TE/mg.

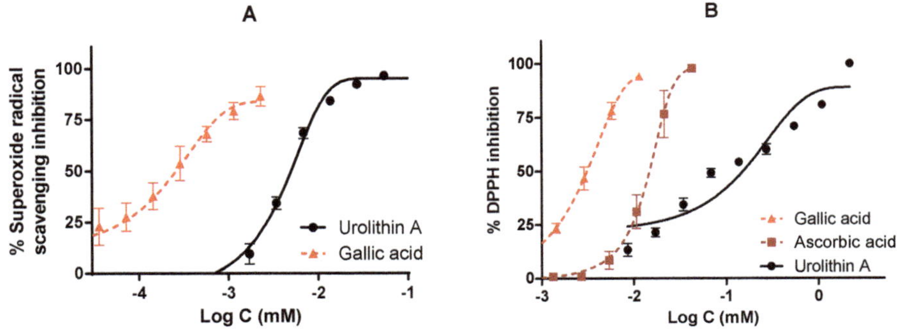

Figure 8. *Cont.*

Compounds	IC$_{50}$ ± SD (μM)	
	Superoxide Radical Scavenging	DPPH Scavenging
Urolithin A	5.01 ± 5.01 μM	152.66 ± 35.01 μM
Gallic acid	0.26 ± 0.21 μM	3.10 ± 3.11 μM
Ascorbic acid	-	14.81 ± 14.90 μM

Figure 8. Antioxidant activity of urolithin A against physiological (superoxide) and synthetic (DPPH) radicals. IC$_{50}$ were calculated by non-linear regression. (**A**) Urolithin A scavenges superoxide radicals generated by the xanthine/xanthine oxidase system. (**B**) DPPH inhibition of urolithin A. Gallic acid and ascorbic acid were used as reference antioxidants.

On the other hand, both in the inhibition of superoxide radicals and DPPH, urolithin A showed a dose-dependent curve reaching the maximum inhibition (100%). The reference compounds achieved better activity in terms of potency. IC$_{50}$ values were 0.26 ± 0.21 μM for gallic acid and 5.01 ± 5.01 μM for urolithin A (in the superoxide radical), while they were 3.10 ± 3.11 μM for gallic acid, 14.81 ± 14.90 μM for ascorbic acid, and 152.66 ± 35.01 μM for urolithin A (DPPH radical).

4. Discussion

Most of the studies involving natural products are performed with the original compounds found in the plant or food matrix, and very few are carried out with the metabolites obtained after the biotransformation in the body. Since ellagitannins are poorly absorbed in the gastrointestinal tract, urolithins have been proposed as the bioactive metabolites responsible for the beneficial effects of pomegranates and other plants containing ellagitannins. Pomegranate juice and extracts have exhibited neuroprotective properties, so here we investigate whether urolithin A, one of the major intestinal metabolites of ellagic acid, might act as a neurotherapeutic and antioxidant agent.

Neuro-2a is a neuroblastoma cell line derived from mice, which has been used because of its ability to produce microtubular proteins. Few works have been done with urolithins using this cell line, but it has been demonstrated that urolithin A protects against ischemic neuronal injury by activating autophagy [29]. This research supports our results on mitochondrial activity, demonstrating that urolithin A is not cytotoxic at this range of physiological concentrations (0.5–50 μM).

The antioxidant activity of urolithin A has already been established in different cell lines using in vitro procedures [30–32]. However, we here investigate whether urolithin can be considered as a neuroprotective agent due to its role as a direct free radical scavenger, as an indirect antioxidant improving the physiological antioxidant defense system of the cells, and as an inhibitor of oxidases such as monoamine oxidase A and tyrosinase.

Frequently, when the ROS and MDA levels decrease, the activity of the cytosolic enzymes (superoxide dismutase, catalase) increases; previous experiments in HepG2 cells treated with urolithin A and H$_2$O$_2$ have demonstrated that cells significantly increased SOD activity compared with cells subjected to oxidative stress with H$_2$O$_2$ [30]; in fact, SOD activity was ameliorated by urolithin in kidney cells from mice [33]. Computational studies performed with different pomegranate juice constituents revealed that they can act as pro-oxidants or antioxidants. Molecular docking studies seem to be controversial, as they have determined that urolithin A may inhibit cytosolic enzymes such as catalase, superoxide dismutase, and glutathione [34]; however, urolithin A seems to improve the activity of these enzymes in our present study. With the aim of elucidating molecular mechanisms involved in the cytoprotective and antioxidant properties of urolithin A, peroxiredoxins expression (Prx) was quantified in this work. These peroxiredoxins are antioxidant enzymes that catalyze the reduction of hydroperoxide. Prx1 and Prx3 play a great role in response to oxidative stress. In this way, the expression of Prx1 and Prx3 was increased when cells were treated with urolithin A, which translates into an antioxidant effect of urolithin A on Neuro-2a cells. Apart from the role of urolithin A as a free radical scavenger, this phenolic compound seems to act in a more specific way through the modulation of antioxidant enzymes. As observed above, urolithin A increased the expression of peroxiredoxins; this fact may explain the cytoprotective properties of urolithin A improving cell viability, decreasing ROS production, and increasing the activity of other physiological antioxidant defense systems, such as catalase, SOD, or glutathione reductase and peroxidase. Other polyphenols, such as resveratrol, have been described in the literature as antioxidant compounds with the ability to induce these cytoprotective proteins known as peroxiredoxins [35]; nevertheless, this is the first time that urolithin A is reported to increase the expression of peroxiredoxins 1 and 3.

This is also the first time that glutathione reductase and peroxidase activities were quantified in neuronal cells subjected to urolithin A treatments. In order to assess this pro-oxidant or antioxidant relationship of urolithin A, other researchers evaluated the in vitro antioxidant activity of the metabolite in HepG2 cancer cells [36]; unexpectedly, urolithin A exerted a pro-oxidant effect on these hepatoblastoma cells, however greater results were obtained in ORAC assay (6.67 ± 0.11). Regarding oxygen radical absorbance capacity, mice treated with urolithin A exhibited significantly different ORAC values after an hour of oral administration [37]. Nevertheless, urolithins may have different antioxidant potency, as shown in previous studies [38], and the ORAC value of urolithin C was lower than ours; the antioxidant properties derived from its role as a free radical scavenger were also confirmed by the DPPH in other studies [39–41].

ROS generation is directly linked to neurodegenerative diseases, such as Alzheimer's, Parkinson's, or Huntington's diseases. In this way, after demonstrating the potential of this metabolite in Neuro2-a cells, different bioassays were carried out involving CNS enzymatic targets known as acetylcholinesterase (AChE), monoamine oxidase A (MAO-A), and tyrosinase; these enzymes are considered pharmacological targets whose inhibition may lead to neuroprotective effects. Urolithin A acted as an inhibitor of oxidase enzymes such as MAO-A and tyrosinase, thus preventing oxidative damage of certain tissues. A human study in older adults with mild memory complaints suggests that 8 ounces of pomegranate juice taken daily over one-month improves verbal memory and alters neural activity during a visual source memory task [42]. Synthetized urolithins have also demonstrated comparable activity to AChE inhibitors such as rivastigmine, galantamine, and donepezil [43]. In silico computational studies predicted that urolithins can penetrate the blood-brain barrier, preventing β-amyloid fibrillation in a *C. elegans* model [44]. In addition, urolithin A has been identified as a proper anti-inflammatory and anti-ageing compound [38]. Recently, 10 μM urolithin A was demonstrated to possess depigmentation efficacy by suppressing tyrosinase activity, attenuating melanogenesis in B16 melanoma cells [45]. Tyrosinase has been recognised as a potential pharmacological target because an excess of tyrosinase activity or dopamine may lead to neurotoxicity through dopamine quinone formation; in this sense, tyrosinase inhibitors might have protective properties in neuronal cells. MAO-A is responsible for the catalytic oxidative deamination of monoamines generating hydrogen peroxide; for this reason, MAO-A inhibitors have been proposed as neuroprotective agents as well. The inhibitory activity of pomegranate polyphenols on MAO-A has already been tested, which could explain its effects in the CNS [22]; however, to the best of our knowledge this is the first report of urolithin A as an antioxidant agent capable of neutralizing MAO-A oxidative reactions.

5. Conclusions

Urolithin A, a gastrointestinal metabolite from ellagitannins, is a promising therapeutic antioxidant agent with potential pharmaceutical or food applications for the prevention of oxidative stress-associated disorders, probably though the improvement of the cell antioxidant capacity by increasing the expression of thiol-dependent peroxidases.

Author Contributions: Conceptualization, V.L. and M.H.; investigation, G.C., F.L., and C.C.-F., analysis, G.C.; writing—original draft preparation, G.C.; writing—review and editing, V.L. and M.H., funding acquisition, V.L. and M.H. All authors have read and agreed to the published version of the manuscript.

Funding: This work has been partially financed by Universidad San Jorge and by Spanish Government grants (RTI 2018-094203-B-I00 to A.M.R and IJCI-2017-34170 to M.H.).

Acknowledgments: Industrias Químicas del Ebro (IQE) and Universidad San Jorge (USJ) are acknowledged for the PhD grant for Guillermo Cásedas. M.H. holds a Juan de la Cierva-Incorporación contract from the Agencia Estatal de Investigación (Spanish Government), reference number IJCI-2017-34170. We thank Antonio Martínez-Ruiz (Unidad de Investigación, Hospital Universitario Santa Cristina, Instituto de Investigación Sanitaria Princesa (IIS-IP), E-28009 Madrid, Spain) for allowing Guillermo Cásedas to perform the Western blot analysis at his lab.

Conflicts of Interest: The authors declare no conflict of interest. The funders had no role in the design of the study; in the collection, analyses, or interpretation of data; in the writing of the manuscript, or in the decision to publish the results.

References

1. Mämmelä, P.; Savolainen, H.; Lindroos, L.; Kangas, J.; Vartiainen, T. Analysis of oak tannins by liquid chromatography-electrospray ionisation mass spectrometry. *J. Chromatogr. A* **2000**, *891*, 75–83. [CrossRef]
2. Charrier -El Bouhtoury, F.; Charrier, B.; Zahri, S.; Belloncle, C.; Charrier, F.; Pardon, P.; Quideau, S.; Charrier, B. UV light impact on ellagitannins and wood surface color of European oak (*Quercus petraea* and *Quercus robur*). *Appl. Surf. Sci.* **2007**, *253*, 4985–4989.
3. Moharram, F.A.; Marzouk, M.S.; El-Toumy, S.A.A.; Ahmed, A.A.E.; Aboutabl, E.A. Polyphenols of *Melaleuca quinquenervia* leaves – pharmacological studies of grandinin. *Phyther. Res.* **2003**, *17*, 767–773. [CrossRef]
4. González-Sarrías, A.; García-Villalba, R.; Romo-Vaquero, M.; Alasalvar, C.; Örem, A.; Zafrilla, P.; Tomás-Barberán, F.A.; Selma, M.V.; Espín, J.C. Clustering according to urolithin metabotype explains the interindividual variability in the improvement of cardiovascular risk biomarkers in overweight-obese individuals consuming pomegranate: A randomized clinical trial. *Mol. Nutr. Food Res.* **2017**, *61*, 1600830. [CrossRef]
5. García-Villalba, R.; Vissenaekens, H.; Pitart, J.; Romo-Vaquero, M.; Espín, J.C.; Grootaert, C.; Selma, M.V.; Raes, K.; Smagghe, G.; Possemiers, S.; et al. Gastrointestinal Simulation Model TWIN-SHIME Shows Differences between Human Urolithin-Metabotypes in Gut Microbiota Composition, Pomegranate Polyphenol Metabolism, and Transport along the Intestinal Tract. *J. Agric. Food Chem.* **2017**, *65*, 5480–5493. [CrossRef]
6. Heilman, J.; Andreux, P.; Tran, N.; Rinsch, C.; Blanco-Bose, W. Safety assessment of Urolithin A, a metabolite produced by the human gut microbiota upon dietary intake of plant derived ellagitannins and ellagic acid. *Food Chem. Toxicol.* **2017**, *108*, 289–297. [CrossRef]
7. Les, F.; Arbonés-Mainar, J.M.; Valero, M.S.; López, V. Pomegranate polyphenols and urolithin A inhibit α-glucosidase, dipeptidyl peptidase-4, lipase, triglyceride accumulation and adipogenesis related genes in 3T3-L1 adipocyte-like cells. *J. Ethnopharmacol.* **2018**, *220*, 67–74. [CrossRef]
8. Toney, A.; Chung, S. Urolithin A, a Gut Metabolite, Induces Metabolic Reprogramming of Adipose Tissue by Promoting M2 Macrophage Polarization and Mitochondrial Function (OR12-02-19). *Curr. Dev. Nutr.* **2019**, *3*, 3. [CrossRef]
9. Wu, X.; Haytowitz, D.; Pehrsson, P. Preliminary Analysis of Isoflavones in Processed Egg Products (P06-127-19). *Curr. Dev. Nutr.* **2019**, *3*. [CrossRef]
10. Mohammed Saleem, Y.I.; Albassam, H.; Selim, M. Urolithin A induces prostate cancer cell death in p53-dependent and in p53-independent manner. *Eur. J. Nutr.* **2019**. [CrossRef]
11. Liu, F.; Cui, Y.; Yang, F.; Xu, Z.; Da, L.-T.; Zhang, Y. Inhibition of polypeptide N-acetyl-α-galactosaminyltransferases is an underlying mechanism of dietary polyphenols preventing colorectal tumorigenesis. *Bioorg. Med. Chem.* **2019**, *27*, 3372–3382. [CrossRef]
12. Ryu, D.; Mouchiroud, L.; Andreux, P.A.; Katsyuba, E.; Moullan, N.; Nicolet-dit-Félix, A.A.; Williams, E.G.; Jha, P.; Lo Sasso, G.; Huzard, D.; et al. Urolithin A induces mitophagy and prolongs lifespan in *C. elegans* and increases muscle function in rodents. *Nat. Med.* **2016**, *22*, 879–888. [CrossRef] [PubMed]
13. Singh, A.; Andreux, P.; Blanco-Bose, W.; Ryu, D.; Aebischer, P.; Auwerx, J.; Rinsch, C. Orally administered urolithin a is safe and modulates muscle and mitochondrial biomarkers in elderly. *Innov. Aging* **2017**, *1*, 1223–1224. [CrossRef]

14. Cásedas, G.; González-burgos, E.; Smith, C.; López, V.; Pilar, M. Regulation of redox status in neuronal SH-SY5Y cells by blueberry (*Vaccinium myrtillus* L.) juice, cranberry (*Vaccinium macrocarpon* A.) juice and cyanidin. *Food Chem. Toxicol.* **2018**, *118*, 572–580. [CrossRef]

15. Mitsuru Uchiyama, M.M. Determination of malonaldehyde precursor in tissues by thiobarbituric acid test. *Anal. Biochem.* **1978**, *86*, 271–278. [CrossRef]

16. González-Burgos, E.; Carretero, M.E.; Gómez-Serranillos, M.P. Kaurane diterpenes from *Sideritis* spp. exert a cytoprotective effect against oxidative injury that is associated with modulation of the Nrf2 system. *Phytochemistry* **2013**, *93*, 116–123. [CrossRef]

17. Cásedas, G.; González-Burgos, E.; Smith, C.; López, V.; Gómez-Serranillos, M.P. Sour cherry (*Prunus cerasus* L.) juice protects against hydrogen peroxide-induced neurotoxicity by modulating the antioxidant response. *J. Funct. Foods* **2018**, *46*, 243–249. [CrossRef]

18. Staal, G.E.; Helleman, P.W.; De Wael, J.; Veeger, C. Purification and properties of an abnormal glutathione reductase from human erythrocytes. *Biochim. Biophys. Acta* **1969**, *185*, 63–69. [CrossRef]

19. Rotruck, J.T.; Pope, A.L.; Ganther, H.E.; Swanson, A.B.; Hafeman, D.G.; Hoekstra, W.G.; Siggins, G.R.; Battenberg, E.F.; Tioffer, B.J.; Bloom, F.E.; et al. Selenium: Biochemical Role as a Component of Glutathione Peroxidase. *Source Sci. New Ser. J. Biol. Chem.* **1973**, *179*, 588–590. [CrossRef]

20. Wilson, S.R.; Zucker, P.A.; Huang, C.R.; Spector, A. Development of Synthetic Compounds with Glutathione Peroxidase Activity. *J. Am. Chem. Soc.* **1989**, *111*, 5936–5939. [CrossRef]

21. Sezer Senol, F.; Orhan, I.E.; Ozgen, U.; Renda, G.; Bulut, G.; Guven, L.; Karaoglan, E.S.; Sevindik, H.G.; Skalicka-Wozniak, K.; Koca Caliskan, U.; et al. Memory-vitalizing effect of twenty-five medicinal and edible plants and their isolated compounds. *S. Afr. J. Bot.* **2015**, *102*, 102–109. [CrossRef]

22. Les, F.; Prieto, J.M.; Arbonés-Mainar, J.M.; Valero, M.S.; López, V. Bioactive properties of commercialised pomegranate (*Punica granatum*) juice: Antioxidant, antiproliferative and enzyme inhibiting activities. *Food Funct.* **2015**, *6*, 2049–2057. [CrossRef] [PubMed]

23. Olsen, H.T.; Stafford, G.I.; van Staden, J.; Christensen, S.B.; Jäger, A.K. Isolation of the MAO-inhibitor naringenin from Mentha aquatica L. *J. Ethnopharmacol.* **2008**, *117*, 500–502. [CrossRef] [PubMed]

24. Dávalos, A.; Gómez-Cordovés, C.; Bartolomé, B. Extending Applicability of the Oxygen Radical Absorbance Capacity (ORAC-Fluorescein) Assay. *J. Agric. Food Chem.* **2004**, *52*, 48–54. [CrossRef]

25. Rodríguez-Chávez, J.L.; Coballase-Urrutia, E.; Nieto-Camacho, A.; Delgado-Lamas, G. Antioxidant capacity of "mexican arnica" *Heterotheca inuloides* cass natural products and some derivatives: Their anti-inflammatory evaluation and effect on *C. elegans* life span. *Oxidative Med. Cell. Longev.* **2015**, *2015*, 1–11. [CrossRef]

26. López, V.; Akerreta, S.; Casanova, E.; García-Mina, J.M.; Cavero, R.Y.; Calvo, M.I. In vitro antioxidant and anti-rhizopus activities of lamiaceae herbal extracts. *Plant Foods Hum. Nutr.* **2007**, *62*, 151–155. [CrossRef]

27. Manta, B.; Hugo, M.; Ortiz, C.; Ferrer-Sueta, G.; Trujillo, M.; Denicola, A. The peroxidase and peroxynitrite reductase activity of human erythrocyte peroxiredoxin 2. *Arch. Biochem. Biophys.* **2009**, *484*, 146–154. [CrossRef]

28. Randall, L.M.; Ferrer-Sueta, G.; Denicola, A. Peroxiredoxins as preferential targets in H_2O_2-induced signaling. *Methods Enzymol.* **2013**, *527*, 41–63.

29. Ahsan, A.; Zheng, Y.; Wu, X.; Tang, W.; Liu, M.; Ma, S.; Jiang, L.; Hu, W.; Zhang, X.; Chen, Z. Urolithin A-activated autophagy but not mitophagy protects against ischemic neuronal injury by inhibiting ER stress in vitro and in vivo. *CNS Neurosci. Ther.* **2019**, *25*, 976–986. [CrossRef]

30. Wang, Y.; Qiu, Z.; Zhou, B.; Liu, C.; Ruan, J.; Yan, Q.; Liao, J.; Zhu, F. In vitro antiproliferative and antioxidant effects of urolithin A, the colonic metabolite of ellagic acid, on hepatocellular carcinomas HepG2 cells. *Toxicol. In Vitro* **2015**, *29*, 1107–1115. [CrossRef]

31. Dobroslawa, B.; Kasimsetty, S.G.; Khan, S.I.; Daneel, F. Urolithins, intestinal microbial metabolites of pomegranate ellagitannins, exhibit potent antioxidant activity in a cell-based assay. *J. Agric. Food Chem.* **2009**, *57*, 10181–10186.

32. Qiu, Z.; Zhou, B.; Jin, L.; Yu, H.; Liu, L.; Liu, Y.; Qin, C.; Xie, S.; Zhu, F. In vitro antioxidant and antiproliferative effects of ellagic acid and its colonic metabolite, urolithins, on human bladder cancer T24 cells. *Food Chem. Toxicol.* **2013**, *59*, 428–437. [CrossRef] [PubMed]

33. Jing, T.; Liao, J.; Shen, K.; Chen, X.; Xu, Z.; Tian, W.; Wang, Y.; Jin, B.; Pan, H. Protective effect of urolithin a on cisplatin-induced nephrotoxicity in mice via modulation of inflammation and oxidative stress. *Food Chem. Toxicol.* **2019**, *129*, 108–114. [CrossRef] [PubMed]

34. Mazumder, M.K.; Choudhury, S.; Borah, A. An in silico investigation on the inhibitory potential of the constituents of Pomegranate juice on antioxidant defense mechanism: Relevance to neurodegenerative diseases. *IBRO Rep.* **2019**, *6*, 153–159. [CrossRef] [PubMed]
35. Khodaie, N.; Tajuddin, N.; Mitchell, R.M.; Neafsey, E.J.; Collins, M.A. Combinatorial Preconditioning of Rat Brain Cultures with Subprotective Ethanol and Resveratrol Concentrations Promotes Synergistic Neuroprotection. *Neurotox. Res.* **2018**, *34*, 749–756. [CrossRef] [PubMed]
36. Kallio, T.; Kallio, J.; Jaakkola, M.; Mäki, M.; Kilpeläinen, P.; Virtanen, V. Urolithins display both antioxidant and pro-oxidant activities depending on assay system and conditions. *J. Agric. Food Chem.* **2013**, *61*, 10720–10729. [CrossRef]
37. Ishimoto, H.; Tai, A.; Yoshimura, M.; Amakura, Y.; Yoshida, T.; Hatano, T.; Ito, H. Antioxidative Properties of Functional Polyphenols and Their Metabolites Assessed by an ORAC Assay. *Biosci. Biotechnol. Biochem.* **2012**, *76*, 395–399. [CrossRef]
38. Tomás-Barberán, F.A.; González-Sarrías, A.; García-Villalba, R.; Núñez-Sánchez, M.A.; Selma, M.V.; García-Conesa, M.T.; Espín, J.C. Urolithins, the rescue of "old" metabolites to understand a "new" concept: Metabotypes as a nexus among phenolic metabolism, microbiota dysbiosis, and host health status. *Mol. Nutr. Food Res.* **2017**, *61*, 1–35. [CrossRef] [PubMed]
39. Nealmongkol, P.; Tangdenpaisal, K.; Sitthimonchai, S.; Ruchirawat, S.; Thasana, N. Cu(I)-mediated lactone formation in subcritical water: A benign synthesis of benzopyranones and urolithins A–C. *Tetrahedron* **2013**, *69*, 9277–9283. [CrossRef]
40. Hideyuki Ito, A. Metabolites of the Ellagitannin Geraniin and Their Antioxidant Activities Reviews. *Planta Med.* **2011**, *77*, 1110–1115.
41. Cerda, B.; Espin, J.C.; Parra, S.; Martinez, P.; Tomas-Barberan, F.A. The potent in vitro antioxidant ellagitannins from pomegranate juice are metabolised into bioavailable but poor antioxidant hydroxy-6H-dibenzopyran-6-one derivatives by the colonic microflora of healthy humans. *Eur. J. Nutr.* **2004**, *43*, 205–220. [CrossRef]
42. Bookheimer, S.Y.; Renner, B.A.; Ekstrom, A.; Li, Z.; Henning, S.M.; Brown, J.A.; Jones, M.; Moody, T.; Small, G.W. Pomegranate Juice Augments Memory and fMRI Activity in Middle-Aged and Older Adults with Mild Memory Complaints. *Evid. Complement. Altern. Med.* **2013**, *2013*, 1–14. [CrossRef] [PubMed]
43. Gulcan, H.O.; Unlu, S.; Esiringu, I.; Ercetin, T.; Sahin, Y.; Oz, D.; Sahin, M.F. Design, synthesis and biological evaluation of novel 6H-benzo[c]chromen-6-one, and 7,8,9,10-tetrahydro-benzo[c]chromen-6-one derivatives as potential cholinesterase inhibitors. *Bioorg. Med. Chem.* **2014**, *22*, 5141–5154. [CrossRef] [PubMed]
44. Yuan, T.; Ma, H.; Liu, W.; Niesen, D.B.; Shah, N.; Crews, R.; Rose, K.N.; Vattem, D.A.; Seeram, N.P. Pomegranate's Neuroprotective Effects against Alzheimer's Disease Are Mediated by Urolithins, Its Ellagitannin-Gut Microbial Derived Metabolites. *ACS Chem. Neurosci.* **2016**, *7*, 26–33. [CrossRef] [PubMed]
45. Wang, S.-T.; Chang, W.-C.; Hsu, C.; Su, N.-W. Antimelanogenic Effect of Urolithin A and Urolithin B, the Colonic Metabolites of Ellagic Acid, in B16 Melanoma Cells. *J. Agric. Food Chem.* **2017**, *65*, 6870–6876. [CrossRef] [PubMed]

MDPI

St. Alban-Anlage 66

4052 Basel

Switzerland

Tel. +41 61 683 77 34

Fax +41 61 302 89 18

www.mdpi.com

Antioxidants Editorial Office

E-mail: antioxidants@mdpi.com

www.mdpi.com/journal/antioxidants